Biopolymers

edited by

Assoc. Prof. Dr. Magdy M. Elnashar

Biopolymers

Edited by Assoc. Prof. Dr. Magdy M. Elnashar

This is a reprint of the book published in 2010 by Sciyo

Republished by InTech

Janeza Trdine 9, 51000 Rijeka, Croatia

Publishing Process Manager Iva Lipovic

Technical Editor Goran Bajac

Cover Designer Martina Sirotic

Additional hard copies can be obtained from orders@intechopen.com

Biopolymers, Edited by Assoc. Prof. Dr. Magdy M. Elnashar
p. cm.
ISBN 978-953-307-109-1

Contents

Preface

Biopolymers are polymers produced by living organisms. Cellulose, starch and chitin, proteins and peptides, DNA and RNA are all examples of biopolymers, in which the monomeric units are sugars, amino acids, and nucleotides.

Biopolymers and their derivatives are diverse, abundant and important for life. They exhibit fascinating properties and are of increasing importance for different applications. Living matter is able to synthesize an overwhelming variety of polymers, which can be divided into proteins and poly(amino acids), polysaccharides such as cellulose, starch and xanthan, chitosan, alginate, carrageenan, organic polyoxoesters such as poly(hydroxyalkanoic acids), poly(malic acid) and cutin.

This book comprehensively reviews and compiles information on biopolymers in 30 chapters which cover occurrence, synthesis, isolation and production, properties and applications, modification, and the relevant analysis methods to reveal the structures and properties of some biopolymers.

This book is written by authors from the USA, Europe, Asia and Africa, yet, the editor has tried to arrange the book chapters in a subject order to make it easier for the readers to find what they need. However, the reader can still find information on the same subject in more than one Section.
Section A, which includes chapters 1-10, is mainly on chitin/chitosan and, sometimes, other biopolymers. It includes chitosan production, chitosan grafting and chitosan applications.
Section B, which includes chapters 11-26, refers to some other biopolymers.
Section C, which consists of chapter 27, deals with theoretical / mathematical / spectroscopical characterization of biopolymers.
Section D, which includes chapters 28-30, is on applications of biopolymers/hydrogels in drug delivery systems and for military use, as well as molecularly imprinted polymers for biomedical applications.

The publishing of this book was accomplished by choosing individual scientists due to their recognized expertise and their excellent contributions to various fields of research. I am very grateful to these scientists for their willingness to contribute to this reference work as well as their engagement. Without them and without their commitment and enthusiasm it would have not been possible to compile such a book.

This book will hopefully be of help to many scientists, physicians, pharmacists, engineers and other experts in a variety of disciplines, both academic and industrial. It may not only support research and development, but also be suitable for teaching.

Last but not least, I would like to thank my family for their patience. I extend my apologies for many hours spent on the preparation of my chapter and the editing of this book, which kept me away from them.

Editor

Assoc. Prof. Dr. Magdy M. Elnashar
Laboratory of Advanced Materials & Nanotechnology,
Polymers and Pigments Department,
Centre of Excellence for Advanced Sciences,
National Research Centre,
12622 Dokki, Cairo,
Egypt

Section A

1

Low-cost Foods and Drugs Using Immobilized Enzymes on Biopolymers

Assoc. Prof. Dr. Magdy M. Elnashar
Laboratory of Advanced Materials & Nanotechnology,
Polymers and Pigments Department,
Centre of Excellence for Advanced Sciences,
National Research Centre,
12622 Dokki, Cairo,
Egypt

1. Introduction

The cost of food and drugs is dramatically increasing on a universal scale, as a result, the production of low-cost foods and drugs is of major national concern. This abstract aims to summarize some of the authors research work on low-cost biopolymers, as an alternative to the expensive carriers, such as *Eupergit C®*, for the immobilization of industrial enzymes to produce more economical foods and drugs. The novel carriers are 15-25 times cheaper compared to the commercial ones in the market (Agaroses and *Eupergit C®*). The present work has been internationally *reported* and appreciated by specialized scientific websites.
From a preliminary study of the world markets and needs, three enzymes, namely, β-galactosdiase, inulinase and penicillin acylase, have been chosen for immobilization. β-galactosdiase will ***solve 70% of the worldwide population*** suffering from lactose intolerance, as well as, ***protecting the environment*** by converting wastes, as whey, to lactose free syrup; inulinase produces fructose, which is ***4 times sweeter than glucose*** and has, additionally, the advantage of being recommended to ***diabetics and people on a diet***. Penicillin acylase produces 6-aminopenicillanic acid (6-APA), which is the precursor for the production of ***19% of the worldwide antibiotics,*** such as Ampicillin and Amoxicillin.
Immobilization of enzymes on natural biopolymers, such as grafted carrageenan, chitosan and alginate will, practically, enable the separation and reusability of the enzymes for tens of times, which will, consequently, reduce both the enzyme and the product costs. Accordingly, we are expecting the prices of foods and drugs to be significantly reduced. One of the major advantages of the immobilization of enzymes, that we have achieved, is the improvement of the enzymes' thermal, as well as, operational stabilities.

2. Worldwide food and drug crisis

According to an assessment made by *US AID,* between March 2007 and 2008, global food prices were increased by an average of 43 percent, according to the International Monetary Fund. [(1)] During this time period, wheat, soybean, corn, and rice prices increased by 146%, 71%, 41%, and 29%, respectively, according to the U.S. Department of Agriculture. The rise in food prices contributed to a significant increase in food insecurity worldwide, particularly among poorer populations. Similarly, the prices of drugs are dramatically increasing. (2) One of the main solutions available is to reduce the price of certain foods and drugs, through the manipulation of immobilized enzymes. Accordingly, in this chapter, we will shed the light on:

a. nature of enzymes,
b. uses of enzymes in different fields,
c. immobilized enzymes and immobilization techniques,
d. carriers for enzyme immobilization (including gel disks and bead formation), and finally, we will re-present some of the distinguished results we achieved in the production of low-cost food and drugs using some industrially immobilized enzymes.

2.1 Nature of enzymes

Enzymes are biological catalysts used for biotransformations, consisting of a protein component (biopolymer) and, in many cases, cofactors or prosthetic groups. Almost, all enzymes are proteins, however, some recent works reported the presence of catalytic activity in RNA. A catalyst, by its turn, is a molecule, which speeds up the chemical reaction, at which equilibrium is achieved without altering its position, and undergoing no insignificant chemical change in itself.(3)

Enzymes have the advantage of catalysing reactions under mild conditions, such as aqueous solutions, moderate temperatures and pressures, while operating with high specificity. In biological processes, enzymes are preferred to conventional chemical catalysts, for the following reasons: (3-7)

1. Enzymes increase the rate of reaction about 10^5 - 10^{14} over that of an uncatalysed reaction, e.g. a catalase increases the rate of decomposition of H_2O_2 into water and oxygen, as much as 10^{11} fold greater than an uncatalysed reaction and 10^7 faster than the manipulation of a platinum based chemical catalyst.(8) Another example is carbonic anhydrase, which catalyses the hydration of up to 600,000 molecules of CO_2 per second of carbonic acid.(4)
2. Enzymes reduce the activation energy barrier (energy difference between ground state and transition state), **Figure 1**.

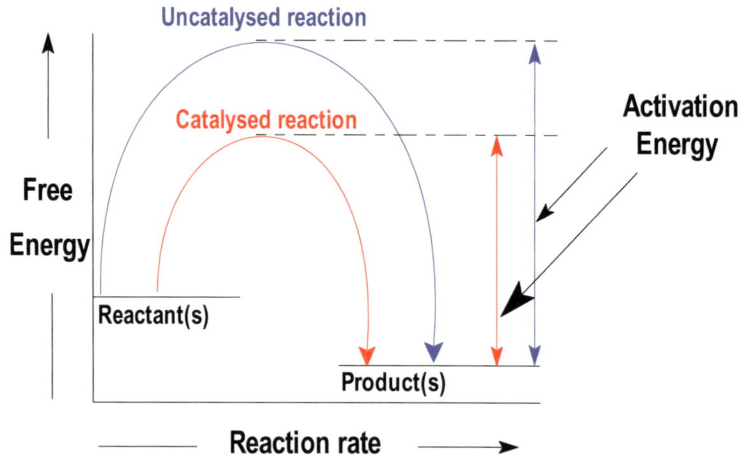

Fig. 1. Effect of enzyme catalysis on the activation energy of a system. The blue and red arrows indicate the uncatalysed and catalysed reactions respectively.

3. Enzymic catalysis can take place at ambient temperatures, neutral pH and at atmospheric pressure, whilst efficient chemical catalysis generally requires an increase in one or more of these variables.
4. Enzyme catalysed reactions are specific and rarely give side products as a consequence of their reactions, while chemical catalysis are limited in specificity and may lead to a number of side products.

2.2 Uses of enzymes in different fields

The use of purified enzymes has been increasing in many fields, e.g. in the detergent, starch, brewing, and baking industries. Enzymes are also used in pharmaceutical, chemical and clinical applications. Biochemical methods are used in clinical diagnosis, e.g. for the measurement of protein, urea, pH and cholesterol.[1, 9] Common uses of enzymes in different fields (analytical, medical and industrial) are indicated in **Table 1**.

2.3 Immobilized enzymes

The worldwide figure of enzymes used in industries (except in medicines) is about € 1.5 billion. In food and drug industries, enzymes are not preferred to be used in a free liquid state, since they can not be easily separated from the product, consequently, they are lost after the first use and some enzymes are very expensive **(Fig. 2a)**. The immobilization technique would enable the reusability of enzymes **(Fig. 2b)** for tens of times, thus reducing the enzyme and product costs significantly.

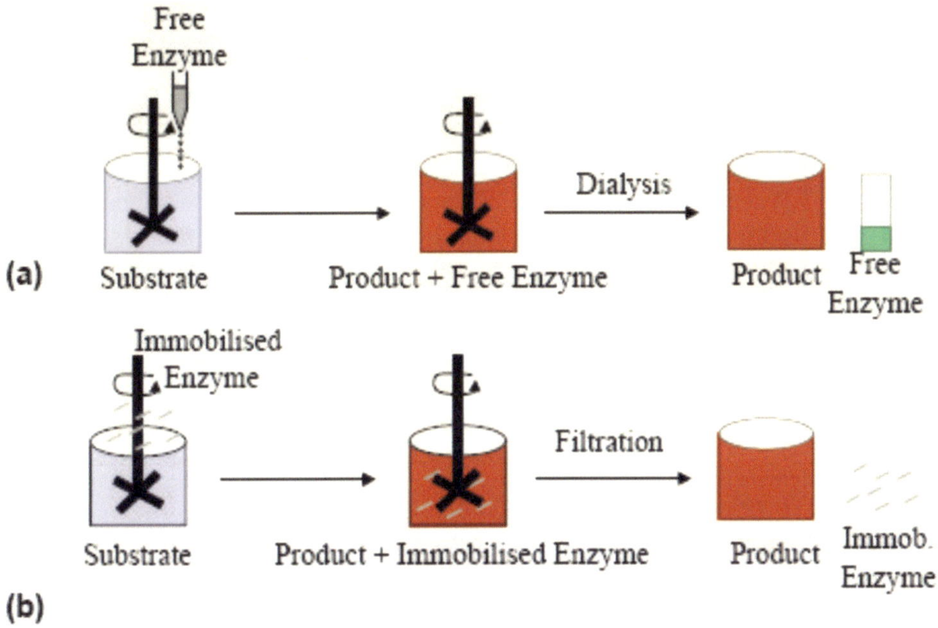

Fig. 2. a&b Schematic diagram of free and immobilized enzyme reactions. (a), Reaction of free enzyme with substrate and formation of product, which has to be separated via dialysis; (b), Reaction of immobilized enzyme with substrate and formation of product, which can be separated via filtration.

Field	Enzymes	Uses
Analytical	Glucose oxidase	In glucose biosensors (biological molecule is attached to a transducer).
	Glucose dehydrogenase	One enzyme system: detects the presence of glucose in urine and blood.
	Glucose oxidase / peroxidase	Two enzyme system: detecting the presence of glucose in urine and blood.
	Urease	Measures the concentration of urea in urine samples.
	Trypsin	For protein sequence studies. In tissue and cell dissociation.
	Lactate dehydrogenase	To measure reactions in which adenosine di- and tri- phosphate, pyruvate are products.
Industrial	Penicillinase	In production of semi-synthetic antibiotics such as penicillin, amoxycillin, ampicillin.
	Inulinase	Production of fructose and high fructose syrup
	Lipase	Hydrolyses lipids in food industries.
	β-galactosidase	Hydrolyses lactose to glucose and galactose.
	Chymosin	In the production of cheese.
	Pectinase and cellulase	In fruit juice industries (to hydrolyse pectin and cellulose).
	Xylanase / phytase	In animal feeds, to improve their nutritional value.
	Detergent enzymes	70 – 80 % of purified enzymes are used in detergent industries.
Medical	Catalase	Eliminates hydrogen peroxide (a potentially harmful substance) from the body.
	L-asparaginase	As an anticancer agent.
	Proteases	In treating thrombosis.
	Lactoperoxidase	Prevents dental caries by the production of hypothiocyanate ions.

Table 1. Use of enzymes in different fields.

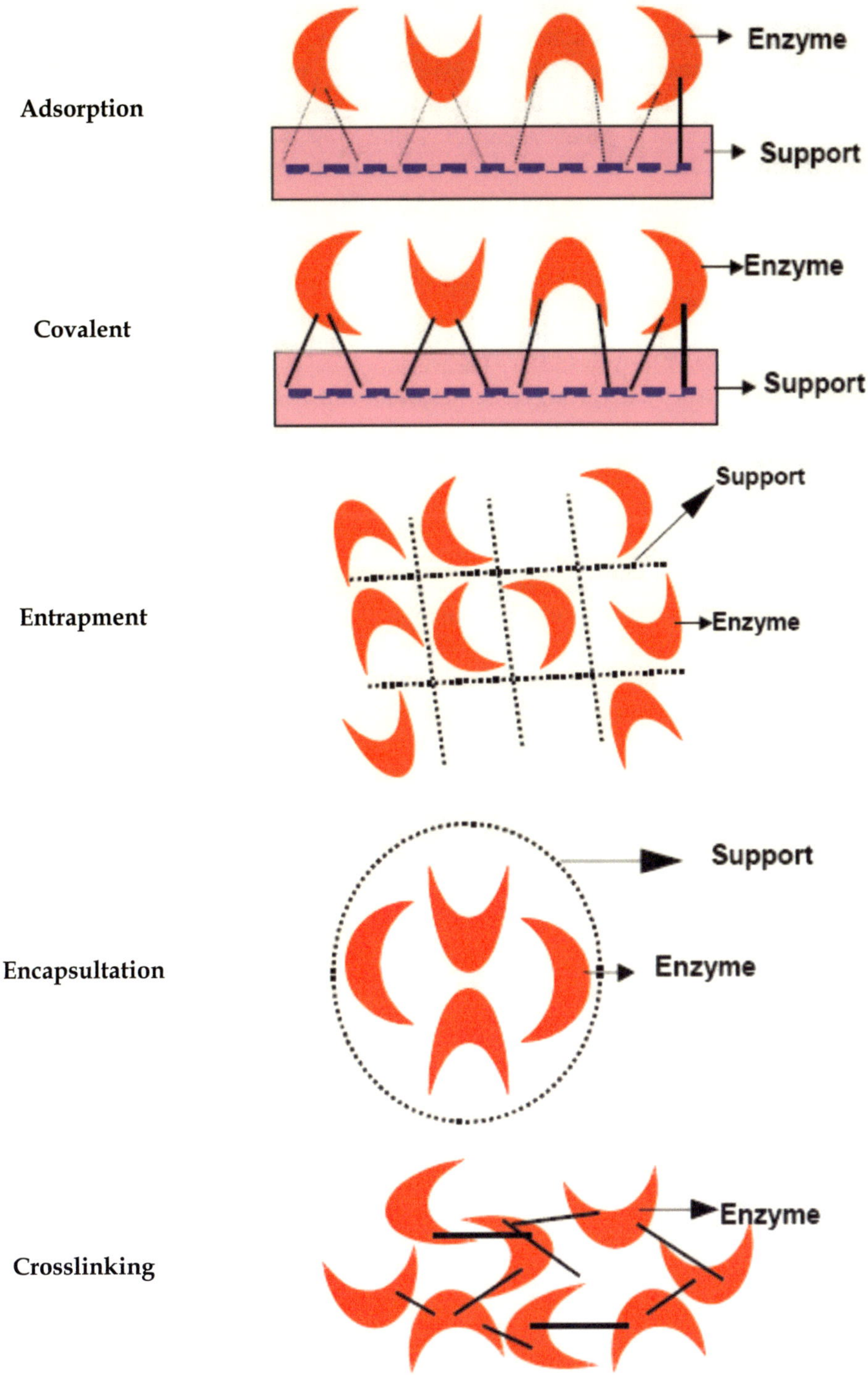

Fig. 3. Immobilization of enzymes using different techniques.

Several techniques have been used for enzyme immobilization **(Fig. 3)**, including entrapment(10), cross-linking(11, 12), adsorption(13), or (an) alternatively a combination of these methodologies.(14) Industrial surfaces such as glass beads(15), nylon-6(16), chitosan(17), Eupergit C® (epoxy-activated acrylic beads) and Agaroses®(18) have been variably used for this purpose. However, efficient commercial carriers suitable for the immobilization of enzymes are fairly expensive(19). For examples, available carriers such as Eupergit C® or Agarose® are sold for € 6,000 and € 3,250 /kg, respectively. (5) Accordingly, the need for a cheaper carrier is still an industrial dream, to compensate the high costs of their entire inputs.

2.4 Carriers for enzyme immobilization

To prepare a new carrier for enzyme immobilization, it is naturally an advantage if substances that are already permitted for use in the pharmaceutical or food industries can be utilized. Hydrogels such as alginates and carrageenans are biopolymers belonging to this commercially available category, have diverse features, and are available at a reasonable cost.(20)

2.4.1 Alginates

Alginates are produced by brown seaweeds (*Phaeophyceae,* mainly *Laminaria*). Alginate consists of monomeric residues of D-mannuronic acid and L-gluronic acid **(Fig. 4).** The gels are formed by ionic network formation in the presence of cations such as calcium ions or other multivalent counter-ions. This method qualifies as safe, mild, fast and cheap.
However, high concentrations of alginate are difficult to work with. Chelating agents such as phosphates and citrates are best avoided as they disrupt the gel structure by binding calcium ions.(21) Microspheres of alginate were produced using an encapsulation technique to immobilize glucoserebrosidase, as an enzyme delivery matrix for treatment of Gaucher's disease.(22) In terms of immobilization of enzymes, the entrapment method is widely used; for example, Tanaka *et al.* (1984) immobilized glucoamylase using Ca-alginate gel, coated with partially quaternized polyethyleneimine 3, whereas, keratinase was entrapped into Ca-alginate gel and 47 % of the total enzyme activity was retained after immobilization.(23)

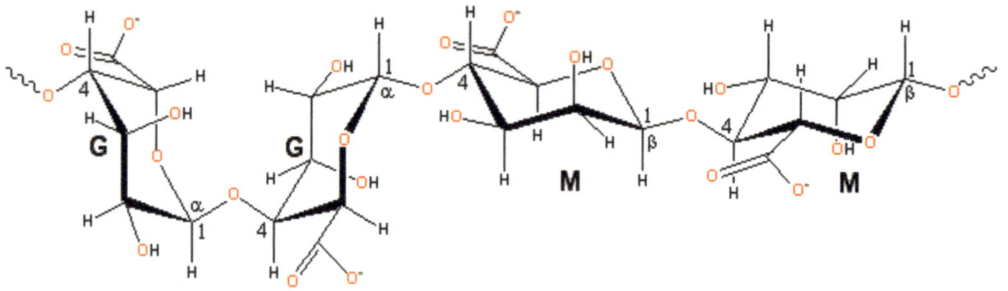

Fig. 4. Structure of Alginates. Alginates are linear unbranched polymers containing β-(14)-linked D-mannuronic acid (M) and α-(14)-linked L-guluronic acid (G) residues.

2.4.2 Carrageenans

Carrageenans are produced from red seaweeds (*Rhodophycae*). They are water-soluble, sulphated galactans that are isolated from red seaweed and contain hydroxyl and sulphate groups. There are three forms of carrageenans: k-carrageenan, ι-carrageenan, and λ-carrageenan and they differ in their ratios of sulphate to hydroxyl groups **(Fig. 5).**(24)

Carrageenan has been used for the immobilization of enzymes and cells using entrapment techniques. It is inexpensive, but suffers from weak mechanical and thermal stability. Some work has been performed to improve its mechanical and thermal stability and it was found that the gels mechanical strength increased with the increased addition of 3,6-anhydro-D-galactose 2-sulphate in the polymer,[25] or after addition of gums.[26] K^+, Al_3^+ were also found to improve the gels characteristics.[27]
In the field of immobilization of enzymes, k-carrageenan is one of the main supports used for cell and enzyme immobilization via entrapment, for example, k-carrageenan was used to immobilize α-chymotrypsin using an encapsulation method.[28] However, one of the main disadvantages of these biopolymers is that they are usually used for immobilization of enzymes using noncovalent bonds (mainly entrapment/encapsulation) due to the lack of functionalities[29] as shown in **Figure 4 & 5**. Unfortunately, the entrapment of enzymes in hydrogels is often characterized by some diffusion of the biocatalyst from the support, particularly for enzymes with molecular weight less than 300 kDa.[30] Recently, a few authors were successful at covalently immobilizing enzymes using hydrogels, and we were fortunate to be among them.[31-35] Some specialized websites highly appreciated our work and distinguishably, published them. [36-38]

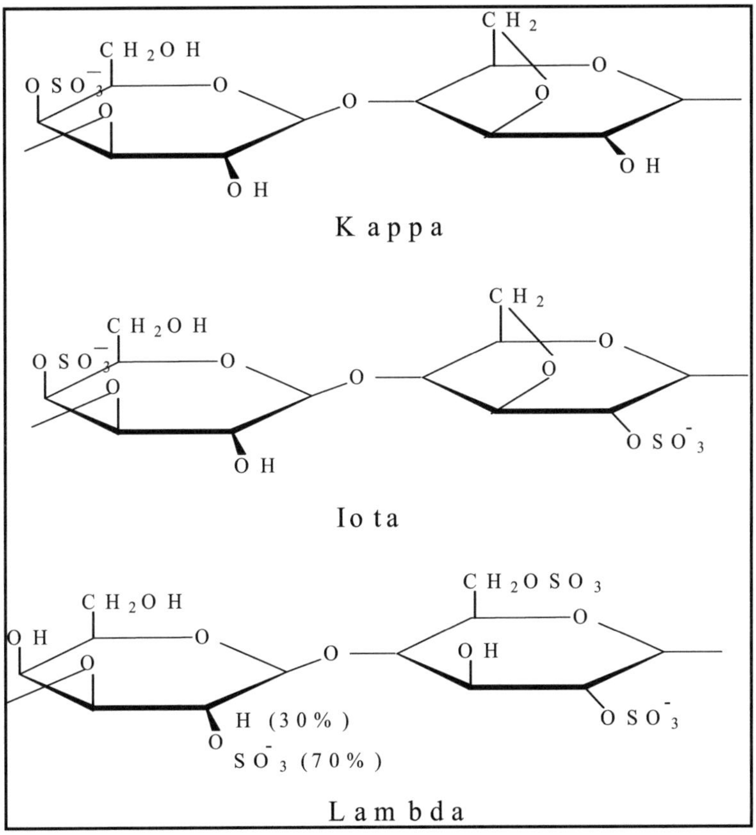

Fig. 5. Structures of repeating units in major carrageenan. Kappa carrageenan (k-) has one sulphate group, iota carrageenan (ι-) has two sulphate groups and lambda carrageenan (λ) has three sulphate groups.

In our laboratories, we succeeded to incorporate a new functionality to alginate and carrageenan gels (aldehyde groups) for covalent immobilization of enzymes. The alginate gels were prepared in uniform beads shape using the "Encapsulator" **(Fig. 6),**(34, 35) whereas k-carrageenan gels were prepared in uniform disks using our invented equipment "parallel plates" **(Fig. 7).** (39)

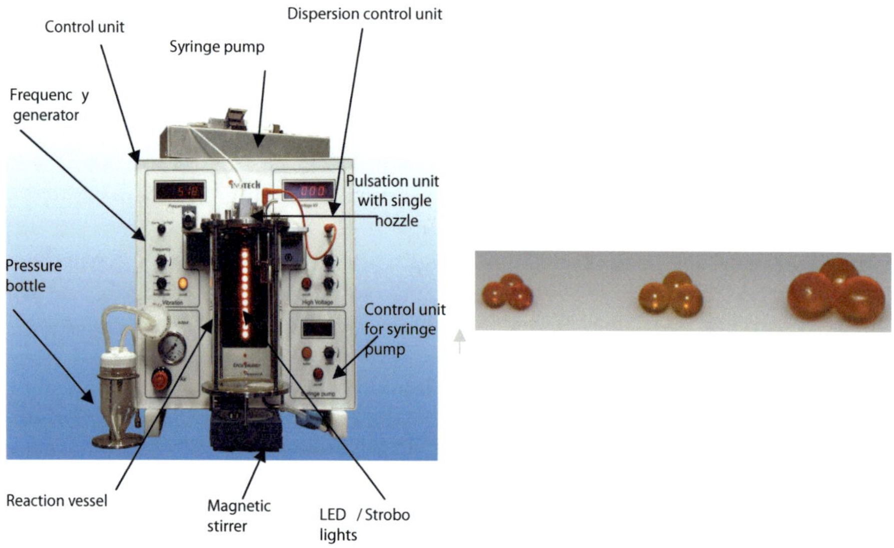

Fig. 6. Encapsulator for making uniform gel beads.

Fig. 7. Parallel plates equipment for making uniform k-carrageenan gel disks.(39)

To overcome the problem of the gels' low thermal stability, both gels were treated with polyamines to form a polyelectrolyte complex. According to Chao, et al., 1986,[40] the thermal stability of k-carrageenan gels could be improved by adding amine compounds and, especially polyamine compounds. For this reason, natural polyamines such as chitosan **(Figure 8)** and synthetic ones such as polyethylenimines **(Figure 9)** were used to improve the carrageenan and alginate gels' thermal stabilities.[30-35]

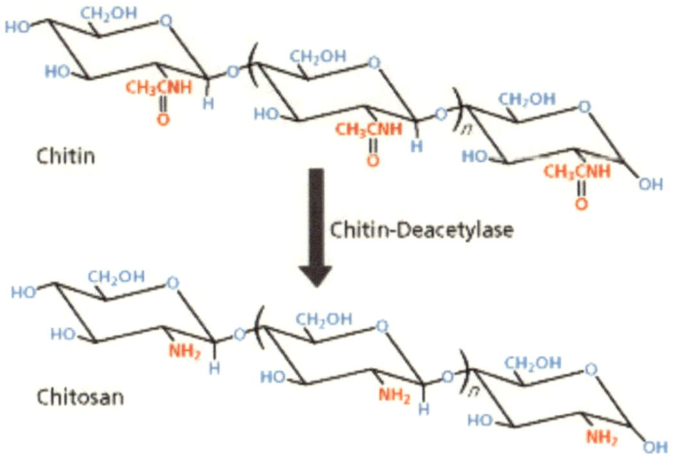

Fig. 8. Structure of chitin and chitosan.

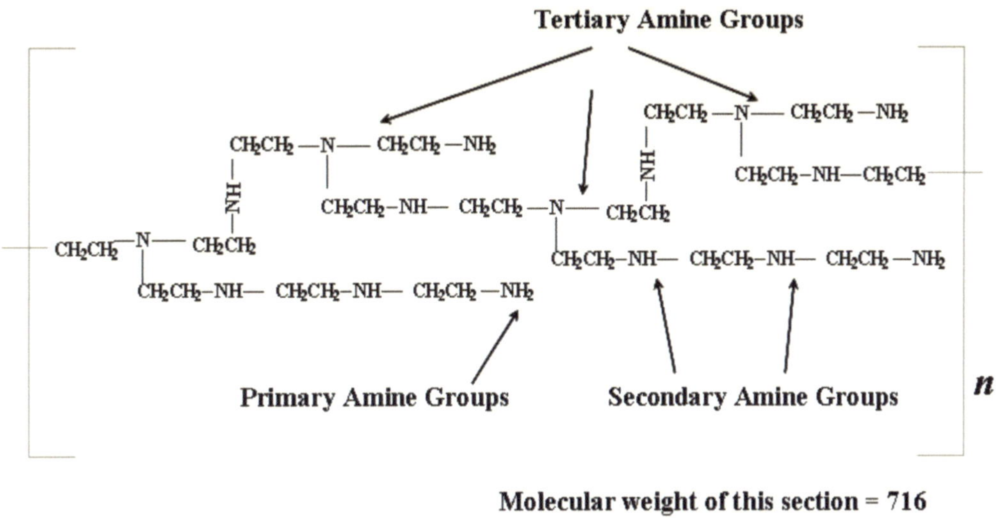

Fig. 9. Polyethyleneimine (PEI): A branched chain cationic polymer. Ratio of primary: secondary: tertiary amine groups is 1:2:1

2.4.3 Chitin and Chitosan

Chitin and chitosan are polysaccharides containing amino groups. They are attractive and widely studied as they are inexpensive. Chitin is an abundant by-product of the fishing and the fermentation industries. It is composed of (1,4)-linked 2-acetamido-2-deoxy-β-Dglucopyranosyl residues. Chitosan is composed of (1,4)-linked 2-amino-2-deoxy-D-glucose and is obtained from chitin by deacetylation, at high pH. It is more soluble than chitin and can be gelled by either crosslinking agents such as glutaraldehyde, or by multivalent anionic counter-ions such as $[Fe(CN)_6]^{4-}$ or polyphosphates. Chitin and chitosan can be prepared in the form of beads and capsules and enzymes can be immobilized using ionotropic gelation methods.(41)

2.4.4 Polyethylenimine

It is a synthetic cationic polymer that contains primary, secondary and tertiary amine groups in its skeleton. Polyethylenimine (PEI) is available in different molecular weights and in linear and branched forms. PEI has been widely used to crosslink the gel surfaces after entrapping enzymes to avoid their leakages, for example, Tanaka *et al* (1984) (42), immobilized glucoamylase using Ca-alginate gel coated with partially quaternized polyethyleneimine.
The novelty of this work resides in the realized production of thermally stable biopolymers of carrageenan and alginates gels treated with polycations. Evidently, this was not only to improve the gels' thermal stabilities, but also to immobilize the enzymes into the modified gels via covalent bonding. Glutaraldehyde was used as both a mediator and as a crosslinker. Three industrial enzymes have been chosen for immobilization due to their industrial importance in the production of low-cost foods and drugs, as follows:

a. penicillin G acylase (PGA),
b. inulinase, and
c. β-galactosidase

a) Penicillin acylase

Penicillins are a group of small compounds with important antibacterial properties. They are the most widely used β-lactam antibiotics, with a share of about 19 % of the estimated world-wide antibiotic market.(43) In particular, they are effective against the micro-organisms responsible for several diseases such as tetanus, diphtheria, syphilis, gonorrhoea and skin boils. Penicillin acylase catalyses the hydrolysis of penicillin G (PG) to phenyl acetic acid (PAA) and 6-aminopenicillanic acid (6-APP) **(Figure 10)**,(44) which is the building block for production of semi-synthetic β-lactam antibiotics,(45) as shown in **Table 2**.

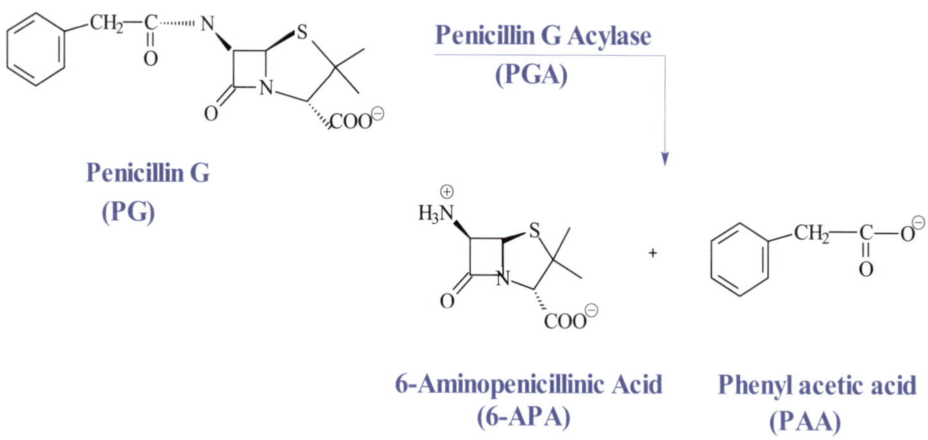

Fig. 10. Hydrolysis of penicillin G by penicillin G acylase.

Modified semisynthetic penicillins have three distinct advantages over native PG. Firstly, some of them, such as *ampicillin* and *cloxacillin,* are acid stable, and are therefore suitable for oral intake. Secondly, *methicillin* and *cloxacillin* are resistant to beta-lactamase activity. Lastly, synthetic penicillins have a significant increased range of susceptible bacteria including many bacteria that are not usually inhibited by PG.

Penicillin core structure	
R =	Penicillin Variant
H -	6-Amino-penicillanic acid (6-APA)
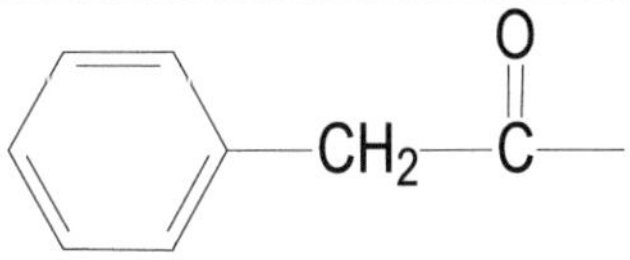	*Penicillin C*
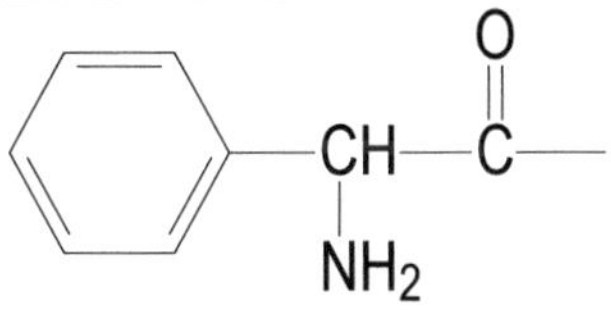	*Ampicillin*
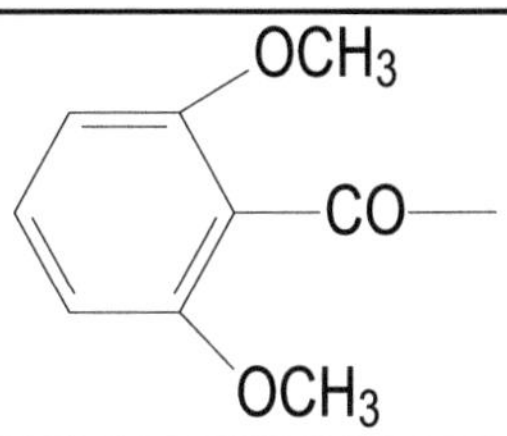	*Methicillin*
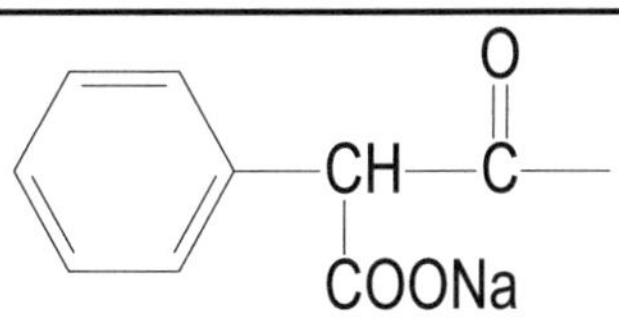	*Carbenecillin*
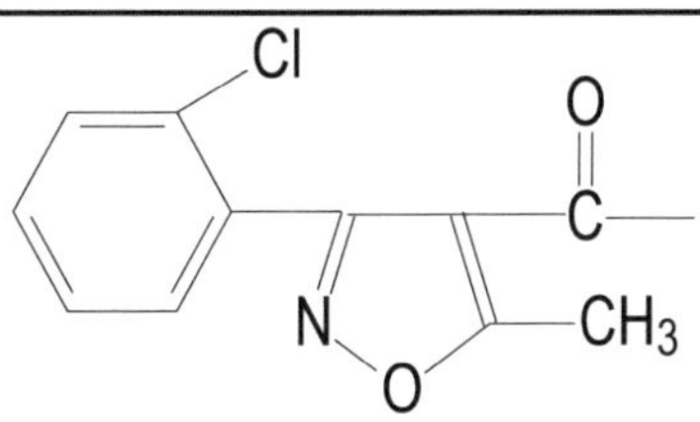	*Cloxacillin*

Table 2. Structural comparison between some semi-synthetic penicillins and naturally produced penicillin G.

b) Inulinases

Inulinases are one of the most important enzymes used in industries. Inulinases are β-fructan fructanohydrolases (EC 3.2.1.7), which hydrolyze inulin, to produce fructose and fructo-oligosaccharides, both of which are important ingredients in food and pharmaceutical industries. Inulin is a polysaccharide of β-(2,1)-linked fructose residues attached to a terminal glucose molecule and is accumulated in the underground organs of chicory, dahlia, and Jerusalem artichoke. Fructose can be produced from starch by enzymatic methods involving R-amylase, amyloglucosidase, and glucose isomerase, resulting in the production of a mixture consisting of oligosaccharides (8%), fructose (45%), and glucose (50%). (46) Separation of fructose from this high fructose syrup is costly and thus makes this method uneconomical. An alternative procedure involves the use of inulinase, which yields 95% of pure fructose after one step of the enzymatic hydrolysis of inulin. Industrial inulin hydrolysis is carried out at 60 °C to prevent microbial contamination and also because it permits the use of higher inulin substrate concentration due to increased solubility. Thus, a thermostable inulinolytic enzyme would be expected to play an important role in food and chemical industries, in which fructose syrup is widely applied. (47)

c) β-galactosidase

β-galactosidase is widely used in milk industries for hydrolysis of lactose to glucose and galactose. Lactose is the main carbohydrate contained in milk at a concentration between 5% and 10% (w/v) depending on the source of milk.(48) Lactose can also be found in whey permeate at higher concentrations. The consumption of foods with a high content of lactose presents a medical problem for approximately 70% of the world population, especially in the developing countries, as the naturally present enzyme in the human intestine, loses its activity during lifetime. (49) Undigested lactose in chyme retains fluid, bacterial fermentation of lactose results in production of gases, diarrhoea, and bloating, abdominal cramps after consumption of milk and other dairy products.

Unfortunately, there is no cure to lactose intolerance. This fact, together with the relatively low solubility and sweetness of lactose, has led to an increasing interest in the development of industrial processes to hydrolyze the lactose contained in dairy products (milk and whey) with both the free and immobilized conditions. (50) Studies have shown that glucose and galactose products hydrolyzed from lactose, are four times sweeter than lactose, more soluble, more digestible,(51) and can be consumed by 'lactose intolerant' people. (48-52) Hydrolysis of lactose present in whey permeate will produce lactose-free syrup, solving an aquatic pollution problem, since whey is commonly thrown in water sources.

3. Some distinguished results

Some of the notable data, we achieved by immobilizing the enzymes are the improvement of the enzymes' optimum temperature, and operational stabilities as a result of immobilization, some examples are herein presented:

3.1 Enzyme's optimum temperature

As shown in **Figures 11-13**, the optimum temperature of the immobilized enzymes is always higher than that of the free enzymes. For example, the optimum temperature for:

- free PGA was 60 °C , whereas for immobilized PGA it was at 70 °C.
- free inulinase was 50 °C, whereas for immobilized inulinase it was at 60 °C.
- free β-galactosidase was 45°C, whereas for immobilized β-galactosidase it was at 55 °C.

This shift of the enzyme's optimum temperature after immobilization could be attributed to the formation of a molecular cage around the protein molecule (enzyme), which the concomitant protection of the enzyme's molecules from the surrounding bulk temperature.(53)

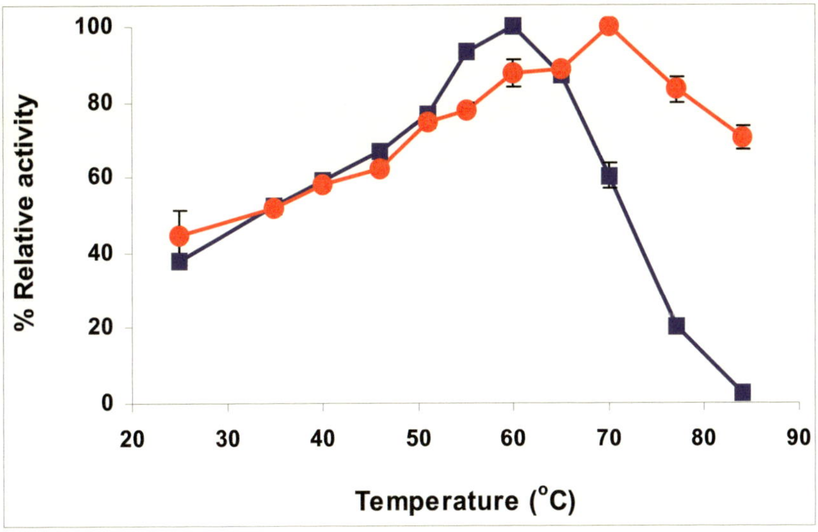

Fig. 11. Optimisation of assay temperature for free and immobilized PGA. Free PGA and immobilized PGA.(31)

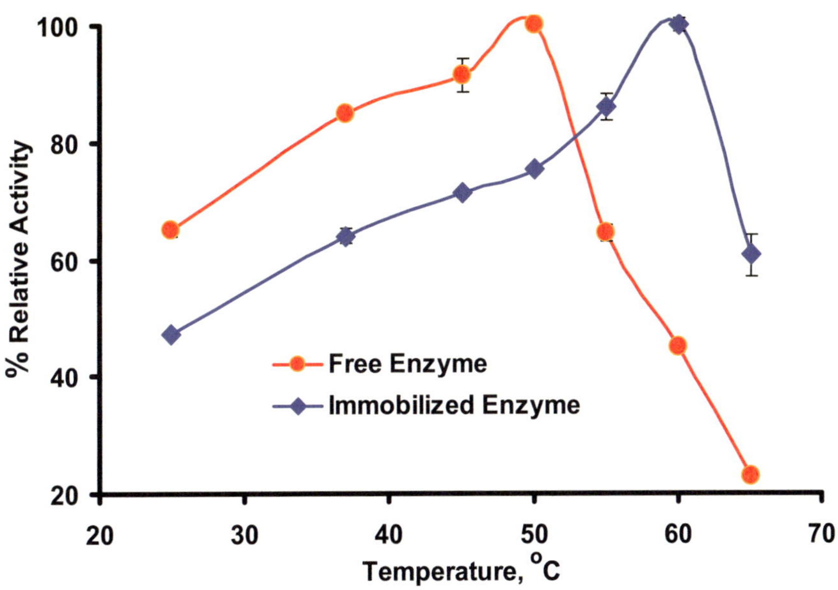

Fig. 12. Optimum temperature profile of free and immobilized inulinase.(34)

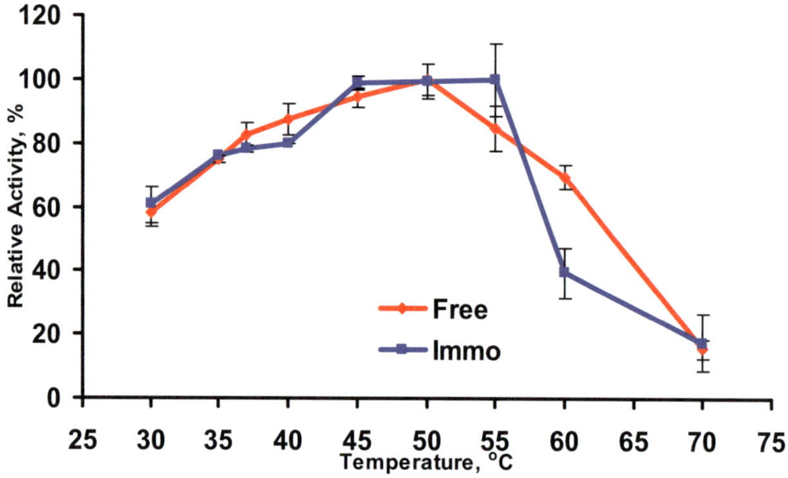

Fig. 13. Temperature-activity profile of free and immobilized *β*- galactosidase.[32,33]

3.2 Enzymes' operational stability

The major advantage of enzyme immobilization is the easy separation and reusability of the starting enzyme. The data shown in **Figure 14-16** indicates that there was almost no decrease in the enzyme activity for the immobilized PGA and inulinase for 20 reuses, and for β-galactosidase for 9 reuses. [54] However, for β-galactosidase, after the ninth use, the relative enzyme activity started to gradually decrease, to attain 97%. The loss in activity could be attributed to the inactivation of enzyme due to continuous use.[55]

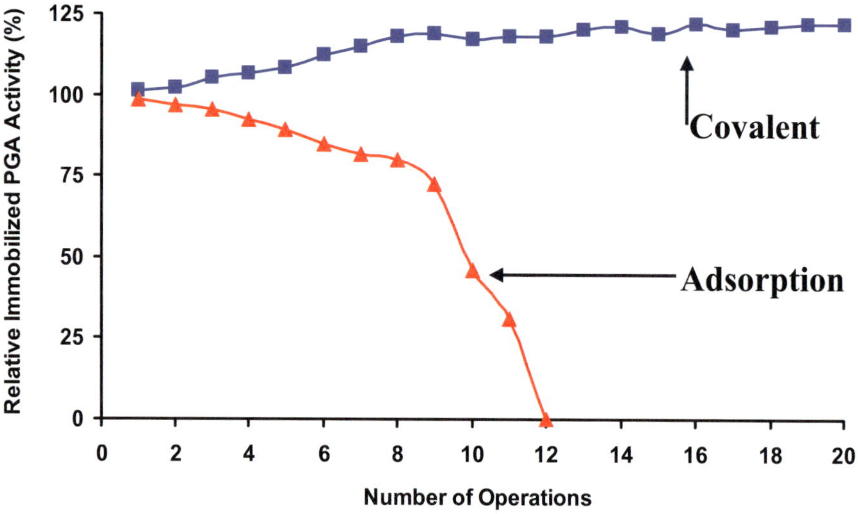

Fig. 14. Operational stability of immobilized PGA. Using modified gel (Carr/PEI/GA) and a control gel (Carr/KCl) for immobilization of PGA.[31]

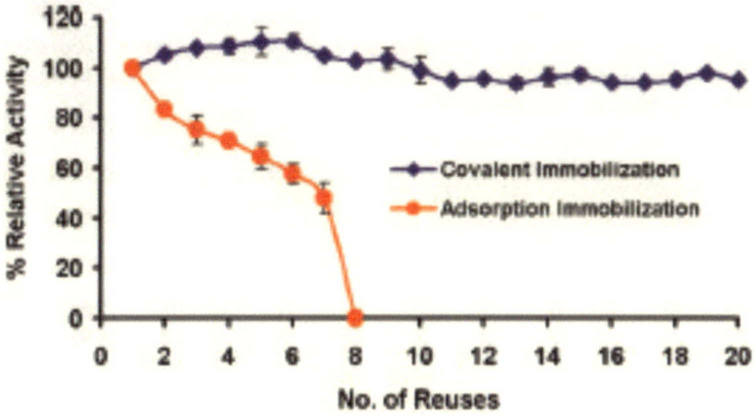

Fig. 15. Operational stability of immobilized inulinase on control gel (Alg/Ca^{2+}) and grafted alginate (Alg/PEI/GA).(34, 35)

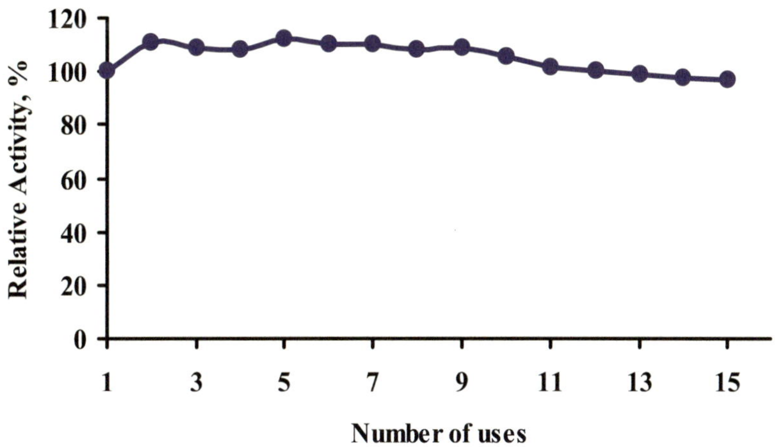

Fig. 16. Operational stability of immobilized β-galactosidase onto the modified gel (carrageenan/chitosan/glutaraldehyde). (32, 33)

4. Conclusion

The modification of carrageenan and alginate biopolymers with chitosan/PEI imparts three extra benefits to these biopolymers. The first is the creation of a new amino groups functionality; the second, is the amelioration of the gel's thermal stability by forming a polyelectrolyte complex (PEC), while the third is the use of the free amino groups to covalently immobilize enzymes, via glutaraldehyde, as a mediator and a crosslinker.

Three industrial enzymes were immobilized using the modified gels to produce ***low-cost foods and drugs***.

1. ***Penicillin G acylase*** (PGA) produces 6-APA, which is the precursor for production of ***19% of the worldwide antibiotics,*** such as *Ampicillin* and *Amoxicillin*.
2. ***Inulinase*** produces fructose, which is ***4 times sweeter than glucose*** (used in high fructose syrups) and has the advantage of being recommended to ***people on a diet and diabetics***;

3. ***β-galactosdiase*** produces glucose from lactose, which will ***solve 70% of the worldwide population*** suffering from lactose intolerance in milk and dairy products as well as ***protecting the environment*** by converting wastes, as whey, to lactose free syrup.

The novel carriers increased the enzymes' optimum temperature due to formation of a cage around the enzymes, protecting the enzyme from the bulk solution's temperature. The most important advantage of immobilization is the operational stability, where the enzymes have been used for up to 15 and 20 cycles with almost no loss of activity. In accordance, the reusability of the immobilized enzyme for tens of times tremendously reduces the enzyme, as well as the enzymatic products cost, e.g. antibiotics and fructose.

In brief, the simplicity and effectiveness of the newly developed methods for covalent immobilization of enzymes, in addition to the realized promising results for the operational stabilities are encouraging to be applicable on the industrial levels, for the prosperity of human beings.

5. Acknowledgement

The author would like to thank the Centre of Excellence, *NRC*, Egypt, for Advanced Sciences, the RDI and the STDF/IMC for supporting this work, and highly appreciates the efforts of Mrs Joanne Yachou and Professor Mohamed Abo-Ghalia for their contribution towards editing.

6. References

[1] http://www.usaid.gov/our_work/humanitarian_assistance/foodcrisis/

[2] http://healthinsurance.about.com/b/2006/06/23/drug-prices-increase-dramatically.htm

[3] Loewy, A. G, Siekevitz, P., Menninger, J. R. & Gallant, J. A. N. (1991). "Cell structure and function: an integrated approach" (3rd edition). Saunders college publishing, Philadelphia. *pp* 133-258.

[4] Bickerstaff, G. F. (1987). Enzymes in industry and medicine. Edward Arnold, London. *pp* 1-95.

[5] *Elnashar, M. M.* (2010). In Biopolymers: Bioapplications of chitosan. Novapublisher, *Accepted and in press.*

[6] Fisher, J. Arnold JPR. (1999). In chemistry for biologists. BIOSS Scientific Publishers Ltd. Oxford. *pp* 64-95, 198-207.

[7] Voet, D. & Voet, J. G. (1995). Biochemistry. New York: John Wiley & Sons. *pp* 29-104 & 331-370.

[8] Hartmeier, W. (1988). Immobilized biocatalysts. Spring-Verlag, New York. *pp*. 4-90.

[9] Boyer, R. F. (1993). In modern experimental biochemistry (2nd edition). The Benjamin / Cumming publishing company, Inc., *pp*. 29 - 58 & 115 – 127.

[10] Betancor, L., Luckarift, R., Seo, H. & Brand, O. (2008). Three-dimensional immobilization of β-galactosidase on a silicon surface. Biotechnol Bioeng, 99, 261.

[11] Wang, Y., Xu, J., Luo, G. & Dai, Y. (2008). Immobilization of lipase by ultrafiltration and cross-linking onto the polysulfone membrane surface. Biores Technol, 99, 2299.

[12] Jozef, S. & Sylwia W. (2006). Immobilization of thermostable β-glucosidase from Sulfolobus shibatae by cross-linking with transglutaminase. Enz Microb Technol, 39, 1417.

[13] Salah, S., Srimathi, S., Gulnara, S., Ikuo, S. & Bengt, D. (2008). Hydroxyapatite as a novel reversible in situ adsorption matrix for enzyme thermistor-based FIA. Talanta, 77, 490.

[14] D'Souza, S. F. (1999). Immobilized enzymes in bioprocess. Current Sci, 77, 69.
[15] Kahraman, V., Bayramoglu, G., Kayaman-Apohan, N. & Atilla, G. (2007). α-Amylase immobilization on functionalized glass beads by covalent attachment. Food Chem, 104, 1385.
[16] Shweta, P., Shamsher, K., Ghanshyam, C. & Reena, G. (2008). Glutaraldehyde activation of polymer Nylon-6 for lipase immobilization: Enzyme characteristics and stability. Biores Technol, 99, 2566.
[17] Emese, B., Ágnes, N., Csaba, S., Tivadar, F. & János, G. (2008). Preparation of chitosan particles suitable for enzyme immobilization. J Biochem Biophys Meth, 70, 1240.
[18] Peppler, H. J., & Reed, G. (1987). Enzymes in food and feed processing. In H. J. Rehm, G. Reed (Eds.), Biotechnology, vol. 7a (*pp*. 578). Weinheim: VCH
[19] Bickerstaff, G. F. (1995). Impact of genetic technology on enzyme technology. Genet Eng Biotechnol, 15, 13.
[20] Hugerth, A., Caram-Lelham, N. & Sundeliir, L. The effect of charge density and conformation on the polyelectrolyte complex formation between carrageenan and chitosan. Carbohydr Polym, 1997, 34, 149.
[21] Woodward, J. (1985). Immobilized enzymes: adsorption and covalent coupling in Immobilized Cells and Enzymes: A practical Approach. Woodward, J., ed., IRL, Oxford. *pp* 3-17.
[22] Ribeiro, C. C., Barrias, C. C. & Barbosa, M. A. (2004). Calcium phosphate-alginate microspheres as enzyme delivery matrices. Biomat, 25, 4363.
[23] Farag, A. & Hassan, M. (2004). Purification, characterisation and immobilization of a keratinase from *Aspergillus oryzae*. Enz Microb Technol, 34, 85.
[24] Gerhard, A. & Brian, R. (1997). Carrageenan biotechnology. Trends Sci Technol, 8, 389.
[25] Ainsworth, P. A. & Blanshard, J. M. V. (1978). The interdependence of molecular structure and strength of carrageenan/carob gels. Part I. Lebensm.-Wiss. U.-Technol., 11, 279.
[26] Audet, P., Paquin, C. & Lacroix, C. (1990). Batch fermentations with a mixed culture of lactic acid bacteria immobilized separetly in k-carrageenan/locust bean gum gel beads. Appl. Microb. Biotechnol., 32, 662.
[27] Chamy, R., Nunez, M. J. & Lema, J. M. (1990). Optimisation of the hardening treatment of S. cereviseae bioparticles. Biotechnol Bioeng, 30, 52.
[28] Belyaeva, E., Della Valle, D. & Poncelet, D. (2004). Immobilization of α-chymotrypsin in k-carrageenan beads prepared with the static mixer. Enz. Microb. Technol, 34, 108.
[29] Moon, S. & Parulekar, S. J. (1991). Characterisation of k-carrageenan gels used for immobilization of *Bacillus firmus*. Biotechnol. Prog, 7, 516.
[30] Tanriseven, A, Dog˘an, S. (2002). A novel method for the immobilization of β-galactosidase. Process Biochem, 38, 27.
[31] *Elnashar, M. M.*, Yassin A. M. & Kahil, T. (2008). Novel thermally and mechanically stable hydrogel for enzyme immobilization of penicillin G acylase via covalent technique. J Appl Polym Sci, 109, 4105.
[32] *Elnashar, M. M.* & Yassin A. M. (2009a). Covalent immobilization of β-galactosidase on carrageenan coated chitosan. J Appl Polym Sci, 114, 17.
[33] *Elnashar, M. M.* & Yassin A. M. (2009b). Lactose hydrolysis by β-galactosidase covalently immobilized to thermally stable biopolymers. Appl Biochem Biotechnol, 159, 426.
[34] *Elnashar, M. M.*, Danial, E. N. & Awad, G. E. (2009). Novel Carrier of Grafted Alginate for Covalent Immobilization of Inulinase. Indus Eng Chem Res, 48, 9781.

[35] Danial, E. N., *Elnashar, M. M.* & Awad, G. E. (2010). Immobilized Inulinase on Grafted Alginate Beads Prepared by the One-Step and the Two-Steps Methods. Indus Eng Chem Res, 49, 3120.

[36] http://goliath.ecnext.com/coms2/gi_0199-8982245/Studies-from-M-M-M.html

[37] http://www.verticalnews.com/premium_newsletters/Journal-of-technology/2008-09-09/66896TE.html

[38] http://www.newsrx.com/newsletters/Medical-Devices-and-Surgical-technology-Week/2009-11-08/2811082009139QW.html

[39] *Elnashar, M. M.*, Millner, P. A., Johnson, A.F. & Gibson, T. D. (2005). Parallel plate equipment for preparation of uniform gel sheets. Biotechnol Lett, 27, 737.

[40] Chao, K. C., Haugen, M. M. & Royer, G. P. (1986). Stabilization of k-Carrageenan Gel with Polymeric Amines: use of immobilized cells as biocatalysts at elevated temperatures. Biotechnol Bioeng, 28, 1289.

[41] Krajewska, B. (2004). Application of chitin- and chitosan-based materials for enzyme immobilizations: a review. Enz Microb Technol, 35, 126.

[42] Tanaka, H, Kurosawa, H, Kokufuta, E. & Veliky, I. (1984). Preparation of immobilized glucoamylase using Ca-alginate gel coated with partially quaternized poly (ethyleneimine). Biotechnol Bioeng, 26, 1393.

[43] Parmar, A., Harish, K., Marwaha, S. S. & Kennedy, J.F. (2000). Advances in enzymatic transformation of penicillins to 6-aminopenicillanic acid (6-APA). Biotechnol Adv, 18, 289.

[44] Balasingham, K., Warburton D., Dunnill P. & Lilly M. D. (1972). Isolation and kinetics of penicillin amidase from Escherichia *coli*. Biochim Biophys Acta, 276, 250.

[45] Cole, M. (1969). Hydrolysis of penicillins and related compounds by the cell-bound penicillin acylase of *Escherichia coli*. Biochem J, 115, 733.

[46] Gill, P. Manhas, R. & Singh, P. (2006). Hydrolysis of inulin by immobilized thermostable extracellular exoinulinase from Aspergillus *fumigatus*. J Food Eng, 76, 369.

[47] Vandamme, E. & Derycke, D. (1983). Microbial Inulinases: Fermentation process, properties, and applications. Adv Appl Microbiol, 29, 139.

[48] Ordoñez, J. A., Cambero, M. A., Fernandez, L., Garcia, M. L., Garcia, G. & Hoz, L. (1998). Tecnologia de los alimentos (II). Madrid, Spain: Editorial Sintesis.

[49] Richmond, M. Gray, J. & Stine, C. (1981). Beta-galactosidase: Review of recent research related to technological application, nutritional concerns, and immobilization. J dairy Sci, 1759, 64.

[50] German, J. H. (1997). Applied enzymology of lactose hydrolysis. In Milk powders for the future, *pp*. 81.

[51] Sungur, S., & Akbulut, U. (1994). Immobilization of β-galactosidase onto gelatin by glutaraldehyde and chromium(III) acetate. J Chem Technol Biotechnol (Oxford, Oxfordshire) 59, 303.

[52] Nijipels, H. H. (1981). Lactases and their applications. In G. G. Birch, H. Blakebrough, & K. J. Parker (Eds.), Enzyme and food processing (*pp*. 89). London: Applied Science Publishers.

[53] Tor, R., Dror, Y. & Freeman, A. (1989). Enzyme stabilization by bilayer "encagement". Enz Microb Technol, 11, 306.

[54] Haider, T. & Husain, Q. (2007). Preparation of lactose-free milk by using salt-fractionated almond (*Amygdalus communis*) β-galactosidase. J. Sci Food Agri, 87, 1278.

[55] Nakane, K., Ogihara, T., Ogata, N. & Kurokawa, Y. (2001). Entrap-immobilization of invertase on composite gel fiber of cellulose acetate and zirconium alkoxide by sol-gel process. J Appl Polym Sci, 81, 2084.

2

Biomedical-Grade Chitosan in Wound Management and Its Biocompatibility *In Vitro*

Chin Keong Lim and Ahmad Sukari Halim
Universiti Sains Malaysia
Malaysia

1. Introduction

Chitin (β-1,4-D-linked polymer of *N*-acetylglucosamine) is a naturally abundant mucopolysaccharide and is second to cellulose in terms of the amount produced annually by biosynthesis. Chitin is visually characterized as a white, hard, inelastic, nitrogenous polysaccharide, and approximately one billion tons are synthesized each year (Peter, 1997). Chitin is a common constituent of the exoskeleton in animals, particularly in crustaceans, mollusks and insects. Commercially sold chitin is usually extracted from shellfish waste (Skjak-Braek et al., 1989; Goosen, 1997). Chitin is structurally similar to cellulose; however less attention has been paid to chitin than cellulose, primarily due to its inertness. Hence, it remains an essentially unutilized resource. Deacetylation of chitin yields chitosan, which is a relatively reactive compound and is produced in numerous forms, such as powder, paste, film and fiber.

Chitosan is a poly-(β-1, 4-D-glucosamine) derived from the *N*-deacetylation of chitin (Figure 1). It is soluble in dilute aqueous acetic, lactic, malic, formic and succinic acids. Chitosan may be fully or partially *N*-deacetylated, but the degree of acetylation is typically less than 0.35. The acetylation ratio is defined by a variety of methods, including pyrolysis gas chromatography, gel permeation chromatography and ultra-violet (UV) spectrophotometry, titration, separation spectrometry and near-infrared spectroscopy (Kumar, 2000). Most commercial chitosans have a degree of deacetylation that is greater than 70% and a molecular weight ranging between 100,000 and 1.2 million Da (Li et al., 1997). Chitosans are of commercial interest due to their high percentage of nitrogen compared to synthetically substituted cellulose, rendering them useful for metal chelation and polyoxysalt and film formulations. Chitosan is polycationic at pH< 6 and it readily interacts with negatively charged molecules, such as proteins, anionic polysaccharides (*e.g.* , alginate and cargeenan), fatty acids, bile acids and phospholipids (Muzzarelli, 1996). Nonetheless, chitosan may also selectively chelate metal ions such as iron, copper, cadmium and magnesium.

Wound healing is defined as a tissue restoration and reparative process that is typically comprised of a continuous sequence of inflammation and repair, in which epithelial, endothelial and inflammatory cells, platelets and fibroblasts interact to resume their normal functions. The wound healing process is regulated by cytokines and growth factors and consists of four phases: the process is initialized by inflammation, followed by granulation, matrix remodeling and re-epithelialization. Research is currently being conducted to discover ways for humans to heal *via* regeneration and the use of a variety of dressing materials to facilitate proper wound management. Significant advancement in wound care was pioneered by moist wound healing theory in the 1960's, which determined that occluded wounds healed faster than dry wounds and a moist wound healing environment increased the healing rates (Winter, 1962).

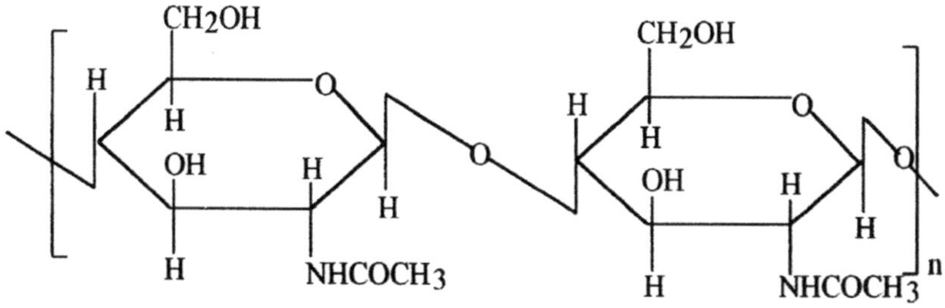

Chitin

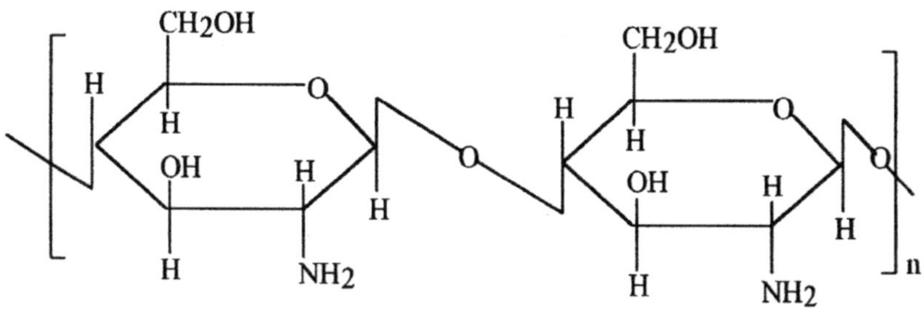

Chitosan

Fig. 1. Chemical structures of chitin and chitosan.

Chitosan has been widely used as a biomedical application due to its excellent biocompatibility (Keong & Halim 2009; Lim et al., 2010). Chitosan is a benefit to wound healing because it stimulates hemostasis and accelerates tissue regeneration (Hoekstra et al., 1998). For a material to be used for biomedical research, it is more preferable a natural product because these materials are more biocompatible than synthetic materials. Chitosan is metabolized by certain human enzymes, such as lysozyme. Thus, chitosan is biodegradable. Chitosan is an attractive material for a tissue engineering scaffold because it has structural similarities to glycosaminoglycans and is hydrophilic. Chitosan's monomeric unit, *N*-acetylglucosamine, occurs in hyaluronic acid, an extracellular macromolecule that is important in wound repair. An effective approach for developing a clinically applicable chitosan is to modify the surface of the material to provide excellent biofunctionality and bulk properties. Surface modification techniques to blend various compound derivatives include coating, oxidation by low-temperature plasma and surfactant addition in order to blend with various derivatives. Furthermore, chitosan can be fabricated into a stable, porous bioscaffold *via* surface modification and lyophilization. However, blending with various additives may affect its biocompatibility. Therefore, evaluation of the biocompatibility of various biomedical-grade chitosan derivatives is necessary to engineer material that is of high quality and biocompatible for human wound management.

2. Chitosan and its derivatives

The practical use of chitosan is restricted to its unmodified forms in wound management, mainly due to its insolubility in water, high viscosity and its tendency to coagulate with proteins at high pH. Recently, modification of chitosan has been found to improve solubility. The introduction of various chemical side chains provides desired properties and expands the potential applications for chitosan use. Alteration of the molecular weight forms water-soluble chitosans, such as the randomly 50% deacetylated and partially depolymerized chitosans. Chitosan purification from proteins, carotenoids and inorganics produces a product of technical, food, pharmaceutical and medical grade, which is approved for use in many countries. Alkali treatment of chitin removes protein and deacetylates the chitin. Some soluble glycans can also be removed depending on the alkali concentration (Madhavan, 1992). In particular, processing of crustacean shells involves the removal of proteins and the dissolution of calcium carbonate, which is abundant in crab shells. Deacetylation of chitosan in 40% sodium hydroxide at 120 °C for 3 hours is approximately 70% efficient. However, it is necessary to perform additional modification on these polymers to improve the chemical properties.

Chitosan has one amino group and two hydroxyl groups in the repeating hexosamide residue. Chemical modification of these groups during a regeneration reaction creates various novel biofunctional macromolecular products that have an original or novel organization. Hence, the bioactivities of chitosan unmodified and in formulation with various drugs may have dual therapeutic effects. In its crystalline form, chitosan is only soluble in an acidic aqueous medium (pH < 6), such as acetic acid, formic acid and lactic acid, in which solubility is conferred by the protonated free amino groups on the glucosamine. Another limitation of sustained release chitosan systems is that they rapidly adsorb water and have a high swelling degree in aqueous environments, which causes rapid drug release to occur. Therefore, several chemically modified chitosan derivatives have been synthesized and examined to improve solubility and versatility (Jayakumar et al., 2005; Prabaharan & Mano, 2007). Chemically modified chitosan structures may result in improved solubility in general organic solvents (Qurashi et al., 1992). For example, phosphorylated chitosan is a water-soluble derivative of chitosan, which is potentially important for drug delivery systems. Non-covalent cross-linking is a useful method to prepare hydrogels from polymers for drug delivery. These gels are likely biocompatible due to the absence of organic solvents. The use of organic solvents may potentially lower drug absorption.

Chitosan derivatives are easily obtained under mild conditions and can be rendered as substituted glucans. The nitrogen content of chitin varies from 5% to 8% depending on the extent of deacetylation, whereas the majority of nitrogen in chitosan is in the form of aliphatic amino groups. Hence, chitosan undergoes reactions that are typical of amines, such as *N*-acylation and the Schiff reaction. *N*-acylation with acid anhydrides or acyl halides introduces amino groups on nitrogens in chitosan. Acetic anhydride fully acetylates chitins. Linear aliphatic *N*-acyl groups above propionyl allow prompt acetylation of hydroxyl groups. At room temperature, chitosan is able to form aldimines and ketimines with aldehydes and ketones, respectively. A reaction with ketoacids, followed by a reaction with sodium borohydride produces glucans that have proteic and non-proteic amino groups. For example, non-proteic amino acid glucans derived from chitosan are the *N*-carboxybenzyl chitosans obtained from *o*- and *p*-phthalaldehydic acids (Madhavan, 1992). *N*-carboxymethyl chitosan is derived from glyocylic acid. Chitosan and simple aldehydes produce *N*-alkyl chitosan upon hydrogenation, where the presence of a bulky substituent can deteriorate the hydrogen bonds of chitosan. This compound swells in water in spite of the presence of hydrophobic alkyl chains (Muzarelli, 1973).

The Schiff reaction between chitosan and aldehydes or ketones yields the corresponding aldimines and ketimines, which can be converted to *N*-alkyl derivatives upon hydrogenation with borohydride. The film-forming ability of *N*-carboxymethyl chitosan imparts a pleasant feeling of smoothness to the skin and protects from adverse environmental conditions and consequences of detergent use. In addition, *N*-carboxymethyl chitosan is superior to hyaluronic acid in terms of its hydrating effects. Chitosan with several molecular designs, is presented when an alkyl or acyl chain is chemically introduced. For example, the introduction of an alkyl chain onto water-soluble modified chitosan (*N*-methylene phosphonic chitosan) introduces both hydrophobic and hydrophilic side chains. The presence of alkyl groups in *N*-lauryl-*N*-methylene phosphonic chitosan weakens the hydrogen bond and provides good solubility in organic solvents (Ramos et al., 2003). In addition, chitosan that is bound with sialic acid using *p*-formylphenyl-a-sialoside by reductive *N*-alkylation is a potent inhibitor of the influenza virus and is used as a blocking agent for acute rejection (Gamian et al., 1991).

Polyethylene glycol (PEG) is a water-soluble polymer that exhibits useful properties, such as protein resistance, low toxicity and immunogenicity. PEG is mixed with chitosan to produce the chitosan derivatives with improved biocompatibility. Chitosan-PEG enhances the protein adsorption, cell adhesion, growth and proliferation (Zhang et al., 2002). *N*-acylation of chitosan with various fatty acid (C_6-C_{16})) chlorides may increase its hydrophobic character and make important structural changes and can be used as a matrix for drug delivery (Tien et al., 2003). Furthermore, chitosan may also be grafted with biomolecules similar to the chitosan derivatives. The conjugation of lipid groups to the chitosan molecule creates an amphiphilic self-aggregate molecule that is useful for drug delivery systems. One example of this is palmitoyl glycol chitosan (GCP), which is prepared by reacting glycol chitosan and sodium bicarbonate with palmitic acid *N*-hydroxysuccinimide in an ethanol solution (Uchegbu et al., 2001). In a different approach, the reaction of the amino group on chitosan and the carboxylic acid group on amino acids with glutaraldehyde may attach various amino acids (lysine, arginine, phenylalanine and aspartic acid) to a chitosan molecule. These amino acid-functionalized chitosan moieties are entrapped on poly L-lactide (PLA) surfaces (Figure 2).

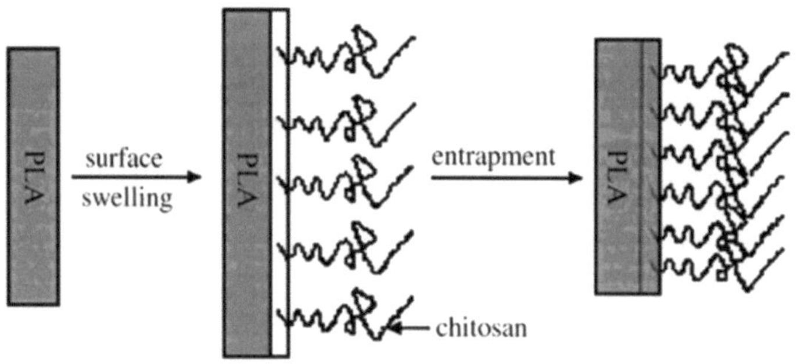

Fig. 2. Entrapment of functionalized chitosan on a PLA. In this process, the solvent swells the surface of the PLA to allow penetration of the amino acid-chitosan derivatives. These adhere upon addition of a non-solvent chitosan solution (Chung et al., 2002).

3. Chitosan in wound management

A wound is defined as the disruption of the anatomic structure and function of a body part. This may be the result of a simple cut, burns and any other injuries. Wounds are generally classified as wounds without tissue loss (*e.g.*, surgical incision) or wounds with tissue loss, such as burn wounds, wounds due to trauma, abrasions or secondary events to chronic ailments (*e.g.*, venous stasis, diabetic ulcers and iatrogenic wounds, such as skin graft donor sites and dermabrasions). In contrast, wound healing is a process of restoration by which tissue repair takes place and usually is comprised of a continuous sequence of inflammation and tissue repair during which epithelial, endothelial, inflammatory cells, platelets and fibroblasts briefly interact to restore normal function. The ordered sequence of healing events is accomplished and regulated by cytokines and growth factors. Soon after the elimination of macrophages, which appear during the inflammatory phase, wound healing is impeded and the tensile strength of the scar is diminished.

The use of chitosan has advantages due to the biocompatibility and biodegradability of the molecules, which does not harm the environment. When chitosan is applied to the body, besides being biocompatible, it is then slowly biodegraded by lysozymes, chitinase and chitosanase to harmless oligomers and monomers (amino sugars), which are completely absorbed by the body. Chitosan embodies analgesic, bacteriostatic and fungistatic properties, which are particularly useful for wound treatment. Additionally, chitosan modulates macrophage function and the secretion of numerous enzymes (*e.g.*, collagenase) and cytokines (*e.g.*, interleukins and tumor necrosis factor) during the wound healing process (Majeti & Ravi, 2000). The degradation of chitosan into monomers and oligomers at a wound site significantly accelerates the wound healing process (Minagawa et al., 2007). In addition, clinical studies have shown an absence of scar formation at the wound site in the presence of chitosan (Okamoto et al., 1993; Okamoto et al., 1995). Chitosan structurally resembles glycosaminoglycans (GAG), which have long-chain, unbranched, repeating disaccharide units and are important for maintaining cell morphology, differentiation and function (Nishikawa et al., 2000). GAG and proteoglycans are widely distributed throughout the human body and may bind and modulate numerous cytokines and growth factors, including heparin and heparan sulfate. Hence, the cell-binding and cell-activating properties of chitosan are crucial for wound healing.

Various chitosan derivatives have been produced for wound management, particularly to enhance wound healing. For example, oligo-chitosan (O-C) and *N, O*- carboxymethyl-chitosan (NO-CMC) derivatives have been fabricated into films for wound dressing (Lim et al., 2007). *N*-carboxybutyl chitosan has also been used in patients undergoing plastic surgery to promote tissue regeneration. The use of *N*-carboxybutyl chitosan improves cutaneous tissue regeneration with good histoarchitecture and vascularization at the wound site (Biagini et al., 1991). Additionally, 5-methylpyrrolidinone chitosan is compatible with other polymer solutions (*e.g.*, gelatin, polyvinyl alcohol, polyvinyl pyrrolidone, and hyaluronic acid), which are beneficial for the treatment of wounded meniscal tissues, decubitus ulcers, depression of capsule formation around prostheses, scar formation and retraction during wound healing (Muzzarelli, 1995).

3.1 Analgesic, antimicrobial and anti-inflammatory effects of chitosan in wound healing

Chitosan treatment reduces inflammatory pain due to intraperitoneal administration of acetic acid in a dose-dependent manner. Studies suggest that chitosan has potent analgesic actions. Bradykinin is one of the main substances related to pain. Okamoto et al. (2002)

reported that the bradykinin concentration during administration of a chitosan-acid acetic solution in the peritoneal lavage fluid was lower than during the administration of a 0.5% acetic acid solution, suggesting that chitosan has analgesic effects. Open wounds are often associated with severe pain in patients. Chitosan that is formulated for wound management may induce analgesia by providing a cool, pleasant and soothing effect when applied to an open wound. Excellent pain relief is conferred by chitosan when it is applied as a topical agent to open wounds, such as burns, skin abrasions, skin ulcers and skin grafted areas (Ohshima et al., 1987).

Chitosan-dependent antimicrobial activity has been observed against various microorganisms, such as fungi, algae and bacteria. These antimicrobial effects are controlled by intrinsic factors, including the type of chitosan, the degree of chitosan polymerization, the host, the natural nutrient constituency, the chemical or nutrient composition of the substrates and the environmental conditions (*e.g.*, substrate water activity or moisture or both). The antimicrobial activity of chitosan differs mainly in live host plants. For example, the fungicidal effects of *N*-carboxymethyl chitosan (NCMC) are different in vegetable and graminea hosts. The antimicrobial activity is more immediate on fungi and algae than on bacteria (Savard et al., 2002). Furthermore, in the presence of more than 0.025% chitosan, the growth of *Excherichia coli*, *Fusarium*, *Alternaria* and *Helminthosporium* is inhibited (Hirano, 1995). The cationic amino groups of chitosan bind to anionic groups in these microorganisms, resulting in growth inhibition. During the infectious period of a burn wound, bacterial infection may delay the healing and probably cause serious complications, such as sepsis. Chitosan that is incorporated with minocycline hydrochloride (CH-MH) was therefore developed to achieve both wound healing enhancement and antibacterial effects (Aoyagi et al., 2007).

Chitosan has anti-inflammatory effects that are beneficial for the treatment of prolonged inflammation at the wound site. Water-soluble chitosan (WSC) significantly suppresses the secretion and expression of proinflammatory cytokines (*e.g.*, tumor necrosis factor-α and interleukin-6) and inducible nitric oxide synthase (iNOS) in astrocytes, the predominant neuroglial cells in the central nervous system, and is actively involved in cytokine-mediated inflammatory events (Kim et al., 2002). Moreover, *N*-acetylglucosamine is an anti-inflammatory drug and is synthesized in the human body from glucose. It is incorporated into glycosaminoglycans and glycoproteins. Chito-oligosaccharides (COS), which have a molecular weight of 5 kDa, are better anti-inflammatory agents than indomethacin, a non-steroidal anti-inflammatory drug (Spindola et al., 2009). Chitosan exerts anti-inflammatory effects by inhibiting prostaglandin E_2 (PGE_2) and cyclooxygenase-2 (COX-2) protein expression and attenuating the pro-inflammatory cytokines (*e.g.*, tumor necrosis factor-α and interleukin-1β). However, chitosan treatment increases the expression of the anti-inflammatory cytokine, interleukin-10 (Chou et al., 2003).

3.2 Chitosan-based wound dressings

Wound dressings are generally classified by their mechanism of action. They are termed passive products, interactive products and bioactive products. Wound dressings before the 1960s were considered passive products minimally affected the wound healing process. Gauze and tulle dressings accounted for the largest market segment. Polymeric films, which are mostly transparent, permeable to water vapor and oxygen but impermeable to bacteria, are commonly recognized as interactive products. Bioactive dressings are important for the delivery of substances for wound healing, for which the delivery of bioactive compounds or dressings is constructed from material having endogenous activity, such as proteoglycans, collagen, non-collagenous proteins, alginates or chitosan. The pioneering research by Winter (1962) initiated the concept of a wound dressing that establishes an optimal environment for wound healing. Therefore, the development of wound dressings from traditional passive materials was replaced by active dressings that create and maintain a moist, healing environment. An ideal wound dressing must be biocompatible, able to protect the wound from bacterial infection and should provide a moist, healing environment (Purna & Babu, 2000).

Some wound dressings are prepared from aqueous solution of 5-methylpyrrolidinone chitosan, which is dialyzed and laminated between stainless steel plates and freeze-dried to yield fleeces. The material can be fabricated into many forms, such as nonwoven fabrics, filaments and so forth. 5-methylpyrrolidinone forms oligomers when applied to a wound site due to lysozyme-derpendent degradation. Flexible, thin, transparent novel chitosan-alginate polyelectrolyte complex (PEC) membranes accelerate the healing of incision wounds in a rat model in comparison to a conventional gauze dressing. The closure rate and appearance of wounds treated with a PEC membrane were comparable with wounds treated with Opsite (Wang et al., 2002). In addition, the chitosan-based Hyphecan cap is useful in the management of deepithelializing fingertip injuries, achieving shorter healing time (Halim et al., 1998). A chitosan bilayer derived from sulfadiazine has excellent oxygen permeability, water vapor transmission rate and water-uptake capability, which benefits the wound dressing (Mi et al., 2001). Chitosan complexed with gelatin has been useful as a surgical dressing. It is prepared by dissolving the chitosan in an acidic solution before addition to gelatin at a ratio of 3:1 chitosan and gelatin (Sparkes & Murray, 1986). The stiffness of the resulting chitosan-gelatin dressing is reduced by the addition of plasticizers such as glycerol and sorbitol. Additionally, chitosan gels may be used in surgery and dentistry as a biological adhesive to seal wounds and to improve wound healing.

3.3 Chitosan as a tissue engineering scaffold for artificial skin

Individuals who suffer from extensive skin loss are in danger of succumbing to either massive infection or severe fluid loss. Patients often cope with problems of rehabilitation arising from deep, disfiguring scars and crippling contractures. Tissue repair requires a complex biological process, where inward cell migration and the proliferation of various types of neighboring cells concertedly restores tissue function.

Tissue engineering is a recent, advanced technology to develop living tissue substitutes and replace diseased or damaged tissues and organs in the human body. Tissue engineering applies the development of polymeric scaffolds, that, among other characteristics, are biodegradable and biocompatible. These scaffolds may be used simultaneously as a carrier matrix for bioactive agents and as a support for primary undifferentiated cells *in vitro*. The three-dimensional (3D) framework of a scaffold must be able to promote adherence, proliferation and differentiation of cells, which ultimately are guided to form the desired tissues. Biological scaffolds are mostly biodegradable and biocompatible, and, with the appropriate growth factors, induce cell growth. In addition, a biological scaffold must also fill space with optimal mechanical strength and control the release of bioactive molecules.

Current tissue engineered systems cover every tissue and organ, with skin and cartilage constructs for repair of skin loss and joints already clinically performed (Pomahac et al.,

1998; Kuo et al., 2006). Acute, chronic, and more extensive wounds or skin loss would be inevitable unless some skin substitutes are applied. The primary role of skin substitutes is to promote wound healing by stimulating the host to produce various cytokines, which may promote the formation of granulation tissue during the wound healing process. Cultured skin from human cells is extremely thin and needs mechanical support from biopolymer complexes, such as collagen, fibrin or chitosan. Hence, skin tissue engineering produces a construct that offers the complete regeneration of functional skin. It restores normal functions, such as barrier formation, pigmentory defence against UV irradiation, thermoregulation and mechanical and aesthetic functions. During the past couple decades, xenografts, allografts and autografts have been used as skin substitutes for wound healing. However, skin substitutes occasionally do not provide skin recovery and cause antigenicity at the donor site. Therefore, these are not widely used.

In general, the substrate material upon which the cells are cultured enhances cellular organization in 3D and provides the initial mechanical integrity for the cell-polymer construct. Chitosan-based scaffolds are of current interest for tissue engineering because these natural products are mostly biocompatible and biodegradable. Moreover, the natural components of living structures have biological and chemical similarities to tissues, in which formation of the native extracellular matrix (ECM) is crucial. One of chitosan's most promising features is its excellent ability to form porous structures for use in tissue transplantation or as template for tissue regeneration. Chitosan scaffolds are commonly porous-structured by freezing and lyophilizing a chitosan solution (Figure 3). Alternatively, the creation of porous chitosan scaffolds may also be achieved through an internal bubbling process (IBP). In this process, calcium carbonate ($CaCO_3$) is added to a chitosan solution to generate a chitosan-$CaCO_3$ gels in a specific shape of a mold (Chow & Khor, 2000). The interconnected porous structure is crucial, and numerous cultured cells can be seeded onto it. Cells proliferate and migrate within the scaffold and ultimately form a tissue or organ.

Fig. 3. A chitosan porous skin regenerating template (CPSRT) produced by lyophilization process in a freeze dryer for 24 hours.

Because chitosan is positively charged, the negatively charged cell surface binds electrostatistically with chitosan and grows in the presence of a medium *in vitro* (Figure 4). For example, a CPSRT that is seeded with skin cells, such as keratinocytes or fibroblasts, may form a skin sheet-like tissue. However, regulation of porosity and pore morphology of a chitosan-based scaffold is particularly important to control angiogenesis, the cellular colonization rate and organization within an engineered tissue *in vitro*. The mechanical properties of chitosan scaffolds formed by the lyophilization technique are primarily dependent on pore size and pore orientation. Tensile testing of hydrated samples shows that porous membranes can greatly reduce elastic moduli compared to non-porous chitosan membranes. Their mean pore size is typically controlled by varying the temperature, whereas the pore orientation can be directed by controlling the geometry of the temperature gradients during freezing and thermal gradients. The freezing and lyophilizing process generates an open microstructure with a high degree of interconnectivity within the inner layer compared with that of the surface layer. In addition, chitosan-gelatin scaffolds have also been used to construct an artificial skin bilayer *in vitro* that consists of co-cultured keratinocytes and fibroblasts (Mao et al., 2003).

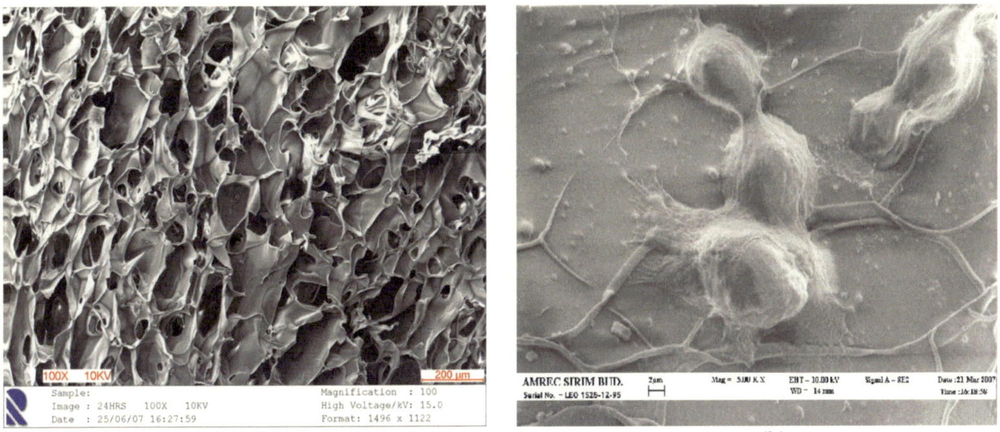

(a) (b)

Fig. 4. CPSRT viewed using a scanning electron microscopy (SEM). (a) Porous structures of a CPSRT without cultured cells. (b) Proliferating cells in the CPSRT.

3.4 Sterilization issues for chitosan as wound dressing

Chitosan products intended for parenteral administration or in contact with bodily fluids (*e.g.*, wounds) must be sterilized before use. Sterilization using dry heat, saturated steam autoclaving, ethylene oxide (EO) and gamma irradiation are among the current methods used for most pharmaceutical and medical products. It is often assumed that the existing sterilization technologies are adequate for chitosan material. A focus on the efficacy of the strerilization process in terms of killing microorganisms, the nature of the residuals formed and the properties of the chitosan must not be ignored. Deleterious effects of a sterilization method on the chitosan material should be minimal.

Before a sterilization method for chitosan products is approved, the effects of sterilization on the properties and performance of the biopolymer must be evaluated and documented. Sterilization methods either chemically or physically result in lethal alteration of the

structure or function of the biomolecule microorganisms. Therefore, various forms of sterilization may also affect the chitosan biopolymers by similar mechanisms, resulting in hydrolysis, oxidation, chain scission, depolymerization or cross-linking of the polymer. For example, saturated steam autoclaving may not be suitable to sterilize chitosan that is complexed with proteins, growth factors or enzymes. This is because the high temperature may completely denature the biomolecules and result in poor biopolymer performance. In addition, heat may alter the physical properties of chitosan, affecting its aqueous solubility, rheology and appearance. Exposure to dry heat resulted in lower chitosan aqueous solubility and insolubility in some acidic aqueous media (Lim et al., 1999). This may be related to the interchain crosslink formation that involves the amino (NH_2) group in chitosan, causing a reduction in the tensile strength and strain at the break point.
Gamma irradiation causes main chain scission events in chitosan (Lim et al., 1998). Irradiation with 2.5 Mrad in air improved the tensile strength of the chitosan film, which is probably due to changes in chain interaction and rearrangement. Additionally, gamma irradiation may have depolymerized chitosan at a radiation dose of 10 kGy (Yang et al., 2007). However, applying anoxic conditions during irradiation did not affect film properties, in part due to the pre-irradiation application of negative pressure that may minimally affect the structure of the chitosan film. Hence, gamma irradiation at 2.5 Mrad under anoxic conditions may provide suitable sterilization for chitosan products. In addition, sterilization using saturated steam autoclaving is recommended for chitosan products becasue it retains the tensile strength of the chitosan film (Rao & Sharma, 1997). Nonetheless, saturated steam autoclaving causes darkening of chitosan to a yellow color, which may result from the Maillard reaction between NH_2 and OH groups (Yang et al., 2007). However, sterilization of chitosan derivatives and porous-structured chitosan scaffolds using EO was also reported to retain the biocompatibility of the porous chitosan (Lim et al., 2007; Lim et al., 2010). Chitosan that is sterilized using EO must be quarantined and saline-irrigated prior to use to remove EO residues. Chitosan sterilized by EO that was quarantined under aeration for 10 days was void of EO residues. Additionally, the chemical properties and structure of chitosan were not affected after EO, as determined by Fourier transform infrared spectroscopy (FTIR) (Yang et al., 2007). Hence, the sterilization method used for chitosan derivatives may depend greatly upon the type of application.

4. *In vitro* biocompatibility evaluations of chitosan as a wound dressing

The application of *in vitro* model systems to evaluate toxicity significantly enhances our understanding of the mechanisms of drug- and chemical-induced toxicity. Biocompatibility of a biomaterial refers to the extent to which the material does not have toxic or injurious effects on biological systems. This means that patient's tissue, that comes into contact with the material does not suffer from any toxic, irritating, inflammatory and genotoxic effects. The United States Food and Drug Administration (FDA), the International Organization for Standardization (ISO) and the Japanese Ministry of Health and Welfare (JMHW) require that manufacturers conduct adequate safety testing of their finished devices through pre-clinical and clinical phases as part of the regulatory clearance process. *In vitro* models for testing the biocompatibility of chitosan and chitosan derivatives are useful to evaluate the toxicity, particularly from the leachability of chitosan as residual monomers or oligomers. Moreover, current *in vitro* toxicity models are preferred to *in vivo* models as the preliminary method to evaluate newly developed dressing materials. These models examine materials outside the body, and data are more reproducible. The use of these methods eliminates concerns about animal ethical issues and subsequently reduces the number of animals used in the *in vivo* biocompatibility tests. Because thousands of drugs and chemical compounds are synthesized every year, the cost of animal testing, which may be expensive, needs to be reduced. The use of animals to evaluate materials may also be very time consuming. Moreover, *in vivo* models are complicated due to the presence of structural and functional heterogeneity, and these models do not clearly define or evaluate drug mechanism.

Analysis of the effects of newly developed chitosan in cell culture systems is useful as a screening tool for their potential activity *in vivo* as wound healing agents. However, it is important to select appropriate cell lines for *in vitro* biocompatibility screening. If chitosan and its derivatives were meant to be used to treat bone injuries, osteoblast or chondrocyte cell cultures would be appropriate for the experiment. In addition, fibroblast and keratinocyte cell culture systems are more reasonable for biocompatibility *in vitro* experiments for chitosan in wound management. Various *in vitro* cell culture systems have been used to investigate and evaluate cellular processes, such as fibroblast and keratinocyte proliferation and cell migration toward growth factors present in a wound (Kawada et al., 1997). However, normal cell or non-transformed cell culture models are of particular interest for in vitro biocompatibility studies of cutaneous toxicity. Studies of new wound dressings, new drugs, cosmetic products and other chemicals require phenotypically normal cell systems. In addition, correlation of *in vitro* biocompatibility testing with *in vivo* irritation potential has been used with normal keratinocyte cultures rather than transformed cell lines (Korting et al., 1994). These *in vitro* models are simple methods to assess material biocompatibility and the ability of chemicals and biomaterials to promote cell proliferation during wound repair. Model compounds, which are known to be toxic (positive control) or non-toxic (negative controls), must be included to determine the validity of the *in vitro* system. The effects of unknown agents are compared to the effects of the controls. For example, organotin-polyvinylchloride (PVC) and high density polyethylene (HDPE) are used as positive and negative controls, respectively, in a direct-contact *in vitro* biocompatibility assay for chitosan wound dressing materials.

Numerous biocompatibility *in vitro* tests must be performed prior to the approval of chitosan products for human use. Otherwise, side effects and tissue toxicity will cause long-term effects, such as alteration of the immune system and development of malignancies, due to genetic damage induced by drug treatment. Optimal *in vitro* biocompatibility must mimic the biological response to materials when they are in contact with tissue. Therefore, *in vitro* biocompatibility testing of newly developed chitosan should measure cellular and molecular responses in cultured cells.

4.1 Cellular assessment: cytotoxicity *in vitro*

In vitro cytotoxicity is considered the most preliminary procedure in the biocompatibility *in vitro* assay. Cultured cells may undergo necrosis, a disruption of membrane integrity, or apoptosis (molecularly-controlled cell death) following treatment with cytotoxic compounds.

Cytotoxicity assays are commonly used to measure the response of cells to toxic substances. These measurements are of either an end-stage event (*e.g.*, permeability of cytoplasmic membranes of dead and dying cells) or some metabolic parameter (*e.g.*, cell division and enzymatic reaction). For example, trypan blue and propidium iodide dye exclusion assays are relatively simple assessment of cell membrane integrity, an end-stage event. These dyes are normally excluded from the interior of healthy cells. Cell membranes become compromised, when exposed to cytotoxic compounds, allowing trypan blue or propidium iodide dyes to cross the membrane and stain intracellular components. The staining is visible under light microscopy. 3-(4,5-Dimethylthiazol-2-yl)-2,5-diphenyltetrazolium bromide (MTT) and 3-(4,5-dimethylthiazol-2-yl)-5-(3-carboxymethoxyphenyl)-2-(4-sulfophenyl)-2H-tetrazolium (MTS) are substrates that are reduced enzymatically only in viable cells, to form formazan crystals, which are either dissolved in organic solvent (MTT assay), or water (MTS assay). The formation of formazan results in a purple color that is used as an indicator of viable cells. The absorbance of the colored solution is measured (*e.g.*, 500 to 600 nm) using a spectrophotometer to generate quantitative data. MTT is a proven, reliable and cost-effective method to measure cell viability *in vitro* (Lim et al., 2007; Keong & Halim, 2009; Lim et al., 2010). In addition, lactate dehydrogenase (LDH) is a stable cytoplasmic enzyme present in most cell types and is instantly released into cell culture medium upon rupture of the cell membrane. The LDH concentration in the medium is proportional to the number of dead or damaged cells. Measurement of LDH is a useful cell-mediated and drug-mediated cytotoxicity assay *in vitro*.

4.2 Genetic assessment: genotoxicity *in vitro*

Even though biocompatibility is typically measured by cytotoxicity, there is a growing concern that newly developed chitosan wound dressings may exert genotoxic effects. Hence, the safety assessment of a newly developed chitosan intended for body contact or permanent implantation would be incomplete without genotoxicity assays. When substances impose a genotoxic effect by damaging and mutating the deoxyribonucleic acid (DNA) of the cells, the growth of viable cells is abnormal or fully retarded.

The Ames test and the *in vitro* micronucleus assay are among the oldest genotoxicity assays and are commonly used. The Ames test, a salmonella point mutation assay that assesses five strains of bacterium *Salmonella typhimurium*, is used to determine the mutagenic potential of chemical compounds. The basis of the Ames test is that the salmonella bacteria cannot reproduce in growth medium unless bacteria undergo mutation. The Ames test is a high throughput genotoxicity screen that requires a small amount of test substance (approximately 2 mg). This test is relatively sensitive and accurate and is an immediate indicator of chemical mutagenic activity. The *in vitro* micronucleus assay, on the other hand, tests the effects of a compound on the induction of chromosomal breakage or clastogenesis. This assay evaluates the induction of micronuclei, which is the product of chromosomal breakage using Chinese hamster ovary (CHO) cells. Micronuclei occur in the cytoplasm following cell division. Therefore, the cellular replication kinetics and the percentage of binucleated cells with micronuclei are measured. This high throughput assay requires relatively little compound (3 mg). These early genotoxic tests may predict the toxicity potential and identify compounds for further tests.

The single-cell gel electrophoresis (SCGE) or Comet assay also has been a useful approach for assessing DNA damage. This technique is relatively more sensitive and is less expensive than other genotoxicity assays, that measure low levels of DNA damage. In addition, it requires few eukaryotic cells and test substance. The term "comet" describes the migration pattern of fragmented or unwound DNA caused by genotoxicity (Figure 5). The first comet assay was originally developed two decades ago using neutral conditions (Ostling & Johanson, 1984). In this method, cells are embedded in agarose and lysed by detergents. The liberated DNA is electrophoresed under neutral conditions. However, the measurement of the breaking potency (*e.g.*, comet tail length and comet tail intensity) is limited to DNA with double-strand breaks (DSB). Therefore, a more sensitive comet assay has been introduced, which uses alkaline electrophoresis conditions (pH > 13) to detect DNA damage in single cells (Singh et al., 1988). This new method not only detects DNA with DSB, but also detects single-strand break (SSB) and alkali-labile sites (ALS) which may result from genotoxic agents. Furthermore, the production of DNA strand breaks correlates with the mutagenic and carcinogenic properties of environmental pollutants (Mitchelmore & Chipman, 1998).

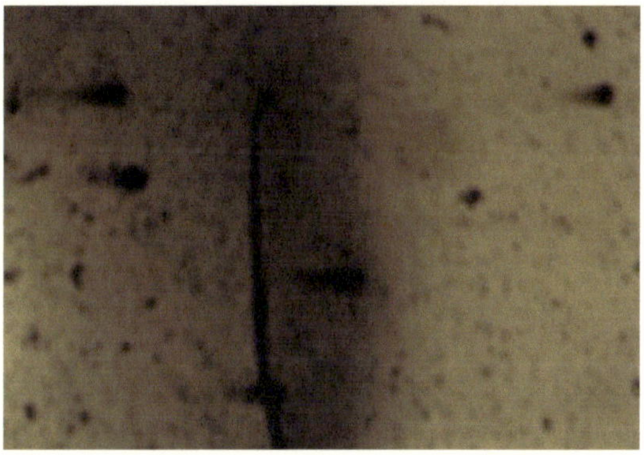

Fig. 5. Cultured human skin keratinocytes treated with a newly developed CPSRT that imposes a genotoxic effect, leading to DNA breakage. Subsequently, a comet-like shape is produced after electrophoressis at a constant voltage.

4.3 Human skin pro-inflammatory cytokine assessment: skin irritation *in vitro*

It is necessary not only to assess the cytotoxicity and genotoxicity of these materials, but also to determine the inflammatory potential *in vitro*. Despite being inert and non-toxic, newly developed chitosan wound dressings trigger adverse foreign body reactions, such as inflammation. Various cells in the dermis and epidermis are involved in these responses and they secrete various cytokines, particularly pro-inflammatory cytokines that cause skin irritation *in vitro*. Cytokines are low molecular weight glycoproteins that are produced by immune and non-immune cells. They are pleiotropic and interact with various receptors expressed on the surface of target cells. The binding of cytokines to cell surface receptors triggers intracellular signaling, protein synthesis and the production of other cytokines. The induction of other cytokines is regulated by autocrine, juxtacrine or paracrine pathways in response to micro-environmental stimuli. Cytokines mediate the interaction between various cells, and cytokine dysregulation indicates disease pathogenesis (Lazutka, 1996). Pro-inflammatory cytokines are detected at low levels in body fluids and in tissues under normal circumstances. Elevated expression may indicate activation of cytokine pathways associated with inflammation or disease progression.

The skin is the primary target tissue for exogenous noxes, which protect against harmful environmental hazards, UV-radiation and endogenous water loss. The skin epidermis consists mainly of keratinocytes, in which the cornified keratinoyctes in the outermost layer is an effective barrier against a vast number of substances. Upon stimulation, keratinocytes conduct immune surveillance of the epidermis and stimulate inflammatory responses (Steinoff et al., 2001). Harvell et al. (1995) defined skin irritation as a local, non-immunogenic inflammatory reaction that appears shortly after stimulation and usually disappears after a few days. Skin irritation is one of the most common adverse responses to cutaneous inflammation. The presence of erythema, oedema, dryness of the skin, fissures, desquamation, itching and pain are attributed to both irritant contact dermatitis and allergic contact dermatitis. Testing for skin irritation in animals can potentially cause them pain and discomfort. The results are not always reflective of effects in humans (Nixon et al., 1975; York et al., 1996). Hence, several alternative *in vitro* tests were developed, of which the *in vitro* reconstructed organotypic skin equivalents is the most favored, because of its resemblance to the structure of human skin. There are two different kinds of reconstituted skin equivalents available: 1) epidermal equivalents that consist of multilayered, differentiated human keratinocytes grown on a synthetic matrix, and 2) full skin equivalents consisting of multilayered, differentiated human keratinocytes grown on fibroblasts containing collagen matrices. Currently, various companies provide reconstituted human epidermal *in vitro* skin equivalents, such as EpiDerm (MatTek, Ashland, MA, USA), Episkin (Episkin, Chaponost, France), Apligraf (Organogenesis Inc., MA, USA) and Skinethic (Skinethic, Nice, France). In addition, because keratinocytes initiate and regulate skin irritation (Coquette et al., 2000), keratinocyte cultures may also serve as indicators for skin irritation *in vitro*.

Keratinocytes, the principle epidermal cell, is also a major contributor to epidermal cytokine production, a fact that is not well recognized by the immunology community (Tizard, 2000). Nevertheless, numerous cytokines are produced by keratinocytes (*e.g.*, Interleukin-1, -6, -7, -8, -10, -12, -15, -18 and -20 and tumor necrosis factor-α), either constitutively or upon induction by stimulants (Grone, 2002). These cytokines trigger multiple biological events, such as the migration of inflammatory cells, systemic effects on the immune system, keratinocyte proliferation and differentiation and the production of other cytokines. Lim et al. (2010) reported that non-biocompatible chitosan wound dressings increase the production of tumor necrosis factor-α and interleukin-8 in an experiment using keratinocyte cultures in an *in vitro* assay. Regardless of the chemical class or mechanism of drug action, the onset of skin irritation by chemicals and newly developed chitosan derivative wound dressings is in accordance with the general principles of toxicology. These biologic effects depend on various factors, such as the concentration of the test substance, the duration and frequency of exposure, the rate of penetration and the intrinsic toxic potential of the substance.

5. Conclusion

Chitosan and chitosan-based derivatives have various medical applications. It is well-known that chitosan possesses medicinal properties that accelerate wound healing and tissue regeneration. Chitosan is a natural product. It is biocompatible and biodegradable, enabling it to be used for wound dressing material. However, the practical use of chitosan is restricted to the unmodified forms, as these are water-insoluble and have high viscosity and the tendency to coagulate with proteins at high pH. Thus, chemical modification of chitosan may ultimately enhance its solubility and potential use for wound dressings. Chitosan has analgesic, antimicrobial and anti-inflammatory effects, which are beneficial for wound treatment. Chitosan is widely applied for the development of various chitosan-based wound dressings and biological scaffolds for tissue engineering. Nevertheless, each chitosan product that is intended for parenteral administration or for wound dressings comes in contact with bodily fluids, must be sterilized prior to application. The most frequently used sterilization methods are autoclaving, EO and gamma irradiation. The choice of sterilization method depends on the intended application. *In vitro* biocompatibility using newly developed chitosan wound dressings should be measured at the cellular and molecular level. These assays measure cytotoxicity, genotoxicity and skin irritation. *In vitro* model systems have made significant contributions to our understanding of the mechanisms of toxicity and carcinogenicity. They are indispensable resources to determine toxicology and identify potentially toxic compounds in chitosan wound dressings for human health risk assessment.

6. Acknowledgments

This work was supported by a grant (No.: 03-03-01-0000-PR0071/ 05) from the Intensification of Research in Priority Area Program (IRPA), Ministry of Science, Technology and Innovation (MOSTI) Malaysia and a Research University grant (1001/PPSP/812037) from Universiti Sains Malaysia.

7. References

Aoyagi, S., Onishi, H. & Machida, Y. (2007). Novel chitosan wound dressing loaded with minocycline for the treatment of severe burn wounds. *International Journal of Pharmaceutics* 330(1): 138-145.

Biagini, G., Bertani, A., Muzzarelli, R., Damadei, A., Di Benedetto, G., Belligolli, A., Riccotti, G., Zucchini, C. & Rizzoli, C. (1991). Wound management with *N*-carboxybutyl chitosan. *Biomaterials* 12(3): 281-286.

Chou, T.Z. C., Fu, E. & Shen, E.C. (2003). Chitosan inhibits prostaglandin E_2 formation and cyclooxygenase-2 induction in lipopolysaccharide-treated RAW 264.7 macrophages. *Biochemical and Biophysical Research communications* 308(2): 403-407.

Chow, K.S. & Khor, E. (2000). Novel fabrication of open-pore chitin matrixes. Biomacromolecules 1: 61-67.

Chung, T.W., Lu, Y.F., Wang, S.S., Lin, Y.S. & Chu, S.H. (2002). Growth of human endothelial cells on photochemically grafted Gly-Arg-Gly-Asp (GRGD) chitosans. *Biomaterials* 23: 4803-4809.

Coquette, A., Berna, N., Poumay, Y. & Pittelkow., M.R. (2000). The keratinocyte in cutaneous irritation and sensitization, *in* Kydonieus, A.F. & Wille, J.J. (ed.), *Biochemical modulation of skin reactions,* CRC Press, Boca Raton, F1, pp. 125-143.

Gamian, A., Chomik, M., Laferriere, C.A. & Roy, R. (1991). Inhibition of influenza A virus hemagglutinin and induction of interferon by synthetic sialylater glycoconjugates. *Canadian Journal of Microbiology* 37: 233-237.

Goosen, M.F.A. (1997). *Application of chitin and chitosan,* Technomic Publishing, Lancaster, USA, pp.320.

Grone, A. (2002). Keratinocytes and cytokines. *Veterinary Immunology and* Immunopathology 88: 1-12.

Halim, A.S., Stone, C.A. & Devaraj, V.S. (1998). The hyphecan cap: A biological fingertip dressing. *Injury* 29(4): 261-263.

Harwell, J.D., Lammintausta, K. & Maibach, H.L. (1995). Irritant contact dermatitis, *in* Guin, J.D. (ed.), *Practical contact dermatitis,* McGraw-Hill, New York, pp. 7-18.

Hirano, S. (1995). *In* Gabelein, C.G. & Carraherjr, C.E. (ed.), *Industrial Biotechnological Polymers*, Technomic Publishing, Lancaster, USA, pp. 189.

Hoekstra, A., Struszczyk, H. & kivekas, O. (1998). Percutaneous micro-crystalline chitosan application for sealing arterial puncture sites. *Biomaterials* 19: 1467-1471.

Jayakumar, R., Prabaharan, M., Reis, R.L. & Manao, J.F. (2005). Graft copolymerized chitosan-present status and applications. *Carbohydrate Polymers* 62(2): 142-158.

Kawada, A., Hiura, N., Shiraiwa, M., Tajima, S., Hiruma, M., Hara, K., Ishibashi, A. & Takahara, M. (1997). Stimulation of human keratinocyte growth by alginate oligosaccharides: A possible co-factor for epidermal growth factor in cell culture. *FEBS Letters* 408: 43-46.

Keong, L.C. & Halim, A.S. (2009). *In vitro* in biocompatibility assessment for biomedical-grade chitosan derivatives in wound management. *International Journal of Molecular Sciences* 10: 1300-1313.

Kim, M.S., Sung, M.J., Seo, S.B., Yoo, S.J., Lim, W.K. & Kim, H.M. (2002). Water-soluble chitosan inhibits the production of proinflammatory cytokine in human astrocytoma cells activated by amyloid-β peptide and interleukin-1β. *Neuroscience Letters* 321: 105-109.

Korting, H.C., Herzinger, T., Hartinger, A., Kersher, M., Angerpointner, T. & Maibach, H.I. (1994). Discrimination of the irritancy potential of surfactants *in vitro* by two cytotoxicity assays using normal human keratinocytes, HaCaT cells and 3T3 mouse fibroblasts: Correlation with *in vivo* data from a soap chamber assay. *Journal of Dermatological Science* 7: 119-129.

Kumar, M.N.V.R. (2000). A review of chitin and chitosan applications. *Reactive and* Functional Polymers 46: 1-27.

Kuo, C.K., Li, W.J., Mauck, R.L. & Tuan, R.S. (2006). Cartilage tissue engineering: Its potential and uses. *Current Opinion in Rheumatology* 18(1): 64-73.

Lazutka, J.R. (1996). Genetic toxicity of cytokine. *Mutation Research* 361 (2-3): 95-105.

Li, Q., Dunn, E.T., Grandmaison, E.W. & Goosen, M.F.A. (1997). Applications and properties of chitosan, *in* Goosen, M.F.A. (ed.), *Applications of chitin and chitosan*, Technomic Publishing, Lancaster, USA, pp. 3-29.

Lim, C.K., Halim, A.S., Lau, H.Y., Ujang, Z. & Hazri, A. (2007). *In vitro* cytotoxicology model of oligo-chitosan and *N, O*-carboxymethyl chitosan using primary normal human epidermal keratinocytes cultures. *Journal of Applied Biomaterials and Biomechanics* 5: 82-87.

Lim, C.K., Yaacob, N.S., Ismail, Z. & Halim, A.S. (2010). *In vitro* biocompatibility of chitosan porous regenerating templates (PSRTs) using primary human skin keratinocytes. *Toxicology In Vitro* 24: 721-727.

Lim, L.Y., Khor, E. & Koo, O. (1998). Gamma irradiation of chitosan. *Journal of Biomedical* Materials Research 43: 282-290.

Lim, L.Y., Khor, E. & Ling, C.E. (1999). Effects of dry heat and saturated steam on the physical properties of chitosan. *Journal of Biomedical Materials Research* 48: 111-116. Madhavan, P. (1992). *Chitin, chitosan and their novel applications*, Science Lecture Series, CIFT, Kochi, pp.1.

Majeti, N.V. & Ravi, K. (2000). A review of chitin and chitosan applications. *Reactive* and Functional Polymers 46: 1-27.

Mao, J., Zhao, L., Yao, K. D., Shang, Q., Yang, G. & Cao, Y. (2003). Study of novel chitosan-gelatin artificial skin in vitro. Journal of Biomedical Materials Research 64: 301-308.

Mi, F.L., Shyu, S.S., Wu, Y.B., Lee, S.T., Shyong, J.Y. & Huang, R.N. (2001). Fabrication and characterization of a sponge-like asymmetric chitosan membrane as a wound dressing. *Biomaterials* 22(2): 165-173.

Minagawa, T., Okamura, Y., Shigemasa, Y., Minami, S. & Okamoto, Y. (2007). Effects of molecular weight and deacetylation degree of chitin/ chitosan on wound healing. *Carbohydrate Polymers* 67: 640-644.

Mitchelmore, C.L. & Chipman, J.K. (1998). DNA strand breakage in aquatic organisms and the potential value of the comet assay in environmental monitoring. *Mutation Research* 399: 135-147.

Muzzarelli, R.A.A. (1973). *Natural chelating polymers*, Pergamon Press, New York, pp. 83.

Muzzarelli, R.A.A. (1995). Methyl pyrrolidinone chitosan, production process and uses thereof. US Patent 5378472.

Muzzarelli, R.A.A. (1996). Chitosan-based dietary foods. *Carbohydrate Polymers* 29: 309-316.

Nishikawa, H., Ueno, A., Nishikawa, S., Kido, J., Ohishi, M., Inoue, H. & Nagata, T. (2000). Sulfated glycosaminoglycan synthesis and its regulation by transforming growth factor-beta in rat clonal dental pulp cells. *Journal of Endodontics* 26(3): 169-171.

Nixon, G.A., Tyson, C.A. & Wertz, W.C. (1975). Interspecies comparisons of skin irritancy. Toxicology and Applied Pharmacology 31: 481-490.

Okamoto, Y., Kawakami, K., Miyatake, K., Morimoto, M., Shigemasa, Y. & Minami, S. (2002). Analgesic effects of chitin and chitosan. *Carbohydrate Polymers* 49: 249-252.

Okamoto, Y., Minami, S., Matsuhashi, A., Sashiwa, H., Saimoto, H., Shigemasa, Y., Tanigawa, T., Tanak, Y. & Tokura, S. (1993). Application of polymeric N-acetyl-D-glucosamine (chitin) to veterinary practice. *Journal of Veterinary Medical Science* 55(5): 743-747.

Okamoto, Y., Tomita, T., Minami, S., Matsuhashi, A., Kumazawa, N.H., Tanioka, S. & Shigemasa, Y. (1995). Effects of chitosan on experimental abscess with *Staphylococcus aureus* in dogs. *Journal of Veterinary Medical Science* 57(4): 765-767.

Ohshima, Y., Nishino, K., Yonekura, Y., Kishimoto, S. & Wakabayashi, S. (1987). Clinical application of chitin non-woven fabrics as wound dressing. *European Journal of Plastic Surgery* 10: 66-69.

Ostling, O. & Johanson, K.J. (1984). Microelectrophoretic study of radiation-induced DNA damages in individual mammalian cells. *Biochemical and Biophysical Research Communications* 123: 291-298.

Peter, M.G. (1997). Introductory remarks. *Carbohydrates in Europe* 19: 9-15.

Pomahac, B., Svensjo, T., Yao, F., Brown, H. & Eriksson, E. (1998). Tissue engineering of skin. Critical Reviews in Oral Biology and Medicine 9(3): 333-344.

Prabaharan, M. & Mano, J.F. (2007). Synthesis and characterization of chitosan-graft-poly (3-trimethoxysilyl) propyl methacrylate initiated by ceric (IV) ion. *Journal of Macromolecular Science Part A: Pure and Applied Chemistry* 44(5): 489-494.

Purna, S.K. & Babu, M. (2000). Collagen based dressings- A review. *Burn* 26: 54-62.

Qurashi, M.T., Blair, H.S. & Allen, S.J. (1992). Studies on modified chitosan membranes. II. Dialysis of low molecular weight metabolites. *Journal of Applied Polymer Science* 46: 263-269.

Rao, S.B. & Sharma, C.P. (1997). Use of chitosan as a biomaterial: Studies on its safety and hemostatic potential. Journal of Biomedical Materials Research 34(1): 21-28.

Ramos, V.M., Rodriguez, N.M., Rodriques, M.S., Heras, A. & Agullo, E. (2003). Modified chitosan carrying phosphonic and alkyl groups. *Carbohydrate Polymers* 51(4): 425-429.

Savard, T., Beaulieu, C., Boucher, I. & Champagne, C.P. (2002). Antimicrobial action of hydrolyzed chitosan against spoilage yeasts and lactic acid bacteria of fermented vegetables. *Journal of Food Protection* 65(5): 828-833.

Singh, N.P., McCoy, M.T., Tice, R.R. & Schneider, E.L. (1988). A simple technique for quantitation of low levels of DNA damage in individual cells. *Experimental Cell research* 175: 184-191.

Skjak-Braek, G., Anthonsen, T. & Sandford, P. (1989). *Chitin and chitosan*, Elsevier Applied Science, London, pp. 560.

Sparke, B.G. & Murray, D.G. (1986). Chitosan based wound dressing materials. *US Patent* 4572906.

Spindola, H., Fernandes, J., De Sousa, V., Tavaria, F., Pintado, M., Malcata, X. & Carvalho, J.E. (2009). Anti-inflammatory effect of chitosan oligomers. *New Biotechnology* 25(1): S9.

Steinhoff, M., Brzoska, T. & Luger, T.A. (2001). Keratinocytes in epidermal immune responses. Current Opinion in Allergy and Clinical immunology 1(5): 469-476.

Tien, C.L., Lacroix, M., Szabo, P.I. & Mateescu, M.A. (2003). *N*-acylated chitosan: Hydrophobic matrices for controlled drug release. *Journal of Controlled Release* 93(1): 1-13.

Tizard, I.R. (2000). Immunity at body surface, *in* Tizard, I.R. (ed.), *Veterinary Immunology*, Saunders, Philadelphia, P.A., pp. 222-234.

Uchegbu, I.F., Sadig, L., Arastoo, M., Gray, A.I., Wang, W., Waigh, R.D. & Schatzleina, A.G. (2001). Quaternary ammonium palmitoyl glycol chitosan. A new polysoap for drug delivery. *International Journal of Pharmaceutics* 224(1): 185-199.

Wang, L., Khor, E., Wee, A. & Lim, L.Y. (2002). Chitosan-alginate PEC membrane as a wound dressing: Assessment of incisional wound healing. *Journal of Biomedical Materials Research* 63(5): 610-618.

Winter, G.D. (1962). Formation of the scab and the rate of epithelialization of superficial wounds in the skin of the young domestic pig. *Nature* 193: 293-294.

Yang, Y.M., Zhao, Y.H., Liu, X.H., Ding, F. & Gu, X.S. (2007). The effect of different sterilization procedures on chitosan dried powder. *Journal of Applied Polymer Science* 104: 1968-1972.

York, M., Griffiths, H.A., Whittle, E. & Basketter, D.A. (1996). Evaluation of a human patch test for the identification and classification of skin irritation potential. *Contact Dermatitis* 34: 204-212.

Zhang, M., Li, X.H., Gong, Y.D., Zhao, N.M. & Zhang, X.F. (2002). Properties and biocompatibility of chitosan films modified by blending with PEG. *Biomaterials* 23: 2641-2648.

3

Biopolymer-Based Stimuli-Responsive Polymeric Systems for Functional Finishing of Textiles

Dragan Jocić, Audrey Tourrette and Pavla Križman Lavrič
University of Twente
The Netherlands

1. Introduction

The technological developments of the first decade of 21st century are slowly changing the way we live by controlling our surroundings and regulating our every day life with intelligent objects. *Smart Materials* and *Intelligent Structures* are novel disciplines which are currently rapidly growing into an interdisciplinary technology. This new technology is being incorporated in contemporary engineering and design with the aim to create the path for materials to gain "intelligent" features. Textile materials also benefit from the rapid advances in a new interdisciplinary technology. Development of textile materials with new advanced functionalities is the perfect example where these base technologies can be brought together by using the knowledge involving *surface science* and *surface engineering* at molecular and atomic level. This knowledge is being responsible for developing and creating a new generation of so-called "smart" textile materials. By redesigning textile material surface, operating at microscopic level, many new possibilities emerge for adapting the macroscopic properties of the material to the present needs of the textile industry and thus fulfil current and future end-user expectations. In this context, this book chapter will focus on an innovative strategy for functional finishing of textile materials by application of *surface modifying systems* (SMS) based on *stimuli-responsive polymers.*

2. Stimuli-responsive surface modifying systems for textiles

Among wide variety of materials that nowadays surround us, textile material is considered unique because we have a natural relationship with it. Textile material for clothing is an example of a material which is personal, comfortable, and used almost anywhere and anytime. Clothing is considered as an extension to our physiological characteristics and is very close to the body, so the eternally existing goal was to make clothing material that possesses the extension of qualities of human skin (protection, breathability, sensing). Therefore, gaining "smart" capability for such a material would be the best example of almost always being surrounded by the object that could sense and response to the changes in the surrounding (human and/or ambient) conditions. The main feature of "smart" textile materials is that they could adapt effectively (significantly and automatically) to their local environment by changing the properties due to defined influences (stimuli) from the immediate surrounding. The functional activity of these materials is an important aspect.

“Smart” textiles are expected to act both as sensors and actuators, so they should not be confused with other existing high-performance or multifunctional textiles that are in fact "passive" materials with advanced properties.

With enormous growth in supporting technologies, primarily in the areas of responsive polymers and surface modification techniques, during the last decade the functionalization of textile material has gone quite a few paces forward. However, by being one of the oldest known technologies, traditional textile technology is rather reluctant to essential technology changes, but it is prepared to accept the modifications being based on a number of production steps. In creating paths for such “readily acceptable” solutions, *functional finishing* is one of the most viable alternatives. Functional finishing approach enables producers to continue to use conventional textile fibres and at the same time, by modifying a very thin surface layer of the material, achieve added-value by implementing “smart” features.

In recent times, an increasing amount of research is being done on functional finishing of textile materials by incorporating stimuli-responsive polymeric systems (Liu & Hu, 2005; Crespy & Rossi, 2007; Jocic, 2008). Through this approach, the new added-value textile material can be created containing fibres that maintains advantageous conventional properties (e.g. mechanical strength, flexibility and wear comfort) but with advanced functionalities and/or environmental responsiveness implemented by the modification of a very thin surface layer of the material.

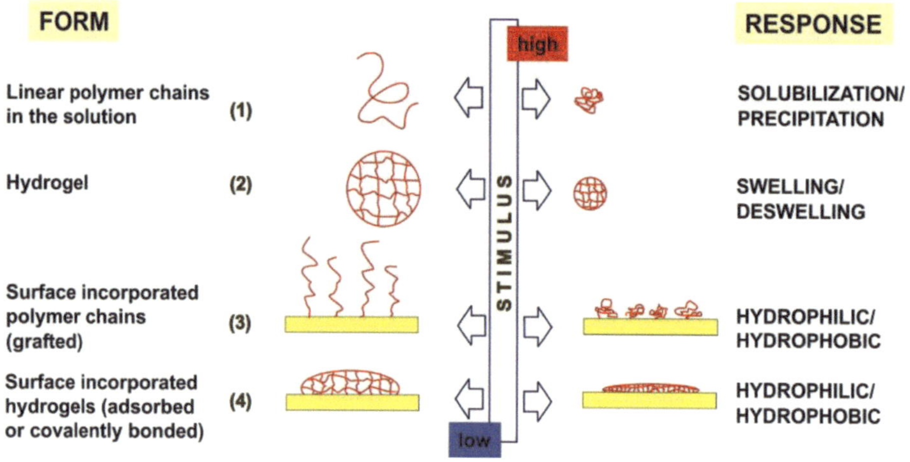

Fig. 1. Different forms of stimuli-responsive polymers and their macroscopic response.

Currently, the most encouraging option for producing efficient surface modifying systems comprises the use of *hydrogels*. This polymeric form exhibits specific volume phase-transition (swelling and shrinking) properties which can be triggered by various stimuli (temperature, pH, humidity, etc.). By incorporation of responsive hydrogels to the surfaces (i.e. textile material), the surface energy or some other property of a material can be switched (on/off). Most often, the switching is between hydrophilic/hydrophobic (Fig. 1).

Hydrogels responsive to temperature and pH have been the most widely studied systems since these two factors have a physiological significance. Versatile dual responsive hydrogels have been reported mainly for biomedical applications and a number of reviews coming up in this area in recent times address the latest developments (Kumar et al., 2007; Mano, 2008; Kopecek & Yang, 2007). However, due to the need for biocompatibility and biodegradability, *biopolymer* based hydrogels are currently of great interest. Such hydrogels can be prepared by combining a thermoresponsive synthetic polymer with a natural based pH-responsive polymeric component, resulting in dual responsive hydrogel systems (Prabaharan & Mano, 2006).

Among the wide choice of natural polymers, biopolymer *chitosan* is a good option for combining with synthetic stimuli-responsive polymers. Chitosan is a typical pH-sensitive polymer (pK_a=6.3) which responds to the changes in the pH of the surrounding medium by protonation/deprotonation that imparts charges on its amino groups (Ravi Kumar, 2000; Rinaudo, 2006). The pH-induced phase transition results in varying dimensions of the hydrogel (swelling and deswelling). Furthermore, the interesting intrinsic properties of chitosan are its biodegradability, antibacterial activity and biocompatibility, thanks to which this biopolymer attracts currently a great deal of interest for biomedical applications.

Among synthetic polymers, *poly(N-isopropylacrylamide)* (poly-NiPAAm) is the most intensively investigated thermoresponsive polymer which exhibits a volume phase-transition (i.e. hydration-dehydration change due to side-chain re-configuration) in response to even slight temperature changes. The coil-to-globule transition, which is a consequence of the rather complex polarity of the molecule, occurs at a temperature around 32°C, named lower critical solution temperature (LCST) (Schild, 1992). Below the LCST, the amide group binds water molecules via hydrogen bonding (i.e. it hydrates to form an expanded structure); above the LCST hydrogen bonds break and the polymer expels water and precipitates (i.e. its chains dehydrate to form a shrunken structure).

3. Biopolymer-based microgel as surface modifying system

The main challenge in achieving the goal of effective functional finishing of textiles with hydrogels lay in the fact that surface modifying system, after its incorporation to the textile material, must exhibit responsive properties without screening the textile's regular performance. Taking in account above consideration, it can be easily concluded that bulk (continuous; macroscopic) hydrogels are not suitable form for functional finishing of textiles, because of slow response time and adverse influence on "textile" properties (flexibility, drapeability, touch, etc.) of the material. Hence, the most modern approach for functional finishing with hydrogels is based on micro- and nano-sized hydrogels, because this polymeric form possesses increased surface area per unit mass and significantly improved response times (Pelton, 2000). Moreover, the submicron particle size of microgels enables their incorporation to textile material surface in a very thin layer, without degrading "textile" properties of the material.

3.1 Synthesis of poly-NiPAAm/chitosan microgel (PNCS)

Microgel is a dispersion of crosslinked hydrogel particles which are swollen by a good solvent. It may also be defined as a disperse phase of discrete polymeric gel particles with sizes ranging between 1 nm and 1 µm. A number of other terms, in addition to the term "microgel", are used in the literature to describe small crosslinked particulate systems, such as "hydrogel microparticles", "hydrogel microspheres", "submicron gel beads", "micro-particulate hydrogels" or "ultrafine microspheres". Microgel has similar properties to both polymers and water-swollen gels (bulk or continuous hydrogels). However, microgel consists of discrete particles whose characteristics are dependent on the method of synthesis, crosslinking density, monomer concentration, monomer composition and solvency conditions. Microgel particles are insoluble and do not form solutions like linear or branched polymers, but they may be considered to form colloidal dispersions.

The preparation of microgel can be achieved by different methods such as: emulsion polymerization; anionic copolymerization; crosslinking of neighbouring polymeric chains; inverse micro-emulsion polymerization or surfactant-free dispersion polymerization (SFDP). Emulsion polymerization is a versatile technique which yields narrow particle size distributions. Conventional emulsion polymerization enables preparation of very small microgel particles (i.e. particle diameters less than 150 nm) and suffers from the difficulty of completely removing the residual surfactant used for emulsion stabilization. Surfactant-free dispersion polymerization (SFDP) yields microgel particles with diameter range between 100 and 1000 nm, and this method does not suffer from residual surfactant contamination. Some authors call this procedure "precipitation polymerization" or "surfactant-free precipitation polymerization (SFPP)". The continuous phase must have a high dielectric constant (e.g. water) and ionic initiators are employed. Generally, the polymerization is conducted at 50-60°C in order to generate free radicals by the decomposition of the initiator (e.g. ammonium persulfate - APS). Also, in case where thermoresponsive polymer chains such as poly-NiPAAm chains are present, elevated temperature is required to ensure that growing chains phase separate to form colloidal particles. The key feature of SFDP is that the particle nucleation period is very short (of the order of minutes), which ensures a narrow particle size distribution.

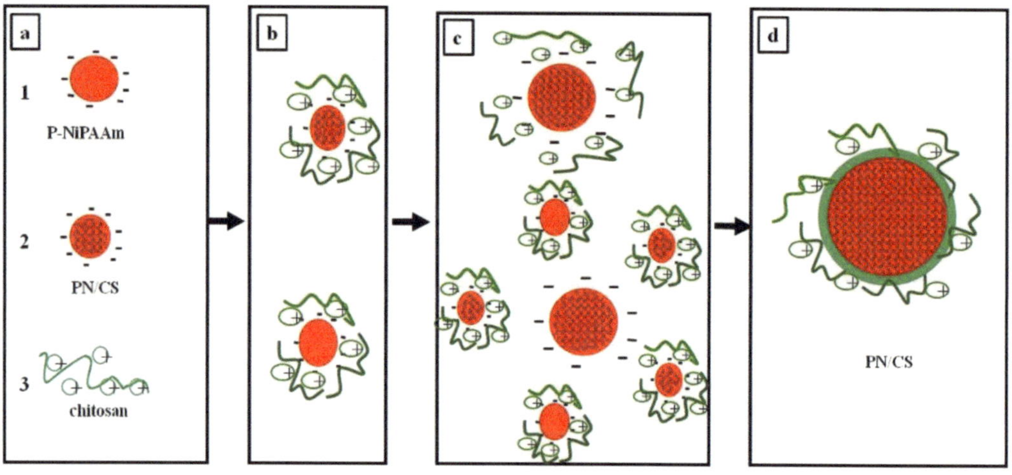

Fig. 2. Synthesis of poly-NiPAAm/chitosan microgel (SFDP mechanism).

The microgel of poly-NiPAAm and chitosan (PNCS), used in our research, was prepared by the surfactant-free dispersion copolymerization method according to the previously reported procedure (Lee et al., 2003; 2004). Three kinds of reactions occur in the reaction system of dispersion copolymerization (Fig. 2). The first reaction is the polymerization of NiPAAm, initiated by APS, which forms negatively charged particles (Pelton & Chibante, 1986) (Fig. 2, a-1). The second reaction is the graft copolymerization of NiPAAm and chitosan, initiated by APS which forms poly-NiPAAm/chitosan complex particles with a negative surrounding charge (Fig. 2, a-2). The last reaction is the ionization of chitosan which results in positively charged chitosan molecules (Fig. 2, a-3). They might surround the negatively charged poly-NiPAAm or PNCS particles and behave like a surfactant to protect the polymer particles and thus avoid the coagulation (Fig. 2, b). If chitosan molecules are not in vicinity of the particles formed, the other possibility is that the particles coagulate to form larger positively charged particles, until the chitosan molecules are close enough to act like surfactant and prevent coagulation (Fig. 2, c-d).

There are several kinds of possible interactions between chitosan and poly-NiPAAm, such as electrostatic and hydrogen interactions (between chitosan amino and poly-NiPAAm amide groups), as well as covalent bonding due to the formation of radicals. The radical degradation of chitosan is the decisive step in the mechanism of copolymerization reaction. Hsu et al. (2002) showed that the thermal decomposition of an initiator similar to APS (potassium persulfate - KPS) produced free radicals which not only initiate the polymerization, but also break the backbone of chitosan molecules. On the other hand, it has been shown (Harish Prashanth & Tharanathan, 2003) that in free radical initiated copolymerization, NH_2 groups of chitosan were involved in microradical formation, which can interact with electrophilic functional group to create covalent bonding. In addition, Liu et al. (2007) stated that in the chemical structure of the grafting copolymers poly-NiPAAm was attached to the C_6-OH reactive group of chitosan and to amino group. In summary, the covalent bonding between chitosan and poly-NiPAAm can be created from different chitosan functional groups: the terminal carbonyl groups of its backbone, its amino groups or its C_6-OH reactive groups. The possibility of anionic persulfate (from the initiator) immobilization inside degraded chitosan chains through electrostatic attraction has been also mentioned (Hsu et al., 2002). Hence, chitosan plays multiple roles in the reaction system: in one way it can increase the polymerization rate by serving as a surfactant; on the other hand, the degraded chitosan chain can inhibit free radicals and slow down the polymerization.

3.2 Physicochemical properties of PNCS microgel

3.2.1 Characterization

A SEM image of dry poly-NiPAAm/chitosan microgel, deposited from a solution onto native oxide layer of a silicon wafer, is shown in Fig. 3. The photomicrograph shows that the microgel particles have a homogeneous spherical shape and that the particle size distribution is quite narrow. The diameter in dry state is estimated at 180 nm.

Since microgel particles are incorporated on textile material not in dry state but in their hydrated state, DLS measurements of diluted PNCS dispersion were performed to determine the particle size (Tourrette et al., 2009). The result was a first proof of the temperature response of the microgel, as the mean microparticle diameter below LCST (426 nm at 25°C) was 2.5 times bigger than above the LCST (166 nm at 38°C). However, the particle size in totally collapsed state was found to be smaller than the one observed in the dry state on SEM picture (Fig. 3). This can be explained by the fact that for SEM analysis the microgel dispersion was deposited onto a native oxide layer of silicon wafer which is hydrophobic. Consequently, the microgel particles interacted strongly with the substrate and did not shrink as much during drying as after collapsing in liquid medium. This finding also shows that the micro-particulate network is not expected to be totally collapsed in dry state, which means that some water remains in the microgel internal structure after drying.

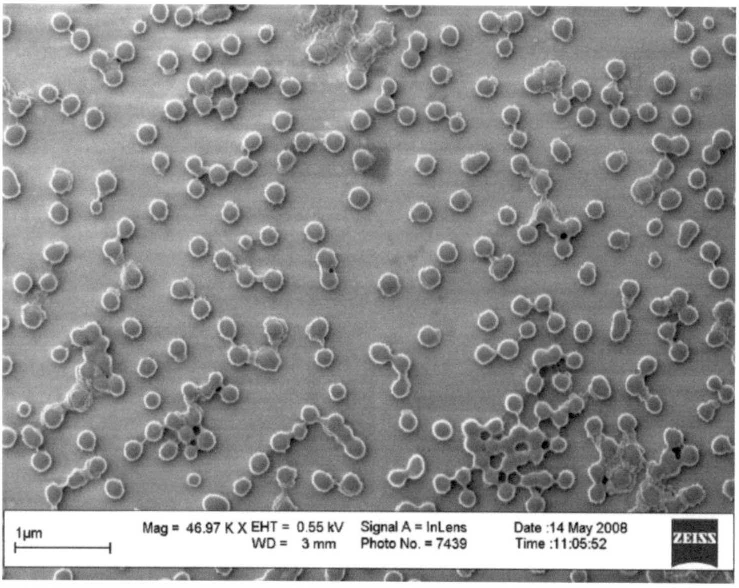

Fig. 3. SEM image of poly-NiPAAm/chitosan microgel particles (PNCS) deposited from a diluted (3.56·10^{-2} g/L) microgel dispersion onto a silicon wafer.

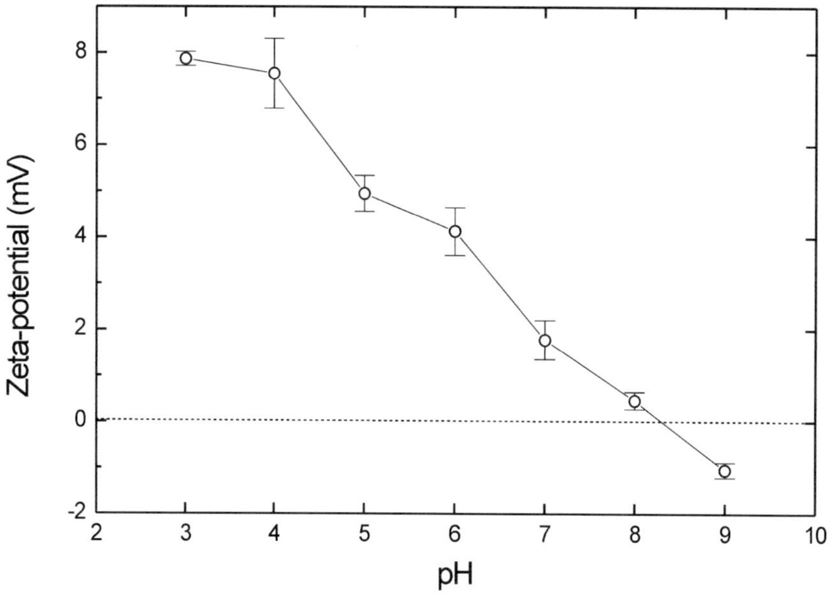

Fig. 4. Variation of zeta-potential with pH for aqueous dispersion of poly-NiPAAm/ chitosan microgel dispersion (PNCS) (1.1·10^{-1} g/L) at 25°C.

The variation of the zeta-potential with the pH change of aqueous dispersion of PNCS microgel is shown in Fig. 4. Zeta-potential measured at low pH values (pH 3) has clearly positive value. With pH increase, zeta-potential decreases and reaches negative values above pH 8. The observed behaviour can be explained through chitosan pH-responsiveness.

In acidic solutions chitosan behaves as a cationic polyelectrolyte due to protonation of the amino groups. The pKa value of chitosan is 6.3 (as already stated before), but it is known that at pH 6.9 the amino groups ($-NH_2$) are still partly (~20%) protonated ($-NH_3^+$) (Muzzarelli, 1977). Therefore, at weak alkaline values (pH 7-8) chitosan is expected to have a very low positive charge. Since chitosan is pH-responsive and the change in its positive charge is detected by zeta-potential analysis, this confirms that chitosan is adsorbed at the surface of microparticles.

3.2.2 Stability

The stability and physical state (hydrophilic or hydrophobic character) of PNCS microgel depends on both temperature and pH. Among all possible interactions between the particles, three of them play a decisive role in microgel stability: electrostatic interaction; van der Waals forces; and steric forces. Hence, the stability of PNCS microgel will depend on the balance of the repulsive and attractive forces that exist between particles as they approach one another. If the particles have a mutual repulsion then they will remain dispersed. However, if the particles have little or no repulsive force, then flocculation, aggregation or coalescence will eventually take place. PNCS microgel stability is controlled by the balance between van der Waals attraction force, which causes aggregation, and steric or electrostatic forces that oppose aggregation. However, below the LCST, thermoresponsive particles are swollen and thus consist mainly of water. Under these conditions, the van der Waals attractive forces are relatively weak. Moreover, polymer tails extend from the gel structure to act as steric stabilizers, further enhancing microgel stability. Charged groups (from chitosan protonated amino groups) incorporated into the particles during polymerization, create electrostatic interactions which contribute to microgel stability. At elevated temperatures the water content of the gels is reduced giving a higher density, thereby increasing the van der Waals forces, which in turn tends to aggregate the microgel.

3.2.3 Stimuli-response

In order to study the stimuli responsive properties (swelling/shrinking behaviour) of PNCS, dynamic light scattering (DLS) analysis was used to study the microgel particle size at different temperatures and pH.

Fig. 5a presents the hydrodynamic diameter of PNCS microgel particles at different pH values. It shows that particle size is very sensitive to pH change as a result of protonation/deprotonation of chitosan. At pH 3 the hydrodynamic diameter is noticeably smaller compared to the values obtained at other pH values, which can be explained by previously reported (Pei et al., 2004) significant pH-dependence of poly-NiPAAm LCST value (it decreases!) in strong acidic medium. Another unexpected result is that the average hydrodynamic diameter of PNCS particles is considerably higher at pH 9, even though this pH value is far above pKa of chitosan (6.3). This result may be attributed to the instrument (DLS) limitation as the consequence of the averaging of a combination of several size distributions which could exist at this particular pH value due to the aggregation of the particles. Indeed, in this case chitosan is not protonated and the particles are slightly negatively charged (Fig. 4). Fig. 5b shows the hydrodynamic diameter of PNCS microgel particles as a function of temperature. The decrease of hydrodynamic diameter of microparticles with an increase of temperature proves their thermoresponsiveness. Polymer network of microgel shrinks gradually with an increase of temperature. Nevertheless, the change of the slope could be noticed around 33°C, which is close to the LCST of poly-NiPAAm.

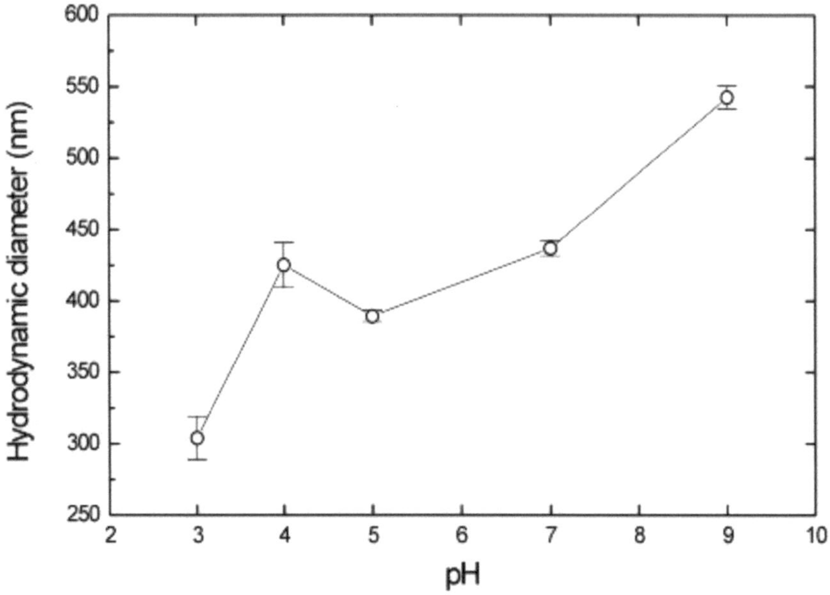

(a)

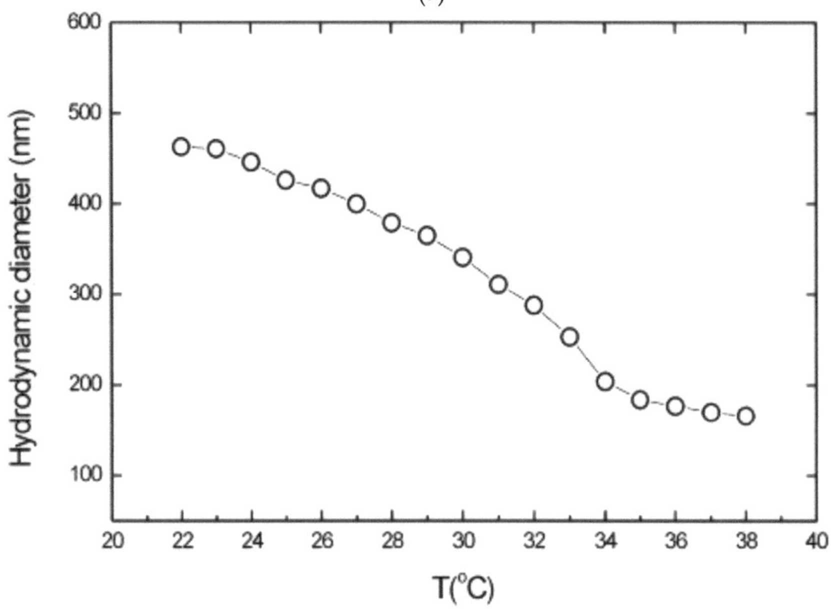

(b)

Fig. 5. Hydrodynamic diameter of PNCS microgel dispersion as a function of pH (a) (25°C; $1.1 \cdot 10^{-1}$ g/L) and temperature (b) (pH 7.3; $2.3 \cdot 10^{-1}$ g/L; NaCl 1 mM).

4. Incorporation of surface modifying system to cotton

The incorporation of the surface modifying system based on poly-NiPAAM/chitosan microgel to cotton can be done from aqueous microgel dispersion by simple *pad-dry* or *pad-*

dry-cure procedure, which is of special interest for the application in industrial conditions. The main challenge during incorporation procedure is to integrate the surface modifying system into textile substrate with sufficient durability while still retaining the effectiveness (responsiveness) of the microgel. Two different approaches are available to achieve this goal. One approach is based on cotton activation before the application of the surface modifying system, and both chemical (Jocic et al., 2009) and physical (Tourrette et al., 2009) methods have been previously reported. Another approach, which does not include previous activation of cotton, is based on the use of the additional film-forming agent to produce three-dimensionally linked network between the microgel particles and the substrate (Kulkarni et al., 2010; Krizman Lavric et al., 2010a; 2010b). Hence, the microgel can be bound to the surface of textile fabric directly (without the need for a separate and/or additional crosslinker) or with a crosslinking agent.

4.1 Microgel incorporation to previously activated cotton

4.1.1 Chemical cotton activation

The aim of textile material (cotton fabric) activation is to impart ionic character to cotton cellulose in order to facilitate the incorporation of micro-particulate system.

Among several possibilities for producing ionic active sites on cotton surface, two methods have been chosen as most convenient (Jocic et al., 2009): *anionic cotton* has been produced by reaction with monochloracetic acid (CAA) to give partially carboxymethylated cellulose (*carboxymethylation*) (Hashem et al., 2003; Racz et al., 1996); *cationic cotton* has been produced by dyeing with reactive dye and subsequent reductive cleavage of the dye attached (*aminization*) (Kim et al., 2007). Thus, aromatic amines are formed on cotton by chemical reduction of covalently attached reactive dye molecules.

The chemical modification of cotton surface has been confirmed by XPS analysis (Jocic et al., 2009). The values for elemental composition and atomic ratio remain similar after carboxymethylation, while after aminization the O/C ratio is significantly increased by a factor 1.6. Both nitrogen and sulphur are detected at aminized cotton, which is the consequence of the presence of the reactive dye's vinyl sulphonic center and the free amino groups on the cotton surface (Fig. 6).

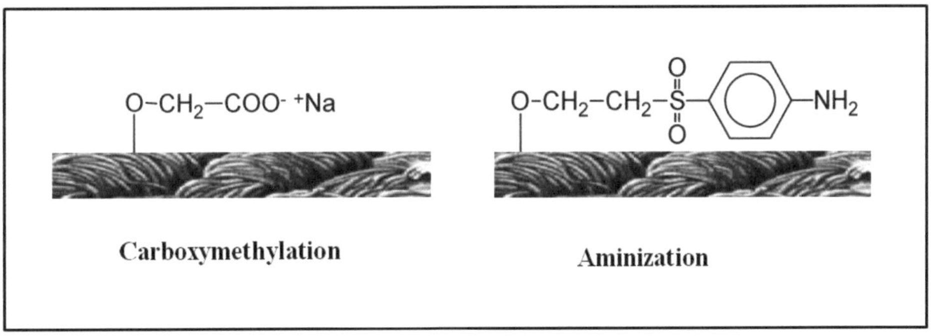

Fig. 6. Chemical activation of cotton.

4.1.2 Physical cotton activation: low-temperature plasma (LTP) treatment

The aim of cotton fabric treatment with LTP is to produce active species on the cotton fabric surface in order to facilitate the incorporation of PNCS microgel. It is well known that plasma treatment produces chemically active species such as radicals, free electrons and ions, which generate cellulosic radicals on the cotton fabric surface (Li et al., 1997). Depending on the nature of the plasma gas used, different modification is expected. Ward et al. (1978) reported that air, nitrogen and argon as plasma gases produce cotton surface with physical and chemical characteristics that differ significantly from untreated cotton. These three most commonly used plasma gases (air, nitrogen and argon) have been used for the activation of cotton surface (Tourrette et al., 2009).

Examination of surface morphology (SEM) of plasma treated and untreated cotton reveals that the untreated cotton fibres have a smoother surface than plasma treated fibres (Tourrette et al., 2009). Micro-cracks are visible on plasma treated cotton fibres surface no matter which gas is used. These micro-cracks and grooves on the plasma treated fibre surface can be attributed to the ablation effect caused by interaction of plasma species with the fabric surface. As a result, the specimen surface becomes rougher; ablation increases fabric surface roughness, producing clearly visible voids and spaces. However, no significant differences are visible by SEM between the three different plasma gases, which means that there is not a big difference in the physical modification of cotton fibres surface. Anyhow, the nature of the plasma gas can lead to different chemical surface modifications, which cannot be analysed by the SEM technique, but with XPS analysis.

It has been shown by XPS analysis that the O/C atomic ratio obtained for all plasma treated samples is significantly higher than the value obtained for the untreated cotton, confirming that some oxidation of the cotton surface occurred after all plasma treatments. The oxidation processes of nitrogen and argon plasma treated cotton fabric can be related to plasma-created free radicals, which initiate oxidation reactions either during plasma treatment (due to: (1) the nature of the gas used (oxygen contained in air); (2) the presence of small amount of oxygen in the inter-electrode of the instrument or due to the vapour desorption from the cotton sample (Osenberg et al., 1999); or (3) after plasma treatment (as the consequence of the exposure of plasma treated fabric to air after treatment). The influence of the nature of the gas used for the treatment has also been shown with: (1) the value of O/C atomic ratio of air plasma treated cotton which was higher compared to cotton treated with other plasma gases; (2) the elemental composition, where nitrogen plasma treated cotton was the only sample that contained nitrogen at the surface (0.9 at.%).

4.1.3 Microgel incorporation method and surface characterization

PNCS microgel was incorporated by a batch method onto untreated and previously activated cotton fabrics. On untreated and chemically activated (carboxymethylated or aminized) cotton, this process is based on adsorption (van der Waals interaction, acid-base interaction, electrostatic interaction and hydrogen bonding) of the microgel particles onto cotton fibre surface without covalent bonding. However, as the consequence of the presence of radicals at the substrate surface, covalent bonding can be additionally expected between the plasma treated substrate and the microgel particles.

Surface features of untreated and both chemically and physically activated cotton samples with subsequently incorporated PNCS microgel were studied by SEM (Fig. 7). The incorporation of PNCS microgel significantly changed the visual aspect of the fibre surface. This was clearly exhibited in case of previously activated fabrics (chemically or physically) where it was easy to locate surface deposition of PNCS microparticles and to distinguish their form, size and amount present at the fibre surface. It is obvious that hydrated microgel particles collapsed after incorporation onto the cotton fibre and deformed their shape to "pancake" structure. This is the consequence of volume shrinking because of water evaporation during drying and, possibly, water absorption from micro-particulate hydrogel into the cotton fibre. Previously untreated cotton (Fig. 7a) does not show clear evidence of microgel particle surface deposition, i.e. there is no significant visual change in comparison with the control sample.

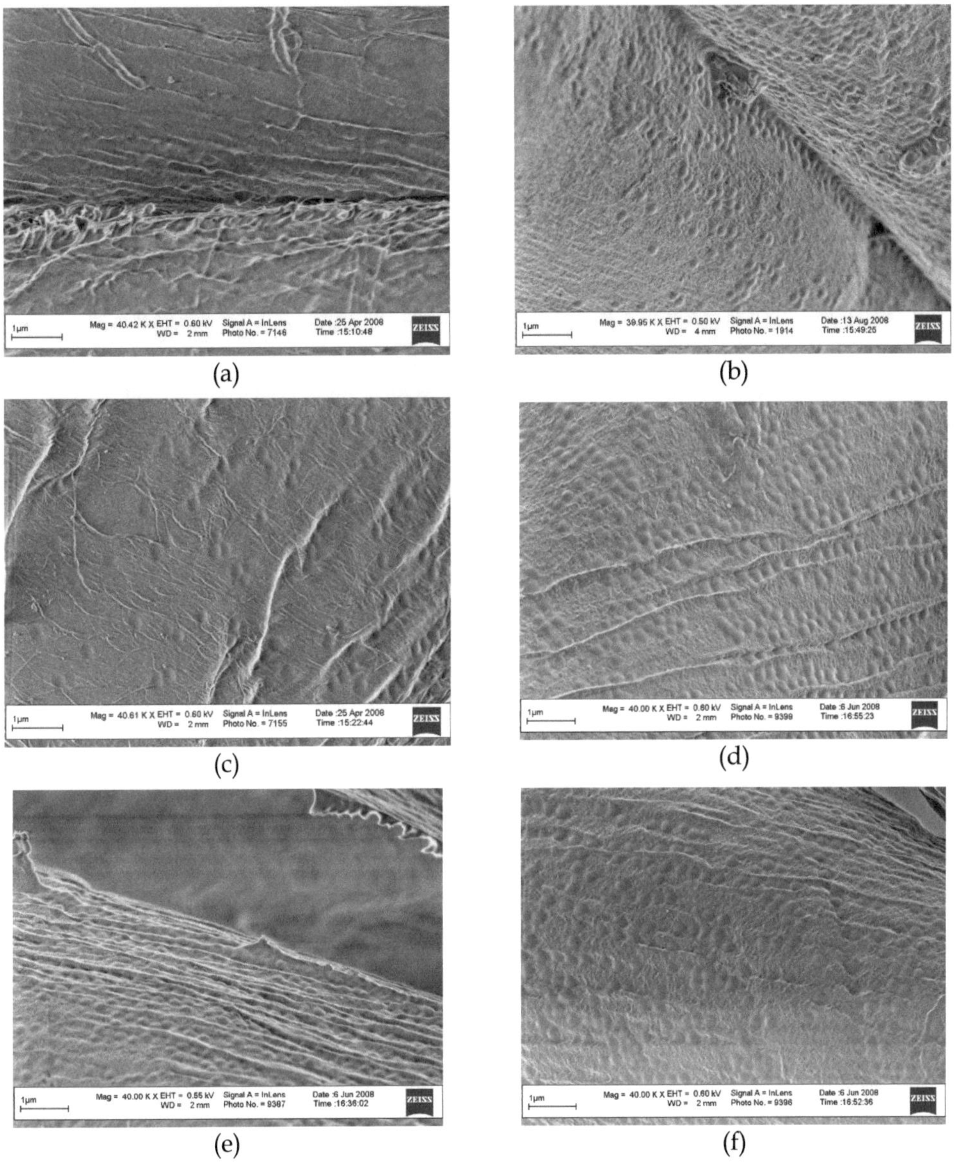

Fig. 7. SEM images after PNCS microgel incorporation: untreated cotton (a); carboxymethylated cotton (b), aminized cotton (c); air plasma treated cotton (d); nitrogen plasma treated cotton (e); and argon plasma treated cotton (e).

XPS analysis was used to determine the surface chemical composition changes of cotton after microgel incorporation (Jocic et al., 2009; Tourrette et al., 2009). After incorporation of PNCS microgel, all samples show the presence of nitrogen in significantly higher amount than in nitrogen plasma treated cotton. This is the consequence of the fact that both poly-NiPAAm and chitosan contain enough nitrogen which can be detected by XPS analysis, thus confirming the incorporation of PNCS microgel to the fibre surface. Therefore, the amount of nitrogen detected can be used as a tool for the estimation of PNCS microgel contents on the fibre surface. However, since previously untreated cotton sample with incorporated microgel shows significant nitrogen content (4.6 at.%), this implies that even without cotton activation some kind of poly-NiPAAm and/or chitosan adsorption occurs. Nevertheless, since SEM analysis did not visually confirm the presence of microgel particles at this sample (Fig. 7a), it can be speculated that a kind of precipitation occurred and polymer(s) have been adsorbed to the nonactivated cotton fibre surface in a thin layer, not in a micro-particulate form. The measured amount of nitrogen is bigger with carboxymethylated sample (5.2 at.%) compared to aminized sample (3.9 at.%), a result which confirms better adsorption of positively charged microparticles (see Fig. 4) on anionic cotton. For plasma treated cotton samples, the highest amount of nitrogen is measured for nitrogen and argon activated samples (6.0 and 6.1 at.%, respectively). Air plasma treated cotton showed nitrogen content similar to previously untreated and chemically activated samples (5.2 at.%). This result implies better microgel incorporation after nitrogen and argon plasma treatment. However, it is difficult to assess at this stage the level of enhancement after nitrogen plasma treatment, concerning the fact that nitrogen plasma treated cotton surface already contains 0.9 at.% of nitrogen as the consequence of the activation step.

These results lead to the conclusion that plasma treatments lead to better incorporation efficiency of PNCS microgel. The enhancement of the adhesion between hydrogel microparticles and plasma treated fibres can be attributed both to physical and chemical modification. The physical modification is the surface roughening of the fibre by the sputtering effect, producing an enlargement of contact area that increases the friction between the fibre and the microgel. The chemical modification increases the functional groups presence at the fibre surface, hence enabling a number of chemical bonds to be formed between the fibre and the microgel. After plasma treatments, there are still a lot of free radicals that remain on the treated fibre surface which can play a role in forming new functional groups and bonds between the fibre and the microgel. In addition, compared to chemical methods, plasma treatment is considered an environmentally friendly process which guarantees the high quality of the material with minimum costs and technological waste (Careiro et al., 2001).

4.2 Microgel incorporation to previously non-activated cotton

In order to obtain satisfying durability of the surface modifying system without previous cotton activation, microgel particles can be covalently bonded to cotton using appropriate crosslinking agent. In textile industry, polycarboxylic acids are well-known crosslinking agents that can serve as formaldehyde-free durable press finishing agents (Welch, 1992; Welch & Andrews, 1990). Among various available polycarboxylic acids, the most effective crosslinking agent for cotton was found to be 1,2,3,4-butanetetracarboxylic acid (BTCA), when combined with catalysts that are inorganic salts of phosphorus-containing acids (Hebeish et al., 2006), among which sodium hypophosphite (SHP) is being the most effective (Gu & Yang, 2000). Polycarboxylic acids have been also used for crosslinking chitosan to cotton, thus adding antimicrobial properties to a durable press finish. It has been reported

(El-tahlawy et al., 2005) that under the action of heat during the curing step, part of BTCA can be consumed for linking chitosan to the cellulosic substrate through esterification. A formation of a covalent bond between the crosslinking agent, chitosan, and cellulosic chains was confirmed (Hsieh et al., 2006; Shin & Ueda, 1999).

Several studies have shown that cellulosic esterification by polycarboxylic acids takes place in two phases. According to Yang & Xu (1998), in the first phase a cyclic anhydride is formed, which forms an ester with the hydroxyl group (-OH) in the cellulose macromolecule (Orhan et al., 2009; Sauperl & Stana-Kleinschek, 2010). This is followed by formation of a second anhydride that subsequently reacts with another cellulose hydroxyl. However, in the presence of SHP, a second mechanism can occur, resulting in formation of a mixed linear anhydride (Gillingham et al., 1999). The number of covalent bonds or linkages which can be formed with cellulose is limited to one less than the number of carboxyl groups in the polycarboxylic acid (Bishof Vukusic & Katovic, 2000). Since BTCA can form at least two ester bonds, two macromolecules can be crosslinked effectively (Xu & Shyr, 2000). Moreover, esterification or amidation with -OH or $-NH_2$ groups of chitosan can occur, if two -COOH groups remain available (El-tahlawy et al., 2005; Hsieh et al., 2006).

4.2.1 Microgel incorporation method

In practice, the incorporation of PNCS microgel onto the cotton surface is possible by a pad (100% WPU) - dry (70°C, 1h) - cure (160°C, 3 min) process. The reaction mixture is prepared by adding BTCA (crosslinker) and SHP (catalyst) into PNCS dispersion. Ratios used are optimised to the following: PNCS:BTCA = 3.75:1 and BTCA:SHP = 2:1. The effectiveness of the crosslinking procedure of cellulose fibre hydroxyl groups with BTCA and/or chitosan, depends on the amount of crosslinking agent and catalyst, temperature, crosslinking reaction time and pH of the finishing bath (Sauperl & Stana-Kleinschek, 2010). The simplest and quickest way to evaluate the crosslinking efficiency is by using a gravimetric method to determine the add-on finish. A sample's weight is taken before and after the treatment and the percentage of add-on is calculated. However, when the mechanism of crosslinking is taken into consideration, then the quantity of accessible BTCA carboxylic groups indicates the efficiency of the cotton cellulose crosslinking (Sauperl et al., 2009; Sauperl & Stana-Kleinschek, 2010). To determine the content of acid carboxylic groups, a staining test with Methylene Blue (MB) dye can be used. Since each molecule of MB is thought to form a single electrostatic bond with each unit of carboxylic acid on the fabric or, in our case, with BTCA, it is possible to calculate the concentration of carboxylic acid groups present on the fabric after being treated either with BTCA and PNCS microgel (Co-PNCS/BTCA) or with BTCA only (Co-BTCA). The degree of crosslinking was studied by increasing the amount of BTCA from 0 to 3% owf. Fig. 8 shows that the content of carboxyl groups increases with increased percentage of BTCA. Up to 1% a slight increase in carboxyl content is noticed, while the biggest increase is noticed from 1 to 3%. The same tendency could be noticed in the case of Co-PNCS/BTCA, the only difference being that less amount of carboxyl groups has been measured, since part of BTCA is consumed in linking chitosan to cellulose substrate. The degree of crosslinking cotton cellulose to BTCA and PNCS microgel particles increased proportionally to the increase in percentage of BTCA applied. By knowing the number of carboxylic groups available, it is possible to determine the proper concentration of BTCA for efficient crosslinking of PNCS to cotton surface. The methylene blue method proved to be suitable for a quick, simple and accurate determination of the degree of crosslinking of cellulose hydroxyl groups with BTCA and/or chitosan.

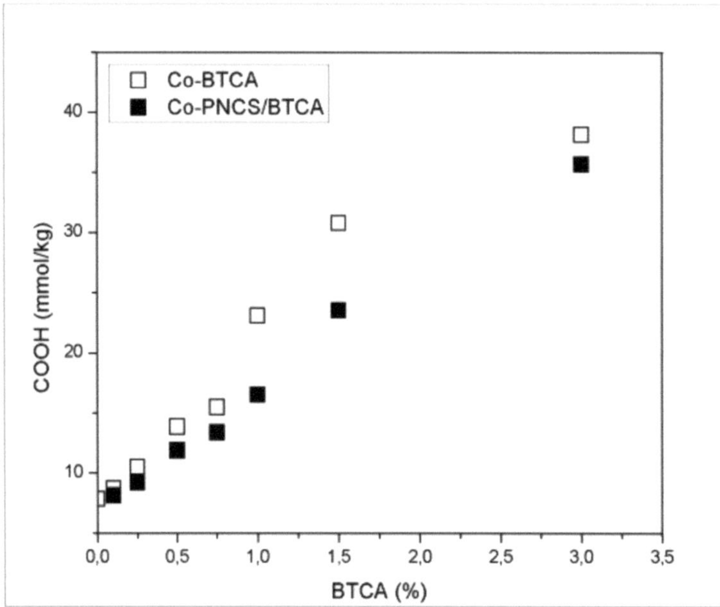

Fig. 8. Carboxyl content (mmol/kg) vs. BTCA concentration (% owf).

4.2.2 Surface characterization

The surface morphology of cotton treated with BTCA only (Co-BTCA) and of cotton with 3% owf of incorporated PNCS microgel (Co-PNCS/BTCA) was assessed by scanning electron microscopy (Fig. 9). Cotton treated with BTCA only (Fig. 9a) shows the presence of continuous film at the fibre surface, without significant visual difference to the well-known surface morphology of untreated cotton. The incorporation of PNCS microgel particles to cotton (Fig. 9b) significantly changes the visual aspect of the fibre surface. The form, size and amount of microgel particles present are clearly noticeable and they seem to be embedded in the crosslinker film.

The surface chemical composition, determined by XPS analysis, confirms that PNCS microgel was successfully incorporated to cotton (Krizman Lavric et al., 2010a). It can be shown that treatment with BTCA only significantly decreases carbon content (from 66.8 at.% to 59.6 at.%) as the consequence of a BTCA film forming at the fibre surface. After PNCS incorporation, even though a BTCA film was formed at the fibre surface, carbon content again reaches the level of untreated cotton (66.1 at.% vs. 66.8 at.%). This was due to the presence of carbon-rich poly-NiPAAm. Nevertheless, with PNCS incorporation, oxygen content decreased to a level that was below even the initial one (29.7 at.% vs. 33.2 at.%). This was again mainly due to the presence of poly-NiPAAm, as well as on behalf of the newly introduced nitrogen, originating from both poly-NiPAAm and chitosan. As expected, nitrogen was present only at PNCS treated cotton which could be considered as the indicator of successful microgel incorporation. However, relatively low observed nitrogen content (4.3 at.%) (compared to theoretical values for chitosan and poly-NiPAAm) could be explained by specific discrete arrangement of microgel particles at the fibre surface which is estimated to be covered not more than 50% by microgel. The SEM analysis gave visual confirmation for this assumption.

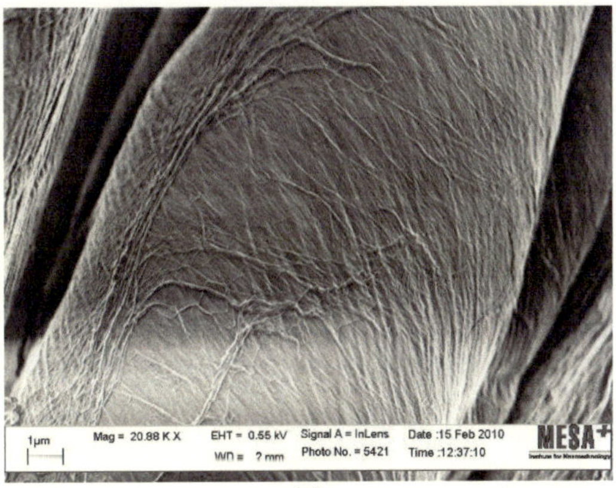

(a)

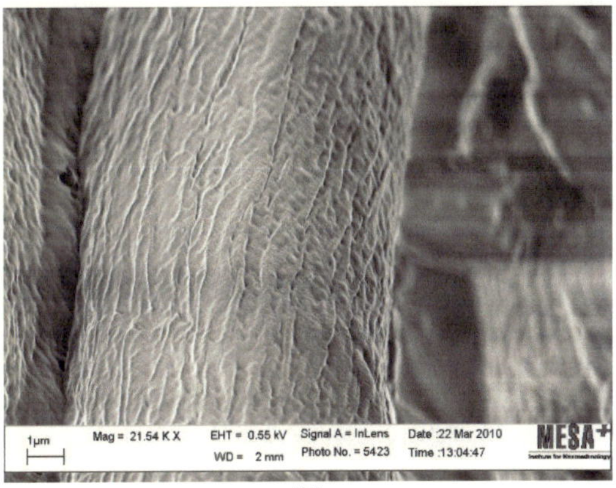

(b)

Fig. 9. Scanning electron microscope images of cotton treated with BTCA only (Co-BTCA) (a) and of cotton with incorporated PNCS microgel (Co-PNCS/BTCA, 3%) (b).

5. The assessment of pH-, temperature- and humidity-responsiveness of functionalized cotton

5.1 Appropriate methods

Even though many test methods are available to characterize fabric functionality, only a few are in use for assessing the responsiveness of new advanced textile materials which are based on pH and temperature responsive polymers. Recently published book chapter (Hu & Babu, 2008) gives an overview of intelligent textile testing with emphasis on shape memory fabrics and phase change materials. The capability of functionalized material to respond to different stimuli (pH, temperature, humidity) is often studied through swelling/shrinking or hydration/dehydration kinetics and equilibrium using a gravimetric method. The most common gravimetric method used for assessing the responsiveness of thermo- and pH-responsive hydrogels, or materials derived thereof, is the determination of *water uptake*. Interactions between water and textiles affect the comfort of apparel directly, therefore measurements of the amount of water absorbed by fabrics such as moisture regain, moisture sorption isotherms, water retention and absorptive capacities are needed (Zeronian, 1984).

Moisture regain can be used as a measure of the hydrophilicity of fibres, which in turn affects their comfort properties. Moisture sorption isotherm gives us basic information on the relation between moisture and fibre. *Water retention values* (WRV) give an estimate of the swelling capacity of fibres, as determined by the amount of water retained (imbibed water) by water-swollen samples after centrifugation. Data obtained on water retention can be used not only to indicate differences between untreated and treated samples in different environmental conditions in terms of water retention, but also for the degree of crosslinking in cellulosic fibres (Zeronian, 1984). In addition to a good moisture absorption performance, the capacity to discharge the absorbed water during drying is important as well, especially in the case of sportswear (Männer et al., 2004). Drying capability is evaluated by the drying rate of the fabric, the drying time being a function of how much water a fabric absorbs, and depends mainly on the thickness of the fabric (Fangueiro et al., 2010; Li, 2001).
A review of standard test methods for moisture in lint cotton, with emphasis on oven-dry techniques, was given by Montalvo & Von Holven (2008), including a list of moisture related terminology, procedural examples and moisture related formulae. All methods that have been mentioned above can be applied for intelligent textile testing.

5.2 Liquid management properties

Water vapour transport properties of textiles fabrics are of considerable importance in determining thermal comfort properties of clothing systems. Textile fabrics with a higher WVT permeability allow the human body to provide cooling due to evaporation of sweat. If the resistance to water vapour diffusion is high, the moisture movement is impeded and discomfort sensation of dampness and clamminess may arise (Keighley, 1985; Huang & Qian, 2007). The main mechanisms of water vapour transport through textile layers are (Indushekar et al., 2005; Das et al., 2007): diffusion of water vapour through the air spaces between the fibres; absorption, transmission and desorption of water vapour by the fibres; adsorption and migration of water vapour along the fibre surface (wicking); and transmission of water vapour by forced convection. The transport through hygroscopic porous materials is complex, due to the tendency of fibres to take up water vapour and experience swelling that can close off the pores in the fabric. This affects the ability of the fabric layer in passing heat and water vapour, as fibre swelling reduces free air volume within the fabric, affecting both the convective and diffusive transport of energy and mass. The size of pores in the fabric is far more important to moisture transfer than the hydrophilicity of the fabric (Gibson & Charmchi, 1997).
Cotton functionalized with the temperature responsive hydrogel which shows specific volume phase-transition (swelling and shrinking), represents a system that is able to control the evaporative rate at which a material can transport moisture vapour. There are several standard test methods available for measuring the water vapour permeability of fabrics, which can also be applied to microgel functionalized materials (Indushekar et al., 2005; Das et al., 2007; Huang & Qian, 2008; McCullough et al., 2003; Ding, 2008). Even though a number of methods are available for the determination of WVT rate of textile materials, the "pot" test method is still widely used. It involves a lot of work in respect to weighing and timing, but it is recognized as a reliable method.
The methods which have been used in our experiments are thoroughly explained in Krizman Lavric et al. (2010a; 2010b). The *moisture content* of the modified fabrics was measured thermo-gravimetrically, using a moisture analyzer. During the drying process the instrument continually measures the weight of the sample and displays the reduction in

moisture. The final moisture content (MC) is expressed in percentage (mass of the material after drying under specified conditions vs. mass of the material before drying). *Water vapour permeability* was measured according to the standard UNI 4818-26. Weight of the pots was recorded before introducing them into the climatic chamber and after 24 h. The difference in weight gives the quantity of water that was transmitted through the fabric as vapour. The water vapour transmission (WVT) rate is expressed in grams per square meter in 24 h. *Water uptake* was measured gravimetrically by soaking the samples either in acidic (pH 3), neutral (pH 6.5) or alkaline (pH 10) solutions at temperature bellow (25°C) and above (40°C) LCST of poly-NiPAAm until the swelling equilibrium was attained. The weight of the wet sample was determined after removing the surface water by blotting with filter paper. Water uptake (WU) is expressed in percentage (mass of the material after taking up water vs. mass of dry material).

It has been previously published that the liquid management properties of microgel functionalized textile depend on the temperature and humidity (Jocic et al., 2009; Krizman Lavric et al., 2010a). Because of its poly-NiPAAm core, each microgel particle acts as a temperature sensor and actuator. Therefore, at temperatures bellow LCST, it is in a swollen state, while above LCST it shrinks, resulting in an increase or decrease of total moisture content of modified cotton, respectively. It is however generally known that ambient moisture concentration is the driving force for moisture absorption, with moisture regain increasing when relative humidity increases (Warner, 1995).

The results obtained on the influence of temperature and ambient humidity on moisture content (MC) and water vapour transport rate (WVT) of untreated cotton (Co-UT) and cotton with incorporated microgel (Co-PNCS/BTCA) are presented in Fig. 10. The humidity values were chosen as low (50% R.H.) and high (80% R.H.), and temperature was chosen to be below (25°C) and above (40°C) 32°C (LCST of poly-NiPAAm). The results show that, when exposed to changes in ambient humidity and temperature, cotton material with incorporated PNCS microgel responds by changes in moisture uptake. The moisture content results (Fig. 10a) generally follow the known fact that the MC of the material decreases with temperature increase or ambient humidity decrease, which can be observed with both untreated and modified material. However, since the water presence is the driving force for the temperature responsiveness of microgels, different behaviour can be observed at low and high R.H. At low R.H. (50%), since there is not enough ambient humidity available, the microgel particles are in a "dry" state and the modified cotton does not show a macroscopically observable temperature response. In fact, it follows the same behaviour as untreated material. However, at high R.H. (80%), when enough humidity is available, there are noticeable differences in the modified cotton behaviour in response to temperature change. Even though the differences are rather small, they clearly imply the responsive property of Co-PNCS/BTCA. At 25°C, where microgel particles are hydrophilic, the moisture content is increased for modified cotton. It is actually the same trend as at low R.H., just with a bigger magnitude of change. At 40°C the microgel particles become hydrophobic, so the moisture content of Co-PNCS/BTCA is consequently significantly decreased (when compared to Co-UT). Therefore, it can be concluded that the responsiveness of the material is more, if not exclusively, pronounced in wet environment.

A similar discussion could be applied to water vapour transport (WVT) rate results (Fig. 10b). At low humidity (50% R.H.) the temperature responsiveness could not be observed. A drop of WVT values is noticed in the Co-PNCS/BTCA sample as a consequence of the microgels presence. It is expected that the microgel at the fibre surface could block water

vapour transport through the material, due to increased roughness. At high humidity (80% R.H.) the responsiveness is clearly evident. At low temperature (25°C), microgel particles swell significantly enough to hinder water vapour passage through the fabric. However, at 40°C (above LCST), hydrophobic microgel particles obstruct moisture absorption (Fig. 10a), which enhances the water vapour transport through the material (Krizman Lavric et al., 2010a). As it was expected, the combined effects of temperature and relative humidity had the biggest impact on water vapour permeability, while relative humidity showed the most significant influence on moisture content (Krizman Lavric et al., 2010b).
Water uptake (WU) results confirm increased water uptake capacity at low pH values at room temperature (Fig. 11a), which is the consequence of chitosan pH-responsiveness. A decreased water uptake capacity was noticed at higher temperature (Fig. 11b), due to the influence of poly-NiPAAm, which adds a hydrophobic character to the hydrogel.

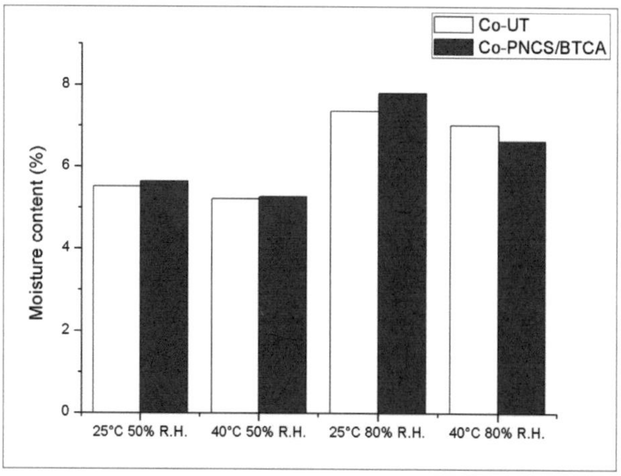

(a)

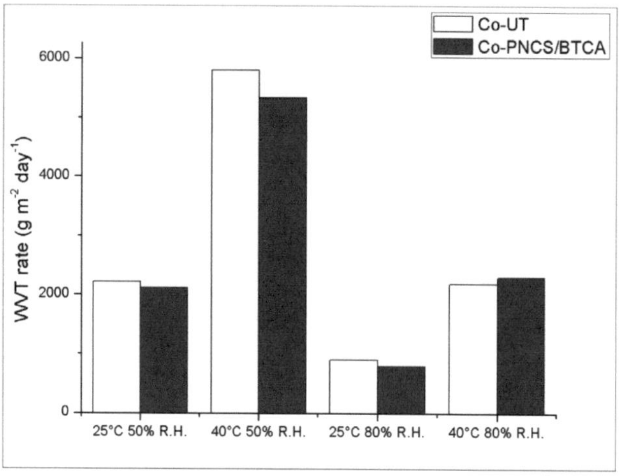

(b)

Fig. 10. Moisture content (a) and water vapour transport rate (b) of Co-UT and Co-PNCS/BTCA at different conditioning parameters (Krizman Lavric et al., 2010a).

6. Expected comfort enhancement

Good thermal sensation and comfort are basic demands for the production of performance apparel. Both are closely linked to fibre hygroscopicity, surface energy of fibres (hydrophilic or hydrophobic), and fabric structure and moisture management abilities, such as wicking, wetting and water vapour transport (Zhou et al., 2007). To satisfy the aforementioned demands, clothing materials require a high moisture retention capacity and a high moisture transportation propensity (Li, 2001; Okubayashi et al., 2005). The comfort perception of the wearer depends on the ability of clothing systems to evaporate the accumulated moisture from the skin to the environment. An important mechanism which the body uses to regulate its temperature is by perspiration, its level depending on the level of activity. Since moisture can derive either from sweat or from the environment (e.g. rain, humidity, etc.) it can have an impact on the comfort perception of both cold and hot conditions. During sweating the vapour is transmitted through the micro-porous structure of functional fabric to the outer layer where it evaporates, which makes the wearer feel dry and comfortable (Männer et al., 2004; Li, 2001; Okubayashi et al., 2005).

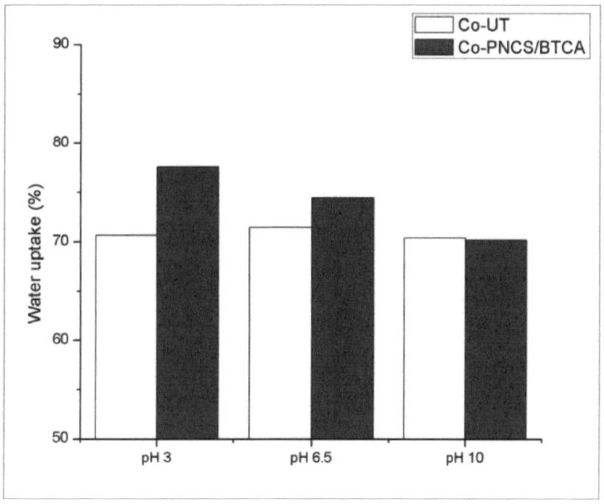

(a)

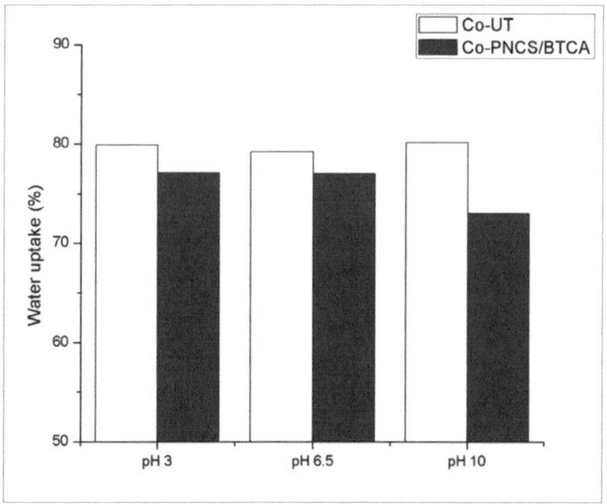

(b)

Fig. 11. Water uptake of Co-UT and Co-PNCS/BTCA at 25°C (a) and at 40°C (b).

In our study comfort does not refer to the psychological comfort but to the physiological comfort such as the moisture vapour transport rate. By exposure of the stimuli-responsive microgel to external stimuli (temperature, pH and humidity), it shows swelling/shrinking or hydration/dehydration properties by causing changes in water vapour transmission rate of modified textile material. As the body temperature exceeds the transition temperature (LCST) of the surface modifying system, PNCS microgel particles collapse enabling the transfer of moisture (water vapour) through the material into the environment. In contrast, when the ambient temperature decreases bellow the body temperature, the microgel tends to swell, preventing perspiration and thereby helping in maintaining the thermal balance of the body. Thus, the controlled expansion or contraction of the surface incorporated microgel particles leads to adjusting the liquid management of textile material and enhances comfort consequently.

7. Conclusions

An encouraging option to meet the goal of achieving "smart" liquid management properties of textiles is the functional finishing by application of microgel based on stimuli-responsive polymers. This surface modifying system shows specific volume phase-transition (swelling and shrinking) which can be triggered by various stimuli (temperature, pH, humidity etc.). The following aspects are reported: the synthesis and characterization of a microgel using the temperature responsive polymer (poly-NiPAAm) and pH-responsive biopolymer (chitosan); the methods for incorporating microgel to cotton, both previously activated and non-activated (with crosslinker); the assessment of pH-, temperature- and humidity-responsiveness of new advanced textile material. The results presented could serve as a proof of the concept and provide a guideline for application of microgel as a surface modifying system in general, with the aim to upgrade the regular textile material quality by providing highly attractive feature of stimuli-responsiveness.

8. Acknowledgement

The authors are thankful for financial support provided by Marie Curie Excellence Grant (EXT) project ADVANBIOTEX (MEXT-CT-2006-042641) funded by the EU's Sixth Framework Programme. (http://www.ctw.utwente.nl/efsm/advanbiotex/)

9. References

Bischof Vukusic, S. & Katovic, D. (2000). Creaseproof finishing using phosphono-based catalyst with polycarboxylic acids. *Colourage Annual*, Vol. 47, 87-94, ISSN 0588-5108

Careiro, N., Souto, A.P., Silva, E., Marimba, A., Tena, B., Ferreira, H. & Magalhães, V. (2001). Dyeability of corona-treated fabrics, *Coloration Technology*, Vol. 117, No. 5, 298-302, ISSN 1472-3581

Crespy, D. & Rossi, M. R. (2007). Temperature-responsive polymers with LCST in the physiological range and their applications in textiles. *Polymer International*, Vol. 56, Issue 12, 1461-1468, ISSN 0959-8103

Das, B.; Das, A.; Kothari, V.K.; Fanguiero, R. & de Araujo, M. (2007). Moisture transmission through textiles, Part I: Processes involved in moisture transmission and the factors at play. *Autex Research Journal*, Vol. 7, No. 2, (June, 2007) 100-110, ISSN 1470-9589

Ding, X. (2008). Fabric permeability testing. In: *Fabric Testing*, Jinlian Hu (Ed.), 189-227, Woodhead Publishing, ISBN 978-1-84569-297-1, Cambridge, UK.

El-tahlawy, K. F., El-bendary, M. A., Elhendawy, A. G., & Hudson, S. M. (2005). The antimicrobial activity of cotton fabrics treated with different crosslinking agents and chitosan. *Carbohydrate Polymers*, Vol. 60, Issue 4, 421-430, ISSN 0144-8617

Fangueiro, R.; Filgueiras, A. & Soutinho, F. (2010). Wicking behaviour and drying capability of functional knitted fabrics. *Textile Research Journal*, (March 18, 2010), Available online: http://trj.sagepub.com, ISSN 0040-5175

Gibson, P.W. & Charmchi, M. (1997). Modeling convection/diffusion processes in porous textiles with inclusion of humidity-dependent air permeability. *International Communications in Heat and Mass Transfer*, Vol. 24, No. 5, 709-724, ISSN 0735-1933

Gillingham, E.L.; Lewis, D.M. & Vončina, B. (1999). An FTIR study of anhydride formation on heating butane tetracarboxylic acid in the presence of various catalysts. *Textile Research Journal*, Vol. 69, No. 12, 949-955, ISSN 0040-5175

Gu, X. & Yang, C.Q. (2000). FTIR spectroscopy study of the formation of cyclic anhydride intermediates polycarboxylic acids catalyzed by sodium hypophosphite. *Textile Research Journal*, Vol. 70, No. 1, 64-70, ISSN 0040-5175

Hashem, M., Hauser, P. & Smith, B. (2003). Wrinkle Recovery for Cellulosic Fabric by Means of Ionic Crosslinking, *Textile Research Journal*, Vol. 73, No. 9, 762-766, ISSN 0040-5175

Harish Prashanth, K. V. & Tharanathan, R. N. (2003). Studies on graft copolymerization of chitosan with synthetic monomers. *Carbohydrate Polymers*, Vol. 54, Issue 3, 343-351, ISSN 0144-8617

Hebeish, A.; Hashem, M.; Abdel-Rahman, A. & El-Hilw, Z.H. (2006). Improving easy care nonformaldehyde finishing performance using polycarboxylic acids via precationization of cotton fabric. *Journal of Applied Polymer Science*, Vol. 100, Issue 4, 2697-2704, ISSN 0021-8995

Hsieh, S. H., Lin, E. S. & Wei, H. C. (2006). Effect of Chitosan Addition to BTCA/CA Processed Cotton Fabrics for Adsorbing Metallic Ions from Waste Water. *Journal of Applied Polymer Science*, Vol. 101, No. 5, 3264–3269, ISSN 0021-8995

Hsu, S.C., Don, T.M. & Chiu, W.Y. (2002). Synthesis of Chitosan-Modified Poly(methyl methacrylate) by Emulsion Polymerization. *Journal of Applied Polymer Science*, Vol. 86, Issue 12, 3047-3056, ISSN 0021-8995

Hu, J & Babu, K.M. (2008). Testing intelligent textiles, In: *Fabric Testing*, Jinlian Hu (Ed.), 275-308, Woodhead Publishing, ISBN 978-1-84569-297-1, Cambridge, UK.

Huang, J. & Qian, X. (2007). A new test method for measuring the water vapour permeability of fabrics. *Measurement Science and Technology*, Vol. 18, No. 9, 3043-3047, ISSN 0957-0233

Huang, J. & Qian, X. (2008). Comparison of test methods for measuring water vapor permeability of fabrics. *Textile Research Journal*, Vol. 78, No. 4, 342-352, ISSN 0040-5175

Indushekar, R.; Awasthi, P. & Gupta, R.K. (2005). Studies on test methods used to measure water vapor transmission of fabrics by DSC and conventional dish techniques. *Journal of Industrial Textiles*, Vol. 34, No. 4, 223-242, ISSN 1528-0837

Jocic, D. (2008). Smart Textile Materials by Surface Modification with Biopolymeric Systems. *Research Journal of Textile and Apparel*, Vol. 12, No. 2, 58-65, ISSN 1560-6074

Jocic, D., Tourrette, A., Glampedaki, P. & Warmoeskerken, M.M.C.G. (2009). Application of temperature and pH responsive microhydrogels for functional finishing of cotton fabric. *Materials Technology: Advanced Performance Materials*, Vol. 24, No. 1, 14-23, ISSN 1066-7857

Keighley, J.H. (1985). Breathable fabrics and comfort in clothing. *Journal of Industrial Textiles*, Vol. 15, No. 2, 89-104, ISSN 1528-0837

Kim, S.Y., Zille, A., Murkovic, M., Guebitz, G. & Cavaco-Paulo, A. (2007). Enzymatic polymerization on the surface of functionalized cellulose fibers, *Enzyme and Microbial Technology*, Vol. 40, Issue 7, 782–787, ISSN 0141-0229

Kopecek, J. & Yang, J. (2007). Hydrogels as smart biomaterials, *Polymer International*, Vol. 56, Issue 9, 1078-1098, ISSN 0959-8103

Krizman Lavric, P, Warmoeskerken, M.M.C.G. & Jocic, D. (2010a). Microgel Functionalized Textiles Responsive To Ambient Conditions, *Proceedings of AUTEX2010 10th WorldTextile Conference*, pp. 1-4, ISBN 978-609-95098-2-2 (CD-ROM), Vilnius, Lithuania, 21-23 June 2010, Kaunas University of Technology, Kaunas.

Krizman Lavric, P., Warmoeskerken, M.M.C.G. & Jocic, D. (2010b). Vapour transmission properties of a surface modified textile material with poly-NiPAAm/chitosan microgel, *Proceedings of 41st International Symposium on Novelties in Textiles (ISNT)*, pp. 104-110, ISBN 978-961-6045-80-3 (CD-ROM), Ljubljana, Slovenia, 27-29 May 2010, University of Ljubljana, Ljubljana.

Kumar, A., Srivastava, A., Galaev, I.Y. & Mattiasson, B. (2007). Smart polymers: Physical forms and bioengineering applications. *Progress in Polymer Scence.*, Vol. 32, 1205-1237, ISSN 0079-6700

Kulkarni, A., Tourrette, A., Warmoeskerken, M.M.C.G. & Jocic., D. (2010). Microgel-based surface modifying system for stimuli-responsive functional finishing of cotton. *Carbohydrate Polymers*, in print, ISSN 0144-8617

Lee C., Wen C. & Chiu, W. (2003). Synthesis of Poly(chitosan-*N*-isopropylacrylamide) Complex particles with the Method of Soapless Dispersion Polymerization. *Journal of Polymer Science Part A: Polymer Chemistry*, Vol. 40, Issue 13, 2053-2063, ISSN 0887-624X

Lee C., Wen C., Lin C. & Chiu W. (2004). Morphology and Temperature Responsiveness–Swelling relationship of Poly(*N*-isopropylamide–chitosan) Copolymers and Their Application to Drug Release. *Journal of Polymer Science Part A: Polymer Chemistry*, Vol. 42, Issue 12, 3029-3037, ISSN 0887-624X

Li, R., Ye, L. & Mai, Y.W. (1997). Application of plasma technologies in fibre-reinforced polymer composites: a review of recent developments. *Composites Part A: Applied Science and Manufacturing*, Vol. 28, Issue 1, 73-86, ISSN 1359-835X

Li, Y. (2001). The science of clothing comfort. *Textile Progress*, Vol. 31, No. 1, 1-135, ISSN 0040-5167

Liu, B. & Hu, Y. (2005). The Application of Temperature-Sensitive Hydrogels to Textiles: A Review of Chinese and Japanese Investigations. *Fibres & Textiles in Eastern Europe*, Vol. 13, No. 6, 45-49, ISSN 1230-3666

Liu, W., Huang, Y., Liu, H. & Hu, Y. (2007) Composite structure of temperature sensitive chitosan microgel and anomalous behavior in alcohol solutions. *Journal of Colloid and Interface Science*, Vol. 313, No. 1, 117-121, ISSN 00219797.

Männer, J.; Schuster, K.C.; Suchomel, F.; Gürtler, A. & Firgo, H. (2004). Higher performance with natural intelligence. *Lenzinger Berichte*, Vol. 83, 99-110, ISSN 0024-0907

Mano, J.F. (2008). Stimuli-Responsive Polymeric Systems for Biomedical Applications. *Advanced Engineering Materials*, Vol. 10, Issue 6, 515-527, ISSN 1527-2648

McCullough, E.A.; Kwon, M. & Shim, H. (2003). A comparison of standard test methods for measuring water vapour permeability of fabrics. *Measurement Science and Technology*, Vol. 14, 1402-1408, ISSN 0957-0233

Montalvo, J.G. & Von Hoven, T.M. (2008). Review of standard test methods for moisture in lint cotton. *The Journal of Cotton Science*, Vol. 12, Issue 1, 33-47, ISSN 1523-6919

Muzzarelli, R.A.A. (1977). *Chitin*, Pergamon Press, ISBN 0080203671, Oxford.

Okubayashi, S.; Griesser, U.J. & Bechtold, T. (2005). Moisture sorption/desorption behaviour of various manmade cellulosic fibers. *Journal of Applied Polymer Science*, Vol. 97, Issue 4, 1621-1625, ISSN 0021-8995

Osenberg, F., Theirich, D., Decker, A. & Engemann, J. (1999). Process control of a plasma treatment of wool by plasma diagnostics, *Surface and Coatings Technology*, Vol. 116-119 , 808-811, ISSN 0257-8972

Orhan, M.; Kut, D. & Gunesoglu, C. (2009). Improving the antibacterial activity of cotton fabrics finished with triclosan by the use of 1,2,3,4-butanetetracarboxylic acid and citric acid. *Journal of Applied Polymer Science*, Vol. 111, Issue 3, 1344-1352, ISSN 0021-8995

Pei, Y., Chen, J., Yang, L., Shi, L., Tao, Q., Hui, B., & Li, J. (2004). The effect of pH on the LCST of poly(N-isopropylacrylamide) and poly(N-isopropylacrylamide-co-acrylic acid). *Journal of Biomaterials Science, Polymer Edition,* Vol. 15, No. 5, 585-594, ISSN 0920-5063.

Pelton, R. (2000). Temperature-sensitive aqueous microgels. *Advances in Colloid and Interface Science*, Vol. 85, No. 1, 1-33, ISSN 0001-8686

Pelton, R.H. & Chibante P. (1986). Preparation of aqueous lattices with N-isopropylacrylamide. *Colloids and Surfaces*, Vol. 20, Issue 3, 247-256, ISSN 0166-6622

Prabaharan, M. & Mano, J.F. (2006). Stimuli-Responsive Hydrogels Based on Polysaccharides Incorporated with Thermo-Responsive Polymers as Novel Biomaterials. *Macromolecular Bioscience*, Vol. 6, Issue 12, 991–1008, ISSN 1616-5187

Racz, I., Borsa, J. & Bodor, G. (1996). Crystallinity and Accessability of Fibrous Carboxymethylcellulose by Pad-Roll Technology, *Journal of Applied Polymer Science*, Vol. 62, Issue 12, 2015-2024, ISSN 0021-8995

Ravi Kumar, M.N.V. (2000). A review of chitin and chitosan applications. *Reactive & Functional Polymers*, Vol. 46, Issue 1, 1–27, ISSN 1381-5148

Rinaudo, M. (2006). Chitin and chitosan: Properties and applications. *Progress in Polymer Science*, Vol. 31, Issue 7, 603–632, ISSN 0079-6700

Šauperl, O.; Stana-Kleinschek, K. & Ribitsch, V. (2009). Cotton cellulose 1,2,3,4 butanetetracarboxylic acid (BTCA) crosslinking monitored by some physical-chemical methods. *Textile Research Journal*, Vol. 79, No. 9, 780-791, ISSN 0040-5175

Šauperl, O. & Stana-Kleinschek, K. (2010). Differences between cotton and viscose fibers crosslinked with BTCA. *Textile Research Journal*, Vol. 80, No. 4, 383-392, ISSN 0040-5175

Schild, H.G. (1992). Poly(N-isopropylacrylamide): experiment, theory and application. *Progress in Polymer Science*, Vol. 17, 163-249, ISSN 0079-6700

Shin, H. & Ueda, M. (1999). Fixation of chitosan on cross-linked cellulose fabrics with polycarboxylic acids. *Fiber*, Vol. 55, No. 1, 42-47, ISSN 0037-9875

Tourrette, A., De Geyter, N., Jocic, D., Morent, R., Warmoeskerken, M. M. C. G. & Leys, C. (2009). Incorporation of poly(N-isopropylacrylamide)/chitosan microgel onto plasma functionalized cotton fibre surface. *Colloids and Surfaces A: Physicochemical and Engineering Aspects*, Vol. 352, Issue 1-3, 126-135, ISSN 0927-7757

Ward, T.L., Jung, H.Z., Hinojosa, O. & Benerito, R.R. (1978). Effect of cold plasmas on polysaccharides, *Surface Science*, Vol. 76, Issue 1, 257-273, ISSN 0039-6028

Warner, S.B. (1995). Environmental Effects: Solvents, Moisture, and Radiation. In: *Fiber Science*, 99-120, Prentice Hall, ISBN 0-02-424541-0, Upper Saddle River, NJ (USA).

Welch, C. M. & Andrews, B. A. K. (1990). Catalysts and Processes for Formaldehyde Free Durable Press Finishing of Cotton Textiles with Polycarboxylic Acids. *U.S. Patent Office*, Pat. No. 4 926 865.

Welch, C. M. (1992). Formaldehyde-Free Durable Press Finishes. *Review of Progress in Coloration and related Topics*, Vol. 22, 32-41, ISSN 0557-9325

Yang, C. Q.; Xu, L.; Li, S. & Jiang, Y. (1998). Nonformaldehyde durable press finishing of cotton fabrics by combining citric acid with polymers of maleic acid. *Textile Research Journal*, Vol. 68, No. 5, 457-464, ISSN 0040-5175

Xu, W. & Shyr, T-W. (2000). Durable press finishing of cotton fabrics by polycarboxylic acids after graft copolymerization with hydroxyethyl methacrylate. *Textile Research Journal*, Vol. 70, No. 1, 8-10, ISSN 0040-5175

Zeronian, S. H. (1984). Analysis of the Interaction between Water and Textiles, In: *Analytical Methods for a Textile Laboratory*, William Weaver (Ed.), 117-127, American Association of Textile Chemists and Colorists, ISBN 978-0-96133-500-7, NC (USA).

Zhou, L.Y.; Li, Y.; Chung, J.; Tokura, H.; Gohel, M.D.I.; Kwok, Y.L. & Feng, X.W. (2007). Effects of fabric surface energy on human thermophysiological responses during exercise and recovery. *Fibers and Polymers*, Vol. 8, No. 3, 319-325, ISSN 1229-9197

4

Production of Biopolymer Composites by Particle Bonding

Sanghoon Kim
United States Department of Agriculture/Agricultural Research Service/ National Center for Agricultural Utilization Research
USA

1. Introduction

With the rising concern for environmental protection, biodegradable polymers and bio-composites have attracted considerable attention as green materials and biocompatible materials that will replace some or all of the synthetic plastics in many applications.

Because petrochemical-based plastic material persists beyond its functional life, a waste disposal problem is facing modern society. Research to alleviate pollution and litter problems includes efforts to develop plastics that degrade more rapidly in the environment. Most of our waste is either stored in landfills or composted. Since most of the natural polymers are biodegradable, the use of natural polymers as a substitute for non-biodegradable synthetic polymers can be environmentally beneficial to some extent. Recently, many research groups have concentrated on the development of biodegradable polymer blends or composites from starch (Ma et al., 2008; Grazuleviciene et al., 2007; Gaspar et al., 2005), corn gluten meal (Samarasinghe et al., 2008; Beg et al., 2005; Wu et al., 2003a), wheat gluten (Zhang et al., 2007; Olabarrieta et al., 2006; Ye et al., 2006; Domenek et al., 2004; Kayseriliolu et al., 2003), and zein (Qu et al., 2008; Corradini et al., 2006; Wu et al., 2003b). In most cases, commercially available biopolymers contain a significant amount of unwanted materials that remain after the extraction/isolation process. Purification of these materials is very costly for the practical utilization of these agricultural products as a component of useful final products. If these agricultural excess products can be used without further purification, it can minimize the use of toxic chemicals, simplify the manufacturing process, save energy, and lower the production cost.

In this chapter, a new fabrication procedure for the production of biopolymer composites will be introduced. In this procedure, a corn protein, zein, is utilized as a binder for matrix materials (Kim, 2008). Matrix material can be chosen from both agricultural products and non-agricultural products. Composite materials fabricated from agricultural products such as starch and protein powders are biodegradable while those fabricated from non-agricultural products such as iron powder and glass spheres are not.

2. Fabrication of composites by particle-bonding

Conventionally, polymer composites had been manufactured by mixing the component materials in an extruder at high temperature. If the raw material is a mixture of various components with impurities, phase-separation, decomposition, and gas generation will interfere with the extrusion process. A new methodology, called particle-bonding technology, was developed for the production of biodegradable/non-degradable polymer composites that will potentially replace existing petroleum-based polymers. This technique requires neither extrusion nor processing at high temperatures. Instead, micrometer scale raw materials (powders) are coated with a corn protein, zein which has a strong adhesive force, and compressed to form a rigid material. Since this technique does not require purification of raw materials, various types of compounds can be used as component materials. During the fabrication process, raw materials in powder form are bound by zein. Zein is utilized because of its good adsorption to hydrophilic surfaces and strong adhesive properties. Both the unusual behavior of zein molecules in the aqueous alcohol solution and its strong adhesive property play a major role in the fabrication process of the particle-bonding technology. Mechanical properties such as compressive yield strength, modulus of elasticity, and yield strength at 0.2% offset of the fabricated polymer composites confirm the validity of the presented process.

3. Characteristics of zein

3.1 Source

Commercial zein is essentially a by-product of the corn wet-milling industry. In the wet milling process, steeped corn in water is milled to separate the hulls and germ from the endosperm, which is further milled into a fine slurry. Centrifugal separation of starch from the endosperm slurry leaves a protein-rich mass, i.e., corn gluten meal, from which zein is extracted (Padua & Wang, 2002). Annual zein production in the U.S. is more than 1 million pounds, mostly used in formulations of specialty food and pharmaceutical coatings (Shukla, 1992; Lawton, 2002).

Biologically, zein is a mixture of proteins varying in molecular size and solubility. These proteins can be separated by differential solubilities and their related structures into four distinct types: α, β, γ, and δ (Coleman and Larkins, 1999). α-Zein is the most abundant, accounting for ca. 70% of the total (Thompson and Larkins, 1989). The next most abundant zein is γ-zein, contributing ca. 20% to the total. α-Zein can be extracted using only aqueous alcohol, whereas the other zeins need a reducing agent in the solvent to be extracted (Lawton, 2002). α-Zein that is extracted with a reducing agent shows up as two bands with apparent migration rates of 19 and 22 kDa on SDS-PAGE (Coleman and Larkins, 1999). Commercial zein is made up of α-zeins (Wilson, 1988).

Zein can be used as a biodegradable material (Shukla & Cheryan, 2001; Lawton, 2002), and as early in 1909, it was used to prepare plastics (Goldsmith, 1909). Later, the related research was delayed due to the emergence of petroleum-based plastics. However, within the last two decades, zein has again attracted attention because of its biodegradability.

3.2 Solution behavior

Zein's defining characteristic is insolubility in water except in the presence of alcohol, high concentrations of urea, high concentrations of alkali (pH 11 or above) or anionic detergents. This is due to its amino acid composition. Zein is particularly rich in glutamic acid (21-26%), leucine (20%), proline (10%) and alanine (10%), but deficient in basic and acidic amino acids. The high proportion of nonpolar amino acid residues and deficiency in basic and acid amino acids is responsible for the solubility behavior of zein. (Shukla, 1992)

Conventionally, aqueous ethanol has been used as a solvent for zein in many experiments (Parris and Coffin, 1997; Fu and Weller, 1999; Lawton, 2002; Dickey et al., 2003; Kim et al., 2004; Guo et al., 2005). It is one of the good solvents for zein (Manley and Evans, 1943). Although it is known that 60–95% ethanol content is adequate for solubilization of zein, the effect of variation of alcohol content on the behavior of individual zein molecules has not been well understood until 2008. Fu and Weller (1999) reported on the rheology of zein solutions in aqueous ethanol. Zein (2–12%) in 50%, 60%, 70%, 80%, and 90% aqueous ethanol was investigated. Regardless of zein concentration, the overall trend showed that the viscosity of the solution decreased as the concentration of ethanol increased from 50% to 90%. This result indicates that the apparent molecular weight of zein decreases at a higher ethanol content. Yamada et al. (1996) reported that zein films prepared from 80% (v/v) aqueous ethanol did not show enough waterproof property compared with those prepared from 70% (v/v) aqueous acetone. In addition, aggregate formation was observed in both films. Dong et al. (2004) reported that SEM revealed that the zein film prepared from 70% ethanol was composed of particles of diameter 100–500 nm. These particles were agglomerated to form a film. The concentration of zein had no significant effect on particle size. Guo et al. (2005) used atomic force microscopy (AFM) for the visualization of zein globules formed from 70% (v/v) aqueous ethanol on the surface of mica. Depending on the concentration of zein in the solution, the diameter of globules ranged between 150 and 550nm, or between 60 and 120nm. If we compare zein with other proteins of similar molecular weight, the diameter of a zein molecule is expected to be ca. 5 nm (Cantor & Schimmel, 1980). Therefore, the data from previous researches clearly indicate that zein forms aggregates with a very large aggregation number in aqueous ethanol.

4. Particle bonding technology

4.1 Aggregation of zein in aqueous ethanol

The developed particle-bonding technology makes use of the aggregate-forming tendency of zein in aqueous ethanol solution. Although the traditional solvents for zein, 70–90% aqueous ethanols, did not show much difference in terms of the capability of making solutions, transmittance measurement showed there is a variation of aggregate size depending on the ethanol content in the solvent mixture (Kim & Xu, 2008). Transmittance of zein solutions was measured at several zein concentrations with a turbidimeter that is composed of a 633nm He–Ne laser, temperature-controlled sample block, stirrer, neutral density filter, and a laser power meter. Zein solutions of 0.2%, 0.5%, 1%, and 2% (w/w) were examined with a turbidimeter to find the best solvent composition (Fig. 1). The data show a peak of transmittance at around 90% aqueous ethanol but the transmittance dropped with a higher concentration of ethanol. However, although there was variation in the transmittance in the range of 70–90%, no precipitation was observed. It is an indication that there is a size variation of zein molecules as the percentage of the ethanol changes. Since the transmittance variation was too dramatic to be considered as due to a change in the individual molecular size, it was expected that there is a growth of particle size by aggregate formation of individual zein molecules. For further understanding of the aggregation behavior of zein in aqueous alcohol solution, the size variation of solute was investigated by a dynamic light scattering (DLS) technique. At greater than 90% ethanol, unlike the case of 70-90% ethanol solution, precipitation was observed and time-dependent transmittance was observed. The reason for the dramatic change in the transmittance in the narrow region at around 90% ethanol could only be explained after obtaining data from DLS experiment.

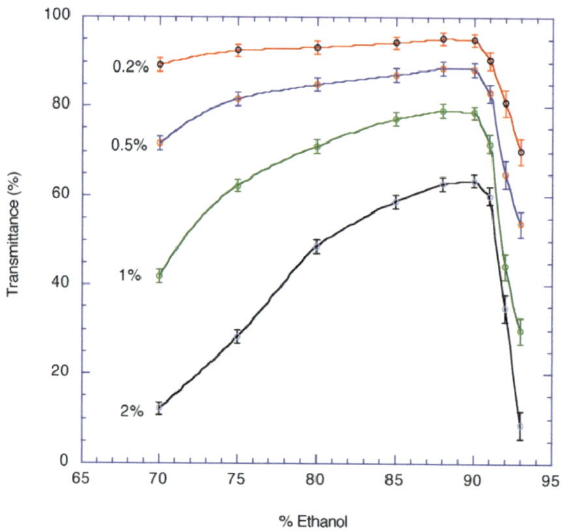

Fig. 1. Transmittance of zein in a series of ethanol solutions. Zein solutions (0.2–2%) were examined at 25.0 ± 0.1 °C. A 10mW He-Ne laser (λ = 633 nm) was used as a light source.

For the measurement of the degree of aggregation of zein in various ethanol/water mixtures, a DLS experiment was performed with an autocorrelator and a goniometer equipped with an argon laser (λ = 514.5 nm). All experiments were carried out with 0.1% (w/w) zein solution at 25.0 ± 0.1 °C. The scattered intensities were determined by using the standard photon counting method. Hydrodynamic radii of zein aggregates were obtained as follows. The DLS experiment yields a correlation function given as

$$g^{(1)}(t) = \exp(-\Gamma t) \quad (1)$$

where $\Gamma = D_t q^2$ (Johnson and Gabriel, 1994). From the slope of q^2 vs. Γ plot, we obtain a diffusion coefficient of a solute molecule, D_t. Here, q is the scattering wave vector. The hydrodynamic radii of aggregates can be calculated by the Stokes-Einstein relationship

$$R_h = (kT)/(6\pi\eta D_t) \quad (2)$$

where k is the Boltzmann constant, T is the absolute temperature (K), η is the viscosity of the solvent (P), and D_t is the diffusion coefficient (cm^2/s).
Fig. 2 shows dimensional variation of hydrodynamic radii of zein aggregates in aqueous ethanol solutions. At around 90% ethanol, a sharp decrease and increase in the hydrodynamic radius of zein was observed. The volume of the aggregates was calculated assuming the aggregates are spherical. As the molecular weight of zein is known from the SDS-PAGE experiment (Cabra et al., 2005; Padua & Wang, 2002), the size of single molecule was estimated from the list of proteins (Canton & Schimmel, 1980). Although it is not practically possible to obtain the data for the hydrodynamic radii for all ethanol/water compositions, it is probable that each zein molecule exists as a freely moving individual particle at around 90% ethanol. The most probable hydrodynamic radius of single zein molecule was assumed at 89.7% ethanol. Including this conjectured data point, the overall trend of the variation of the volume of aggregates is shown by the red line in Fig. 2. The volume of each aggregate and the degree of aggregation were calculated by assuming the

specific volume to be close to one. The degree of aggregation of zein was gradually decreased when the composition was changed from 70% to 90% ethanol. At around 90% ethanol solution, the aggregation number was at the minimum. After that, the aggregation number increased dramatically in the solvent composition higher than 90% ethanol. The overall behavior of aggregate formation of zein suggests that there is a structural change in the conformation of zein aggregates at around 90% ethanol. Since the amphiphilic behavior of zein has been known (Wang et al., 2004a & 2004b), this behavior can be interpreted as the inversion of micellar structure of zein aggregates as the solvent composition changes from a hydrophilic environment to a more hydrophobic one. According to the aforementioned aggregation behavior of zein, the structural inversion of zein can be induced by increasing or decreasing the composition of ethanol from 90%.

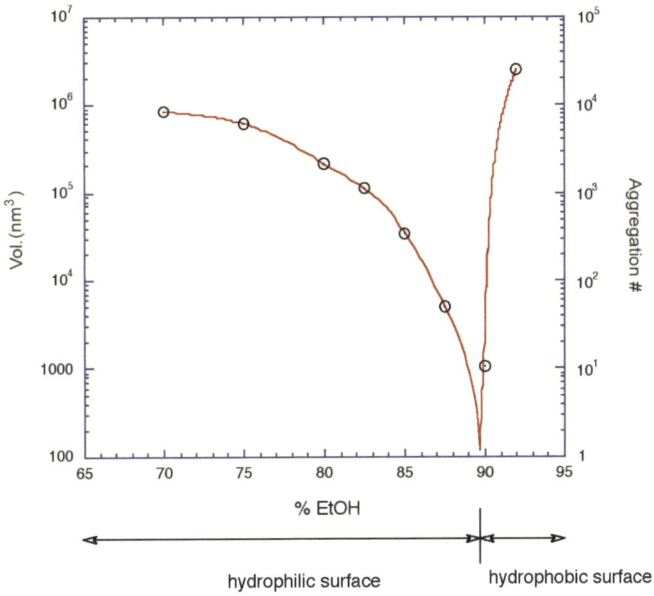

Fig. 2. Dimensional variation of zein aggregates in a series of ethanol solutions. Data were obtained by a dynamic light scattering instrument equipped with an argon ion laser (λ = 514.5 nm). The experiment was carried out with 0.1% (w/w) zein solution at 25.0 ± 0.1 °C.

4.2 Amphiphilic nature of zein

If each zein molecule behaves as an amphiphile, it is expected that zein molecules form a macromolecular micelle—globular aggregates with the hydrophilic moiety exposed to the surface and hydrophobic moieties clumped together in the aqueous ethanol with up to 90% ethanol content. 90% aqueous ethanol should be the best solvent for zein as the aggregation number was at its minimum (Fig. 2). As the percentage of ethanol increases to higher than 90%, the solvent medium turns more hydrophobic and a different type of macromolecular micelle will be formed. This time, the hydrophobic moiety will be exposed to the surface and hydrophilic moieties will clumped together. In other words, the orientation of each molecule in the micelle will be reversed.

The structural inversion of zein molecules or aggregates had been suggested by Yamada et al. (1996) when they found that the hydrophobicity of zein film depends on the type of solvent system used for its preparation. This observation supports our view that the inversion of micelle-like structure of zein aggregates occurs at around 90% ethanol. The micelle-like structures formed in lower than 90% ethanol have hydrophilic moieties that are oriented toward the solvent medium. In this case, the surface charges of particles repel each other without precipitation whereby they form a turbid solution. On the other hand, the micelle-like structures formed in greater than 90% ethanol have hydrophobic moieties that are oriented toward the solvent medium. In this case, the particles attract each other by van der Waals force so that aggregation, followed by precipitation, is induced. This behavior indicates that the micelles formed in this ethanol content have no or negligible charges on the surface whereby van der Waals force between aggregates can play a major role in the formation of precipitates. Micelle formation of zein in aqueous ethanol is schematically illustrated in the top row of Fig. 3 (no added substrate particles). Considering the micelle formation of surfactants at above CMC (Critical Micelle Concentration), the simplified model for the aggregation of zein molecules is conceivable. Inclusion of the experimental result shown in Fig. 2 readily yields inversion of micelle at around 90% aqueous ethanol. This behavior resembles pH-induced inversion of micelle that was formed by diblock copolymer unimers (Wang et al., 2006). It is expected that zein molecules would surround hydrophilic or hydrophobic particles depending on the percentage of ethanol in the solution. The schematic presentation of this behavior is shown in the bottom row of Fig. 3 (with added

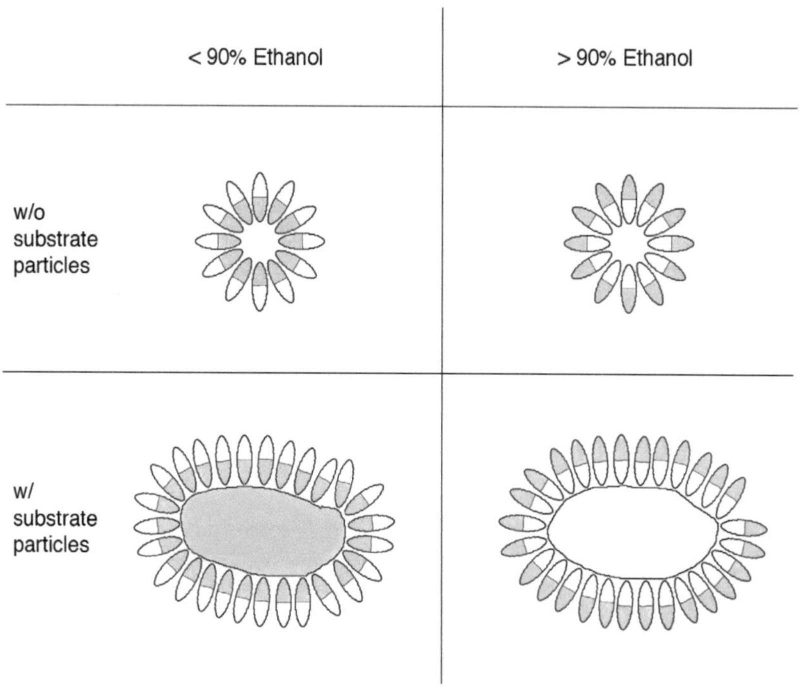

Fig. 3. Schematic diagram of the structural inversion of micelle-like structured zein molecules with and without added substrate particles in aqueous ethanol solutions. Shaded areas represent hydrophobic surfaces. When particles with hydrophilic (hydrophobic) surfaces are mixed with zein molecules, their surfaces will be surrounded by zein molecules with hydrophilic (hydrophobic) moieties oriented toward their surfaces.

substrate particles). This viewpoint is again supported by a simple experiment as follows. According to Fig. 2, addition of ethanol or water to a zein solution in 90% ethanol will induce larger aggregates. If two kinds of particles, one with hydrophilic surface and the other with hydrophobic surface, are dispersed in this system (zein in 90% ethanol solution), depending on the added solvent, ethanol or water, zein molecules would surround only one of the two particles. The aggregate formation of zein was experimentally observed with the above-mentioned procedure where glass spheres were employed as particles with hydrophilic surface and toner with hydrophobic surface. The sizes of the two particles were similar to each other and the color of the glass spheres was white while that of toner was black (Fig. 4).

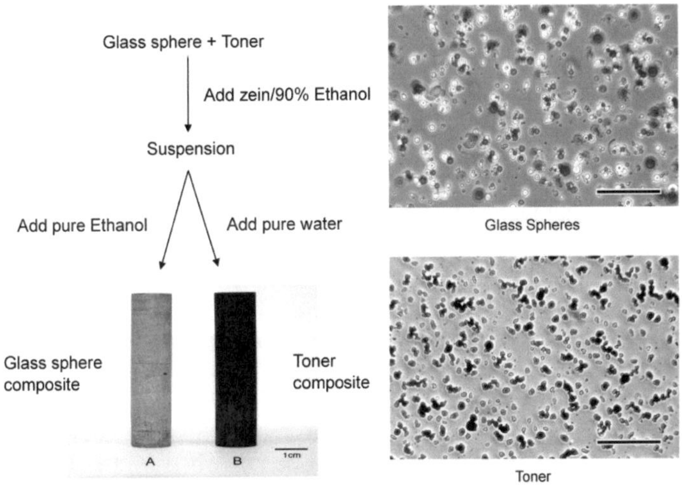

Fig. 4. Composite materials prepared from a mixture of glass spheres and toner. The selective adhesion of zein molecules to the surfaces of hydrophilic (glass spheres) or hydrophobic particles (toner particles) allows us to prepare a glass sphere composite (A) or a toner composite (B) from their mixture. 100 μm bars are shown in the microscopic images.

In this demonstration, 10% zein in 90%ethanol mixed with hydrophilic particles (glass spheres, o.d. = ca. 10 μm) and hydrophobic particles (toner, o.d. = ca. 10 μm) was prepared. When ethanol was gradually added to the solution, zein surrounded the glass spheres and formed a large agglomerate while the toner particles were minimally affected. The surface of precipitated zein is so sticky that the glass spheres coated with zein stick together and form large agglomerates. During this time, some of the toner particles, although they are not coated with zein, are physically embedded in the agglomerates. That is why the color of glass spheres is not pure white. The same reasoning applies to the toner composites. As the agglomerate was compressed in the cylindrical mold, a glass sphere composite was produced (Fig. 4). Since a phase contrast microscope was used for taking images in the inset of Fig. 4, some particles of glass spheres look black. On the other hand, when water was added to the solution, zein surrounded the toner particles and formed small agglomerates while the glass spheres were not affected. Smaller agglomerates than those of the previous case were formed because of the surface charges of particles that repel each other. A toner composite was produced as the agglomerate was compressed in the cylindrical mold (Fig. 4). In this experiment, the behavior of zein in the latter case is similar to that of detergents in a washer. Toner particles are nonpolar and insoluble in water. When detergents and hydrophobic particles (e.g., toner or oily dirt) are mixed in water, the hydrophobic moiety of detergent molecules are attracted to the surface of hydrophobic particles. The micelle-like structures formed by this mechanism contain hydrophobic particles in the center. By the same reasoning, if some amphiphiles (e.g., zein) and hydrophilic particles (e.g., glass spheres) are dispersed in a hydrophobic medium, the hydrophilic moiety of zein molecules will be attracted to the surface of hydrophilic particles. The above experimental result supports the concept for the structural inversion of micelle-like zein molecules. As stated above, addition of more ethanol to a zein solution in 90% ethanol induces large aggregates. It is the basis of the procedure developed in our lab for the production of polymer composite materials.

4.3 Principle of fabrication

The particle bonding technology makes use of the powerful adhesive properties of zein to bind the matrix materials. The adhesive properties of zein had not been recognized for a long time until recent research by Parris and Dickey (2003). The enormous bonding strength of zein could easily be demonstrated by using zein as a glue; zein solution prepared in organic solvent mixture (CH_2Cl_2/MeOH; 60/40) was applied on the surface of wood blocks (Uy, 1998). After 24 h, when the bonded wood blocks were taken apart, bonded parts were more difficult to break apart than the original wood block.

The surface of most matrix materials prepared from agricultural commodities is hydrophilic. At greater than 90% aqueous ethanol solution, the hydrophilic moiety of zein molecules should have a tendency to find hydrophilic materials in its solution and adsorb on its surface to hide the hydrophilic moiety and externally expose the hydrophobic moiety. In this situation, when matrix materials with hydrophilic surfaces, e.g., most of the biopolymers, are mixed with zein, the latter will be adsorbed on the surface of the former. As a result, we obtained matrix materials coated with zein molecules. Consequently, large agglomerates are formed because of the adhesive character of zein in the mixture. When these agglomerates are squeezed in the mold, the solvent is removed and a lump of protein composite is obtained.

There is a fundamental difference in the behavior of aggregates between the particles with hydrophilic surface and those with hydrophobic surface. In the case of the particles with hydrophobic surface, they form larger agglomerates by dispersion force between particles as soon as they are formed. Therefore, these particles are very unstable in the solution and form larger particles until they precipitate. The aggregation process is readily observed every time when the biopolymer composites are fabricated. On the other hand, the particles with hydrophilic surface do not spontaneously form larger agglomerates because of the electrostatic repulsion between particles.

The first step of the fabrication process is to disperse the matrix material and zein in ethanol. Only a small amount of ethanol that will barely immerse the matrix material is enough at this stage. The matrix material must be a powder of micrometer-scale particle size, and most of the spray-dried biopolymers satisfy this condition. The second step is to dissolve zein in the mixture by adding a calculated amount of water to make the composition of the solvent mixture to be ca. 90% aqueous ethanol. Enough time (typically 2-3 min) should be allowed for zein molecules to fully dissolve in the solution. From this point, depending on the hydrophilicity of the matrix material, the fabrication procedure needs to follow a different process.

In the case of hydrophilic powders, ethanol is added to make the ethanol content of the solvent mixture greater than 95%. At this moment, zein molecules are adsorbed to the surface of the matrix particles, and the sticky characteristics of zein induce agglomeration of the particles. The final step is to squeeze out the solvent mixture from the chunk material in the mold.

In the case of hydrophobic powders, water is added to make the ethanol content of the solvent mixture less than 95%. At this moment, zein molecules are adsorbed to the surface of the matrix particles, but the charges on the surface of proteins prevent the particles from binding together. To obtain aggregated form of particles, the suspension needs to be filtrated to remove the solvent mixture. An alternative choice is to add salts in the solvent mixture to screen the charges on the surface of protein molecules that are surrounding particles. The principle of this procedure is the same as that of "salting out" in colloid chemistry. The final step is to squeeze out the solvent from the chunk material in the mold.

4.4 Fabrication procedure - example

4.4.1 Particles with hydrophilic surface

Composites can be prepared by using molds fabricated from Teflon and aluminum. Typical procedure is as follows (example for composites with 20% zein). Both 6 g of matrix material and 1.5 g of zein powder are dispersed and mixed thoroughly in 13.5 g of absolute ethanol in a plastic container. Then 1.5 g of distilled water are added and stirred until the zein is fully dissolved (2-3 min). After that, 50 g of ethanol are added with vigorous stirring. During this time, particles aggregate to form a large chunk. Excess aqueous ethanol is decanted, and the aggregate is poured into a Teflon mold. Compression pressure is maintained at 75 lbs/cm^2 for 2 min. The fabricated specimens are air-dried at least for 72 h. For the measurement of mechanical properties, test specimens are fabricated by the method specified in the above and the content of zein are calculated by (wt. of zein)/(wt. of both matrix and zein) x 100. If the density of matrix material is very low, more than 15 g (ethanol 13.5 g + water 1.5 g) is needed for dispersion because of its large volume. In that case, the amount of the second addition of ethanol (50 g) needs to be adjusted accordingly. If n times larger composites need to be fabricated, the above procedure needs to be modified by simply multiplying the amount of the materials and chemicals by n times.

4.4.2 Particles with hydrophobic surface

In the case of particles with hydrophobic surface, both 6 g of matrix material and 1.5 g of zein powder are dispersed and mixed thoroughly in 13.5 g of absolute ethanol (example for composites with 20% zein). Then 1.5 g of distilled water are added and stirred until the zein is fully dissolved. Fabrication procedure up to this stage is the same as that of hydrophilic particles. After that, 50 g of water are added with vigorous stirring. During this time, unlike as in the hydrophilic particle case, particles do not spontaneously aggregate to form a large chunk because of the surface charges of each zein-coated particle. By adding a small amount of sodium chloride solution, agglomeration/precipitation will be induced. The rest of the procedure is the same as that of hydrophilic particle case.

4.5 Mechanical properties

As far as the size of the particle that will be suspended is not too large, any type of particles can be used for the base materials of polymer composites. As a test system, a wheat protein mixture, gluten, was used as a matrix material for the fabrication of biopolymer composite. From the preliminary examination of the procedure, it was found that too little or too great of an amount of zein yields a mechanically weak product. To know the optimum amount of zein necessary to produce the sturdiest material, the percentage of zein was varied in 8 compositions (4%, 8%, 12%, 16%, 20%, 22%, 24%, and 32%), and the compressive strengths of the product specimens were examined.

A typical compressive stress-strain diagram of gluten composite is shown in Fig. 5. In general, maximum compressive stress increased monotonously until the content of zein reached 20-22%. Repeated experiment revealed that there is a variation in the overall profile of the compressive strain vs. stress curves, but the maximum strength was always obtained with 20-22% zein, at which composition the compressive strength was comparable to that of polypropylene (Northolt, 1981). Further increase in zein content lowered the maximum compressive stress.

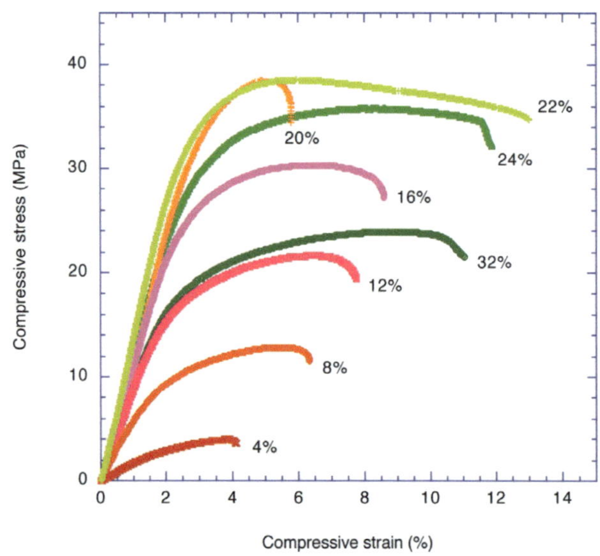

Fig. 5. Mechanical property of gluten (wheat protein) composites. Effect of the percentage amount of zein was demonstrated. Test was performed by ASTM standard, D695.

The compressive yield strength, modulus of elasticity, and yield strength at 0.2% offset are plotted as a function of percentage of zein to find the optimum composition of zein (Fig. 6). These three mechanical properties show the same trend, i.e., mechanical strength reaches its maximum when around 22% of zein is included in the composite. This experimental result clearly demonstrates that 20-22% is the optimum composition of zein in the case of gluten composite.

Processing pressure is another variable that determines the strength of the final product. Its effect was investigated by varying the pressure that is applied to the mold during the process. Fig. 7a shows that the overall stress-strain curve profile does not change at various processing pressures. To obtain an overview of the pressure effect, compressive stress at 10% strain was plotted as a function of processing pressure in Fig. 7b. This graph demonstrates that the mechanical strength of the composites monotonously increases as greater pressure is applied to the mold in the tested pressure range. In addition, it shows that the composite has a base strength even without applied pressure to the mold. According to this data, the strength of the composite will be doubled at a processing pressure of 150 lb/cm^2.

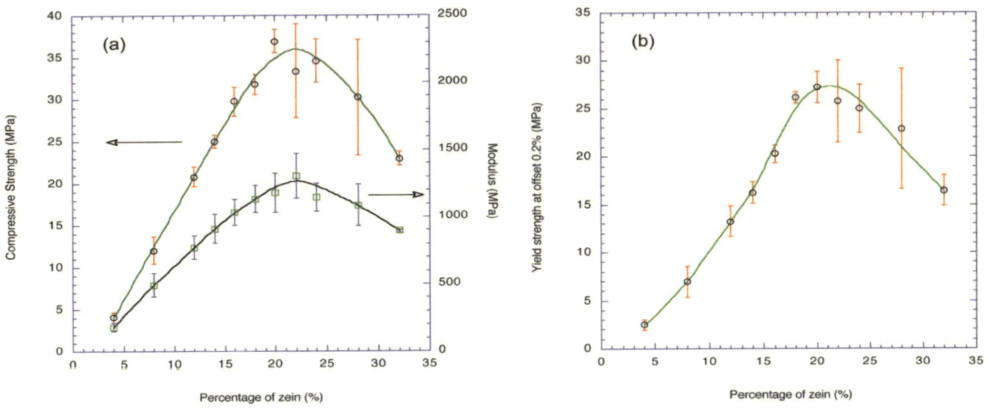

Fig. 6. Effect of the percentage of protein adhesive on compressive modulus, compressive strength, maximum load, and yield strength at 0.2% offset. (a) Composition vs. compressive modulus and compressive strength. (b) Composition vs. yield strength at 0.2% offset.

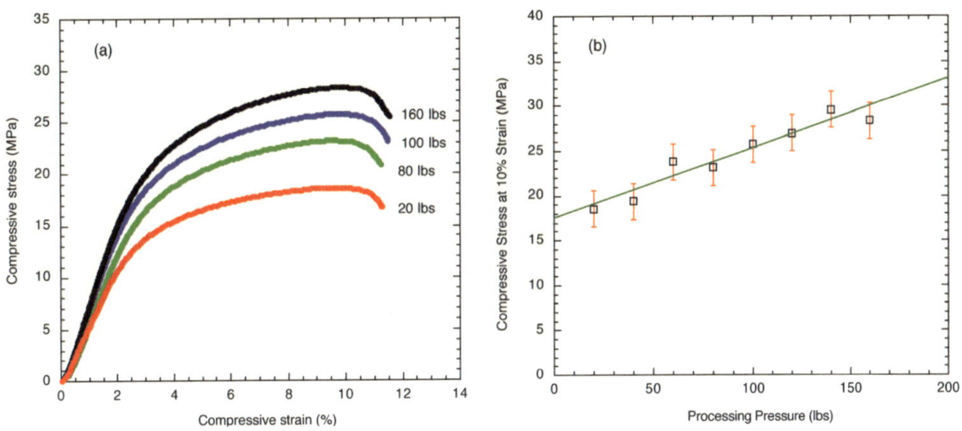

Fig. 7. Processing pressure effect. (a) Stress-strain curve at various processing pressure. (b) Compressive stress at 10% strain was plotted as a function of processing pressure. The cross section of the mold was 1.33 cm^2. For this experiment, the composition of zein was fixed to 15%.

In order to show that the fabrication of polymer composite from hydrophilic particle and zein is a universal behavior, glass spheres were used as a matrix material. The percentage of zein was varied from 4% to 28% and the mechanical strengths of the fabricated glass sphere composites were measured. Fig. 8 shows the compressive strain vs. stress curves obtained from glass sphere composites. The strength of the composites increased until the composition reached 24% and was weakened thereafter. It is noticeable that the maximum strength of glass sphere composites is reached in less than 3% compressive strain, while that of gluten composites is as high as 20% (Fig. 5). This discrepancy should be caused by the difference in the hardness of the constituent particles.

The current processing technique uses controlled precipitation to coat particles of biopolymers with a binder, zein, to form a microcomposite, and then consolidating the material in a compression molding process. Therefore, it can be used with any type of materials that does not dissolve in ethanol. Most proteins and carbohydrates belong to this category. Mixtures of two or more compounds can also be used for the production of polymer composites.

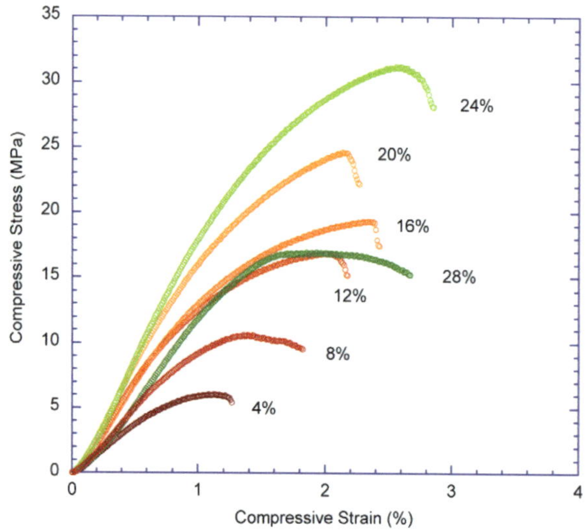

Fig. 8. Mechanical property of glass sphere composites. Effect of the percentage amount of zein was demonstrated.

The compressive strength of the composite produced from cellulose is a little weaker than that from gluten. When composites are manufactured from the mixture of these two materials, it is expected that the strength of the composite becomes weaker as the content of cellulose is increased in the mixture. Fig. 9(a) demonstrates that the incorporation of weaker matrix material weakens the strength of the composite. For this experiment, sample composites were manufactured from 80 wt.% matrix material mixture and 20 wt.% zein. The weight fraction of cellulose in the abscissa of Fig. 9(a) denotes wt.% of cellulose in the mixture of the two matrix materials. This observation indicates that the strength of the composite relies on the hardness of the constituent particles. In this example, the sizes of the matrix particles are not much different from each other, as shown in the inset of Fig. 9. On the other hand, when the particle sizes are very much different from each other as in the case of the mixture of gluten and cornstarch, somewhat different result was observed. The compressive strength of composite produced from cornstarch was a little weaker than that from gluten as well. As these two materials, gluten and cornstarch, were mixed together, the highest strength was observed when 40% starch was incorporated in the mixture (Fig. 9(b)). This behavior was interpreted as follows. Gluten powders are composed of larger particles than those of cornstarch. The strength of the composites originates from the strength of the individual particles. When the composites were fabricated at the same condition, the compressive strength of gluten composite (30 MPa) was higher than that of cornstarch composite (23 MPa). When the two materials are mixed together, the small cornstarch particles will fill the gaps between large gluten particles. For up to 40% cornstarch composition, the reinforcement offered by cornstarch plays a role. Once all the gaps are filled with cornstarch, the excessive cornstarch which is weaker than gluten degrades the strength of the final product.

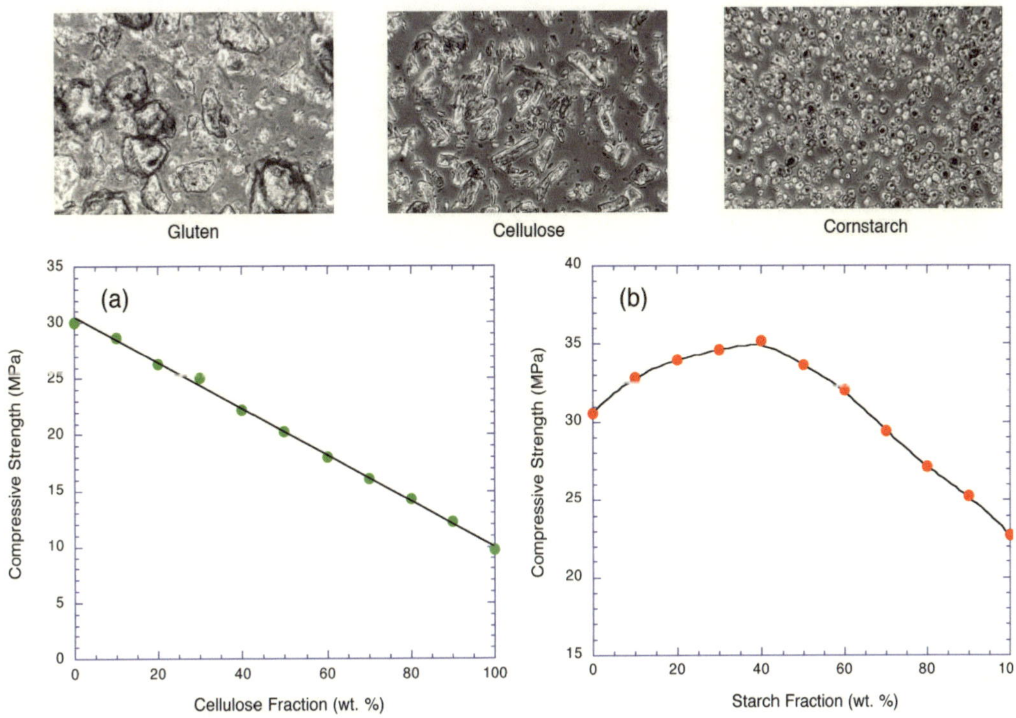

Fig. 9. Effect of incorporation of second component (cellulose (a) and starch (b)) into the gluten composites. Compressive strength was monitored as the content of second component was increased. Each composite contains 20% (w/w) zein. Microscopic images of gluten, cellulose and cornstarch and are also shown.

To show that the aforementioned technology can be applied to most of the biopolymer particles, various type of materials were used to fabricate the composites (Fig. 10). Some of

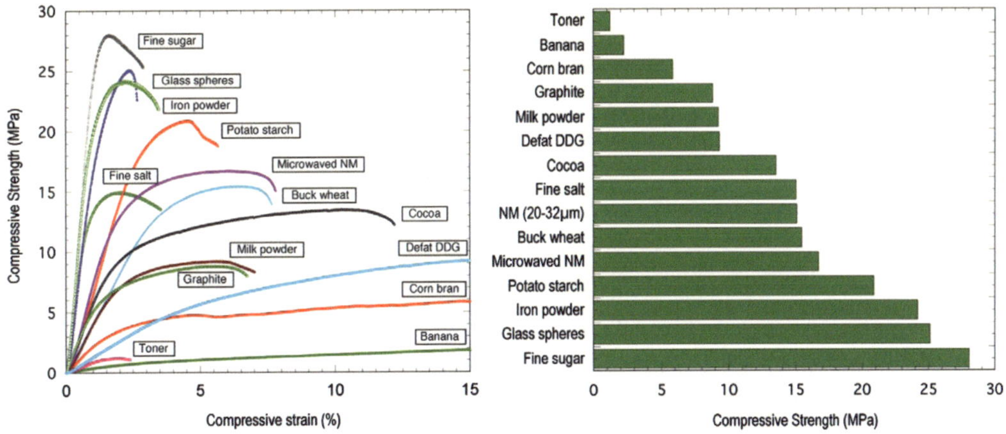

Fig. 10. Compressive stress–strain diagram of various biopolymer samples. Each composite contains 20% (w/w) zein. This illustration shows that various materials can be used for the production of polymer composites by using the particle-bonding technology.

the materials such as fine sugar, glass spheres, iron powder, fine salt, and graphite are not polymers. Since the suspension medium of the matrix material(s) contains equal or higher than 90% ethanol, matrix materials that dissolve in ethanol cannot be used. Other requirements of the particles to be used for fabricating composites are the micrometer-scale size and hydrophilicity of the particle surface. As particles with hydrophobic surface are rare, only toner was used as an example. In this case, unlike the case with hydrophilic surface, the particles were coated with zein by lowering the concentration of ethanol.

4.6 Conductive composites

As any type of particles can be used for the production of composites with the proposed technology, a conductive filler, graphite, was incorporated into the gluten composite to fabricate conducting polymer composites (Kim et al., 2010). For the presentation of the data, resistivity which is the inverse of conductivity was used as a unit for comparing the conductance of fabricated composites. It is defined as,

$$\rho = RA/L \tag{3}$$

where ρ is the resistivity (Ω-cm), R is the electrical resistance (Ω), L is the length (cm), and A is the cross-sectional area (cm^2).

By common sense, it is predictable that the more added conducting fillers, the higher the conductance of the produced composites. Fig. 11a clearly shows this trend, i.e., as more graphite was incorporated in the gluten composite, lower resistivity was observed. On the other hand, the amount of zein did not affect the resistivity of the composites. Fig. 11b is the same plot as Fig. 11a but showed the resistivity as a function of the percentage of incorporated graphite. The resistivity decreases as the percentages of graphite gets higher, but the variation of the resistivity at high percentages of graphite is not so dramatic. This decrease in the resistivity can be explained by the percolation theory (Pike & Seager, 1974; Vilcakova et al. 2002). The mechanical strength of the conducting composites was measured with the same samples as used for Fig. 11a and b. Although the percentage of zein does not affect the conductance of the composites, it does with mechanical strength as is expected. As

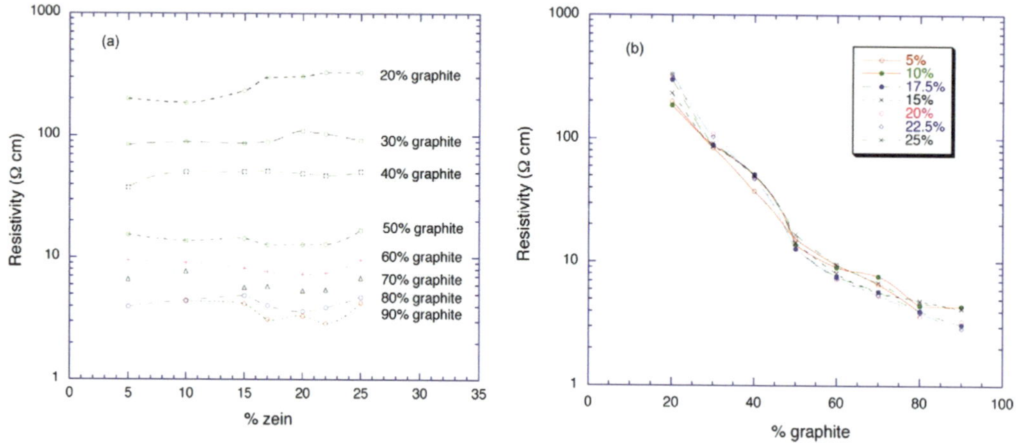

Fig. 11. Resistivity of conducting polymer composites. (a) Effect of the percentage of zein. (b) Effect of the percentage of incorporated graphite.

in the case of previous gluten composites, the strength increases until zein content increases to 22%. After that, a decrease in the mechanical strength was observed (Fig. 12a). Therefore, 22% zein is the optimum composition for this system.

Also, it was found that the inclusion of graphite weakened the mechanical strength. This trend was clearly shown again in the replot version (Fig. 12b). According to this data, the surface of graphite is not hydrophilic enough to be bound firmly with gluten particles. For this reason, the mechanical strength and conductance of graphite/gluten composite have to be compromised.

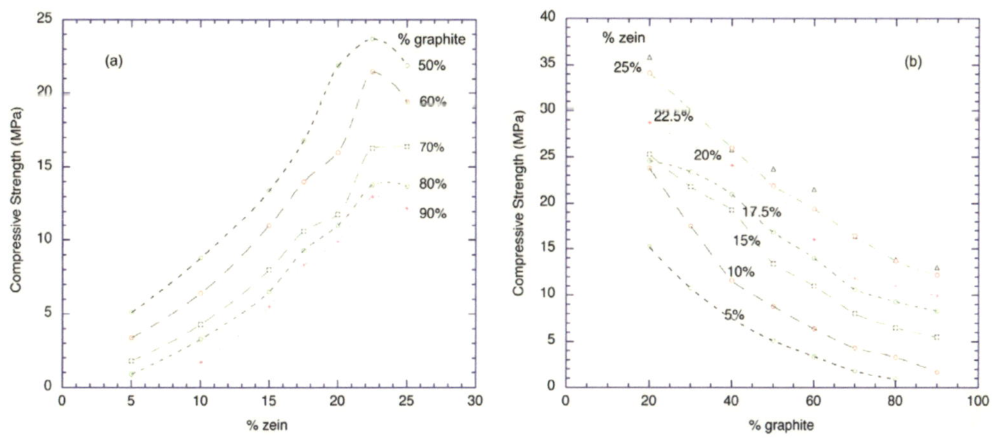

Fig. 12. Mechanical strength of conducting polymer composites. (a) Effect of the percentage of zein. (b) Effect of the percentage of incorporated graphite.

5. Future research

Although the fabrication of the composites makes use of the amphiphilic nature of zein molecules, the detailed structure of the individual molecule is not known yet. Several models have been proposed for the 3D shape of α-zein (Argus et al., 1982; Garrett et al., 1993; Tatham et al., 1993; Matsushima et al., 1997; Bugs et al., 2004; Forato et al., 2004). Despite all these trials, a generally accepted axial ratio of the individual molecule has not even been obtained yet. In this situation, discussion on the local hydrophilicity of the surface of each molecule is premature. Crystallography of zein, however, would give us a clear picture on the distribution of both polar and non-polar amino acids on the surface of each zein molecule.

Like zein, gliadin in wheat belongs to the characteristic class of proteins known as prolamines, which occur specifically in cereals (Shukla & Cheryan, 2001). Therefore, the solvent behaviour of gliadin is expected to be similar to that of zein. This point of view was examined with several instruments and supported by preliminary data. The only major difference was that the best ethanol composition for the solubilization of gliadin is 65 wt% ethanol while that of zein is 90 wt% ethanol. Preliminary data from the measurement of mechanical strength at various gliadin content is shown in Fig. 13. It is clearly shown that composite fabricated with ca. 20% gliadin is the sturdiest and the mechanical strength of gliadin composite is compatible with that of zein composite. Since the production cost for gliadin is much lower than that of zein ($0.50-1.00/lb vs. $5-10/lb), gliadin is a promising material.

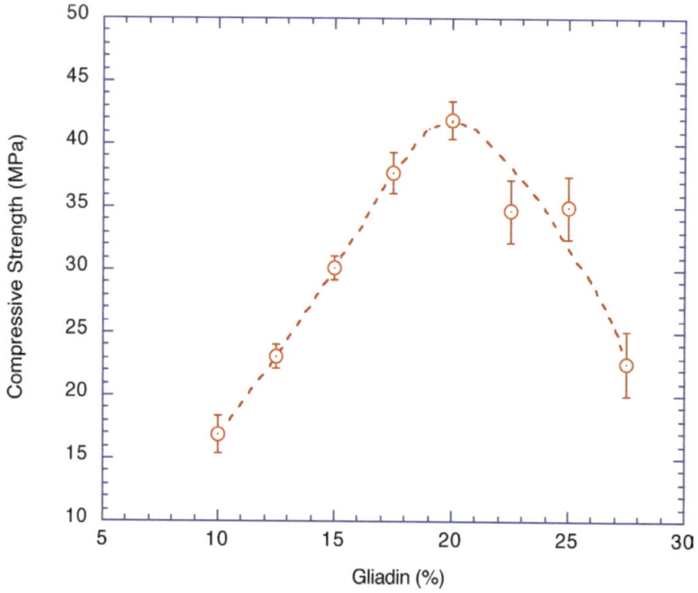

Fig. 13. Mechanical strength of composites fabricated with glass spheres and gliadin.

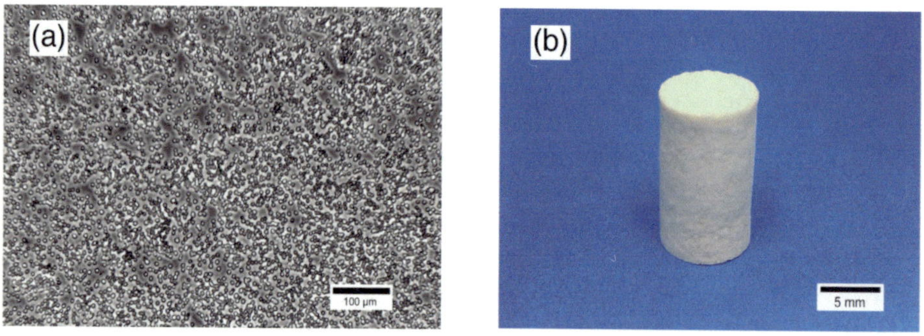

Fig. 14. Fabrication of composite with a living microorganism, yeast. (a) yeasts bearing toxin. (b) fabricated yeast pellet before cutting.

6. Conclusion

The particle bonding technology described in this chapter has the following merits.

a. It does not require purification of raw materials, except zein, thus reducing the use of chemicals and processing time.
b. Since the fabricating process is performed at room temperature, it does not require heating.
c. Biodegradable composites can be fabricated when biopolymers are used as raw material.
d. Products with compressive yield strength comparable to that of petroleum-based polymers can be fabricated.
e. In the case of the composite made with gluten, the compressive strength was comparable to that of polypropylene.

f. Any type of micrometer-scale powder that does not dissolve in alcohol can be processed.
g. Two or more types of raw materials can be used at any mixing ratio.
h. It allows for easy coloring of the product by addition of food dyes during the process.
i. Compressive strength can be adjusted by varying the processing pressure.
j. It allows the incorporation of fibers to improve the strength of the product. Since this process makes use of the adhesive properties of zein, any type of fiber with a hydrophilic surface can be incorporated into the matrix materials.

The developed composites can be used as substitutes for existing commercial products where biodegradation is desirable, such as plant pots, golf tees, clay targets (skeets), lightweight construction material or as a binder for foods and drugs. Recently, it is proposed that the composite can be used as a carrier for the pesticides. As an example, a composite has been fabricated with intoxicated yeast (Fig. 14). Since the matrix material is a living microorganism, the processing pressure needed to be maintained at a much lower value (15 lb/cm^2). Other than that, the fabrication procedure is exactly the same as those of other composites. As the test result with living insects was satisfactory, filing a patent is in progress.

7. References

Argos, P.; Pedersen, K.; Marks, M. D. & Larkin, B. A. (1982). A structural model for maize zein proteins, *Journal of Biological Chemistry*, 257: 9984-9990.

Beg, M.; Pickering, K. & Weal, S. (2005). Corn gluten meal as a biodegradable matrix material in wood fibre reinforced composites, *Materials Science and Engineering A*, 412: 7–11.

Bugs, M. R.; Forato, L. A.; Bortoleto-Bugs, R. K.; Fischer, H.; Mascarenhas, Y. P.; Ward, R. J. & Colnago, L. A. (2004). Spectroscopic characterization and structural modeling of prolamin from maize and pearl millet, *European Biophysics Journal*, 33: 335-343.

Cabra, V.; Arreguin, R.; Galvez, A.; Quirasco, M.; Vazquez-Duhalt, R. & Farres, A. (2005). Characterization of a 19 kDa a-zein of high purity, *Journal of Agricultural and Food Chemistry*, 53: 725–729.

Cantor, C. & Schimmel, P. (1980). Chapter 10: Size and Shapes of Macromolecules, In: *Biophysical chemistry, Part II: Techniques for the study of biological structure and function*, pp. 539-590, W.H. Freeman and Company, New York.

Coleman, C. E., & Larkins, B. A. (1999). The Prolamins of Maize, In: *Seed Proteins*. P. R. Shewry and R. Casey, eds., pp. 109-139, Kluwer Academic Publishers, The Netherlands.

Corradini, E.; de Medeiros, E.; Carvalho, A.; Curvelo, A. & Mattoso L. (2006). Mechanical and morphological characterization of starch/zein blends plasticized with glycerol, *Journal of Applied Polymer Science*, 101:4133–4139.

Dickey, L.; McAloon, A. & Parris, N. (2003). Minimizing entrainment of extract liquid by settling maize particles, *Industrial Crops and Products*, 18: 77–84.

Domenek, S.; Feuilloley, P.; Gratraud, J.; Morel, M.-H. & Guilbert, S. (2004). Biodegradability of wheat gluten based bioplastics, *Chemosphere*, 54 (4): 551–559.

Dong, J.; Sun, Q. & Wang, J.-Y. (2004). Basic study of corn protein, zein, as a biomaterial in tissue engineering, surface morphology and biocompatibility, *Biomaterials*, 25 (19): 4691–4697.

Forato, L. A.; Doriguetto, A. C.; Fischer, H.; Mascarenhas, Y. P.; Craievich, A. F. & Colnago, L. A. (2004). Conformation of the Z19 prolamin by FTIR, NMR, and SAXS, *Journal of Agricultural and Food Chemistry*, 52: 2382-2385.

Fu, D. & Weller, C. (1999). Rheology of zein solutions in aqueous ethanol, *Journal of Agricultural and Food Chemistry*, 47:2103–2108.

Garratt, R.; Oliva, G.; Caracelli, I.; Leite, A. & Arruda, P. (1993). Studies of the zein-like R-prolamins based on an analysis of amino acid sequences: implications for their evolution and three-dimensional structure, *Proteins: Structure, Function, and Genetics*, 15: 88-99.

Gaspar, M.; Benko, Z.; Dogossy, G.; Reczey, K. & Czigany, T. (2005). Reducing water absorption in compostable starch-based plastics, *Polymer Degradation and Stability*, 90 (3): 563–569.

Goldsmith, B. (1909). Thermoplastic Compound of Proteins and Amins. US Patent 922,133.

Grazuleviciene, V.; Augulis, L.; Grazulevicius, J.; Kapitanovas, P. & Vedegyte, J. (2007). Biodegradable starch, PVA, and peat composites for agricultural use. *Russian Journal of Applied Chemistry*, 80(11): 1928–1930.

Guo, Y.; Liu, Z.; An, H.; Li, M. & Hu, J. (2005). Nano-structure and properties of maize zein studied by atomic force microscopy, *Journal of Cereal Science*, 41: 277–281.

Johnson Jr., C. & Gabriel, D. (1994). Dynamic Light Scattering, In: *Laser Light Scattering*, pp. 22-38, Dover Publications, Mineola, New York.

Kayseriliolu, B.; Bakir, U.; Yilmaz, L. & Akka, N. (2003). Use of xylan, an agricultural by-product, in wheat gluten based biodegradable films: mechanical, solubility and water vapor transfer rate properties, *Bioresource Technology*, 87 (3): 239–246.

Kim, S.; Sessa, D. & Lawton, J. (2004). Characterization of zein modified with a mild cross-linking agent, *Industrial Crops and Products*, 20: 291–300.

Kim, S. & Xu, J. (2008). Aggregate formation of zein and its structural inversion in aqueous ethanol, *Journal of Cereal Science*, 47:1–5.

Kim, S. (2008). Processing and properties of gluten/zein composite, *Bioresource Technology*, 99: 2032–2036.

Kim, S., Xu, J. & Liu, S. (2010). Production of biopolymer composites by particle bonding, *Composites: Part A*, 41:146–153.

Lawton, J. (2002). Zein: a history of processing and use, *Cereal Chemistry*, 79 (1): 1–18.

Ma, X.; Chang P. & Yu, J. (2008). Properties of biodegradable thermoplastic pea starch/carboxymethyl cellulose and pea starch/microcrystalline cellulose composites, *Carbohydrate Polymers*, 72: 369–375.

Manley, R. & Evans, C. (1943). Binary solvents for zein, *Industrial Engineering Chemistry*, 35 (6): 661–665.

Matsushima, N.; Danno, G.; Takezawa, H. & Izumi, Y. (1997). Threedimensional structure of maize R-zein proteins studied by smallangle X-ray scattering, *Biochimica et Biophysica Acta*, 1339: 14-22.

Northolt, M. (1981). Compressive strength and glass transition temperature, *Journal of Materials Science*, 16: 2025–2028.

Olabarrieta, I.; Llstedt, M.; Ispizua, I.; Sarasua, J.-R. & Hedenqvist, M. (2006). Properties of aged montmorillonite-wheat gluten composite films, *Journal of Agricultural and Food Chemistry*, 54: 1283–1288.

Padua, G. W. and Wang, Q. (2002). Chapter 2: Formation and properties of corn zein films and coatings, In: *Protein-based films and coatings*, A. Gennadios, ed., pp. 43-68, CRC Press, USA.

Parris, N. & Coffin, D. (1997). Composition factors affecting the water vapor permeability and tensile properties of hydrophilic zein films, *Journal of Agricultural and Food Chemistry*, 45: 1596–1599.

Parris, N. & Dickey, L. (2003). Adhesive properties of corn zein formulations on glass surfaces, *Journal of Agricultural and Food Chemistry*, 51 (13): 3892–3894.

Pike, G. & Seager, C. (1974). Percolation and conductivity: a computer study I, *Physical Review B: Condensed Matter*, 10: 1421–1434.

Qu, Z.-H.; Wang, H.-J.; Tang, T.-T.; Zhang, X.-L.; Wang, J.-Y. & Dai, K.-R. (2008). Evaluation of the zein/inorganics composite on biocompatibility and osteoblastic differentiation, *Acta Biomaterialia*, 4: 1360–1368.

Samarasinghe, S.; Easteal, A. & Edmonds, N. (2008). Biodegradable plastic composites from corn gluten meal, *Polymer International*, 57(2): 359–64.

Sessa, D.; Eller, F.; Palmquist, D. & Lawton, J. (2003). Improvedmethods for decolorizing corn zein. Ind. Crop. Prod. 18, 55–65.

Shukla, R & Cheryan, M. (2001). Zein: the industrial protein from corn. *Industrial Crops and Products*, 13:171–192.

Tatham, A. S.; Field, J. M.; Morris, V. J.; Anson, K. J. I.; Cardle, L.; Dufton, M. J. & Shewry, P. R. (1993). Solution conformational analysis of the R-zein proteins of maize, *Journal of Biological Chemistry*, 268: 26253-26259.

Thompson, G. A. & Larkins, B. A. (1989). Structural elements regulating zein gene expression, *BioEssays*, 10: 108-113.

Uy, W. (1998). Dry spinning process for producing zein fibers. US Patent 5,750,064.

Vilcakova, J.; Saha, P. & Quadrat, O. (2002). Electrical conductivity of carbon fibres/polyester resin composites in the percolation threshold region, *European Polymer Journal*, 38 (12): 2343–2347.

Wang, D.; Yin, J.; Zhu, Z.; Ge, Z.; Liu, H.; Armes, S. P. & and Liu, S. (2006). Micelle Formation and Inversion Kinetics of a Schizophrenic Diblock Copolymer, *Macromolecules*, 39 (21): 7378–7385.

Wang, Q.; Wang, J.-F.; Geil, P. & Padua, G. (2004a). Zein adsorption to hydrophilic and hydrophobic surfaces investigated by surface plasmon resonance, *Biomacromolecules*, 5 (4): 1356–1361.

Wang, Q.; Geil, P. & Padua, G. (2004b). Role of hydrophilic and hydrophobic interactions in structure development of zein films, *Journal of Polymers and the Environment*, 12 (3): 197–202.

Wilson, C. M. (1988). Electrophoretic analysis of various commercial and laboratory-prepared zeins, *Cereal Chemistry*, 65: 72-73.

Wu, Q., Sakabe, H., Isobe, S. (2003a). Processing and properties of low cost corn gluten meal/wood fiber composite, *Industrial Engineering and Chemistry Research*, 42: 6765–6773.

Wu, Q.; Yoshino, T.; Sakabe, H.; Zhang, H. & Isobe, S. (2003b). Chemical modification of zein by bifunctional polycaprolactone (PCL), *Polymer*, 44 (14): 3909–3919.

Yamada, K.; Noguchi, A. & Takahashi, H. (1996). Effects of the solvents on properties of zein, *Nippon Shokuhin Kagaku Kogaku Kaishi*, 43 (3): 306–312.

Ye, P.; Reitz, L.; Horan, C. & Parnas, R. (2006). Manufacture and biodegradation of wheat gluten/basalt composite material, *Journal of Polymers and the Environment*, 14 (1): 1–7.

Zhang, X.; Do, M.; Dean, K.; Hoobin, P. & Burgar, I. (2007). Wheat-gluten-based natural polymer nanoparticle composites, *Biomacromolecules*, 8: 345–353.

5

Life Span of Biopolymer Sequestering Agents for Contaminant Removal and Erosion Resistance

Anna Sophia Knox[1], Ioana G. Petrisor[2], Charles E. Turick[1], Jesse Roberts[3], Michael. H. Paller[1], Danny. D. Reible[4], and Casey R. Forrest[4]
[1]Savannah River National Laboratory
[2]Haley&Aldrich, Inc.
[3]Sandia National Laboratories
[4]University of Texas
United States of America

1. Introduction

The objective of this paper is to report the development and life span of cross-linked biopolymers that remove contaminants, resist biodegradation over long periods of time, and resist erosion in dynamic aquatic environments.

Biopolymers are polymeric compounds produced by living organisms (e.g., microorganisms, plants, crustaceans). They have repeated sequences that vary broadly in chemical composition including a variety of repeating functional groups (such as carboxyl, hydroxyl, amino, etc.). This makes them reactive and subject to cross-linking. Therefore, biopolymers, a great molecular weight compounds with repeated sequences, may have high opportunity for chemical interaction with other compounds. Depending on their functional groups, biopolymers can bind metals, organic contaminants, or soil particles and form interpenetrating cross-linking networks with other polymers. The ability of biopolymers (cross-linked or not) to bind a large variety of metals is supported by many studies (Chen et al., 1993; Etemadi et al., 2003; Knox et al., 2008 a, b). The capacity of alginate as a cross-linked product (calcium alginate) for Cr(VI) uptake was demonstrated by Fiol et al. (2004), who obtained an uptake of 86.42 mmol of Cr(VI) per L of wet sorbent volume using grape stalk wastes encapsulated into calcium alginate. The Cr(VI) removal ability of cross-linked calcium alginate was also shown by Araujo and Teixeira (1997), and its ability to bind Cu was shown by Chen et al. (1990 and 1993) and Wan et al. (2004). The removal of Cu, Cr, and As from treated wood onto the biopolymers, chitin and chitosan, was shown by Kartal and Iamamura (2004). The use of biopolymers based on elastine-like polypeptides for the selective removal of Hg was reported by Kostal et al. (2003), who also reported their potential for binding and removal of other metals such as As and Cr. Recently, the use of a similar elastin-like polypeptide composed of a polyhistidine tail was exploited as a metal-binding biopolymer with high affinity toward Cd by Prabhukumar et al. (2004). Knox et al. (2007 and 2008 a, b) showed that biopolymers (with and without cross-linking) have the ability to sequester a large variety of metals (e.g., Cu, Pb, Cd, As, Cr, Zn, and Ni) and organic contaminants (e.g., phenanthrene and pyrene).

Apart from their contaminant binding ability, the use of biopolymers as plugging agents is well known. They are easily introduced in subsurface environments by injection under pressure using drilling equipment similar to that in the oil industry. Several studies (Yen et al., 1996; Stewart and Fogler, 2001; Khachatoorian et al., 2004) reported the application of biopolymers and associated microorganisms as plugging agents to construct a range of impervious barriers. Apart from their plugging effect, biopolymers can bind metals, soil/sediment particles and other biopolymers with the added ability to create cross-linking interpenetrating networks that may encapsulate contaminants into very stable forms as geopolymers (Kim et al, 2004 and 2005). Subsequently, the application of biopolymers to soils or sediments may result in the formation of barriers that isolate the contaminants, with possible permanent encapsulation of some contaminants and fixation of soils or sediments thereafter.

Cross-linking is the process of chemically joining two or more molecules by a covalent or ionic bond using a cross-linking agent (or cross-linker). Cross-linkers contain at least two reactive groups (identical or different). Functional groups that can be targeted for cross-linking include primary amines, sulfhydryls, carbonyls, carbohydrates, and carboxylic acids (most of these groups are possessed by biopolymers). Coupling also can be nonselective using a photoreactive phenyl azide cross-linker. Cross-linked biopolymers should become resistant to biodegradation and have the potential for remedial applications. As a result of the cross-linking process the biopolymer molecular mass increases. Many additional bonds are created between the biopolymer chains inside the cross-linked network, which makes an enzymatic attack (brake in the chain) less or even non- effective in terms of changes in the molecular weight of the whole cross-linked product. Cross-linking agents are used to enhance the strength of polymers and to decrease their biodegradability. Cross-linking agents are chosen based on the functional groups of polymers. The interpenetrating polymer networks (IPNs) developed by cross-linking may stop or slow the migration of contaminants as a result of increased viscosity, reduced permeability of porous media, and greater stablility at lower pH values.

In this study commercially available biopolymers were treated with cross-linking agents to produce cross-linked biopolymers that stabilize contaminants in soil or sediments while improving soil/sediment structure to reduce physical processes (e.g., erosion) that result in contaminant dispersal. Sediment and aqueous environments contaminated with heavy metals and organic contaminants remain a serious concern worldwide. New technologies focusing on permanent methods of in situ enclosure of contaminants effectively reducing risk levels to acceptable levels are currently being developed; one of them is capping technology. Capping technology is one of the few in-situ technologies available to environmental scientists and engineers to remediate polluted sediments in aquatic environments. Conventional capping technology usually employs sand as a cap, which may provide inadequate risk reduction at some sites. An alternative to a passive plain-sand cap is to directly amend the sand to provide a more sorptive medium to retain contaminants and further retard their transport from the sediment into the benthic zone and water column (Palermo et al., 1998). Retardation of the migrating contaminants serves a two-fold benefit. Not only are contaminants withheld from the overlying water column but their increased retention time in the sediment allows for natural attenuation processes to more fully degrade and reduce the associated hazards. Increased retention time allows for the deposition of clean sediment on top of the cap as well as more effective progress by the slower biodegradation processes.

Active caps, or permeable adsorptive barriers, are being developed and implemented as an alternative, cost effective remediation technology. Active capping involves the use of capping materials that are reactive and not only isolate but also sequester contaminants, further preventing their mobility, toxicity, and bioavailability (Knox et al., 2006, 2007, and 2008 a and b). Active capping technologies currently lay at the forefront of contaminated river sediment remediation techniques and research. Innovative capping technologies need to be explored to enhance the capabilities of a sand cap (Reible et al., 2006). Biopolymer materials composed, for example, of chitosan, xanthan gum, and guar gum may be a promising addition to materials for active capping technology due to greater organic content than in a conventional sand cap and therefore better sorption-related retardation. Also, the repeated sequences exhibited by biopolymers provide ample opportunity for chemical reactions with metals and sediment particles, allowing for the effective containment of contaminants.

Previous research suggests that cross-linked biopolymers are stable in soil, and that stability may increase over time, in some cases entrapping contaminants in stable geological structures such as geopolymers. However, the stability of biopolymers and their anticipated life span may be affected by environmental conditions indicating a need for further evaluation before this technology is deployed for remediation of contaminated soils or sediments. Biopolymers such as xanthan gum and chitosan are used extensively in industry resulting in substantial information concerning their biogdegradaability. Xanthan gum is not easily degraded by microorganisms (Cadmus et al., 1982), and chitosan polymers have antimicrobial activities. Interestingly, partially degraded chitosan demonstrates enhanced antimicrobial activities (Rhodes and Roller 2000). While biopolymer biodegradation is enhanced by elevated temperature and salt concentrations, metals sorbed to organic chelators can significantly decrease their rate of biodegradation (Francis and Dodge, 1993).

In this study biopolymers were evaluated for metal and organic contaminant removal, alone and in combination with soil/sediment, as potential candidate materials in active capping applications. This evaluation included the assessment of stability over time under varying environmental conditions (e.g., elevated temperature and moisture). Also evaluated were microbial effects on the properties of cross-linked biopolymers, and long term effects of selected cross-linked biopolymers on the mobility and retention of contaminants in sediments. Commercially available biopolymers that were evaluated in these laboratory studies included chitosan, xanthan, guar gum, and alginate. Cross-linking agents for these biopolymers included borax, xanthan, and calcium chloride. Since erosion control is a big part of successful remedial technology for contaminated sediments, e.g., active capping, the selected biopolymers were tested (as a slurry or coated on sand particles) for erosion resistance in the lab.

2. Materials and methods

Development of cross-linked biopolymer and biopolymer coated sand or sand/amendments is presented in Tab. 1 and 2. The most promising products, which had high carbon fractions (indicating greater coverage of biopolymer) and high viscosity were evaluated further for metal and organic sorption, biodegradability, and resistance to physical disturbance/erosion.

2.1 Sorption of metals on biopolymers

Xanthan crossed linked with guar gum and xanthan cross-linked with chitosan were evaluated for metal removal in a sorption study (Tab. 1). The experiments were conducted in 50 mL centrifuge tubes for one week. Each treatment had three replicates. The spike solution used in the experiment was obtained from Inorganic Ventures, Lakewood, NJ. The metal concentrations in the spike solution were 5 mg L^{-1} of As, Cd, Cr, Co, Cu, Pb, Ni, Se, and Zn. Suspensions composed of 0.2 gram of solid and 15 mL of spike solution were shaken for one week, phase separated by centrifugation, and then analyzed for metals by inductively coupled plasma - mass spectrometry (ICP-MS).

Major Product Name	**Primary Biopolymers**	**Cross-link Agent**	**Modified Product name**	**Biopolymer Sand ratio**	**Additives**			
					5% HCl	**Glutar-aldehyde**	**1N NaOH**	**water**
					mL	mL	mL	mL
CGB	Chitosan Guar gum	Borax	CGB1	0.05	475			500
			CGB2	0.05	200	5		300
			CGB3	0.05	200		20	400
GB	Guar gum	Borax	GB1	0.005	100			500
			GB2	0.005			20	600
			GB3	0.025		5	20	600
GX	Guar gum	Xanthan	GX1	0.05	100			500
			GX2	0.05			20	600
			GX3	0.05				500
XCc	Xanthan Chitosan	Calcium chloride	XCc	0.025	100	5		500
XC	Xanthan	Chitosan	XC					500
XG	Xanthan	Guar gum	XG					500

Table 1. Biopolymer products that were used for sand coating and contaminant sorption.

2.2 Sorption of organic contaminants on biopolymers

Sand samples coated with chitosan/guar gum cross-linked with borax (CGB3) and with xanthan/chitosan cross-linked with calcium chloride (XCc) were evaluated for their capacity to sorb organic contaminants (Tab. 1). PAHs used in this study were purchased from a commercial supplier (Sigma Aldrich, MO). They included 5000 μg L^{-1} phenanthrene in methanol, 1000 μg L^{-1} pyrene in methanol, and 200 μg L^{-1} benzo(a)pyrene in methylene chloride. These solutions were diluted in electrolyte solutions (0.01M NaCl, 0.01M $CaCL_2.2H_2O$) to prepare a mixture of 20 μg L^{-1} phenanthrene and 100 μg L^{-1} pyrene. Exact concentrations of these compounds were determined by high performance liquid chromatography (HPLC) affiliated with a Waters 2475 Multi λ Fluorescence Detector and Waters 996 Photodiode Array Detector. Sodium azide (0.05M) was added to the electrolyte solution to inhibit bacterial degradation of the PAHs.

2.3 Stability of cross-linked biopolymers under varying environmental conditions

2.3.1 Biodegradability of biopolymers

Biopolymer products were evaluated for biodegradability by microorganisms associated with the polymers. One gram of polymer material was mixed with and without 10 ml sterile basal salts medium (BSM) (Turick et al. 2002) and sealed in sterile test tubes with airtight

Product Number	Coated Sand	Biopolymers/ Cross-link Agent	Preparation Method
1	CGB	Chitosan / Guar Gum/ Borax	2 kg sand + 50 g guar gum + 50 g chitosan + 25 g borax + 300 mL 1N NOH + 6 L tap water Sand, biopolymer powders, and cross-link agent were well mixed as solids. One N NaOH was added to create a basic pH for cross-linking of guar with borax. Water was added in small amounts under continuous shaking, followed by the addition of acid under continuous shaking. The prepared material was placed on a rotary shaker (at about 30-40 rpm) overnight (12 h), then neutralized by the addition of 1N NaOH. The coated sand (as slurry) was collected wet and stored wet for erosion testing.
4*	XCc	Xanthan/ Chitosan / Calcium chloride/ glutaral-dehyde	2 kg sand + 50 g xanthan + 50 g chitosan + 15 g $CaCl_2$ + 75 mL glutaraldehyde + 6 L tap water Sand, biopolymer powders, and cross-link agent $CaCl_2$ were well mixed as solids, then 50 mL glutaraldehyde was added. Additional mixing was performed mechanically. Water was added in small amounts under continuous shaking. The prepared material was left on a rotary shaker (at about 30-40 rpm) overnight (12 h), after which the pH was adjusted to neutral and the mixture filtered through a sieve. The coated sand (as slurry) was collected wet and stored for erosion testing.
5*	XG	Guar Gum/ Xanthan	2 kg sand + 25 g guar gum + 25 g xanthan + 6 L tap water Sand and biopolymer powders were well mixed as solids. Water was then added under continuous shaking. The prepared material was placed on a rotary shaker (at about 30-40 rpm) overnight (12 h). The coated sand (as slurry) was collected wet and stored wet for erosion testing.
6*	XG		2 kg sand + 50 g guar gum + 50 g xanthan + 6 L tap water The same procedure as product #5.
7*	AXG	Guar Gum/ Xanthan/ NC apatite	1.75 kg sand + 0.25 kg apatite + 50 g guar gum + 50 g xanthan + 6 L tap water The same procedure as product #5.
8*	OXG	Guar Gum/ Xanthan/ PM-199 organoclay	1.75 kg of sand + 0.25 kg of organoclay + 50 g guar gum + 50 g xanthan + 6 L tap water The same procedure as product #5.
9*	XG/AO	Guar Gum/ Xanthan/ NC apatite/PM-199 Organoclay	1.5 kg of sand + 0.25 apatite + 0.25 kg of organoclay + 50 g guar gum + 50 g xanthan + 6 L tap water The same procedure as product #5.

Table 2. Preparation methods for biopolymers and biopolymer coated sand materials. Materials selected for erosion tests in an Adjustable Shear Stress Erosion and Transport (ASSET) flume are marked by an asterisk.

butyl rubber stoppers. Uncoated sand was used as a control for comparison with the biopolymer coated sand. Static incubation in the dark was at 0°C and 35°C. Low temperature (0°C) and high temperature (35°C) and wet and dry moisture regimes simulated a broad range of environmental conditions and seasonal changes. Microbial activity (biopolymer degradation) was measured by CO_2 release with a Hewlett Packard 5890 series 2 gas chromatograph (GC) with a mass spectrometer. A 250 μl sample of the headspace gas was injected into the GC using a gas tight syringe with a side-hole needle. A carrier gas of helium was used to move the sample through the column into the mass spectrometer. An internal standard of argon was used to calculate CO_2 production in the samples. The release of CO_2 from the biopolymer coated sands was measured for ten weeks. Additionally, metal concentrations for biopolymers from the sorption experiment were also evaluated (ICP/MS) upon termination of this experiment in an effort to correlate biopolymer breakdown with metal release. Microbial density on biopolymers was characterized at the termination of the study by direct microscopic counts. Filtered biopolymers and cells were stained with 496-diamidino-2-phenylindole (DAPI), and stained cells were detected using epifluorescent illumination (Lehman et. al 2001). The heterogeneous nature of the biopolymers did not permit quantitative enumeration of cells, but results generally correlated with CO_2 evolution.

2.3.2 Microbial effects on the properties of biopolymers

Microbial degradation of the biopolymers was further evaluated by sampling polymers that appear to be biodegradable (as indicated by increased CO_2 evolution) by soil microbes. This method included addition of sterile biopolymers to BSM or BSM solidified with 1.5% Noble agar, thus providing the biopolymers as the sole source of carbon for microbial isolates associated with the biopolymers (above). As a source of bacterial inocula, fresh sediment was treated following the methods of Lehman et al. (2001), which included addition of 0.1% sodium pyrophophate, and blended followed by sedimentation (for 24 hrs) of sediment particles. Disorbed bacteria, in suspension, were used as inocula for degradation studies following washing in phosphate buffer. A 1% inoculun (wt/vol) was added to sterile biopolymers in gas tight vials as above. Carbon dioxide evolution and oxygen utilization were monitored over time to determine the rate (if any) of degradation. Controls consisted of uninoculated biopolymers as well as inoculum without biopolymers. The samples were incubated statically in the dark for 2 weeks at 25°C to evaluate microbial growth resulting from biopolymer breakdown. Gas analyses were conducted either through periodic GC headspace analysis or via respirometry. Following the incubation period (when CO_2 concentrations level off), biopolymer material was dried and weighed to determine loss due to biodegradation, and a carbon balance was attempted with the CO_2 data. Microbial isolates were grown with specific polymers (i.e., guar gum, chitosan, glutaraldehyde, etc.) on BSM and Nobel agar plates to obtain a gross characterization of microbial community changes resulting from polymer biodegradation. Growth did not occur on Nobel agar plates without supplemental carbon. Biopolymers were also evaluated before and after degradation using electron microscopy to determine if biodegradation affected polymer size. Polymer breakdown products were evaluated by ion chromatography and/or GC/MS.

2.4 Evaluation of biopolymer resistance to physical disturbance

The resistance of biopolymers to physical disturbance was evaluated by shaker tests and by an Adjustable Shear Stress Erosion and Transport (ASSET) flume.

2.4.1 Evaluation of suspension by shaker tests

The biopolymer materials tested for erosion resistance and methods for preparing biopolymer products are described in Tab. 2. These products were prepared and kept wet (as a slurry) for testing as a wet product. Two -5 kg of each product were produced. The main objective of this study was to identify biopolymer products that could resist erosion.

A standard shaker test (Tsai and Lick, 1986) was used to assess the ability of sand and amendments with and without biopolymers to resist physical disturbance. The shaker allowed a standardized assessment of the shear stress needed to suspend a sample. The shear stress needed to suspend a sample was measured using a sampling port (Fig. 1). Seven cm of sample was placed inside of the cylindrical chamber, and sufficient distilled water was added to the cylinder to bring the depth to 12.7 cm. Suspended sample particles were allowed to settle, the cylinder was reattached to the shaker, and the motor speed was recorded when the following conditions were visually observed: fine top particles disturbed, motion of top particles, cloudiness, full re-suspension of top layer, and full re-suspension of bottom layer. The motor speed of the shaker drive disc was measured with a tachometer in meters per minute. The circumference of the drive disc (17 cm) was measured and used to convert the m/min measurements into revolutions per minute (0.17 rpm), which was further converted to oscillations by the following equation (equation 1)

$$oscillationperiod = \frac{1}{60 \times RPM} \tag{1}$$

Fig. 1. Shaker for simulating erosion.

2.4.2 Evaluation of erosion by Adjustable Shear Stress Erosion and Transport (ASSET) flume

2.4.2.1 Description of the ASSET flume

The ASSET flume is considered a next generation SEDflume in that it maintains all capabilities of its predecessor while also quantifying the transport modes of the sediments after erosion. The erosion test section of the ASSET flume is similar in design, with a slightly taller channel height, and identical in erosion testing operation to the SEDflume, which has been described extensively in the literature (McNeil et al., 1996; Jepsen et al., 1996; Roberts et al., 1998; Jepsen et al., 1997; Roberts and Jepsen, 2001). It consists of eight primary components (Fig. 2): a 120 gallon reservoir, a 200 gpm centrifugal pump, a motor controlled screw jack, an erosion channel including erosion test section, a transport channel including bedload traps, a three way valve, a magnetic flow meter, and connective plumbing. Water is pumped from the reservoir through the three-way valve, which either sends water directly back to the reservoir or through the flow meter to the erosion and transport channels (and then back to the reservoir). A manually controlled screw jack is used to push the sediments through the core tube to keep the sediment surface flush with the channel floor such that, as closely as possible, the sediments are exposed only to an applied shear stress.

The ASSET Flume's enclosed (internal flow) erosion and transport channels are 5 cm tall and 10.5 cm wide (Fig. 2). The erosion test section is preceded by 180 cm of enclosed rectangular channel to ensure fully developed turbulent flow over the sediment core. The cylindrical sediment core tube is 10 cm in diameter. The first bedload trap is located 1m from the center of the erosion test section, and the center of each successive trap is 1 m from the center of the preceding one. Based on the theoretical definition of bedload in combination with fluid velocities and particle/aggregate settling speeds, a bedload particle/aggregate should contact the flume floor at least once every 15 cm of downstream travel. Consequently, the traps are 15 cm long and span the width of the channel (10.5 cm). Capture basins that are 10 cm deep and have a 2 L volume are located below the traps, each with a baffle system that reduces recirculation and minimizes the resuspension of trapped sediments. As the sediment core is eroded upstream, some of the material is suspended and some is transported as bedload. All sediment that falls into the traps is considered bedload.

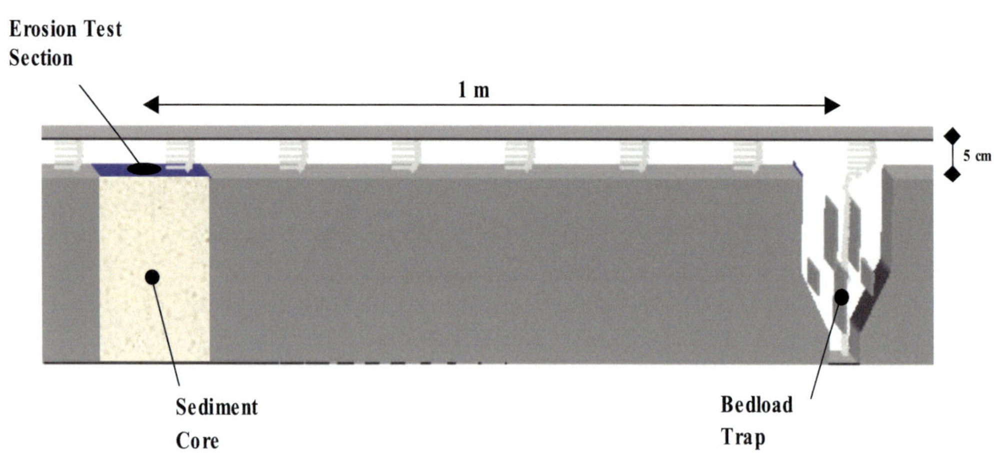

Fig. 2. Schematic of the ASSET flume.

2.4.2.2 Hydrodynamics

The hydrodynamics within the flow channel of the ASSET Flume are equivalent to those of the SEDflume (McNeil et al., 1996); however, the increase in duct height necessitated a change to the system inlet. To achieve fully developed turbulent flow over the sediment core, the flume inlet was lengthened to 180 cm and preceded by a 20 cm circular-to-rectangular flow converter and several meters of inlet pipe. Turbulent flow through pipes has been studied extensively, and empirical functions have been developed that relate the mean flow rate to the boundary shear stress. In general, flow in circular cross-section pipes has been investigated. However, the relations developed for flow through circular pipes can be extended to non-circular cross-sections by means of a shape factor. An implicit formula relating the boundary shear stress to the mean flow in a pipe of arbitrary cross-section can be obtained from Prandtl's Universal Law of Friction (Schlichting, 1979). For a pipe with a smooth surface, this formula is

$$\frac{1}{\sqrt{\lambda}} = 2.0\log\left[\frac{UD\sqrt{\lambda}}{\nu}\right] - 0.8 \qquad (2)$$

where U is the mean flow speed, ν is the kinematic viscosity, λ is the friction factor, and D is the hydraulic diameter. For a duct with a rectangular cross-section the hydraulic diameter is

$$D = 2hw/(h + w) \qquad (3)$$

where w is the duct width and h is the duct height. The friction factor is defined as

$$\lambda = \frac{8\tau}{\rho U^2} \qquad (4)$$

where ρ is the density of water and τ is the wall shear stress. Substituting Eqs. (3) and (4) into Eq. (2) yields the boundary shear stress as an implicit function of the mean flow speed. The mean flow speed and hence the boundary shear stress are controlled by the pump speed. For flow in a circular pipe, turbulent flow theory suggests that the transition from laminar to fully turbulent flow occurs within 25 to 40 diameters from the entrance to the pipe. Because the hydraulic diameter of the duct is 6.8 cm, this indicates a necessary entry length between 170 and 270 cm, which is supplied by the inlet piping, converter, and ducting. Furthermore, for shear stresses in the range of 0.1 to 10 N/m^2, the Reynolds numbers, UD/ν, are on the order of 10^4 to 10^5 implying that turbulent flow existed in all experiments performed for this study. These arguments along with direct observations indicate that the flow is fully turbulent in the test section.

2.4.2.3 Sample collection and preparation for the ASSET flume Test

Samples tested in the ASSET flume were prepared following the method described in Tab. 2. All materials were stored in a refrigerator at 4°C. The samples were stirred and poured into erosion core tubes to a depth of 10 cm. Five cm of water was gently poured on top of each material, which was returned to a refrigerator. Each sample remained in the refrigerator until the day of the erosion test for consolidation times of 2, 5, 10, and 175 days.

2.4.2.4 Measurements of sediment erosion rates and critical shear stress

To measure the erosion rates of the samples as a function of shear stress and depth, the samples were placed upward into the test section until the sample surface was even with the bottom of the test section. A measurement was made of the depth to the bottom of the sample in the tube. The flume was then run at a specific flow rate corresponding to a particular shear stress. Erosion rates were obtained by measuring the remaining sample depth at different time intervals, taking the difference between each successive measurement, and dividing by the time interval.

To measure erosion rates at several different shear stresses using only one sample, the flume was run sequentially at higher shear stresses with each succeeding shear stress being 1.33, 1.5 or 2 times the previous one. Generally between three and five shear stresses were run sequentially. Each shear stress was run until at least 0.5 mm but no more than 10 mm were eroded. The time interval was recorded for each run with a stop watch. The flow was then increased to the next shear stress, and so on until the highest shear stress was run.

A critical shear stress can be quantitatively defined as the shear stress at which a very small, but accurately measurable rate of erosion occurs. In the present study, this rate of erosion was chosen to be 10^{-4} cm/s; this represents 1 mm of erosion in approximately 15 minutes. Since it would be difficult to measure all critical shear stresses at exactly 10^{-4} cm/s, erosion rates were generally measured above and below 10^{-4} cm/s at shear stresses which differ by a factor of 1.33, 1.5 or 2. The critical shear stress was then linearly interpolated to an erosion rate of 10^{-4} cm/s. This gave results with 20% accuracy for determination of critical shear stress.

2.4.2.5 Erosion rate ratio analysis

The erosion rate data collected for each sample is generally plotted as erosion rate as a function of depth from the sediment surface and applied shear stress. The non-linear relationship between erosion rate and bed shear stress can make it difficult to quantify variability in the erosion behavior within a single core and between many cores. In order to overcome this limitation, the data can be analyzed to determine an erosion rate ratio that produces a single numerical value for a particular erosion rate data series that accounts for this non-linear relationship (Jones et. al, 2008). The erosion rate ratio is used to make direct comparisons between erodibility within a single core (i.e. to identify changes with depth), between similar cores, and between all tested cores to aid in the identification of the most erosion resistant cap material.

In this analysis, each core was sub-sampled into separate depth intervals. Following the methods of Roberts et al (1998), the erosion rate for each depth interval can be approximated by a power law function of sediment density and applied shear stress. For a particular depth interval, density is assumed to remain relatively constant, therefore the density term is dropped. For each depth interval, the measured erosion rates (E) and applied shear stresses (τ) are used to develop the following equation.

$$E = A\tau^{n} \qquad (5)$$

Where E is the erosion rate (cm/s) and τ is shear stress (Pa). The parameters A and n are determined using a log-linear regression analysis. From this analysis an average erosion rate for the entire core can also be determined, and the erosion rate at each depth interval can be directly compared to this average. The result is an erosion rate ratio which provides an estimation of the erosion susceptibility of each depth interval relative to the core average. In addition, an average erosion rate of similar core and for all cores can be determined. The erosion rate for each depth interval within a core as well as each cores average erosion rate can be compared to the specified average and a graph of the erosion rate ratios for all of the cores can be created and compared to the average erosion behavior of all cores.

3. Results and discussion

3.1 Sorption of metals and organic contaminants by biopolymers

The most promising materials for metal removal were xanthan crossed linked with guar gum and xanthan cross-linked with chitosan (Fig. 3). These cross-linked biopolymers sequestered a variety of metals from a spike solution; e.g., As, Cd, Co, Cr, Cu, Ni, Pb, and Zn (Fig. 3). Removal of most metals by xanthan crossed linked with guar gum and xanthan cross-linked with chitosan exceeded 90% (Fig. 3). The tested biopolymers were as effective at removing metals from the spike solution as apatite (rock phosphate from North Carolina), which has proven ability to immobilize metals such as Pb, Cd, and Zn in soils and sediments (Knox et al, 2008a) (Fig. 3).

Various processes such as adsorption, ion exchange, and chelation dominate the mechanisms responsible for complex formation between biopolymers and metal ions. The interactions of metal ions with biopolymers (e.g., chitosan) also are influenced by the degree of polymerization and deacetylation, as well as by the distribution of acetyl groups along the polymer chains (Bassi et al., 1999). The evidence currently available supports the concept that chitosan-metal ion complex formation occurs primarily through the amino groups functioning as ligands (Udaybhaskar et al., 1990; Randall et al., 1979).

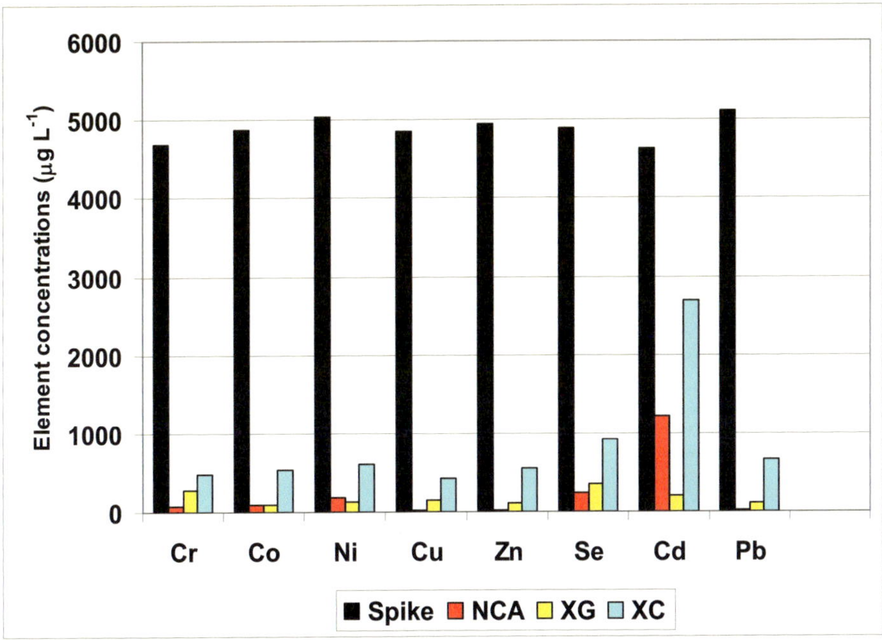

Fig. 3. Removal of metals by biopolymers from a spike solution with an initial concentration of 4800 µg L^{-1} of As, Cd, Co, Cr, Cu, Ni, Pb, Se, and Zn; NCA - North Carolina apatite, XG - xanthan cross-linked with guar gum, XC - xanthan cross-linked with chitosan.

Organic carbon content was measured as an indicator of the efficiency of the procedure for coating the sand with biopolymers and as an indicator of the potential for the sorption of organic contaminants on the coated sand. The measured carbon fractions are presented in Fig. 4. The carbon fractions of coated sand with one wash, two washes and three washes did not differ substantially from the unwashed sand, indicating that the coated sand was resistant to washing.
Sorption capacities of sand samples coated with chitosan/guar gum cross-linked with borax (CGB) and with xanthan/chitosan cross-linked with calcium chloride (XCc) are presented in Tab. 3. Both biopolymer coated sand materials had a significantly higher soption capability than sand for pyrene (Tab. 3).

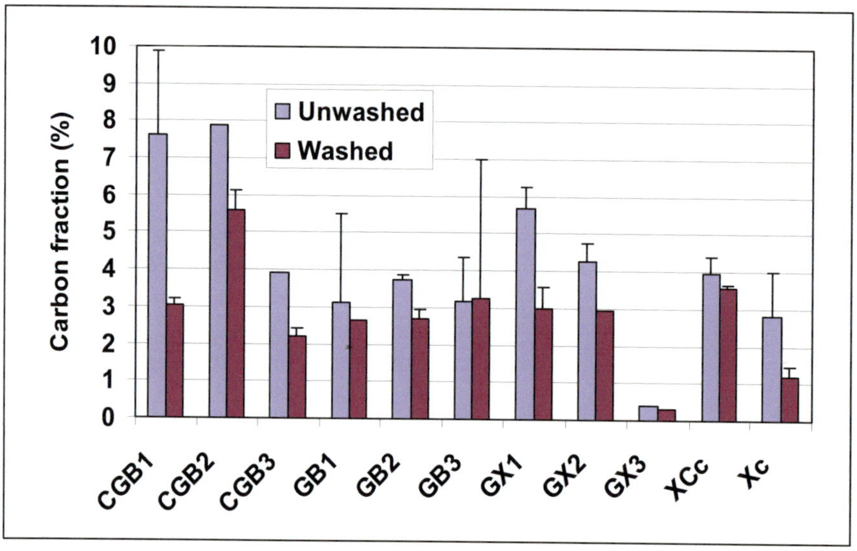

Fig. 4. Carbon fraction of biopolymer coated sand; C – chitosan, G – guar gum, B - borax, X – xanthan, c – calcium chloride.

	Phenanthrene L kg^{-1}	**Pyrene L kg^{-1}**
Sand	**3.19** (1.87)	**27.01** (5.34)
CGB1	**0.4** (-)	**29.54** (4.19)
CGB2	**27.72** (3.07)	**127.1** (23.26)
CGB3	**40.64** (24.32)	**118.3** **(17.15)**
GB2	**13.18** (2.32)	**68.82** (14.4)
XCc	**12.8** (3.42)	**106.7** (15.08)

Table 3. Average sorption coefficients of sand and sand coated by biopolymers (standard deviation in parentheses); B - borax, C - chitosan, G - guar gum, X - xanthan, c - calcium chloride.

3.2 Stability of cross-linked biopolymers under varying environmental conditions

Cross-linking agents added to biopolymers enhance their strength and decrease their biodegradability. However, the stability of biopolymers and their life span in the field may be affected by environmental conditions. The stabilities of selected cross-linked biopolymers were evaluated over extended periods (10 weeks to 6 months) in the laboratory using temperature and leaching studies to simulate accelerated weathering. Indications of degradation or loss of effectiveness served to identify potential concerns with long-term stability.

3.2.1 Biodegradability of biopolymers

A ten week evaluation of several biopolymers (Fig. 5) showed that chitosan cross-liked with guar gum and borax (CGB) and xanthan cross-linked with chitosan and calcium chloride (XCc) had the lowest evolution of CO_2; i.e., the lowest degradability. Biopolymers, especially xanthan cross-linked by guar gum, degraded faster under wet conditions and high temperatures (35°C) than under dry conditions (Fig. 6).

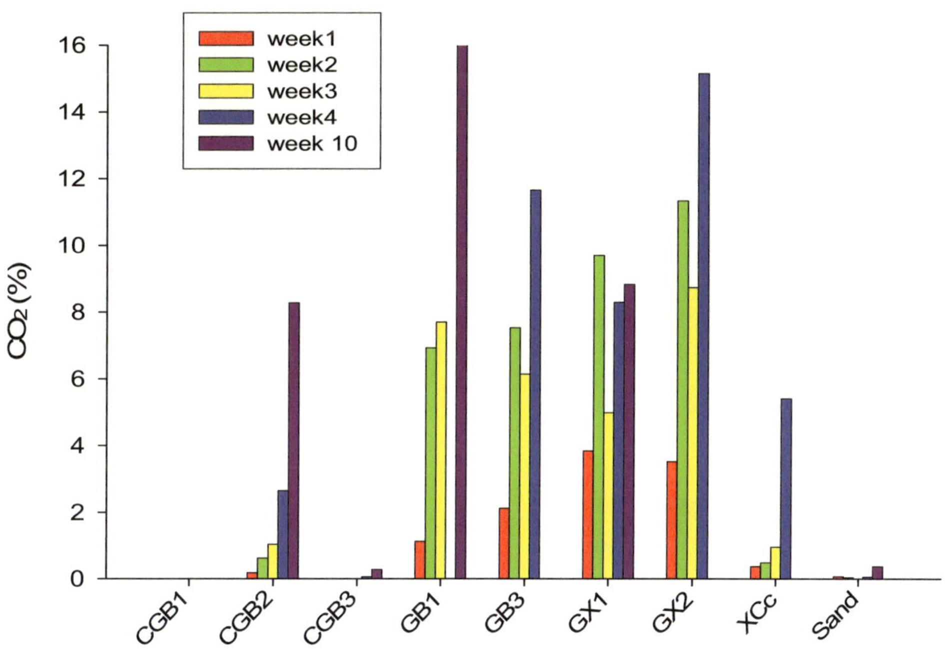

Fig. 5. Release of CO_2 (measured by GC-MS) from several cross-linked biopolymers: B - borax, C - chitosan, G - guar gum, X - xanthan, c - calcium chloride, 1 & 3 - without glutaraldehyde, 2 - with glutaraldehyde, 3 – with NaOH

Microbial densities associated with the biopolymers likely were a result of bacteria present during manufacture of the biopolymers. Minimal increases in bacterial densities and CO_2 release over 6 months and under various conditions indicated that biopolymer-associated microbes did not contribute significantly to the degradation of some biopolymers (Fig. 7). Biopolymers with sorbed metals demonstrated decreased CO_2 release and likely minimal biodegradation compared to biopolymers without sorbed metals (Fig. 8 and 9). Obvious morphological differences in bacteria isolated from biopolymers further indicated that different microbial consortia were associated with biopolymers as a function of metal concentration. Biodegradation of biopolymers resulted in minimal release of metal contaminants (Fig. 10).

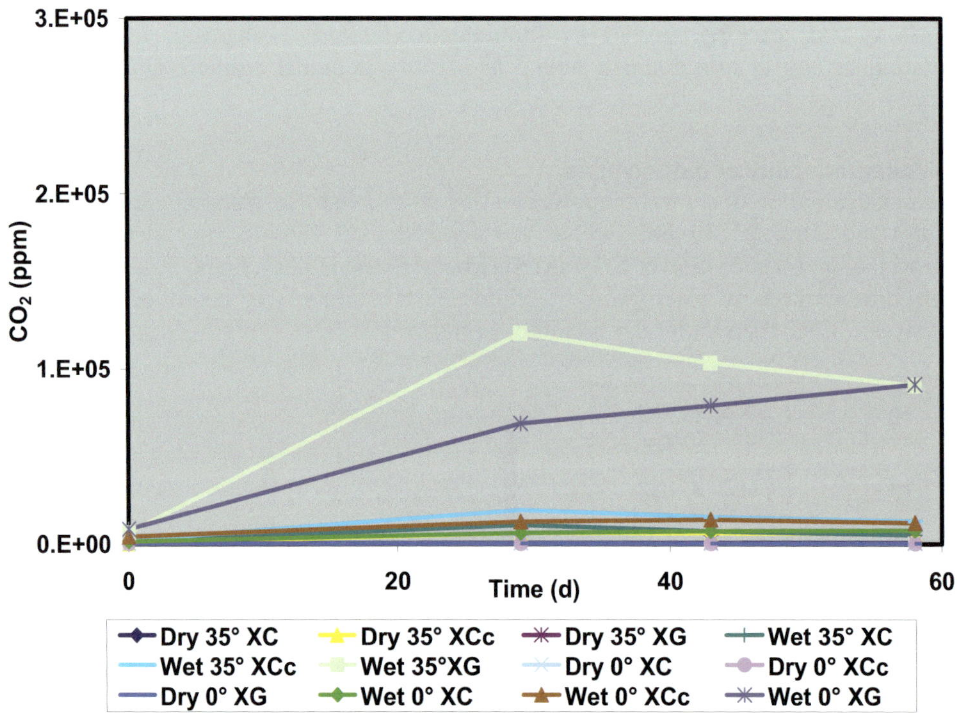

Fig. 6. Evaluation of biopolymer degradation under wet/dry conditions and different temperatures; X - xanthan, G - guar gum, C- chitosan, c - calcium chloride.

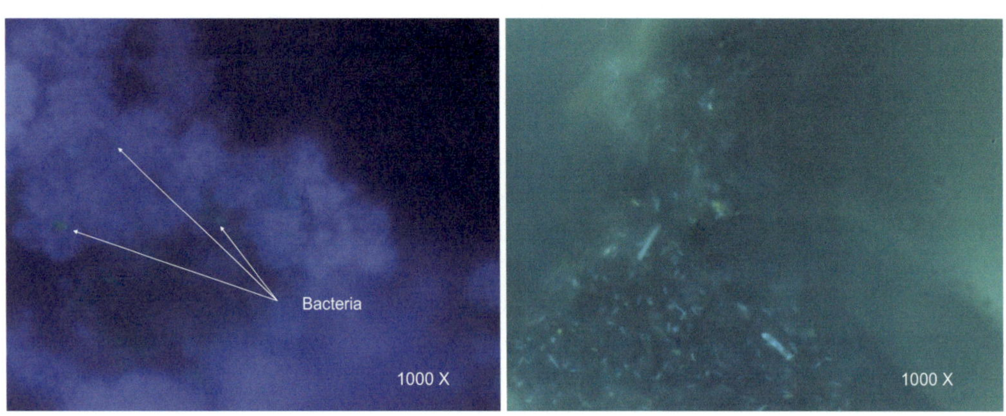

Fig. 7. Microscopic analyses of biopolymer surfaces using 4',6-diamidino-2-phenylindole (DAPI) and epifluorescence microscopy. Biopolymer XCc (left) contained fewer bacteria than CGB (right) after 6 months of contact with sediment suggesting limited biodegradation.

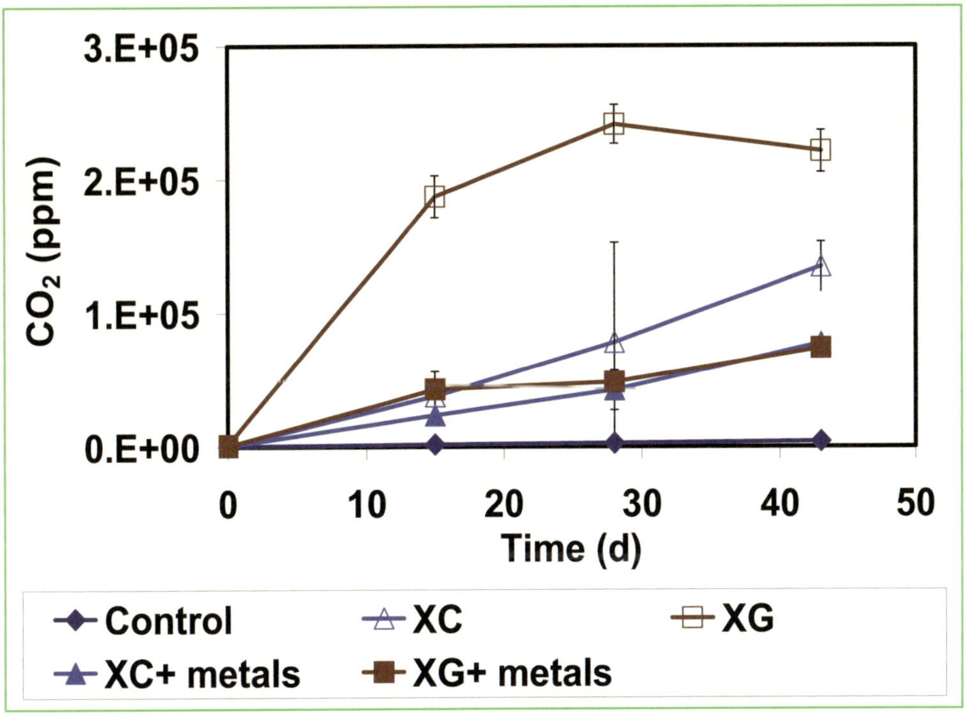

Fig. 8. Evaluation of biopolymer degradation for 45 days; X - xanthan, G - guar gum, and C-chitosan. Metal sorption of biopolymers inhibited bacterial activity.

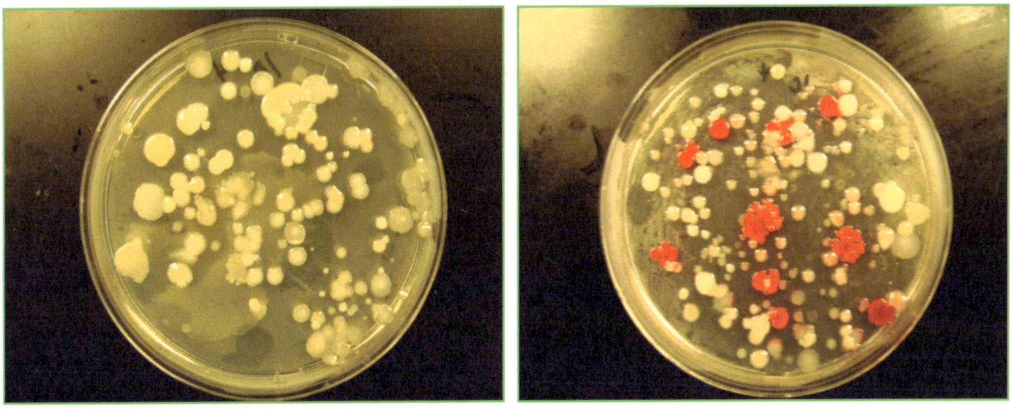

Fig. 9. Morphological differences in bacterial populations after exposure to Xanthan biopolymers without (left) and with (right) sorbed metals.

This study showed that cross-linked biopolymers have the potential to remove contaminants from the aqueous phase and to stabilize contaminants in soils/sediments. Cross-linked biopolymers vary in their susceptibility to biodegradation, with some being resistant for several months. Biopolymer degradation did not result in contaminant release during the test period. Our research showed that cross-linked biopolymers are promising for remediation, but longer periods of evaluation under field conditions are still needed.

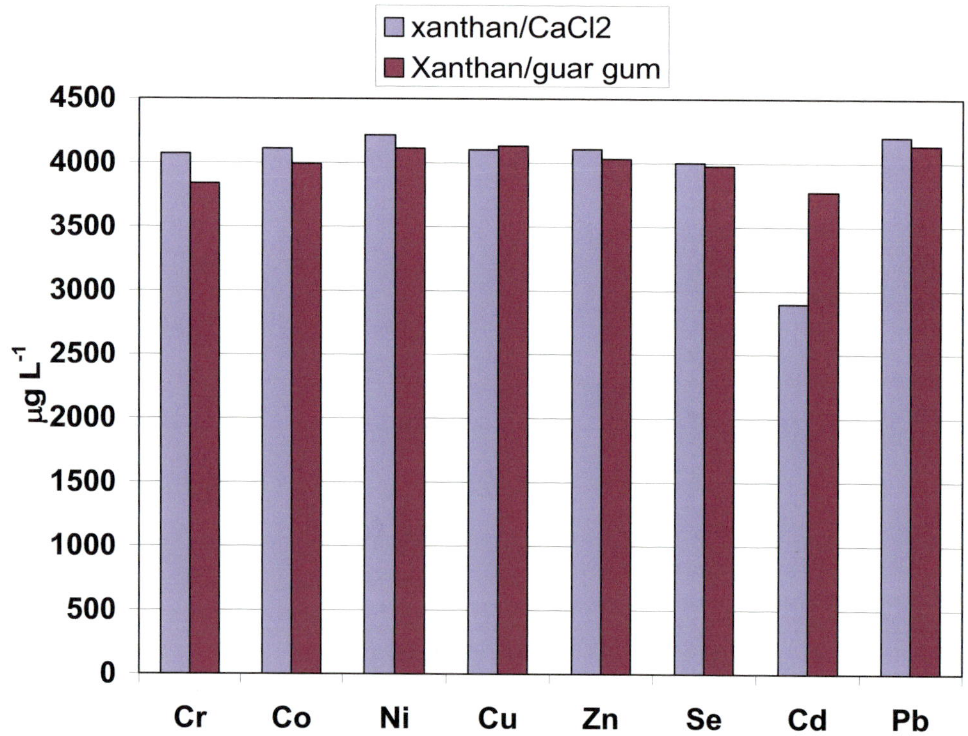

Fig. 10. Metals remaining in biopolymers after 3 months of biodegradation (initial concentration of the spike solution was 5000 μgL^{-1}).

3.3 Evaluation of biopolymer resistance to physical disturbance

Various biopolymers were cross-linked with and without coating on sand or amendments (Tab. 2). The cross-linked biopolymer products had increased viscosity and shear strength (results not shown here). They also had an evident cohesiveness, some of them looking and acting literally like "glues" when wet (Fig. 11). These physical characteristics indicate that some biopolymers have potential for use in active caps as an adhesive, binding cap materials together, and for removal of contaminants.

3.3.1 Shaker tests

The shaker suspension-simulation device was used to test the suspension resistance of sand, biopolymer coated sands, and organoclay. Five suspension thresholds were established: fine top particles disturbed, motion of top particles, cloudiness, full re-suspension of top layer, and full re-suspension of the bottom layer. The oscillations of the grid used to produce these motions were converted into shear stresses. These stresses were compared with those calculated with Shield's curve equations, which are indicators of the stability of non-cohesive particles in a bed. Full re-suspension of the bottom layer of sand was not achieved (Fig. 12). The deepest penetration into the 7cm high sand layer was 1.5 cm. The maximum speed that the motor was able to achieve was approximately 650 rpm. In the paper by Tsai and Lick, maximum speed derived from the given oscillation periods was 750 rpm (Tsai and

Lick 1986). Full re-suspension of the bottom layer may have been achieved if the motor would have reached higher speeds.

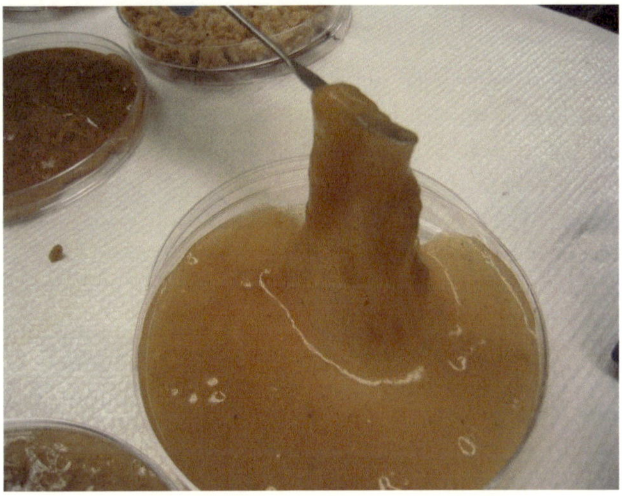

Fig. 11. Adhesive product of sand coated with guar gum cross-linked by borax.

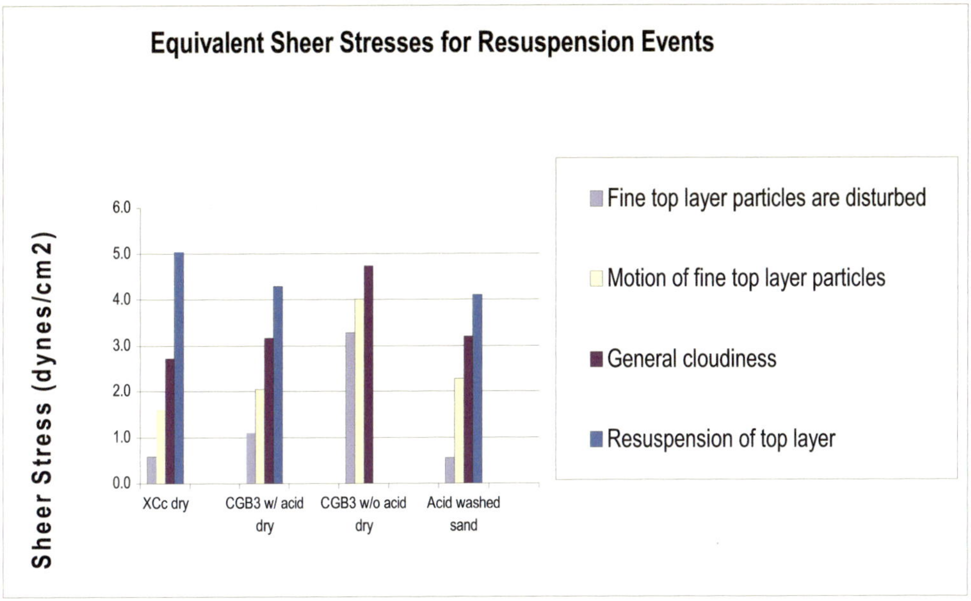

Fig. 12. Effects of equivalent sheer stresses on resuspension of sand and three types of dry and rewetted biopolymer coated sand.

Dried biopolymer coated sands CGB3 and XCc treated with addition of HCl in the preparation process performed similarly to plain sand. The biopolymer coatings in these products did not become viscous after rewetting and did not aid in preventing suspension of the sediment columns (Fig. 12). However, the dry coated sand CGB3 prepared without

HCl (Table 2) produced a viscous gel immediately upon rewetting and was resistant to suspension (Fig. 12). The gel properties of the rewetted biopolymer significantly increased the sheer stresses required for resuspension of sediments (Fig. 12). Even at maximum rotational speed very little disturbance of CGB3 was observed and the top layer was never resuspended (Fig. 12).

Additional suspension experiments were conducted to test the stability of multiple biopolymer materials when placed in viscous slurry rather than first drying and then rewetting. The results of these tests are displayed in Fig. 13. The slurry products performed very well in the suspension experiments. Initial oscillations of the grid caused the uppermost portions of the slurries to pulse vertically but no sloughing of the samples occurred. Increased sheer stresses resulted in minor sloughing of small particles but no resuspension occurred. Sand with biopolymers, e.g., xanthan cross linked with guar gum (XG), was suspension resistance even at a speed of 11 m/s (Figs. 14 & 15). Mixtures of biopolymer XG, sand, and other amendments such as organoclay and/or apatite were also suspension resistant at a speed 11 m/s (Fig. 16). The significant resistance of the slurry products to suspension shows promise for future applications as a stand-alone active cap or as armament for other amendments.

Organoclay (PM-199) without biopolymers was not suspension resistant. If placed in a flowing aquatic environment, a cap of organoclay would erode like a plain sand cap (Fig. 17).

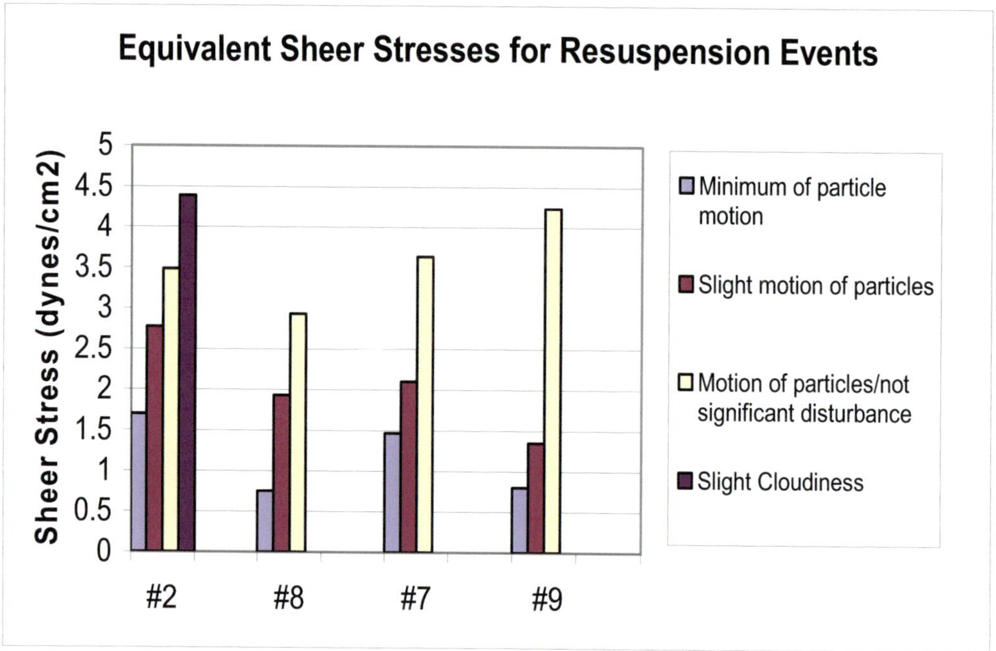

Fig. 13. Effects of equivalent sheer stresses on sand and slurries of biopolymer coated sand: 2 - sand with chitosan/guar gum/borax, 7 - sand with xanthan/guar gum and apatite, 8 - sand with xanthan/guar gum and organoclay (PM-199), 9 - sand with xanthan/guar gum and apatite, and organoclay.

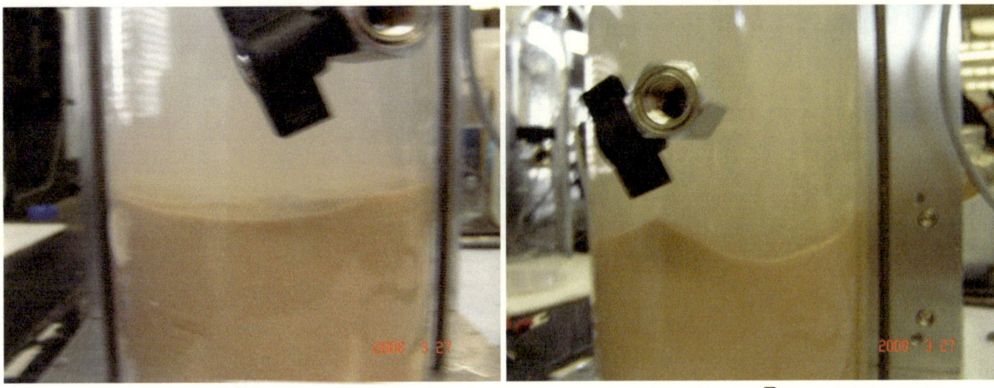

A B

Fig. 14. Plain sand was easily resuspended at very low speed (3.7 m/s) (A) and exhibited marked erosion at higher speed (11 m/s) (B).

A B

Fig. 15. Coated sand with xanthan and guar gum did not exhibit erosion at speeds of 3.7 m/s (A) or 11m/s (B).

A B

Fig. 16. Sand and organoclay (PM-199) mixed with the biopolymers xanthan and guar gum did not show erosion at speeds of 3.7m/s (A) or 11m/s (B).

A B

Fig. 17. Resuspension of organoclay (PM-199) leading to cloudiness (A) and surface suspension (B).

3.3.2 Adjustable Shear Stress Erosion and Transport (ASSET) flume

The materials evaluated in the shaker and ASSET flume tests are listed in Tab. 4 where they have been assigned product numbers. The products included:

- #4 XCc - sand and xanthan/chitosan cross-linked with calcium chloride and glutaraldehyde,
- #5 XG - sand and 2.5% guar gum cross-linked with xanthan (Kelzan brand),
- #6 XG Coyote - sand and 5% guar gum cross-linked with xanthan (Coyote brand),
- #6 XG Kelzan - sand and 5% guar gum cross-linked with xanthan (Kelzan brand),
- #7 AXG - sand, 12.5% apatite and 5% guar gum cross-linked with xanthan (Kelzan brand),
- #8 OXG - sand,12.5% organoclay, and 5% guar gum cross-linked with xanthan (Kelzan brand); and
- #9 XG/AO - sand, 12.5% organclay, 12.5% apatite, and 5% guar gum cross-linked with xanthan (Kelzan brand).

Erosion rates as a function of shear stress and depth were obtained for six of the seven materials after 2, 10, and 175 days of consolidation at shear stresses of 0.25, 0.5, 0.75, 1.0, 1.5, 2.0, 3.0 and 4.0 N/m^2. After preliminary tests of the #7 AXG material at 2 and 5 days, the sample did not show promise as an erosion resistant cap material and was dropped from further testing. The ASSET flume erosion tests enabled stability evaluation of nearly the entire thickness of the cap (~10 cm) under simulated flow conditions that ranged from quiescent to the shear environment in extreme storm events. The erosion rate ratio analysis is used to compare average erosion behavior at distinct intervals within a core. An example of how the two methods correlate is shown for the #9 XG/AO and #8 OXG samples after 10 days of consolidation (Fig 18). Graphics A and B in Fig. 18 show that erosion decreases with increasing depth in the # 9XG/OA cap material while graphics C and D show increased erosion in the center of the core with the most erosion resistant layer at the bottom.

The individual erosion behavior of the six dominant cap materials is shown in Fig. 19. The six graphics compare the erosion behavior at 2, 10 and 175 days of consolidation at each erosion inverval as well as the core average. This enables the evaluation of the erosion behavior of each cap material as a function of consolidation time and depth within the core sample. For example, the #4 XCC and #9 XG/OA samples display a general hardening or resistance to erosion with increasing core age. Both samples also show that each core became more stable at depth at all ages except for the second depth interval in the 2 day #9 XG/OA sample. It is important to note that the scale on the #9 XG/OA plot is expanded by two orders of magnitude and shows that the oldest, 175 days of consolidation, sample is the most stable or erosion resistant core of all. The #6 XG Kelzan sample shows the opposite trend in that it becomes softer or less erosionaly stable as the sample ages. This same sample shows that the surface layer at all ages is always the easiest to erode with general, but not consistent trends, of hardening at depth. The remaining cores, #5 XG, #6 XG Coyote, and #8 OXG, show inconsistent erosion behavior with sample age and depth within the core.

The core average erosion rates for the six primary materials are compared at 2, 10, and 175 days of consolidation along with the time average erosion rate for each material (Fig 20). Sand mixed with XG (xanthan gum) Coyote and XG Kelzan generally became more difficult to erode with depth for all shear stresses. Erosion was in the form of small aggregates (~0.5-2 mm) that often formed small runnel-like features in the surface of the core parallel to the flow path and left a fairly uniform and smooth surface layer. The material was very cohesive and exhibited behaviors consistent with naturally cohesive sediments (Roberts et al, 1998).

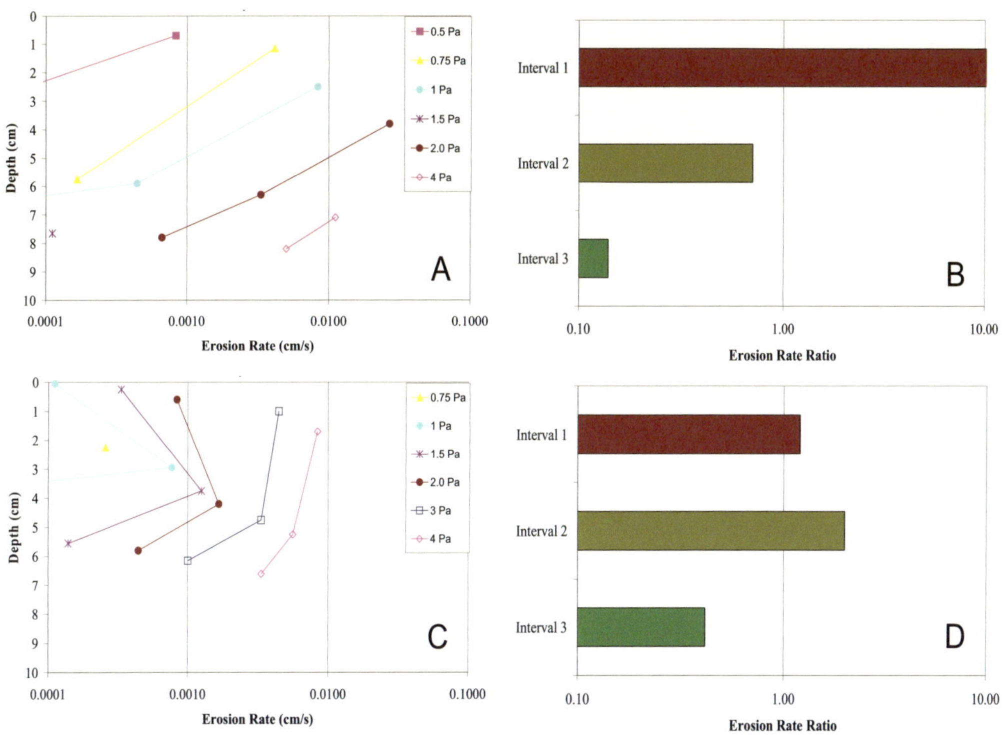

Fig. 18. (A) #9 XG/AO 10 day consolidation erosion rate as a function of depth at shear stresses of 0.5, 0.75, 1.0, 1.5, 2.0, and 4.0 Pa. (B) #9 XG/AO 10 day consolidation erosion rate ratio for the 3 erosion intervals. (C) #8 OXG 10 day consolidation erosion rate as a function of depth at shear stresses of 0.75, 1.0, 1.5, 2.0, 3.0, and 4.0 Pa. (D) #8 OXG 10 day consolidation erosion rate ratio for the 3 erosion intervals.

Sample ID	Critical Shear Stress Range (Pa)	Erosion Rate Range at 1.0 Pa (cm/s)	Erosion Rate Generally Decreases with Depth
2 Day Consolidation			
#4 XCC	0.17 - 0.51	0.0083 - 0.013	Yes
#5 XG	0.73 – 0.99	0.00011 - 0.00067	Yes
#6 XG Coyote	0.86 – 1.90	$<10^{-4}$ - 0.00022	Yes
#6 XG Kelzan	0.73 - 1.73	$<10^{-4}$ - 0.00033	Yes
#7 AXG	0.28 – 0.75	0.00067 – 0.0083*	No, center layer easiest to erode
#8 OXG	0.30 – 0.73	0.00017 – 0.0033	No, easier with depth except for second depth interval
#9 XG/AO	0.30 – 1.2	$<10^{-4}$ - 0.0033	No, center layer easiest to erode although bottom layer was most erosion resistant
10 Day Consolidation			
#4 XCC	0.28 – 0.71	0.00095 - 0.00278	Yes
#5 XG	0.73 – 1.45	$<10^{-4}$ - 0.00022	Yes
#6 XG Coyote	0.28 – 0.73	0.00017 - 0.002	No, surface and bottom layers easier to erode
#6 XG Kelzan	0.73 – 1.65	$<10^{-4}$ - 0.00067	Yes
#7 AXG (5-days)	0.25 – 0.75	0.00067 – 0.0035	No, center layer easiest to erode
#8 OXG	0.60 – 1.36	$<10^{-4}$ - 0.00077	No, center layer easiest to erode
#9 XG/AO	0.28 – 1.45	$<10^{-4}$ - 0.00083	Yes
175 Day Consolidation			
#4 XCC	0.98 – 1.06	$<10^{-4}$ - 0.00011	Yes
#5 XG	0.65 – 1.61	$<10^{-4}$ - 0.0005	Yes
#6 XG Coyote	0.61 – 0.98	0.00011 - .000056	Yes
#6 XG Kelzan	0.73 – 0.98	0.00011 - .000044	Yes, but second depth interval was most erosion resistant
#8 OXG	0.24 – 0.86	0.00022 - .00303	Yes
#9 XG/AO	0.73 – 3.3	$<10^{-4}$ - 0.00033	Yes

Table 4. Summary of erosion properties for 2, 10 and 175 days of consolidation.

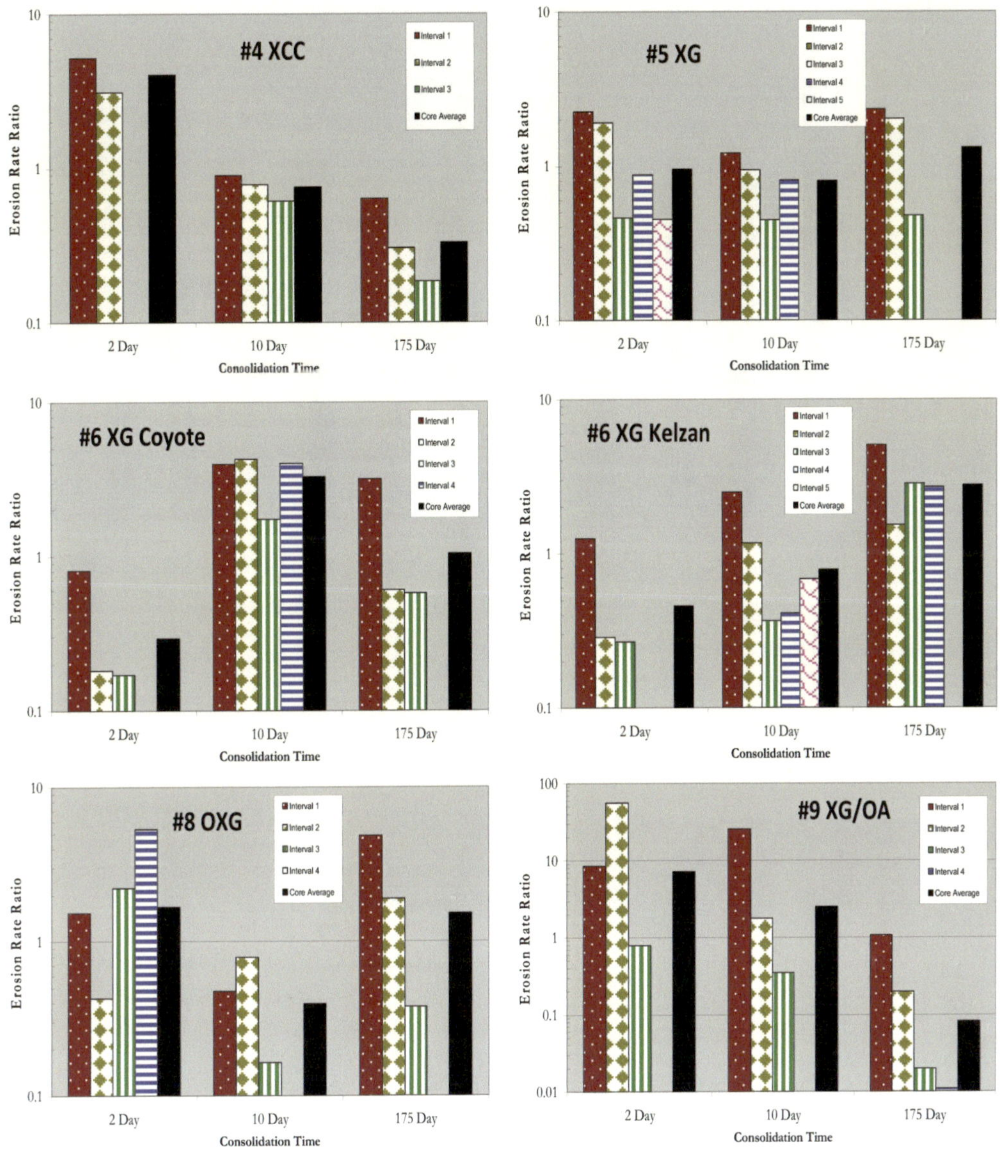

Fig. 19. Erosion rate ratio for six primary cap materials that compares all erosion intervals and the core average erosion at consolidation times of 2, 10, and 175 days.

The erosive behavior of AXG differed from that of the other materials (Tab. 4). The erosion resistance of some materials increased with time as indicated by core average critical shear stress shown in Fig. 20. Comparisons between the two and 10 day consolidation periods showed that XCC, XG (2.5% biopolymer), XG Kelzan, OXG Kelzan, and XG/AO Kelzan became harder to erode as they became more consolidated (Tab. 4 and Fig. 20 and 21). Xanthan/guar gum mixed with apatite and organoclay showed long term (175 day) physical stability (Fig. 20 and 21). The critical shear stress for this material exceeded 2Pa in the 175 day consolidation test indicating its promise as a cap material (Fig. 21).

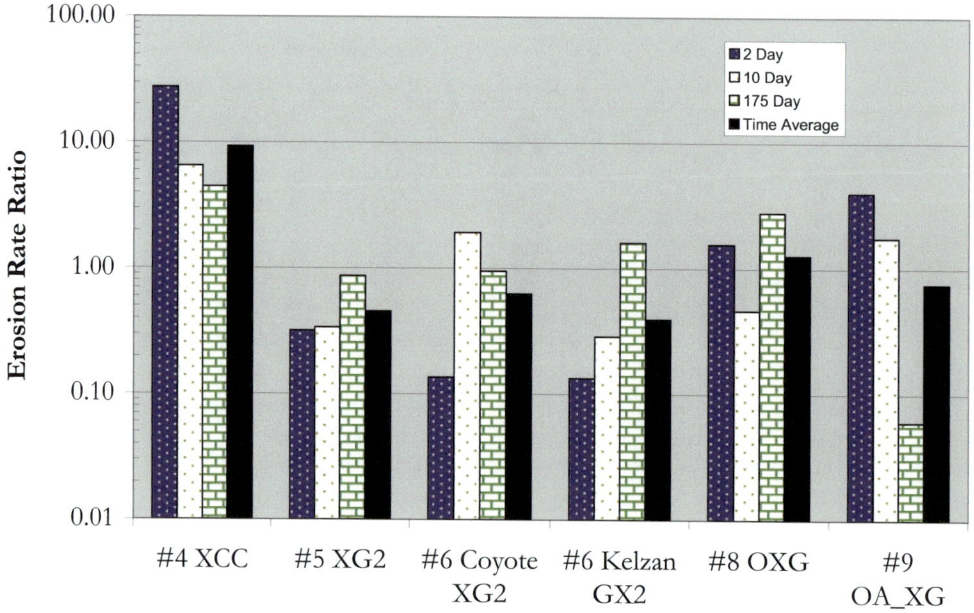

Fig. 20. Comparison of core average erosion rates at 2, 10, and 175 day consolidation including the time average of all three.

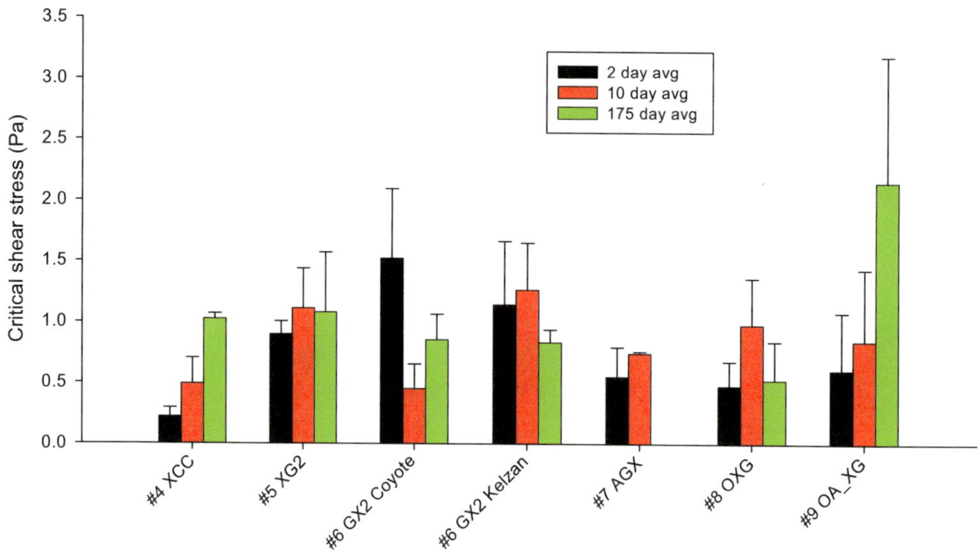

Fig. 21. Critical shear stress comparison among biopolymer materials at 2, 10, and 175 days. Each value is an average of measurements taken at two to five different depths in a core sample. Error bars represent standard deviations.

4. Conclusions

The biopolymer materials tested for possible application in remediation of contaminated soils or sediments showed potential for immobilization of metals and some organic contaminants. Most of the tested products demonstrated good erosion resistance capabilities in the laboratory. Slurry mixtures consisting of xanthan and guar gum with entrained amendments and sand showed the greatest resistance to erosion. Addition of highly metal or organic sorptive amendments (e.g., apatite or organoclays) to biopolymer products should significantly augment the sequestering capabilities of the products and increase their remedial applicability and efficiency. The results from this laboratory evaluation of biopolymers were consistent with a recently conducted pilot study in which xanthan and guar gum were evaluated in the field (Knox et al., 2009). Both laboratory and field studies showed that guar gum cross-linked with xanthan (Kelzan) became less resistant to erosion after two months. The application of xanthan/guar gum in the field as the top layer of active caps is beneficial for a short time for erosion resistance. This mixture also reduced sediment suspension during cap construction and caused the rapid settling of other amendments that were placed below the biopolymer layer. Biopolymers can also increase the pool of carbon in the sediment beneath caps and lower the release of metals and other elements, especially P, in comparison with apatite only (Knox et al., 2009). However, more research is needed on the type of biopolymers applied to caps and methods for delivering biopolymers to the cap. Further research is especially needed on the biodegradability of biopolymers under extreme aquatic conditions (e.g., high summer temperature, changing ratios of Fe-S-P in sediment pore and surface water, and other factors). Finally, the applicability of biopolymer products in active capping technology is also dependent on the effects of these products on benthic organisms. Although biopolymers are nontoxic, the viscous matrix produced by biopolymers may have the potential to physically entrap or suffocate burrowing organisms (Paller & Knox, in press).

5. Acknowledgment

This work was sponsored by the DoD Strategic Environmental Research and Development Program (SERDP) under project ER 1501 and the Savannah River National Laboratory (SRNL) Lab Directed Research Development and Mini-Sabbatical Program. The SRNL is operated by Savannah River Nuclear Solutions, LLC for the U.S. Department of Energy under Contract DE-AC09-798861048.

6. References

Araujo, M.M. & Teixeira, J.A. (1997). Trivalent Chromium Sorption on Alginate Beads. *Int. Biodeter. Biodegr.* 40(1): 63-74.

Cadmus, MC., Jackson, J.K, Burton, K.A. Plattner, R.D. & Slodki, M.E. (1982). Biodegradation of xanthan gum by *Bacillus* sp. *Appl. Enviorn. Microbiol.* 44-5-11.

Bassi, R.; Prasher, S.O. & Simpson, B.K. (1999). Remediation of metal-contaminated leacgate using chitosan flakes. *Eviron. Tech.* 20:1177-1182.

Chen, D., Lewandowski, Z., Roe, F. & Surapanemi, P. (1993). Diffusivity of Cu^{2+} in calcium alginate gel beads. *Biotechnol. Bioeng.* 41: 755-760.

Chen, J. & Yen, T. F. (1990). Transport of Microorganisms to Enhance Soil and Groundwater Bioremediation, Proceedings, Hazmacon1990 (T. Burszynski, ed.) ABAG, San Francisco, Vol. II, pp. 95-100.

Deans, J.R. & Dixon, B.G. (1992). Uptake of Pb^{2+} and Cu^{2+} by novel biopolymers. *Water Res.* 26 (4): 469-472.

Etemadi, O.; Petrisor, I.G.; Kim, D.; Wan, M.-W. & Yen, T.F. (2003). Stabilization of Metals in Subsurface by Biopolymers: Laboratory Drainage Flow Studies. *Soil & Sediment Contamination: An International Journal* 12 (5): 647-661

Fiol, N.; Poch, J. & Villaescusa, I. (2004). Chromium (VI) uptake by grape stalks wastes encapsulated in calcium alginate beads: equilibrium and kinetics studies. *Chemical Speciation and Bioavailability* 16(1/2): 25-33.

Francis, A.J. & Dodge, C.J. (1993). Influence of complex structure on biodegradation of iron-citrate complexes. *Appl. Environ. Microbiol.* 59: 109-113.

Jepsen, R..; Roberts, J. & Lick, W. (1996). Effects of Bulk Density on Sediment Erosion Rates, *Water, Air, and Soil Pollution,* 99: 21-31.

Jepsen, R.; Roberts, J. & Lick, W. (1997). Long Beach Harbor Sediment Study, Report Submitted to the U.S. Army Corps of Engineers, DACW09-97-M-0068

Khachatoorian, R.; Petrisor, I.G. & Yen, T.F. (2004). Prediction of Plugging Effect of Biopolymers Using Their Glass Transition Temperatures. *Journal of Petroleum Science & Engineering* 41: 243-251.

Kartal, S.N. & Imamura, Y. (2004). Removal of copper, chromium, and arsenic from CCA-treated wood onto chitin and chitosan. *Bioresource Technology* 96(3): 389-392.

Kim, D.; Petrisor, I.G. & Yen, T.F. (2004). Geopolymerization of biopolymers; a preliminary inquiry. *Carbohydrate Polymers,* 56: 213-217.

Kim, D.; Petrisor, I.G. & Yen, T.F. (2005). Evaluation of Biopolymer-Modified Concrete Systems for Disposal of Cathode Ray Tube Glass. *J. Air & Waste Manage. Assoc.,* 55: 961-969.

Knox, A.S.; Paller, M.H.; Reible, D.D. & Petrisor, I.G. (2006). Innovative in-situ remediation of contaminated sediments for simultaneous control of contamination and erosion. Annual Report 2007, WSRC-RP-2006-01149.

Knox, A.S.; Paller, M.H.; Petrisor, I.G. & Reible, D.D. (2007). Sequestering agents for metal immobilization - application to the development of active caps in fresh and salt water sediments. Conference Proceedings of the Fourth International Conference on Remediation of Contaminated Sediments, Savannah, Georgia, January 22-25, 2007, Battelle Press: ISBN 978-1-57477-159-6.

Knox, A.S.; Dixon, K.L.; Paller, M.H.; Reible, D.D.; Roberts, J. & Petrisor, I.G. (2008a). Innovative in-situ remediation of contaminated sediments for simultaneous control of contamination and erosion. Annual Report 2008, SRNL-RP-2008-01216.

Knox, A.S.; Paller, M.H.; Reible, D.D.; Ma, X. & Petrisor, I.G. (2008b). Sequestering Agents for Active Caps - Remediation of Metals and Organics, Soil and Sediment Contamination: An International Journal, 17(5): 516-532.

Knox, A.S.; Paller, M.H.; Dixon, K.L.; Reible, D.D. & Roberts, J. (2009). Innovative in-situ remediation of contaminated sediments for simultaneous control of contamination and erosion. Annual Report 2009, SRNL-RP-2009-01497.

Kostal, J.; Mulchandani, A.; Gropp, K.E.; & Chen, W. (2003). A Temperature Responsive Biopolymer for Mercury Remediation. *Environ. Sci.Technol.* 37: 4457-4462.

Lehman, R.M.; Colwell, F.S. & Bala, G.L. (2001). Attached and Unattached Microbial Communities in a Simulated Basalt Aquifer under Fracture- and Porous-Flow Conditions. Appl. Environ. Microbiol. 67: 2799-2809.

McNeil, J.; Taylor, C. & Lick, W. (1996). Measurements of the Erosion of Undisturbed Bottom Sediments with Depth, *Journal of Hydraulic Engineering*. 122 (6): 316-324.

Palermo, M.R.; Maynord, S.; Miller, J. & Reible, D.D. (1998). Guidance for in-situ subaqueous capping of contaminated sediments. Chicago, Great Lakes National Program Office.

Paller, M.H. & Knox, A.S. (2010). Amendments for the remediation of contaminated sediments: Evaluation of potential environmental impacts. *Sci. Total Environ.* (in press).

Prabhukumar, G.; Matsumoto, M.; Mulchandani, A. & Chen W. (2004). Cadmium Removal from Contaminated Soil by Tunable Biopolymers. *Environ Sci. Technol.* 38: 3148-3152.

Randall, J.M.; Randall, V.G; McDonald, G.M.; Young, R.N. &Marsi, M.S. (1979). Removal of trace quantities of nickel from solution. *J. Appl. Polym. Sci.* 23: 727-732.

Reible, D.D.; Lampert, D.; Constant, W.D.; Mutch, R.D. & Zhu, Y. (2006). Active Capping Demonstration in the Anacostia River, Washington, DC, Remediation: The Journal of Environmental Cleanup Costs, Technologies and Techniques, 17 (1): 39-53.

Rhodes, J. & Roller, S. (2000). Antimicrobial actions of degraded and native chitosan against spoilage organisms in laboratory media and foods. *Appl. Environ. Microbiol.* 66: 80-86.

Roberts, J.; Jepsen, R.; Gotthard, D. & Lick, W. (1998). Effects of Particle Size and Bulk Density on Erosion of Quartz Particles. *Journal of Hydraulic Engineering.* 124(12): 1261-1267.

Roberts, J. & Jepsen, R. (2001). Development for the Optional Use of Circular Core Tubes with the High Shear Stress Flume. SNL report to the US Army Corps of Engineers, Waterways Experiment Station.

Roberts, J.; Jepsen, R. & James, S. (2003). Measurement of Sediment Erosion and Transport with the Adjustable Shear Stress Erosion and Transport Flume. *Journal of Hydraulic Engineering.* 129(11): 862-871.

Schlichting, H. (1979). Boundary-Layer Theory. Seventh Edition, McGraw-Hill. Stewart, T.L., Fogler, H.S. 2001. Biomass Plug Development and Propagation in Porous Media. Biotechnology and Bioengineering, 5: 353-363.

Tsai, C-H. & Lick, W. (1986). A portable device for measuring sediment resuspension. J. Great Lakes Res. 12(4): 314-321.

Turick, C.E.; Tisa, L.S. & Caccavo, F. Jr. (2002). Melanin production and use as a soluble electron shuttle for Fe(III) oxide reduction and as a terminal electron acceptor by *Shewanella algae* BrY. *Appl. Environ. Microbiol.* 68:2436–2444.

Udaybhaskar, P.; Iyengar, L. & Prabhakara Rao, A.V.S. (1990). Hexavalent chromium interaction with chitosan. *J. Appl. Polym. Sci.* 39: 739-747.

Wan, M.-W.; Petrisor, I.G.; Lai, H.-T.; Kim, D. & Yen, T.F. (2004). Copper adsorption through chitosan immobilized on sand to demonstrate the feasibility for in situ soil decontamination. *Carbohydrates Polymers* 55: 249-254

6

Microfoams of Biopolymers by Laser-Induced Stretching: Mechanisms and Applications

Lazare Sylvain
Institut des Sciences Moléculaires (ISM) UMR 5255, Université Bordeaux 1
351 cours de la Libération, 33405 Talence
France

1. Introduction

Biopolymers and polymers derived from the biomass (Gandini, 2008; Mooney, 2009) are abundant natural chemical substances that find applications in many fields like agriculture, agro-food industry, textile and medicine, etc.. . A vast majority of them does not cause problems to the environment (Pérez et al., 2002), contrary to synthetic polymers (Moore, 2008). Their natural resource is continuously renewed and they are traditionally used since the prehistoric times. The recent problems like climate change, fossil fuels depletion and growing costs of recycling due to strictly observed modern environmental concerns, have renewed our interest for a modern form of bio-based economy. This results in an increased use of natural substances and a search for new energy efficient engineering processes. The laser-based technologies have emerged (Bäuerle, 2000) since the early sixties date of the laser birth, that is to say very recently, and are now explored for new types of processing, like micro-precise laser characterization, ablation and machining, by various appealing experimental approaches (Fig.1). Laser processing and characterization of materials in general offer definite advantages like the cost-efficiency due to rapidity and the submicro-precision (Lazare & Tokarev 2005a). They are considered as techniques lending themselves easily to future automation and industrialization. Moreover it is generally recognized that laser beams make it possible to perform a remote characterization and treatment of the target with a large working distance, which in the extreme case can easily be up to the kilometer range (as in LIDAR) depending on the laser beam quality (Telle et al., 2007). A new micro-foaming transition of biopolymer films was discovered (Lazare et al., 2005b) during the study of laser ablation of collagen, a biopolymer extracted from rat tails tendon. In these experiments the absorption of one KrF excimer laser pulse produces a layer of foamy material on the surface of the collagen film (Fig.1). For the first time, with a unique pulse a clearly visible and highly interesting transformation of a material was produced, and which was not the simple removal of matter like in regular laser ablation (Figs.1.1 for surface ablation, 1.2 for multi-pulse micro-drilling, and 1.3 for surface modification), often called "clean ablation" as known from previous studies (Weisbuch, 1999; Tokarev, 2003; Lippert & Dickinson, 2003; Elaboudi, 2008). Usually thermal and photochemical decomposition of the polymer are considered to account for the ejection of the gas products.

In the foaming process, whose mechanism has been studied and modeled (Lazare, 2009) since the discovery, a more selective combination of laser-induced forces is involved to explain the sudden expansion of the laser-excited polymer. This expansion approximatively occurs in a timescale of the order of 50 ns after the end the laser pulse, when the high pressure wave induced by the laser excitation energy propagates away from the surface and gives rise to a tensile wave. It results in a sudden and dense cavitation phenomenon yielding a micro-foam composed of closed or open microscopic cells, quite similarly to the chemically engineered polymer foams, although the timescale is comparatively quite smaller. It is shown that, for the laser-induced foaming to occur, the target polymer must be moderately absorbing in order to generate a strong pressure wave and the laser absorption must provide enough gas pressure produced by the ablation products. Biopolymers tend to display coefficients of absorption situated in the adequate range centered around ~10^3 cm^{-1}. The obtained microfoams of biopolymers display micro-features, microscopic cells and submicron-size filaments, having some similarities with that of familiar biomaterials e.g. wood, tissue matrix, etc.. . They are therefore expected to find applications in various fields of biomaterial sciences e.g. in biomedical. The purposes of this paper are mainly to review the results, to discuss the mechanisms and speculate about the future applications. Among the laser ablation regimes studied up to now the photomechanical regime, which allows ablation by using high pressure rather than high temperature, is still recent and mostly unknown. The main questions to be answered are 1) how matter is moved from the surface, by which mechanisms 2) is part of the surface material expulsed from the target and in which form, 3) is molecular decomposition necessary, to which extend the polymer is decomposed into gas molecules and 4) what are the laser and material parameters that best control the process and that can be adjusted to select the sub-regime of interest?. The experimental results and the explanatory theoretical modeling are presented in the following text.

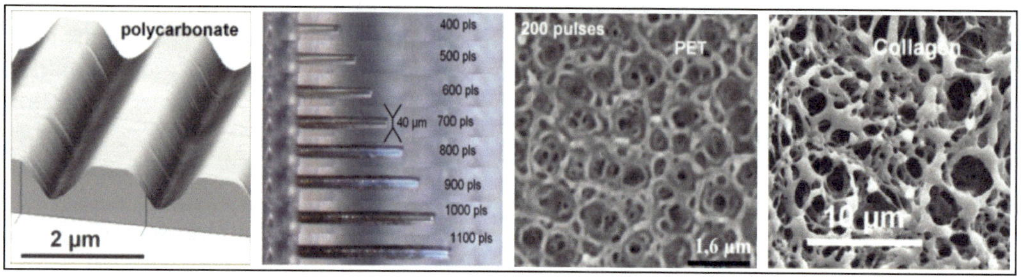

Fig. 1. Various laser (KrF laser radiation) processing of polymers studied in the recent years. Left to right: 1) Micro-precise surface ablation of polycarbonate (one pulse), 2) Micro-drilling in polycarbonate, 3) Surface modification of poly(ethylene terephthalate), 4) Micro-foam induced by a single laser pulse (0.9 J/cm^2) in the surface of a collagen film.

2. Lasers systems

Lasers used in this work are either nanosecond KrF ultraviolet lasers or ultrashort pulses femtosecond lasers for the time resolved experiments. KrF excimer lasers are classical excimer lasers which have been described in details in previous works. The KrF laser radiation at 248 nm is a good compromise since biopolymers are in general moderately absorbing at this wavelength. The main qualities are the large energy per pulse (~500 mJ) and the long-lasting gas mixture ($Kr/F_2/ Ne$) filling which makes it cost efficient. The beam is patterned with homemade optical setups and attenuated variably and at constant spot area (either 0.1×0.1 mm^2 or 1.0×1.0 mm^2) in order to control over a large range the energy density (0.001 to ~12 J/cm^2) delivered to the target. The energy profile on the spot area was near homogeneous in all cases, or measured as approximately Gaussian. The energy of each pulse is carefully monitored with a detector, a precaution made necessary by the single-pulse experiments. The femtosecond laser system has a pulse width of 90 fs (Gaspard, 2008a) and is composed of a Ti:sapphire amplifier delivering linearly polarized fs pulses centered at 800 nm at a maximum repetition rate of 1 kHz. For some experiments isolation of single pulses was carried out by a Pockels cell in the regenerative cavity. Wavelength of the final pulse could be tuned by frequency conversion to 2 and 3 ω (resp. 400 and 266 nm) with BBO crystals. The absorption mechanisms of the femtosecond pulses are different from the 1 photon-induced electronic transition involved in the ultraviolet nanosecond laser interaction. More multiphoton type of electronic excitation takes place in the fs case and makes it possible to yield ablation with materials that cannot be excited by nanosecond laser pulse due to the absence of absorption.

3. Materials

3.1 Biopolymers of Interest

It happens that most biopolymers have optical features that are well suited to obtain the present laser-induced foaming phenomena. First they must have a moderate absorption coefficient, approximately α<3000 cm^{-1}, in order to generate a strong pressure wave. Second enough gas pressure must be produced during their ablation in order to support the cavitation phenomenon that occurs upon the laser-induced stretching. Third their viscosity must decrease rapidly upon the laser heating (low T_g) for providing a high nucleation rate giving rise to a dense cavitation and bubbling. If these conditions are met at the same time, foaming is produced when one pulse is absorbed at the surface.

Polymers made with the glucose repeat unit obtained by the photosynthesis are the most abundantly produced by the biomass. Depending on the linkage between the glucose unit (α or β), cellulose or starch molecules are obtained. The two molecules have different properties. Cellulose is a straight and rigid molecule maintained in a linear configuration by the side groups ($-CH_2OH$) and the network of hydrogen bonds as in Fig.3 (Pinkert et al., 2009). It is linked to the neighbour cellulose molecules in a very stable way, so that elaborated 3D structures known in plants can be formed. In starch molecules the glucose repeat units are linked differently and the side groups are evenly distributed on the same side. Their steric effect induces a curvature in the polymer backbone which results in the formation of helical molecules. Because of this, starch molecules cannot be so strongly self-associated than cellulose molecules.

Owing to the high insolubility of cellulose in common solvents (Liu et al., 2007), many derivatives have been studied for making it more industrially tractable. Some industrially important cellulose derivatives, like cellulose triacetate (CTA), cellulose acetate phthalate (CAP), cellulose dinitrate (CDN) (Fig.2), have been developed long ago but have been since then supplanted by synthetic polymers made by the petrochemistry. However they are regaining interest as polymers derived from sustainable resources. Other important biopolymers are proteinic molecules made of amino-acid chains. They are structurally very

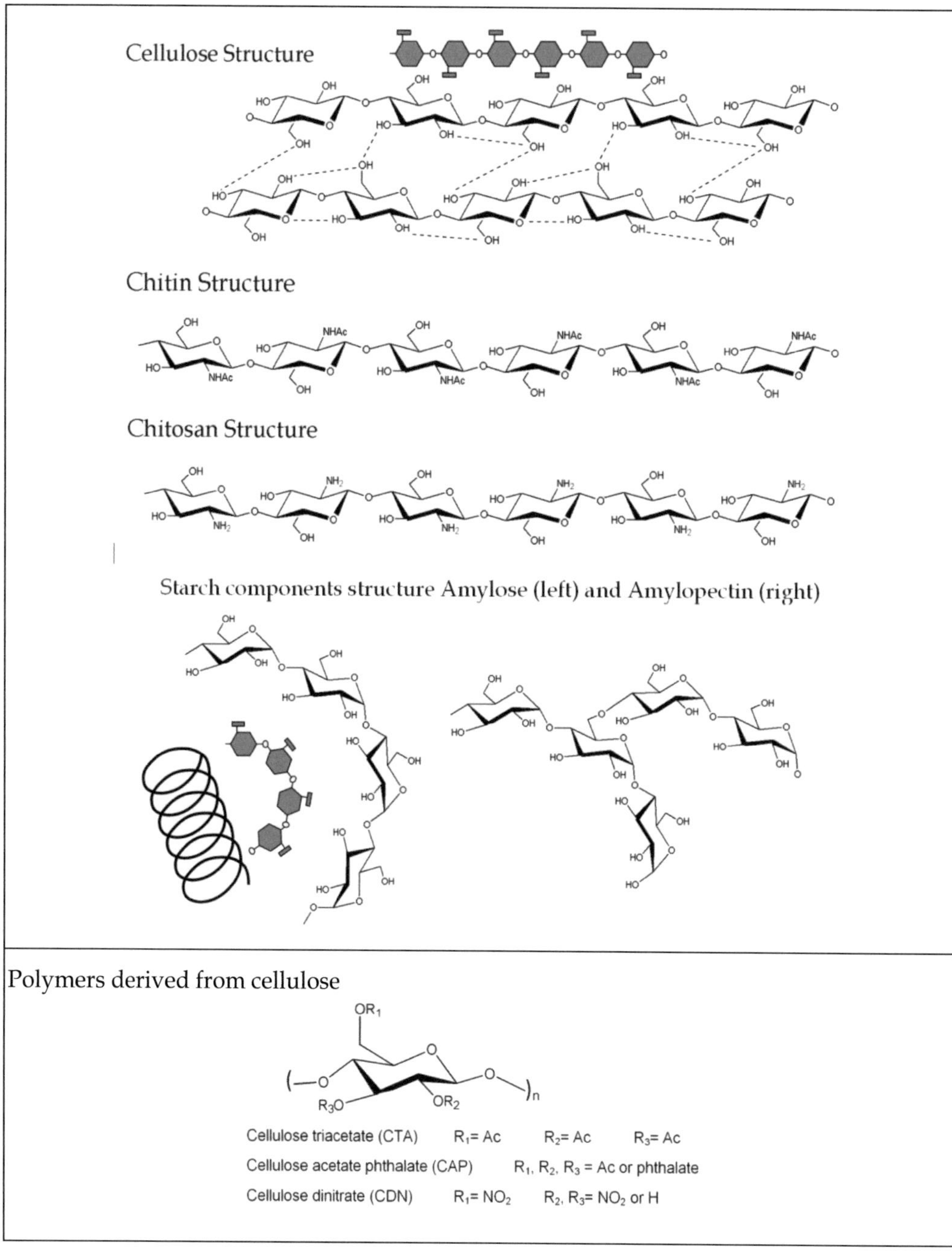

Fig. 2. Natural polymers studied in this work are based on glucose repeat unit (cellulose) or are protein type like collagen which is made of glycine, proline and hydroxyproline mainly. Cellulose is transformed into chitin by animals (insects, crustaceans) and mushrooms which is industrially derived in more tractable chitosan. Other industrially important cellulose derivatives include cellulose triacetate (CTA), cellulose acetate phthalate (CAP), cellulose dinitrate (CDN).

diverse and display a variety of molecular functions in natural systems as well. They are for a long time used as commodity polymers. Some of the most important practical interest include gelatin, casein, keratin, etc.. Here we present collagen (Gordon, 2010) from which gelatine is extracted. Collagen single strand is a polymer mostly composed of a repeating tri-peptide (proline, hydroxyproline, glycine POG) which provides the capability of forming a triple helix by the winding of three strands (Fig.3). Triplet molecules can in turn self-assemble into fibrils and fibers which are integrated into the extracellular matrix. Apart being a structural material for cellular tissues the collagen network has a variety of functions like the presence of cryptic RGD (Arg-Gly-Asp)-containing peptides (Taubenberger et al., 2010) which can eventually be recognised by osteoblast cells in damaged tissue in need of reconstruction. Therefore collagen molecules are always engaged in a molecular assembly unless chemically treated by heat or aqueous basic solution. In that case the triple helix unwinds to yield gelatine, a random coil polymer having same amino-acid composition. It is the result of the so-called "melting" of the collagen assemblies.

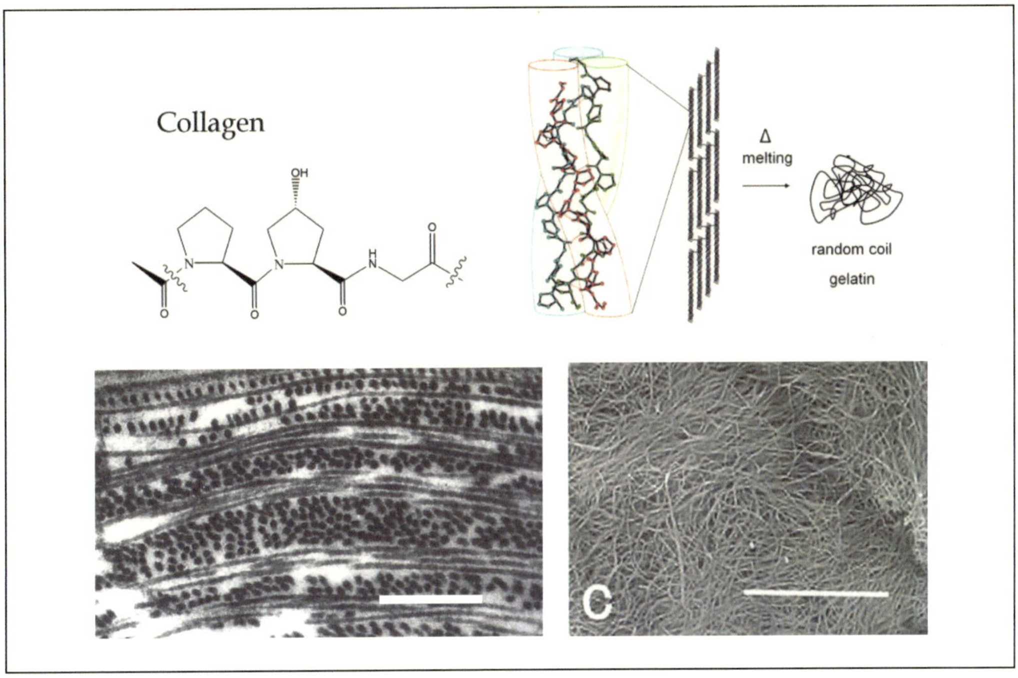

Fig. 3. Collagen single strand has a repeat unit mainly composed of tripeptide (proline, hydroxyproline and glycine POG). It is a structural protein whose molecule is a triple helix capable of easy self assembling. It is present in the form of fibrils in many tissues like cornea (picture on the left, scale bar is 1 μm) or lungs (picture on the right, scale bar is 100 μm). Collagen can be extracted in the triple helix form or can be denatured into a random coil form called gelatine under the action of a basic solution or hot water.

As mentioned previously and demonstrated below by the model, a polymer is a good candidate to laser-induced foaming if its absorption at the laser wavelength places it in the situation of pressure confinement, that is to say $\frac{l_p}{l_\alpha} = \frac{c_s\tau}{\alpha^{-1}} = \alpha c_s\tau < 1$. Pressure propagation length $l_p=c_s\tau$, during the laser pulse duration τ must be much smaller than the laser absorption length $l_a=1/\alpha$ in order to observe an elevation of pressure. The model shows that the optimum of absorption coefficient is 1000 cm^{-1} if the assumed speed of sound is 1 µm/s. Further short pulses favour pressure elevation and changing to smaller coefficients polymer necessitates increasing incident fluence to yield same elevation. The pressure model below brings more precise explanation.

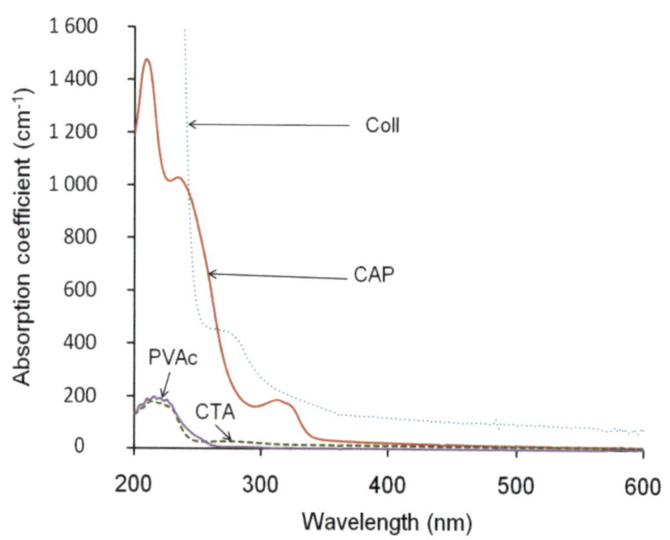

Fig. 4. Absorption spectra of polymer films. The absorption coefficients at the KrF laser wavelength 248 nm are (in cm^{-1}): CAP/1000; CTA/50; Chi/325; Coll/600; Gel/600.

Absorption of the KrF laser light at 248 nm (Fig.4) in the considered polymers can be due to molecular valence transitions of electrons located in chromophores like unsaturations ($\pi^*\leftarrow\pi$ in C=C double bonds) or free doublets located on the oxygen atoms ($\pi^*\leftarrow n_O$ and $\sigma^*\leftarrow n_O$ in C=O carbonyl group) or can be due to Rydberg series in which the excited electron has a more diffuse final orbital. Aromatic systems generally give high absorption and are in high concentration in synthetic polymers which display high absorption coefficients of the order of 10^5 cm^{-1}. On the contrary these unstaturations and aromatic systems are in small concentration in natural polymers, even absent in cellulose derivatives except in CAP in which phthalate groups have been introduced, in low concentration however. In proteins, like collagen, residual aromatic amino-acids include tyrosine, phenylalanine and tryptophan and occurrence is approximately ~1 % of the residues. So, as expected from this consideration, absorption coefficient of collagen (600 cm^{-1} in fact) is two orders of magnitude smaller than that of a synthetic aromatic polymer (10^5 cm^{-1} at 248 nm). In cellulose derivatives non-aromatic esters like acetate yield a rather small absorption which is of the order of 50 cm^{-1} enough for the foaming experiments with however much larger fluence thresholds compared to higher coefficients. As a consequence the absorption depth $l_a=\alpha^{-1}$ is large and becomes larger than the pressure transit distance during the laser pulse duration. This results in the accumulation of pressure in the absorption depth which makes possible the appearance of the dense cavitation effect.

3.2 Experimental results

In Fig.5 are displayed a few examples of laser-induced nanofoams observed at various magnifications with either the optical or electron microscope. They are all formed with only one laser pulse whose energy density is above the threshold of foaming of the considered polymer. The thickness of the foam layer depends on the target polymer and is of the order of a few tens of microns. It is clearly visible in Fig.5c where a cross-section on a gelatine film is shown. Obviously there is a relationship between energy density of the laser pulse, the absorption coefficient of the material, the foaming threshold and the thickness of the foam layer. With laser pulse energy close to threshold, the onset of the phenomenon, that is to say the formation of a network of micro-bubbles can be observed. Even it has been measured by fast light transmission measurements that some bubbles or cavities are only transiently formed and live only a few microseconds. It is an indication that the gas formed by laser ablation may play an important role in the pressure equilibration inside the bubbles. High fluences are capable of material expulsion and yield more open cells type foamy structure. On the contrary low fluences near threshold tend to give foams with closed cells and probably do not expulse a significant amount of the surface material.

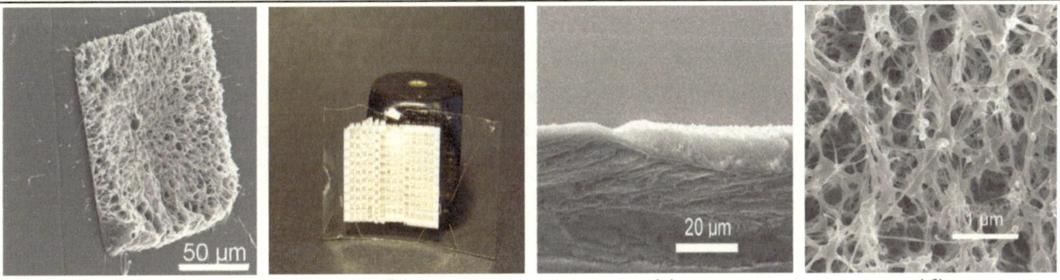

(a) (b) (c) (d)

Fig. 5. Examples of KrF laser induced foams on the surface of (a) collagen film (b) gelatine, macroscopic view (c) gelatine, cross-section (d) chitosan, fine nanofibers network

Poly(vinyl acetate) (PVAc) is studied as a model polymer. Its absorption coefficient at 248 nm is 30 cm^{-1} and is well suited to give rise to a strong pressure wave upon absorption of a strong laser pulse. Therefore similar foaming (Fig.6) is obtained although at a higher fluence threshold of 3 J/cm^2. Interestingly PVAc yields the same foaming phenomenon but lends itself more easily to modelling owing to the large body of physical data available in the literature on this synthetic polymer. For all polymers studied so far some general features can be observed like bump formation with a height equivalent to a fraction of the absorption depth, for instance 25 µm on PVAc at 6 J/cm^2, a tendency to give closed bubbles near threshold fluence and to give open cell structure at a fluence well above threshold. Moreover the ejected matter which is still poorly known can be solid particles and gases and is probably more abundant at high fluence. Results obtained during experiments with fs laser are reported in details in (Sionkowska et al, 2007; Gaspard et al., 2007, 2008a and 2008b). As shown below they show strong similarity with the ns laser results. Owing to the multi-photon absorption mechanism strongly favoured with fs pulses the radiation wavelength does not need to match the absorption spectrum of the target polymer therefore focusing inside the material makes it possible to create voids in the bulk (Oujja et al., 2009).

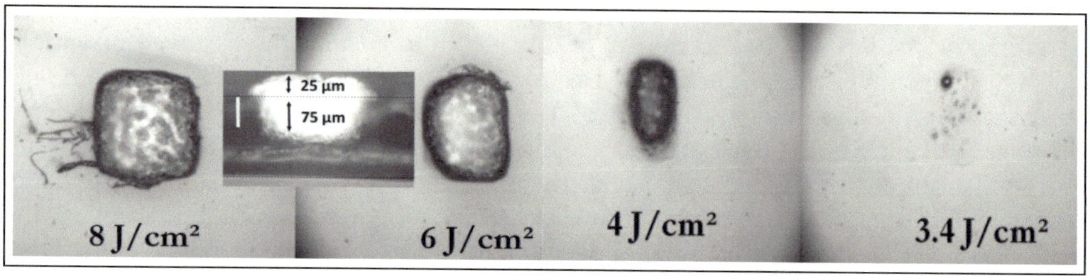

Fig. 6. Ablation spots on a film of poly(vinyl acetate) with one KrF laser pulse, showing the formation of a microfoam due to the tensile breaking of the polymer. The central inset shows a cross section of the foamed part upon irradiation at 6 J/cm^2 with maximum spot size of 145 µm.

4. Fast transmission measurements

Upon absorption of a laser pulse of energy above threshold, transparent polymer film of this study becomes instantaneously opaque due to the rapid formation of the foamy layer. The rate of appearance of the foam was measured by a specific optical setup (Fig.7) designed to record the data with a time resolution of 1 ns. The light of a probe HeNe laser at 632.8 nm is normally transmitted without loss before the foam. This transmission of 100 % drops very rapidly within the first 100 ns after the KrF ablation laser pulse absorption.

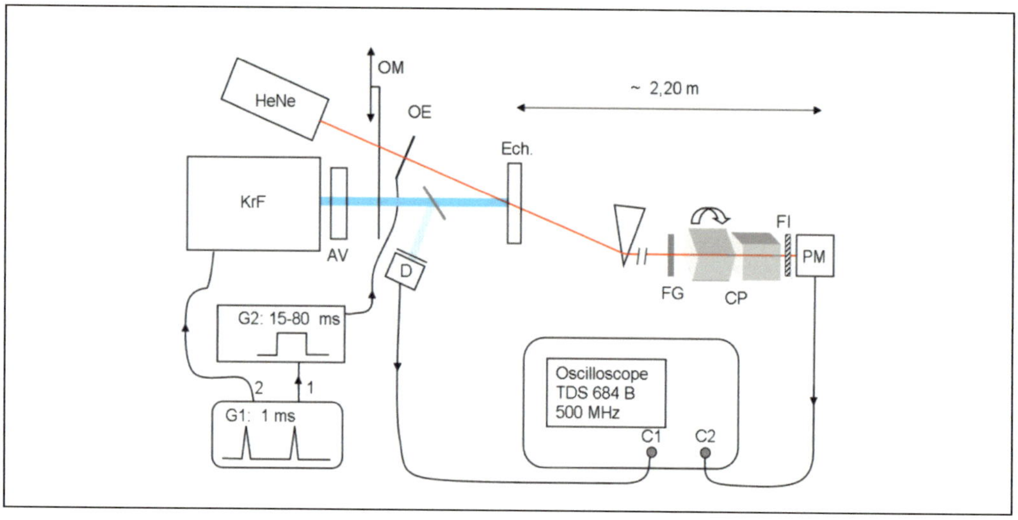

Fig. 7. Setup to measure the fast transmittance drop produced by the laser-induced foam. **KrF**: Excimer laser 248 nm, **HeNe**: Laser HeNe 632,8 nm, 5 mW, **AV**: Variable attenuator, **G1**: Pulse delay generator (1-5 ms, SRS DG535), **G2**: Square pulse generator (home made), **OE**: Electric shutter (Uniblitz, Vincent Associates), **OM**: Manual shutter, **D**: Fast Diode Thorlabs DET210 (monitoring of KrF pulse and triggering), **Ech.** Sample, **FG:** Grey Filter, **FI,** Interference filter allowing 632.8 nm, **CP**: 2 Polarizing cubes making adjustable attenuation, **PM**: Photomultiplier 1R446, **Oscilloscope**: Tektronix digital TDS684 B, 500 MHz, 2,5 Gsamples/s

To understand this experimental fact one may consider three important data of the present system. The propagation speed of the pressure in the polymer surface of the order of 1 µm/ns, the absorption depth below the polymer surface, 16 µm e.g. for collagen and the ablation KrF laser pulse duration ~30 ns. Before the more detailed model in sect.5 and 6 the following simple reasoning can be of valuable help. During the laser pulse absorption and because the volume can be considered as constant the pressure accumulates and does not relaxes before the end of the absorption. In the meantime, the reflexion of the pressure wave on the surface occurs with change of sign, giving rise to a new growing tension wave propagating into the bulk. The cavitation phenomenon does not occur when pressure is high but its nucleation becomes very fast once the compression wave reverses to tension, as will be shown by the model. As a consequence and since tension appears only after a delay time necessary to evacuate the high pressure component of the acoustic wave, the onset of cavitation is observed only ~55 ns after the beginning of the laser pulse absorption. By using a shorter UV laser pulse with a 90 fs duration, it was shown that this delay is reduced to a lower value of 18 ns and in full agreement with the model as shown below. The pressure wave dynamics does not depend much on the pulse duration. Changing the duration from 30 ns to 90 fs, that is to say a reduction of $\sim 10^6$ only alters the delay time of appearance of cavitation by less of a factor 2 (Lazare at al., 2009).

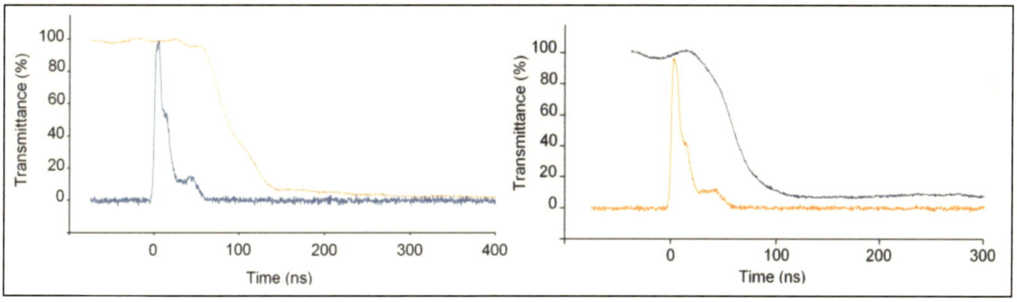

Fig. 8. Example of transmittance fast recording during the KrF laser-induced foaming phenomenon. Left: for gelatin at 1.7 J/cm² and right: for PVAc at 4.5 J/cm².

Polymer	Abbrev.	α_{248nm} (cm^{-1})	F_t(J/cm^2)	$\Delta T_{max}(F_t)$	$\Delta P(1J/cm^2)$	T_g (K)
Gelatin	GEL	600	0.5	144	90 %	434
Collagen	COL	600	0.5	144	90 %	
Chitosan	CHI	325	1.5	260	67 %	
Cellulose triacetate	CTA	50	4	100	19 %	378
Cellulose dinitrate	CDN	430				333
Cellulose Acetate Phthalate	CAP	1000	0.15	74	100 %	348
Poly(vinyl acetate)	PVAc	30	3	60	10 %	305
Poly(vinyl pyrrolidone)	PVP	120	1.5	90	43 %	448

Table 1. Physical data of the polymers considered for the laser-induced foaming phenomenon. Target temperature is 25 °C just before laser pulse.

5. Theory of laser-induced strong pressure wave

Three parameters define the state of the matter excited by the laser pulse during and after the absorption step. Temperature, pressure and volume are time dependent as a characteristic of the non-equilibrium due to the short pulse laser excitation and can be estimated by the following theoretical approach. The dynamics of temperature and pressure are very different because heat diffusion is slow in polymers and pressure relaxes by propagation at the speed of sound. Therefore for the time window needed to be considered, the temperature profile is stationary and given by:

$$T(z,t) = \frac{\alpha A_0}{C_p \rho} F_0 e^{-\alpha z} + T_0 = \Delta T_{\max} e^{-\alpha z} + T_0 \tag{1}$$

with the signification of the symbols given in the nomenclature below. This temperature elevation is the origin of the pressure rise and its profile is not expected to vary in the timescale of the foaming transition (100 ns) since the appropriate model shows that 10 μs are necessary to display a significant evolution of the profile by diffusion, a time much longer than the foaming appearance. As seen in Table 1 the maximum temperature elevation at threshold, as estimated by the model, is for most polymers very low and is very rapidly averaged and applies for durations of the order of milliseconds. It suggests that temperature alone does not account for the observed transition. Several other factors have to be considered like pressure elevation and ablation products. Ablation products are usually very diverse since they can be gases, polymers, particles and resulting of thermal or photochemical reactions. To consider them is important because they are the driving force that makes the foaming process possible once the tension wave comes into play. The pressure is considered now as the result of the temperature elevation before the volume expansion that occurs at finite speed, the speed of sound c_S. As soon as volume expansion is possible and occurs, pressure decreases. The dynamics of the pressure wave is governed by its propagation equation:

$$\frac{\partial^2 P(z,t)}{\partial^2 t} - c_s^2 \nabla^2 P(z,t) = -\frac{\beta}{C_P} \frac{\partial [A\alpha F(z,t)]}{\partial t} \tag{2}$$

A one dimension model is assumed since the spot size of diameter larger than 500 μm implies a diffusion along the z-axis. This type of equation does not take into account the z-attenuation of the pressure wave due to bulk and shear viscosity because only the short distance propagation is considered. One is interested in the pressure developed within the absorption length and not more. This approximation has to be reconsidered for a larger propagation. The solution to the propagation equation (2) can be easily described for an ultrashort (e.g. τ=90 fs) pulse of incident fluence f_0 for which the confinement is total, that is to say no pressure relaxation at the end of the pulse. In that case almost all the absorbed energy contributes proportionally to the pressure elevation according to the Mie-Grüneisen equation of state of the target material (Lazare et al., 2009).

$$\Delta p(z,t=\tau) = \Gamma \cdot \alpha \cdot f_0 \cdot e^{-\alpha \cdot z} = p_{\max} \cdot e^{-\alpha \cdot z} \tag{3}$$

$$\Gamma = -\left(\frac{\partial \ln P}{\partial \ln(U/V)}\right)_V = \frac{c_s^2 \beta}{c_p} \tag{4}$$

Nomenclature

A	target absorptivity
$A_v(t)$	avalanche function
α	absorption coefficient
$l_a=\alpha^{-1}$	laser absorption depth
c_s	speed of sound in the polymer
C_p	heat capacity of the polymer
d	depth in which foaming occurs
ρ	density of the polymer
β	thermal expansion coefficient
$F(z,t)$	instantaneous fluence
$g(t)$	normalized time profile of laser pulse
τ	pulse width
τ_0	pulse width parameter in $g(t)$
F_0	total fluence for a ns pulse
f_0	fluence of a "Dirac" or fs laser pulse
$\alpha c_s \tau < 1$	pressure confinement condition
t	time
z	depth
$J(t,z)$	nucleation rate
σ	surface tension of the material
Z	Zeldovich frequency factor
n_0	molecular density factor
v	volume of a free volume hole in PVAc
$J_0=Zn_0$	pre-exponential factor
P_i	pressure inside a cavitation bubble
P_o	pressure outside the nucleating bubbles
P_v	vapor pressure inside bubbles
$T(z,t)$	temperature at depth z and time t
$P(z,t)$	pressure at depth z and time t
T_c	critical temperature of the polymer
T_0	ambient temperature
T_{max}	maximum surface temperature
Γ	Grüneisen constant
$R_s=-1$	surface reflection coefficient of sound
S	sum of critical bubbles formed
$p_o(z,t)$	pressure for a fs pulse (ultrashort)
$P_o(z,t)$	pressure for a ns pulse (long)
$p_1(z,t)$, $p_2(z,t)$ and $p_3(z,t)$	pressure of the 3 superimposed subwaves
δT, δP, T_R, P_R	parameters of the Avramov model

The proportionality constant Γ is called the Grüneisen coefficient and is approximated to 1 in the case of polymers.
In fact the considered pressure elevations are of the order of ~10-10^5 bar, high values compared to ambient pressure which can be neglected in the formulation. The physical solution to (2) can be explicited as a sum of three sub-waves p_1, p_2, p_3, with one p_2 being reflected on the surface of the sample to yield p_3 with a reflection coefficient R_s=-1.

$$p_0(z,t) = \Delta p(z,t) = p_1(z,t) + p_2(z,t) + p_3(z,t) \tag{5}$$

$$p_1(z,t) = 0.5 p_{\max} e^{-\alpha(z - c_S t)}, \text{ if } z > c_s t, \text{ otherwise } 0 \tag{6}$$

$$p_2(z,t) = 0.5 p_{\max} e^{-\alpha(z + c_S t)}, \text{ if } z > 0, \text{ otherwise } 0 \tag{7}$$

$$p_3(z,t) = 0.5 p_{\max} R_s e^{-\alpha(c_S t - z)}, \text{ if } c_s t > z > 0, \text{ otherwise } 0 \tag{8}$$

This sub-wave number 3 has a negative sign due to R_s=-1 and is associated with the tension wave that can be observed in the experiments. The absorption of the laser ultrashort pulse launches two counter-propagating compression sub-waves and one is reflected on the surface with the formation of a retro-propagating tension wave. As the surface compression is converted into tension with momentum conservation, in the theoretical absence of energy dissipation due to viscosity. It appears progressively and after relaxation of the compression sub-waves. The tension wave develops enough intensity to induce cavitation after some propagation time which can be evaluated in the case of an ultra-short pulse by:

$$\Delta t \cong \frac{1.5}{\alpha c_s} \tag{5}$$

The applicability of this consideration has proved to be successfull by comparing the ultra-short laser pulse (90 fs at 266 nm) case with the ns KrF (25 ns at 248 nm) laser pulse case. The absorption depths are assumed to be close if the non-linear absorption of the ultra-short pulse is negligible (it is the case at low fluence). Now the pressure wave $P_0(z,t)$ due to the absorption of a long KrF pulse is evaluated as the result of the absorption of discrete series of many ultra-short laser pulses equally spaced and with allowance for relaxation between two elementary pulses. It is therefore the convolution product which is developed below. Let us consider a theoretical KrF laser pulse time profile of the following form:

$$F(t) = F_0 \int_0^t g(t')dt' = F_0 \left(\int_0^\infty t^2 e^{-t/\tau} dt \right)^{-1} \int_0^t t'^2 e^{-t'/\tau} dt' \tag{6}$$

It is normalized to be equal to the incident fluence at the end of the pulse and its duration is adjusted to the experimental one with the parameter τ=7.5 ns. The function g(t) is the time envelop or profile of the KrF laser pulse and of the successive theoretical elementary ultrashort pulses. If one considers an isolated "Dirac delta pulse" delivered at time t_1, at time t_2 after relaxation during t_2-t_1 the pressure elevation due to this pulse is $p_0(z,t_2-t_1)$. The sum of all pressure elevations is the given by the convolution product:

$$P_0(z,t) = p(z,t) \otimes g(z,t) = \int_0^t g(t_1) p_0(t - t_1) dt_1 \tag{7}$$

This model of pressure wave accounts well for the confinement effect which occurs when the absorption depth $l_a=\alpha^{-1}$ increases up to an optimum above which it decreases again at constant fluence because the absorbed energy density becomes too low. Two competing factors are therefore to be considered in the experiment, one varies like α and the other like $e^{-\alpha}$. This sensitivity of polymers to develop pressure confinement is expressed in the Table 1 by ΔP the pressure increase as a function of α for an incident pulse of 1 J/cm^2. It is predicted that the best value of absorption coefficient giving the most intense acoustic pressure is for α=1000 cm^{-1} which is that of CAP, for instance. Examples of pressure waves are displayed in Fig.9 for two different polymers with significantly different absorption coefficients. It is seen that PVAc develops a stronger acoustic wave because the fluence needed for the ablative foaming is one order of magnitude larger.

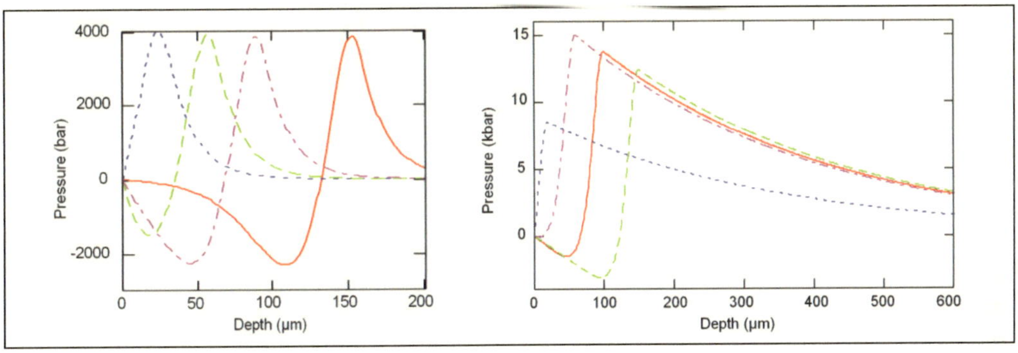

Fig. 9. Examples pressure depth profiles of acoustic waves propagation for two significantly different polymers: ***left*** is collagen or gelatine with same α= 600 cm^{-1} at t= 20, 40, 60 and 100 ns, fluence is Fo=0.5 J/cm^2; ***right*** is PVAc with lower α= 30 cm^{-1} at t=20, 60, 100 and 150 ns for Fo=6 J/cm^2.

The important feature of the acoustic profile is the negative component that develops with time after some propagation in depth. It corresponds to the tensile state of the material and its increasing intensity with depth is at some critical point capable of creating the cavitation phenomenon which is the topic of this work. In the approximations of this model the parameter c_s can be adjusted by fitting the experimental data of Fig.8 so that the appearance time of the foam is predicted by the cavitation kinetic theory. However since the data do not allow a high precision c_s is normalized to the realistic value of 1 µm/ns in the following. Further comments can be made on the transition fall from high compression peak to the tensile component maximum. It is called the rarefaction wave since a fast density decrease is associated to it. In the absence of absorption it is propagating into the bulk, like the acoustic pressure wave without change in profile. Its width or steepness is characteristic of the laser pulse duration which can be varied dramatically by changing the laser type and pulse duration. Taking an ultrafast laser pulse with a 90 fs duration makes it very narrow compared to the KrF ns laser which has a 10^6 time larger pulse width.

6. Kinetic theory of dense cavitation

The foaming process develops in the absorption zone because of the slight temperature elevation and is triggered by the acoustic wave. Its kinetics distributed in space and time $J(z,t)$ can be expressed by the classical theory of nucleation (Zeldovich, 1942).

$$J(z,t) = J_0(z,t)e^{\frac{\Delta G(z,t)}{kT(z,t)}} = n_0 \frac{\sigma(z,t)}{\eta(z,t)} \sqrt{\frac{\sigma(z,t)}{kT(z,t)}} \exp(-\frac{16\pi\sigma(z,t)^3}{3kT(z,t)[P_i(z,t)-P_0(z,t)]^2}) \tag{8}$$

This rate expression in (8) has a classical exponential factor with the free-energy barrier ΔG to form the critical bubble depending on liquid surface tension and the inner and outer pressure difference P_i-P_o. The pre-exponential factors are n_0 a molecular concentration and the Zeldovich frequency factor Z which includes surface tension and viscosity of the liquid in this case. Although various forms of Z have been introduced by several authors, this particular one taken in (Baidakov, 2007) has been chosen because it shows an interesting dependence on viscosity. Almost all material related quantities like viscosity and surface tension are temperature and pressure dependent, consequently time and depth dependent, a difficulty which imposes large approximations. The evaluation of the molecular concentration n_0, called sometime number density, is a real problem in the present case of voids concentration contrary to the usual kinetic theory in an elemental liquid like the rare gas liquids. However since the rate J quantifies bubbles or cavities formed upon the stretching situation created by the laser pulse absorption one can look for the concentration of their possible precursors in the starting polymer. The possible candidates for yielding bigger bubbles after critical expansion are the free volume holes that are well known and present in the condensed matter in general. In molecular liquids their presence and concentration is tightly related to the concentration and shape of the molecule. In polymers the free volume holes distribution reflects the complexity of the chain molecules. Below we discuss the case of PVAc for which extensive literature data is available and which provides numerical data of rate. In a previous work (Lazare et al. 2009) the ratio J/J_0 was used to explore the dynamics capable of interpreting the transmission curves of Fig.8. By doing so, only the exponential factor is considered in the evaluation of the avalanche function $A_v(t)$ which was used to compare long ns pulse to ultrashort fs foaming of Fig.10.

$$A_\upsilon(t) = 100(1 - S(t)/S_\infty) \tag{9}$$

$S(t)$ is the sum of the critical bubbles formed at time t in the considered depth and S_∞ at t=∞.

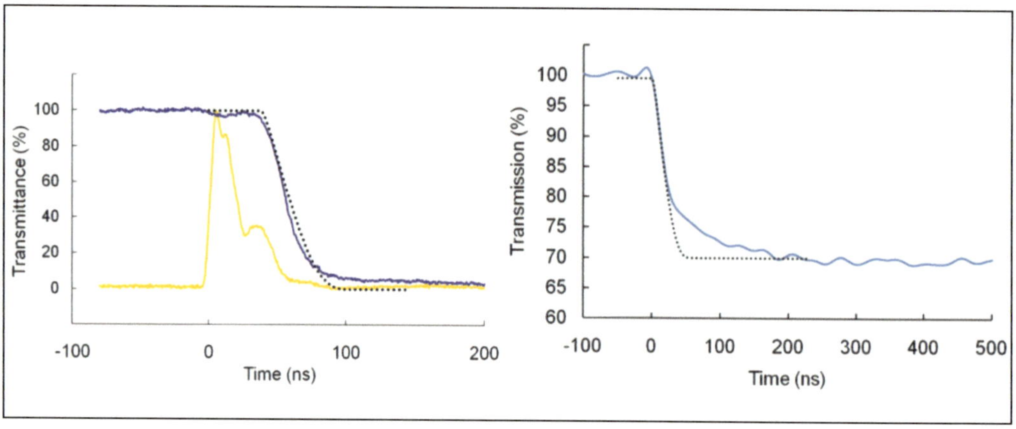

Fig. 10. Transmittances and predicted theoretical avalanche function $A_v(t)$ for a KrF ns pulse (left) and an ultrashort fs pulse (right).

The accumulated bubble since the beginning of the laser pulse absorption is given by:

$$S(t) = \int_0^t \int_0^d J(t', z, F_0)\, dz\, dt' \quad (10)$$

In which J is explicitly given by a combination of (1), (7) and (8). Necessarily the depth of integration d is limited to a realistic value usually taken slightly larger than l_a, for example d=20 µm for collagen having l_a=16 µm.

Collagen	ρ=1.3 g/cm^3	σ=30 mJ/m^2	C_p=1.5 J/g.K	C_s=1 µm/ns

Table 2. Data used for the collagen model

The model predicts a sigmoid curve that can be fitted to the experimental data as in Fig.10 if the speed of sound is c_s=1 µm. The other major approximation brought to the model concerns the pressure of ablation gas P_i that definitely contributes to the increase of the bubble growth. However it can be neglected in some part of the time window without significant loss of prediction power, but may play a role in the mechanism depending on the experimental parameters. In these conditions, if it is assumed that the tensile state alone is enough to reach a critical state that triggers the intense foaming, so a good prediction of the experimental data is obtained like in Fig.10 for both ns and fs regimes. It is observed that foaming does not occur in the high pressure compressive first component of the acoustic wave but during the rarefaction component, that is to say in the tensile tail. The nucleation wave propagation coincides with that of the tensile component as shown by the model for collagen (Fig.11). Furthermore the experimental value of the threshold fluence is well predicted by the present model. The main governing data of the threshold prediction are in table 2 and can be refined if a better fitting is needed. Among them probably the most determinant one is the surface tension σ. The lack of reliable data on viscosity of the biopolymer in order to be able to fully explicit the pre-exponential factor in (8) led us to consider the case of a new model polymer PVAc (see next section). In section 7 it is shown that the introduction of a model of viscosity with T and P dependence opens a full access to the numerical values of the voids nucleation rates

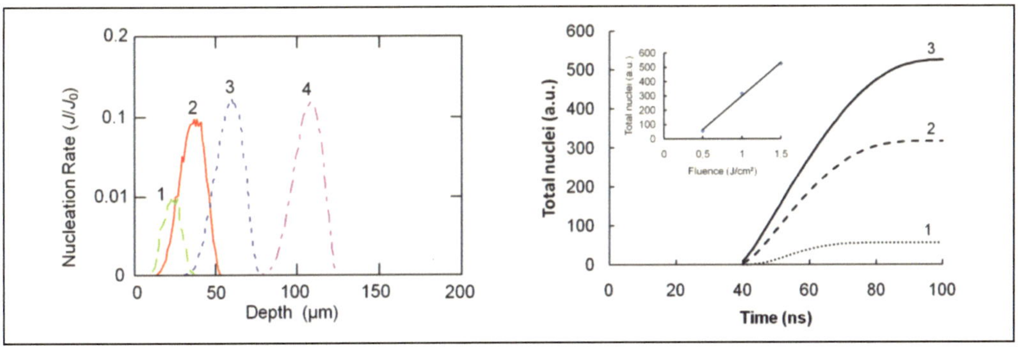

Fig. 11. (*left*) Nucleation wave in collagen for a KrF laser pulse of F_0=0.5 J/cm^2 for various time (1) t=45 (2) 55 (3) 70 (4) 100 ns. (*right*) Total number of critical bubbles as a function of time at various fluences (1) F_0= 0.5 (2) 1 (3) 1.5 J/cm^2. The inset shows the ablation curve predicting the experimental value.

7. Case of a model polymer (PVAc)

The data on laser-induced foaming of PVAc are presented in Fig.6, 12 and table 1. Further details are given in (Lazare et al., 2010). The case of this synthetic polymer simplifies the modeling owing to its molecular structure made of a simpler repeat unit. Therefore it has lent itself to many theoretical and physical studies, concerning viscosity of interest in this work and specially the temperature and pressure dependence.

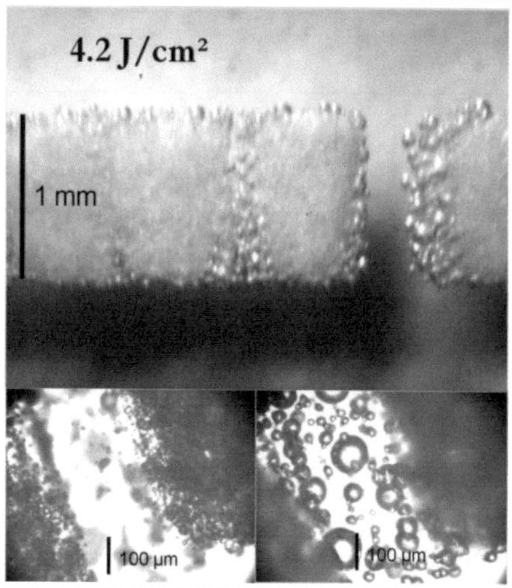

Fig. 12. Poly(vinyl acetate) PVAc film foaming with single pulses of the KrF laser at 4 J/cm². Several spots show small bubbles on surface and larger bubbles deeper in the film.

Examples of large KrF laser spots are shown in Fig.12 main picture which reveals some characteristic features like bubbles bigger than in the case of biopolymers. This may be explained by the larger absorption depth and the lower glass transition temperature T_g which allows to reach lower viscosity and higher bubble diffusion at lower temperature. In the bottom two-pictures which focus on the same spot but at two different depths in the polymer, it can be seen that the surface bubbles are smaller and conversely larger more in depth (right picture). The model predicts that the tensile component becomes larger with propagation in depth, therefore the viscosity does the same and since the diffusion varies like the reciprocal of the viscosity, it is thought that bubble collisions and growths are increased. Polymer melt viscosity is an important topic for the applications like thermal processing and molding. Therefore the literature is wide and reports many models of temperature-"only" dependence like Williams-Landel-Ferry (WLF) (Williams et al., 1955) and the (Angell, 1985) (AM). Although the P dependence alone is discussed for a long time since (Barus, 1891), the development of simultaneous T and P dependent viscosity models $\eta(T,P)$ is more recent and can be provided by (Avramov & Milchev, 1988), later adapted to PVAc by (Roland & Casalini, 2003) in equation (11) where δ_T, δ_P, T_R, P_R are parameters characteristic of the material.

$$\eta(T,P) = \exp[30\left(\frac{T_R}{T}\right)^{\delta T}\left(1+\frac{P}{P_R}\right)^{\delta P}] \tag{11}$$

ρ=1.19 g/cm^3	C_p=1.5 J/g	σ_0=36.5 mJ/m^2	c_s=1 µm/s	n_0=0.77×10^{21}cm^{-3}
δ_T=7.11	δ_P=1.62	T_R=302.2 K	P_R=2710 bar	T_c=1000 K

Table 3. PVAc physical data used in the modelling

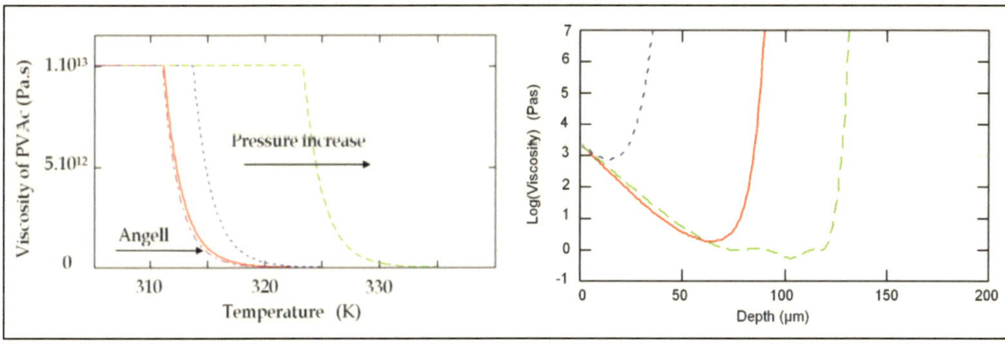

Fig. 13. (left) Poly(vinyl acetate) PVAc viscosity dependence on temperature and pressure (1, 10 and 50 bar) with the Avramov model equation (11) and comparison with Angell model at 1 bar. (right) KrF one pulse (F_0= 6 J/cm^2) laser-induced viscosity change at various times 70, 120, 160 ns.

The numerical model of viscosity with temperature and pressure dependence is presented in Fig.13 with the physical data of PVAc shown in table 3. As seen an increase of pressure results in an increased viscosity so that during the laser pulse absorption the ΔT,ΔP couple can be so that the absorbing layer remains solid up to the rarefaction wave time. This explains why in the transmission recording experiments of Fig.8 and 10, the blackening reflecting the appearance of the avalanche of bubbles occurs well after the end of the laser pulse. As suggested in the simulation Fig.13 (right) of the time evolution of the depth viscosity profile in the PVAc film the polymer of low viscosity appears only at t~100 ns and in a layer located at z>25 µm below the surface. In the same line the rate of nucleation must be in agreement with this picture. The model of surface tension of the polymer is provided by the (Guggenheim, 1945) equation (12) which is only temperature dependent since the pressure dependence is not known.

$$\sigma(T) = \sigma_0 (1 - T / T_c)^{11/9} \tag{12}$$

Before going further it important to give some background next section on the considerations on the Free Volume Theory, because it provides some numerical value of n_0, the concentration of free volume holes, in equation (8) and also it plays a role in the variation of viscosity with temperature and pressure.

7.1 Free volume theory

The free volume existence is common to any condense phase. For a long time, its two components, the interstitial free volume and the free volume holes are known and studied. Both types can be found in polymers (Budd et al., 2005) like any solid or liquids, however free volume holes are nanometer size and more frequent in polymer phases for statistical reasons due to the macromolecular character of the molecule. It is well known that at the solid to liquid transition of polymers the hole free volume fraction increases faster with temperature than the molecular volume and the interstitial free volume (Fig.14). This means that either their number, or their size, or both increase by thermal expansion and confers new properties to the polymer like decrease of viscosity, increase of diffusion and permeability, etc.. The interstitial free volume is the unoccupied space between atoms, which is necessarily present even at 0 K and increases with temperature like the molecular volume expansion.

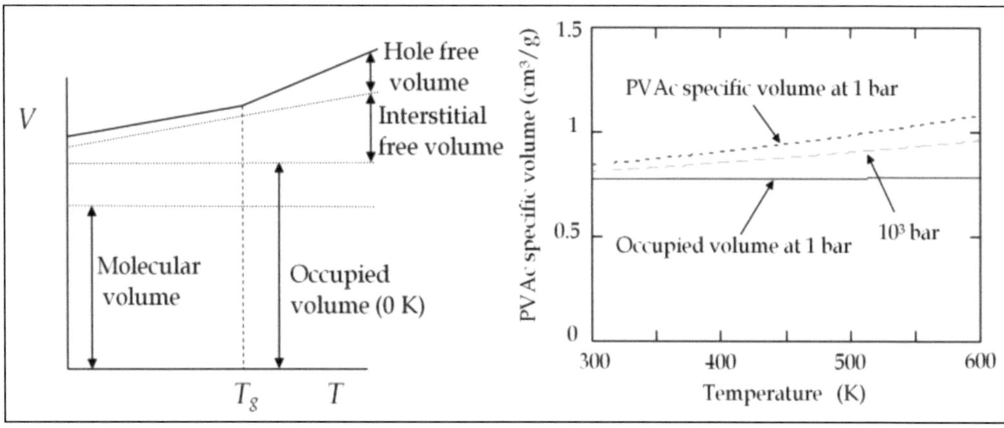

Fig. 14. (left) Temperature variation of the specific volume of polymers around Tg and its contributive components. (right) PVAc theoretical specific volumes variations with temperature at 1 and 10^3 bar calculated with the Simha equation of state (Dlubek et al., 2000). The difference between occupied and specific volumes at 1 bar is due to the free volume holes.

Holes of free volume (fvh) can be reduced in size and number by decreasing temperature and by processes like physical aging. Their concentration and volume are a fraction of that of the constitutive molecules and are known experimentally from dilatometry (McKinney & Belcher, 1963), X-ray diffraction (Wendorf & Fischer, 1973), *ortho*-positronium anhilation lifetime spectroscopy (*o*-PALS) (Tao, 1971; Dlubek et al., 2000; Hong-ling et al., 2008). This experimental knowledge is predicted by the theoretical equation of state (SS-EOS) developed by (Simha & Somcynski, 1969) using the cellular method of (Lennard-Jones & Devonshire, 1937). In the work of (Dlubek et al., 2000) a numerical solution of the theoretical SS-EOS can be found and used to compute the specific and hole free volumes of PVAc in Fig.14 (right). The plot shows the importance of the hole free volume that increases with temperature, whereas the occupied volume increases only slightly. At higher applied pressure 1000 bar, a normal compression phenomenon occurs. These data are for an equilibrium situation only and have to be used with caution when interpreting the laser experiment since the timescale, the dynamic windows and the fast evolution are placing the system in a region out of equilibrium. Nevertheless SS-EOS data are useful to characterise the starting polymer system and its free volume holes. The precise o-PALS results provide the concentration free volume holes (n_0=0.77×10^{21}cm^{-3}) in PVAc and the mean hole volume (0.1 nm^3) at room temperature. This provides a satisfactory value of n_0 in our calculation of nucleation rate (8) for PVAc (see next section). Furthermore it appears very likely that at the rarefaction wave the ablation gas can diffuse to the inner of these fvh. Just before rarefaction the high pressure maintains the system in a solid state with high viscosity and one may consider that ablation gas are blocked in the polymer or are not yet formed. For PVAc ablation gas products obtained with a CO_2 laser have been reported to be mainly H_2, CO, CO_2, CH_3COOH, $CH_3COOCHCH_2$ (Kupcik et al., 2005). Most of these molecules have sizes that are small enough to be contained in the original fvh, however at rarefaction an expansion of the system increases dramatically the diameter of the holes and amplifies the diffusion phenomenon which builds up the inner gas pressure in the nucleating bubbles.

7.2 Stretched state of the polymer

Evaluating the nucleation rate (8) requires some approximation to be considered as far as ablation gas pressure P_i inside the growing bubbles is concerned. As in our previous work one may consider that P_i can be neglected and that the tensile state of the polymer is enough to create the new voids. As shown this yields a polymer in a stretched state which is well capable of high nucleation rate in the condition of the experiments. The case of PVAc is shown in Fig.15 and reveals the dependence of the nucleation rate on time and fluence. The onset of nucleation occurs at a well defined depth of 27 μm and starts at a time of ~60 ns. The calculation further shows that pre- and exponential factors in (8) grow simultaneously with time but with different rates. In the present case of PVAc the pre-exponential factor is the limiting one. This seems to indicate that viscosity is the controlling parameter of the system when T and P vary in a large range during the pulsed heating. The inset of Fig.15 indicates that above the predicted threshold which fits well the experimental one of 3 J/cm^2 (table 1), there is a strong accumulation of the post-nucleation bubbles.

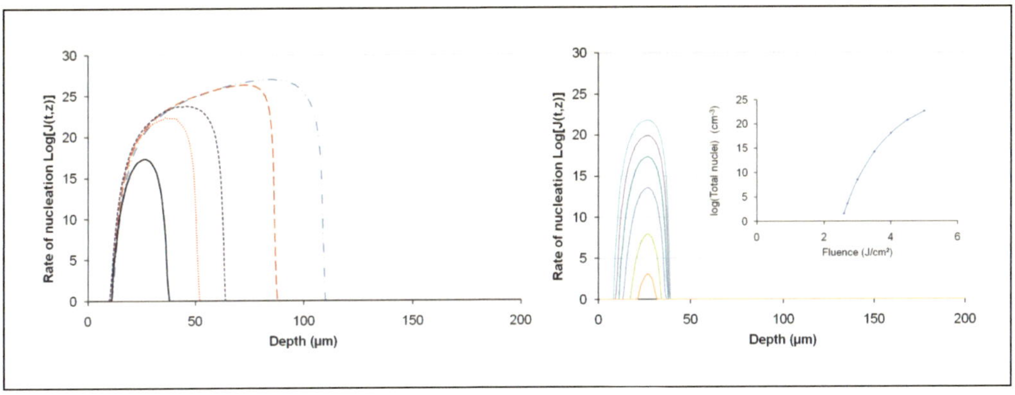

Fig. 15. (left) Theoretical rate of nucleation (in nucl.cm^{-3}s^{-1}) as a function of depth in PVAc at various times 80, 90, 100, 120, 140 ns after the laser pulse absorption (F_0=4 J/cm^2). (right) Theoretical rate of nucleation (in nucl.cm^{-3}s^{-1}) as a function of fluence (F_0=2.7, 3.0, 3.5, 4.0, 4.5, 5 J/cm^2) in PVAc at a time of 80 ns after the pulse absorption. The inset is the total number of nuclei formed at 80 ns as a function of fluence.

These results also show that cavitation does not occur selectively at the extreme surface of the polymer as would be suggested by the laser temperature or excitation profile but rather at the depth where the tension wave reaches a maximum intensity. Moreover cavities, free volume holes and nanobubbles formed in the stretching phase of the mechanism are expected to facilitate the molecular diffusion according to the principle of (Cohen & Turnbull, 1959). As expressed with the Stoke-Einstein equation (13) the diffusion coefficient D varies like the reciprocal viscosity in a system with spherical molecules of radius r.

$$D = \frac{kT}{6\pi\eta r} \tag{13}$$

It follows that nano-bubbles collisions and merging is activated by the "in-depth" viscosity drop giving rise to the birth of larger bubbles like that seen in Fig.12 deeper below the surface. Nano-cavities merging is also a mechanism which can help the system reach the critical state faster at some points. Now if some ablation gas pressure model $P_g(z,t)$ is added to the present system, then equation (8) giving the nucleation rate is accelerated versus time due to the increase of difference $P_i(z,t) - P_0(z,t)$ in which P_i is more or less proportional to P_g. A possible tentative model is equation (14) which depends strongly on fluence.

$$P_g(z,t) = P_g(0,t)e^{-\alpha z} \tag{14}$$

7.3 Limits of prediction

The approximations of the model should be recalled to estimate the future applicability of the results to other systems, especially to natural biopolymers like collagens which are good candidates for artificial type of cellular matrix materials. To obtain the pressure wave, the Grüeneisen function was used as a constant and equal to one for polymers. This has no incidence on the dynamics but only on the shape of the waveform. The nucleation rate is a product of exponential factor, Zeldovich frequency factor and molecular concentration factor n_0. All of them require slight approximations because T and P dependence of the parameters are most of the time not known. Among them viscosity η and surface tension σ play an important role. If viscosity dependence on T and P is known for some synthetic polymers like PVAc, it is not the case of surface tension and unfortunately biopolymers are not well documented with that respect. Also the molecular concentration factor n_0 has been tentatively attributed to the concentration of free volume holes of PVAc. Again biopolymers literature does not report such value of n_0 due to the more complex molecular structures of the biological systems. The growing importance of the new present field is hopefully going to boost new studies.

8. Properties of the obtained polymer foams

The laser-induced foams obtained so far have shown that the molecular structure of the polymer or biopolymer is conserved by the laser treatment which then appears as a very mild method for local processing. This has been proved by spectroscopic methods like Raman micro-spectrometry and fluorescence (Wiesniewski et al., 2007; Gaspard et al., 2008c). In most cases the post-treatment material has spectral signature similar to the starting polymers. This is true for a large range of fluence although for values much larger than the threshold slight traces of carbon particles can be detected on the foam samples. An appealing feature of this laser method to obtain foamy material is the high density of nucleation which can be evaluated at 1 μm^{-3}. Such high densities are not achieved by classical methods and confer to the final laser-induced polymer foam a large potential of new applications. The use of high values of laser fluence can achieve nano-structures like filaments in high concentration (Fig.5d). The obtained 3D network made of nano-sized filaments have valuable properties, for future experiments in biomedical applications like artificial cellular matrices for tissue repair, for chemical or biosensors, for gas or nano-filtration.

9. Future experiments and applications

Little is known on the part of target which is ejected from the surface during the development of the foam layer. However we are in the presence of a mechanism which does not require a high temperature but rather relies on the force field created by the pressure developed during the confinement. The surface material in a fluid state is accelerated non uniformely along the normal of the surface due to the cavitation phenomenon, stronger below the surface as indicated by the model. As a result an extensive fragmentation of this portion of polymer occurs and is sometime visible in the form of a small cloud of white smoke ejected at high speed from the laser spot. When the laser irradiation is done trough a silica window placed near the sample surface, this cloud of nanoparticles is deposited on its surface as seen in Fig.16. The experiment is performed in air and does not require a vacuum, and can therefore be easily adapted to the deposition of nanoparticles onto substrates that do not stand the vacuum, like cell cultures in biological research or other polymer science experiments. Similar results have been repeated for the polymers studied in this work. As suggested by Raman micro-spectrometry the chemical structure of the polymer is not destroyed by the laser treatment and the atomic force microscopy indicates that the nanodroplets are in a molten state when they hit the silica surface. This preparation of nanoparticles by laser ablation in the photomechanical regime is new and will probably open up further ideas of future applications.

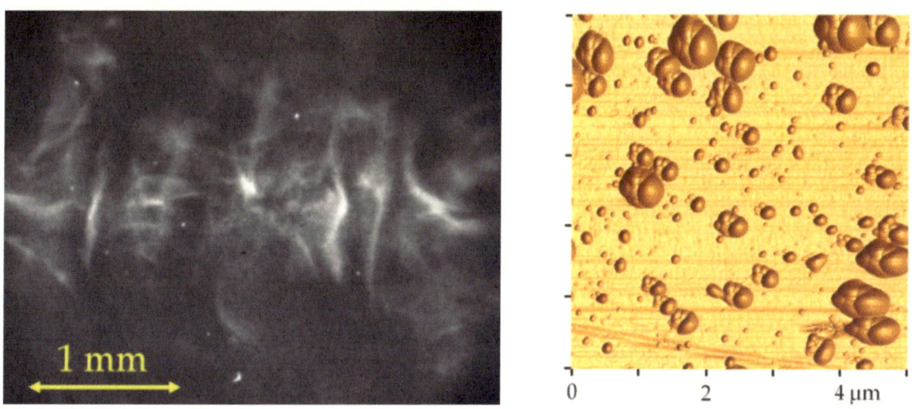

Fig. 16. Nanoparticles of gelatin deposited by KrF laser ablation in air (F_0=1.7 J/cm^2) on a silica window placed 300 μm in front the target. Several square spots of 1 mm were used with a 1 mm shifting between each. (left) Optical microscope image with scattered light. (right) Atomic force microscope image.

The obtained foams are attractive substrates for building new cell culture media because of their interesting properties that are approaching that of the natural extracellular matrices. Based on the small numbers of already studied polymers it can be said that with a laser pulse of high fluence, several times the threshold fluence, the foam structure is forced to develop open cells configuration due to large amount of ablation gas formed. This means that the new material is permeable to fluids like in the natural tissue. However, due to the small characteristic dimensions of the laser-induced foam, usually in the nanometer or sub-micrometer range, the stability in the contact with liquids is a critical parameter which needs to be carefully studied. To circumvent this difficulty cross-linking the polymer before or after laser step may be an interesting treatment. The model has shown how the rate of cavitation is sensitive to viscosity which in turn can be modified by cross-linking reagents. There are some biopolymers or materials which are naturally cross-linked. For instance this is the case of keratin which is stable in liquids due to the significant amount of SS cross-links formed by dimerisation of cystein. Fig.17 is the illustration that two different synthetic polymers do not behave identically. CAP tends to give nano-droplets and CTA preferably forms nano-filaments for fluencies much larger than threshold. Future mechanistic investigation will bring more insight to explain these peculiarities.

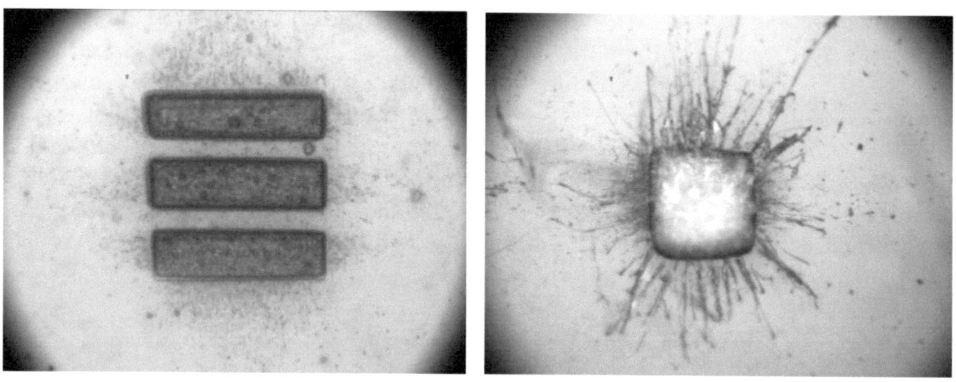

Fig. 17. KrF laser ablation with a single pulse on a 150 μm spot. (left) CAP ablated at F_0=1.5 J/cm^2; the spot is divided in three parts for the purpose of focusing. (right) CTA ablated at 8 J/cm^2 tends to produce many small filaments by ejection of molten polymer whereas lower fluence yields only foam.

10. Conclusions

A new type of polymer foam is obtained with the aid of laser ablation. Biopolymers and synthetic polymers concerned by the laser-induced foaming have absorption coefficient of the order of 10-10^3 cm^{-1} in order to allow the pressure to increase during the radiation absorption by pressure confinement. So far the experiments have been performed with one pulse of ns lasers in the condition of pressure confinement and single fs pulses for which this condition is necessarily met and no polymer one-photon absorption is needed. The mechanisms of foaming are studied experimentally and by modelling with a set of several biopolymers. The dynamics of foam formation was monitored and showed an appearance time of the order of 50-100 ns after the start of the laser pulse. This characteristic time is explained by the necessity of the high pressure component of the laser-induced acoustic wave to propagate away from the absorption depth, a phenomenon which occurs the speed of sound, and to be reflected off at the target surface to launch a tension wave back to the bulk of the polymer. When the tension intensity rises and combines with the action of the ablation gases, it overpasses the critical tensile strength of the polymer. The phenomenon of cavitation occurs giving rise to the formation of critical bubbles which further grow, merge and stabilize to form a dense nanofoam. The proposed model of nucleation rate based on the classical theory of homogeneous nucleation accounts well for the experimental ns and fs laser dynamics. A new numerical development was possible for poly(vinyl acetate), a model synthetic polymer rich in literature data. The rate of nucleation is expressed in three factors which are each discussed and explicited numerically. The importance of the temperature and pressure dependence of the viscosity is revealed by the results. Also the model brings some information on the depth profile of the foaming kinetics which has a maximum below the surface at a well defined depth, e.g. 27 μm for PVAc and the KrF laser. It is proposed that the holes of free volume, known in polymers by *o*-PALS and theoretical equation of state, can be at the inception of the critical nucleus involved in the cavitation step. The laser-induced modifications, foams with closed or open cells, filaments, ejected nano-droplets, etc have appealing properties and unprecedented by competing and more classical techniques. Many new and promising experiments are made possible by these results concerning laser-material microprocessing based on a photomechanical regime of ablation. Future detailed studies of the specificities of new polymers and biopolymers may open up more applications.

11. References

Angell, C.A. (1985). Spectroscopy simulation and scattering, and the medium range order problem in glass. *J.Non-Cryst.Solids*, Vol 73., No 1-3., (August 1985), (1-17), ISSN 00223093

Avramov, I. & Milchev, A. (1988). Effect of disorder on diffusion and viscosity in condensed systems. *J.Non-Cryst.Solids*, Vol 104., No 2-3., (September 1988), (253-260), ISSN 00223093

Baidakov,V.G. (2007). Explosive Boiling of Superheated Cryogenic Liquids, Wiley-VCH*Springer*, ISBN 3527405755, 9783527405756, *Berlin*, Germany.

Barus, C., (1891). The dependence of viscosity on pressure and temperature. *Proc.Amer.Acad.*, Vol 27., (1892), (13)

Bäuerle, D. (2000). Laser Processing and Chemistry, 3rd edn. *Springer*, ISBN 978-3-540-66891-6, *Berlin*,. Heidelberg, Germany.

Budd, P.; McKeon, N.B. & Fritsch, D. (2005). Free volume and intrinsic microporosity in polymers. *J.Mat.Chem.*, Vol 15., No 20., (May 2005), (1977-1986), ISSN 09599428

Cohen, M.H. & Turnbull, D. (1959). Molecular Transport in Liquid and Glasses. *J.Chem.Phys.*, Vol 31., No 5., (November 1959), (1164-1169), ISSN 00219606

Dlubek, G.; Lüpke, T.; Stejny, J.; Alam, M. & Arnold, M. (2000). Local Free Volume in Ethylene-Vinyl Acetate Copolymers: A Positron Lifetime Study. *Macromol.*, Vol 33., No 3., (February 2000), (990-996), ISSN 00249297

Elaboudi, I.; Lazare, S.; Belin, C.; Talaga, D. & Labrugère, C. (2008). Underwater excimer laser ablation of polymers. *Appl.Phys.A*, Vol 92., No 4., (September 2008), (743-748), ISSN 0947-8396

Gandini, A. (2008). Polymers from Renewable Resources: A Chalenge for the future of Macromolecular Materials. *Macromol.*, Vol 41., No 24., (December 2008), (9491-9504), ISSN 0024-9297

Gaspard, S.; Oujja, M.; DeNalda, R.; Abrusci, C.; Catalina, F.; Banares, L.; Lazare, S. & Castillejo, M. (2007). Nanofoaming in the surface of biopolymers by femtosecond pulsed laser irradiation. *Appl.Surf.Sci.*, Vol.254, No 4, (December 2007), (1179-1184) ISSN 01694332

Gaspard, S.; Oujja, M.; de Nalda, R.; Castillejo, M.; Bañares, S.; Lazare, S. & Bonneau, R. (2008a). Nanofoaming dynamics in biopolymers by femtosecond laser irradiation. *Appl.Phys.A*, Vol.93, No.1, (October 2008), (209-213), ISSN 0947-8396

Gaspard, S.; Forster, M.; Huber, C.; Zafiu, C.; Trettenhahn, G.; Kautek, W. & Castillejo, M. (2008b). Femtosecond laser processing of biopolymers at high repetition rate. *Phys.Chem.Chem.Phys.*, Vol.10, No.40, (October 2008), (6174-6181), ISSN 14639076

Gaspard, S.; Oujja, M.; Abrusci, C.; Catalina, F.; Lazare, S.; Desvergne, J.P. & Castillejo, M. (2008c). Laser induced foaming and chemical modifications of gelatine films. *J.Photochem.Photobiol.A:Chem.*, Vol 193., No 2-3., (January 2008), (187-192), ISSN 10106030.

Gordon, M.K. & Hahn, R.A. (2010). Collagens. *Cell.Tissue Res.*, Vol.339, No.1, (247-257), ISSN 0302766X

Guggenheim, E.A. (1945). The principle of the corresponding states. *J.Chem.Phys.*, Vol 13., No 7., (July1945), (253-261), ISSN 00219606

Hong-ling Lv; Bao-guo Wang & Ji-chu Yang (2008). Correlations between PALS and diffusivities in amorphous polymer-solvent systems. *Desalinisation*, Vol.234 No.1-3, (December 2008), (33-41), ISSN 00119164

Kupcik, J.; Blasvska-Gilev, J. & Pola, J. (2005). IR Laser-Induced Degradation of Poly(vinyl acetate): Novel Thermal Reactions in Solid. *Macromol.Rap.Commun.*, Vol.26 No.5, (March 2005), (386-389), ISSN 10221336

Lazare, S. & Tokarev, V. (2005a). Ultraviolet Laser Ablation of Polymers and the Role of Liquid Formation and Expulsion. In: *Recent Advances in Laser Processing of Materials,* Elsevier, ISBN 0750660791, Oxford England

Lazare, S.; Tokarev, V.; Sionkowska, A. & Wisniewski, M. (2005b). Surface foaming of collagen, chitosan and other biopolymer films by KrF excimer laser ablation in the photomechanical regime. *Appl.Phys.A*, Vol 81., No 3., (May 2005), (465-470), ISSN 0947-8396

Lazare, S.; Bonneau, R.; Gaspard, S.; Oujja, M.; De Nalda, R.; Castillejo, M. & Sionkowska, A.; (2009). Modeling the dynamics of one laser pulse surface nanofoaming of biopolymers. *Appl.Phys.A*, Vol 94., No 4., (March 2009), (719-729), ISSN 0947-8396

Lazare, S.; Elaboudi, I.; Castillejo M. & Sionkowska, A., (2010). Model Properties relevant to Laser Ablation of moderately absorbing polymer. *Appl.Phys.A*, Vol 9X., No X., (Xxx 2010), (ppp-ppp), ISSN 0947-8396

Lennard-Jones, J. E. & Devonshire, A. F. (1937), Critical Phenomena in Gases I, *Proc.R.Soc.Lond.* A, Vol.163, No.912, (November 1937), (53-70), ISSN 1364-5021

Lippert, T. & Dickinson, J. T. (2003). Chemical and Spectroscopic Aspects of Polymer Ablation: Special Features and Novel Directions. *Chem.Rev.*, Vol.103, No.2, (January 2004), (453-486), ISSN 0009-2665

Liu, C-F; Sun, R-C; Zhang, A-P; Qin, M-H; Ren, J-L & Wang, X-I, (2007). Preparation and Characterization of Phthalated Cellulose Derivatives in Room-Temperature Ionic Liquid without Catalysts. *J.Agric.Food Chem.*, Vol. 55, No 6, (March 2007), (2399-2406), ISSN 00218561

McKinney, J.E. & Belcher, H.V. (1963). Dynamic compressibility of poly(vinyl acetate) and ist relation to free. *J.Res. oft he NIST*, Vol.67A, No.1, (January-February 1963), (43-53), ISSN 1044-677X

Mooney, B.P. (2009). The second green revolution? Production of plant-based biodegradable plastics. *Biochem.J.*, Vol.418, No.2, (March 2009), (219-232), ISSN 02646021

Moore, C.J. (2008). Synthetic polymers in the marine environment: a rapidly increasing, long-term threat. *Environ.Res.*, Vol 108, No.2, (Oct 2008), (131-139), ISSN 0013-9351

Pérez, J. ; Muñoz-Dorado, J. ; de la Robia, T. & Martinez, J. (2002). Biodegradation and biological treatments of cellulose, hemicellulose and lignin : an overview, *Int.Microbiolo.*, Vol.5, No.2 , (June 2002), (53-63), ISSN 1139-6709

Pinkert, A.; Marsh, K.N.; Pang, S. & Staiger, M.P. (2009). Ionic liquids and their interaction with cellulose, *Chem.Rev.*, Vol.109, No.12 , (December 2009), (6712-6728), ISSN 00092665

Moore, C.J. (2008). Synthetic polymers in the marine environment: a rapidly increasing, long-term threat. *Environ.Res.*, Vol 108, No.2, (Oct 2008), (131-139), ISSN 0013-9351

Oujja, M. ; Pérez, S., Fadeeva, E. ; Koch, J.; Chichkov, B.N. & Castillejo, M., (2009). Three dimensional microstructuring of biopolymers by femtosecond laser irradiation. *Appl.Phys.Lett.*, Vol.95, No.26, (December 2009), (263703-3p), ISSN 00036951

Roland, C.M. & Casalini, R., (2003). Temperature and volume effects on local segmental relaxation in poly(vinyl acetate). *Macromol.*, Vol.36,No.4, (February 2003), (1361-1367), ISSN 00249297

Sionkowska, A. ; Wisniewski, M.; Lazare, S.; Lopez, J.; Hernandez, M-C., Guillemot, F. & Durrieu, M-C. (2008). Surface properties of femtosecond laser irradiated collagen films. *Mol.Cryst.Liq.Cryst.*, Vol.486, (January 2008), (250-256), ISSN 15421406

Telle, H.U. ; Urena, A.G. & Donovan, R.J. (2007). *Laser Chemistry*, J Wiley & Sons, ISBN 978-0-48570-4 (HB), England

Tao, S.J. (1972). Positronium annihilation in molecular substances. *J.Chem.Phys.*, Vol 72., No 11., (June 1972), (5499-5510), ISSN 00219606

Tokarev, V.; Lopez, J.; Lazare, S. & Weisbuch, F. (2003). High aspect ratio microdrilling in polymers with UV laser ablation: experiment and model. *Appl.Phys.A*, Vol 76., No 3., (March 2003), (385-393), ISSN 0947-8396

Weisbuch, F.; Lazare, S.; Goodall, F.N. & Débarre, D. (1999). Submicron-resolution ablation with a KrF excimer laser beam patterned with a projection. *Appl.Phys.A*, Vol 69., Suppl. No 1., (December 1999), (S413-S417), ISSN 0947-8396

Wendorf, J.H. & Fisher, E.W. (1973). Thermal density fluctuations in amorphous polymers as revealed by small angle X-ray diffraction. *Coll.&Pol.Sci.*, Vol 251., No 11., (November 1973), (876-883), ISSN 0303-0402

Williams, M.L.; Landel, R.F. & Ferry,D (1955). The temperature dependence of relaxation mechanisms in amorphous polymers and other glass-forming liquids. *J.Amer.Chem.Soc.*, Vol 77., No 14., (1942), (3701-3707), ISSN 27863.

Wisniewski, M.; Sionkowska, A.; Kaczmarek, H; Lazare, S.; Tokarev, V. & Belin, C. (2007). Spectroscopic study of a KrF ecximer laser treated surface oft hin collagen films. *J.Photochem.Photobiol.A:Chem.*, Vol 188., No 2-3., (May 2007), (192-199), ISSN 10106030.

Zeldovich, Y., (1942). Theory of Formation of New Phase, Cavitation. *Sov.J.Exp.Theor.Phys.*, Vol 12., No 11., (1942), (525-538), ISSN 0044-4510

7

Production of Fungal Chitosan by Enzymatic Method and Applications in Plant Tissue Culture and Tissue Engineering: 11 Years of Our Progress, Present Situation and Future Prospects

Nitar Nwe, Tetsuya Furuike and Hiroshi Tamura
Faculty of Chemistry, Materials and Bioengineering, Kansai university
Suita, Osaka 564-8680
Japan

1. Introduction

Chitin is a copolymer of *N*-acetyl-D-glucosamine and D-glucosamine units linked with β-(1-4)-glycosidic bond, here *N*-acetyl-D-glucosamine units are predominant in that polymer chain. The deacetylated form of chitin refers to chitosan. Chitin usually refers to a copolymer with a degree of acetylation (DA) of more than 40% [i.e., number of *N*-acetyl-D-glucosamine more than 40% and number of D-glucosamine less than 60%] and insoluble in dilute acids. The name chitosan is used for a copolymer with less than 40% DA [i.e., more than 60% DD (degree of deacetylation), number of *N*-acetyl-D-glucosamine less than 40% and number of D-glucosamine more than 60%] that, in most cases, will be soluble in dilute acid (Nwe & Stevens, 2008). Chitin and chitosan are found as supporting materials in many aquatic organisms (shells of shrimps and crabs and bone plates of squids and cuttlefishes), in many insects (mosquitoes, cockroaches, honey bees, silkworms, Drosophila *melanogaster, Extatosoma tiaratum* and *Sipyloidea sipylus*), in terrestrial crustaceans (*Armadillidium vulgare, Porcellio scaber*), in nematode, in mushrooms (*Agaricus bisporus, Auricularia auriculajudae, Lentinula edodes, Trametes versicolor, Armillaria mellea, Pleurotus ostreatus, Pleurotus sajo-caju* and *Pleurotus eryngii)* and in some of microorganisms (yeast, fungus, and algae) (Carlberg, 1982; Nemtsev et al., 2004; Veronico et al., 2001; Paulino et al., 2006; Moussian et al., 2005; Tauber, 2005; Hild et al., 2008; Anantaraman & Ravindranath, 1976; Pochanavanich & Suntornsuk, 2002; Mario et al., 2008; Yen & Mau, 2007 cited in Nwe et al., 2010).

Nowadays, commercially, chitins and chitosans are produced from biowastes obtained from aquatic organisms. The production of those products in industrial scale for medical and agriculture applications appears in inconsistent physico-chemical characteristics of products because of seasonal and variable supply of raw materials, as well as variability and difficulties of process conditions (Ashford et al., 1977 cited in White et al., 1979). Moreover, in 1996, main argument on the production of chitosan from shells of shrimps was the present of shrimp antigen in the final product of chitosan (Tan et al., 1996, *Proceedings of 2nd Asia Pacific Chitin and Chitosan Symposium*, Bangkok, pp 50-57). These problems on production of chitosan may be circumvented by extracting chitosan from fungal mycelia.

The production of chitosan from fungal mycelia has a lot of advantages over crustacean chitosans such as the degree of acetylation, molecular weight, viscosity and charge distribution of the fungal chitosan are more stable than that of crustacean chitosans; the production of chitosan by fungus in a bioreactor at a technical scale offers also additional opportunities to obtain identical material throughout the year and to obtain chitosans with a radioactive label or with specific changes in its polymeric composition; and the fungal chitosan is free of heavy metal contents such as nickel, copper (Tan et al., 1996, Arcidiacono & Kaplan, 1992, Nwe & Stevens, 2002a). Moreover the production of chitosan from fungal mycelia give medium-low molecular weight chitosans (1–12 × 10^4 Da), whereas the molecular weight of chitosans obtained from crustacean sources is high (about 1.5 ×10^6Da) (Nwe & stevens, 2002). Chitosan with a medium-low molecular weight has been used as a powder in cholesterol absorption (Ikeda et al., 1993) and as thread or membrane in many medical-technical applications. For these reasons, there is an increasing interest in the production of fungal chitosan. However, so far, the extraction of high yield pure chitosan production from fungal cell wall material has not been accomplished upto 2001 (Stevens, 2001).

In our research, we have investigated the bond between chitosan and glucan in fungal cell wall and developed a powerful enzymatic method for the production of excellent quality chitosan from the fungal mycelia in a high yield. In this process a heat-stable alpha-amylase is used to cleave the covalent link between chitosan and glucan in cell wall of *Gongronella butleri* (Nwe & Stevens, 2002 & 2002a; Nwe et al., 2008). In 2006, the effect of shrimp and fungal chitosan on the growth and development of orchid plant meristemic tissue in culture was investigated in liquid and on solid medium (Nge et al., 2006). In recent years, the scaffold and membrane were prepared using the fungal chitosan and their mechanical and biological properties were evaluated together with the scaffolds and membranes of chitosans obtained from crustacean sources to use as a tissue regeneration template (Nwe et al., 2009; Nwe et al., 2010a). Among the tested chitosans in here, fungal chitosan showed as an excellent growth stimulator to apply in agriculture sector and as an excellent scaffolding material to construct a biodegradable tissue regeneration template. In this chapter, we summarize all our research contributions on last 11 years in 4 sections: (1) Investigation of a method to produce high quality and quantity of fungal chitosan and study the bond between chitosan and glucan in the cell wall of fungi, *Gongronella butleri* and *Absidia coerulea,* (2) Optimization of fermentation conditions for production of high yield chitosan by solid substrate fermentation, (3) Application of fungal chitosan in plant tissue culture, and (4) Application of fungal chitosan in tissue engineering.

2. Investigation of a method to produce high quality and quantity of fungal chitosan and study the bond between chitosan and glucan in the cell wall of fungi, *Gongronella butleri* and *Absidia coerulea*

2.1 Growth of fungus and extraction of chitosan by traditional method

Chitosan is a substantial component of cell wall of certain fungi, particularly those belonging to the class Zygomycetes (Bartniki-Garcia, 1968). Tan et al., 1996 evaluated the yield of chitosan from several Zygomycetes fungi including *Absidia, Gongronella, Mucor* and *Rhizopus* and concluded that *G.butleri* gave the highest yield of chitosan. At the same time, Crestini et al., 1996 reported that the yield of chitosan produced from *Lentinus edodes* grown in solid state fermentation, 6.18 g/kg was higher than that in submerged fermentation, 0.12

g/l. In 1998, fungus *Gongronella butleri* was selected to produce chitosan in our research. Firstly, a comparison was made between the yield of chitosan from fungal mycelia grown in solid substrate fermentation (SSF) and in submerged fermentation (SMF) using various nitrogen sources. The Termamyl assayed extraction method was not discovered yet at that time. The chitosan was extracted using vacuum filtration and β-glucanase treatment method. It was observed that the yield of chitosan obtained from fungal mycelia grown in SSF (3.7 g chitosan/kg of solid substrate) was higher than that in SMF (0.6 g chitosan/L of fermentation medium) due to the low amount of mycelia produced in SMF (Nwe et al., 2002). Based on the results obtained from our research and Crestini et al., 1996, solid substrate fermentation was selected as the best fermentation method to produce chitosan by fungus *Gongronella butleri* (Figure 1).

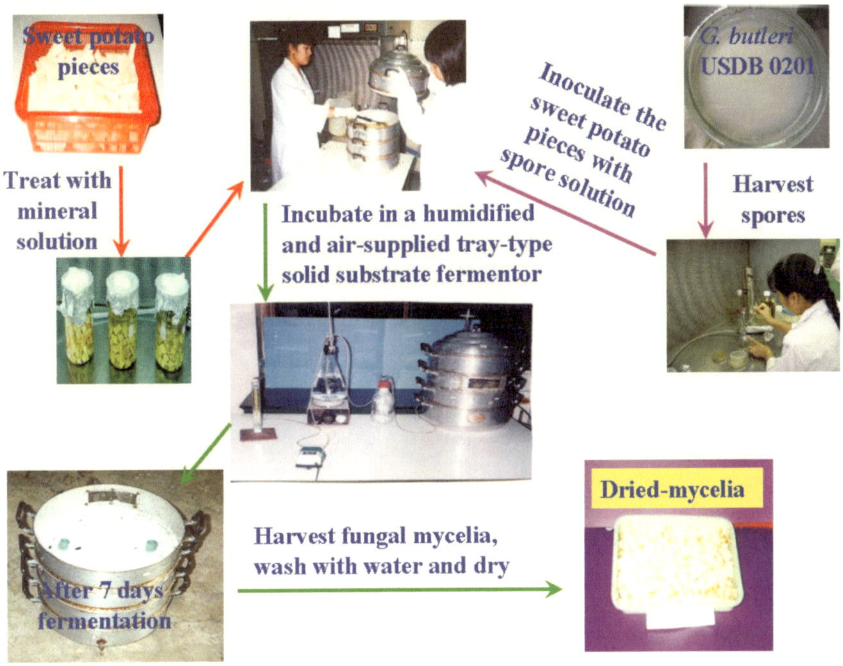

Fig. 1. Fungus *Gongronella butleri* USDB 0201 was grown on sweet potato pieces in a tray-type solid substrate fermentor. Sweet potato pieces were used as solid support and as carbon source. The dried fungal mycelia were used to extract chitosan.

The history of the development of chitosan extraction procedure by enzymatic method started with the work of Mr. Su Ching Tan from the National University of Singapore, Singapore. In his method, mycelia were treated with 1 M NaOH and the resultant alkaline insoluble material (AIM) was treated with 0.35 M acetic acid at 25°C for 2 h (Tan et al., 1996). In our research the extraction of chitosan from fungal mycelia was started according to the method described by Tan et al., 1996 with minor modification (Figure 2). The yield of chitosan extracted from fungal mycelia grown in solid substrate fermentation was 2-3 g/100 g of mycelia. An effective chitosan extraction procedure is essential for an economical production of fungal chitosan.

Several methods have been developed for the extraction of chitosan from the fungal mycelia (Table 1). Most methods used 1 M NaOH to remove protein and other cell wall materials and then the chitosan was extracted with 2 % acetic acid. The yield of chitosan produced from the fungal mycelia treated in this way is very low. The extraction procedure for high yield production of pure chitosan from the fungal cell wall material has not yet been accomplished up to 2001 (Stevens, 2001).

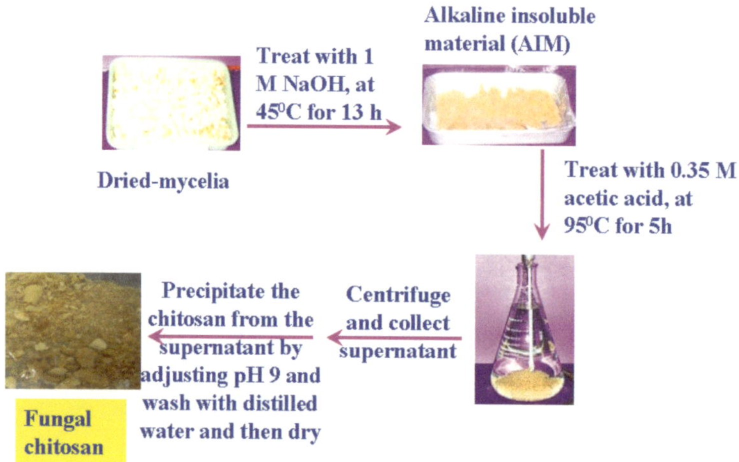

Fig. 2. Extraction of chitosan from mycelia of fungus *G.butleri* grown in solid substrate fermentation

Reference	Strain	Deproteination (NaOH)			Acetic acid extraction			Chitosan yield
		Conc:	Temp:	Time	Conc:	Temp:	Time	
Jaworska and Konieczna, 2001	*Absidia orchidis*[a]	1M	121°C	10 min	1%	-	-	1.84 g/L
Crestini *et al.*, 1996	*Lentinus Edodes*[b]	1M	121°C	15 min	2%	95°C	14h	6.18 g/kg of wheat straw
Crestini *et al.*, 1996	*Lentinus Edodes*[a]	1M	121°C	15 min	2%	95°C	14h	120 mg/L
Tan *et al.*, 1996	*G. butleri*[a]	1M	121°C	15 min	2%	25°C	1 h	467 mg/L
Muzzarelli *et al.*, 1994	*A.Coerulea*[a]	25%	boil	3 h	0.1%	25°C	16h	1,800 mg/L
Rane and Hoover, 1993	*G.butleri*[a]	1M	121°C	15 min	2%	95°C	12h	230 mg/L
Rane and Hoover, 1993	*A.Coerulea*[a]	1M	121°C	15 min	2%	95°C	12h	480 mg/L
Arcidiacono and Kaplan, 1992	*M. rouxii*[a]	1 M	121°C	15 min	2%	95°C	-	250 mg/L
Hang, 1990	*R. oryzae*[b]	2%	121°C	15 min	2%	95°C	12h	700 mg/L

[a]Submerged fermentation, [b]Solid state/substrate fermentation

Table 1. Extraction of chitosan from fungal mycelia by different methods

Glucose joined through α-1,3-, α-1,4-, α-1,6-, β-1,3-, β-1,4- or β-1,6-linkages is called glucan (Carbonero et al., 2005; Schmid et al., 2001; Hochstenbach et al., 1998; Wessels et al., 1990; Wolski et al., 2005; Marchessault & Deslandes, 1979 cited in Nwe et al., 2008). Chitin/chitosan and glucan are the main fungal skeletal polysaccharides. In the fungal cell wall, chitin/chitosan occurs in two forms, as free aminoglucoside and covalently bonded to β-glucan (Bartnicki-Garcia, 1968; Gooday, 1995; Robson, 1999; Wessels et al., 1990). In 1990, Wessels et al. proposed that initially chitin and β-glucan chains accumulate individually in the fungal cell wall and thereafter form the interpolymer linkage. The formation of the chitin/chitosan–glucan complex chains results in a rigid cross-linked network in the cell wall (Gooday, 1995; Robson, 1999) and causes a considerable problem for the extraction of intact chitosan and glucan. It does not break down easily under mild extraction condition (Muzzarelli et al., 1980). Under the above mentioned conditions only free chitosan, that is chitosan unbounded to other cell wall components is extracted (Nwe & Stevens, 2002). Chitosan bounded to insoluble cell wall components will not be extracted (Figure 2 and 3).

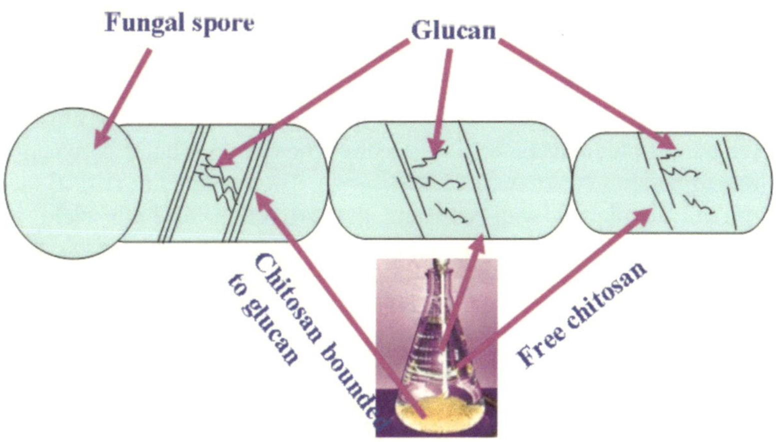

Fig. 3. A proposed model for the synthesis of chitosan (straight line) and glucan (wave line) in fungal hyphae

To extract the high quality and quantity of chitosan and glucan from cell wall of fungi, the bond between chitosan and glucan in cell wall of fungi must be investigated. Most of the researchers are trying to find the linkage between the chitosan and glucan in the fungal cell wall by digestion with glucanase, chitinase and amylase. In 1979, Sietsma and Wessels reported that 90%of β-glucan obtained from the chitin-glucan complex by digestion with (1-3)-β-glucanase and *N*-acetyl-glucosamine, lysine and/or citrulline were identified as products after digestion with chitinase. Therefore they proposed that the bridge linking the glucan chain with the chitin contains lysine, citrulline, glucose and *N*-acetyl-glucosamine. Similar evidence was obtained by Gopal et al., 1984 for the degradation of chitin-glucan complex by (1,3)-β- and (1,6)-β-glucanase, and subsequently by chitinase. Carbohydrate expressed as glucose and *N*-acetyl-glucosamine monomers was detectable in equivalent amounts in the hydrolysate. The residue after chitinase treatment was further treated with α-amylase but additional release of glucose could not be detected colorimetrically. Surarit et al., 1988 suggested a glycosidic linkage between position 6 of *N*-acetyl-glucosamine in chitin and position 1 of glucose in β-(1-6)-glucan in the cell wall of *Candida albicans.* In 1990, Wessels et al., proposed that a direct link between free amino groups in the glucosaminoglycan and the reducing end of the glucan chains forming the inter-polymer linkages in the chitin-glucan complex. Up to 1992, no cleare evidence of the identity of the chemical link between the chitosan and glucan polymer chains had been uncovered (Roberts, 1992). After 1995, Kollar et al., 1995 and Fontaine et al. 2000 digested the cell wall of *Saccharomyces cerevisae* and *Aspergillus fumigatus* with 1%SDS and 1 M NaOH respectively and the insoluble fractions were digested with (1,3)-β-endoglucanase and chitinase. The soluble fractions were analyzed. Based on their results, they concluded that the terminal reducing residue of a chitin chain is attached to the non-reducing end of a β-(1,3)-glucan chain by a β-1,4 linkage. An insoluble residue was remained at the end of both extraction processes.

Importantly, cell wall matrix must be broken down far enough in order to study the linkage between the chitosan and glucan in the fungal cell wall and to extract total chitosan (i.e., free chitosan plus chitosan bounded to glucan). In 1994, Muzzarelli et al., reported that the chitosan-glucan complex can be split by 25 % NaOH (Table 1). Therefore further treatment of cell wall matrix was investigated.

2.2 Decomposition of fungal cell wall materials by acid and alkaline treatment

In order to decompose the AIM matrix into molecular level of chitosan-glucan complex, chitosan and glucan, the cell wall matrix has to be weakened and the chitosan from the complex has to be soluble in acetic acid. Thereafter the covalent bond between the chitosan and glucan has to be split by enzymatic hydrolysis. There steps are critical factors for the decomposition of fungal cell wall and for the extraction of high yield chitosan. To investigate the condition for the maximum solubility of chitosan in acetic acid, the duration and temperature of the acetic acid extraction step were varied (Figure 2). It was found that

1 M NaOH,
45^{0}C for 13 h

11 M NaOH,
45^{0}C for 13 h

Treatment	Temperature (°C)	NaOH concentration (M)	Time (h)	Decomposition of AIM in acetic acid
1	79	1.77	13.50	-
2	46	1.77	13.50	-
3	79	5.08	4.93	-
4	46	5.08	4.93	-
5	95	6.75	13.50	-
6	63	6.75	13.50	-
7	30	6.75	13.50	-
8	79	8.41	22.07	+
9	46	8.41	22.07	+
10	79	11.73	13.50	++++
11	46	11.73	13.50	++++

Table 2. Effect of NaOH treatment conditions on decomposition of the AIM in 0.35M acetic acid at 95^{0}C for 5h (Reproduced from Nwe et al., 2008a, International Journal of Biological Macromolecules, 43, 2-7)

the yield of chitosan increased to 3.14 g/100g of mycelia when the AIM was treated with 0.35 M acetic acid at 95°C for 5 h (Nwe et al., 2008a). However the AIM matrix did not break down at that condition (Figure 2, Table 2). The condition of acetic acid treatment at 95°C for 5h was selected and the NaOH treatment step was carried out with a randomized experimental design (Figure 2 and Table 2).

An experimental condition to break the rigid cell wall matrix into a suspension form was investigated without degradation of extracted chitosan during acid and alkaline treatment (Nwe et al., 2008a). The best decomposition condition was the mycelia treated with 11 M NaOH at 45°C for 13 h and then the AIM was treated wtih 0.35 M acetic acid at 95°C for 5 h. In this situation, if the chitin present in cell wall of fungus, probably all chitin will be deacetylated to chitosan using 11 M NaOH instead of 1 M NaOH (Nwe & Stevens, 2002 and 2002a). Upto now it is not clear yet the formation of chitin and chitosan in cell wall of fungi. Here it appeared very difficult to isolate pure chitosan from that suspended solution (Table 2, Figure 4). In order to isolate pure chitosan, the AIM suspended solution was treated under different conditions.

2.3 Isolation of chitosan from the AIM suspended solution

In the process of fungal mycelia treated with 11 M NaOH, nearly all proteins, soluble glucan, other polysaccharides, DNA and RNA presented in the fungal hyphae might wash out from alkaline insoluble materials (AIM) of the fungal hyphae during washing step of the AIM with distilled water upto neutral pH. Here it was assumed that chitosan, insoluble glucan and chitosan-glucan complex might remain in the AIM material (Nwe & Stevens, 2002a). These polymers suspended in 0.35M acetic acid during treatment of the AIM with 0.35M acetic acid at 95°C for 5 h (Table 2, Figure 4).

pH of solution	Solubility of chitosan and glucan		
	CTS	GLU	CTS-GLU
4	Soluble	Soluble	Clear solution
		Insoluble	Turbid suspension
9	Insoluble	Soluble	Precipitate
		Insoluble	Precipitate

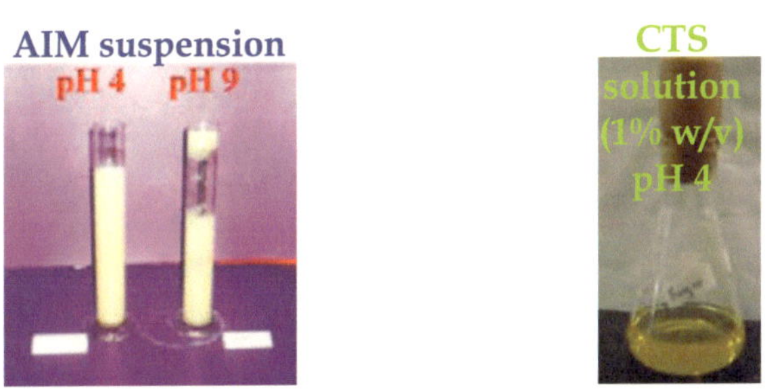

Fig. 4. Nature of chitosan (CTS), glucan (GLU), and chitosan-glucan complex (CTS-GLU) in 0.35 M acetic acid at pH 4 and 9

According to the polycationic nature of chitosans, chitosans dissolve in organic acid at pH 4 and they precipitate at pH 9. Most of the glucans are insoluble at pH 4 and 9. At pH 4, the AIM components in 0.35 M acetic acid form a stable suspension and the turbid components precipitate at pH 9 (Figure 4, Nwe & Stevens, 2002a). As a hypothesis it is assumed that chitosan and glucan are bound to each other and show suspension/precipitation behavior according to the physico-chemical nature of the chitosan component (Nwe & Stevens, 2002a). The chitosan component of the material is heavily protonated at pH 4 and supports the complex to stay in suspension (Nwe & Stevens, 2002a). At pH 9, chitosan lost its charge and does not longer sustain the complex to remain in suspension and the whole complex precipitate. To confirm this assumption, the turbidity of the AIM suspended solution was measured at pH 4 and the suspension was twice-centrifuged and than turbidity of the suspension was measured again. It was found that the turbidity of the suspended solution decreased from more than 1000 NTU (above detection limit) to about 600 NTU after twice centrifugation (Nwe & Stevens, 2002a). This observation confirmed that the turbid material in the suspended solution is linked to the chitosan chains. In order to isolate chitosan from the suspended solution, the suspended solution was treated with glucanase, promozyme and amylase (Nwe & Stevens, 2002a).

It was found that the turbidity of the suspended solution reduced to >300 NTU when the AIM suspended solution was treated with glucanase or promozyme (Table 3). Most glucans in the fungal cell wall synthesize as linear form with β-1,3 linkages and as branched chain with β-1,6 linkages (Herrera, 1991). Here it can be assumed that glucanase and promozyme could not be cleaved the glucan component from the complex when the glucan chains were stable in the suspension as a triple helix (Case II, III, and IV, Figure 5). However the turbidity of the solution decreased sharply from >1000 to <30 NTU when the suspension was treated with Termamyl Type LS, Termamyl type L or α-amylase (Table 3). These enzymes cleave the α-(1,4) glycosidic bond. The turbidity of suspension decreased to 10 NTU at the optimum conditions of Termamyl Type LS: pH 4.5, 4% (v/v) enzyme, at 65 °C, 225 rpm for 3 h (Table 4, Nwe & Stevens, 2002a). The maximum shaking speed for the water bath in laboratory of Bioprocess Technology, Asian Institute of Technology, Thailand used was 200 rpm. Therefore shaking speed, 200 rpm was used in the later experiments.

Enzyme	Enzyme % (v/v)	Temp (°C)	RPM	Time (h)	Turbidity
Glucanase (β-1,3)	5	25	100	2	>300 NTU
Promozyme (α-1,6)	5	50	175	2	>500 NTU
Diazyme (α-1,4)	5	50	100	2	<30 NTU
Termamyl Type LS (α-1,4)	5	50	175	2	<30 NTU
Termamyl Type L (α-1,4)	5	50	175	2	<30 NTU
α-amylase (α-1,4)	5	50	175	2	<30 NTU

Table 3. Turbidity of chitosan solutions after treatment of the AIM suspended solutions with glucanase, promozyme and amylase

After Termamyl treatment, a large amount of precipitate and clear supernatant was obtained (Figure 6). This precipitate is supposed to be glucan, freed from the chitosan-glucan complex. After removal of this precipitate by centrifugation (1600g, 15min), the supernatant was collected. After adjusting the pH of supernatant to 9, a heavy precipitate

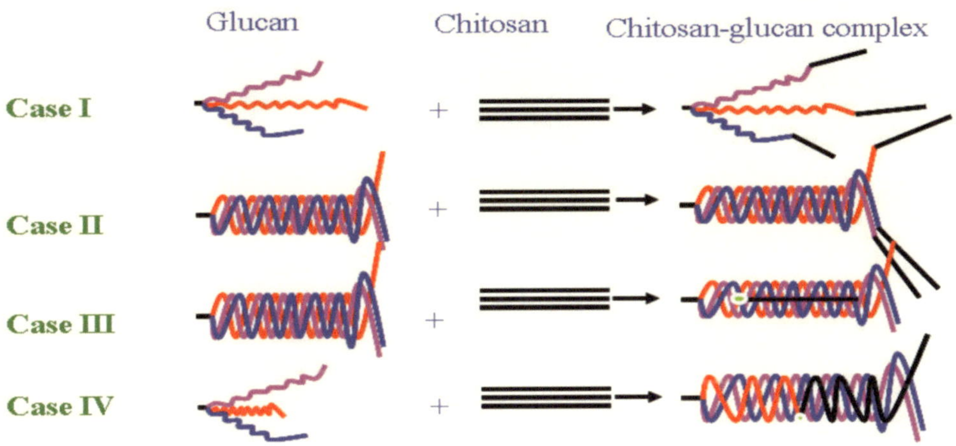

Fig. 5. Proposed model for the formation of chitosan-glucan complex in cell wall of fungus *Gongronella butleri*

Parameter to be optimized	Experimental conditions					Conditions for variable parameter				
	pH	Enz (%)	Time (h)	Temp (°C)	rpm	A	B	C	D	E
pH	V	5	3	65	200	3.5	4	4.5	5	-
Enzyme volume (% v/v)	4.5	V	3	65	200	1	2	3	4	5
Time (h)	4.5	4	V	65	200	0.5	1	1.5	2	3
Temperature (°C)	4.5	4	3	V	200	45	55	65	75	85
Shaking (rpm)	4.5	4	3	65	V	125	150	175	200	225

Parameter to be optimized	Turbidity of chitosan after enzymatic treatment at specified condition (NTU)				
	A	B	C	D	E
pH	139 ±3	169 ± 3	31 ± 3	197 ± 5	
Enzyme volume	455 ± 5	307 ± 16	53 ± 2	15 ± 3	31 ± 3
Time	35 ± 1	30 ± 1	32 ± 1	26 ± 2	15 ± 3
Temperature	313 ± 6	71 ± 3	15 ± 3	25 ± 0	58 ± 3
Shaking	62 ± 1	23 ± 0	17 ± 2	15 ± 3	10 ± 0

Table 4. Turbidity of chitosan solutions after treatment of the AIM suspended solutions with Termamyl, type LS under different experimental conditions (Reproduced from Nwe & Stevens, 2002a, Biotechnology Letters, 24, 1461-1464).

was obtained. This material easily can be redissolved in 0.35 M acetic acid and is apparently glucan-free chitosan (Figure 4). The Termamyl treatment is obviously highly efficient to separate the chitosan from the glucan and offers possibilities for the isolation of purified chitosan. Based on these observations, it was proposed that the bond between the two polysaccharides, chitosan and glucan is linked by α-(1-4)-glycosidic bond (Nwe & Stevens, 2002a).

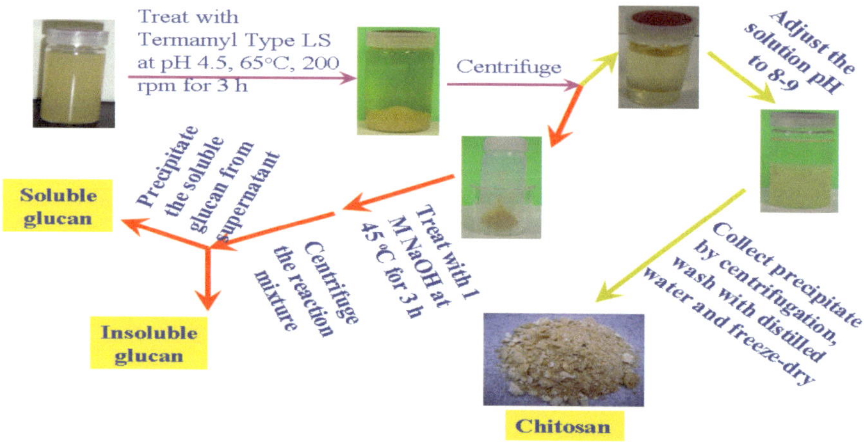

Fig. 6. Extraction of chitosan and glucan from the AIM suspended in 0.35M acetic acid by treatment with Termamyl Type LS (Nwe & Stevens, 2002 and Nwe et al., 2008)

2.4 Confirmation the bond between the chitosan and glucan

To confirm the bond between the chitosan and glucan, the suspension was centrifuged and filtered with GF/C filter paper and then the turbidity of the filtrate was recorded as 39 NTU (Figure 7). In this process the glucan unbounded to chitosan and glucan bounded to chitosan larger than 1.2 µm might remain on the GF/C filter paper and these materials smaller than 1.2 µm might pass through the GF/C filter paper (Nwe et al., 2008). The resultant filtrate was treated with β-glucanase, the turbidity of solution decreased to 32 NTU. This result suggested that at least a part of the turbidly is caused by glucan bounded to the chitosan. After glucanase treatment, the reaction mixture was adjusted to pH 8-9 to precipitate the chitosan. The precipitate was washed with distilled water upto neutral pH and dried and then weighted (Figure 7 and Table 5, product **A**). In order to observe the purity of product A, the product A was redissolved again in dilute acetic acid and the solution A was digested with Termamyl Type LS at optimal conditions (Nwe et al., 2008). Chitosan, product B was purified from the solution B after treatment with Termamyl Type LS enzyme. The concentration of reducing end in the chitosan solution A and B was measured (Figure 7, Table 6). The results of product A and B was compared with the yield of product C, which extracted from the AIM suspension treated with Termamyl Type LS directly (Figure 6 and 7 and Table 5).

Although IR spectrum of product A was very similar to that of shrimp chitosan (Figure 8), product A, 8.2 g composed with 5 g glucan (3.2 g chitosan, product B). This result clearly showed that β-glucanase did not completely remove glucan from the product A and once the bond between chitosan and glucan was broken by Termamyl, Type LS, this glucan precipitated in chitosan solution (Figure 7). In case Termamyl was directly applied to the AIM suspended in acetic acid, chitosan, 5.9 g was obtained (Table 5). This was 2.7 g of chitosan more than the yield of product B. Apparently this chitosan was part of a cell wall component that did not pass through the GF/C filter paper in absence of Termamyl treatment (Figure 7 and Table 5).

Moreover the amount of reducing ends in the chitosan solution B was about 40 times higher than that in the chitosan solution A (Figure 7, Table 6). The enzyme is effective in the

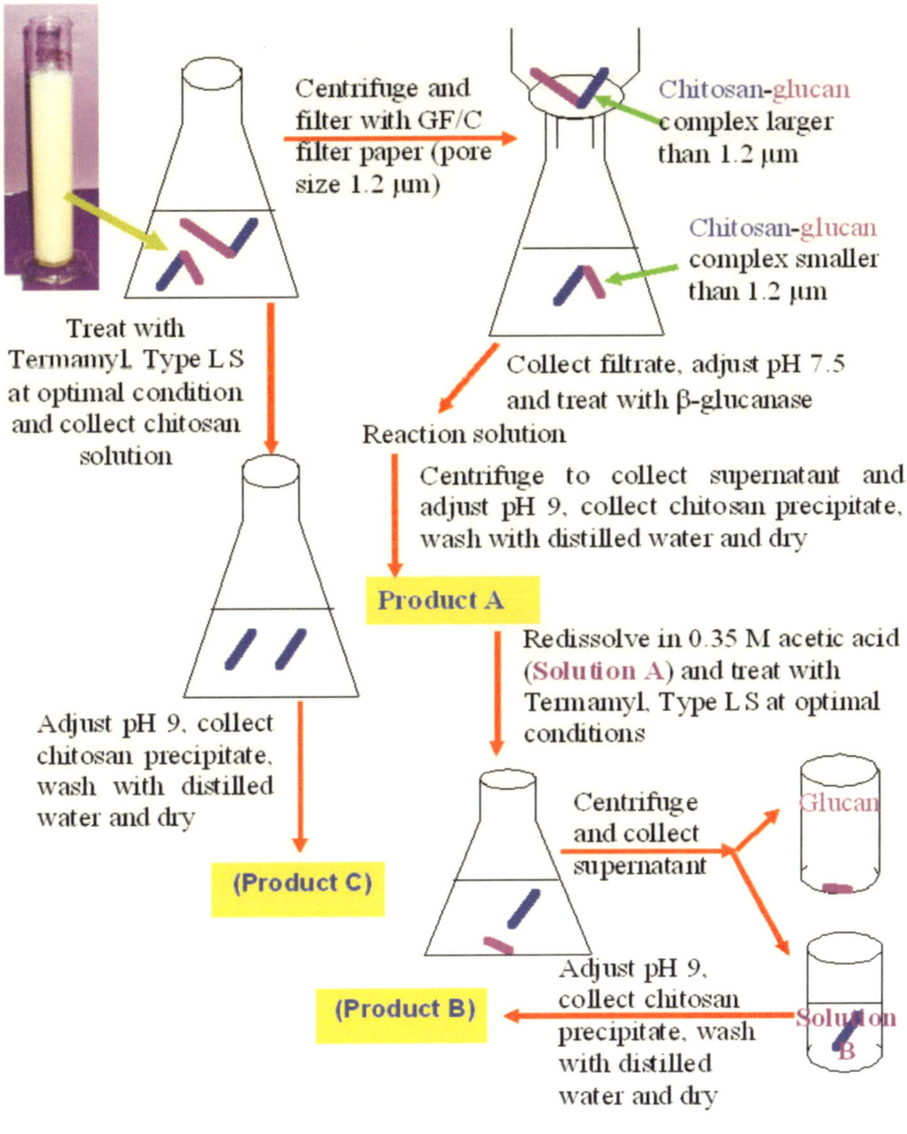

Fig. 7. Extraction of chitosan from the AIM suspended in 0.35M acetic acid by different methods (Nwe et al., 2002, 2002b & 2008, Nwe & Stevens, 2002)

hydrolysis of α-(1-4)-glycosidic linkages, leading to a sharp increase in 1-glucose end groups in the solution (Nwe et al., 2008). Since Termamyl does not split the β-glycosidic bonds in chitosan, the increase must originate from hydrolysis of 1-4 bonds between chitosan and glucan and on α-1-4 bonds within the glucan molecule. Based on the results presented in this study it is obvious that reducing end of chitosan chain and non-reducing end of glucan chain are bound by α-(1-4)-glycosidic bond and that probably glucan has also internal α-(1-4)-glycosidic bond, as concluded from the overwhelming increase in the concentration of reducing ends. This evidence is one of the confirmations of α-(1-4)-glycosidic linkages between chitosan and glucan in cell wall of fungus *G. Butleri* (Nwe et al., 2008). Pure α-amylase (Sigma) was used to confirm the above results. This enzyme gave same result as Termamyl, Type LS enzyme (Nwe & Stevens, 2002a). Here the application of Termamyl, Type LS in production of fungal chitosan are (1) glucan is removed from the extracted chitosan solution more effectively than by filtering with GF/C filter paper (2) high purity of fungal chitosan is obtained (3) less time consuming and (4) status of enzyme is food grade (Nwe et al., 2001).

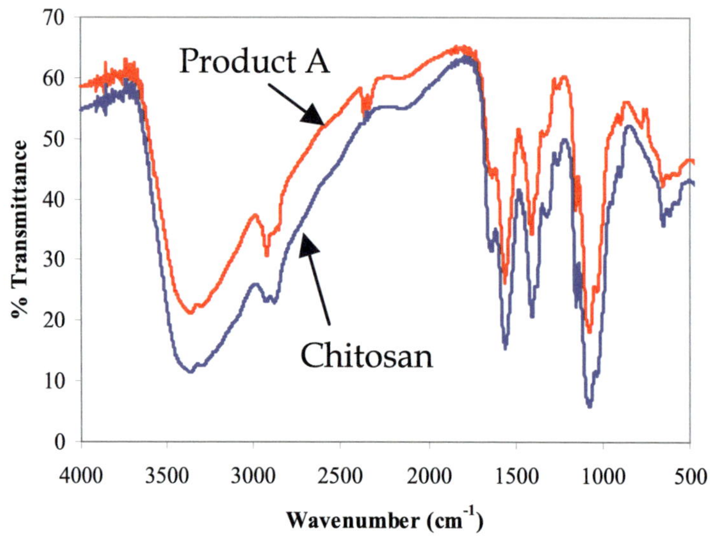

Fig. 8. IR-spectra of chitosans from shells of shrimps and from cell wall of *G. butleri* grown on solid substrate supplied with mineral solution, basic medium (product A)

Growth condition	Biomass (g/kg sweet potato, w.b)	Number average molecular weight (kDa)	Yield of chitosan (g chitosan/100 g of mycelia)		
			Product A	Product B	Product C
Sweet potato	30.1 ± 3.2	-	-	-	2.29 ± 0.49
Basic medium	24.6 ± 2.0	32 ± 3	8.2 ± 0.4	3.2 ± 1.3	5.91 ± 0.11
Basic medium & peptone	28.5 ± 1.3	33 ± 3	10.5 ± 0.7	6.4 ± 1.0	-
Basic medium & $(NH_4)_2SO_4$	19.8 ± 3.2	36 ± 1	11.7 ± 2.0	4.8 ± 2.4	-
Basic medium & $NaNO_3$	25.6 ± 0.5	32 ± 3	10.2 ± 0.7	5.8 ± 1.0	-
Basic medium & urea	29.3 ± 1.8	33 ± 1	12.7 ± 1.1	7.8 ± 0.2	-

Table 5. Biomass (dry weight) and chitosan yield obtained by solid substrate fermentation with different nitrogen sources (Basic medium: Sweet potato pieces impregnated with mineral solution) (Nwe & Stevens, 2002a & 2002b, Nwe et al., 2002 & 2008)

Condition	Reducing ends concentration (mg/ml)
Before Termamyl treatment (A)	0.043 ± 0.001
After Termamyl treatment (B)	1.85 ± 0.085

Table 6. The concentration of reducing ends in the chitosan solution A and B (Reproduced from Nwe et al., 2008, Enzyme and Microbial Technology, 42, 242-251)

For further confirmation, the extracted chitosan and glucan fractions from chitosan-glucan complex (Figure 6) were analyzed by IR and ^{13}C-NMR spectroscopy and elementary analysis (Nwe et al., 2008). The IR spectra of alkaline soluble and insoluble glucan, and chitosan showed the peak at 1550 cm^{-1}, amide II band (Figure 9A). This observation indicated that some chitosan polymers might be entrapped in the glucan triple helices and heat stable alpha-amylase enzyme could not cleave the chitosan from the glucan polymers (case III and IV in figure 5). In the ^{13}C-NMR spectrum of fungal chitosan (Figure 9B), the

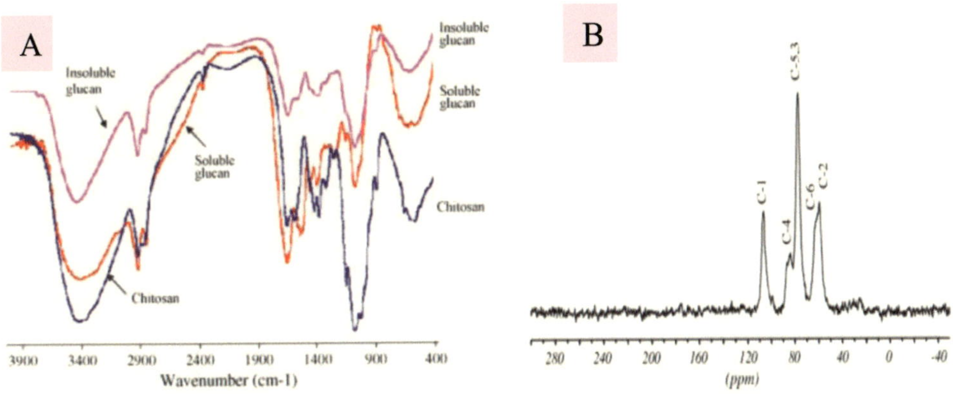

Fig. 9. (A) IR spectra of fungal chitosan and alkaline soluble and insoluble glucan (B) ^{13}C-NMR spectrum of purified fungal chitosan (Reproduced from Nwe et al., 2008, Enzyme and Microbial Technology, 42, 242-251)

chemical shift on the fungal chitosan spectrum at approximately δ 68 was absent. This result suggested that there is no chain branching in the major monomer units in the fungal chitosan molecules and lack of the free non-reducing end in the major monomer units of the polymers. The analytical result from determination of nitrogen content in the fungal chitosan (8%) strongly proved that 95% purified chitosan was obtained from Termamyl treatment. Taken together, these investigations proved that chitosan and glucan in the cell wall of fungus *G.butleri* are linked by α-(1-4)-glycosidic bond (Figure 10, Nwe & Stevens, 2002a, Nwe et al., 2008).

After proved the bond between chitosan and glucan by Termamyl treatment, chitosan was extracted from the fungal mycelia grown on solid substrate fermentation medium supplied with urea by method I and II (Table 7). It was found that the amount of chitosan extracted by method II leads to a four times higher than that by method I (Table 7). The similar result has been reported from the fungal mycelia grown in the basic medium (Nwe & Stevens, 2002). These results confirmed that total chitosan in the mycelia of fungus *G. Butleri* is extracted using method II and free chitosan from the fungal mycelia is extracted using method I.

By applying these two methods, method I and II, the bond between chitosan and glucan in cell wall of fungus *Absidia coerulea* ATCC 14076 was investigated. Fungus *Absidia coerulea* was grown in submerged fermentation medium (Figure 11) and free chitosan and total chitosan (free chitosan plus chitosan bounded to glucan) were extracted from the fungal mycelia using method I and II (Table 7). The free chitosn, 6.5 g/100 g of mycelia and total chitosan 8-9 g/100g of mycelia were isolated from fungal mycelia of *A.coerulea* (Nwe et al., 2008a).

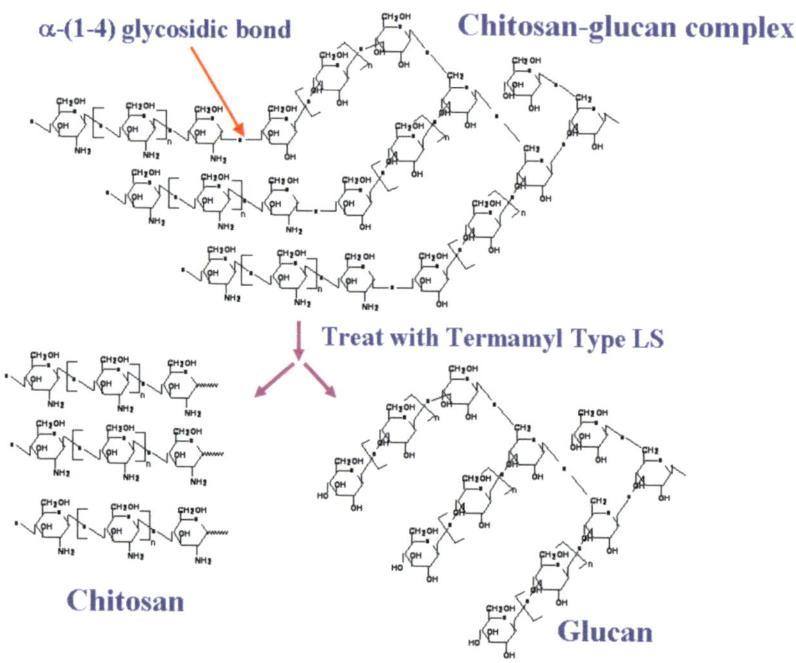

Fig. 10. Possible molecular structure of chitosan and glucan obtained from the treatment of chitosan-glucan complex by Termamyl, Type LS (Reproduced from Nwe et al., 2008, Enzyme and Microbial Technology, 42, 242-251)

Chitosan extraction Method	Yield of chitosan (g/ 100 g mycelia)	Number average molecular weight (kDa)
Method I	1.74	35
Method II	8.09	55

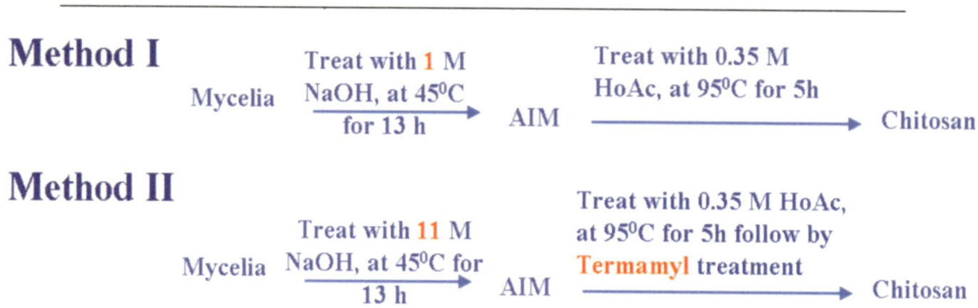

Table 7. Yield and number average molecular weight of chitosan extracted from fungal mycelia using method I and II (Nwe et al., 2008a)

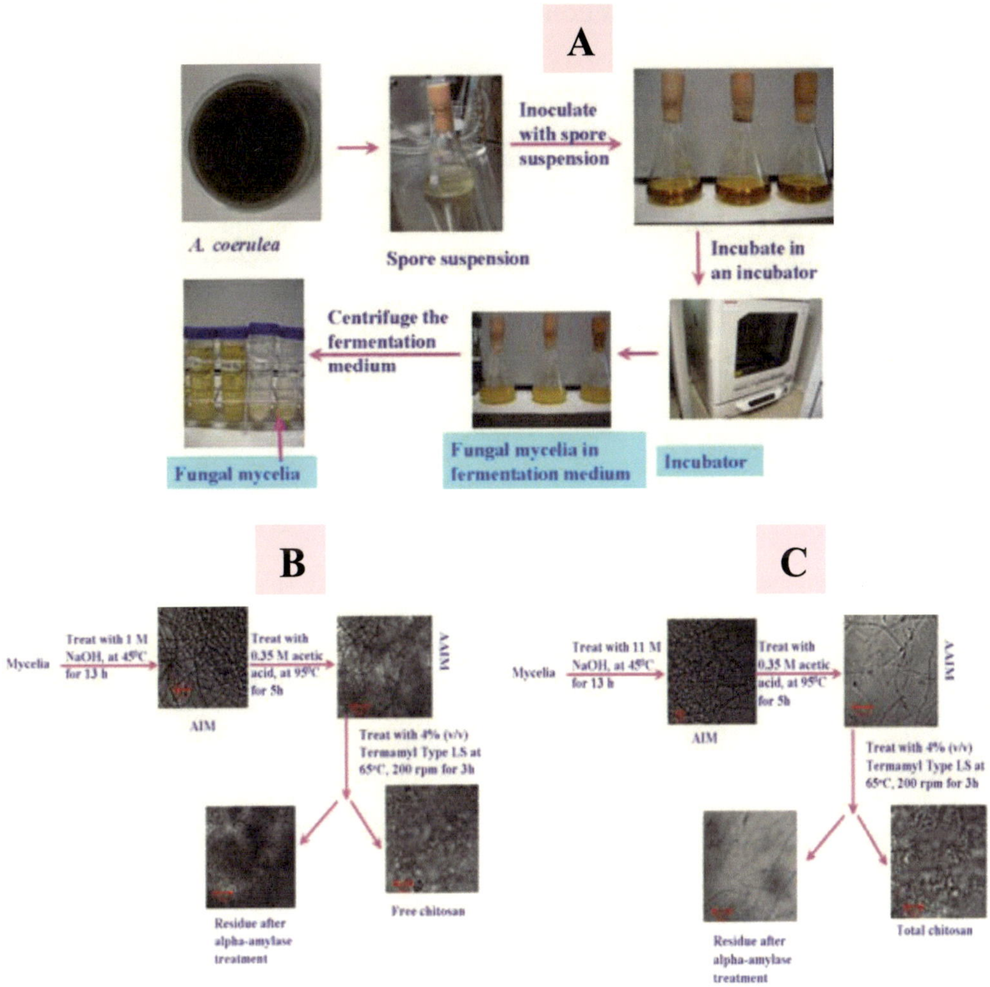

Fig. 11. Cultivation of fungus *Absidia coerulea* in submerged fermentation medium (A) and Laser scanning micrographs of fungal cell wall materials obtained from treatment of fungal mycelia with NaOH solution (1M NaOH, B and 11 M NaOH, C) for extraction of free chitosan and total chitosan (Nwe et al., 2008a)

In this research free chitosans and total chitosans have been isolated from fungi, *G. butleri* and *A. coerulea*. These observations indicated that fungi, *G. butleri* and *A. coerulea* separately synthesize chitosan and glucan in their cell wall and these two polymers are linked together afterwards to synthesize the chitosan-glucan complex (Nwe et al., 2008a). These observations agreed with finding of Sietsma et al., 1996. Moreover it can be assumed that the chitosan-glucan complex may be growing polymer since the number average molecular weight (42 kDa) and yield of chitosan product C, 5.9g/100g of mycelia gave higher result than that of chitosan product B (32 kDa and 3.2 g/100g of mycelia). These two results are compared with number average molecular weight of free chitosan, 35 kDa (Table 7 and Nwe & Stevens, 2002 & 2002b). These evidences confirmed that the bond between chitosan and glucan chains in fungi, *A.coerulea* and *G. butleri* is α-(1,4)-glycosidic bond.

4. Optimization of fermentation conditions for production of chitosan in a high yield

The quality and quantity of chitosan extracted from the fungal mycelia depend on the fungal strain: *Absidia, Gongronells, Rhizopus,* etc; fermentation type: solid substrate/state fermentation and submerged fermentation (batch fermentation, fed batch feremementation and continuous fermentation); fermentation medium composition: carbon source and concentration, nitrogen source and concentration, and metal ions and their concentration; fermentation conditions: inoculum size, harvesting time, fermentation temperature; and chitosan extraction procedure (Arcidiacono & Kaplan, 1992; Crestini et al., 1996; Rane & Hoover, 1993; Nwe & Stevens, 2002, 2002b, & 2006; Tan et al., 1996; Jaworska & Konieczna, 2001; Nwe et al., 2002 & 2008a). Among the different fermentation methods, SSF has shown as the best fermentation method for the production of chitosan from fungal mycelia (Nwe et al., 2002; Crestini et al., 1996). Therefore it was decided to design a solid substrate fermentor for the growth of fungus *Gongronella butleri* USDB 0201. Most researchers have been designed various types of soli substrate/state fermentors (Durand et al., 1996; Laukevics et al., 1984). Among them, tray type fermentor is practical for the production of chitosan from fungal mycelia. Based on these published solid state/substrate fermentor designs, a try-type solid substrate fermentor was designed to use in this research (Figure 1, Nwe & Stevens, 2002).

The fermentor was designed containing 3 trays of perforated aluminium with small holes (a tray volume of about 9.6 liters and a tray-working volume of about 1.6 liters). Each tray mounted on a steamer pot and the top tray was covered by an aluminium lid (fermentor cover). The steamer pot has two holes in the wall connected with a pipe for humidified air inlet (upper pipe) and to take out condense water after completion of the sterilization (lower pipe). The air outlet pipe is mounted on the fermentor cover. Two PVC (polyvinylchloride) tubes with the diameter of 2.5 cm and length 5.5 cm were fixed vertically on each of the perforated trays to pass the humidified air through the mycelia layer. Because the mycelia layer is covering all the holes in the aluminium trays after 3 days of growth, the inlet air can pass through these PVC tubes to the next compartment above. The fermentor was inoculated while applying a laminar flow of sterilized air and then wrapped with sterilized tape on the vim between the trays to prevent contamination and to avoid leakage of the inlet air. An air pump was used to pump fresh air and pass through the air-filter. Sterilized air was first humidified in heated and sterilized water and then cooled to reach the desired inlet air temperature. The capacity of each tray for solid substrate was 850 g of peeled raw sweet potato. The cost of fermenter is low. The materials are widely available. It does not need extensive maintenance since it is very simple. Taking into consideration all these factors, this fermentor is the most suitable for production of fungal chitosan.

Sweet potato was selected to use as a cheap solid carbon source for the growth of fungus *G. butleri* USDB 0201 in small-scale fermentation. The fungus *G. butleri* USDB 0201 was grown on sterilized sweet potato pieces under sterilized and humidified air supply in the solid substrate fermentor for 7 days. The mycelium is formed on the surface and does not penetrate into the sweet potato. Therefore it is easy to separate the mycelia from the substrate at the end of the fermentation (Nwe & Stevens, 2002). Mycelia were detached from the solid substrate by flotation in water. The detached mycelia were washed with water for several times to remove remaining sweet potato pieces. Finally, the mycelia were dried to assay for the amounts of mycelia and to extract chitosan.

Firstly, fungus *G. butleri* was grown in SSF using only sweet potato without additional mineral solution for 7 days (Nwe & Stevens, 2002b). The amount of mycelia yield was 30.1 ± 3.2 g/kg of solid substrate. Chitosan was extracted from dried mycelia using method II (Table 7). The production yield of chitosan extracted from fungus grown on sweet potato pieces without further supplementation was low, only 0.69 g per kg sweet potato. Sweet potato contains 2.63 ± 0.47 elementary N/kg (Nwe et al., 2006). Therefore, this endogenous nitrogen is probably not available for chitosan production because it is part of existing sweet potato macromolecules (protein, DNA). The solid substrate was treated with a mineral solution as first step to improve the fermentation conditions for a better yield of mycelia and chitosan. In this treatment 1kg potato pieces was sterilized in the presence of 1L mineral solution. After sterilization, free fluid was decanted and spores of fungus *G. butleri* was inoculated and incubated in the fermentor for 0 - 8 days (Nwe & Stevens, 2002). It was observed that the amount of chitosan did not increase after 6 days of fermentation. However the production yield of mycelia increased up to 7-8 days of fermentation. The maximum production yield of chitosan, 1.54 g per kg of solid substrate was obtained from the mycelia grown on solid substrate for 7 days (Figure 12). According to these results, the best time to harvest the mycelia from solid substrate fermentation medium was 7days of fermentation.

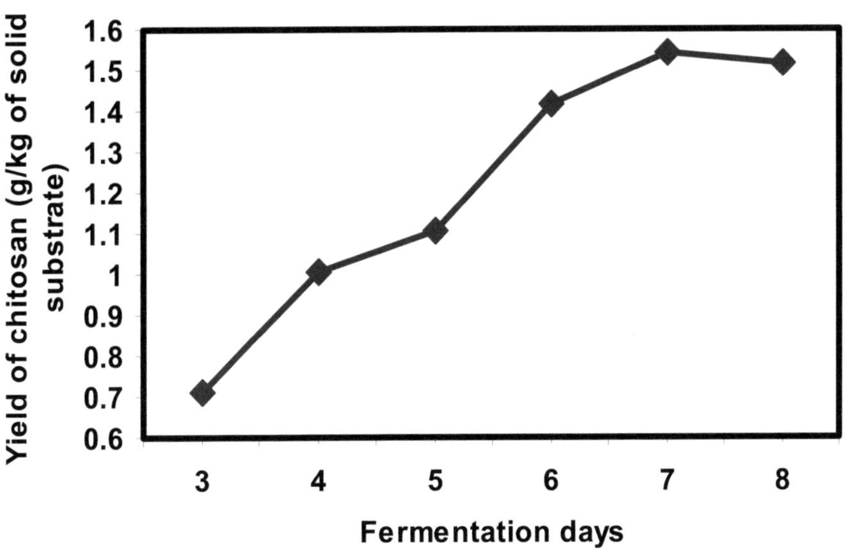

Fig. 12. Production yield of chitosan from mycelia of fungus *Gongronella butleri* USDB 0201 grown in solid substrate fermentation medium for 3–8 days (Nwe & Stevens, 2002)

The chitosan polymer has a high nitrogen content, 8% (Nwe et al., 2008). Sufficient amount of nitrogen should be available during synthesis of chitosan in the cell wall of fungus (Nwe et al., 2006). Therefore *G. butleri* USDB 0201 was grown on peeled sweet potato pieces supplied with mineral solution and different nitrogen sources: peptone, ammonium sulfate, sodium nitrate and urea (Nwe et al, 2002). It was found that the yield of mycelia and chitosan obtained from fungus grown on solid substrate fermentation medium supplied with urea was higher than that from fungus grown on solid substrate fermentation medium supplied with other nitrogen sources (Table 5). Thereafter solid substrate fermentation media were supplied with different amounts of urea and fungus was grown on these media to produce chitosan. The yields of chitosans obtained from mycelia of fungus grown on 1 kg of solid substrate supplied with 7.2 or 14.3 g urea, 3.59 ± 0.29 and 4.31 ± 0.65 g respectively was not significant difference. The best fermentation medium to grow fungal mycelia was the solid substrate supplied with 7.2 g urea for 1 kg of solid substrate at pH 4.5 (Nwe & Stevens, 2004).

After that, conditions for solid substrate fermentation was optimized to improve yield of chitosan from fungus mycelia grown in solid substrate fermentation. Five fermentation parameters were varied to study the yield of chitosan and growth of mycelia on the sweet potato pieces supplied with 7.2 g urea for 1 kg of solid substrate at pH 4.5 (Nwe & Stevens, 2006). Those parameters are inoculum size (0.15 -24 x 10^8 spores/kg SS), inlet airflow rate (0.4-2.3 l min^{-1} $(kgDM)^{-1}$), humidity of inlet air (70 - 95 %), solid substrate moisture content (53 - 80 %) and outside fermentor temperature (26-35°C) (Table 8). The chitosan yield was

A

No	Parameter to be optimized	Conditions for variable parameters					Experimental conditions for the other parameters				
		A	B	C	D	E	1	2	3	4	5
1	Inoculum size (x 10^8 spores/kg of SS)	0.15	1.5	10	18	24	V	1.2	95	80	26
2	Inlet air flow rate (L min^{-1} kg DM^{-1})	0.4	0.9	1.2	1.5	2.3	0.15*	V	95	80	26
3	Humidity of inlet air (% Humidity)	70	90	95	-	-	0.15	1.2*	V	80	26
4	Solid substrate moisture content (%)	80	73	53	-	-	0.15	1.2	95*	V	26
5	Outside fermentor temperature (°C)	26	29	35	-	-	0.15	1.2	95	80*	V

V, Variable conditions in that experiment;

*Selected condition, as concluded from the previous experiment

B

No	Parameter to be optimized	Production yield of chitosan (g/kg of solid substrate)				
		A	B	C	D	E
1	Inoculum size (x 10^8 spores/kg of SS)	3.28 ± 0.22	3.48 ± 0.15	3.52 ± 0.14	3.59 ± 0.29	3.38 ± 0.23
2	Inlet air flow rate (L min^{-1} kg DM^{-1})	3.42 ± 0.44	3.21 ± 0.14	3.28 ± 0.22	3.41 ± 0.26	3.74 ± 0.19
3	Humidity of inlet air (% Humidity)	4.06 ± 0.04	3.95 ± 0.27	3.28 ± 0.22	-	-
4	Solid substrate moisture content (%)	3.28 ± 0.22	2.74 ± 0.12	1.65 ± 0.22	-	-
5	Outside fermentor temperature (°C)	3.28 ± 0.22	4.60 ± 0.16	0.00 ± 0.00	-	-

Table 8. (A) Experimental conditions for growth of fungus *G.butleri* USDB 0201 in the solid substrate fermentation and (B) production yield of chitosan from the fungal mycelia obtained from each experimental condition (Reproduced from Nwe & Stevens, 2006, Journal of Chitin and Chitosan, 11, 11-15)

increased to maximal 4.6 g per kg sweet potato under the fermentation conditions of inoculum size, 1.5 x 10^7 spores/kg solid substrate; inlet air flow rate, 1.2 l min^{-1} $(kgDM)^{-1}$; humidity of inlet air, 95 %; outside fermentor temperature, 29°C; and moisture content of the solid substrate 80 % (Nwe & Stevens, 2006). Based on these results, it is concluded that fermentation medium compositions, fermentation conditions, harvesting time of mycelia and chitosan extraction procedure are critical factors to obtain high yield of chitosan from the fungal mycelia.

The characteristics of fungal chitosan have been determined and reported (Nwe & Stevens, 2002 & 200b; Nwe & Stevens, 2006; Nwe et al., 2006a). All chitosans have the degree of deacetylation 87 ~ 90 % (measured by UV spectrophotometry method and HPLC method) and the number average molecular weight 20-70 kDa (GPC method). The nitrogen content in the chitosan sample is 7.91% (measured by Micro-Kjeldahl method), moisture and ash content are 8.59% and 0.73% respectively (AOAC method, 1991). The viscosity of 1% (w/v) chitosan solution is 13.2 cps (measured by Brookfield viscometer). The turbidity of chitosan solution is 10-20 NTU (measured by Turbidity meter). Fungal chitosan is soluble in dilute organic acids such as 4% citric acid, 25 % formic acid, 1% lactic acid and 1 % acetic acid but insoluble in 0.2 % benzoic acid, 0.2 % cinnamic acid and 6.3 % oxalic acid. It is soluble in 0.6 % HCl and 96% H_2SO_4 but insoluble in 3% HCl and 5% H_2SO_4 solution (Nwe et al., 2001). The fungal chitosan has 95% purity, which has been confirmed by elementary analysis, IR, ^{13}C-NMR spectroscopy and UV spectroscopy (Nwe et al., 2008). In a number of applications of chitosan in agriculture and medical sectors, chitosan with low molecular weight and DA is more powerful than chitosan with high molecular weight and low DA (Nwe & Stevens, 2008). According to these properties, fungal chitosan is proposed to use in agriculture, food, cosmetic and medical sectors. In our research, the fungal chitosan has been applied to stimulate the growth of orchit tissue in solid and liquid tissue culture media (Nge et al., 2006) and to prepare scaffold and membrane for tissue regeneration template (Nwe et al., 2009 & Nwe et al., 2010a).

5. Application of fungal chitosan in plant tissue culture

In agriculture, chitosan has been used in seed, leaf, fruit and vegetable coating; as fertilizer to control release of agrochemical; to increase plant product; to stimulate the immunity of plants; to protect plants against microorganisms; and to stimulate plant growth (Devlieghere et al., 2004; Sukwattanasinitt et al., 2001; Wanichpongpan et al., 2001; Chandrkrachang, 2002; Nwe et al., 2004; Hadwiger et al., 2002; Pospieszny et al., 1991; Struszczyk & Pospieszny, 1997; Bautista-Baňos et al., 2003 cited in Nge et al., 2006). However the effect of chitosan depend on its molecular weight and degree of acetylation. The effect of chito-oligosaccharide (CTS-O, Kitto Life Co. Ltd., Korea) has been tested in the paddy field (Nwe et al., 2004). The Myanmar rice variety Manaw Thukha was used as seed sample and experimental paddy field was in Thanlyin, Yangon, Myanmar. The CTS-O, 1.5 g was dissolved in 100 ml of 1% acetic acid and the solution was used as CTS-O stock solution for making dilutions into different dosage. The rice seeds were soaked with water for 1 day and cleaned with water to remove the sand and mud. The cleaned seed was soaked in 500 times diluted CTS-O solution for 1 day. The CTS-O coated seed was taken out from the solution and put in a bamboo sieve and then covered with a wet gunny bag for 2 days. The germinated seeds were planted in the field after pudding. In week 4, the rice plants were sprayed with 1000 times diluted CTS-O solution. In week 7, urea, one bag (50 kg) per acre was applied. The 1000 times diluted chitosan solution was sprayed again to the paddy field in week 9. Plants were harvested in week 18. Non chitosan treated rice seeds served as control. It was found that the leaves of the CTS-O supplied plants were greener than those of control plants. After one month, the CTS-O supplied plant had a length of 22.5 inches, compared to 15 inches for the control. After three months, the CTS-O supplied plants had 8 crops and the control plants had 4 crops (Figure 13). The rice yield from CTS-O supplemented field was 1.6 times higher per acre than control field (Nwe et al., 2004).

Fig. 13. Effect of chito-oligosaccharide on yield of rice

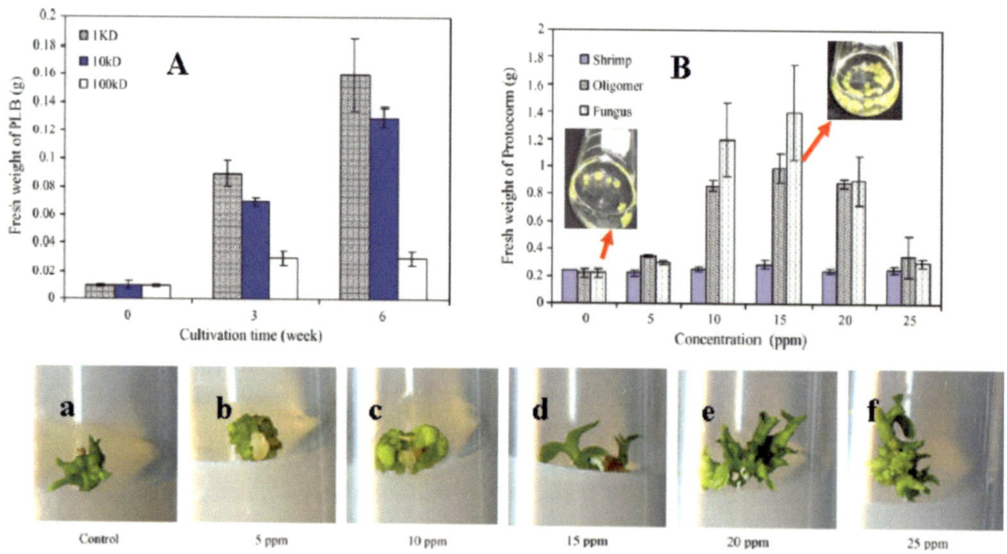

Fig. 14. (A) Effect of molecular weight of chitosan on growth of orchid protocorm-like bodies (PLB's) in VW liquid medium supplemented with 15 ppm of shrimp chitosan with molecular weights of 1 kDa (chito-oligosaccharide), 10 kDa and 100 kDa. (B) Effect of various concentrations of shrimp chitosan (10 kDa), shrimp chito-oligosaccharide (1kDa), and fungal chitosan (10 kDa) on the propagation of total fresh weight of PLB's after six weeks cultivation in liquid VW medium. (a-f) Effect of various concentrations of fungal chitosan (10 kDa) on the production of plantlets by the fractionated PLB's explant of *Dendrobium phalaenopsis* cultivated on VW agar media at 12 weeks. (Reproduced from Nge et al., 2006, Plant Science, 170, 1185-1190)

In addition, the effect of chitosan on chili plants and orchid plants had been studied in Thailand. It was found that the application of chitosan improved the yield of plant products (unpublished data). The effect of molecular weight of chitosan on germination rate of mung bean, yard long bean and soy bean seeds has been studied. The beans showed a higher

germination rate in the presence of chitosan with a molecular weight of 40 kDa (Nwe et al., 2006b). Fathermore the effect of different concentration of fungal chitosan on green gram bean production was studied at Government Technological College, Kyaukse, Myanmar. The field trial was completed in 10 weeks and non-chitosan treated field served as control. The results showed that the number of pod per plant (30 pods/plant), number of seed per pod (12 seeds/pod) and production yield (3500 kg/ha) increased by soaking the seeds with 20 ppm and by spraying the field with 100 ppm chitosan solution when compared with other fields supplied with 0, 50, 150 and 200 ppm chitosan solution (Han et al., 2006).
In order to confirm these results, chito-oligosaccharide, fungal chitosan (10 kDa) and shrimp chitosan (10^6 Da) were used to propagate the plantlet from orchid protocorm in liquid and on solid tissue culture medium (Nge et al., 2006). The highest number of plantlets was observed in the presence of 20 ppm using either 10 kDa fungal chitosan or 1 kDa chito-oligosaccharide. High molecular weight chitosan (100 kDa) had no stimulating effect. The 10 kDa fungal chitosan was more effective compared with 1 kDa oligomer. The data in tissue culture and in the field proved that low molecular weight chitosan in the dosage of 15–100 ppm to plant results in a considerable positive effect on plant growth (Figure 14).

6. Application of fungal chitosan in tissue engineering

Pure chitosan is non-toxic, polar compound and bioadhesive and bioactive polymer; it has free of antigenic effects, biocompatible and biodegradable properties, and antimicrobial and hemostatic activities; and it can accelerate tissue regeneration and wound healing (Vandevord et al., 2002; Onishi etal., 1994; Kim et al., 2001; Maslova & Krasavtsev, 1998; Shigemasa et al., 1998; Tokura et al., 1994 cited in Nwe & Stevens, 2008 & 2008a). It has been used to prepare a variety of forms such as powders, hydrogels, fibers, membranes, beads and porous scaffolds that have been tested in many medical and biological applications (Tamura et al., 2006; Madihally & Mattew, 1999 cited in Nwe & Stevens., 2008 & 2008a). For tissue engineering applications, chitosan scaffolds have been prepared by the freeze-drying and freeze-gelation method and by a 3-axis robotic arm dispensing system and chitosan membranes have been prepared by solvent-casting method (Gravel et al., 2006; Ng et al., 2004; Chupa et al., 2000; Huang et al., 2005; Vandevord et al., 2002; Madihally & Mattew, 1999; Thein-Han & Kitiyanant, 2006 cited in Nwe & Stevens, 2008a).
In our research, scaffolds were prepared using fungal chitosan by the freeze-drying method and membranes of fungal chitosan were prepared using a solvent-casting method. The mechanical and biological properties of the fungal chitosan scaffolds and membranes were evaluated and compared with those of scaffolds and membranes prepared using chitosans obtained from shells of shrimps and crabs and bone plates of squids (MW 10^5-10^6 Da and DA 10-20%) (Nwe et al., 2009 & 2010a). It was observed that the fungal chitosan scaffold had excellent mechanical and biological properties than shrimp, crab and squid chitosan scaffolds (Figure 15). However the fungal chitosan membrane had lower mechanical properties and higher degradation rate than membranes of shrimp and squid chitosan (Nwe et al., 2010a).
For wound therapy, nowadays artificial skin is available, but it is very expensive. The biopolymers such as chitosan, gelatin, alginate and bacterial cellulose are cheap and can be obtained easily. Aim of our present study is to develop an artificial skin using biopolymers that are low in production costs and in more than one beneficial way for wound healing. In which, there needs to be solved the problems in lowering the mechanical and water holding properties and cellular activities of the fungal chitosan membrane.

A Parameters		Chitosans from various sources			
		Shells of Shrimp SHCTS 10^6 Da	Shells of Crabs CRCTS 10^5 Da	Bone plates of squids SQCTS 10^5Da	Fungal mycelia FCTS 10^4 Da
Pore size of chitosan scaffold (μm)		64 ± 20	77 ± 22	54 ± 17	84 ± 21
Porosity of water absorbed scaffold (%)		88 ± 4	90 ± 3	96 ± 5	97 ± 1
Amount of absorbed water in scaffold (g/g of scaffold)		36 ± 4	43 ± 3	46 ± 0	53 ± 2
Fibroblast cell no. (x 10^{-6} cell / cm^3 of scaffold	0 day	0.38 ± 0.02	0.35 ± 0.04	0.33 ± 0.07	0.37 ± 0.02
	14 days	1.3 ± 0.2	2.1 ± 0.3	1.9 ± 0.3	1.8 ± 0.2
Degradation of scaffold (%)		2 ± 0.8	5 ± 0.7	5 ± 0.5	11 ± 1.1
Tensile strength	Force (cN)	141 ± 37	88 ± 09	151 ± 31	209 ± 20
	Elongation (%)	15.1 ± 4.4	20.3 ± 4.2	5.4 ± 1.7	7.2 ± 1.5

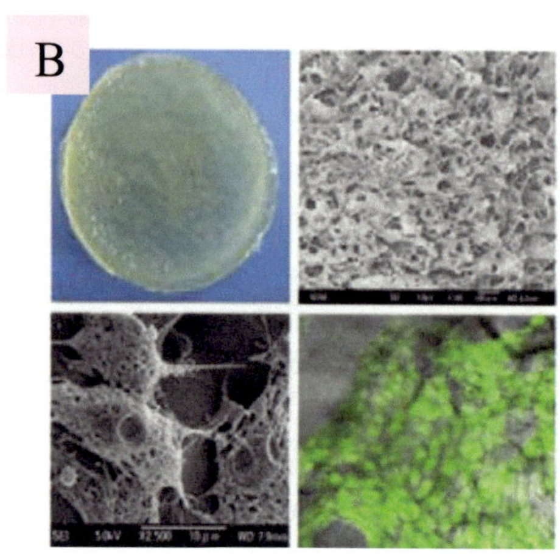

Fig. 15. (A) Characteristics of scaffolds prepared using chitosan from shells of shrimps and crabs, bone plates of squids and fungal mycelia, (B) Photo and SEM micrograph of FCTS scaffold and SEM micrograph and confocal laser microscopic image of fibroblast NIH/3T3 cells on FCTS scaffold (Reproduced from Nwe et al., 2009, Materials, 2, 374-398)

To solve these problems, alginate, bacterial cellulose (BC) and gelatin were selected based on available material and published data and the membranes were prepared. The water absorption properties and cellular activities on these membranes were examined to select the best material for the preparation of a novel membrane by incorporation with fungal chitosan for tissue regeneration template (Nwe et al., 2010b). The cells on alginate membrane crosslinked with Ca^{2+} (AGM_Ca) showed a spherical morphology and the NIH/3T3 cells grown on the BC membrane (BC_M) and membrane of glutaraldehyde (GTA) crosslinked gelatin (GTA_GM) had a polygonal morphology. The proliferation rate of fibroblasts on the GTA_GM was faster than that on the BC_M (Figure 16). However cytotoxic effects of GTA were observed on the GTA_GMs prepared with molar ratio higher than 0.033, except pork skin gelatin membrane with a molar ratio, 0.033, which showed the cytotoxic effects on fibroblast cells. In which the similar amounts of water absorbed in the AGM_Ca and BC_M that was lower than the amount of water absorbed in GTA_GM.

Parameters		AGM_Ca	BC_M	GTA_GM_6
Thickness of the dried membrane (μm)		20	5	85
Area of membrane before and after water absorption (mm^2)	Before	10	10	10
	After	10.7 ± 1.2	11.4 ± 1.2	24.6 ± 3.6
Dry weight of membrane (mg/cm^2)		6.45 ± 1.9	1.05 ± 0.07	12.35 ± 0.78
Amount of absorbed water (g/g of membrane)		2.5 ± 0.6	3.1 ± 0.5	15.9 ± 1.1

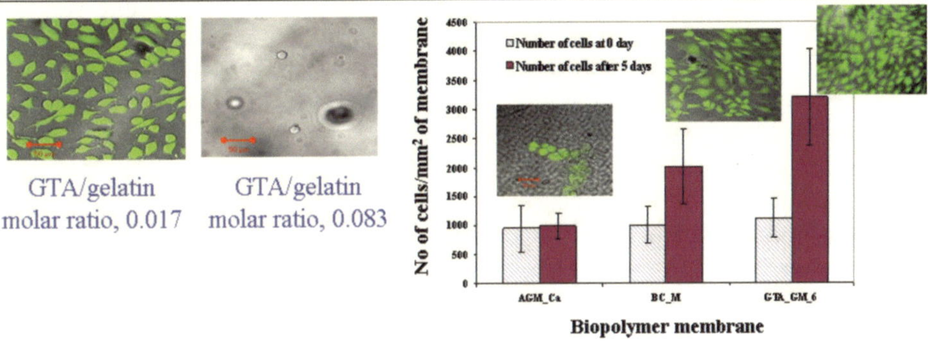

Fig. 16. Characteristics of membranes of alginate crosslinked with Ca^{2+}, bacterial cellulose and glutaraldehyde crosslinked gelatin and attachment, morphology and proliferation of fibroblast NIH/3T3 cells on these membranes (Reproduced from Nwe et al., 2010b, Process Biochemistry, 45, 457-466)

7. Present situation and future prospects

We are working on fungal chitosan production and its application in agricultural and medical sectors enters to 12 years. In this research, fungal mycelia production and enzymatic chitosan extraction method have been developed to obtain high yield fungal chitosan in very easy way. This new enzymatic method is proposed to apply in large-scale production of low molecular weight chitosan from fungal source. The resultant chitosan has been tested in plant tissue culture for agriculture applications and has been tested as matrix for tissue regeneration template. The data obtained from this research point out that the selection of chitosan is one of the important factors to apply chitosan in agriculture and medical sectors. Among the tested chitosans in here, fungal chitosan shows as an excellent plant growth

stimulator to apply in agriculture sector and as an excellent scaffolding material to construct a tissue regeneration template. For further research it needs to study the effect of chitosan and its degradation products in plant and animal systems. At that moment, ^{13}C labeled chitosan has been synthesized by fungus *A.coerulea* in submerged fermentation to study metabolic pathway of chitosan and its degradation pathway (Nwe et al., 2010c).
For application of fungal chitosan in tissue engineering, there need to be improve mechanical properties and water holding capacity of chitosan membranes. In recent years, significant progress has been made on the characterization of biopolymers for tissue regeneration template, gelatin has been found to be the material that can enhance the mechanical and water holding properties of membrane. Taking together, it is expected that there will be reach to the final goal on the development of a double-layered biodegradable scaffold for skin regeneration template by constructing fungal chitosan and gelatin composite membrane/scaffold in near future.

8. Acknowledgements

The authors would like to extend their thanks to the late Prince Leo de Lignac (The Netherlands); "High-Tech Research Center" Project for Private Universities: matching fund subsidy from MEXT (Ministry of Education, Culture, Sports, Science and Technology, Japan), 2005–2009; the Japan Society for Promotion of Science (JSPS), Japan (H19:07068); and the Japan Scientific Technology Agency, Japan for providing funds for this research, to Siam Modified Starch Co. LTD (Thailand) and Novo Nordisk Bioindustrial (Thailand, Japan and Demark) for providing Termamyl, Type LS, to the members of Bioprocess Technology, Asian Institute of Technology, Thailand; Temasek Life Science Laboratory, The National University of Singapore, Singapore; Faculty of Chemistry, Materials and Bioengineering and HRC, Kansai University, Japan and Koyo Chemical Co., Japan for their technical supports. Specially, author, Nitar Nwe would like to express her heartful appreciation to her advisor, prof. Willem F. Stevens for his guidance, encouragement, and critical comments during her Ph.D research work. Moreover sincere thanks are also extended to Porf. Seiichi Tokura, Prof. Suwalee Chandrkrachang, Dr. Eugene Khor, Prof. Tan Teck Koon, Dr. Didier Montet, Dr. Malgorzata M. Jaworska and Prof. George A. F. Roberts for their invaluable discussion and support in this research work.

9. References

Arcidiacono, S.; Kaplan, D. L. (1992). Molecular weight distribution of chitosan isolated from *Mucor rouxii* under different culture and processing conditions. *Biotechnology and Bioengineering*, 39, 281–6.

Bartnicki-Garcia, S. (1968). Cell wall chemistry, morphogenesis and taxonomy of fungi. *Annual Review of Microbiology*, 22, 87–108.

Crestini, C.; Kovac, B. & Giovannozzi-Sermanni, G. (1996). Production and isolation of chitosan by submerged and solid state fermentation from *Lentinus edodes*. *Biotechnology and Bioengineering*, 50, 207–10.

Durand, A.; Renaud, R.; Maratray, J.; Almanza, S. & Diez, M. (1996). INRA-Dijon Reactors for solid state fermentation: designs and applications. *Journal of Scientific and Industrial Research*, 55, 317-332.

Fontaine, T.; Simenel, C.; Dubreucq, G.; Adam, O. ; Delepierre, M.; Lemoine, J.; et al. (2000). Molecular organization of the alkali-insoluble fraction of *Aspergillus fumigatus* cell wall. *Journal of Biological Chemistry*, 275, 27594–607.

Gooday, G. W. (1995). Cell walls, In: *The growing fungus.* pp.3–62, Chapmanan and Hall.

Gopal, P.; Sullivan, P. A. & Shepherd, M. G. (1984). Isolation and structure of glucan from regenerating spheroplasts of *Candida albicans. Journal of General Microbiology,* 130, 1217–25.

Han, T.; Nwe, N.; Oo, M. M. & Aye, K.N. (2006). Chitosan production from yeast and fungus and application in green gram production, *Proceeding of 10th International Chitin and Chitosan Conference and 7th European Chitin and Chitosan Conference,* PS8, September 6th -9th, Montpellier.

Hang, y. D. (1990). Chitosan production from *Rhizopus* mycelia. *Biotechnology Letters,* 12, 911-913.

Herrera, J. R. (1991). Fungal glucans, In: *Fungal cell wall: Structure, Synthesis, and Assembly,* pp. 59-88, Mexico.

Ikeda, I.; Sugano, M.; Yoshida, K.; Sasaki, E.; Iwamoto, Y. & Hatano, K. (1993). Effect of chitosan hydrolysates on lipid absorption and on serum and liver lipid concentration in rats. *Journal of Agricultural and Food Chemistry,* 41, 431–435.

Jaworska, M. M. & Konieczna, E. (2001). The influence of supplemental components in nutrient medium on chitosan formation by fungus *Absidia orchidis. Apply Microbiology and Biotechnology*, 56, 220-224.

Kollar, R.; Petrakova, E.; Ashwell, G.; Robbins, P. W. & Cabib, E. (1995). Architecture of The yeast cell wall, the linkage between chitin and β-1,3-glucan. *Journal of Biological Chemistry*, 270, 1170–8.

Laukevics, J. J.; Apsite, A. F. & Viesturs, U. E. (1984). Solid substrate fermentation of wheat straw to fungal protein. *Biotechnology and Bioengineering,* 26, 1465-1474.

Muzzarelli, R. A. A. ; Ilari, P. ; Tarsi, R. ; Dubini, B. & Xia, W. (1994). Chitosan from *Absidia coerulea. Carbohydrate Polymers*, 25, 45-50.

Muzzarelli, R. A. A.; Tanfani, F. & Scarpini, G. (1980). Chelating, film-forming and coagulating ability of the chitosan–glucan complex from *Aspergillus niger* industrial wastes. *Biotechnology and Bioengineering*, 22, 885–96.

Nge, K. L.; Nwe, N.; Chandrkrachang, S. & Stevens, W. F. (2006). Chitosan as growth stimulator in orchid tissue culture. *Plant Science*, 170, 1185-1190.

Nwe, N.; Chandrkrachang, S. & Stevens, W. F. (2001). Fungal chitosan from solid state fermentation, *Proceeding of 8th International Conference on Chitin and Chitosan and 4th Asia Pacific Chitin and Chitosan Symposium,* pp 482-484, ISBN4-906464-43-0, Yamaguchi.

Nwe, N.; Chandrkrachang, S.; Stevens, W. F.; Maw, T.; Tan, T. K.; Khor, E. & Wong, S. M. (2002). Production of fungal chitosan by solid state and submerged fermentation. *Carbohydrate Polymers*, 49, 235-237.

Nwe, N. & Stevens, W. F. (2002). Production of fungal chitosan by solid substrate fermentation followed by enzymatic extraction. *Biotechnology Letters*, 24, 131-134.

Nwe, N. & Stevens, W. F. (2002a). Chitosan isolation from the chitosan-glucan complex of fungal cell wall using amylolytic enzymes. *Biotechnology Letters*, 24, 1461 -1464.

Nwe, N. & Stevens, W. F. (2002b). Increased yield of chitosan from fungal mycelia, *Proceeding of 5th Asia Pacific Chitin and Chitosan Symposium*, pp 36 – 39, ISBN 974-229-412-7, Bangkok.

Nwe, N.; Chandrkrachang, S. & Stevens, W.F. (2004). Application of chitosan in Myanmar's agriculture sector, *Proceedings of the Sixth Asia Pacific Chitin and Chitosan Symposium*, pp S6-2, May 23–26, Singapore.

Nwe, N. & Stevens, W. F. (2004). Effect of urea on fungal chitosan production in solid substrate fermentation. *Process Biochemistry*, 39, 1639-1642.

Nwe, N.; San, N. M. M.; Nge, K. L.; Chandrkrachang, S.; Stevens, W. F. & Tamura, H. (2006b). Application of chitosan in the agriculture sector. *Chitin Chitosan Research*, 12, 239.

Nwe, N. & Stevens, W. F. (2006). Optimization of solid substrate fermentation for the production of fungal chitosan. *Journal of Chitin and Chitosan.*, 11, 11-15,.

Nwe, N.; Stevens, W. F.; Nagahama, H.; Tokura, S. & Tamura, H. (2006). Nitrogen balance of fungal chitosan synthesis in solid substrate fermentation, *Proceeding of 10th International Chitin and Chitosan Conference and 7th European Chitin and Chitosan Conference*, pp PS7, September 6th -9th, Montpellier.

Nwe, N.; Stevens, W. F.; Tokura, S. & Tamura, H. (2006a). Fungal chitosan production on sweet potato. *Chitin and Chitosan Research*, 12, 118-119.

Nwe, N. & Stevens, W. F. (2008). Production of chitin and chitosan and their applications in the medical and biological sector, In: *Recent Research in Biomedical Aspects of Chitin and Chitosan*, Tamura, H. (Ed.), pp161-176, Research Signpost, ISBN 978-81-308-0254-1, India.

Nwe, N. & Stevens, W. F. (2008a). Preparation and characteristics of chitosan scaffolds for tissue engineering, In: *Recent Research in Biomedical Aspects of Chitin and Chitosan*, Tamura, H. (Ed.) pp. 57-68, Research Signpost, ISBN 978-81-308-0254-1, India.

Nwe, N.; Stevens, W. F.; Tokura, S. & Tamura, H. (2008). Characterization of chitin and chitosan-glucan complex extracted from cell wall of fungus *Gongronella butleri* USDB 0201 by enzymatic method. *Enzyme and Microbial Technology*, 42, 242-251.

Nwe, N.; Stevens, W. F.; Montet, D.; Tokura, S. & Tamura, H. (2008a). Decomposition of myceliar matrix and extraction of chitosan from *Gongronella butleri* USDB 0201 and *Absidia coerulea* ATCC 14076. *International Journal of Biological Macromolecules*, 43, 2-7.

Nwe, N.; Furuike, T. & Tamura, H. (2009). The mechanical and biological properties of chitosan scaffolds for tissue regeneration templates are significantly enhanced by chitosan from *Gongronella butleri*. *Materials*, 2, 374-398.

Nwe, N.; Furuike, T. & Tamura, H. (2010). Chitin and chitosan from terrestrial organisms, In: *Chitin, Chitosan, Oligosaccharides and Their Derivatives: Biological Activities and Applications*, Kim, S. K. (Ed), pp. 3-10, Taylor & Francis Group LLC, ISBN 978-14-398-1603-5, Florida.

Nwe, N.; Furuike, T.; Tokura, S. & Tamura, H. (2010a). *In vitro* characterization of fungal chitosan scaffolds and membranes of chitosan from various sources and study the behavior of fibroblast NIH/3T3 cells on these matrices. (Submitted to publish in Carbohydrate polymers).

Nwe, N.; Furuike, T. & Tamura, H. (2010b). Selection of biopolymer base on attachment, morphology and viability of fibroblast NIH/3T3 cells for the development of a biodegradable tissue regeneration template: alginate, bacterial cellulose and gelatin. *Process Biochemistry*, 45, 457-466.

Nwe, N.; Furuike, T.; Osaka, I.; Fujimori, H.; Kawasaki, H.; Arakawa, R.; Tokura, S.; Stevens, W. F.; Kurozumi, S.; Takamori, Y.; Fukuda, M.; & Tamura, H. (2010c). Laboratory scale production of ^{13}C labeled chitosan by fungi *Absidia coerulea* and *Gongronella butleri* grown in solid substrate and submerged fermentation, Carbohydrate Polymers, doi:10.1016/j.carbpol.2010.06.023.

Rane, K. D. & Hoover, D. G. (1993). Production of chitosan by fungi. *Food Biotechnology*, 7, 11–33.

Roberts, G. A. F. (1992). Preparation of chitin and chitosan, In: *Chitin chemistry*. pp.55–831, Macmillan Press, London.

Robson, G. (1999). Hyphal cell biology, In: *Molecular fungal biology*. pp.164–84, University Press, Cambridge.

Sietsma, J. H.; Vermeulen, C. A. & Wessels, J. G. H. (1996). The role of chitin in hyphal morphogenesis, *Advances in Chitin Science, Proceedings of the 5th International Conference on Chitin and Chitosan*, pp. 40-47, Lyon.

Sietsma, J. H. & Wessels, J. G. H. (1979). Evidence for covalent linkages between chitin and beta-glucan in a fungal wall. *Journal of General Microbiology*, 114, 99–108.

Stevens, W. F. (2001). Production of chitin and chitosan: Refinement and sustainability of chemical and biological processing, *Chitin and Chitosan in Life science, Proc. 8th Int. Conf. on Chitin and Chitosan and 4th Asia Pacific Chitin and Chitosan Symp*, pp. 293-300, ISBN4-906464-13-0, Yamaguchi.

Surarit, R.; Gopal, P. K. & Shepherd, M. G. (1988). Evidence for a glycosidic linkage between chitin and glucan in the cell wall of *Candida albicans*. *Journal of General Microbiology*, 134, 1723–30.

Tan, S. C.; Tan, T. K.; Wong, S. M. & Khor, E. (1996). The chitosan yield of Zygomycetes at their optimum harvesting time. *Carbohydrate Polymers*, 30, 239–242.

Wessels, J. G. H.; Mol. P. C.; Sietsma, J. H. & Vermeulen, C. A. (1990). Wall structure, wall growth, and fungal cell morphologenesis, In: *Biochemistry of cell walls and Membranes in fungi*, Kuhn, P. J.; Trinci, A. P. J.; Jung, M. J.;

Goosey, M. W.& Copping, L. G.; (Eds.) pp. 81-95, Springer-Verlag, Berlin.

White, S. A.; Farina, P.R. & Fulton, I. (1979). Production and isolation of chitosan from Mucor rouxii. *Apply Environmental Microbiology*, 38, 323-328.

8

Chitin Based Biocomposites for Removal of Contaminants from Water: A Case Study of Fluoride Adsorption

Jose R. Rangel-Mendez, Vladimir A. Escobar-Barrios and Jose L. Davila-Rodriguez
Instituto Potosino de Investigacion Cientifica y Tecnologica, A.C.
Mexico

1. Introduction

Around the world, safe water for human consumption and use is more and more scarce, due to its increasing demand and to the natural or anthropogenic contamination of the water resources (Jackson et al., 2001). World population grew 15% only in a decade from 1990 to 2000, that is to say 789 million people, which means an equal quantity of new water consumers (WHO, 2000). In addition, an annual mortality rate of 2.2 million of people associated to contaminated water consumption was reported by the World Health Organization (WHO) in 2000 (WHO, 2000). Such contamination, either natural or anthropogenic, affects lakes, rivers or groundwater; from these reservoirs, the last represents one of the most important water sources for human consumption.

On one hand, the natural contamination of groundwater occurs due to geochemical phenomena, which take place during the natural cycle of water, when this comes into contact with underground rocks, reaching chemical equilibrium which causes in turn dissolution of part of such rocks. Due to these facts, groundwater contains diverse chemical species, and some of them can be considered as contaminants, based on the negative effects of each substance in water on the human health and the integrity of other living beings.

On the other hand, the anthropogenic contamination of groundwater is mainly due to emissions of some industries e.g. mining, textile and metal-mechanic. Other human activities as agriculture are also sources of water contamination with fluoride due to the intensive use of fertilizers.

People from arid regions around the world use groundwater as main source to satisfy their water necessities. In Mexico, for example, 37% of the total water consumption is groundwater, according to data generated in 2008 (CONAGUA, 2010). Nevertheless, the use of groundwater for human consumption in humid regions is also important, due to the increasing contamination of surface water. These facts have increased the need to eliminate contaminants from water before its use by human beings, which can be achieved by means of diverse methods.

2. Typical methods to remove contaminants from water

The selection of the method to remove a specific water contaminant depends on: chemical speciation of such contaminant, concentration, presence of suspended solids, required system size, influent flow variability, etc. (Clifford et al., 1986). Also, the treatment cost must be evaluated in order to minimize it and have a good cost-benefice balance. The most used methods for water contaminants removal are chemical precipitation, membrane separation, ion exchange, electrochemical methods and adsorption (Clifford et al., 1986). One of the preferred methods, especially at low concentrations of contaminants, is adsorption, due to its low cost, easy operation and little use of chemical additives, as well as the possibility of reusing the adsorbent materials. At industrial scale, the adsorption process is carried out in packed columns. Figure 1 shows a typical scheme of an adsorption system, as well as an example of a real adsorption system that uses activated carbon as adsorbent.

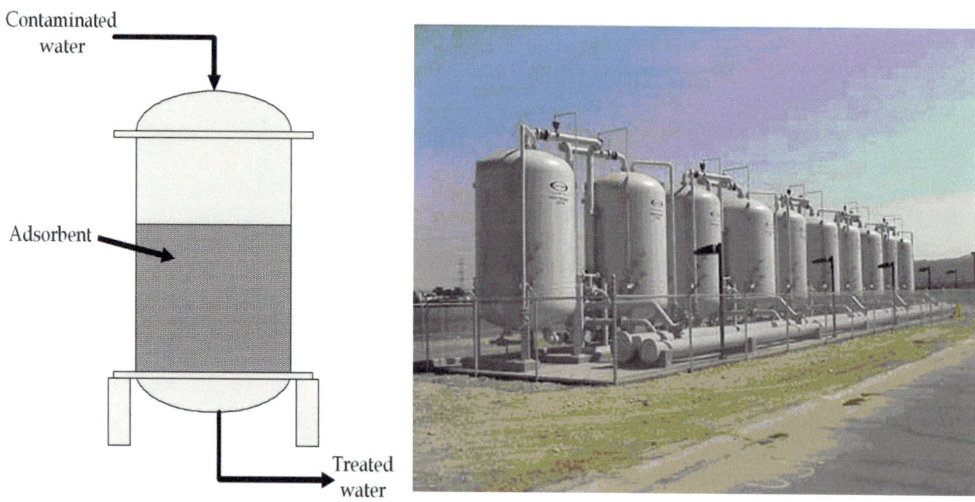

a) b)

Fig. 1. a) Scheme of a continuous adsorption system (USEPA, 2002). b) Example of an adsorption system consisting of activated carbon columns (Alcan, 2005).

3. Biosorption

The list of typical adsorbent media includes materials produced by means of processes which consume great quantities of energy and/or chemical substances (e.g. activated carbon) involving high environmental costs. Therefore, there is a need of cheaper and more sustainable materials. This need has promoted the increasingly scientific study of biosorbents during the last decades. Biosorbents are adsorbent materials obtained from biomass by relatively simple processes that use considerably lower quantities of chemical substances and energy, compared to the production of typical adsorbents. Besides, most biosorbents are obtained from materials typically considered like byproducts or wastes.

The biomass is naturally composed by very diverse substances as proteins, polysaccharides, pigments, carboxylic acids, etc., which have many functional groups as hydroxyl, carboxyl, carbonyl, tiol, sulfhydryl, sulfonate, phosphonate, amine, amide, etc. (Volesky, 2003). Under determined conditions, such groups can interact physically or chemically with contaminants present in water, which causes their retention on the surface of the biosorbent, that is to say, the biosorption occurs (AWWA, 1999; Volesky, 2003).

4. Biosorbents

4.1 Sources and applications

Only certain types of biomass have the capacity and selectivity to remove an appreciable quantity of a contaminant from an aqueous medium and, thus, can be considered for use in high scale biosorption processes. Nevertheless, at present, many biomass-derived materials have been studied as biosorbents, as it can be seen in Table 1.

Source	Examples
Vegetable matter	Aquatic plants, canola, moss, peat moss, cellulosic materials, seaweed.
Agricultural residues	Apple residues, banana residues, coconut husk, palm fiber, corncob, olive residues, peanut husk, rice husk, tea leafs, coffee residues, beet residues.
Other	Starch, human hair, eggshell, residual water sludge, scallop shell, crustaceans' residues, chitin, chitosan.

Table 1. Example of the variety of sources from which have been obtained the biosorbents studied at present (Volesky, 2003).

Biosorbents have been studied as ion removers, being heavy metal (cations) or anions. The removal of cations by biosorbents from water has been considered as more viable than the removal of anions, since most chemical groups of biomass act as ligands, i.e. as electron donors, which enable them to form coordination complexes with heavy metals. Only some chemical groups, specifically amine groups (primary, secondary and tertiary) are capable to acquire positive electrical charge at acid pH values, because the protonation phenomenon takes place. Protonation occurs due to the chemical nature of amine groups, which act as Brönsted-Lowry bases regarding that these accept hydrogen ions, or as Lewis bases regarding they donate their nitrogen free electron pair to the hydrogen ions. Hence, nitrogen acquires positive charge and the amine group becomes an anion adsorption site. Two of the biosorbents that have high density of amine groups are chitin and its main byproduct, chitosan.

4.2 Chitin and chitosan

4.2.1 Nature and purification

Chitin is the second natural polymer in abundance, exceeded only by cellulose, and it is found in nature mainly as constituent of invertebrates' exoskeleton (Pastor, 2004). This biopolymer was discovered in 1811 by H. Braconnot, who purified it from fungi. In 1823, A. Odier obtained the chitin from beetles and named the biopolymer from the greek word "χιτωμγ", which means cover or wrapper. In 1859, C. Rouget obtained "modified chitin" from alkaline chitin treatment with sodium hydroxide. However, in 1894 F. Hoppe-Seyler (who ignored Rouget's studies) subjected chitin to a new alkaline treatment with potassium hydroxide at 180°C; the product was very soluble in acetic and hydrochloric acids and Hoppe-Seyler named it "chitosan" (Pastor, 2004).

The chemical name of chitin is poly-[β-(1-4)-2-acetamide-2-deoxy-D-glucopyranose], which is conformed by repetitive units of saccharide rings linked by β-(1-4) bonds, with presence

of an acetamide group jointed at C2 position (Figure 2a), composed in turn by acetyl and secondary amine groups.
Chitin is obtained by alkaline deproteinization, acid demineralization and decoloration by organic solvents contact of crustaceans' wastes (Pastor, 2004). Although this chemical method is the most known and used, a more sustainable method has been recently developed in Mexico, specifically in the Universidad Autonoma Metropolitana Iztapalapa (UAMI). Such method consists of a biological step, where crustacean wastes are partially demineralized and deproteinized by lactic fermentation. The second step consists of chemical elimination of residual pigments, minerals and proteins from the solid produced in the first step (Cira et al., 2002). This novel process produces chitin with similar composition compared to the chitin obtained by the purely-chemical method, but with considerable savings of chemical substances and energy. The developed technology at the UAMI has been transferred to a local company, which will allow to exploit the high potential of Mexico as chitin producer. This country produces, for example, about 100,000 tons of shrimp per year (CONAPESCA, 2007), from which around 40% are chitinous wastes (Cira et al., 2002). At present, only 5% of such wastes are used in farming activities. The rest of the chitinous wastes from shrimp are discarded and represent an important environmental problem (Cira et al., 2002).

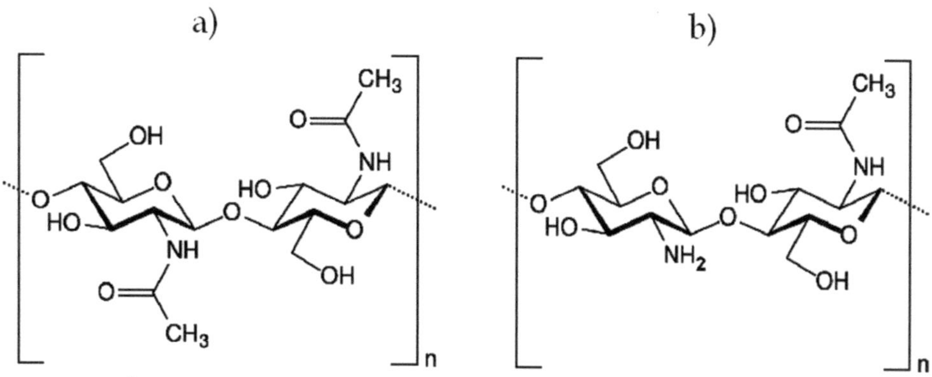

Fig. 2. Chemical structures of a) chitin and b) chitosan.

Chitosan, poly-[β-(1-4)-2-amine-2-deoxy-D-glucopyranose], has a very similar chemical structure to that of chitin (Figure 2b). However, a part of its repetitive units does not have the acetyl group or, in other words, such units have primary amine groups. Thus, chitosan is a product of the deacetylation process of chitin, which can be partially or totally deacetylated. In fact, the typical process to obtain chitosan consists on the chitin deacetylation by contacting the biopolymer with a concentrated alkaline solution at temperatures over 60°C (Pastor, 2004). Nevertheless, the degree of deacetylation (DD) of chitin is variable and depends on the process conditions (alkali concentration, contact time, temperature, etc.), which produces degree of deacetylation values from 0 to 100%. Because of this, chitin is known as the biopolymer which has a DD from 0 to 60%; likewise, when chitin has been deacetylated over 60%, the biopolymer is named chitosan. It is important to say that such convention is still in debate in the scientific community; in fact, some researchers consider chitin only between 0 and 50% of deacetylation (Sengupta, 2007). Besides, the use of the degree of acetylation (DA) rather than the degree of deacetylation as a way to distinguish between chitin and chitosan is common. Both parameters, DD and DA, are complementary, since a DD of 0% is equivalent to a DA of 100% and a DD of 100% is equivalent to a DA of 0%. However, owing to its greater use in cited literature, only DA will be used hereafter.

The importance to know the DA is based on the great difference of chitin and chitosan properties, which are evident by the greater reactivity and, in general, greater sensibility of chitosan to the variability of environment physicochemical conditions, especially the pH. Such sensibility is due to the higher density of primary amine groups of chitosan respect to that of the chitin.

Other important difference between chitin and chitosan is the greater crystallinity of the former. The lower crystallinity of chitosan increases the accessibility to the adsorption sites of the biopolymer. A chitin with DA of 100% has a highly crystalline structure, which is arranged in one of three molecular forms, named α, β y γ-chitin, respectively (Pastor, 2004). The α-chitin is the most common form of this biopolymer (Carlstrom, 1957), present in crustaceans as shrimp, from which nowadays is obtained the most commercial chitin (Peesan et al., 2003). The α-chitin is characterized by an antiparallel arrange of its chains; in other words, adjoining chitin chains run in opposite directions (Pastor, 2004). This type of arrange produces a greater union between biopolymeric chains due to created forces by hydrogen bonds, which causes in turn the formation of orthorhombic units (Peesan et al., 2003; Pastor, 2004; Zhou et al., 2005), with axis longitude of 0.5, 1.0 and 1.8 nm, respectively (Carlstrom, 1957), and right angles between each pair of axis. Such crystallinity is decreased during the deacetylation process, due to the removal of a portion of the acetyl groups from the chitin structure and, thus, to the greater presence of primary amine groups. These facts change the intra and intermolecular interactions, conformed mainly by hydrogen bonds (Ramírez-Coutiño et al., 2006), and generate amorphous zones in the biopolymer, with swelling capacity higher than that of crystalline zones, owing in part to the great affinity of primary amine groups to water. Summarizing, chitosan has a lower crystallinity than chitin and, therefore, is more susceptible to hydration and dissolution in aqueous medium, especially at low pH values.

4.2.2 Application of chitin and chitosan in water treatment

Since chitin was discovered, it has been object of numerous studies in order to understand its properties and find its application in very diverse fields. Nowadays, chitin and chitosan have application in pharmaceutical, biomedical, food, agricultural, paper, cosmetic and textile industries, as well as in chromatography and water treatment (Pastor, 2004). At present, the application of chitin and chitosan in water treatment focuses mainly in coagulation-flocculation processes to remove organic residues, suspended solids, amino acids and dyes (Pastor, 2004). Nevertheless, this way of application of the biopolymers implies the loss of the material, because the added material to the process is not reused, which contributes to the generation of sludge which must be after processed or confined. Due to this disadvantage, scientific studies have focused on other applications of chitin and chitosan in water treatment, especially as biosorbents in continuous adsorption processes, where these materials can be regenerated and reused.

4.2.3 Biosorption onto chitin and chitosan

Chitin and chitosan have been widely studied as biosorbents of cations (mainly heavy metals) and to a lesser extent as biosorbents of anions (metals, metalloids and organics) and organic compounds. Table 2 shows a series of examples of studies regarding this.

Chitin has been little studied as biosorbent compared to chitosan, due to the concentration of primary amine groups in chitosan. Such chemical groups are the main responsible of adsorption either of cations (by chelation), or anions (by electrostatic attraction) or polar molecules (by interactions as ion-dipole, dipole-dipole, van der Waals forces, etc.). However, chitin also can have a considerable density of primary amine groups even at high DA values; such density can be enough to make chitin a competitive biosorbent compared to other materials. Theoretically, 100% acetylated chitin has about 5 mmol/g of nitrogen. For example, if a certain mass of chitin is 20% deacetylated, the density of primary amine groups would be 1 mmol/g, which would be equivalent to a lead removal potential of 207 mg/g, value greater than the greatest reported capacity for typical adsorbents as activated carbon, around 100 mg/g (Sekar et al., 2004; Issabayeva et al., 2006). Besides, chitin has advantages compared to chitosan as a lower solubility in acid medium and lower cost.

Removed species	Contaminant / biopolymer	Reference
cation	Al/ Qs	(Septhum et al., 2007)
cation	Cd / Qs	(Evans et al., 2002)
cation	Co / Qs	(Minamisawa et al., 1999)
anion	As / Qs	(Kwok et al., 2009)
anion	Au, Se, Cr, V / Q	(Niu & Volesky, 2003)
anion	Remazol black 13 reactive / Qs	(Annadurai et al., 2008)
anion and polar molecule	palm oil residues / Qs	(Ahmad et al., 2005)

Table 2. Examples of contaminants removal studies by biosorption onto chitin and chitosan. The abbreviations Q and Qs mean chitin and chitosan, respectively.

4.3 Advantages and disadvantages of biosorbents

Some of the advantages of the biosorbentes are the following: (Volesky, 2003):

- Their obtaining sources are abundant in most of cases.
- Their cost is low, due to the use of byproducts or wastes.
- Their adsorption capacities can be competitive with commercial grade absorbents.
- The improvement of their adsorption capacity by diverse treatments is feasible.
- Their regeneration and reuse is possible.

The main disadvantages of the biosorbents are the following:

- Extreme variation of pH and temperature denaturalize them.
- Their physical and morphological properties restrict their application in packed columns, since the obstruction of these is possible.

The first of these disadvantages can be overcome in an adsorption process because pH and temperature are controllable variables that can be set at not extreme values, especially in water treatment for human consumption.

Finally, both of the mentioned disadvantages of the biosorbents can be overcome by supporting and/or reinforcing them.

4.4 Biocomposites

The physical and chemical resistance of the biosorbents can be improved by producing biocomposites i.e. the addition or mixing of the biosorbents into or with a supporting and/or reinforcing matrix. In addition, the morphological drawbacks of the biosorbents can be also overcome by producing biocomposites, since during their obtaining process it is possible to give hydrodynamic shapes to the particles of the material. In other words, the biocomposites can be produced in such way that, when they are placed inside of an adsorption column, a porous bed is obtained. Such bed must allow water to flow through it. However, despite the great quantity of scientific studies about the biosorbents-based biocomposites, at present there is not commercial biocomposites for water treatment. This situation can be mainly attributed to the low physical and chemical stability of the produced materials, which does not allow their application under real process conditions.

The biosorbents are supported/reinforced by means of matrixes which are frequently of synthetic nature (organic or inorganic) such as polymers and minerals, or the biosorbents are crosslinked with itself, by adding adequate chemicals. Table 3 shows the major methods to produce biocomposites, their description, and the most important drawbacks.

It is important to stand out that the adsorption capacity and the intraparticle diffusion of the biocomposites are lower than that of the pure biosorbents, regardless the method used to support/reinforce the biosorbents. Therefore, one of the most important aims of the study of the biocomposites is to maximize the reinforcing of the biosorbents and keeping, at the same time, their original adsorption capacity as much as possible.

One of the methods that have been used to overcome the loss of adsorption capacity of the biocomposites with regard to the biosorbents is the use of additives. An additive should be physically or chemically affine to the biosorbent and, at the same time, should have chemical groups that can act as adsorption sites of the contaminant of interest, or remains inert toward the biosorbent, at least.

4.5 Biosorption onto chitin/chitosan-based biocomposites

As the rest of the biosorbents, chitin and chitosan show the typical aforementioned disadvantages of its nature. Thus, the support/reinforcement of both biopolymers by means of biocomposites synthesis has been a matter of many studies. Table 4 shows a series of examples on the biosorption onto chitin/chitosan-based biocomposites. As can be seen in Table 4 and as it was aforementioned for the pure biosorbent, chitosan has also been preferred by researchers in order to obtain biocomposites to remove contaminants from water. Nevertheless, chitin has a great potential as biosorbent, above all if it is considered its greater chemical resistance compared to chitosan.

One reason used to justify the use of chitosan instead of chitin is its greater solubility, especially because many of researchers look for the maximum adsorption potential of these biopolymers by separating and spreading out the polymeric chains by dissolution processes. Such processes are expected to enhance the accessibility toward the adsorption sites and make more easy-to-use the chitin/chitosan. However, the dissolution eliminates the original crystallinity of the biopolymer, which decreases its physical and chemical resistance and, thus, the support/reinforcement of the biopolymer becomes essential. These facts make necessary to use additional chemical substances which represent an increase of the economic and environmental cost to produce the chitin/chitosan-based biocomposites.

Consequently, the study of novel ways to support/reinforce of chitin and chitosan that does not necessarily depend of their dissolution is important. In this manner, the inherent physical and chemical resistance of both biopolymers, particularly chitin, could be utilized. A possible way to synthesize chitin/chitosan-based biocomposites without the need of dissolving the biopolymer is by mixing of chitin with the support/reinforcement matrix.

Method	Description	Restrictions
Entrapment	The biosorbent particles keep fixed by means of other material e.g. a gel in which they are inserted.	• The matrix causes problems of mass transport of the contaminants towards inside of the biosorbent particles. • Part of the effective treatment volume of the biosorbent is lost. • If the proportion of biosorbent is excessive, the biocomposite becomes fragile.
Encapsulation	The biosorbent particles are confined inside of a wrapper conformed by a permeable membrane.	• The development of adequate membranes for this application is still incipient. For example, the existent membranes at the present are easily broken.
Adhesion	The biosorbent particles are joined by means of a sticky substance.	• The sticky substance can be an additional limitation to the contaminants diffusion. • The adhesive could deteriorate the mechanical properties of the biocomposite compared to the pure biosorbent.
Crosslinking	A crosslinker chemically reinforce the biosorbent, by creating bonds between adjoining molecules or polymeric chains of the biosorbent.	• Macro-particles are not produced. • Part of the potential adsorption sites is obstructed. • Additional costs by the crosslinking process. • Possible generation of undesirable residues.
Anchorage	The biosorbent chemically adhere on a pre-conditioned surface.	• Chemical substances are used intensively. • Dissolution of the biosorbent is required. • The biosorbent loss its original properties, which decreases its mechanical and chemical resistance.
Grafting	The biosorbent is reinforced by means of compounds that are inserted in the original chemical structure of the biosorbent.	• Chemical substances are used intensively. • Dissolution of the biosorbent is required. • The biosorbent loss its original properties, which decreases its mechanical and chemical resistance.

Table 3. Common methods, their description and restrictions reported in literature for support/reinforcement of the biosorbents (Volesky, 2003).

The mixture of chitin and the supporting/reinforcement matrix permits the union of both phases in terms of their exposed surface. Such matrix should allow the diffusion of the contaminants towards the supported biopolymer particles and, at the same time, chemically and physically stabilize the biopolymer. In fact, the most studied supporting/reinforcement mediums of chitin/chitosan are organic compounds and polymers (see Table 4).

Contaminant/ biopolymer	Method of support and/or reinforcement	Medium of support and/or reinforcement	Additive	Reference
Ag/Qs	crosslinking	TU [1], glutaraldehyde	Fe_3O_4	(Donia et al., 2007)
Cd, Cr (III)/Qs	crosslinking	silicate	----------	(Copello et al., 2008)
Cd, Ni/Q	grafting	poly-vinylpyrrolidone	----------	(Filho et al., 2004)
Cu/Qs	crosslinking	tripolyphosphate	----------	(Lee et al., 2001)
Cu/Qs	crosslinking	polyvinyl alcohol	----------	(Wan Ngah et al., 2004)
Cu/Qs	anchorage	perlite	----------	(Hasan et al., 2008)
Cu, Ni/Qs	----------	----------	alginate	(Huang et al., 1996)
Hg/Qs	crosslinking	glutaraldehyde	----------	(Jeon & Höll, 2003)
Hg/Qs	grafting	polyacrylamide	----------	(Li et al., 2005)
Hg/Qs	crosslinking	EDGE [2]	PEI [3]	(Kawamura et al., 1997)
Ni/Qs	anchorage	alginate, silica	----------	(Vijaya et al., 2008)
Pb/Q	crosslinking	thiourea, cellulose	----------	(Zhou et al., 2004)
U/Qs	grafting	glutaric acid	----------	(Guibal et al., 2002)
As/Qs	----------	----------	MoO_4^{2-}	(Dambies et al., 2002)
As/Qs	anchorage	alumina	----------	(Boddu et al., 2008)
As/Q, Qs	crosslinking	glutaraldehyde	----------	(Elson et al., 1980)
F/Qs (1.7 mg/g)*	crosslinking	glutaraldehyde	----------	(Viswanathan et al., 2009)
F/Q, Qs (3.8 mg/g)*	----------	----------	La	(Kamble et al., 2007)
F/Qs (4.2 mg/g)*	crosslinking	glutaraldehyde	HClAc [4], Fe	(Viswanathan & Meenakshi, 2008)
Mo, V/Qs	crosslinking	glutaraldehyde	----------	(Guibal et al., 1998)
humic acid/Qs	anchorage	PET [5]	----------	(Zhang & Bai, 2003)
dyes: PaMX [6], Rvbr [7], ArH5G [8]/Qs	grafting	poly methylmethacrylate	----------	(Singh et al., 2009)
phenol, chlorophenol/ Qs	----------	----------	alginate	(Nadavala et al., 2009)

[1] thiourea; [2] ethyleneglycoldiglycidylether; [3] polyethylenimine; [4] chloroacetic acid;
[5] polyethyleneterephthalate; [6] Procion Yellow MX; [7] Remazol Brilliant Violet;
[8] Reactive Blue H5G

* Maximum reported adsorption capacity

Table 4. Examples of the study of contaminants removal by biosorption onto chitin/chitosan-based biocomposites. The abbreviations Q and Qs mean chitin and chitosan, respectively.

This preference can be attributed to the affinity of such substances to the biopolymers i.e. the formation of stable chemical links (covalent bonds) and/or physical interactions (electrostatic attraction) between the chitin/chitosan and such substances. The use of polymers as supporting/reinforcement matrix has given encouraging results with biosorbents as seaweed, moss, sunflower waste and maize plant mixed with polyurethane (Alhakawati & Banks, 2004; Zhang & Banks, 2006).
Polymers as polyurethane show advantages as: easy handling, versatility and the possibility to obtain homogeneous, porous, malleable and resistant biocomposites. The polyurethane includes a group of polymers derived from the isocyanates, organic compounds which are very reactive. When chemical reaction between the isocyanate and a primary amine group occur, the urea group is produced; on the other hand, when the reaction occurs with hydroxyl group instead of amine, the urethane group is constituted. Both urea and urethane are very stable chemical groups, which make the polyurethane a very useful polymer in applications that must resist extreme conditions of temperature, friction and UV radiation (Hepburn, 1982). Figure 3 shows the main steps of the polyurethane polymerization from its precursors (poliol and diisocyanate) to the stable polymer. Due to the high reactivity of the isocyanate with the primary amine and hydroxyl groups, it would be possible to suppose that the polyurethane establish strong unions with a partial or totally deacetylated chitin, particularly over the contact surface of both phases during polyurethane polymerization, since chitin and chitosan have the mentioned groups on its structure.

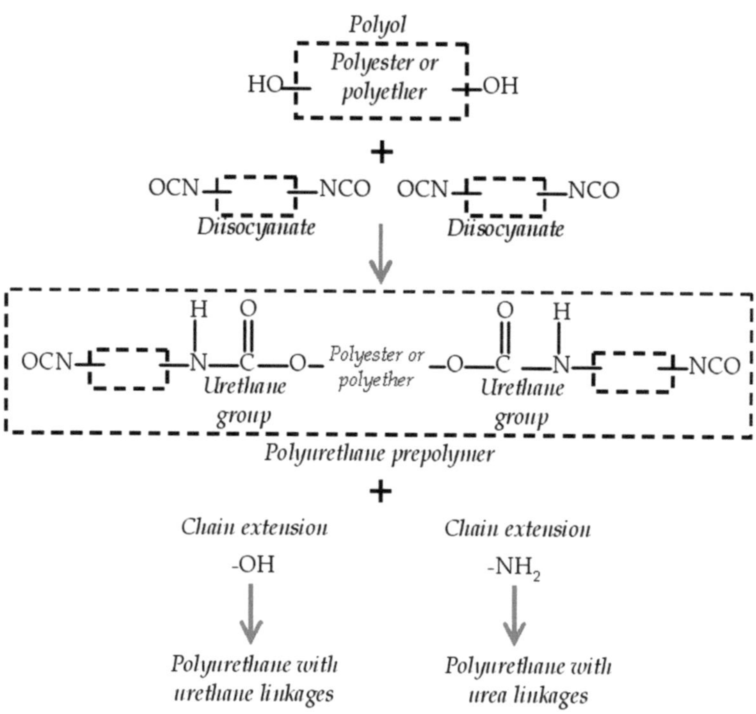

Fig. 3. Schematic representation of polyurethane synthesis (Hepburn, 1982).

In addition, Table 4 shows that the use of additives in the production of chitin/chitosan-based biocomposites has been little studied, particularly during cations biosorption. As mentioned, an additive should establish chemical or physical links with either the biopolymer or the support/reinforcement matrix, as well as to provide adequate adsorption sites. For example, a molybdenum oxyanion (molybdate) has been used as additive of chitosan for arsenic removal from water (Dambies et al., 2002); such oxyanion establishes covalent bonds with the nitrogen and oxygen of the biopolymer and, at the same time, has a strong affinity for arsenic anions. Nevertheless, in studies where molybdate has been applied, it has been demonstrated that this oxyanion is desorbed during the arsenic removal (Dambies et al., 2000), which is a serious technical drawback.

Regarding the regeneration of the chitin/chitosan-based biocomposites, some studies reported its feasibility by chemical processes which are based on a sudden pH change. For example, citric acid, tartaric acid and phosphate ions were studied to desorb arsenic (V) previously adsorbed onto molybdate impregnated chitosan (Dambies et al., 2000). In batch tests, phosphate was the best eluent, achieving regeneration values over 95% with a 0.1 M phosphate solution. In other work, hexavalent chromium previously adsorbed onto a chitosan-alumina biocomposite was desorbed in a packed column, using a 0.1 M NaOH solution as eluent (Boddu et al., 2003); the total desorption was carried out using 15 bed volumes, without appreciable decrease of the adsorption capacity of the biocomposite during the first operation cycles. In other study, desorption of 85% of fluoride initially adsorbed onto chitosan spheres by means of a 0.1 M NaOH solution was reported (Viswanathan et al., 2009). Therefore, the regeneration of chitin/chitosan-based materials is feasible, which is suitable in terms of a lower operation cost.

Finally, it is important to point out that, since a few years ago, the study of chitin and chitosan as biosorbents has been focused towards the inorganic anions removal. Among these anions, the arsenic oxyanions have been the most studied in the last decade, due to the reduction of the maximum permitted level in water for human consumption (from 50 to 10 $\mu g_{arsenic}/L$), established in many countries from the WHO recommendations in 1993 (WHO, 2006). It is important to mention that other inorganic anions e.g. fluoride, are also considered by the WHO as relevant water contaminants, since these affect the health of million of people around the world (WHO, 2006). That is why in recent years fluoride biosorption has started to be studied in a more intense manner.

5. Fluoride

5.1 Nature and removal methods

Fluoride (F^-) is one of the most ubiquitous inorganic water contaminants in groundwater around the world, due to its easy dissolution from underground rocks that contain it; consequently, fluoride can be found in natural water in concentrations that reach 25 mg/L (AWWA, 1999). Presence of fluoride in groundwater is considered an endemic problem in at least 25 countries including México (Gupta et al., 2007).

Besides, there are also anthropogenic sources of water contamination by fluoride around the world, as superphosphate industry, aluminum and zinc smelters, brickworks, ceramic works, uranium enrichment facilities, coal fired power plants, and oil refineries (Gupta et al., 2007). Fluoride ingestion in low concentration (between 0.5 and 1.5 mg/L) is recommended as prevention of dental caries. However, when fluoride is ingested in quantities that exceeds the recommendation of the WHO i.e. >1.5 mg/L (WHO, 2006) it permanently deposits on teeth and bones, which causes calcium deficiency and abnormal growth of bones. The most frequent diseases due to the excessive ingestion of fluoride are dental fluorosis (ingestion of water containing 1.5 to 4 mg/L) and skeletal fluorosis (ingestion of water containing > 4 mg/L) (BGS, 2000). However, the possible relationship between fluoride ingestion, and other health problems such as osteosarcoma, and damages to the immune and reproductive systems has been discussed by researchers for the last decades (Harrison, 2005).

Regarding chemical aspects, fluoride is the reduced state of fluorine, which is the most electronegative and reactive of the chemical elements. Due to these properties, fluoride is the only natural form of fluorine. In the solid state, fluoride is always associated with metals; in aqueous solution, fluoride can be found as either free ion or joined to metals/ metalloids in complexes forms. At pH values lower than 5, fluoride associates with the hydrogen ion (H^+) to form the HF molecule and the complex HF_2^-. The relative proportion of each of these chemical species in pure water is showed in Figure 4.

The study of fluoride removal from water started in the 30's, by means of coagulation and batch adsorption processes, using diverse aluminum-based compounds (Boruff, 1934). Subsequently different methods as chemical precipitation (Maier, 1947; Sorg, 1978; Clifford et al., 1986), inverse osmosis (Clifford et al., 1986; Cohen & Conrad, 1998) and electrodialysis (Annouar et al., 2004) were implemented. However, adsorption has remained as the most used technique (Maier, 1947; Bishop & Sansoucy, 1978; Clifford et al., 1986; WHO, 2006) due to its operative and cost advantages. The most used fluoride adsorbent at present is the activated alumina, with a maximum adsorption capacity that can vary between 1 and 3 mg/g (Ghorai & Pant, 2004; Medellín, 2006), depending of the type of alumina and the experimental conditions, particularly pH.

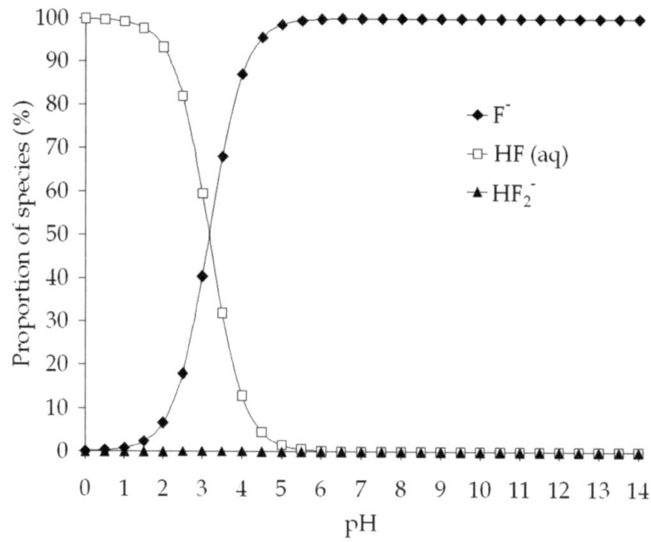

Fig. 4. Diagram of chemical species of fluoride in aqueous solution.

5.2 Fluoride biosorption

The biosorption of fluoride has been recently explored. In fact, only since 2007 the biosorption of fluoride has been reported; fluoride has been adsorbed on algae (Venkata Mohan et al., 2007), fungus (Ramanaiah et al., 2007), pure chitosan (Viswanathan et al., 2009) and added with iron (Viswanathan & Meenakshi, 2008), lanthanum (Kamble et al., 2007; Bansiwal et al., 2009), neodymium (Yao et al., 2009), titanium (Jagtap et al., 2009), hydroxyapatite (Sairam-Sundaram et al., 2008) and magnesium oxide (Sairam-Sundaram et al., 2009). The reported results are encouraging, since the maximum fluoride adsorption capacities has been reported around 20 mg/g, which is in agreement with the high density of amine groups contained on the biological structures, especially due to the presence of proteins. Amine groups work as either adsorption sites or anchorage sites of the substances used as additives. Nevertheless, the biocomposites chemical resistance has not been reported, which is a very important aspect of adsorbent materials in order to be used in real adsorption processes. In addition, some of the additives used in the aforementioned studies would considerably increase the biocomposites production cost.

Chitin has been little studied as biosorbent, and fluoride biosorption is not an exception; however, properties of chitin as chemical resistance and lower cost compared to other biosorbents, particularly to chitosan, could represent important advantages to its application in adsorption processes of fluoride or other contaminants. Based on these advantages, a biocomposite based on chitin and a polymeric matrix was recently obtained (Davila-Rodriguez et al., 2009), capable of adsorbing fluoride from aqueous solutions. Such biocomposite was composed mainly by chitin and polyurethane, which were mixed during the polymerization reaction of the last. This method promoted the formation of physical and chemical interactions between both polymers, which caused in turn a decrease of the exposure level of chitin (surface area decreased from 0.32 to 0.08 m^2/g). Such decrease was due to a lower accessibility to the adsorption sites measured as a decrease of the ion exchange capacity (IEC) of the biocomposite compared to pure chitin (from 2.12 to 0.38 mmol/g). However, the biocomposite showed a greater chemical resistance, measured as a decrease of around 10 times in the loss of mass when the material was submerged in an acid aqueous medium at pH 5 (from 19.6 to 1.5%). Besides, the chitin adsorption properties were retained and optimized by improving the biocomposite composition. The optimum composition of the chitin-based biocomposite was: 58% chitin (DA=79%), 40% polymer and 2% catalyst. Such biocomposite had a fluoride adsorption capacity of 0.29 mg/g at pH 5, being the initial fluoride concentration of 15 mg/L at 25°C, which was around 2 times lower than that of pure chitin (0.70 mg/g) at the same experimental conditions. Although such decrease of the adsorption capacity was considerable, the aforementioned increase of the chemical resistance was even more important.

The study of chitin/polyurethane biocomposite showed that a decrease in the degree of acetylation of chitin promoted a greater fluoride adsorption capacity, which was due to the greater availability of adsorption sites (primary amine groups) as the degree of acetylation of chitin decreased. Besides, the pH of the aqueous medium was a critical fluoride adsorption parameter, since fluoride adsorption notably increased as pH decreased from 8 to 5. This fact was caused by the greater protonation of the primary amine groups as pH decreased.

In addition, chitin crystallinity was not affected when mixed with the polymeric matrix, since the interaction chitin-polyurethane was mainly on the contact surface. In fact, the X-ray diffraction pattern of the chitin-based biocomposite showed an intermediate behavior between chitin and polyurethane diffraction patterns, due to the dilution effect of the mixture biopolymer-polymer.

Additional work has shown that the optimum chitin-based biocomposite has a swelling capacity which permits that the fluoride ions can enter to the hydrated structure of the

material and, thus, access to the adsorption sites contained mainly on the chitin structure. Besides, the fluoride adsorption capacity of both chitin and biocomposite is considerably great, if they are compared to other chitinous materials, as reported in Table 4.
The authors consider that future work should include studies about the adsorption kinetics of the fluoride adsorption process, as well as packed columns tests in order to determine the behavior of the respective breakthrough curves. Such knowledge would permit to scale the fluoride biosorption process to real processes and so achieve the application of the chitin-based biocomposites, which in turn would contribute to resolve one of the most important water contamination problems around the world, by means of more sustainable technologies.

6. Conclusion

Although chitin is an abundant biopolymer in nature, its utilization in the daily activities of people is even little. Chitin and chitosan have a great potential to be used in many fields because of the particular properties of these biopolymers. In fact, the possibility to use chitin and chitosan as biosorbents has been widely demonstrated; however, the low mechanical and chemical resistance of both biopolymers is a disadvantage which restricts their application in real scale adsorption processes. The production of chitin/chitosan-based biocomposites has shown encouraging results in terms of improved adsorption capacity and resistance. Regarding chitin, this has been studied as biosorbent to a lesser extent than chitosan; however, the natural greater resistance of the former compared to the last, due to its greater crystallinity, could mean a great advantage. Besides, the possibility to control the degree of acetylation of chitin permits to enhance its adsorption potential by increasing its primary amine group density. Recent studies regarding the production of chitin-based biocomposites and its application as fluoride biosorbents have demonstrated the potential of these materials to be used in continuous adsorption processes. Moreover, these biocomposites could remove many different contaminants, including cations, organic compounds and anions. Future work will permit to advance towards the application of biosorption processes, especially those based on the use of chitin and its derivatives.

7. Acknowledgements

The authors are grateful to the Instituto Potosino de Investigacion Cientifica y Tecnologica, A.C. (IPICyT) for the support to carry out this work. In addition, the authors appreciate the technical support of M.C. Dulce Partida-Gutierrez, M.C. Guillermo Vidriales Escobar, M.C. Grisel Ramirez-Manzanares and Ing. Daniel Ramirez-Gonzalez.

8. References

Ahmad, A. L.; Sumathi, S. & Hameed, B. H. (2005). Adsorption of residue oil from palm oil mill effluent using powder and flake chitosan: Equilibrium and kinetic studies. *Water Research*, 39, 12, 2483-2494, 0043-1354

Alcan (2005). *Arsenic and fluoride removal adsorptive process: technical bulletin*, Alcan Primary Products Corp., Cleveland, OH, U.S.A.

Alhakawati, M. S. & Banks, C. J. (2004). Removal of copper from aqueous solution by Ascophyllum nodosum immobilised in hydrophilic polyurethane foam. *Journal of Environmental Management*, 72, 4, 195-204, 0301-4797

Annadurai, G.; Ling, L. Y. & Lee, J.-F. (2008). Adsorption of reactive dye from an aqueous solution by chitosan: isotherm, kinetic and thermodynamic analysis. *Journal of Hazardous Materials*, 152, 1, 337-346, 0304-3894

Annouar, S.; Mountadar, M.; Soufiane, A.; Elmidaoui, A. & Menkouchi Sahli, M. A. (2004). Defluoridation of underground water by adsorption on the chitosan and by electrodialysis. *Desalination*, 165, 437, 0011-9164

AWWA (1999). *Water Quality and Treatment—A Handbook of Community Water Supplies*, McGraw-Hill, 0070016593, New York, N.Y.

Bansiwal, A.; Thakre, D.; Labhshetwar, N.; Meshram, S. & Rayalu, S. (2009). Fluoride removal using lanthanum incorporated chitosan beads. *Colloids and Surfaces B-Biointerfaces*, 74, 1, 216-224, 0927-7765

BGS (2000). *Water Quality Fact Sheet: Fluoride*, British Geological Survey, Natural Environment Research Council, U.K.

Bishop, P. L. & Sansoucy, G. (1978). Fluoride removal from drinking water by fluidized activated alumina adsorption. *Journal - American Water Works Association*, 70, 10, 554-559, 1551-8833

Boddu, V. M.; Abburi, K.; Talbott, J. L. & Smith, E. D. (2003). Removal of hexavalent chromium from wastewater using a new composite chitosan biosorbent. *Environmental Science & Technology*, 37, 19, 4449-4456, 0013-936X

Boddu, V. M.; Abburi, K.; Talbott, J. L.; Smith, E. D. & Haasch, R. (2008). Removal of arsenic (III) and arsenic (V) from aqueous medium using chitosan-coated biosorbent. *Water Research*, 42, 3, 633-642, 0043-1354

Boruff, C. S. (1934). Removal of Fluorides from Drinking Waters. *Industrial & Engineering Chemistry*, 26, 1, 69-71, ----

Carlstrom, D. (1957). The Crystal Structure of α-Chitin (Poly-N-Acetyl-D-Glucosamine). *The Journal of Biophysical and Biochemical Cytology*, 3, 5, 669-683, 0142-9418

Cira, L. A.; Huerta, S.; Hall, G. M. & Shirai, K. (2002). Pilot scale lactic acid fermentation of shrimp wastes for chitin recovery. *Process Biochemistry*, 37, 12, 1359-1366, 0032-9592

Clifford, D.; Subramonian, S. & Sorg, T. J. (1986). Water treatment processes. III. Removing dissolved inorganic contaminants from water. *Environmental Science & Technology*, 20, 11, 1072-1080, 1520-5851

Cohen, D. & Conrad, H. M. (1998). 65000 GPD fluoride removal membrane system in Lakeland, California, USA. *Desalination*, 117, 19-35, 0011-9164

CONAGUA (2010). *Estadísticas del Agua en México Edición 2010*, Comisión Nacional del Agua, Mexico City

CONAPESCA (2007). *Plan de Manejo para la Pesquería de Camarón en el Litoral del Océano Pacífico Mexicano*, Comisión Nacional de Acuacultura y Pesca, Mexico City

Copello, G. J.; Varela, F.; Vivot, R. M. & Díaz, L. E. (2008). Immobilized chitosan as biosorbent for the removal of Cd(II), Cr(III) and Cr(VI) from aqueous solutions. *Bioresource Technology*, 99, 14, 6538-6544, 0960-8524

Dambies, L.; Guibal, E. & Roze, A. (2000). Arsenic(V) sorption on molybdate-impregnated chitosan beads. *Colloids and Surfaces A: Physicochemical and Engineering Aspects*, 170, 1, 19-31, 0927-7757

Dambies, L.; Vincent, T. & Guibal, E. (2002). Treatment of arsenic-containing solutions using chitosan derivatives: uptake mechanism and sorption performances. *Water Research*, 36, 15, 3699-3710, 0043-1354

Davila-Rodriguez, J. L.; Escobar-Barrios, V. A.; Shirai, K. & Rangel-Mendez, J. R. (2009). Synthesis of a chitin-based biocomposite for water treatment: Optimization for fluoride removal. *Journal of Fluorine Chemistry,* 130, 8, 718-726, 0022-1139

Donia, A. M.; Atia, A. A. & Elwakeel, K. Z. (2007). Recovery of gold(III) and silver(I) on a chemically modified chitosan with magnetic properties. *Hydrometallurgy,* 87, 3-4, 197-206, 0304-386X

Elson, C. M.; Davies, D. H. & Hayes, E. R. (1980). Removal of arsenic from contaminated drinking water by a chitosan/chitin mixture. *Water Research,* 14, 9, 1307-1311, 0043-1354

Evans, J. R.; Davids, W. G.; MacRae, J. D. & Amirbahman, A. (2002). Kinetics of cadmium uptake by chitosan-based crab shells. *Water Research,* 36, 13, 3219-3226, 0043-1354

Filho, J. A. R.; Bach, E. E.; Vargas, R. R.; Soares, D. A. W. & Queiroz, A. A. A. d. (2004). An investigation of cadmium(II) and nickel(II) adsorption by chitin graft copolymer. *Journal of Applied Polymer Science,* 92, 2, 1310-1318, 1097-4628

Ghorai, S. & Pant, K. K. (2004). Investigations on the column performance of fluoride adsorption by activated alumina in a fixed-bed. *Chemical Engineering Journal,* 98, 1-2, 165-173, 1385-8947

Guibal, E.; Jansson-Charrier, M.; Saucedo, I. & Cloirec, P. L. (2002). Enhancement of Metal Ion Sorption Performances of Chitosan: Effect of the Structure on the Diffusion Properties. *Langmuir,* 11, 2, 591-598, 0743-7463

Guibal, E.; Milot, C. & Tobin, J. M. (1998). Metal-Anion Sorption by Chitosan Beads: Equilibrium and Kinetic Studies. *Industrial & Engineering Chemistry Research,* 37, 4, 1454-1463, 0888-5885

Gupta, V. K.; Ali, I. & Saini, V. K. (2007). Defluoridation of wastewaters using waste carbon slurry. *Water Research,* 41, 15, 3307-3316, 0043-1354

Harrison, P. T. C. (2005). Fluoride in water: A UK perspective. *Journal of Fluorine Chemistry,* 126, 11-12, 1448-1456, 0022-1139

Hasan, S.; Ghosh, T. K.; Viswanath, D. S. & Boddu, V. M. (2008). Dispersion of chitosan on perlite for enhancement of copper(II) adsorption capacity. *Journal of Hazardous Materials,* 152, 2, 826-837, 0304-3894

Hepburn, C. (1982). *Polyurethane Elastomers,* Applied Science Publishers, London and New York

Huang, C.; Chung, Y.-C. & Liou, M.-R. (1996). Adsorption of Cu(II) and Ni(II) by pelletized biopolymer. *Journal of Hazardous Materials,* 45, 2-3, 265-277, 0304-3894

Issabayeva, G.; Aroua, M. K. & Sulaiman, N. M. N. (2006). Removal of lead from aqueous solutions on palm shell activated carbon. *Bioresource Technology,* 97, 18, 2350-2355, 0960-8524

Jackson, R. B.; Carpenter, S. R.; Dahm, C. N.; McKnight, D. M.; Naiman, R. J.; Postel, S. L. & Running, S. W. (2001). Water in a changing world. *Ecological Applications,* 11, 4, 1027-1045, 1051-0761

Jagtap, S.; Thakre, D.; Wanjari, S.; Kamble, S.; Labhsetwar, N. & Rayalu, S. (2009). New modified chitosan-based adsorbent for defluoridation of water. *Journal of Colloid and Interface Science,* 332, 2, 280-290, 0021-9797

Jeon, C. & Höll, W. H. (2003). Chemical modification of chitosan and equilibrium study for mercury ion removal. *Water Research,* 37, 19, 4770-4780, 0043-1354

Kamble, S. P.; Jagtap, S.; Labhsetwar, N. K.; Thakare, D.; Godfrey, S.; Devotta, S. & Rayalu, S. S. (2007). Defluoridation of drinking water using chitin, chitosan and lanthanum-modified chitosan. *Chemical Engineering Journal,* 129, 1-3, 173-180, 1385-8947

Kawamura, Y.; Yoshida, H.; Asai, S. & Tanibe, H. (1997). Breakthrough curve for adsorption of mercury (II) on polyaminated highly porous chitosan beads. *Water Science and Technology,* 35, 7, 97-105, 0273-1223

Kwok, K. C. M.; Lee, V. K. C.; Gerente, C. & McKay, G. (2009). Novel model development for sorption of arsenate on chitosan. *Chemical Engineering Journal,* 151, 1-3, 122-133, 1385-8947

Lee, S.-T.; Mi, F.-L.; Shen, Y.-J. & Shyu, S.-S. (2001). Equilibrium and kinetic studies of copper(II) ion uptake by chitosan-tripolyphosphate chelating resin. *Polymer,* 42, 5, 1879-1892, 0032-3861

Li, N.; Bai, R. & Liu, C. (2005). Enhanced and Selective Adsorption of Mercury Ions on Chitosan Beads Grafted with Polyacrylamide via Surface-Initiated Atom Transfer Radical Polymerization. *Langmuir,* 21, 25, 11780-11787, 0743-7463

Maier, F. J. (1947). Methods of removing fluorides from water. *American Journal of Public Health,* 37, 12, 1559-1565, ----

Medellín, N. (2006). *Remoción de fluoruros en solución acuosa por medio de adsorción sobre varios materiales (Tesis de Maestría),* CIEP, Facultad de Ciencias Químicas, UASLP, San Luis Potosí, México

Minamisawa, H.; Iwanami, H.; Arai, N. & Okutani, T. (1999). Adsorption behavior of cobalt(II) on chitosan and its determination by tungsten metal furnace atomic absorption spectrometry. *Analytica Chimica Acta,* 378, 1-3, 279-285, 0003-2670

Nadavala, S. K.; Swayampakula, K.; Boddu, V. M. & Abburi, K. (2009). Biosorption of phenol and o-chlorophenol from aqueous solutions on to chitosan-calcium alginate blended beads. *Journal of Hazardous Materials,* 162, 1, 482-489, 0304-3894

Niu, H. & Volesky, B. (2003). Characteristics of anionic metal species biosorption with waste crab shells. *Hydrometallurgy,* 71, 1-2, 209-215, 0304-386X

Pastor, A. (2004). *Quitina y Quitosano: obtención, caracterización y aplicaciones.,* Programa CYTED, CIAD, A.C., Fondo Editorial de la Pontificia Universidad Católica del Perú, Lima, Perú

Peesan, M.; Rujiravanit, R. & Supaphol, P. (2003). Characterisation of beta-chitin/poly(vinyl alcohol) blend films. *Polymer Testing,* 22, 4, 381-387, 0142-9418

Ramanaiah, S. V.; Venkata Mohan, S. & Sarma, P. N. (2007). Adsorptive removal of fluoride from aqueous phase using waste fungus (Pleurotus ostreatus 1804) biosorbent: Kinetics evaluation. *Ecological Engineering,* 31, 1, 47-56, 0925-8574

Ramírez-Coutiño, L.; Marín-Cervantes, M. d. C.; Huerta, S.; Revah, S. & Shirai, K. (2006). Enzymatic hydrolysis of chitin in the production of oligosaccharides using Lecanicillium fungicola chitinases. *Process Biochemistry,* 41, 5, 1106-1110, 1359-5113

Sairam-Sundaram, C.; Viswanathan, N. & Meenakshi, S. (2008). Uptake of fluoride by nano-hydroxyapatite/chitosan, a bioinorganic composite. *Bioresource Technology,* 99, 17, 8226-8230, 0960-8524

Sairam-Sundaram, C.; Viswanathan, N. & Meenakshi, S. (2009). Defluoridation of water using magnesia/chitosan composite. *Journal of Hazardous Materials,* 163, 2-3, 618-624, 0304-3894

Sekar, M.; Sakthi, V. & Rengaraj, S. (2004). Kinetics and equilibrium adsorption study of lead(II) onto activated carbon prepared from coconut shell. *Journal of Colloid and Interface Science,* 279, 2, 307-313, 0021-9797

Sengupta, A. K. (2007). *Ion Exchange and Solvent Extraction. A series of advances,* CRC Press, Taylor and Francis Group, 978-0-8493-7397-8, U.S.A.

Septhum, C.; Rattanaphani, S.; Bremner, J. B. & Rattanaphani, V. (2007). An adsorption study of Al(III) ions onto chitosan. *Journal of Hazardous Materials,* 148, 1-2, 185-191, 0304-3894

Singh, V.; Sharma, A. K.; Tripathi, D. N. & Sanghi, R. (2009). Poly(methylmethacrylate) grafted chitosan: An efficient adsorbent for anionic azo dyes. *Journal of Hazardous Materials,* 161, 2-3, 955-966, 0304-3894

Sorg, T. J. (1978). Treatment technology to meet the interim primary drinking water regulations for inorganics. *Journal - American Water Works Association,* 70, 105-112, 1551-8833

USEPA (2002). *Arsenic treatment technologies for soil, waste, and water,* United States Environmental Protection Agency, Cincinnati, OH, U.S.A.

Venkata Mohan, S.; Ramanaiah, S. V.; Rajkumar, B. & Sarma, P. N. (2007). Removal of fluoride from aqueous phase by biosorption onto algal biosorbent Spirogyra sp.-IO2: Sorption mechanism elucidation. *Journal of Hazardous Materials,* 141, 3, 465-474, 0304-3894

Vijaya, Y.; Popuri, S. R.; Boddu, V. M. & Krishnaiah, A. (2008). Modified chitosan and calcium alginate biopolymer sorbents for removal of nickel (II) through adsorption. *Carbohydrate Polymers,* 72, 2, 261-271, 0144-8617

Viswanathan, N. & Meenakshi, S. (2008). Selective sorption of fluoride using Fe(III) loaded carboxylated chitosan beads. *Journal of Fluorine Chemistry,* 129, 6, 503-509, 0022-1139

Viswanathan, N.; Sairam-Sundaram, C. & Meenakshi, S. (2009). Removal of fluoride from aqueous solution using protonated chitosan beads. *Journal of Hazardous Materials,* 161, 1, 423-430, 0304-3894

Volesky, B. (2003). *Sorption and Biosorption,* BV Sorbex, Inc., 0-9732983-0-8, Montreal-St. Lambert, Quebec, Canada

Wan Ngah, W. S.; Kamari, A. & Koay, Y. J. (2004). Equilibrium and kinetics studies of adsorption of copper (II) on chitosan and chitosan/PVA beads. *International Journal of Biological Macromolecules,* 34, 3, 155-161, 0141-8130

WHO (2000). *Global Water Supply and Sanitation Assessment 2000 Report,* World Health Organization, 92 4 156202 1, Geneva, Switzerland

WHO (2006). *Guidelines for drinking-water quality [electronic resource]: incorporating first addendum. Vol. 1, Recommendations,* World Health Organization, 92 4 154696 4, Geneva, Switzerland

Yao, R.; Meng, F.; Zhang, L.; Ma, D. & Wang, M. (2009). Defluoridation of water using neodymium-modified chitosan. *Journal of Hazardous Materials,* 165, 1-3, 454-460, 0304-3894

Zhang, X. & Bai, R. (2003). Mechanisms and kinetics of humic acid adsorption onto chitosan-coated granules. *Journal of Colloid and Interface Science,* 264, 1, 30-38, 0021-9797

Zhang, Y. & Banks, C. (2006). A comparison of the properties of polyurethane immobilised Sphagnum moss, seaweed, sunflower waste and maize for the biosorption of Cu, Pb, Zn and Ni in continuous flow packed columns. *Water Research,* 40, 4, 788-798, 0043-1354

Zhou, D.; Zhang, L. & Guo, S. (2005). Mechanisms of lead biosorption on cellulose/chitin beads. *Water Research,* 39, 16, 3755-3762, 0043-1354

Zhou, D.; Zhang, L. N.; Zhou, J. P. & Guo, S. L. (2004). Cellulose/chitin beads for adsorption of heavy metals in aqueous solution. *Water Research,* 38, 11, 2643-2650, 0043-1354

9

Cellulose Fibres Functionalised by Chitosan: Characterization and Application

Simona Strnad, Olivera Šauperl and Lidija Fras-Zemljič
University of Maribor, Faculty of Mechanical Engineering
Slovenia SI-2000

1. Introduction

Fibrous (textile) materials, such as used in the medical field or health services can be divided into two basic groups, according to whether they are used 1) inside organic tissues (internal/implantable): vascular grafts, meshes, stents, tendons and ligament implants, surgical threads, etc.) or 2) on their surface (external/non-implantable): gauzes, bandages, surgical covers, nappies, tampons, etc. The use of natural cellulose fibers, such as cotton and flax, goes back in medical applications to ancient times and still today, in some medical applications, cellulose fibrous materials represent materials that can not be exchanged with any other. In the more recent past new procedures and technologies enabled the production of various chemical cellulose fibers such as viscose, modal and lyocell, which are cleaner and even more hygroscopic than cotton, and as such highly applicable within hygiene and medical fields. The great potential of cellulose fibers lies in their molecular structure, which offers excellent possibilities as a matrix for the design of bioactive, biocompatible, and intelligent materials.

Over the last twenty years the increase in the number of microbially caused diseases and hospital infections has led to intensive research into new materials and procedures, which would assure permanent bioactive effects together with complete safety for the people (Chung et al. 1998, Liu et al. 2000, Lee et al. 1999). Fungi, as well as gram positive and gram negative bacteria are commonly found in textile materials, especially in bedclothes. Many of these microorganisms are pathogens or potential pathogens and quite often related to nosocomial infections (Takai et al. 2002). Fibrous materials used in the medical and hygiene fields are usually in contact with extremely contaminated environments. *Escherichia Coli* and *Penicillium Crysogenum* very often cause diseases, as well as the disintegration and foul smell of fibrous materials (Chang et al. 1996). Therefore, the most important biomedical characteristics of a fibrous material for external use are its' antimicrobial properties. In textiles these are usually achieved by treatments with silvers salts, quaternary ammonium chloride, metals, aromatic, halogen compouns, etc. (Kenawy et al. 2007, Takai et al. 2002). Antimicrobial agents act either as growth inhibitors (bacterio- or fungi-static) or they kill the microorganisms (bio-cidal). Almost all antimicrobial agents used in convetional textile treatments are biocidal, as they damage the cell wall or the membrane permeability, denature proteins, inhibit enzyme activity or lipid synthesis, which are all essential for cell survival (Gao et. al. 2008). The most important issue regarding biocides used on commercial textiles is the potential induction of bacterial resistance to these substances, which could also lead to increased resistance to certain antibiotics in clinical use, especially because of large quantities of these substances, which are needed for conventional textile treatments. Furthermore, some of these substances produce unpleasant odour and/or discoloration of textiles (N-halamine), need to be regenerated in bleaching process (N-halamine, peroxyacids), or can be depleted (silver) (Gao et al. 2008).

All the mentioned factors have resulted in a high interest in researching and developing technologies, which are based on the use of alternative natural materials. One of such natural polymers is chitosan, which has received lots of attention due to its non-toxicity, antimicrobial properties and wide possibilities for chemical derivations (Berger 2004, Uragami 2006, Muzzarelli 2005, Alonso et al. 2009). The antimicrobial activity of chitosan is assigned to its amino groups, which in diluted acids form ammonium salts. The manipulation of chitosan's binding strategies onto cellulose surfaces i.e.: (i) reversible binding - which enables the release of a bioactive substance from a fibre's surface and, (ii) irreversible binding resulting in the permanent bioactivity of a fibrous (textile) surface, determines the applicabilities of such surfaces.

Efficient and permanent fixation of chitosan is of major importance in the cases of conventionally applied (washable) textiles. Many different approaches have been used in order to permanently bind chitosan to cotton fibers (Fras-Zemljič et al. 2004, El-ahlavy 2005, Strnad et al. 2008). Successful permanent chitosan binding with a large number of accessible amino groups requires a certain amount of anchoring sites on the fibrous substrate. However, interactions with the anchoring groups lower the free amino groups' amount and as such decreases the surface antimicrobial activity of the treated fibres. In order to achieve satisfactory results, it is extremely important to balance between treatment résistance, materials' mechanical properties, and the number of free amino groups in/on the treated material.

2. Cellulose fibers in medical applications

The most important cellulose fibres in medical applications are natural cotton fibers and different regenerated (man-made) cellulose fibres such as viscose, modal, and lyocell. These fibres find medical application in two somewhat different fields: *1) hygene and healthcare* (bed linen, mattress covers, incontinence care pads, nappies, tampons, etc.) and *2)* so-called *external* textile materials (wound dressings, bandages, gauzes, etc.) The most important characteristics, which determine the medical applications of these fibres, are: moisture and liquids' adsorption, antistatic behaviour, low content of dyes and other chemicals, high mechanical stability, and ease of laundering and/or sterilisation (Rajesh 2006).

A cotton fibre is a single cell, which grows on the seed's surface of a plant from the genus Gossipium. At the growth beginning the outer primary wall of a cell grows in length for about 16 - 17 days and after that the secondary wall begins to form with the deposition of concentric layers of cellulose lamella from the outside towards the lumen. The secondary wall in a mature fibre represents the body of the fibre. The primary-wall is composed of cellulose, fats, waxes, pectic substances, and proteinaceous matter, while the secondary-wall is considered to be pure cellulose (De Gruy et al. 1973). During growth, the internal central canal (lumen) is large and filled with plant juices from the protoplasm of the cell. When the growing period ends, the fibers dry out and collapse into shrivelled, twisted, and flattened tubes (Fig. 1). Cotton fibres are around 10 - 55 mm long and with a diameter between 10 and 40 μm. The cross-sections of cotton fibres are elliptical or bean-shaped, with the narrow lumen in the middle.

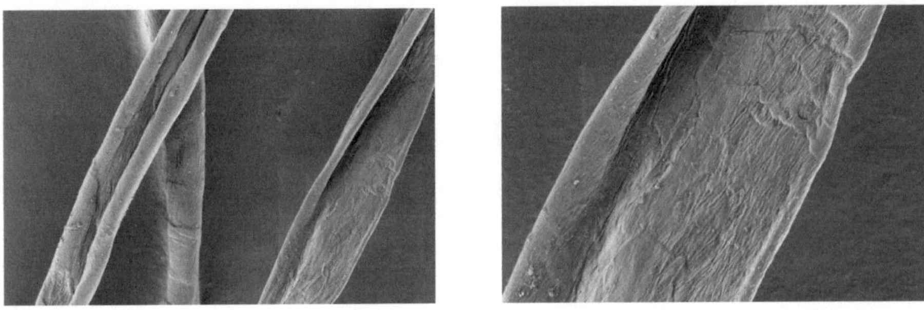

Fig. 1. SEM micrograf of cotton fibre: a) magnification 10 000X and b) magnification 30 000 X

Cellulose macromolecules organise themselves into well-ordered crystalline regions and less ordered amorphous regions. These differently ordering levels alternate along fibrils, which compose fibril bundles and larger fibril structures up to the cell walls. The crystalline unit of natural cellulose (cellulose I) has monoclinic symmetry (Krässig 1993). The main intermolecular conections are H-bonds. During special alkaline treatments, natural cellulose converts into cellulose II, which is also characteristic for regenerated cellulose fibres.

Owing to hemicelluloses, waxes, and other impurities in the primary-wall, raw cotton fibres have rather hydrophobic characteristics. Some pretretament procedures of fibres are needed in order to purify and to increase the hydrophilicity of cotton fibres. Scouring is an alkaline treatment in NaOH solutions with concentration < 10 %, by which impurities and waxes are removed and fibre hydrophilicity is increased. Mercerization, however, is an alkaline treatment at NaOH conc. > 10 %, by which the supermolecular structure and fibre morphology are changed and, owing to these changes, the mechanical properties, accessibility to aqueous media, affinity to dyes, and lustre are increased (Šauperl 2009). Besides this, cotton usually needs to be bleached and treated in order to improve its wrinkle resistance.

Another group of cellulose fibres applied in medicine is regenerated (man-made) cellulose fibres (Fig. 2). The main sources of the cellulose for regenerated fibre production are highly refined wood pulps and, to a lesser extent, cotton (Krässig 1993). Conventional regenerated cellulose viscose fibres, are generally produced by the indirect viscose process, based on deriving cellulose from carbon bisulphite, and a modal fibre is produced using a modification of this basic procedure (Cook 1984). Owing to considerable environmental problems connected to viscose process, several decades ago lyocell fibres were produced on the basis of an environmentally-friendly procedure, where the cellulose was dissolved directly in the organic solvent N-methylmorpholine-N-oxide, without the formation of derivatives (Albrecht et al. 1997, Berger 1994, Cole 1994, Krüger 1994). Different production processes regarding viscose, modal and lyocell cause differences in the molecular and supermolecular arrangements of the fibres despite the same chemical composition. The degree of crystallinity of lyocell fibres is appoximately 43 % higher than of viscose and about 16 % higher when compared to modal (Kreze & Malej 2003). The same trend can be observed regarding molecular orientation. Fibre sorption properties are, in addition to size and orientation of amorphous regions, influenced by void fracture (diameter, volume, and specific inner surface). Owing to the similar void structures in lyocell and viscose fibres, the sorption, swelling and dyeing properties of these two fibres are similar (Kreze et al. 2001). Lyocell fibres are superior in comparison to viscose in regard mechanical properties. The wet state mechanical properties of fibres are of crucial importance for medical applications.

In the wet-state, lyocell fibres keep over 90 % of their tensile strength in a conditioned state, viscose about 60 %, and modal about 50 % (Kreže & Malej 2003).

a) b) c)

Fig. 2. SEM micrographs of a) viscose, b) modal and c) lyocell fibre at magnification 10 000X

3. Chitosan

Over the last 30 years chitosan has gone through unimaginable development and has been successfully infiltrated into different areas of our lives. It is a natural product gained from chitin which is, after cellulose, the second more common biopolymer on earth (Majeti et al. 2000). Besides crayfish, molluscs and insects, chitin is also present in vegetal species, mainly in the cell-walls of fungi and moulds, where the nutritional cycle requires nitrogen (Table 1). Chitin structures are mainly derived from the ectoderm of multicellular animals and the typically built skeletons of invertebrates, while the collagen structures originate mainly from mesoderm cells (Table 2). Crayfish, molluscs, insects and fungi produce around 100 milliards tons of chitin a year, which is one of the least exploited sources of biomass on earth (Tharanathan 2003).

Type	Content of chitin [%]	Type	Content of chitin [%]
CRAYFISH		**MOLLUSCA**	
Crabs	72.1[c]	Mussels (shell)	6.1
Crayfish	69.1[c]	Oysters (shell)	3.6
Alaska crayfish	28.0[d]	Squid (skeleton)	41.0
Lobster	60-75[c]		
INSECTS		**MOULDS, FUNGI**	
Cockroaches (blatela)	18.4[d]	aspergillus niger	42.0[e]
Beetles	27-35[c]	penicillium notatum	18.5[e]
Flies (true flies)	54.8[c]	penicillium crysogenum	20.1[e]
Larva of silkworm (bombyx)	44.2[c]	mucor rouxii	44.5[a]
		Fungi (lactarius vellereus)	19.0[b]

Table 1. The content of chitin in some of the organisms (Tharanathan 2003): [a]on the weight of the body; [b]on the dry weight of the body; [c]on the weight of the organic part of the cuticle; [d] on the joint dry weight of the cuticle; [e]on the dry weight of the cell wall

Chitin is a polysaccharide, a combination of 2-acetamino-2-deoksi-ß-D-glucosic units, connected with ß-1,4 linkages, and chitosan is the common name for a large group of chitins, deacetylated to different degrees. It is mainly a combination of 2-amino-2-deoksi-D-glycopyranose units, connected by ß (1-4) linkages (Fig. 3).

The source of chitin	Proteins	Chitin	Ashes	Fats
Crabs (Collinectes sapidus)	25.1	13.5	58.6	2.1
Crabs (Chinoecetes opilio)	29.2	26.6	40.6	1.3
Crayfish (Pandalus borealis)	41.9	17.0	34.2	5.2
Crayfish (Cragon cragon)	40.6	17.8	27.5	9.9
Crayfish (Penaeus monodon)	47.4	40.4	23.0	1.3
Lobsters (Procamborus clarkii)	29.8	13.2	46.6	5.6
Common snipe	61.6	33.0	29.4	1.4

Table 2. The approximate composition of crayfish skeletons' offal in % according to the dry weight (Synowiecky & Al-Katheeb 2003).

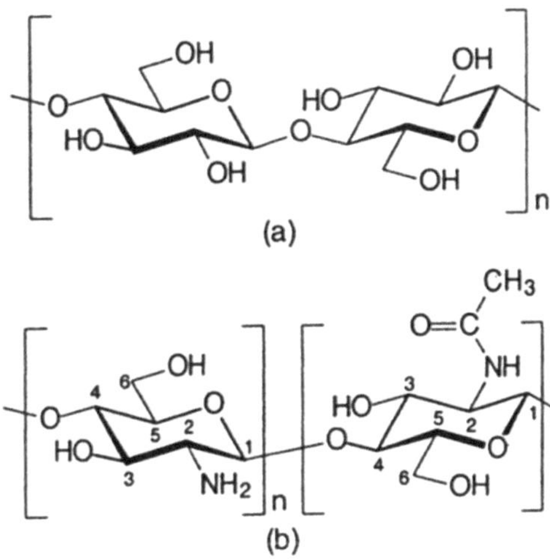

Fig. 3. Chemical structure of celluloses (a), chitin and chitosan (b) – chitin mainly consists of monomers in the shape of »m« (N-acetyl form), while chitosan, according to the degree of deacetylation, consists mainly of monomers in the shape of »n« (amine form) (Lim 2003).

Chitosan is a weak acid and, as such, is subject to reactions of neutralization in alkaline mediums. The free electron pairs on the primary amine group delegate chitosan as a potential nucleofil, which easily reacts with most aldehydes and forms imines. The molecules of chitosan have a highly positive polarity and attract negative molecules. Although most reactions with chitosan occur primarly on amino groups, it is also possible that hydroxyl groups selectively change. A hydroxyl group on the atom C6 is more reactive than the group of atom C3.

During the procedure of acquiring chitosan (Fig. 4), rough conditions usually lead to the degradation of products. The molecular mass of natural chitin is usually higher than one million, while the products of commercial chitosan have a molecular mass between 100, 000 and 1, 200, 000. According to the Horowitz method of deacetylation (Goosen 1997), which assumes a 30 minute treatment of chitosan at 180 °C, it is possible to gain chitosan with ta 95% degree of deacetylation, but such a kind of chitosan has an average chain length of only about twenty units. Generally speaking, factors such as the presence of oxygen, high temperature, and cutting charge can cause the degradation of a chitosan product.

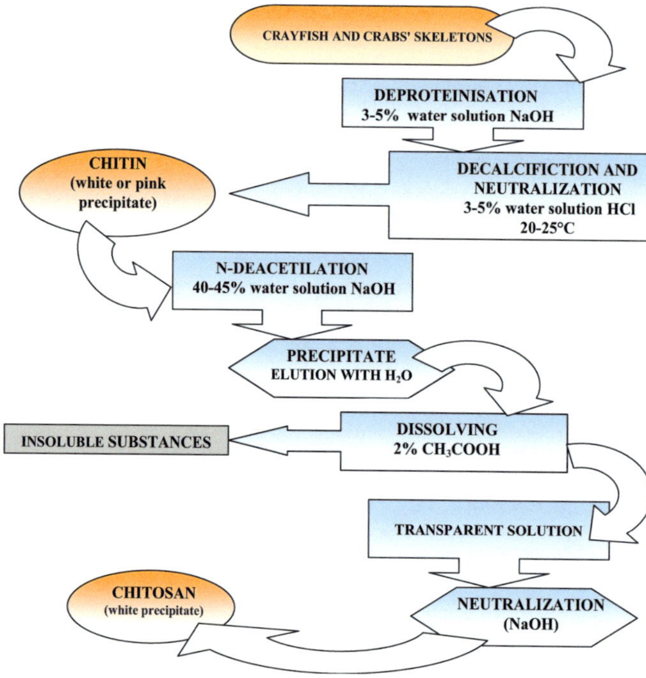

Fig. 4. The ways of chitin and chitosan acquisition from the skeletons of arthropods (Strnad et al. 2008)

The degree of deacetylation has a significant affect on the degree of chitosan dissolution in water solutions, since the chitosan with a degree of deacetylation of 40 % can be dissolved in water solutions to pH 9, while at 85 % degree of deacetylation it can be dissolved only in solutions to pH 6,5. Chitosan in the form of free amine is insoluble in water near its neutral pH, in concentrated acids' (except sulphuric acid) the bases, and organic solvents. On the other hand, chitosan is soluble in dilute HCl, HI, HBr, HNO_3 and $HClO_4$. There is a possibility that, in hydrogen acid after dissolution of the polymer, precipitation occurs in the form of a white-coloured gel. This biopolymer is also slightly soluble in dilute H_3PO_4 (Goosen 1997, Knnittel et al. 1998) The dissociation constant of chitosan is variable depending on the degree of dissociation at which it is determined. The Katchalsky equation is appropriate when calculating the variation of the pk_a value (Roberts, 1992). There are many factors that affect the viscosity of chitosan solutions, such as the degree of the polymer's deacetylation, the average, and the organisation of the molecular masses, concentration, ionic strength, pH value and temperature. The change in pH value of the polymer solution affects the viscosity of the solution differently depending on the acid used. The viscosity of the chitosan solution in an acidulous medium of ethan acid generally increases with any reduction in pH value of the solution, while the viscosity of the solutions in HCl with any reduction of pH value also reduces (Goosen 1997, Domzsy et al. 1983).
Since the discovery of chitin, the use of chitin and chitosan has expanded over almost all fields of research (Muzzarelli 2005). Chitin and chitosan can be modified using different

chemical modifications e.g., nitration, acetilation, sulphonation, phosphorylation, etc (De Smedt et al. 2000). A broad spectre of derivates from chitosan can be gained using chemical modification which can then be used in various areas (Fig. 5).

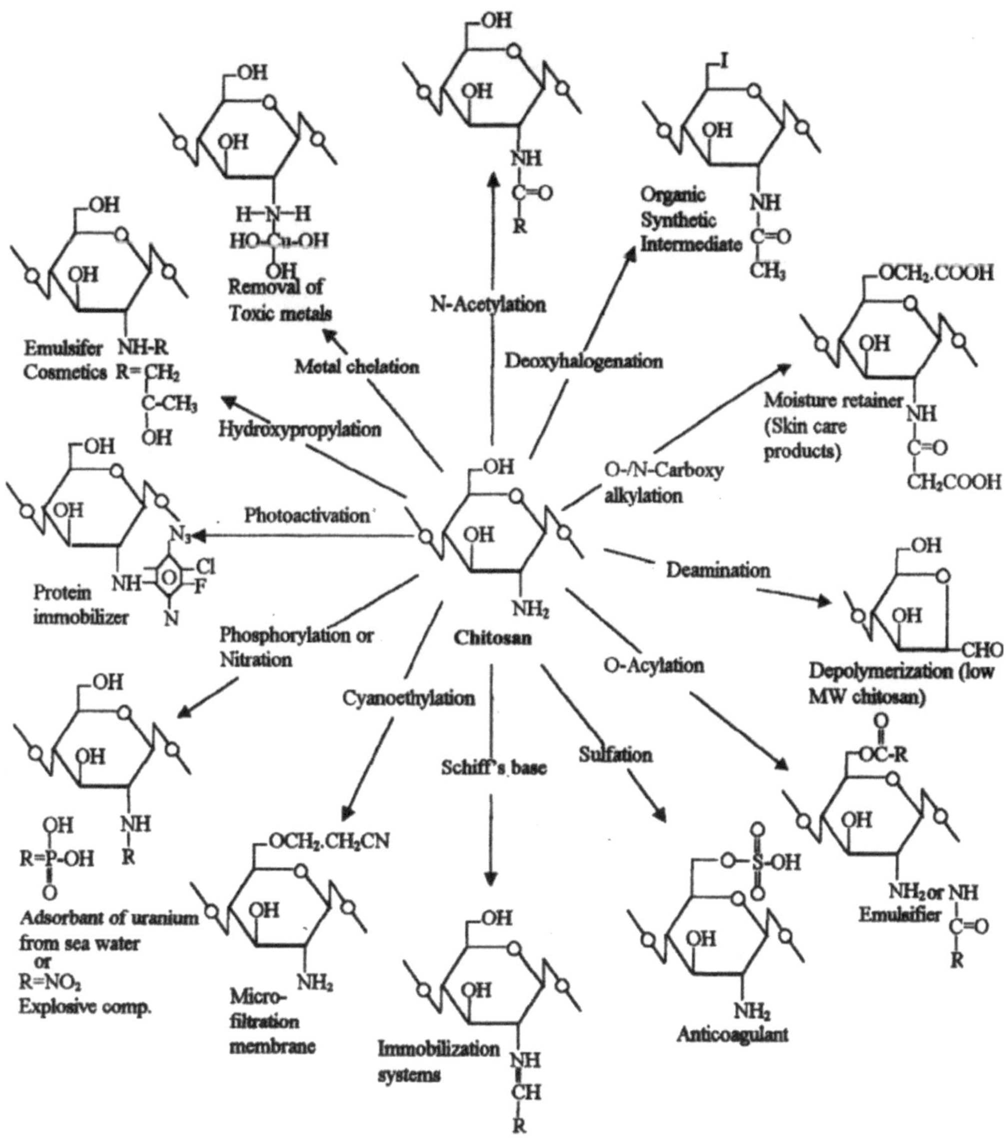

Fig. 5. Chitosan derivates and their potential use (Tharanathan 2003)

Biologically important characteristics have been found using some of these derivates, such as specifically immune, anticoagulant, spermicidal, fungistatic, bacteriostatic as well as flocculation properties (Rinaudo 2006, Smart et al. 2006, Chi et al. 2006, An at al. 2009, Kumar et al. 2010). As such chitosan derivatives can be used in ecology (for cleansing waste waters) as well as in nutritional, medical and pharmaceutical areas.

5. Cellulose fibres' surface functionalization

The use of natural cellulose fibres for developing medical materials has recently gained considerable attention, as emphasized by the numerous publications on that topic. Several methods to modify cellulose fibres for medical applications are described in literature: (i) oxidation procedures, (ii) synthesis of microbial cellulose, (iii) incorporation of metallic nanoparticles, and iv) various coating strategies at the finishing stages using quaternary ammonium compounds, PHMB (polyhexamethylene biguanides), triclosans, regenerable N-halamine and peroxyacid, some synthetic dyes, etc. (Edward et al. 2000, Browning 1967, Kaputskii et al. 2005, Kotel'nikova et al. 2003, Czaja et al. 2006, Hoenich 2006, Ravi 2000, Lim 2004). Besides the various fibrous, hydrogel (i.e. carboxymethyl cellulose) and alginate - based products, chitin and its derivate chitosan are currently the most popular as antimicrobial coatings for cellulose fibres. Chitosan offers many advantages over traditional cotton and regenerated cellulose fibres treatments because of its non-toxicity, biodegradability and biocompatibility (Fras Zemljič et al. 2009). Chitosan can be used for the introduction of active amino groups onto/into cellulose fibers, and in this way, for cellulose fibers' surface functionalisation. Similarities between the chemical and molecular structures of cellulose and chitosan enable high affinity between both polymers. The more possible cellulose-chitosan intermolecular interactions are based on H-bonds and Van der Waals forces, however, ionic and/or covalent bonds can be formed under special conditions and cellulose treatments.

The manipulation of chitosan binding strategies onto cellulose surfaces i.e. a) reversible binding, which enables the release of a chitosan from a fibre's surface and, b) irreversible binding resulting in the permanent bioactivity of a fibrous (textile) surface, enable the achieving of functional fiber surfaces for different applications.

5.1 Irreversible binding

For conventional textile applications, irreversible binding means that the substance-fiber binding is permeable to washing procedures. It is extremely important for irreversible attachment to introduce carboxyl or aldehyde groups into/onto the fibres which are then potential anchoring sites for chitosan molecules. In this way, electrostatic attraction is assured between cellulose fibres as host material, and chitosan as adsorbent. It has been confirmed that carboxyl groups in the cellulose have different effects on chitosan binding in comparison with aldehyde groups (Browning 1967).

In order to achieve permanent antimicrobial properties, the procedures for textile treatments using chitosan have to fulfill two rather contradictory conditions,: 1) a satisfactory amount of anchoring sites in/on the fibers, onto which chitosan could permanently bind and, at the same time 2) a satisfactory amount of free amino groups of chitosan, which would be responsible for antimicrobial activity. The more appropriate activation procedures of cellulose fibres for permanent chitosan binding are those which introduce aldehyde groups. In this way, chitosan could be chemically attached onto cellulose. There are several procedures for obtaining aldehyde groups in cellulose, but the most useful are oxidation procedures using different specific oxidising agents (Browning 1967).

According to previous studies oxidation procedures with potassium or sodium iodatum (VII) (KIO_4) under special conditions lead to an increase in the cellulose reductive (aldehyde) group's amount on C2, C3 and C6 glycoside atoms (Browning 1967). There are also small amounts of aldehyde end-groups on C1, which can be further oxidized to carboxyles (Fig. 6). The contents of the reducing groups in cellulose fibres increase with increasing oxidation time, and with KIO_4 solution concentration.

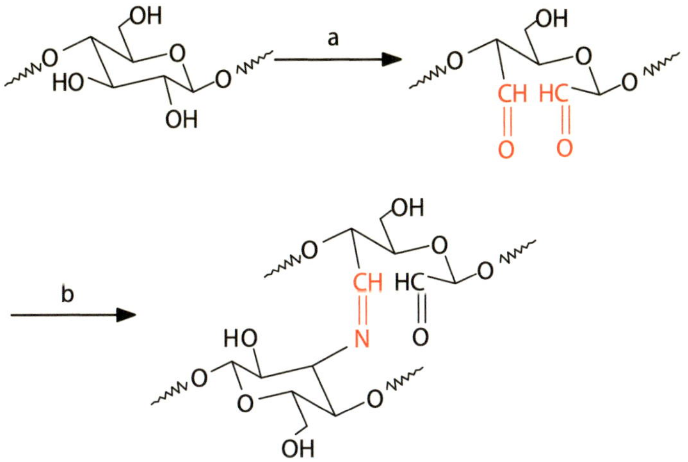

Fig. 6. Scheme of oxidation (a) and chitosan binding (b) onto cellulose

During the binding of chitosan onto oxidised fibres, the aldehyde groups on the fibres surface were reduced, due to the formation of Schiff's base between the chitosan and oxidised cellulose (Fras Zemljič et al. 2009). However, an increase of accessible amino groups was expected in cotton samples (non-oxidised and oxidised) coated with chitosan. A variety of techniques can be used for determination of accessible amino groups' amounts in cellulose fibres. One of them is spectroscopic determination of dye adsorption for C.I. Acid Orange 7 into cotton fibres when coated with chitosan, and the second very accurate method Polyelectrolyte titration and Acid Orange 7 methods showed that cotton samples previously activated by oxidation using potassium or sodium iodatum after chitosan treatment possessed lower amount of amino groups than non-activated fibres. This is expected since presence of aldehyde groups in fibres leads to the formation of electrostatic bonds with accessible amino groups of chitosan. Moreover, both techniques also showed that the amounts of the amino groups in the cotton sample, functionalised by a higher molecular weight of chitosan, were higher than in the cotton sample functionalised with low molecular weight chitosan. These results were in accordance with the zeta potentials of the fibres' surfaces which were calculated from measurements of streaming potential, as a function of pH (Fig. 7).

Pretreated cotton fibre showed typical ZP=f(pH) function for cotton, which was negative in practically the whole pH region, with a ZP plateau value -8 mV. The higher amount of aldehyde groups in the oxidised samples influenced the higher negative values of ZP in the plateau of these samples. After chitosan treatment, the functions $\zeta = f(pH)$ showed typical amphoteric characteristics. At higher pH regions, the negative streaming potential is due to the dissociation of cotton carboxyl groups. Protonation of amino groups in acidic region causes the sign of the streaming potential to change into positive streaming potential at the isoelectric point.

After chitosan treatment the shifting of isoelectric point into higher pH region was observed. This is expectable since chitosan coatings of fibres lower the acidity of the cellulose fibre surface. On another hand, adsorbed chitosan compacts the fiber surface and thereby reduces the hydration of the fibre interior (Čakara 2009) and consequently fibre swelling, which increase the negative ZP values in plateau values (see Figure 7).

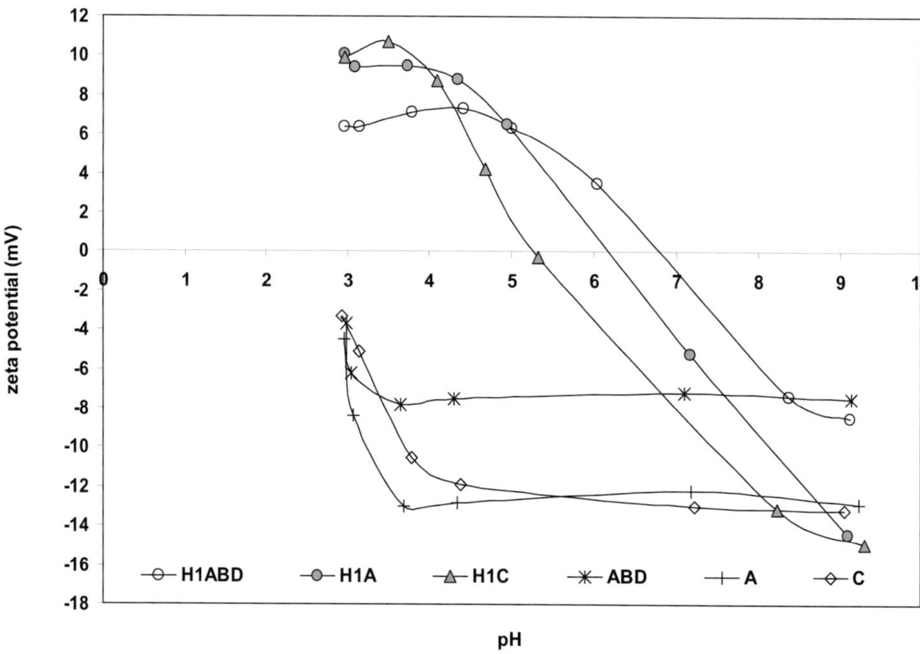

Fig. 7. Zeta potential (ζ) as a function of pH (ζ = f(pH)) for pretreated (ABD) and KIO_4 oxidized (A - 24 h, with 2,3 mg/mL KIO_4 and C - 72 h, with 2,3 mg/mL KIO_4) cellulose samples and the same samples, treated with chitosan H1

It was found out in the same research, that oxidation procedures using potassium iodatum had a large influence on fibres' mechanical properties. Even after a relatively short time of oxidation at room temperature the breaking force of the fibres decreased by about 34,5 % in comparison with non-oxidised samples (Strnad et al. 2008).

The second more appropriate way of permanent chitosan binding onto cellulose fibres are the reactions between chitosan amino groups and cellulose carboxyl groups. Several methods available in literature describe how cellulose fibres are modified in order to obtain carboxyl groups: oxidation procedures, adsorption of anionic polymers and dyes, plasma activation, etc. (Browning 1967, Kaputskii et al. 2006, Kotel'nikova et al. 2003, Fras Zemljič et al. 2006, Vesel et al. 2010).

In our previous research work, fibres were oxidized selectively as well as non-selectively with the aim of obtaining a different content of carboxyl groups (Fras Zemljič et al. 2004). The Methylene Blue method and Complexometric titration, which are based on the ion-exchange capacity of the cellulose, were used for determining the contents of the carboxyl groups. It was found that carboxyl groups' contents for oxidized fibers depend on the oxidation procedure conditions (concentration of oxidizing agent, time, pH, etc). Increasing the carboxyl groups' contents was more expressive using a non-selective oxidation in comparison to selective oxidation, where the oxidation mechanism is known. During non-selective oxidation, any possible oxidation products may occur because the course of the oxidation process is unpredictable. A mixture of products is present oxidized to different degrees and according to different mechanisms. The DP (degree of polymerization) of cellulose fibres was drastically reduced. On the basis of extensive studies it was established that, due to extreme worsening of fibre mechanical properties, oxidation is an inappropriate tool for fibre activation. It has become clear that different standard chemical processes such as oxidation procedures usually modify the fibres both structurally and chemically and, in addition to frequently being ecologically less desirable, are detrimental to the fibres' mechanical properties (Liu 2007).

Therefore, over the last decade, the tendency of fibre functionalisation has been aimed at using natural, biodegradable, and non-aggressive chemicals. Polysaccharides are currently the most promising among them as antimicrobial coatings for cellulose fibres (Dimitriu 2002).

It has been shown that modification of fibres by the adsorption of carboxymethyl cellulose (CMC) introduces new carboxyl groups onto the fibres' surfaces. The increase of charge was determined indirectly by phenol-sulphuric acid method, and directly by conductometric titration. It was discovered that the total charge of cotton increased by more than 50 % in the case of all used CMC products. Even more, the modified fibres appeared to have better mechanical properties (Fras Zemljič et al. 2006) (Table 3).

Sample	**Titer** dtex	**Tenacity** cN/tex	**Elongation** %	**Force** cN	**Young modulus** cN/tex
Non-modified fibres	1.84	31.94	11.61	5.66	105
CMC-modified fibers	1.94	33.24	13.57	6.39	82

Table 3. Mechanical properties of CMC-functionalized fibres

The titer and tenacity of CMC-modified fibers can be increased by about 5 %, while with respect to elongation and force, about a 15 % increase is obtained in comparison with non-modified fibers (reference sample). Moreover, because of the capacity of cotton fibres to adsorb cationic surfactants, the rate of adsorption is increased by the previously adsorbed carboxymethyl cellulose (CMC) on the fibres' surfaces (Fras Zemljič et al. 2006), see Fig. 8.

The adsorption of cationic surfactant onto fibres increases as the anionic charge on the fibres increases (due to CMC adsorption). Due to the attraction between surfactant cations and dissociated fibre carboxyl groups, a higher charge density on the fibres leads to a higher and faster adsorption of surfactant (Fras Zemljič et al. 2006).

This procedure of fibre anionization may be due to extremely anionic charge increase and improved mechanical fibre properties at the same time, very promising for irreversible chitosan binding. This work is presently in progress.

Another possibility of introducing carboxyle groups into cotton fibres is treatment with 1, 2, 3, 4 buthanetetracarboxylic acid (BTCA) (El-ahlavy 2005). Our research results showed that an adequate number of free carboxyl groups can be assured with the treatment of cellulose fibres by this anionic crosslinking agent. BTCA treatment played an important role in chitosan fixation and chitosan treatment durability. The amount of carboxyl groups introduced into cellulose fibers strongly depends on the BTCA concentration in the impregnation solution. In combination with an appropriate catalyst (e.g. $NaH_2PO_2.H_2O$) a satisfactory amount of accessible carboxyl groups is achieved. The carboxyl groups' amounts in the BTCA treated samples may be determined using the Sobue-Okubo modified complexometric titration method (Fras Zemljič 2004). The results of complexometric titrations showed that after treatment with 7 % BTCA solution, the cellulose contained a

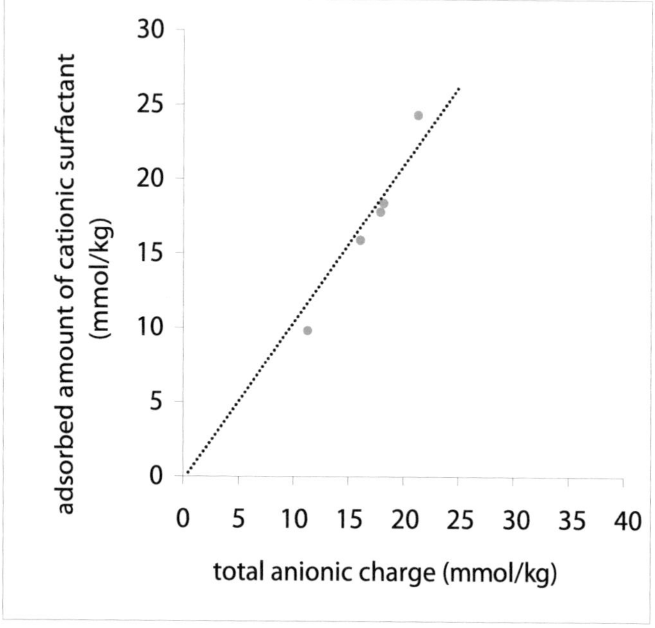

Fig. 8. The adsorbed amount of cationic surfactant (determined by spectroscopy) as a function of the total charge of the fibres, as determined by conductometric titration.

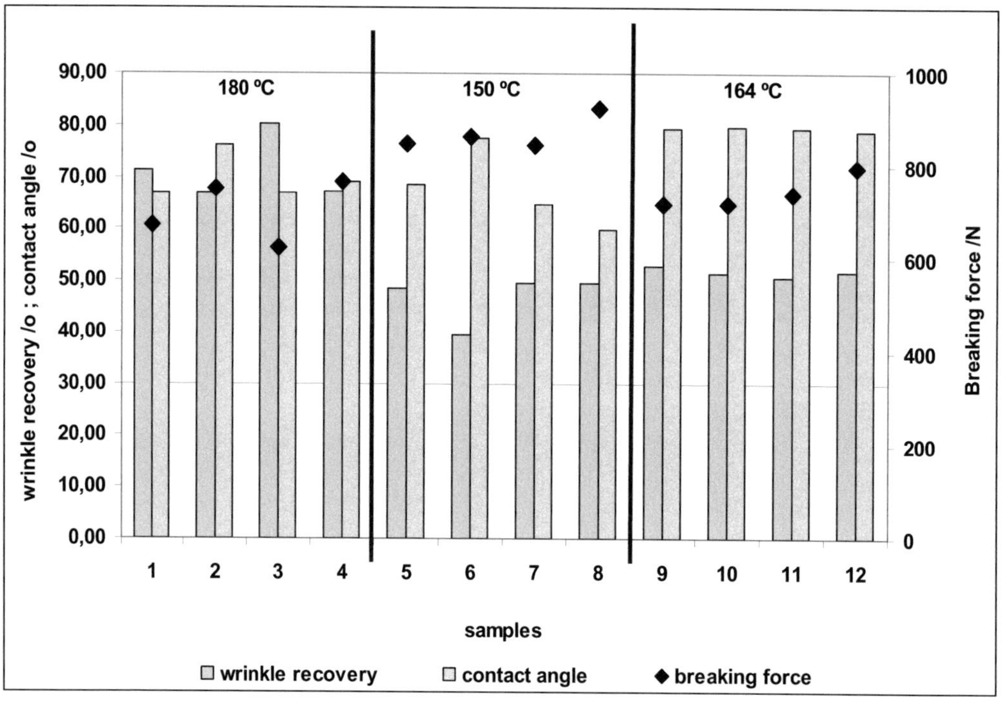

Fig. 9. The influence of process conditions on the wrinkle recovery, contact angles and breaking force of BTCA and chitosan treated cotton fibers

higher amount of carboxyl groups when compared to the 3 % BTCA solution. The BTCA treatment, on the other hand, improved the mechanical properties of the samples (Fig. 9). Chitosan was adsorbed onto previously BTCA-treated fibres, which gave the same tendency regarding results as in the case of oxydised fibres i.e. reduction of the amounts of accessible amino groups on the chitosan treated fibers due to interactions with carboxyle groups of fibres. However, it has been discovered, that BTCA- treated samples, showed in comparison with non treated material, a much better washing resistency of chitosan.

Gas plasma treatment of fibres is the next possibility of cellulose fibre surface activation for later chitosan adsorption. Cold plasma treatment is an extremely versatile technique for modifying the polymer surfaces of totally different shapes (Clark 1979 et al., Wertheimer et al. 1996, Kunaver et al. 2004, Mozetic 2004, Morales et al. 2006, Sharnina et al. 1996, Tu et al. 1994, Prabaharan et al. 2005). It has been reported that plasma treatment can improve polymer-polymer adhesion (Clark et al. 1979, Wertheimer et al. 1996), the best results being obtained when using oxygen plasma (Carlsson 1991, Felix et al. 1994, Couto et al. 2002). The purpose of our investigation (Fras Zemljič et al. 2008) was evaluating to what extent oxygen plasma treatment could improve the chitosan adsorption, and in this way, the antimicrobial activities of chitosan covered fabrics. The result of surface activation using oxygen plasma is the formation of different oxygen-containing polar functional groups such as C-O, C=O, and O-C=O, which act as nucleophilic centres to which adsorbent atoms can bind. The influence of low pressure oxygen plasma treatment on the functionalization of cellulose material using chitosan was investigated by conductometric titration and XPS. The results were supported by conventional Kjeldahl analysis. The effect of plasma treatment on the antimicrobial capacity of chitosan treated fabrics was studied according to the ASTM E2149-01), which is a quantitative antimicrobial test method performed under dynamic contact conditions (Fras Zemljič et al. 2008). Both conductometric titration and XPS indicated that chitosan adsorbed onto the viscose fabric, irrespective of whether it had been activated with plasma or not. However, the amounts adsorbed on plasma-treated samples were significantly higher than those on non-treated samples (Fig. 10), and increased with increasing chitosan concentration in the solution used for treatment.

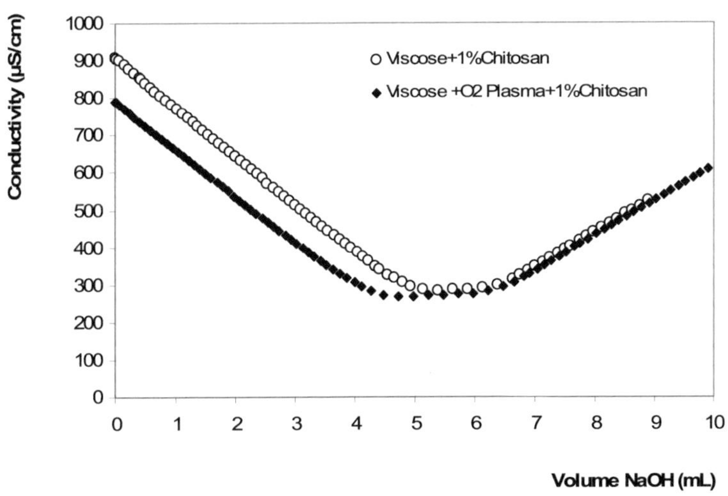

Fig. 10. Conductometric titration curve of viscose fibers (activated and non-activated by oxygen plasma), modified with 1 % chitosan solution

Increasing the number of active functional groups due to plasma activation in the cellulose surface enabled the adsorption of higher amounts of chitosan and, consequently, a higher amount of amino groups responsible for antimicrobial activity (Lim 2004, Ravi Kumar 2000, Liu et al. 2007, Zitao et al. 2003). The higher amount of amino groups in plasma-activated–chitosan-treated samples increased the probability that a protonated amino group met the bioplasm of the bacteria and, in this way, resulted in a greater bacterial reduction capacity (Zitao et al. 2003).

Sample treatment	*R*, %				
	Pathogenic bacteria			**Pathogenic fungi**	
	Staphylococcus aureus	*Escherichia coli*	*Streptococcus agalactiae*	*Candida albicans*	*Candida glabrata*
Non-treated viscose fabric	0	0	99	20	34
Plasma treated fabric (24 h after treatment)	0	0	80	51	21
Non-treated fabric impregnated by 1 % chitosan solution for 24 h	70	0	100	0	98
Plasma treated fabric impregnated by 1 % chitosan solution for 24 h	91	0	100	100	100

Table 3. Antimicrobial properties of fibres impregnated by chitosan; plasma pre-activated and withaout plasma pre-activation

Table 3 shows that the plasma activation for sample treated using 1 % of chitosan solution, affected the degree of bacterial reduction R. *R* for *Staphylococcus aureus* increased from 70 % to 91 % and for *Candida Glabrata* from 98 % to 100 %. Even more, the plasma-activated-chitosan-treated samples showed a total reduction in *Candida Albicans*, for which, in the case of the non-activated sample $R = 0$ % (Fras Zemljič et al. 2008).

It has to be pointed out that several past researches by our group showed that, in some cases, when chitosan interacts with non-activated cellulose, adsorption could also be irreversible. This is somehow expected due to the fact that, cotton fibres especially are weakly acidic due to pre-treatments such as scouring and bleaching (Fras Zemljič et al. 2008) and cationic polymers may be readily adsorbed onto fibres by electrostatic attraction. However, Čakara et al. 2010 showed that irreversible adsorption of chitosan onto weakly acidic cotton fabric is, under present conditions, predominately driven by a non-electrostatic attraction. Myllitye et al. 2009 evidenced a non-electrostatic interaction between chitosan and cellulose at low pH. This may be attributed to specific structural interaction between chitosan and cellulose (H-bonds and hydrophobic interactions). In the same way, several neutral or even anionic polysaccharides, including xyloglucan and carboxymethy cellulose are irreversibly adsorbed onto cellulosic fibres as well (Laine et al. 1994).

For cellulose fibres functionalised by chitosan (irreversible binding), the charging isotherms exhibit a charge reversal around pH≈6, which is identified as the point of zero charge (PZC). These fibres protonate according to the one-p*K* model with two determined p*K* values such as $pK_{(chitosan)}=6.3$ and $pK_{(acidic\ cotton\ groups)}= 4.7$. p*K* is not only important for classifying acid strength but also determines the properties of a substance in nature or its possible use as a drug. Determination of the p*K* value is therefore of great importance for biomedical applications as, for example, medical textiles in wound healing, gynaecological treatment, etc. (Čakara 2009).

5.2 Reversible binding of chitosan

Chitosan solubility is pH-dependent, therefore, changes in pH could be applied for a tuning of adsorption/desorption of chitosan by changes from soluble to insoluble forms of it.

One of the latest researches of our group is based on comparing the adsorption of totally soluble chitosan (from acidic solution) against the adsorption of precipitated chitosan, onto cellulose fibres. This topic was investigated by ATR-FTIR, potentiometric titration, and the conventional spectrophotometric method using C.I. Acid Orange 7 adsorption. The binding of precipitated chitosan leads to a min. 70 % of chitosan desorption. However, the functionalization of viscose fibres using precipitated chitosan is, in a contrast to the functionalization of fibres with acidic chitosan solution, more efficient in the sense of active (accessible) amino groups' amounts and, in this way, in antimicrobial activity.

Cellulose fibres, treated using chitosan precipitation showed a slightly increased reduction of pathogenic bacteria (*S. aureus)* and fungus *(C. albicans)* in comparison with the fibres treated by the adsorption of totally dissolved chitosan from acidic solution.

The most popular approach for reversible chitosan/fibre binding is currently the preparation of chitosan micro/nanoparticles. These can act as an antimicrobial agent itself or as drug carrier in delivery systems. There are several ways for synthesising chitosan nanoparticles, such as emulsification, oppositely-charged polymers' precipitation, the ionic gelation process etc. (Wan Ajun et al. 2008).

Chitosan nanoparticles showed high cytotoxic activity toward tumor cells, while low toxicity against normal human liver cells (L-02) (Qi, et al. 2005). Furtehrmore, chitosan nanoparticles showed high sorption capacity and anti-bacterial activity (Qi & Xu, 2004). The unique cationic character of chitosan nanoparticles could provide higher affinity to negatively-charged biological membranes and site-specific targeting in vivo (Qi et al., 2004). Particle size had an substantial influence on increasing their anti-tumor efficacy when chitosan nanoparticles are applied by intravenous injection (Qi et al., 2005). The unique character of the positive charge and small particle sizes of chitosan nanoparticles is responsible for their in vivo efficacy (Qi & Xu 2006).

If certain preparatory conditions are used, the resulting nanoparticles prove to be highly effective for providing new functionalities to the host material, such as with cellulose fibres. The treatment of cotton with poly(n-butyl acrylate) cores and chitosan particles confers the fabric with excellent antibacterial property. Other characteristics of particle-coated fabrics including mechanical properties, air permeability, hand feel and antibacterial durability over repeated launderings are reported as well. Using chitosan microcapsules/nanocapsules containing several extracts oil were performed to introduce antimicrobial properties. Moreover, the pleasant fragrance of aromatic oils was retained on fibre surfaces at least six months (Alonso et al. 2010, Ye et al. 2005). It has been shown that cellulose coated by chitosan nanoparticles describe a cost effective and non-toxic methodology for new textiles developing. There are several possible interactive mechanisms (Fig. 11) between cellulose and chitosan nanoparticles.

In the future, this work will have implications in the design of medical textiles adsorbed by vehicles for the loading of several drugs as targeted drug delivery, protection from enzymatic degradation, and reduced drug toxicity or side effects. This may present therapy for solving several skin and several other mucous membranes problems (vaginal, mouth, etc).

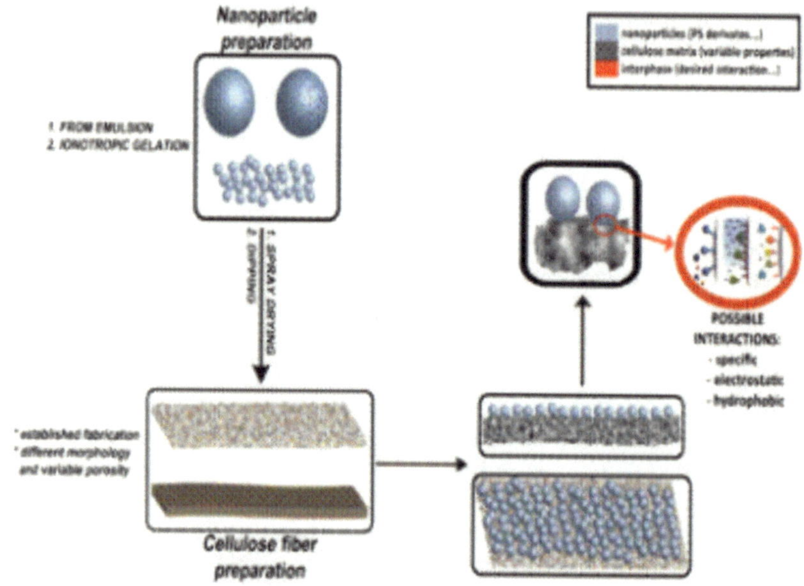

Fig. 11. Possible interaction mechanisms between cellulose and chitosan nanoparticles

5. References

Albrecht, W., Reintjes, M., Wulfhorst, B.; (1997). Lyocel Fibres. Chem Fibres Int. 47, 289-304

Alonso, D., Gimeno, M., Olayo, R., Vasquez-Torres, H., Sepulveda-Sanchez, JD, Shirai, K., (2009). Cross-linking chitosan into UV-irradiated cellulose fibers for the preparation of antimicrobial-finished textiles. *Carbohydrate Polymers*, 77, 536-543

An, NT; Thien, DT; Dong, NT; Le Dung, P, (2009). Water-soluble N-carboxymethylchitosan derivatives: Preparation, characteristics and its application. *Carbohydrate Polym*, 75, 3: 489-497

ASTM E2149 - 01 *Standard Test Method for Determining the Antimicrobial Activity of Immobilized Antimicrobial Agents under Dynamic Contact Conditions*

Berger, W., (1994). Möglichkeiten und Grenzen alternativer Verfahren zur Celluloseauflösung und Verformung. *Chemifas. Textilind.* 44, 747-750

Berger, J., Reist, M., Mayer, J. M., Felt, O., Peppas, N. A. and Gurny R., (2004). Structure and interactions in covalently and ionically crosslinked chitosan hydrogels for biomedical applications. *European Journal of Pharmaceutics and Biopharmaceutics*, 57,1, 19-34

Browning, B. L., (1967). *In Methods of Wood Chemistry*, Interscience Publisher, New York

Carlsson, G.; Ström, G., (1991). *Langmuir*, 7, 2492-2497

Chang, J. C. S., Foardee, K. K., Van Osdel D., W.; (1996) Assessment of fungal (*Penicillium Crysogenum)* growth on three HVAC duct materials; *Environmental International*, 22; 4; 425-431

Chi FH, Cheng WP, (2006), Use of chitosan as anticoagulant to treat wastewater from milk processing plant, *J Polym Environ*, 14, 4 , 411-417

Chung, Y. S.; Lee, K.K.; Kim, J.W.; (1998) Durable Press and Antimicrobial Finishing of Cotton Fabrics with a Citric Acid and Chitosan Treatment. *Textile Research Journal*, 68, 10, 722-755

Clark, D. T.; Shuttleworth, D., (1979). *Journal of Polymer Science: Polymer Chemistry Edition, 17*, 957-976.

Cole, D.J.; (1994). Courtaulds Tencel Fibre in Apparel fabrics. Lenz Ber, 74, 45-48

Cook, G., (1984) Handbook of Textile Fibres, Man-made Fibres, Merrow, Durham, UK

Couto, E., Tan I. H., Demarquette, N., Caraschi J. C., Leao A., (2002). Oxygen plasma treatment of sisal fibres and polypropylene: Effects on mechanical properties of composites. *Polymer Engineering and Science*, 42, 4, *790-797*

Czaja, W., Krystynowicz, A., Bielecki Sand Brown, R. M., (2006). Microbial cellulose- the natural power to heal wounds. *Jr. Biomaterials*, 27, 2, 145 - 151

Čakara, D., Fras Zemljič, .L., Bračič, M., Stana-Kleinschek, K. (2009). Protonation behavior of cotton fabric with irreversibly adsorbed chitosan : A potentiometric titration study. *Carbohydr. polym.*, 78, 1, 36-40

De Gruy, I. V.; Carra, J. H.; Goynes, W.R.; (1973). An atlas of cotton microscopy, The fine structure of cotton, O'Connor, R.T. (Ed.), 1-224, Marcel Dekker Incorporation, ISBN: 0-8247-6109-X, New York

De Smedt, S.C., Demeester, J., Hennink, W. E., (2000). Cationic polymer based gene delivery systems. *Pharmaceut. Res.*,17, 113-126

Domszy, JG., Moore GK., Roberts GAF, (1983) The Adsorption of Chitosan on Cellulose, Department of Physical Sciences, Trend Polytechnic, Notthingham England

Domitriu, S. (2002). *Polymeric Biomaterials*. Second edition, Revised and Expanded, Marcel Dekker, Inc. New York, 1-213

Edward, J. V., Buschle-Diller, G.; Goheen, S. C., (2006). *In Modified Fibres with Medical and Specialty Application,* Springer, The Netherlands

El-ahlavy, K. F., El-bendary, M. A., El-hendavy, A.G., Hudson, S.M., (2005). The antimicrobial activity of cotton fabrics treated with different crosslinking agents and chitosan. *Carbohydrate Polymers*, 60, 4,. 421-430

Felix, J. M., Carlsson, C. M. G., Gatenholm, P., (1994), *Journal of Adhesion Science and Technology*, 8, 2, 163-180

Fras Zemljič, L., Stana-Kleinschek, K., Ribitsch, V., Sfiligoj-Smole, M., Kreže, T. (2004). Quantitative determination of carboxyl groups in cellulose polymers utilizing their ion exchange capacity and using a complexometric titration. *Mater. res. innov.*,8, 3, 145-146

Fras Zemljič, L., Stenius, P., Laine, J., Stana-Kleinschek, K., Ribitsch, V., (2006). Characterization of cotton fibres modified by carboxylethyl cellulose. Lenzing. Ber., 85, 68-76

Fras Zemljič, L., Peršin, Z., Stenius, P., Stana-Kleinschek, K. (2008). Carboxyl groups in pre-treated regenerated cellulose fibres. *Cellulose*, 15, 5, 681-690

Fras Zemljič, L., Strnad, S., Šauperl, O., Stana-Kleinschek, K., (2009). Characterization of amino groups for cotton fibers coated with chitosan. *Textile Research journal*, 79, 3, 219-226

Gao, Y., Cranston, R., (2008). Recent advances in antimicrobial treatments of textiles. *Textile research Journal*, 78, 1, 60-72

Goosen, M. F. A., (1997). Applications of Chitin and Chitosan. Technomic publishing company Inc. ISBN 1-56676-449-1, Lancaster Pennsylvania, USA

Hoenich, N., (2006). Cellulose for medical applications. *BioResources*, 1, 2, 270-280

Kaputskii, F. N., Gert, E. V., Torgashov, V. I.; Zubets, O. V., (2005). *Fibre Chemistry*, *37*, 485-489

Kaputskii, F. N., Gert, E. V., Torgashov, V. I., Zubets , O.V., (2006). Combination of oxidative and hydrolytic functions of nitric acid in production of enterosorbents based on carboxylated microcrystalline cellulose. *Fibre Chemistry*, 37, 417-504

Kenawy, E. R., Worley, S. D., Broughton, R. (2007) The chemistry and applications of antimicrobial polymers: A state-of-the-art review; Biomacromolecules, 8, 5, 1359-1384

Knittel, D., Schollmeyer, E. (1998). Chitosan und seine Derivate für die Textilveredlung. *Textilveredlung*, 33, 3-4, 67-71

Kotel'nikova, N. E., Wegener, G., Paakkari, T., Serimaa, R., Demidov, V. N., Serebriakov, A. S., Shchukarevand, A. V., Gribanov, A. V., (2003). Silver Intercalation into Cellulose Matrix. An X-Ray Scattering, Solid-State ^{13}C NMR, IR, X-Ray Photoelectron, and Raman Study, *Russian Journal of General Chemistry*, 73, 3, 418-426

Krässig, H.A.; (1993) Cellulose Structure, Accessibility and reactivity, Polymer Monographs, Huglin M.B. (Ed.), 1-375, Gordon and Breach Science Publisher, ISBN: 2-88124-798-9, Yverdon, Switzerland

Kreže, T., Jeler, S., Strnad, S. (2002) Correlation between structure characteristics and adsorption properties of regenerated cellulose fibres. *Mater. res. innov.*, May, 5, 6, 277-283

Kreže, T., Malej-Kveder, S. (2003). Structural characteristics of new and conventional regenerated cellulosic fibres. *Tex. res. j.*, 73, 8, 675-684

Krüger, R.; (1994). Cellulosic filament yarn from the NMMO Process. Lenz Ber 74, 49-52

Kumar, S.; Nigama, N.; Ghosh, T.; Dutta, P.K., Singh, S.P., Datta, P.K., An, L.J.; Shi, T.F., (2010). Preparation, characterization and optical properties of a novel azo-based chitosan biopolymer. *Materials Chemistry And Physics*, 120, 2-3, 361-370

Kunaver, M., Mozetic, M., Klanjsek-Gunde, M., (2004). Selective plasma etching of powder coatings. *Thin Solid Films*, 459, 1-2, 115-117

Laine, J., (1994). In *dissertation*, Department of Forest Products Technology; Helsinki University of Technology, Espoo, Finland, 208-214

Lee, S., Cho, J.S., Cho, G., (1999) Antimicrobial and Blood Repllent Finishes for cotton and Nonwoven Fabrics Based in Chitosan and Fluoropolymers, *Textile Res. Journal*, 69, 2, 104-112

Lim, SH., Hudson, SM. Review of chitosan and its derivatives as antimicrobial agents and their uses as textile chemicals. *Journal of macromolecular science*, part C – polymer reviews, 2003, C43, 2, 223-269

Lim, S. H., Hudson, S. M., (2004). Synthesis and antimicrobial activity of a water-soluble chitosan derivative with a fiber-reactive group . *Carbohydrate Research*, *339*, 2, 313-319

Liu, X.F., Guan, Y.L., Yang, D.Z., Li, Z., and Yao, K.D., (2000) Antibacterial action of chitosan and carboxymethylated chitosan. *J. Appl. Polym. Sci.* 79 7, 1324–1335,

Liu, J., Wang, Q., Wang, A., (2007). Synthesis and characterization of chitosan-g-poly(acrylic acid)/sodium humate superabsorbent. *Carbohydrate Polymers*, 70, 166-173

Majeti, NV., Ravi Kumar, MNV. A review of chitin and chitosan applications. *Reactive & Functional Polymers*, 2000, vol. 46, no. 1, p. 1-27

Morales, J., Olayo, M. G., Cruz, G.J., Herrera-Franco, P., Olayo, R., (2006). Plasma modification of cellulose fibers for composite materials. *Journal of Applied Polymer Science*, 101, 6, 3821-3828

Muzzarelli, R. A. A., Muzzarelli, C., (2005). Chitosan Chemistry: Relevance to the Biomedical Sciences; in: Polysaccharides I, Structure, Characterisation and Use, T. Heinze (Ed.), Springer-Verlag Berlin Heidelberg,

Myllyte, P., Salmi, J., Laine j., (2009). The influence of pH on the adsorption and interaction of chitosan with cellulose. *BioR esurces*, 4, 4, 1674.

Prabaharan, M., Carneiro, N., (2005). *Indian Journal of Fibre & Textile Research*, 68-75

Qi, L. F., Xu, Z. R. (2004). Lead sorption from aqueous solutions and chitosan nanoparticles. *Colloids Surfces*, 251, 183-190

Qi, L. F., Xu, Z. R. (2006). In vivo antitumor activity of chitosan nanoparticles. *Bioorganic and Medical Chemistry letters*, 16, 4243-4245

Qi, L. F., Xu, Z. R., Jiang X., Hu, C. H., Zou, X. F., (2004). Preparation and antibacterial acitivity of chitosan nanoparticles. *Carbohydrate Research*, 339, 2693-2700

Qi, L. F., Xu, Z. R., Jiang X., Li, Y., Wang, M. Q., (2005). Cytotoxic activities of chitosan nanoparticles and copper-loaded nanoparticles. , *Bioorganic and Medical Chemistry letters* 15, 1397-1399

Rajendran, S., Anand, S. C., Harrison, P. W., (2002) Developments in medical textiles: A critical appreciation of recent developments. *Textile Progress*. 32, 4,

Rajesh D. A., (2006). Role of advanced textile materials in healthcare, Medical Textiles and Biomaterials for Healthcare, Anand, S.C.; Kenndy, J.F.; Miraftab, M.; Rajendran, S.; (Eds.), 90-98, Woodhead Publ Ltd, ISBN 1-85573-683-7, Cambridge, England

Ravi Kumar, M. N. V., (2000). *Reactive &Functional Polymers*, 46, 1-27

Rinaudo, M., (2006). Chitin and chitosan: Properties and applications, Prog In Polym Sci, 31, 7: 603-632

Roberts, A. F., Baxter, A., Dillon M.l, Anthony Taylor, K. D., (1992). Improved method for I.R. determination of the degree of *N*-acetylation of chitosan. *International Journal of Biological Macromolecules*, 4, 3, 166-169

Smart, G; Miraftab, M; Kennedy, JF; Groocock, MR (2006). Chitosan: Crawling from crab shells to wound dressings, In: Medical Textiles and Biomaterials for Healthcare, Anand, S.C.; Kenndy, J.F.; Miraftab, M.; Rajendran, S.; (Eds.), 67-72, Woodhead Publ Ltd, ISBN 1-85573-683-7, Cambridge, England

Sharnina, L. V., Mel'nikov, B. N., Blinicheva, I. B., (1996). *Fibre Chemistry, 28, 4,* 8-51

Strnad, S., Šauperl, O., Fras Zemljič, L., Jazbec, A., (2007), Chitosan - universally applicable polymer, *Tekstilec*, 50, 10-12,

Strnad, S., Šauperl, O., Jazbec, A., Stana-Kleinschek, K., (2008). Influence of chemical modification on sorption and mechanical properties of cotton fibers treated with chitosan. *Textile Research Journal*, 78, 5, 390-398

Synowiecky, J., Al-Khateeb, A. N. (2003) Production, Properlincages, and some new applications of chitin and chitosan, *Critical Reviews in Food Science and Nutrition*, 43, 2, 145-171

Šauperl. O., Stana-Kleinschek, K., Ribitsch, V., (2009). Cotton cellulose 1, 2, 3, 4 buthanetetracarboxylic acid (BTCA) crosslinking monitored by some physical-chemical methods. *Tex. res. journal*, 79, 9, 780-791

Takai, K., Ohtsuka, T., Senda, Y., Nakao, M., Yamamoto, K., Matsuoka, J., et al. (2002) Antibacterail properties of antimicrobial-finished textile products, *Microbiology and Immunology*, 46, 2, 75-81

Tharanathan, RN., Kittur, FS. (2003). Chitin - the undisputed Biomolecule of great potential. Critical reviews in food science and nutrition, , vol. 43, no. 1, p. 61-87

Tu, X., Young, R. A., Denes F., (1994). *Cellulose,* 1, 1, 87-106

Uragami, T., Okura, S. (Eds.); (2006). Material Science of Chitin and Chitosan, Springer-Verlag Berlin

Vesel, A., Mozetič, M., Strnad, S., Stana-Kleinschek, K., Hauptman, N., Peršin, Z., (2010). Plasma modification of viscose textile. *Vacuum*, 84, 1, 79-82

Wan, A., Sun, Y., Gao, L. Li, H., (2009). Preparation of aspirin and probucol in combination loaded chitosan nanoparticles and in vitro release study, *Carbohydrate Polymers*, 75, 4, 566-574

Wertheimer, M. R.; Martinu, L., Liston, E. M., (1996). In *Handbook of Thin Film Process Technolog*, ED.; D. A. Glocker and S. I. Shah, Bristol, Inst. Of Physics Publishing: Bristol

Zitao, Z., Cheng, L., Jinmin, J., Yanliu, H., Donghui, C., (2003). Antibacterial properties of

10

Chitosan Based Membranes for Separation, Pervaporation and Fuel Cell Applications: Recent Developments

Tina Chakrabarty, Mahendra Kumar and Vinod K. Shahi
Electro-Membrane Processes Division,
Central Salt and Marine Chemicals Research Institute, Council of Scientific & Industrial Research (CSIR), G. B. Marg, Bhavnagar-364002 (Gujarat)
India

1. Introduction

Growing public health and environmental awareness accompanied by increasing number of stricter environmental regulations on wastes discharge. Attention has been focused on the use of biopolymers from renewable resources as alternatives to synthetic polymers [1]. Biopolymers are produced in nature by living organisms and plants, participate in the natural biocycle and are eventually degraded and reabsorbed in nature. The most widespread biopolymers are polysaccharides among them chitosan(CS) is most valuable, whose swellability in water and viscous solution/gel-forming properties were utilized by manufacture for number of industrial and consumer products. CS (primary derivative of chitin) is commercially available basic polysaccharide [2,3]. The basicity of CS responsible for singular chemical and biological characteristics, biocompatibility, antibacterial properties, heavy metal ion chelation ability, gel-forming properties, hydrophilicity, affinity to proteins and good membrane forming capability. In this chapter we will discuss for modification of CS and its exploitation for advance membrane separation applications. The membrane processes were classified by Howell includes [4]:

1. Cleaner industrial process: adsorption, ultrafiltration and electro-ultrafiltration.
2. Energy: fuel cell applications
3. Pervaporation: separation of organic solvents from their azeotropic mixtures.
4. Water: virus–free supply, water reuse and micro-pollutant-free water

Chitosan is obtained by varied extent *N*-deacetylation and characterized by degree of deacetylation (Fig. 1). It is a copolymer of *N*-acetyl glucosamine and glucosamine and insoluble in water. CS readily dissolves in acidic solutions due to the presence of amino groups and 80–85% degree of deacetylation is necessary to obtain a soluble product. Commercially, CS is obtained from low cost shells of shellfish (mainly crabs, shrimps, lobsters and krills), the wastes of the seafood processing industry [2-6]. Chemical and biological properties of CS attributable to the presence of amino and hydroxyl groups [2,3,5-8]. These groups allow chemical modifications of chitosan: acylation, *N*-phthaloylation, tosylation, alkylation, Schiff base formation, reductive alkylation, *O*-carboxymethylation, *N*-carboxyalkylation, silylation, and graft copolymerization [3,9]. Modifications of CS will help for tailor-made materials of specific applications such as adsorption, polymer electrolyte membrane (PEM) for the separation of metal ions, amino acids and protein by adsorption, ultrafiltration, electro-ultrafiltration, fuel cell application and pervaporation. Additionally, amino groups made CS a cationic polyelectrolyte ($pK_a \approx 6.5$), due to its solubility in aqueous acidic media at $pH < 6.5$. When dissolved, CS possesses a high positive charge on $-NH_3^+$ groups and makes aggregates with polyanionic compounds, and chelates with heavy metal ions. CS has drawn particular attention as effective biosorbent, due to its low cost compared to activated carbon and its high amino and hydroxyl functional group contents responsible for high adsorption potential for various aquatic pollutants [9-14]. This biopolymer represents an attractive alternative over other biomaterials because of its physico-chemical characteristics, chemical stability, high reactivity, excellent chelation behavior and high selectivity toward pollutants [15-16]. Natural chitosan was modified by several methods (either physically or chemically) to enhance the adsorption capacity for various types of pollutants. Different shapes of chitosan, e.g. membranes, microspheres, gel beads and films have been prepared and examined for the removal of various pollutants from water and wastewater. CS has been employed as a promising polymeric matrix for DMFC application considering its low cost, desirable alcohol barrier property and proton conductivity as well as adequate thermal stability after cross-linking [17–19]. Reports are available, where chitosan has been used as membrane forming material in pervaporation for the dehydration of alcohols, because of its hydrophilic nature and very high affinity towards water [20-22].

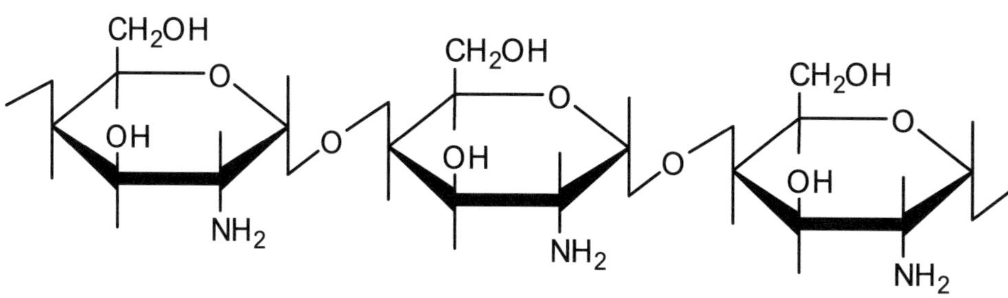

Fig. 1. Chemical structure of Chitosan.

2. Adsorption

Water is one of the basic necessities required for the sustenance and continuation of life. It is therefore important to supply good quality of water for various activities. However, this is becoming increasingly difficult in view of large scale pollution caused by industrial, agricultural and domestic activities. These activities generate wastewater which contains both inorganic and organic pollutants. Some of the common pollutants are phenols, dyes, detergents, insecticides, pesticides and heavy metals [23,24].

2.1 Chitosan based adsorbents for the separation and recovery of pollutants from wastewater

During recent years, CS derivatives were successfully utilized for removal of anions from wastewater. The adsorption of nitrate by CS hydrobeads was examined by Chatterjee and Woo [25]. Desorption of nitrate from the loaded beads was accomplished by increasing the pH of the solution to the alkaline range, and 87% desorption ratio of nitrate was achieved at pH 12.0. The applicability of neodymium-modified chitosan as adsorbents for removal of

excess fluoride ions was studied by Yao et al [26]. Three cross-linked CS derivatives were used for the removal of Cr(VI) from aqueous solutions: (i) CS, without grafting; (ii) CS grafted with acrylamide (CS-g-Aam) and (iii) CS grafted with acrylic acid (CS-g-Aa). CS-g-Aam exhibited the highest sorption capacity for Cr(VI) removal (935 mg g^{-1}) at pH: 4.0 [27]. The metal-binding property of chitosan was used to incorporate titanium (Ti) metal and applied as an adsorbent for fluoride adsorption by Ti loading. It was observed that an increase in titanium loading from 5 to 15% improved fluoride removal from 61 to 89% [28]. CS intercalated montmorillonite (CS-MMT) was prepared by dispersing sodium montmorillonite (Na+-MMT) into CS solution at 60 °C for 24 h. CS-MMT showed the highest 46–49 mg g^{-1}adsorption capacity, which was attributed to the existence of intercalate-CS, and enlarged its pore structure and facilitated the penetration of macromolecular dyes bound electro-statically [29]. N-benzyl mono- and disulfonate derivatives of CS were also used for the removal of dyes in aqueous media. Results confirmed the strong cation exchange character of the sulfonated derivatives with 121.9 mg g^{-1} adsorption capacity at pH: 3.0 [30]. Four kinds of phenol derivatives: 4-hydroxybenzoic acid (BA), 3,4-dihydroxybenzoic acid (DBA), 3,4-dihydroxyphenyl-acetic acid (PA), and hydrocaffeic acid (CA), were grafted onto CS substrate, individually. The modified CS derivatives were used for removal of cationic dyes (crystal violet (CV) and bismarck brown Y (BB)). The optimum adsorptive uptake for CV and BB occurred at pH: 7.0 and 9.0 at 30 °C, respectively [31].
Simultaneous removal of various metal ions (zinc, copper, cadmium, and lead) in aqueous media was achieved by commercially available CS flakes [32]. Cross-linked CS gel beads were also used for molybdate sorption. Authors reported 700 mg g^{-1} adsorption capacity in batch studies [33]. Cross-linking of CS particles with glutaraldehyde, epichlorhydrine, or EGDE (ethylene glycol glycidyl ether) enhanced the resistance of sorbent beads against acids, alkali or chemicals [34]. Contamination of natural waters by arsenic in is a worldwide problem, which is a health hazard. Extensive research has been conducted to control/minimize the arsenic contamination in drinking water. Sorption of As(V) on molybdate impregnated CS gel beads was investigated [35]. CS derivatives were also found to be efficient for removing gold from dilute acidic solutions, with 600 mg g^{-1} (3 mmol g^{-1}) uptake capacity. However, 102 mg g^{-1} adsorption capacity was recorded as high as for 100 mgl^{-1} initial Cr(VI) concentration. Results showed that both monolayer adsorption and intraparticle diffusion mechanisms limited the rate of Cr(VI) adsorption [36]. Quaternary ammonium salt of chitosan (QCS) was synthesized via reaction of a quaternary\trimethyl ammonium, glycidyl chloride and examined for Cr(VI) removal [37]. Cross-linked CS/PVA beads with low water content were prepared by suspension of aqueous solution of chitosan/PVA in toluene–chlorobenzene medium using glutaraldehyde as cross-linking agent in acidic conditions [38]. The developed beads exhibited quite low degree of swelling of prepared beads due to water loving nature of beads and considerable hydrogen bonding with hydrophilic groups. Both nitrogen and oxygen atom posses lone pair of electrons, can bind a positively charged ion through the electron pair sharing. The easy release of lone pair from nitrogen atom makes it the main binding site and forms stable metal complex. With the above consideration, the following chemical reactions are proposed to account the mechanism of Cd(II) adsorption and desorption:

$$\text{R-NH}_2\text{+H}^+ \rightarrow \text{R-NH}_3^+ \quad (1)$$

$$\text{R-NH}_2\text{+Cd}^{2+} \rightarrow \text{R-NH2Cd}^{2+} \quad (2)$$

$$\text{R-NH}_3^+ + \text{Cd}^{2+} \rightarrow \text{R-NH2Cd}^{2+} + \text{H}^+ \tag{3}$$

$$\text{R-NH}_2^+\text{Cd}^{2+} + \text{H}_2\text{O} \rightarrow \text{CdOH}^+ + \text{R-NH}_3^+ \tag{4}$$

The maximum adsorption capacity of Cd(II) on chitosan-coated perlite beads at pH: 6 was found to be 78 and 178.6 mg g^{-1} of beads from a solution containing 1000 and 5000 mg l^{-1} of Cd(II), respectively, at 298 K. Breakthrough data from a column were used to calculate the diffusion coefficients for Cd(II) into the chitosan coated perlite beads, and the diffusion coefficient was found to be 8×10^{-13} m^2s^{-1} [39]. The modification of –OH groups on CS chemical reactions using ehtylenediamine and carbodiimide (CR) were also studied [40]. Hg(II) uptake by CS cross-linked with gluteraldehyde (GLA) and (CR) subsequently treated with EPI (CR-Cl) followed by reaction with ethylene diamine (CR-amine) or 3-amino-1,2,4-thiazole-5 thiol (CR-azole) was studied [41]. Hg(II) adsorption behavior of cross-linked carboxymethyl chitosan was reported with 124.4 mg g^{-1} maximum adsorption capacity (pH: 5.0) [42]. Donia et al. modified magnetic chitosan by use of a Schiff's base cross-linker (GLA+thiourea) and studied the removal of Hg(II) solutions by the obtained resin [43]. Observed adsorption capacity was attributed to the presence of the free lone pair of electrons on N and S atoms suitable for coordination with soft acid i.e. Hg(II) ions. Novel CS based chelating resin (CCTS-TAA) was also synthesized for Hg(II) adsorption from EDGE-cross-linked chitosan (CCTS) followed by chemical modification with tris(2-aminoethyl)amine (TAA) moiety The CCTS-TAA resin possesses tertiary and primary amine groups, presenting the tertiary amine higher affinity and selectivity for Hg(II) [44]. Gen et al. reported the use of Procion Brown MX 5BR immobilized poly(hydroxyethylmethacrylate/CS) composite membrane (pHEMA/CS) for removal of Hg(II) from aquatic systems.

The CS and *N*-carboxymethyl CS (NCMCS) beads were used as an adsorbent for the gold (III) (Au^{3+}) removal from wastewater. Adsorption of Au^{3+} ions occurred only at the optimum pH and kinetic studies showed a rapid adsorption of Au^{3+} from aqueous solution. The experimental data of the adsorption equilibrium from Au^{3+} solutions correlated well with the Langmuir adsorption isotherm. The maximum desorption of Au^{3+} was obtained from CS/ NCMCS beads by the treatment with EDTA solution of known concentration. The Au^{3+} ions can be easily removed from CS or NCMCS by treatment with an aqueous EDTA solution, therefore making it possible to collect the Au^{3+} ions and recycle the CS and NCMCS beads [45]. The adsorption of Al(III) from aqueous solutions onto CS was studied in a batch system [46].

The cross-linked N,O-carboxymethyl- CS resin with Pb(II) as template ions (cross-linked CMCS template) were synthesized and cross-linked with glutaraldehyde to improve adsorption selectivity for heavy metals. The adsorption experiments demonstrated high adsorption selectivity for Pb(II) of cross-linked CMCS template. Furthermore, although the adsorption capacities for Cu(II), Co(II), Ni(II), Zn(II) decreased in compare with CMCS for adsorption of Pb(II) from the mixture solution of Pb(II)–Cu(II)–Zn(II) solutions. This was an effective way to separate Cu(II), Zn(II) and Pb(II) ions selectively. The sorption of CMC and the cross-linked CMC template for Pb(II) was a chelate process through both oxygen and nitrogen atoms co-ordinated with Pb(II) in the polymer chain [47] The PAAc-functionalized chitosan granules (CS-PAAc) showed significantly greater adsorption capacities for lead ions than CTS, and the performance improved with the increase of pH: 1-6. The excellent adsorption performance of CS-PAAc for Pb^{2+} was attributed to the many carboxyl groups grafted on CS-PAAc beads. Desorption study showed regeneration of adsorbent without

any loss of adsorption capacity and confirmed utilization of CTS-PAAc to remove Pb^{2+} ions from the wastewater [48].
ions from aqueous solution. The amount of adsorbed Cu^{2+} per gram of the beads decreased with the increase of CS and NaOH concentration in the beads [49]. CS was chemically modified with 2[-bis(pyridylmethyl)aminomethyl]-4-methyl-6-formylphenol (BPMAMF) and employed to study the kinetics and the equilibrium adsorption of Cu(II), Cd(II), and Ni(II) metal ions as functions of the solution pH [50]. The kinetics was evaluated for the utilization for the pseudo-first-order and pseudo-second-order equation models and the equilibrium data were analyzed by Langmuir and Freundlich isotherms models. The adsorption kinetics follows the mechanism of the pseudo-second-order equation for all studied systems and this mechanism suggests that the adsorption rate of metal ions by CHS–BPMAMF dependent on the number of ions on the adsorbent surface, as well as on their number at equilibrium. Modified CS, 2[-bis-(pyridylmethyl) aminomethyl]-4-methyl-6-formyl-phenol (HL by Schiff's base reaction), was identified as new adsorbent materials for the removal of Cu(II) in aqueous media [51].

2.2 Chitosan based membranes for the adsorption/separation of amino acids and proteins

Introduction of silica into the CS created a dense and uniform hybrid network and reduced the degree of swelling of the materials in aqueous system, which ensured the formation and maintenance of imprinting sites. The specific interactions of imprinting cavities in the hybrid membrane with templates resulted in a significantly improved chiral resolution of the imprinted membranes by strengthening the binding ability of the imprinting molecules, hindering their diffusion and facilitating the transport of the other isomers. The chiral separation ability of CS/ glycidyl propyl trimethoxysilane (GPTMS) hybrid molecularly imprinted membranes towards the underivatized d,l-Phe aqueous mixture was evaluated by permeation and binding experiments. The selectivity of the developed membrane was increased along with a 4.5 separation factor [51,52]. Porous structured CS–β–cyclodextrin (CD) membranes were synthesized by incorporating β–CD polymers as the functional moiety and GPTMS used as cross-linking agent [53]. The developed membrane exhibited the significant changes in molecular constitutions, crystalline and thermal properties.
Modifications of CS were carried out by either introducing phosphonic acid group or by quaternization of existing primary amine groups in order to make it water soluble [54]. The functionalized materials were used to prepare charged nanoporous composite membranes of different compositions of *N*-methylene phosphonic chitosan (NMPC) and quaternized chitosan (QC) by taking required modified materials and poly(vinyl alcohol) 50% (w/w) in aqueous media and later the membranes were gelated in methanol at 10°C to tailor pore structure in the membrane matrix. *FTIR* and *IEC* studies confirmed the introduction of phosphonic acid groups and quaternary ammonium groups, resulting in formation of cross-linking or covalent bonding with PVA that resulted the composite membranes. Thermodynamic properties of dilute aqueous solutions of NMPC, QC, and their mixture were also studied and it was concluded that both individual materials have good hydrophilic interactions with water and mixed solution exhibited electrostatic interactions between negatively charged –PO3H2 and positively charged –N+(CH3)3 groups. Membrane conductivity studies revealed that PC/QC–30 was a relatively dense membrane and may be able to discriminate between different types of ions based on their ionic radii. The values of

diffusion coefficient of ions across the membranes (Dm^i) for different electrolytes follow the trend NaCl > CaCl2 > MgCl2 in the case of PC/QC-30 membrane which suggested the applicability of the developed membrane for the electrolyte separations, because of their varied diffusional migration. Membrane permeation studies also confirmed the lowest equivalent pore radius for PC/QC-30 membrane, and the relative permeability of these membranes for different types of ions suggests the suitability of PC/QC-30 membrane for the separation of Na+ from Ca2+ and Mg2+ mixture.

Cross-linked CS/ poly(methacrylic acid) (PMAA) particles was obtained via template polymerization. The resultant particles were characterized for different charge density regions. Albumin was adsorbed on these particles (after cross-linking with glutaraldehyde), and pH was controlled for: (i) adsorption of positively charged albumin and (ii) adsorption of albumin at its iso-electric point. Adsorption isotherms and zeta potential (ζ) measurements showed that albumin adsorption was controlled by hydrogen bonding/van der Waals interactions and the brush-like structures may enhance the adsorption of albumin on these particles. It was also found that shearing can induce desorption of albumin from PEC surface, depending on the continuous phase albumin concentration [55]. Semi-interpenetrating networks (IPN)-structured CS/β-cyclodextrin (CS/CD) composite membranes were prepared and used for the enantiomeric separation of tryptophan (Trp) racemate mixture. The diffusion selectivity as compared to sorption selectivity was responsible for the enantiomeric separation of Trp. Thus the overall sorption–diffusion process could be condensed to a mechanism which is schematically illustrated in Fig. 2. Both CS and β-CDP exhibited l-Trp selectivity in the sorption process, and the concentration gradient built by sorption process between donor side and receptor side within membrane thickness contributed to the following diffusion process. The diffusion process was predominant for the fact that the sorption process showed a l-Trp selectivity while the overall sorption–diffusion process favored d-Trp. Due to lower complexation selectivity of β-CDP which narrow the difference between diffusion resistances of d and l-Trp, the

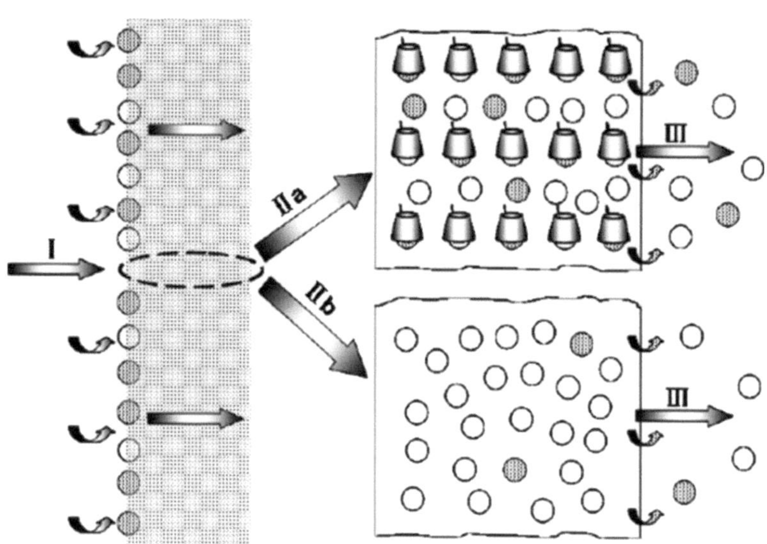

Fig. 2. Schematic illustration of the overall solution-diffusion process within CS/CD membrane.

diffusion rate difference between d- and l-Trp diminished that accounts for a decreased permselectivity in the presence of β-CDP. Meanwhile, permeate flux for both d- and l-Trp was found to increase considerably with the increase of β-CDP content within the membrane, which could be attributable to the facilitated mass transport of Trp enantiomers through β-CDP within the membrane. Both CS and β CDP exhibited l-sorption selectivity in the sorption experiment, and while the complexation selectivity of β-CDP with Trp was relatively lower, which would reduce. The obtained results provided useful information for the development of composite membranes for enantiomers separations [56].The surface of CS films was modified to introduce the positive and negative charges. The positively charged CS surface was prepared by a reaction between the amine group of CS and methyl iodide (CH_3I) to form the quaternary ammonium-functionalized CS surface (QAC film) as depicted in Scheme 1[57].

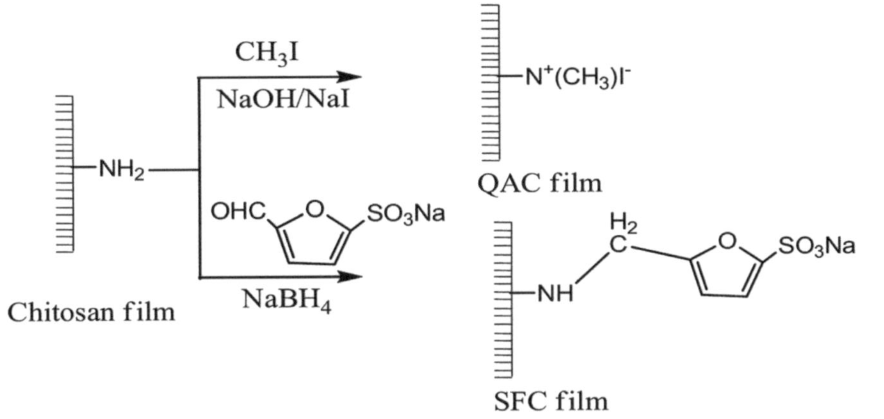

Scheme 1. Introduction of charged fuctional groups to the surface of chitosan.

The CS/CMC blend membranes were prepared by a simple solution-blending method with glutaraldehyde as a cross-linking agent for CS and with silica particles as progens [58]. Shi et al. oworkers developed affinity membranes by immobilizing Cu^{2+} ions onto CS modified disc alumina to purify hemoglobin from a hemolysate and achieved a binding capacity of ~17.5 mg of hemoglobin per gram of membrane [59]. CS-Cibacron blue F3GA affinity membranes were prepared in three steps: (i) dissolving CS in a dilute acetic acid solution; (ii) pouring the CS solution containing PEG as progen over a microporous polyether sulfone (PES) membrane; and (iii) coupling the membrane with the dye under the mild alkaline condition [60]. Adsorption studies of human serum albumin (HSA) on developed substrate under the equilibrium and dynamic conditions revealed as affinity membrane having the good adsorption capacity for HAS. The surface of CS films modified with amine groups of the glucosamine units affected the hydrophobicity. The improved surface hydrophobicity affected by the stearoyl groups promoted protein adsorption. In contrast, selective adsorption behavior was observed in the case of CS films modified with anhydride derivatives. Lysozyme adsorption was enhanced by H-bonding and charge attraction with the hydrophilic surface. While the amount of albumin adsorbed was decreased possibly due to negative charges that gave rise to repulsion between the modified surface of CS and albumin [61]. The adsorption of human IgG on the immobilized protein on a composite membrane was further studied under both equilibrium and dynamic conditions. The equilibrium adsorption did not follow the Langmuir model, but well fitted the Freundlich-Langmuir adsorption isotherms. The results indicated that the affinity of CS/cellulose membrane has a great potential in fast purification of biomolecules [62]. The concept of chiral ligand exchange was employed to achieve the chiral resolution of tryptophan (Trp) enantiomers by using CS membrane in a sorption resolution mode and Cu(II) ion acted as the complexing ion. Porous CS membranes were prepared by freeze-drying method (CS–LT) and sol–gel process at high temperature (CS–HT), respectively, to investigate their sorption resolution characteristics. The CS chiral ligand exchange membranes exhibited good chiral resolution capability. Meanwhile the sorption selectivity of the CS membranes was found to be reversed from L-selectivity at low Cu(II) ion concentration to D-selectivity at high copper(II) ion concentration, which was attributable to the stability difference between the copper(II)–L-Trp and Cu(II)–D-Trp complexes. Moreover, CS-HT membrane showed better performance with respect to both sorption selectivity and sorption capability than CS-LT membrane, which mainly resulted from its more amorphous structures as compared with the more crystalline structures of CS-LT membrane. The superiority of sorption capability of CS-HT membrane is also attributable to its larger specific surface area than that of CS-LT membrane. The results obtained were conducive to the design and fabrication of chiral ligand exchange membranes for enantiomeric separation in sorption mode [63]. Affinity membranes based on a Ni^{2+}–chelating CS surface were fabricated to immobilize C-terminus hexa histidine-tagged green fluorescent protein (His-GFP) [64]. The ability of histidine-tagged proteins to chelate with Ni^{2+} ions that were coordinated to functionalized polymeric matrices. Both membranes had comparable amounts of his-GFP immobilized on the surface. However, the amount of anti GFP bound to the Ni^{2+}-chelated his-GFP complex at saturation was higher than that bound to the glutaraldehyde-immobilized his-GFP by a factor of five. Furthermore, fitting the data to a single-site Langmuir model resulted in an affinity constant for the Ni^{2+} chelated his-GFP complex towards anti GFP was 14 times higher than the glutaraldehyde-immobilized his-GFP. The higher affinity suggested that immobilizing a protein at its C-terminus resulted in the proper orientation for subsequent binding of antibodies. At low antibodies concentrations, the sensitivity of the affinity membrane was 70 times than that of the control. The ability to capture specific proteins with their respective antibodies would be highly advantageous to any assay based on the identification of biomarker proteins. Beppu et al. [65] prepared CS membrane and cross-linked with glutaraldehyde in order to see the effect on ion permeability and water adsorption property. Heterogeneous cross-linking with glutaraldehyde showed to produce more CS structures in CS membranes, which then interfere in the interaction with water and ions and changes their mechanical characteristics. Thus the developed membrane can be easily used for the separation of cations from their mixture solution [66]. The permeability coefficients of the metal ions obeyed the models, those for CS membranes were found lower than predicted values and found the largest deviation than the transition metal ions. The diffusive properties of CS membranes towards metal ions offer potential for protection of CS based biological systems against destructive effects of heavy metal ions. A new kind of metal affinity membrane based on a ceramic support was prepared by: (i) deposition of a CS layer in order to functionalize the ceramic support; (ii) cross-linking with epichlorohydrin to stabilize the polymer layer and to enable for further grafting, and (iii) grafting of iminodiacetic acid [67]. Schematic structure of affinity membrane is presented in Fig. 3. Due to the ceramic support, this membrane is highly resistant and CS layer brings

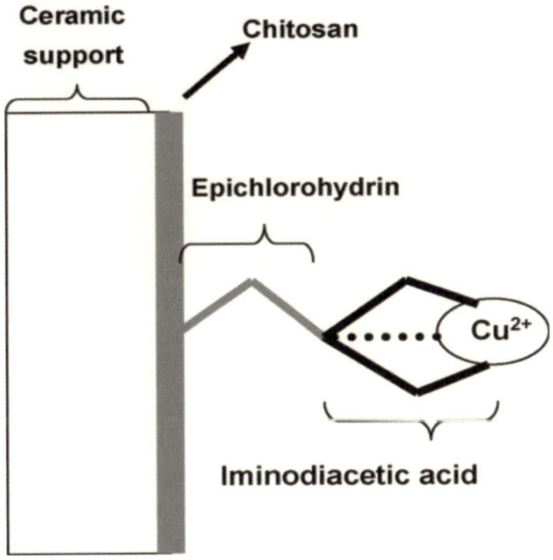

Fig. 3. Schematic representation of the affinity membrane.

biocompatibility properties. The commercially available membranes were modified with cationic layer of quaternized CS to enhance their selectivity towards the metal ions [68]. Highly porous CS/ cellulose acetate (CA) blend hollow fibers as adsorptive membranes were successfully fabricated through a wet spinning process with CA and CS. The developed hollow fiber based membranes can be efficiently used for affinity-based separations of metal ions, amino acids and proteins [69]. The hollow fiber membranes were prepared from CA acted as a matrix polymer and CS functional polymer to provide the membrane with coupling or active sites for affinity-based separations. Formic acid was used as the co-solvent for both CA and CS to prepare the dope solution and NaOH solution was used as the external and internal coagulants in the wet spinning fabrication process. The properties of the blend hollow fibers membranes were characterized through water flux measurements, surface and cross-section examinations. Adsorption performances of the composite membranes for Cu^{2+} ions and BSA on the surface were compared with CA hollow fibers membranes. The blend hollow fibers membranes achieved significantly better adsorption performance as compared to CA hollow fibers, indicating the benefit of adding CS into CA to make novel blend hollow fibers membranes in improving the performance of the traditional CA hollow fibers, especially for the affinity-based separations [70]. Compared to chromatographic methods, membrane separation techniques offer advantages of lower cost and ease to scale-up for commercial production. However, the lack of membrane selectivity and its fouling due to protein absorption during filtration has severely restricted their ultrafiltration (UF) applications [71]. Nowadays, UF has been widely used as preferred method for protein concentration and buffer exchange that replaced size exclusion chromatography in these applications [72]. UF membranes prepared from a variety of synthetic polymers have high thermal stability, chemical resistivity, and restricted the use of fairly harsh cleaning chemicals [73]. The choice of membrane was usually guided by its molecular weight cut-off (MWCO), which is defined as the equivalent molecular weight of the smallest protein that would exhibit above 90% rejection. However, the experimental conditions and systems used to evaluate 90% MWCO have not been standardized [74]. Fractionation of dairy wastewater into lactose-enriched and protein-enriched streams using

UF membrane technique was also studied. Three regenerated cellulose membranes of 3, 5 and 10 kDa MWCO were used to determine the process efficiency. The performance was determined under various processing conditions that include the operating temperature, trans-membrane pressure across the membranes and the concentration of lactose in the feed solution [75].

CS/polystyrenesulfonate (PSS) multilayers membrane on PES supports was prepared by casting method. The developed membrane was used in ultrafiltration for the separation of ternary mixture of proteins (BSA, ovalbumin and Lysozyme (LYS) mixture [76]. BSA was rejected by the multilayer membrane at all the studied concentrations (0.25, 0.5, 1.0 and 2.0 mg ml^{-1}). BSA solution flux was found to decrease with an increase in BSA concentration. BSA–lysozyme separation may also be achieved using the with multilayer composite charged ultrafilter membrane. Hydrophobic poly (vinylene diflouride) (PVDF) membrane was modified with CS in order to reduce the protein fouling in ultrafiltration. The hydrophilicity of the modified membrane increased (decline in contact angle) with increasing CS concentration and modification time. However, the water permeate flux was decreased with the increase in CS concentration and modification time. Moreover, the modification method significantly affected the deposition of CS on membrane surface and pore wall. The protein fouling study was carried out and found that the normalized flux during the BSA filtration slightly declined when modified membrane having the high CS concentration. The modification of PVDF membrane with CS could effectively prevent protein fouling on membrane. Protein adsorption on the modified membrane was highest at BSA iso-electric point (IEP) while the adsorption decreased at BSA solution pH was far from the IEP [77].

The fabrication of a novel composite membrane with high flux and good selectivity was carried out by modifying the surface of a microfiltration membrane with a uniform coating of CS/ PSS polyelectrolyte multilayer, which may find application in protein separations under ultrafiltration conditions. The individual transport studies of ovalbumin and lysozyme indicated that these egg white proteins could be separated using multilayer deposited charged ultrafilter membrane [78]. Experimental data collected includes total flux as a function of time, as well as final protein deposition dependant on feed volume flow, pH and protein concentration [79]. The Poly(acrylonitrile) (PAN)/CS composite ultrafiltration membranes were prepared by filtration of CS solution through PAN base membrane and subsequent curing and treatment with NaOH. The formation of CS layer on the surface of PAN base membrane was confirmed by Fourier transformer infrared (FTIR), X-ray photoelectron spectroscopy (XPS) and scanning electron microscopy (SEM). While the deposition of CS on the pore walls was confirmed by determination of various parameters such as pure water permeation, pore size distribution and MWCO and it was found that decline in studied parameters compared with pristine membrane. The composite membranes were stable in aqueous medium and showed reduction in pure water fluxes after the filtration of aqueous acidic (pH: 3.0) and basic (pH: 11.0) solution, which was attributed to the swelling of CS layer. The composite ultrafiltration membranes can be efficiently used for the separation of proteins from their mixture [80].

The use of electric field in UF goes back to the first study carried out by Bechhold by imposing electric field in UF and utilized a combination of electro-osmosis and electrophoresis to purify colloids in an apparatus he called an 'electro-ultrafiltration' (EUF) [81]. EUF is an effective method to decrease gel layer formation on the membrane surface

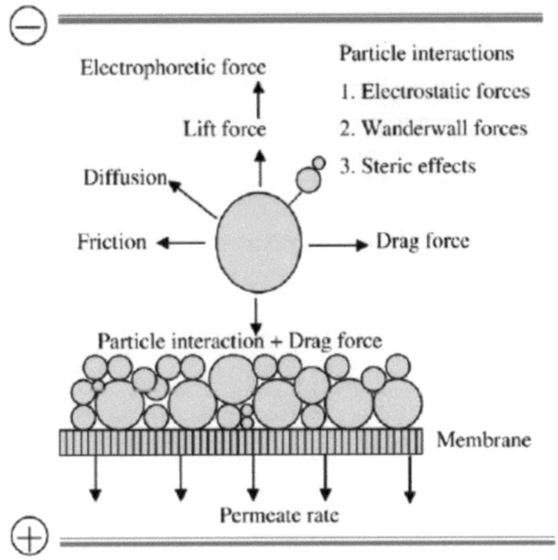

Fig. 4. Principles of electro-ultrafiltration.

and to increase the filtration flux, owing to electro-kinetic phenomena such as electrophoresis and electro-osmosis [81-86]. Basic principle of EUF is presented in Fig.4. This process aroused from a combination of number of mechanisms, including ion association, ion adsorption or ion dissolution. The electrochemical properties of the membrane surface and the dispersed materials or solutes can have a significant influence on the nature and magnitude of the interactions between the membrane and the used substances and their separation characteristics. The utilization of such properties by the application of external electric fields improved substantially the membrane performance. The accumulation of the solutes on the membrane surface is limited by the imposed electro-phoretic force. In addition, the filtration rate through the filter cake is dramatically enhanced due to electro-osmosis as a secondary electro-kinetic phenomenon. This method is best suited for the separation of protein since its surface charge changes according to the solution pH [87-103]. As biological products like proteins and peptides are sensitive to shear stress and temperature, the coupled effects of electric field and pressure served as an additional driving force for the separation, which is an interesting way to improve the membrane permeates flux without increasing the shear stress [81,102,103]. Since proteins carry a net electrical charge, an electrical field may be used to reduce the influence of polarized layer. By applying a suitable external dc electric field, the protein molecules were transmitted through the membrane due to the electrostatic attraction and decline in concentration polarization. EUF was also studied as a membrane cleaning procedure, in which electrical fields attracted the particles with opposite charge from the membrane surface, and initial permeate flux was restored by eliminating fouling up to large extent. It was observed that as the voltage increased, the cake layers deposited on the membrane surface became thinner. Eventually when the critical voltage, at which the foulants were stationary reached no particle deposition on the membrane surface was observed [104]. Functionalizations of CS were carried out by either introducing phosphonic acid group or by quarternization of existing primary ammonium groups for making it water soluble biopolymer material. Negatively charged PC–Si and positively charged QC–Si membranes were prepared using

TEOS and PVA as plasticizer by acid catalyzed sol–gel reaction in aqueous media. These membranes were gelated in the methanol under optimized conditions for tailoring the pore structure of the membrane. FTIR spectra and IEC study of the membranes confirmed the membrane conductivity, surface charge density and counter-ion transport number of the membranes also revealed mild charged nature. These membranes were employed for the separation of proteins from their mixture under coupled driving forces. It was concluded that separation of LYS from the mixture of BSA–LYS at pH: 4.8 (p*I* of BSA) using negatively charged PC–Si membrane or BSA from the mixture of BSA–LYS at pH 10.7 (p*I* of LYS) using positively charged PC–Si membrane, was possible with high selectivity. Also in all cases due to coupling of driving forces, filtrate flux and protein transmission was enhanced. Furthermore, applied electric gradient further progressively enhanced the separation factor (SF) suggested highly selective separation of protein under coupled driving forces. Protein transmission (selectivity) and membrane throughput across both membranes were studied using binary mixture of protein under different gradients at pH: 2.0, 4.8, 10.7, and 13.0. Furthermore, applied electric gradient progressively increased the SF, which was close to 10 for PC–Si and 15 for QC–Si membranes. Relatively high SF of individual protein from binary mixture and filtrate velocity suggests the practical usefulness of this novel process. BSA transmissions across both types of membranes and its effect on the pH of the feed protein solution were studied under coupled driving forces (pressure and electric gradient) in order to investigate the effect of nature of charge on the protein molecule (BSA), membrane matrix and direction of electric polarity applied. Schematic diagram for the different modes of protein transmission such as protein transmission using uncharged ultrafilter membrane without any electric gradient is presented in Fig. 5A. While Fig. 5B represents protein transmission under coupled driving forces using negatively charged ultrafilter membrane (PC–Si) and same for positively charged membrane (QC–Si) is depicted in Fig. 5C [105].

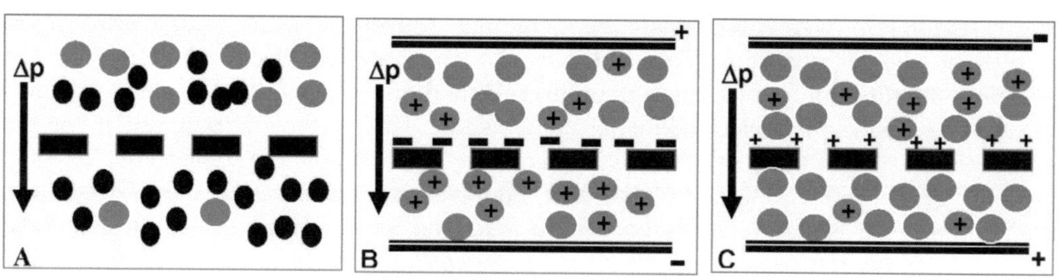

Fig. 5. Schematic presentation of: 6(A) Ultrafiltration; (B) ultrafiltration under applied potential gradient for CEM and (C) for AEM.

3. Polymer electrolyte membranes for fuel cell applications

The fuel cell (FC) is an electro-catalytic membrane reactor where the chemical energy of a fuel and an oxidant is continuously converted to electrical energy. The basic principle of a FC is not different than that of electrochemical batteries. However, in batteries the chemical energy is stored inside and conversion to electrical energy requires either disposal or recharge. In a FC, however, the chemical energy is provided by the fuel and an oxidant that stored outside the cell. Electrical energy can then be continuously produced as long as the fuel and the oxidant are provided. In a FC, the chemical energy is directly converted to electrical energy with efficiencies much higher than the conventional thermo-mechanical systems. In addition, FC operates without combustion and it is pollution free [106]. The FC is basically composed of two electrodes i.e. (anode and cathode) and an electrolyte. The energy conversion reaction occurs at the electrodes. The fuel is oxidized at the anode, the oxidant moves through the electrolyte and is reduced at the cathode. Hydrogen-oxygen (H_2-O_2) FC, one of the most important fuel cell in which the hydrogen releases electrons and converted into the hydrogen ions (H^+). The H^+ ions permeate through the electrolyte and the electrons flow through an external load circuit to reach the air-electrode (cathode). On the cathode, H^+ ions, electrons and oxygen react to form water. The reaction take place at the electrode in FC is presented as:

$$\text{Anode: } H_2 \rightarrow sH^+ + 2e^- \tag{5}$$

$$\text{Cathode: } 2H^+ + 1/2O_2 + 2e^- \rightarrow H_2O \tag{6}$$

$$\text{Over all reaction: } H_2 + O_2 \rightarrow H_2O \tag{7}$$

The nature of the electrolyte (liquid or solid) determines the operating temperature of a FC. The electrolyte blocks the electrons and prevents the electrical contact between the electrodes. It can either be a proton contactor or an oxygen ion contactor. The major difference between the two is the side of FC where the water is produced. When a proton conductor is used, the water is produced in the cathode side of FC. The FC systems are commonly distinguished by the type of electrolyte applied, in the following categories: alkaline fuel cells (AFC), phosphoric acid fuel cells (PAFC), molten carbonate fuel cells (MCFC), solid oxide fuel cells (SOFC) and polymer-electrolyte fuel cells (PEFCs). In this chapter, we focus on the PEFCs where a polymeric membrane is used as electrolyte. The H_2/O_2 fuel cell, commonly referred to as polymer electrolyte fuel cell (PEFC) and the direct methanol fuel cell (DMFC) are the two types of fuel cells which use polymer electrolytes. DMFCs have higher energy density but exhibit shortcomings such as (a) slower oxidation kinetics than PEFC below 100°C and (b) significant permeation of the fuel from the anode to the cathode resulting in a drop in efficiency of fuel utilization up to 50% [107]. In order to improve fuel cell performance, it is essential to understand the effect of operating parameters on fuel cell performance. Membrane is the core component of polymer electrolyte membrane (PEM) fuel cell. To achieve high efficiency, membrane must possess the following desirable properties: (i) high proton conductivity to support high currents, (ii) resistive losses and zero electronic conductivity, (iii) adequate mechanical strength and stability, (iv) chemical and electrochemical stability under operating conditions, and (v) moisture control in stack.

A series of cross-linked CS sulfate membranes have been developed by grafting CS monomers with -SO_3H groups and then cross-linked the polymers by the reaction between -SO_3H groups in CS sulfate and the amido groups in pure CS [108]. Although these types of membranes do not offer significant advantages over Nafion, as far as proton conductivity concerned and a significant reduction in methanol permeability make these polymers appeared to be suitable for DMFCs applications [109].

The ratio between PVA and CS in the substrate membrane varied in order to determine an optimal substrate for the reduction in methanol crossover. The pore filling developed membranes were characterized in terms of methanol permeability and conductivity. The composition ratio between PVA and CS can affect the proton conductivity. The developed composite membranes were also tested for methanol permeability [110]. The ionic cross-linking occurred on blending the polyelectrolyte that excludes the need of using other cross-linking agents. The study thus reveals the possibility of preparing low cost acid-base polyelectrolyte blend membranes having IEC comparable to that of Nafion but relatively lower methanol permeabilities. New polymer electrolyte composite membranes were prepared by using CS as the matrices and incorporating potassium hydroxide for ionic functionality [111].

Hybrid membranes were prepared from CS as organic matrix and surface-modified Y zeolite as inorganic filler. Membranes were tested for their applicability for use in DMFC by methanol permeability, proton conductivity and swelling property [112].

Two series of novel cross-linked composite anion exchange membranes were prepared using different mass of QAPVA and HACC and using GA as the cross-linking reagent [113]. The quaternary ammonium group grafted onto the matrix of quaternized poly(vinyl alcohol) (QAPVA) and quaternized CS (2-hydroxypropyltrimethyl ammonium chloride chitosan, (HACC) composite membranes have exchangeable anions. Novel cross-linked composite membranes were investigated for their applicability in alkaline membrane fuel cells.

The composite membranes showed a high conductivity (10^{-3} to 10^{-2} Scm^{-1} along with low methanol permeability (from 5.68×10^{-7} to 4.42×10^{-6} cm^2 s^{-1}) at 30°C. It was concluded that membrane structure was the principal factor for affecting the conductivity and methanol permeability of membranes. A series of quaternized-chitosan derivatives (QCDs) with various degrees of quarternization was synthesized using glycidyltrimethylammonium chloride as a main quaternized reagent [114]. These QCDs were then processed into hydroxide by dipping the resultant membranes in aqueous potassium hydroxide solutions. The resultant anion-exchange membranes in hydroxide form were further cross-linked by using ethylene glycol diglycidyl ether as a cross–linker. An anhydrous proton conducting membrane was prepared from CS, which possesses a large proton exchange capacity. The composite membrane (CS–200 wt% MP) showed the high proton conductivity (5×10^{-3} S cm^{-1}) at 150°C under anhydrous conditions [115]. Additionally, the proton conduction mechanism of the composite membrane was attributable due to proton transfer into the proton defect site without the assistance of diffusible vehicle molecules. The utilization of a biopolymer, such as CS for PEMFC technologies is novel and challenging where biological products are usually considered as waste, non-hazardous, and environmentally benign. Especially, the low production cost of the biopolymer is an attractive feature.

Zeolite beta particles with different sizes and narrow size distribution were hydrothermally synthesized and incorporated into CS matrix to prepare CS/zeolite beta hybrid membranes for DMFC [116]. Interfacial interaction between modified-zeolite-filled chitosan membranes has been illustrated in Fig. 6.

Sultone precursor (1,3-propane sultone (PS)) directly endowed the sulfonic acid groups from ring opening of sultone and did not need further treatment after the surface functionalization of zeolite beta particles. In the case of phenyl group precursor phenyl trimethoxy silane (PTMS), the phenyl groups grafted onto zeolite beta further reacted with concentrated sulfuric acid at 80 °C. As for thiol precursor i.e. mercaptopropyl trimethoxy silane (MPTMS), the mercapto groups was grafted onto zeolite beta and then oxidized into $-SO_3H$ groups with H2O2 solution at 25°C. Then sulfonated zeolite beta particles were incorporated into CS membrane matrix. Then the membranes were characterized and found

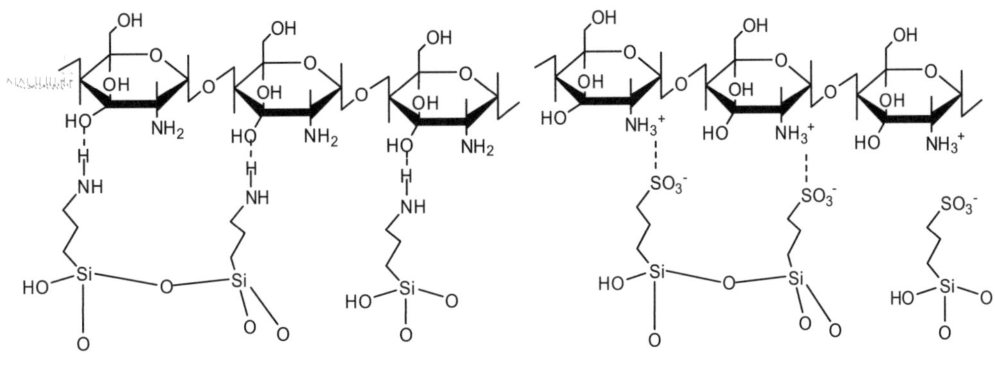

Fig. 6. Schematic presentation of interfacial interaction between modified-zeolite-filled chitosan membranes (a) CS–H_2NY membranes and (b) CS–HO_3SY membranes

that the introduction of –SO_3H groups could reduce the methanol permeability further as a result of the enhanced interfacial interaction between zeolite beta particles and CS matrix. Further sulfonation of zeolite beta particles improved their compatibility with CS matrix through increasing the ionic interaction between -SO_3H groups and -NH_2 groups. Thereby, most of CS/beta–SO_3H hybrid membranes showed lower methanol permeability and higher selectivity than CS/beta hybrid membranes. However, their proton conductivity changed only a little, indicating that the vehicle mechanism was dominant in proton conduction. Considering low methanol permeability, moderate proton conductivity, high selectivity, environmental benignity, low cost as well as facile fabrication, CS membranes filled by zeolite beta, especially sulfonated zeolite beta, showed a promising potential for DMFC applications [117]. CS served as catalyst support for Pt particles. LBL films of CS may also be used in hybrid cells with a Nafion membrane, in which synergism is sought between the properties of Pt in self-assembled films and Nafion membranes.

The CS membranes were prepared from CS flakes extracted from the exoskeleton of Cape rock lobsters, and were cross-linked with H_2SO_4 [118]. The four composite membranes i.e. CS/ heteropolyacids (HPAs), CS/ phosphomolybdic acid (PMA), CS/phosphotungstic acid (PWA) and CS/silicotungstic acid (SiWA)) were prepared by an infiltration and self-assembly method [119]. The methanol permeability of CS/PMA composite membrane was approximately two times higher than Nafion 117. Among the all membranes, CS/PMA membrane was identified as ideal candidate for DMFC applications as it exhibited low methanol permeability (2.7×10^{-7} cm^2 s^{-1}) and comparatively high proton conductivity (0.015 S cm^{-1} at 25°C). Thus, CS/HPAs composite membranes were promising proton-conducting materials and have great potential for use in DMFCs application [120]. The effects of zeolite pore size, particle size, hydrophilic/hydrophobic nature, and zeolite content on the membrane performance were explored. The transport of methanol in all the as-prepared membranes was mainly controlled by the diffusivity [121]. Hydrophobic zeolite with a relatively high Si/Al ratio preferentially adsorbed methanol by London force, leading to decreased water uptake, swelling, free volume cavity size and methanol permeability but increased methanol uptake. Meanwhile, the proton conductivity was slightly decreased with zeolite content due to the decrease of water content in the membranes. The hybrid membranes displayed desirable thermal and mechanical stabilities within the working temperature range of DMFC. Solid superacid ($STiO_2$) particles into CS-intermolecular interactions between $STiO_2$ and CS segmental chains were confirmed by FTIR image, suggesting that the addition of sulfate ions into TiO_2 particles might be retard the crystallization during the calcinations procedure. The results confirmed that $STiO_2$ particles (10 nm) as inorganic filler were successfully incorporated into the CS membrane matrix. Meanwhile, due to the interfacial interactions between $STiO_2$ and CS chains, the hybrid membranes exhibited an enhanced mechanical strength and adequate thermal stability as verified by mechanical strength characterization and thermo-gravimetric analysis. Consequently, CS/$STiO_2$ hybrid membranes acquired higher selectivity compared with control CS membrane [121]. New polymer electrolyte composite membranes were prepared by using CS as the matrices and incorporating KOH for ionic functionality. These membranes had a three-layer structure, which consists of a porous intermediate layer and two cross-linked solid surface layers. Their ionic-conductive properties were investigated using impedance spectroscopy. All fuel cells showed an open-circuit potential around some possible improvements on the performance of the resultant fuel cells were also found. Proton conducting cross-linked complex membranes was prepared by blending of a cationic polyelectrolyte, i.e. CS and an anionic polyelectrolyte, i.e. poly (styrenesulfonic acid-co-maleic acid) (PSSA-MA). The synthetic procedure for cross-linked CS/PSSA-MA membranes is illustrated in Scheme 2. CS contain two functional groups, i.e. hydroxyl (-OH) and amine (–NH_2) whereas PSSA-MA contains two acids, i.e. strong acid (–SO_3H) and weak acid (–COOH). Thus, two interactions occurred in blend membranes. The other was the complex formation between –NH_2 of CS and –SO_3H of PSSA-MA. Thus the PSSA-MA acted as a cross-linker and a proton conductor in the blend membranes. The interaction between CS and PSSA-MA was confirmed by FTIR study. The minimum water uptake and conductivity were attributed to the competitive effect between the formation of cross-linking/complex and the concentration of ionic groups in the membrane matrix. All the membranes exhibited temperature dependent proton conductivity, i.e. the increase of conductivity with temperature. Furthermore, CS is an extremely cheap, nonhazardous, and environmentally polymer which is advantageous over other polymer electrolyte membranes in terms of the green technology fuel cells [122]. Polymer electrolyte complexes of CS and poly vinyl phosphonic acid (PVPA) were prepared by in situ polymerization of vinyl phosphonic acid (VPA) in CS matrix [123]. Polybenzamidazole (PBI) and sulfuric acid (H_2SO4) composite membranes were prepared and the exclusion problem was observed especially in PBI/H_3PO_4 composite membranes [124]. Sulfonated polysulphone (SPSF) /CS exhibited high IEC and proton conductivity higher than Nafion 117 at temperature above 100°C. The membrane also showed adequate geometrical and thermal stabilities and can therefore be considered as a potential alternative fuel cell membrane especially for high temperature operations. The modification of CS was carried out by introducing -PO_3H_2 group and composite membranes were prepared by blending the phosphonated CS and PVA of different compositions [125]. Then the membranes were cross-linked by HCHO as cross-linker by formal reaction. All these bonds exhibited affinity towards water and repel organic molecules such as methanol, which leads to the hydration of polymer matrix with enhanced water retention. NMPC/PVA composite biomaterial was used for the membrane preparations with known composition.

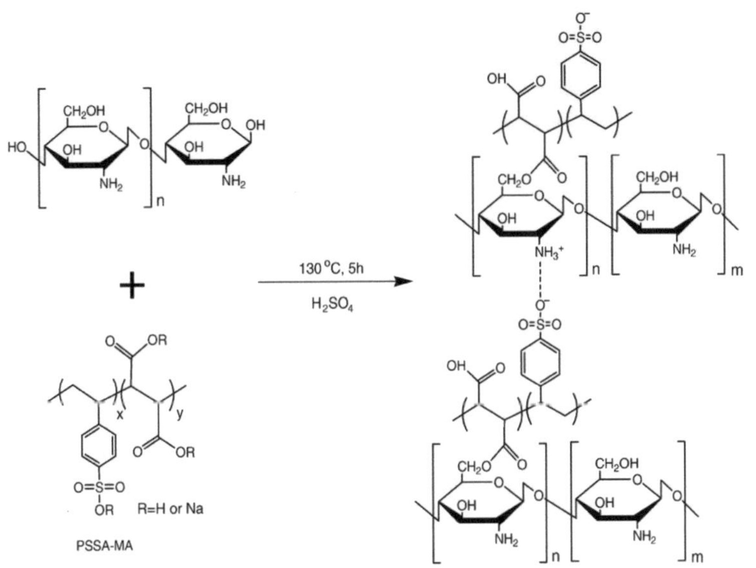

Scheme 2. Synthetic procedure for the cross-linked membranes consisting of CS and PSSA-MA.

Novel cross-linked composite membranes were synthesized to investigate their applicability in AEM fuel cells [126]. These membranes consist of QAPVA and HACC with glutaraldehyde as the cross-linking reagent. With the quaternary ammonium group grafted onto the matrix of QAPVA and HACC, the composite membranes possessed exchangeable anions [127]. PEC membrane for DMFCs was also prepared by blending a cationic polyelectrolyte, CS with an anionic polyelectrolyte, acrylic acid-2-acrylamido-2-methylpropane sulfonic acid copolymer (P(AA-AMPS)) [128]. The CS/P(AA-AMPS) membrane with a P(AA-AMPS) content of 41 wt.% exhibited a methanol permeability (P) of 2.41×10^{-7} cm^2 s^{-1} which was fifteen times lower than that of Nafion®117 where as its proton conductivity (σ) was comparatively high (3.59×10^{-2} S cm^{-1}). In terms of the overall selectivity index ($\beta = \sigma/P$), the PEC membrane showed a remarkably higher selectivity than Nafion®117 and, furthermore, the overall selectivity index increased with the increase of P(AA-AMPS) content. This kind of PEC membranes jointly utilized the methanol-rejecting property of one polyelectrolyte and the proton-conducting property of the other polyelectrolyte showed a promising application for DMFCs [129-132]. Organic-inorganic nanostructured composites constitute an emerging research field, which has opened the possibility of tailoring new materials because they combine in a single solid both the attractive properties of a mechanically and thermally stable inorganic backbone and the specific chemical reactivity and flexibility of the organo functional groups. Reports are available for diversified applications of CS inorganic hybrid nanostructured material. Blended PEM of CS with zeolite was also reported [133–139]. But mechanical stability and leaching out are serious problems for their prolonged use. It was expected that grafting of aromatic ring and less acidic –COOH groups on CS moiety will balance its hydrophilic-hydrophobic nature and enhance proton conducting properties [140]. The distribution of silica particles within the membrane matrix was assessed by TEM study. This provides evidence that these hybrid membranes have nanosized silica particles homogeneously distributed within the polymer matrix. The main advantage of preparing hybrid membrane by the sol-gel method was uniform homogeneous distribution. These results suggest that the uniform hybrid membrane with nanosized silica and $-SO_3H$ clusters in the membrane matrix. The activation energy (Ea) increased significantly with the increase in NCBC-silica content in the membrane phase. In all cases, 2.0 h cross-linked membranes with the same compositions showed maximum *E*a. The E_a of the prepared membranes were slightly higher than that of Nafion117. Thus, the thermal activated conduction process for prepared composite membranes was higher in comparison with Nafion117 membrane. At higher temperature, fast proton and water molecule diffusion resulted in a rapid conduction process, due to the more continuous pathway because of interlinking of hydrophilic channels. Furthermore, comparable *E*a values of prepared and Nafion117 membranes indicated a Grotthus-type conduction mechanism. The performance of PCS-3-3 membrane was tested by recording current-voltage polarization curves in DMFCs under different experimental conditions with varying methanol concentrations (20, 30, and 50% MeOH) in feed at 70 °C (Fig. 7A). The variations of power density with current density are also presented in Fig. 7(B) under similar experimental conditions.

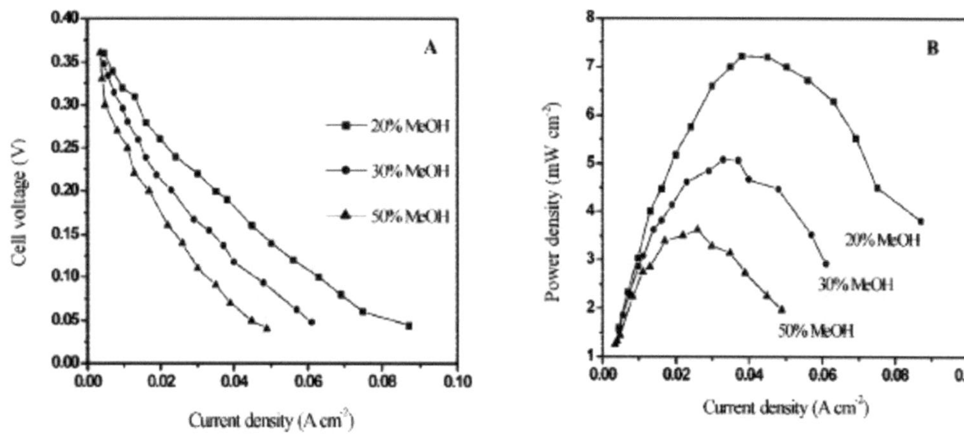

Fig. 7. Current-voltage polarization curves for PCS-3-3 membrane with varying methanol concentration at 70 °C in air mode.

At low methanol concentration, low methanol crossover increased the open circuit voltage (OCV) and significantly improved the performance. In this study, all polyelectrolyte membranes showed almost comparable OCV (0.35 V). The current density for PCS-3-3 at the different methanol concentrations was measured at about 38, 35, and 26 mA cm^{-2}.

The nanostructure hybrid material was prepared from sepiolite with the cationic biopolymer CS. The strong interaction occurred between them after blending to each other and enhanced the thermal stability of the CS-sepiolite nanocomposite. Thus, the new CS-sepiolite hybrid materials are processed as self-supporting films that could potentially be used in different applications such as separation processes of gas mixtures, components for electrochemical sensors, or membranes in fuel-cell devices [141]. New PEM composite membrane comprises of CS, PVA and calcium oxide (CaO) were fabricated. The effect of CaO particle addition into CS-PVA on conductivity and the feasibility of membranes for DMFC application were also carried out. The CaO particles have the profound effect on the CS–PVA membrane conductivity. These results indicated that CS-PVA-CaO composite membrane have excellent methanol barrier properties and found to be feasible for DMFC applications [142]. CS complex membranes were prepared and characterized at room temperature [143]. Fluorescence measurements were correlating the improved activity to alterations in the amphiphilic nature of the chemical microenvironment (immediately surrounding the enzyme) that the modified CS affords [144]. Phosphorylated CS membranes were prepared from the reaction of orthophosphoric acid and urea on the surface of CS membranes in N,N-dimethylformamide [143]. Membrane based on stabilized phosphotungstic acid (PTA) incorporated to chitosan (CS)-hydroxy ethyl cellulose (HEC) for application in direct methanol fuel cells (DMFCs) was reported. Membranes were characterized using FTIR, TGA, SEM and their mechanical properties were also evaluated. The PTA content in the CS-HEC blend membrane and its influence on proton conductivity, water/methanol sorption, and methanol cross-over in the DMFC was studied. The PTA-CS-HEC mixed matrix 3 (wt. %) stabilized membrane used in DMFC testing delivered peak power density of 58 mWcm2 at a load current [145].

4. Chitosan based polymer electrolyte membranes for pervaporation (PV)

PV is a relatively new membrane separation process that has elements in common with reverse osmosis and membrane gas separation. In pervaporation, the liquid mixture to be separated (feed) is placed in contact with one side of a membrane and the permeated product (permeate) is removed as a low pressure vapor from the other side. The permeate vapor can be condensed and collected or released as desired. The chemical potential gradient across the membrane is the driving force for the mass transport. Pervaporation transport is usually described to be a three-step process: solution-diffusion-evaporation. The separation is based on the selective solution and diffusion. In developing pervaporation membranes, three issues must be addressed: (i) membrane productivity, (ii) membrane selectivity, and (iii) membrane stability. Membrane productivity is a measure of the quantity of a component that permeates through a specific area of membrane surface in a given unit of time. Membrane productivity is frequently characterized by permeation flux, *J*, which relates the product rate to the membrane area required to achieve the separation. Note that permeation flux depends on both the intrinsic permeability and the effective thickness of a membrane. The commercialization of the pervaporation technique is, to a large extent, attributed to the engineering approach of making thin membranes in asymmetric and composite forms. When the separation factor is unity, no separation occurs [146]. The silica CS membrane modified with functionalized silica (CSM-5 and CSM-10) were more suitable for the pervaporation and dehydration of ethanol/ aqueous mixture solution. Organic-inorganic hybrid membranes with various TEOS contents from quaternized CS - tetra ethoxy silane (*q*-Chito- TEOS) were prepared by sol-gel reaction [147]. This study suggested that the preparation of organic-inorganic hybrid membranes for the dehydration of an azeotropic mixture of ethanol/water was possible by minimizing membranes swelling which was accomplished by controlling TEOS content in the hybrid membrane matrix (Fig.8.). The CS membrane was modified with oppositely charged surfactants such as sodium dodecyl sulfate (SDS), sodium laurate (SL), sodium stearate (SS), dioctyl sodium sulfosuccinate (DSS) and amphoteric sodium *N*-lauroyl sarcosinate (SLS) [148]. The developed membranes were used for the separation of water–isopropanol mixtures at different temperatures. The highest separation selectivity was found to be 31,648 with a flux of 4.4×10^{-2} kg m^{-2} h^{-1} for membrane (M-3) at 30°C for 5 mass% of water in the feed.

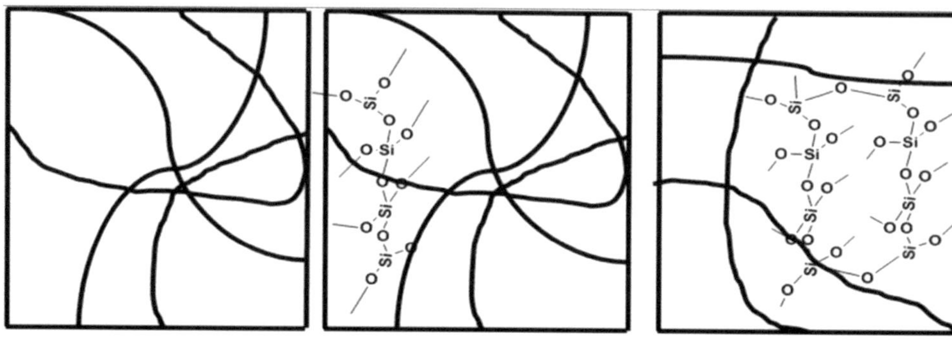

Fig. 8. Change of cross-linked structure in q-chitosan/TEOS hybrid membranes.

Significantly activation energy for water (E_{pw}) lower than that of isopropanol (E_{pIPA}) suggested high separation ability for developed PECs membranes. Huang et al. prepared composite CS membranes on porous polyetherimide substrate support for the pervaporation separation of ethanol/toluene and methanol/toluene mixtures [149]. The CS membranes were acetylated in various concentrations of acetic anhydride solutions. The developed membranes were mechanically robust and stable to withstand the corrosive nature of ethanol/toluene mixture during the PV experiments. Developed composite membranes were considered as good candidate for PV separation of organic/organic liquid mixture. Homogeneous polyelectrolyte complex membranes (HPECMs) were fabricated by poly electrolyte complexes (PECs): CS/ sodium carboxymethyl cellulose (CMCNa) in aqueous NaOH solution [150]. TiO_2 particles were generated at the nanosized level, and dispersed homogeneously within the polymer matrix without aggregation even at high TiO_2 content. CS/TiO_2 nanocomposite membranes were considered as good candidate for the separation of ethanol from ethanol and water mixture.

CS/PSF composite membrane was prepared in order to enhance the permeation rate [151]. CS composite membranes were prepared by casting CS solution onto a porous poly(ether sulfone) (PES) based ultra-filtration membrane with various surface cross-linking densities. The surface cross-linked CS composite membranes exhibited a high selectivity with a low permeation flux for the dehydration of water-alcohol mixtures by PV. Permeate flux and water concentration in the permeate decreased drastically with feed of ethanol concentration greater than 97 wt%. Permeation rate of CS composite membranes was less temperature dependent than that of PVA. IPA-water mixture was similar tendency as that of ethanol-water mixture in pervaporative dehydration performances [152]. Nwawi et al. prepared homogeneous and composite CS membranes by solution casting technology [153]. The composite CS/PSF composite membrane exhibited an advantage over the homogeneous when the isopropanol concentration in feed solution was increased more than 70%. CS–silica hybrid membranes (CSHMs) were prepared by cross-linking of CS by using 3-aminopropyl-triethoxysilane (APTEOS) as cross–linker to enhance flux and selectivity [154]. PV separation performance of water-ethanol by using the CS membrane was clearly correlated with the crystal property aroused from the preparation temperature. The prepared membranes showed the best separation index at 323K temperature. CS membranes were converted into its free amine form [155]. Devi et al. prepared novel cross-linked CS/poly(vinylpyrrolidone)(PVP) blend membranes for the dehydration of tetrahydrofuran (THF) by PV. Blend membrane was cross-linked with glutaraldehyde, followed by further cross-linking with sulfuric acid to improve the membrane resistivity towards water-rich media [156]. Won et al. prepared CS membranes and cross-linked by sulfuric acid. Dimethyl carbonate (DMC) an environmentally benign chemical is commercially produced by oxidative carbonylation of methanol, and the separation and purification of DMC is critical for the production of DMC due to azeotropic nature of the reaction mixtures. The membranes were tested for separation of binary DMC/water, DMC/methanol and methanol/water mixtures and ternary DMC/methanol/water mixtures in a temperature range of 25-55°C. It was found that coupling effect was important in PV separation of multi-component mixtures due to interactions among the permeating species [157]. CS homogeneous membrane, PVA-poly (acrylonitrile) (PVA-PAN) and CS-PVA/poly (acrylonitrile) (CS-PVA/PAN) composite membranes were prepared for the separation of ethanol–water solutions [158].

Bacterial cellulose membrane (BCM) impregnated with CS based membrane was prepared and designated as CS–BCM. Then the modified membrane was used for the separation of ethanol/water (EtOH/H_2O) azeotrope by PV. The effect of temperature on selectivity and flux through CTSN–BCM was also investigated. It was concluded that the developed CS-BCM based membrane can be an alternate for use in PV dehydration of EtOH/H_2O azeotrope mixture [159]. Sodium zeolite-Y (NaY zeolite) filled CS polymeric membranes were developed and characterized for the PV dehydration performances [160].

5. Conclusions

Interest in chitin was initially cultivated mainly by zoologists, marine entomologists, and physiologists, but in the late 1970s, chemists all over the world devoted attention to chitin. In particular, it was immediately realized that chitin was an abundant source of chitosan, the unique cationic polysaccharide (as opposed to a variety of easily accessible anionic polysaccharides), and, as such, was superior to man-made cationic derivatives of cellulose and starch. Chitin and its primary derivative chitosan, biopolymers from renewable resources (shells of shellfish, the wastes of the seafood industry), offer a distinct set of advantageous biological and physico-chemical characteristics that qualify them for a multiplicity of industrial, biotechnological, medical, food industry, separations and applications for energy devices such as fuel cell. These membrane processes are contributing to cleaner industries, a cleaner environment, water and food. The effective wide applications of chitosan based materials or membranes; however, becoming possible once the manufacture of chitinous materials of standardized characteristics has been worked out.

The distinctive advantages of chitosan include availability, biocompatibility, biodegradability, non-toxicity, antimicrobial properties, heavy metal ions chelation, gel forming properties, ease of chemical modification, and high affinity to proteins. These characteristics of the polymer (chitosan) can be widely exploited in the near future in environmentally benign membrane based applications for environment, bio-technology and energy devices. This Chapter looks at how chitin/chitosan materials can contribute to the development of membrane-based processes.

6. Acknowledgements

Authors are thankful to Dr. P. K. Ghosh, Director, CSMCRI, Bhavnagar-364002, India for his encouragements and supports.

7. References

[1] E. Chiellini, F. Chiellini, P. Cinelli, in: G. Scott (Ed.), Kluwer, Dordrecht, 2002, p. 163.
[2] S.M. Hudson, C. Smith, in: D.L. Kaplan (Ed, Springer, Berlin, 1998, p. 96.
[3] K. Kurita, Prog. Polym. Sci. 26 (2001) 1921.
[4] J.A. Howell, Desalination 144 (2002) 127.
[5] Q. Li, E.T. Dunn, E.W. Grandmaison, M.F.A. Goosen, J. Bioact. Comp. Polym. 7 (1992) 379.
[6] M. Peter, J. Macromol. Sci. Pure Appl. Chem. A32 (1995) 629.
[7] W. Paul, C.P. Sharma, STP Pharma Sci. 10 (2000) 5.
[8] A.K. Singal, M. Chawla, Pharm. Pharmacol. 53 (2001) 1047.
[9] M. Minoru, S. Hiroyuki, S. Yoshihiro, Trends Glycosci. Glycotechnol. 14 (2002) 205.
[10] M.N.V.R. Kumar, React. Funct. Polym. 46(2000) 1.
[11] E. Guibal, Sep. Purif. Technol. 38(2004) 43.
[12] A.J. Varma, S.V. Deshpande, J.F.Kennedy, Carbohydr. Polym. 55 (2004) 77.
[13] C. Gerente, V.K.C. Lee, P.L. Cloirec, G. McKay, Crit. Rev. Environ. Sci. Technol. 2007;37:41.
[14] G. Crini, P.M. Badot, Prog. Polym. Sci. 33 (2008) 399.
[15] P. Miretzky, A. C. Fernandez, J. Hazard. Mater. 167 (2009) 10.
[16] G. Crini, Prog. Polym. Sci. 30 (2005) 38.
[17] J. Ramirez-Salgado, Electrochim. Acta 52 (2007) 3766.
[18] B.P. Tripathi, V.K. Shahi, J. Phys. Chem. B 112 (2008) 15678–15690.
[19] P. Mukoma, B.R. Jooste, H.C.M. Vosloo, J. Power Sources 136 (2004) 16–23.
[20] T. Uragami, Polysaccharides, Marcel Dekker, New York, 1998, p. 887.
[21] T. Uragami, K. Takigawa, Polymer 31 (1990) 668.
[22] Y.-L. Liu, C.-Y. Hsu, Y.-H. Su, J.-Y. Lai, Biomacromolecules 6 (2005) 368.
[23] N.L. Nemerrow, Liquid waste of industry. California: Addison Wesley Publishing Company; 1971.
[24] F.W. Pontius, Water quality and treatment. 4th Ed. New York: McGraw-Hill, Inc.; 1990.
[25] S. Chatterjee, S.H.Woo, J. Hazard. Mater. 164(2009) 1012.
[26] R. Yao, F. Meng, L. Zhang, D. Ma, M.Wang, J. Hazard. Mater. 165(2009) 454.
[27] G.Z. Kyzas, M. Kostoglou, N.K .Lazaridis, Chem. Eng. J. 152(2009) 440.
[28] S. Jagtap, D. Thakre, S. Wanjari, S. Kamble, J. Colloid Interface Sci. 332 (2009) 280.
[29] P. Monvisade, P. Siriphannon, Appl. Clay. Sci. 42 (2009) 427.
[30] G. Crini, B. Martel, G. Torri, Int. J. Environ. Pollut. 34 (2008) 451.
[31] A.C. Chao, S.S. Shyu, Y.C. Lin, F.L. Mi, Bioresour. Technol. 91(2004) 157.
[32] R. Bassi, S.O. Prasher B.K. Simpson, Sep. Sci. Technol. 35 (2000) 547.
[33] E. Guibal, C. Milot, J.Roussy, Water Environ. Res. 71(1999)10.
[34] E. Guibal C. Milot, J.M. Tobin, Ind. Eng. Chem. Res. 37(1998) 1454.
[35] L. Dambies, A. Roze, E. Guibal Adv. Chitin. Sci. 4 (2000) 302..
[36] M.L. Arrascue, H.M. Garcia, O. Horna, E.Guibal Hydrometallurgy 71 (2003)191.
[37] V.A. Spinelli, M.C.M. Laranjeira, V.T. Favere, React. Funct. Polym. 61(2004) 347..
[38] M. Kumar, B. P. Tripathi, V. K. Shahi, J. Hazard. Mater. 172 (2009) 1041.
[39] S. Hasan, A. Krishnaiah, T. K. Ghosh, D. S. Viswanath, Ind. Eng. Chem. Res. 45 (2006) 5066.
[40] C. Jeon, W.H.H. Holl, 37 (2003) 4770.
[41]A. Atia, Hydrometallurgy 80 (2005) 13.
[42] J. Song, J. Li, D. Wang, Y. Wang, Ion Exchange Adsorption 24 (2008) 175.

[43]A.M. Donia, A. Atia, K.Z. Elwakeel, J. Hazard. Mater. 151 (2008) 372.
[44] L. Hakim, A. Sabarudin, K. Oshita, M. Oshima, S. Motomizu, Talanta 76 (2008) 1256.
[45] W. S. W. Ngah, K. H. Liang, Ind. Eng. Chem. Res. 38 (1999) 1411.
[46] C. Septhum, S. Rattanaphani, J.B. Bremner, V. Rattanaphani, J. Hazard. Mater. 148 (2007) 185.
[47] S. Sun, L. Wang, A. Wang, J. Hazard. Mater. B136 (2006) 930.
[48] N. Li, R. Bai, Ind. Eng. Chem. Res. 45(2006)7897.
[49] F. Zhao, B. Yu, Z. Yue, T. Wang, X. Wen, Z. Liu, C. Zhao J. Hazard. Mater.147 (2007) 67.
[50] K. C. Justi, V. T. Favere, M. C.M. Laranjeira, A. Neves, R. A. Peralta, J. Colloid Interface Sci. 291 (2005) 369.
[51] K. C. Justi, M. C.M. Laranjeira, A. Neves, A. S. Mangrich, V. T. Favere, Polymer 45 (2004) 6285.
[52] R. L. Machado, E. J. de Arruda, C. C. Santana, S. M. A. Bueno, Process Biochemistry 41 (2006) 2252.
[53] F. Zha, S. Li, Y. Chang, J. Yan, J. Membr. Sci. 321 (2008) 316
[54]A. Saxena , A. Kumar , V. K. Shahi, J. Colloid Interface Sci. 303 (2006) 484.
[55]C.L.de Vasconcelos, P.M. Bezerril, T.N.C. Dantas, M.R. Pereira, and J.L.C. Fonseca, Langmuir 23(2007) 7687.
[56] H.-D. Wang, L.-Y. Chu, H. Song, J.-P. Yang, R. Xie, J. Membr. Sci.297 (2007) 262.
[57]V.P. Hoven, V.Tangpasuthadol, Y. Angkitpaiboon, N. Vallapa, S. Kiatkamjornwong, Carbohydrate Polymers 68 (2007) 44.
[58] X. Chen, J. Liu, Z. Feng, Z. Shao, J. Appl. Polym. Sci. 96(2005) 1267.
[59] W. Shi, Y.Q. Shen, D.T. Ge, M.Q. Xue, H.H. Cao, S.Q. Huang, J.X. Wang., G.L. Zhang, F. B. Zhang, J. Membr. Sci. 325 (2008) 801.
[60] V. Tangpasuthadol, N. Pongchaisirikul, V. P. Hoven, Carbohydrate Res. 338 (2003) 937.
[61] L. Yang, P. Chen, J. Membr. Sci. 205 (2002) 141.
[62] H.-D. Wang, R. Xi, C.H. Niu, H. Song, M. Yang, S. Liu, L.-Y. Chua, Chem. Eng. Sci. 64 (2009) 1462.
[63] S.R. Ahmed, A.B. Kelly, T.A. Barbari, J. Membr. Sci. 280 (2006) 553.
[64] M.M. Beppu, R.S. Vieira, C.G. Aimoli, C.C. Santana, J. Membr. Sci. 301 (2007) 126.
[65] B. Krajewska, React. Funct. Polym. 47 (2001) 37.
[66] C.J.M. Nova, D. Paolucci-Jeanjean, M.-P. Belleville, M.B.M. Rivallin, G. Rios, J. Membr. Sci. 321 (2008) 81.
[67] Y. Hu, M. Wang, D. Wang, X. Gao, C. Gao, J. Membr. Sci. 319 (2008) 5.
[68] C. Liu, R. Bai, J. Membr. Sci. 279 (2006) 336.
[69]C. Liu, R. Bai, J. Membr. Sci. 267 (2005) 68.
[70] C.K. Larive, S.M. Lunte, M. Zhong, M.D. Perkins, S. George, G.S. Wilson, Anal. Chem. 71 (1999) 389R.
[71] R.T. Kurnik, A.W. Yu, G.S. Blank, A.R. Burton, D. Smith, A.M. Athalye, Biotechnol. Bioeng. 45 (1995) 149.
[72] R.V. Reis, A.L. Zydney. J. Membr. Sci. 297 (2007) 16.
[73] A. Chollangi, M.M. Hossain, Chem. Eng. Process 46 (2007) 398.
[74] J. Mathew, C.T. A. Kumar, U. K. Aravind, J. Membr. Sci. 325 (2008) 625.
[75] S. Boributh, A. Chanachai, R. Jiraratananon, J. Membr. Sci. 342 (2009) 97.
[76] U. K. Aravind, J. Mathew, C.T. Aravindakumar, J. Membr. Sci. 299 (2007) 146.
[77] P. Schausberger, N. Norazman, H. Li, V. Chen, A. Fried, J. Membr. Sci. 337 (2009) 1.

[78] D.A. Musale, A. Kumar, G. Pleizier, J. Membr. Sci. 154 (1999) 163.
[79] H. Bechhold New York: Chemical Catloge Company New York; 1926.
[80] H. Yukawa, K. Shimura, A. Suda, J. Chem. Eng. Japan 16 (1983) 305.
[81] K. Weber, W. Stahl, Chem. Eng. Technol. 26 (2003) 44.
[82] H. Yoshida, K. Kitajyo, M. Nakayama, Drying. Technol. 17(1999) 539.
[83] H. Yukawa, K. Shimura, A. Suda, A. Maniwa, J. Chem. Eng. Japan 16(1983) 246.
[84] J.W. Yu, I. Neretnieks, Chem. Eng. Sci. 51(1996) 4355.
[85] J.D. Henry, L.F. Lawler, C.H. A. Kuo, AIChE J. 23(1977) 851.
[86] A.D. Enevoldsen, E.B. Hansen, G. Jonsson, J. Membr. Sci. 299 (2007) 28.
[87] C.W. Robinson, M.H. Siegel, A. Condemine, C. Fee, T.Z. Fahidy, B.R. Glick, J. Membr. Sci. 80 (1993) 209.
[88] G. Bargeman, G.H. Koops, J. Houwing I. Breebaart H.C. Van der Horst, M. Wessling Desalination 149 (2002) 369.
[89] M. Sung, C.P. Huang, Y.H. Weng, Y.T. Lin, K.C. Li, Sep. Purif. Technol. 54(2007) 170.
[90] B.D. Fair, A.M. Jamiesson, J. Colloid Interface Sci. 73 (1980) 130.
[91] R.J. Wakeman, E.S. Tarleton, Filtr. Sep. 23 (1986) 174.
[92] C. Guizard, F. Legault, N. Idrissi, A. Larbot, C. Gavach, J. Membr. Sci. 38 (1988) 147.
[93] G.M. Rios, H. Rakatoarisoa, B.T. Fuente. J. Membr. Sci. 38 (1988) 147.
[94] E. Iritani, K.Ohashi, T. Murase J. Chem. Eng. Japan 14 (1998) 79.
[95] D. Zhou, H. Zhao, W.E. Price, G.G. Wallace. J. Membr. Sci. 98(1995) 173.
[96] H. Nakakura, A. Yamashita, M. Sambuichi, K. Osasa, J. Chem. Eng. Japan 30(1997) 1020.
[97] R.J.Wakeman, Trans I Chem. E Part C Food. Bioprod. Proc. 76 (1998) 53.
[98] S. Kimura, T. Nomura J. Membr. Sci. 7(1982) 245.
[99] V. Karthik, S D. Gupta, S. De. J. Membr. Sci. 199(2002) 29.
[100] A.K. Turkson, J.A. Mikhlin, M.E. Weber, Sep. Sci. Technol. 24 (1990) 1261.
[101] B.M.Verdegan, Sep. Sci. Technol. 21(1986) 603.
[102] C. Visvanathan R.B. Aim Sep. Sci. Technol. 24(1989) 383.
[103] S.P. Moulik, F.C. Cooper, M. Bier J. Colloid Interface Sci. 24 (1967) 427.
[104] J.M. Radovitch, N.S. Mason, R.E. Spark. Sep. Sci. Technol. 15 (1980) 1491.
[105] A. Saxena, B. P. Tripathi, V.K. Shahi, J. Colloid Interface Sci. 319 (2008) 252.
[106] K. Kordesch, G. Simader, VCH Publishers Inc., New York, 1996.
[107] A. B. Stambouli, E. Traversa, Fuel cells, Renew. Sustain. Energy Rev. 6 (2002) 297.
[108] A. B. Stambouli, E. Traversa, Renew. Sustain. Energy Rev. 6 (2002) 297.
[109] Y. Xiang, M. Yang, Z. Guo, Z. Cui, J. Membr. Sci. 337 (2009) 318.
[110] Y. Zhang, Z. Cui, C. Liu, W. Xing, J. Zhang J. Power Sources 194 (2009) 730.
[111] B. Smitha, S. Sridhar, A.A. Khan, Macromolecules, 37 (2004) 2233.
[112] Y. Wan, B. Peppley, K. A.M. Creber, V. T. Bui, E. Halliop, J. Power Sources 162 (2006) 105.
[113] H. Wu, B. Zheng, X. Zheng, J. Wang, W. Yuan, Z. Jiang, J. Power Sources 173 (2007) 842.
[114] Y. Xiong, Q. L. Liu, Q. G. Zhang, A. M. Zhu, J. Power Sources 183 (2008) 447.
[115] P. Mukoma, B.R. Jooste, H.C.M. Vosloo, J. Membr. Sci. 243 (2004) 293.
[116] M. Yamada, I. Honma, Electrochim. Acta 50 (2005) 2837.
[117] P.O. Osifo, A. Masala, J. Power Sources 195 (2010) 4915.
[118] Y. Wan, K. A.M. Creber, B. Peppley, V. Tam Bui, J. Membr. Sci. 280 (2006) 666.

[119] J. Wang, X. Zheng, H. Wu, B. Zheng, Z. Jiang, X. Haob, B. Wang J. Power Sources 178 (2008) 9.
[120] J. Wang, Y. Zhang, H. Wu, L. Xiao, Z. Jiang, J. Power Sources 195 (2010) 2526.
[121] J.A. Seo, J. H. Koh, D. K. Roh, J. H. Kim, Solid State Ionics 180 (2009) 998.
[122] F. Goktepe, S.U. Celik, A. Bozkurt, J. Non-Crystalline Solids 354 (2008) 3637.
[123] B. Smitha, D. A. Devi, S. Sridhar, J. Hyd. Energy, 33 (2008) 4138.
[124] V.V. Binsu, R.K. Nagarale, V.K. Shahi, P.K. Ghosh, React. Funct. Polym. 66 (2006) 1619.
[125] J. Ramırez-Salgado, Electrochim. Acta 52 (2007) 3766.
[126] P. Mukoma, B.R. Jooste, H.C.M. Vosloo, J. Power Sources 136 (2004) 16.
[127] B. Smitha, S. Sridhar, A.A. Khan, Europ. Polymer J. 41 (2005) 1859.
[128] Y. Wang, D. Yang, X. Zheng, Z. Jiang, J. Li, J. Power Sources 183 (2008) 454.
[129] L.C. Cogo, M.V. Batisti, M.A. Pereira-da-Silva, O.N. Oliveira Jr, F.C. Nart, F. Huguenin, J. Power Sources 158 (2006) 16
[130] H. Wu, W. Hou, J. Wang, L. Xiao, Z. Jiang, J. Power Sources 195 (2010) 4104.
[131] Z. Cui, C. Liu, T. Lu, W. Xing, J. Power Sources 167 (2007) 94.
[132] Y. Xiong, Q.L. Liu, Q.G. Zhang, A.M. Zhu, J. Power Sources 183 (2008) 447.
[133] Z. Jiang, X. Zheng, H. Wu, J. Wang, Y. Wang, J. Power Sources 180 (2008) 143.
[134] L. Depre, M. Ingram, P.M. Poinsignon, Electrochim. Acta 45 (2000) 1377.
[135] V.V. Binsu, R.K.Nagarale, V.K. Shahi, J. Mater. Chem. 15 (2005) 4823.
[136] A. Saxena, B.P. Tripathi, V.K. Shahi, J. Phys. Chem. B 111(2007) 12454.
[137] V. Ganesan, A.Walcarius, Langmuir, 20 (2004) 3632.
[138] R. K.Nagarale, G. S.Gohil, V. K. Shahi, R. Rangarajan, Macromolecules 37(2004) 10023.
[139] M. Schuster, W.H. Meyer, G.Wenger, H.G. Herz, M. Ise, M. Schuster, K. D. Kreuer, J. Solid State Ionics 145 (2001) 85.
[140] S. Jacob, S. Cohet, C. Poinsignon, M. Popall, Electrochim. Acta 48 (2003) 2181.
[141] H.Wu, B. Zheng, X. Zheng, J. Wang, W. Yuan, Z. Jiang, J. Power Sources 173 (2007) 842.
[142] K. Soontarapa, U. Intra, Chem. Eng. Comm. 193(2006)855.
[143] C. Lau, G. Martin, S. D. Minteer, M.J. Cooney, Electroanalysis 22 (2010) 793.
[144] S. Mohanapriya, S.D. Bhat, A.K. Sahu, S. Pitchumani, P. Sridhar, A.K. Shukla, Energy Environ. Sci. 2 (2009) 1210.
[145] M.Y. Kariduraganavar, J.G. Varghese, S.K. Choudhari, R.H. Olley, Ind. Eng. Chem. Res. 48 (2009) 4002.
[146] T. Uragami, T. Katayama, T. Miyata, H. Tamura, T. Shiraiwa, A. Higuchi, Biomacromolecules 5(2004) 1567.
[147] R.Y.M. Huang, G.Y. Moon, R. Pal, J. Membr. Sci. 184 (2001) 1.
[148] R.Y.M. Huang, G.Y. Moon, R. Pal, J. Membr. Sci. 176 (2000) 223.
[149] Q. Zhao, J. Qian, Q. An, C. Gao, Z. Gui, H. Jin, J. Membr. Sci. 333 (2009) 68.
[150] X. Feng, R.Y. H. Huang, J. Membr. Sci. 116 (1996) 67.
[151] Y.M. Lee, S. Y. Nam, D. J. Woo, J. Membr. Sci. 133 (1997) 103.
[152] M.G.M. Nawawi, R.Y.M Huang, J. Membr. Sci. 124 (1997) 53.
[153] J. H. Chen, Q. L. Liu, X. H. Zhang, Q. G. Zhang, J. Membr. Sci. 292 (2007) 125.
[154] W. Won, X. Feng, D. Lawless, J. Membr. Sci. 209 (2002) 493.
[155] D.A. Devi, B. Smitha, S. Sridhar, T.M. Aminabhavi, J. Membr. Sci. 280 (2006) 45.
[156] W. Won, X. Feng, D. Lawless, Sep. Purif. Technol. 31 (2003) 129.
[157] B.-B. Li, Z.-L. Xu, F. A. Qusay, R. Li, 193 (2006) 171.
[158] V. Dubey, L. K. Pandey, C. Saxena, J. Membr. Sci. 251 (2005) 131.

[159] A.L. Ahmad, M.G. M. Nawawi, L.K. So, J. Appl. Polymer Sci. 99, (2006) 1740
[160] K.S.V.K. Rao, M.C.S. Subha, M. Sairam, N.N. Mallikarjuna, T.M. Aminabhavi, J. Appl. Polymer Sci. 103 (2007) 1918.

Section B

11

Fabrication of HA/PLLA Composite Scaffolds for Bone Tissue Engineering Using Additive Manufacturing Technologies

Fernando Cruz
Polytechnic Institute of Setubal
Portugal

1. Introduction

This chapter presents an overview of a research work carried out for the rapid manufacture of bioceramics (HA)/biopolymers (PLLA) composite scaffolds by means of Selective Laser Sintering (SLS) technique, to be used for bone tissue implantation aiming to replace and/or repair bone defects due to traumatised, damaged or lost bone.

SLS is an Additive Manufacturing Technology (AMT) that selectively sinters powders of engineering materials, from solid or surface models created by a CAD - 3D file, by means of a CO_2 laser and in a layer-by-layer basis.

HA (hydroxyapatite) is a bioceramic used since several years for medical applications although being mainly processed by conventional methods (cast, machined or manually produced).

In recent years ceramics and their composites are being used to augment or replace various parts of the body, particularly bone. Their relative inertness to the body fluids, high compressive strength, and aesthetically pleasing appearance led to the use of ceramics in dentistry as dental crowns. Some carbons have found use as implants, especially for blood-interface applications, such as heart valves. Due to their high specific strength as fibbers and their biocompatibility, ceramics are also being used as reinforcing components of composite implants and for tensile loading applications such as artificial tendons and ligaments.

Calcium phosphate based bioceramics have been in use in medicine and dentistry in the last thirty years, in several applications: dental implants, periodontal treatment, alveolar ridge augmentation, orthopaedics, maxillofacial surgery, and otolaryngology. Different phases of calcium phosphate ceramics are used depending upon whether a resorbable or bioactive material is desired. Porous hydroxyapatite has been studied for use in repairing large defects in bone. Implanted HA is slowly resorbed by the body over several years and replaced by bone. HA as the capability to form a direct chemical bond with hard tissues. On implantation of HA particles or porous blocks in bone, new lamellar cancellous bone forms within 4 to 8 weeks. Porous materials are used on bone compatible implants to encourage bony ingrowths. Therefore, pore size is of considerable biological importance.

PLLA (Poly-L-lactic acid) is a biodegradable polymer. The interest in biodegradable polymeric biomaterials for biomedical engineering has increased dramatically during the

past decade, because this class of biomaterials has two major advantages that non-biodegradable biomaterials do not have.
First, these biomaterials don't elicit permanent chronic foreign-body reaction due to the fact that the human body would gradually absorb them, and they do not permanently retain trace of residual in the implantation sites. Second, some of them have recently been found to be able to regenerate tissues (*tissue engineering*), through the interaction of their biodegradation with immunologic cells. Hence, surgical implants made from biodegradable biomaterials could be used as temporary scaffold for tissue regeneration.
This approach toward the reconstruction of injured, diseased, or aged tissues is considered one of the most promising fields for this century. One of the bone formation strategies is to construct bio-scaffold, which allows bone reformation once it is surgically placed in bone-defect. The bio-scaffold has to combine mechanical strength with biocompatibility and osteoinductiveness (induction of bone cells) and osteoconductiveness (cells attachment).
A key requirement in many biomedical applications, including artificial bones, is a property called connected porosity. Green parts produced by the chosen SLS technology inherently have this property. While HA has been used in bone replacement and grafting for more than 20 years, this material was a logical choice for this work. PLLA was chosen as a binder because it offers a higher degradation time and a low melting temperature when compared with other acid lactic biodegradable derivates. This biopolymer acts as a binder during the sintering operation, and afterwards improves the mechanical properties of the implants.
In this research work CAD/CAM/CAE systems, AMT, biomaterials and medical imaging techniques were integrated and this enabled the production of anatomical models. This approach provides a direct method of producing HA-based scaffolds suitable for bone replacement or bone tissue engineering.

2. Biomaterials

Biomaterials are natural or synthetic materials, used to replace part of a living system or to function in intimate contact with living tissue.
This group of materials includes:

- Metals (e.g. titanium and its alloys, Co-Cr alloys, stainless steels, etc.);
- Ceramics (e.g. alumina, zirconia, calcium phosphates, marine coral, nacre etc.);
- Polymers (e.g. polyethylene, polypropylene, polyvinyl, poly(lactide acid); poly(glycolic acid), etc.);
- Composites (e.g. carbon-carbon, wire or fibre reinforced bone cement, etc.).

Some of these biomaterials are *biodegradable or resorbable* (e.g. hydroxyapatite, polylactides, or composites of both).
The required properties of biomaterials, for artificial tissue engineering are:

- Biocompatibility (does not induces immune response or inflammation);
- Biodegradability (biodegrade for non-toxic components);
- Sterilizability (able to be sterilisable);
- Stability over timescales, allowing the growth of new tissues;
- Adequate mechanical and physical properties;
- Adequate manufacturing.

A description of materials for use in the human body is shown in Table 1:

Materials	Advantages	Disadvantages	Examples
Polymers (nylon, silicone rubber, polyester, polytetrafuoroethylene, etc.)	Resilient, easy to fabricate	Not strong, deforms with time, may degrade	Sutures, blood vessels, hip socket, ear, nose, other soft tissues
Metals (Ti and its alloys, Co-Cr alloys, stainless steels, Au, Ag, Pt, etc.)	Strong, tough, ductile	May corrode, dense, difficult to make	Joint replacements, bone plates and screws, dental root implants, pacer and suture wires
Ceramics (aluminium oxide, calcium phosphates including hydroxyapatite, carbon, etc.)	Very biocompatible, inert, strong in compression	Brittle, not resilient	Dental, joint replacements, coating of dental and orthopaedic implants
Composites (carbon-carbon, wire or fibre reinforced bone cement)	Strong, tailor-made	Difficult to make	Joint implants, heart valves

Table 1. Materials for use in the body (After Bronzino et al., 1995)

This section will mainly focuses on the biomaterials used in the experimental work of the research (bioceramics and biopolymers).

Calcium phosphate based bioceramics have been in use in medicine and dentistry in the last thirty years, in several applications: dental implants, periodontal treatment, alveolar ridge augmentation, orthopaedics, maxillofacial surgery, and otolaryngology. Different phases of calcium phosphate ceramics are used depending upon whether a resorbable or bioactive material is desired.

Apatite is a natural calcium phosphate containing a little fluorine (Ca_4 (Ca F) $(PO_4)_3$) or chlorine (Ca_4 (Ca Cl) $(PO_2)_3$). It is found in granular limestone, igneous rocks and metalliferous ores.

Hydroxyapatite (HA or HAp) is a complex phosphate of calcium (Ca $(PO_4)_3OH$) that occurs as a mineral and is the chief structural element of vertebrate bone. It is a very important material for bioceramics.

One of the important research issues concerning HA is its mechanical behaviour limitation. In fact, brittleness, poor fatigue resistance, low tensile strength, and low fracture toughness value precludes HA from use in load bearing situations.

Dense, porous, or particulate forms have been prepared. However, porous HA as an implant material, is preferred. The pores (100-300μm) allow bone to grow into implant, promoting mechanical fixation with the natural bone. Nevertheless, porosity and pore size can reduce the mechanical properties of HA ceramic. The minimum pore size of approximately 100-150μm has been established as necessary for the continued health of bony ingrowths.

Porous hydroxyapatite has been studied for use in repairing large defects in bone. Implanted HA is slowly reabsorbed by the body over several years and replaced by bone.

HA has the capability to form a direct chemical bond with hard tissues. On implantation of HA particles or porous blocks in bone, new lamellar cancellous (soft) bone forms within 4 to 8 weeks (Zan et al., 2010).

Porous materials are used on bone compatible implants to encourage bony ingrowths. Pore size can be of considerable biological importance. Studies have shown (Hench, 1991; Chelule, 2002) that with pore sizes ≥ 150μm, in orthopaedic implants, bony ingrowths into the pores occur, and this is useful to anchor the implant. It was found experimentally that pores smaller than 75μm did not permit the ingrowths of bone tissue. Moreover, it was difficult to maintain fully viable osteons within pores in the 75-150μm size range. This large pore size is needed so that capillaries can provide a blood supply to the in-growth of connective tissues. Vascular tissues do not appear if pores < 150μm. If the micro-movements occur at the interface of a porous implant, the capillaries can be cut off, leading to tissue death, inflammation and destruction of interfacial stability (Hench, 1991).

When a porous material is implanted in bone, the pores become filled first with blood, which clots, then with osteoprogenitor mesenchymal cells, then after about 4 weeks, bony trabeculae. The ingrowths bones then become remodelled in response to mechanical stress. The bony ingrowths process depends on a degree of mechanical stability in the early stages. If too much motion occurs, the ingrowths tissue will be collagenous scar tissue, not bone.

The porous and dense HA materials show excellent biocompatibility after implantation. The porous material can only be used to replace those places in the skeleton which are not loaded or are loaded mainly in compression. The dense sintered material does not convert to natural bone after implantation (Chelule, 2002).

Synthetic polymeric materials have been widely used in medical disposable supplies, prosthetic materials, dental materials, implants, dressings, extracorporeal devices encapsulates, polymeric drug delivery systems, and orthopaedic devices such as metal and ceramics substitutes.

The most natural polymer is demineralised bone matrix (DBM), which is produced by decalcifying pieces of bone. Other popular synthetic biodegradable polymers are poly (lactide acid) (PLLA) and poly(glycolic acid) (PGA) (Gibson, 2003).

The required properties of polymeric biomaterials are similar to other biomaterials, which are given in Table 2:

Properties	*Description*
Biocompatibility	Noncarcinogenesis, nonpyrogenicity, nontoxicity, nonallergic response.
Sterilizability	Autoclave, dry heating, ethylenoxide gas, radiation.
Physical property	Strength, elasticity, durability.
Manufacturability	Machining, moulding, extruding, fibre forming.

Table 2. Requirements for biomedical polymers (After Bronzino et al., 1995)

The biomedical applications of polymeric biomaterials are very large, as shown in Table 3.

For implant applications special attention have be given to polyethylene (PE) and polymethylmetacrylate (PMMA).

The effect of implantation on the behaviour of biocompatible polymers is given in Table 4.

The interest in biodegradable polymeric biomaterials for biomedical engineering has increased dramatically during the past decade, because this class of materials has two major

Synthetic Polymers	I. Applications
Polyvinylchloride (PVC)	Blood and solution bag, surgical packaging, dialysis devices, catheter bottles, connectors, and cannulae.
Polyethylene (PE)	Pharmaceutical bottle, no woven fabric, catheter, pouch, flexible container, and orthopaedic implants.
Polypropylene (PP)	Disposable syringes, blood oxygenator membrane, suture, no woven fabric, and artificial vascular grafts.
Polymethylmetacrylate (PMMA)	Blood pump and reservoirs, membrane for blood dialyser, implantable ocular lens, and bone cement.
Polystyrene (PS)	Tissue culture flasks, roller bottles, and filter wares.
Polyethylenterephthalate (PET)	Implantable sutures, mesh, artificial vascular grafts, and heart valve.
Polytetrafluoroethylene (PTFE)	Catheter and artificial vascular grafts.
Polyurethane (PU)	Film, tubing, and components.
Polyamide (nylon)	Packaging film, catheters, sutures, and mould parts.

Table 3. Biomedical applications of polymeric materials (After Bronzino et al., 1995)

Polymers	II. Effects of Implantation
Polyvinylchloride (rigid)	Tissue reaction, plasticizers may leach out and become brittle.
Polyethylene (PE)	Low-density ones absorb some lipids and lose tensile strength; high-density ones are inert, and no deterioration occurs.
Polypropylene (PP)	Generally no deterioration.
Polymethylmetacrylate (PMMA)	Rigid form: crazing, abrasion, and loss of strength by heat sterilization. Cement form: high heat generation, unreacted monomers during and after polymerisation may damage tissues.
Polyethylenterephthalate (PET)	Susceptible to hydrolysis and loss of tensile strength. Solid specimens inert; if fragmented, irritation will occur.
Polytetrafluoroethylene (PTFE)	No tissue reaction, very little deterioration.
Silicone rubber	Absorb water and irritate tissue, lose tensile strength rapidly.
Polyamide (nylon)	

Table 4. Effect of implantation on polymers (After Bronzino et al., 1995)

advantages that non-biodegradable biomaterials do not have. First, these biomaterials don't elicit permanent chronic foreign-body reaction due to the fact that the human body would gradually absorb them, and they do not permanently retain trace of residuals in the implantation sites. Second, some of them have recently been found to be able to regenerate tissues (tissue engineering), through the interaction of their biodegradation with immunologic cells. Hence, surgical implants made from biodegradable biomaterials could be used as temporary scaffold for tissue regeneration.

This approach toward the reconstruction of injured, diseased, or aged tissues is considered one of the most promising fields of this century (Stevens, 2002).

The most commercially significant biodegradable polymeric biomaterials are originated from:

- Glycolic acid (e.g. polyglycolide (PG), polyglycolic acid (PGA));
- Lactic acid (e.g. polylactide (PLA), poly-L-lactide (PLLA)).

Their biomedical applications have been limited to mainly orthopaedic surgery, drug control/release devices, coating materials for suture, vascular grafts, and surgical meshes to facilitate wound healing after dental extraction.

PLLA is a biodegradable semi-crystalline polymer derived from lactic acid ($C_3H_6O_3$).

Lactic acid is a natural organic acid. Long before it became commercially available lactic acid was formed by natural fermentation in products such as cheese, yoghurt, soy sauce, meat products, pickled vegetables, beer and wine. Animal and human bodies also produce significant amounts of L (+) lactic acid during daily activities such as walking and running.

Today, lactic acid, its salts and esters are extensively used in food, industrial, cosmetic, medical and pharmaceutical industries. It is produced by fermentation of sugar and water or by chemical process and is commercially sold as a liquid. Lactic acids are optically active compounds including L-and D- forms.

Applications of this polymer in medicine and pharmacy include wound closure products, drug delivery systems and surgical implant devices. The main advantages of these polymers are their mechanical strength, combined with biodegradability and biocompatibility. The biodegradation of PLLA takes place by random hydrolysis, being the final degradation products eliminated by natural way.

Fig. 1. PLLA image (Designinsite, 2001)

PLLA presents an extremely slow biodegradation rate at body temperature (> 24 months). This biopolymer has a Tm = 170°C and a Tg = 56°C. The high Tg is mainly responsible for the mentioned slow degradation rate. It resembles clear polystyrene as can be seen in Figure 1.

Recently, biodegradable polymers, e.g. polylactic acid (PLA) have been used to treat minimally loaded fractures, thereby eliminating the need for a second surgery for implant removal.

Because metals are too stiff to prevent stress protection, polymers tend to be too flexible and to weak to meet the mechanical demands for an internal fixation device, and bioceramics (e. g. HA) are to brittle and have unfavourable mechanical properties where weight bearing is concerned, composite materials should be considered for this purpose.

Composite materials are solids that contain two or more distinct constituent materials or phases, on a scale larger than the atomic.

Natural composites include bone, wood, dentin, cartilage and skin. In biomaterials, e.g. HA/PLLA composites, each constituent of the composite must be biocompatible. Moreover, the body environment should not degrade the interface between constituents.
Composite materials can be made having carefully tailored properties for specific engineering performance and design requirements whether these are mechanical, biological, chemical, or electrical.
The property of the composite is related to the properties and proportions of the composite constituents and the interface between those constituents. Composites can be produced with a high specific modulus and an advantageous strength-to-weight ratio. Composites can also often have superior toughness since the filler fibres or particles can deflect or stop the process of a crack. This is reflected in a higher strength of the composite compared to that of the matrix material alone. In addition, the filler can confer certain biological properties, such as bioactivity, when incorporated into the composite in appropriate amounts. Bioactive materials, e.g. hydroxyapatite and various bioactive glasses and glass-ceramics, are generally renowned for low toughness and tensile strength.
The properties of the composites (density, stiffness, chemical activity, and possibly biological activity) are determined by the Rule of Mixtures.
Many of the synthetic biocomposites are used in orthopaedics to reconstruct or augment living bone.
In Table 5 the classes of materials that can be used in biocomposites are shown:

Possible matrices	Possible fillers
Thermoplastics	Glasses
Thermosets	Carbon
Ceramics	Ceramics
Metals	Metals
	Polymers

Table 5. Classes of materials that could either be used as fillers or matrices (After Yamamuro et al., 1990)

For bone reconstruction different composites can be used, as shown in Table 6.
Because metals are to stiff to prevent stress protection, polymers tend to be too flexible and to weak to meet the mechanical demands for an internal fixation device, and bioceramics (e. g. HA) are to brittle and have unfavourable mechanical properties where weight bearing is concerned, composite materials should be considered for this purpose. Recently attention has been paid to the application of HA in combination with a polymeric substance. The use is still confined to the field of the filling of bony defects and as drug carrier. Polyethylene, polybyturate and PLLA are the most frequently used polymers in such composite materials.
Recently attention has been paid to the application of HA in combination with a polymeric substance. The use is still confined to the field of the filling of bony defects and as drug carrier. Polyethylene, polybyturate and PLLA are the most frequently used polymers in such composite materials.
Composites of HA / PLLA combines bioresorption (PLLA) with bone bonding potentials (HA) resulting in a potentially bioactive and bioresorbable composite with higher strengths and stiffness than the unfilled polymer (Ignjatovic et al., 1999; Kesenci et al., 2000).

Type of system	Composite
Stable systems	Epoxy resin / carbon fibre
	Polysulfone / carbon fibre
	Epoxy resin / alumina / stainless steel
	Polymethylmethacrilate / polyaramid
	Polyethylene / carbon fibre
	Carbon fibre-reinforced carbon
Bioactive systems	Epoxy resin / Bioglass®
	Collagen / hydroxyapatite
	Polyethylene / hydroxyapatite
	Polymethylmethacrilate / phosphate / silicate / apatite (CPSA) glass fibre
	Polymer / phosphate glass
	Hydroxyapatite / collagen + gelatin - resorcinol - formaldehyde
Degradable systems	Polylactic acid / glycolic acid / with self-reinforcement
	Polyhydroxybutirate / hydroxyapatite
	Polylactic acid / glycolic acid / hydroxyapatite

Table 6. Composites for bone reconstruction (After Yamamuro et al., 1990)

Figure 2 shows new bone formation surrounding an implant of HA/PLLA after 32 weeks:

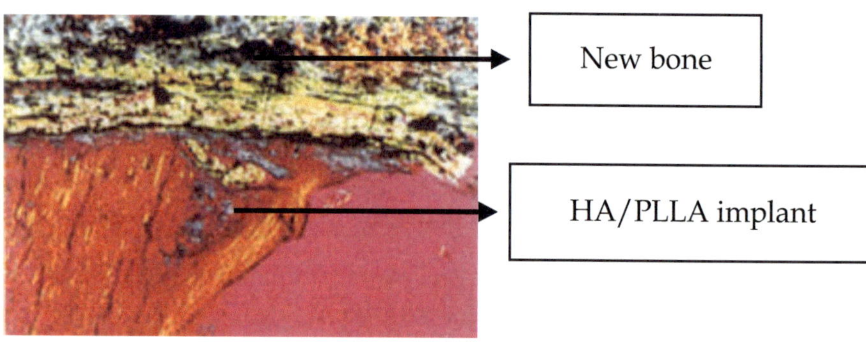

Fig. 2. HA/PLLA implant (Biocomposites, 2001)

3. Selective Laser Sintering

The Selective Laser Sintering (SLS) process belongs to the solid-based Additive Manufacturing Technologies, which primarily use powder as the basic medium for manufacture. The objects are built layer by layer by sintering / interfusing thin layers of powder. Each layer fuses to the previous one creating a physical object.

Thus, the Selective Laser Sintering process creates three-dimensional objects, layer by layer, from powdered materials, with the heat generated by a CO_2 laser within the apparatus (Sinterstation System). First, three dimensional CAD data must be output in the industry-standard .STL (Standard Triangulation Language) format.

The process comprises 4 phases:

- As the selective laser sintering process begins, a thin layer of the heat-fusible powder is deposited into the part build chamber;

- An initial cross-section of the object under fabrication is selectively "drawn" (or scanned) on the layer of powder by a heat-generating CO_2 laser. The interaction of the laser beam with the powder elevates the temperature to the point of melting, fusing the powder particles and forming a solid mass, i.e., sinters the powder particles (heats and bonds selected portions of each layer). The intensity of the laser beam is modulated to melt the powder only in areas defined by the object's design geometry;
- An additional layer of powder is deposited, via a roller mechanism, on top of the previously scanned layer;
- The process is repeated, with each layer fusing to the layer below it. Successive layers of powder are deposited and the process is repeated until the part is complete.

After the building the part is removed from the build chamber and the loose powder falls away. Parts may then require some post-processing, such as sanding, depending upon the intended application.

There is no need to create support structures with the CAD design prior to or during processing and, therefore, no support removal is needed when the part is complete.

The software components of the Sinterstation System include a Unix operating system and proprietary application software.

In Figure 3 a schematic view of the Sinterstation Process Chamber is present:

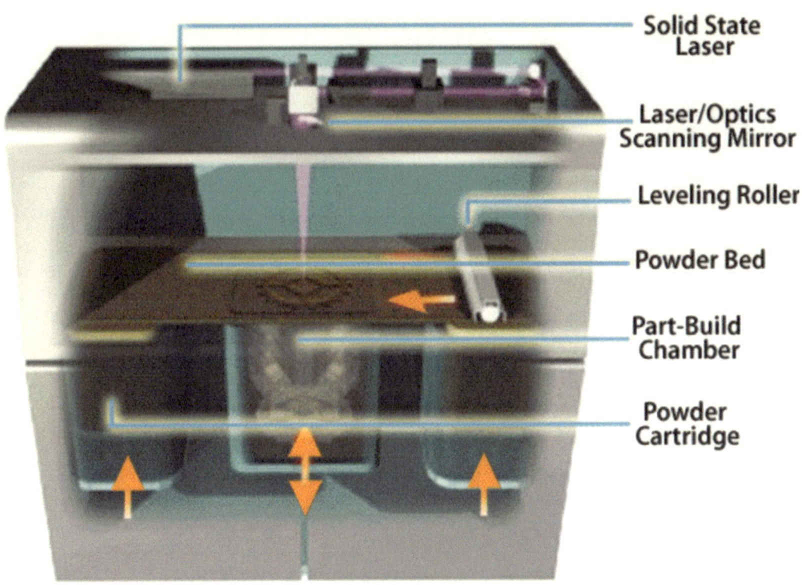

Fig. 3. Process Chamber of the SLS Sinterstation System

The main aspects to be considered in the SLS technology are the properties of the powders used in the process and the fabrication parameters. The fabrication parameters depend strongly on the materials used, and have significant influence in the mechanical properties (e.g. tensile strength, surface hardness and density), dimensional accuracy and surface quality (i.e. roughness) of the parts produced. With SLS it is not possible to achieve the best quality in appearance with the best mechanical properties. Maximum density of the parts is only achieved with parameters that result in excess powder sticking to the surface and low speed of construction (which is time-consuming). Using the correct combination of coating and finishing in post-processing, the mechanical properties and surface of SLS parts can be improved.

A large range of materials is available for this process. Basically, any substance that can be grounded to a fine powder may be employed. At present there are 12 commercially available powders that can be used in the SLS process. The most used are nylon (polyamides), nylon composites (glass fibre nylon), polystyrene and polycarbonate. They are non-toxic, safe, and can be sintered with relatively low power lasers (10-20W).

The main SLS process variables are: part bed temperature, laser power, scan size, scan spacing, slice thickness, and part position and orientation (3D Systems, 2003).

4. Experimental work

In this section an overall view of the experimental work is presented.

Based on previous screening trials, a choice of 40% ratio of PLLA by weight has been made. This choice was based on the fact that adding lower than 30% wt. of PLLA to HA will not add significantly for the biodegradation or to enhance the ductility properties, and higher than 50 % wt. the implant could be too plastic and therefore not behave as bone tissue (Cruz, 2005).

4.1 Materials

The materials used in this work (in powder form) were as follows:

- HA: Captal® 120 grade - sintering powder, supplied by Plasma Biotal Limited (UK), with mean particle size (D50) of 111±5μm, melting point of 1250 (-0 +50) °C and 1.30 g/cm^3 of density (~£300 /Kg);
- PLLA: Purasorb® L, supplied by PURAC (The Netherlands), with mean particle size (D50) of 163±5μm (after sieving), melting range of 182.4 - 192.3°C and 0.47 g/cm^3 of density (after sieving) (~£1700 /Kg).

4.2 Equipments

The experiments were performed on a SLS Sinterstation 2000 (3D Systems™, USA) apparatus installed at CRDM (Centre for Rapid Design and Manufacture), High Wycombe, UK.

This apparatus was modified, namely the feeding envelopes and the part platform, to allow operating with non-standard materials (as the standard material is nylon), and with small amounts of HA/PLLA (1Kg instead of the standard 20Kg) due to cost control of testing.

In Figure 4 a general view of the SLS apparatus used in the experimental work is presented.

4.3 Method

The aim of the first phase of the experimental work was to identify those factors (process parameters) that have large effects on the response variables (physical properties of the parts produced). In this phase a screening experimental design and analysis, based in the fractional factorial design technique, was applied to determine the most influential factors of interest on fabrication of HA/PLLA models by SLS. A DoE (Design of Experiments) methodology was used (Montgomery, 1997).

The results obtained in the tests were focused on three (3) response variables (density, geometric accuracy and surface quality of the parts produced).

The results obtained in the screening analysis show that the three factors with most influence on the physical properties of the parts produced by SLS were laser power, scan speed and scan space, i.e., the Applied Energy Density (AED). In fact, these three factors together constitute the Andrew's Equation (Cruz, 2005), corresponding to the energy density applied to the powder bed, which determines the scanning strategy during the SLS operation.

Fig. 4. The SLS Sintersation 2000 apparatus (Courtesy of CRDM)

After this preliminary stage of the experimental work, the next step was focused on the measurement of the ultimate compressive strength and the elastic modulus in compression, as well as the bending strength and modulus of the parts produced, by means of a 2^k full factorial design and analysis, in order to achieve more detailed information about the most influent factors selected in the previous phase.

A 2^3 full factorial design was applied (Montgomery, 1997). The objective was to obtain a mathematical equation to express the effects of the AED in the ultimate compressing strength, bending and density of HA/PLLA specimens.

A design consisting of sixteen (16=8+8) experiments (i.e., a 2^3 experimental design: 3 independent variables at 2 levels each, with one (1) genuine replicate added to the design) was conducted in a randomised sequence.

The experimental design was conducted according with the data presented in Table 7 (Cruz, 2005):

Factors	*Levels* Low	High
Coding	–1	1
Laser power (W)	5	7.5
Laser scan speed (mm sec^{-1})	200	300
Laser scan spacing (mm)	0.10	0.15
Applied energy density (cal cm^{-2})	2.66	8.96

Table 7. Data for the 2^3 factorial design

To investigate the degradation behaviour of the HA/PLLA parts produced by SLS, namely its PLLA component, in vitro tests were performed. To study the in vitro degradation mechanism of the HA/PLLA specimens, an aqueous media (physiological fluid) - saline phosphate buffer (SPB, pH=7.4) was used. This media has been taken as a model of biological fluids (Taddei et al., 2002). The study was conducted on their physical properties, namely their mass loss along the time (six months).

5. Results

In the following Table 8 and Figures 5 to 7, the main results obtained are summarized (Cruz, 2005):

	Compression		Bending		Density
	σ_c (MPa)	E_c	σ_c (MPa)	E_c	(g cm^{-3})
Cortical bone (hard bone)	200 ± 36	23 ± 4.8 GPa	35–283	5–23 GPa	1.7–2.1
Cancellous bone (soft bone)	1.5–38	10–1570 MPa	-	<1 GPa	0.14–1.1
HA/PLLA (laser sintered)	2.4–4.6	17.2–40.8 MPa	1.6–4	140–257 MPa	0.78–1.1

Table 8. Comparison of properties between HA/PLLA parts and human bone

Fig. 5. SEM image of a HA/PLLA part with measurement of internal porosity (Example in the Z direction - magnification 100x)

Fig. 6. HA / PLLA scaffolds as SLS produced

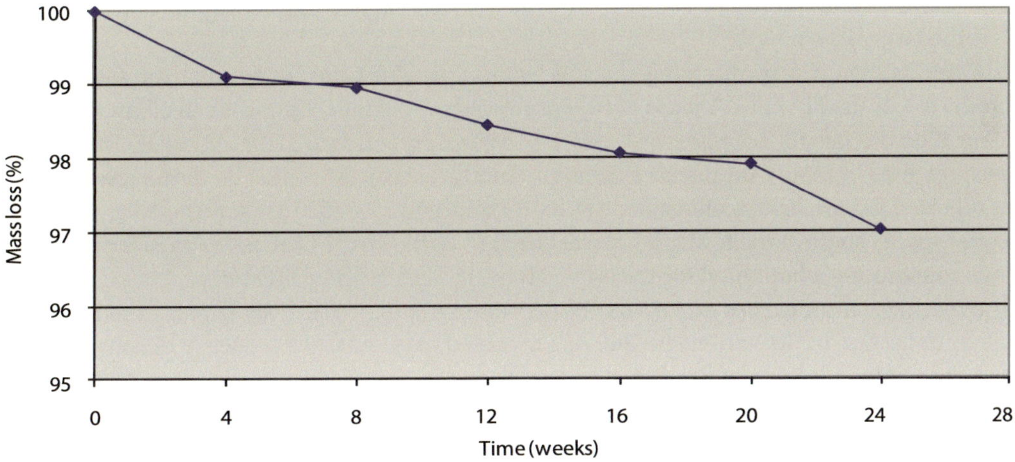

Fig. 7. Percentage of mass loss versus time

6. Discussion

The main goal of the experimental work was to demonstrate the feasibility of producing HA/PLLA parts by selective laser sintering. The results were successful in proving that feasibility.

In fact, the main results of the trials performed, have shown that:

- It is possible to produce bioceramic parts (HA based) by means of AMT - Additive Manufacturing Technologies, namely the Selective Laser Sintering (SLS) process, using a biocompatible/biodegradable polymer as binder;
- For that purpose it is necessary to use a polymeric binder (40% wt.) to promote the agglomeration of the ceramic particles during the laser operation (sintering). The component, in this phase, is the so-called "green body";

- In the case of using a biocompatible binder (e.g. PLLA), the blend HA/PLLA produces a part that can be used directly in the body without removal of the binder (green body). This technique is a direct route to fabricate laser sintered HA based ceramic parts for medical purposes;
- Thus, this route allows the production of parts directly, without the need of post-process operations, therefore reducing the time and cost of production;
- In such a case the mechanical properties of the parts produced constitutes a major limitation, allowing their application only in non-load bearing situations or supported load bearing uses;
- The internal porosity of the HA/PLLA parts produced (mean value >150μm) is suitable to promote the growth of new tissues in the implant. In fact, this condition is essential to allow extensive blood supply of the new bodies, providing a rich supply of cells, growth factors and other nutrients needed to make bone grow. Moreover, these small holes allow access of bone marrow elements from the host tissues to the bone graft, providing nutrients for bone healing;
- The density of the HA/PLLA sintered parts (mean value = 0.883 g/cm^3) is within the density range values for the cancellous bone (0.14 to 1.10 g/cm^3) as previously referred. It can be seen that the green density in nearly the same of the powder bed.
  This result is important because, in principle, the higher the density of bone implant, the larger its strength.

The results obtained by applying the full factorial design have shown that the mechanical properties of the HA/PLLA parts as SLS produced are relatively poor when compared with other biomaterials (σ_c between 2.44 and 4.57 MPa / E_c between 17.19 and 40.80 MPa), but, even so, the Ultimate Compressive Strength and the Elastic Modulus lie in the lower limits of reported values for cancellous bone (1.5 - 38 MPa and 10 - 1570 MPa, respectively).
These results confirm the low mechanical strength of the HA/PLLA parts, as SLS produced, thus constituting a limitation for the applications in load-bearing situations.
The mechanical properties of the HA/PLLA sintered parts by SLS are expected to rise with the optimisation of the various parameters involved, particularly the energy density applied and the grain dimension of the starting powder.

7. Recommendations

It is believed that the investigation carried out can give guidance for other researchers interested in this subject. For successful SLS with HA powder and binder, the following care must be taken:

- Particle size is very important as it decides good powder spread with dense structure for component strength;
- When using coarse particles the friction between successive layers is higher, not allowing the correct spreading of the powder in the part platform during the build. In some cases a kind of "powder wave" can originate the displacement of the part being built;
- Also, the higher the particle size the higher the surface roughness and the lower the surface definition and the strength of the components;
- To avoid those phenomena it is recommended that powders be used with grain size in the range 80–120 μm, though the grinding of ductile polymers is problematic.

Therefore, an investigation into methods for reducing PLLA raw granules in order to obtain particles in that size range is necessary. The use of a ball mill for ground blends of HA coated with PLLA (produced using emulsions) is a possible solution;

- Although no references have been found in literature, and in testing, the possible existence of carbon in the HA matrix, as a result of the burning from the laser of the organic portion of the PLLA binder, is a concern. The effect of carbon on the behaviour of the implant is not fully understood, but some researchers suggest that it can have a possible negative influence either on the living tissues around the implantation area (eventual inflammation) or by the generation of hard spots in the implant matrix (eventually promoting a fragile structure). The thermal analysis performed didn't indicate any signs of carbon. The use of high temperature DSC (Differential Scanning Calorimeter) is therefore recommended for that purpose. The use of such equipment was not possible during the time of this research.

8. Conclusion

The major contribution of this research is the establishment of data and formulation of guidelines for the rapid manufacture of hydroxyapatite based components, by means of the SLS technology, to be used for bone tissue implantation, to replace and/or repair bone defects, due to traumatised, damaged or lost bone.

The overall feasibility of producing hydroxyapatite based bone shapes, from SLS, has been established.

While potentially very biocompatible, the porous calcium phosphate/poly-lactide implants are probably not strong enough for load bearing applications, such as artificial hips, bone screws, or dental implants. This is a limitation of the SLS technology, due to the fact that this process cannot produce directly full-density implants because it always yields porous parts.

This bioactive and resorbable composite (HA/PLLA) can be a potential improvement of the currently used internal fixation devices in orthopaedic surgery, because bone healing is a dynamic process and such a material should be resorbable to insure a progressive stress transfer to the bone.

This approach will allow the promotion of autograft (i.e., the patient's bone reconstruction), bone reconstruction, eliminating the need of using allograft bone graft substitutes (i.e., cadaver bone obtained from a tissue bank), thus reducing the chance of remote risk of disease transmission that all forms of allograft carry. A synthetic replacement is a major contribution for eliminating such problems.

The use of biodegradable polymers is, nowadays, still quite expensive, which can constitute a major drawback for the application of this type of products in bone reconstruction, although the good potentialities presented for that purpose.

The main advantages of RM (Rapid Manufacturing) of medical implants are: to reduce the number of patient visits prior to the surgery, remove time consuming operations, and reduce prostheses construction times.

The research has shown that the additive fabrication of biomaterials is possible with an adequate accuracy, strength and geometry of the models produced. The main advantages of using AMT to produce medical implants include the reduction of patient's visits prior to the surgery, remove time consuming operations, and reduce prostheses construction times. The mechanical properties (i.e., the compressive strength, compressive elastic modulus, bending strength and bending modulus) of the sintered SLS specimens produced in this research are

lower than that of cortical bone and pure solid sintered HA, but acceptable when compared with the mechanical properties of cancellous bone. *In vitro* studies have shown a slow degradation rate, and a good biocompatibility of the parts, as SLS produced, meaning that the implants can act as a good scaffold for bone ingrowths.
Thus, this bioactive and resorbable composite (HA/PLLA) can be a potential improvement of the currently used internal fixation devices in orthopaedic surgery, because when bone healing is a dynamic process such a material should be resorbable to insure a progressive stress transfer to the bone.

9. References

3D Systems Inc. (2003). In: *3D Systems Web Site,* http://www.3dsystems.com, USA.

Biocomposites (2001). Bone Graft Substitutes - Orthopaedic Sports Medicine, In: *Biocomposites Brochure,* www.biocomposites.com, March 2001, UK.

Bronzino, J. D. (1995). *The Biomedical Engineering Handbook,* CRC Press Inc. ISBN 0-8493-8346-3, Boca Raton, USA.

Chelule, K. L. (2002). *Characterisation of Hydroxyapatite for Direct Manufacture of Bone Implants using CAD/CAM Techniques,* Ph.D. Dissertation, Staffordshire University, Stafford, UK.

Cruz, F. (2005). *Direct Manufacture of Hydroxyapatite Based Bone Implants Using Selective Laser Sintering,* Doctoral Dissertation, Brunel University, West London, UK.

Designinsite (2001). In: *Designinsite Web Page,* http://designinsite.dk, Denmark.

Gibson, I. R. (2003). Current Status of Calcium Phosphate-based Biomedical Implant Materials in the USA, *Medical Devices Technology Watch,* Article 4, August 2003, pp. 1-4.

Hench, L. L. (1991). Medical and Scientific Products, In: *Ceramics and Glasses of the Engineered Materials Handbook,* pp. 1007-1013, Section 13, Vol. 4, ASM International, ISBN 0-87170-282-7, USA.

Ignjatovic, N.; Tomic, S.; Dakic, M.; Miljkovic, M.; Plavsic, M. & Uskokovic, D. (1999). Synthesis and properties of hydroxyapatite/poly-L-lactide composite biomaterials, In: *Biomaterials,* Vol. 20, pp. 809-816, ISSN 0412-9612.

Kesenci, K.; Fambri, L.; Mibliaresi, C. & Piskin, E. (2000). Preparation and properties of poly (L-lactide) / hydroxyapatite composites, *Journal of Biomaterials Society, Polymer Edition,* Vol.11, No.6, pp. 617-632.

Montgomery, D. C. (1997). *Design and Analysis of Experiments - Fourth Edition,* John Wiley & Sons, Inc., USA.

Stevens, E. S. (2000). *Green Plastics: An Introduction to the New Science of Biodegradable Plastics,* Princeton University Press, USA.

Taddei, P.; Monti, P. & Simoni, R. (2002). Vibrational and thermal study on the in vitro and in vivo degradation of a poly (lactic acid) - based bioabsorbable periodontal membrane, *Journal of Materials Science: Materials in Medicine,* Vol. 13, No. 5, pp. 469-475.

Yamamuro, T.; Hench, L. L. & Willson, J. (1990). Calcium Phosphate and Hydroxyapatite Ceramics, In: *Handbook of Bioactive Ceramics,* Volume II, CRC Press, ISBN 0-8493-3242-7, Boca Raton, USA.

Zan, Q., Wang, C., Dong, L., Cheng, P., & Tian, J. (2010). Design and preparation of chitosan/HA composite scaffolds for tissue enginerring with long-bone-like strucuture. *International Journal of Materials and Product Technology,* Vol. 37, No. ¾.

12

PEGylation and BioPEGylation of Polyhydroxyalkanoates: Synthesis, Characterisation and Applications

John Foster
*Bio/Polymer Research Group, University of New South Wales
Australia*

1. Introduction

Chemical conjugations with poly(ethylene glycols) (PEGs) are established procedures to facilitate solubilisation of hydrophobic compounds. Techniques for PEGylation are currently applied to various hydrophobic pharmaceutical agents for drug delivery. More recently PEGylation has been applied to members of a novel family of microbial biopolyesters: the 'Polyhydroxyalkanoates' (PHAs). Complementing these chemical techniques, PEG modulation of bioprocessing protocols for the production of PHAs has shown that the strategic addition of certain PEGs not only supports hybrid synthesis, but may provide a technique for control of PHA composition and molecular mass. Furthermore, the addition of PEGs to the microbial cultivation systems can also result in end-capping of the hydrophobic polymer chains; a process termed 'bioPEGylation'. In this chapter we review the processes of PEGylation and bioPEGylation of PHAs as well as the influence they exert on the physiochemical, material and biological properties of the resulting natural-synthetic hybrid copolymer materials.

2. Polyhydroxyalkanoates (PHAs)

Polyhydroxyalkanoates (PHAs) are a family of microbial polyesters synthesized under conditions of environmental stress with excess of available carbon and one or more limiting nutrients (Doi, 1990; Steinbüchel & Fuchtenbusch, 1998; Lenz & Marchessault, 2005;). These biopolyesters are intracellularly sequestered as readily visible white refractile inclusion bodies and are considered to act as carbon reserves and possibly also serve as ion sinks; (Figure 1), (Anderson and Dawes, 1990; Foster, 2000).

First reported by Lemoigne in 1925, Poly(3-hydroxybutyrate), (PHB) is the most studied member of the PHA family and has a track record of commercial success. PHB is a homopolymer comprised of monomer 3-hydroxybutyric acid units (HBA). As one of the ketone bodies, HBA plays the vital role of carbon supply in mammalian systems responding to periods of prolonged starvation or suffering from diabetes mellitus (Bondy et al. 1949). Foster and Tighe have shown that microbial HBA is chemically identical to its mammalian counterpart and is recognised and processed by mammalian enzymes (Foster and Tighe 1995). Martin et al. subsequently patented its use, and that of its oligomers, as therapeutic agents for the treatment of insulin resistant states and appetite control as well as neurodegenerative disorders and epilepsy (Martins et al. 1999). In addition, Reusch et al. have reported that mammalian cells also synthesise trace quantities of PHB which is proposed to serve a membrane transport function (Reusch et al. 1992). In fact, such low molecular weight PHB is widely distributed in cells and found in representative organism of nearly all phyla. Complexation of this PHB with a range of other macromolecules modifies the physical and chemical properties and as a consequence, this complexed PHB can also be found in cytoplasm (Reusch et al. 1992).

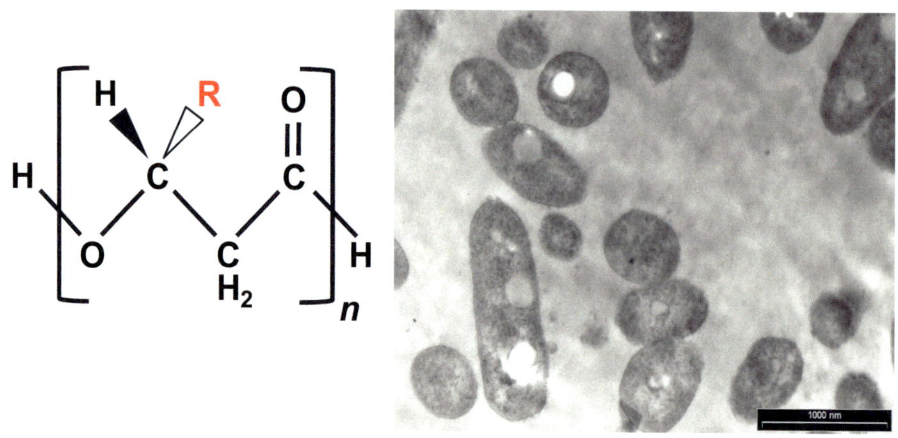

Fig. 1. Chemical formula for poly(3-hydroxyalkanoate)s and transmission electron micrograph PHA refractile inclusion bodies in *Pseudomonas oleovorans* (bar = 1,000nm).

As a consequence of its biocompatibility, PHB was initially pursued as a commercial biomaterial for various medical devices (Foster, 1996). However, its ability to successfully fulfil such a role is hampered by its relatively high crystallinity and brittle nature which also influence in degradation behaviour *in vivo* (Wu & Chen, 2009). These properties can be modified through copolymerisation with hydroxyvalerate to yield P(HB-*co*-HV) (Lenz, 1993; Yasin & Tighe 1993). The physiochemical and material properties of P(HB-*co*-HV) are more commercially appropriate and it is this copolymer which is the focus of much attention as a 'bioplastic'; a biodegradable replacement to conventional thermoplastics (Hänggi, 1995; Steinbüchel, 1996; Lenz & Marchessault, 2005).

Copolymerisation of PHB with random units of HV occurs through manipulation of bioprocessing systems, usually through variations in carbon feedstock. Further investigations in bioprocessing parameters have shown significant metabolic flexibility in a wide range of microorganisms for the synthesis of a spectrum of PHAs with over 105 different monomeric components. The properties of these PHAs range from the brittle and crystalline, as with PHB, to the flexible and elastomeric found in Poly(3-hydroyoctanoate-*co*-undecylenoate), (PHOU), (Steinbüchel, 1996; Steinbüchel, 2001). PHA synthesis is species dependent and while some species such as *Ralstonia eutropha* synthesise PHAs with comparatively short chian length substituents in the side chain (*scl*-PHA), many of the pseudomonads belonging to RNA homology Group I, produce PHAs with medium chain length alkyl groups in the side chain (*mcl*-PHAs) such as Poly(3-hydroxyoctanoate), (PHO). An additional group may be also be classified as PHAs possessing interesting chemically functional groups in their side chains (*fcl*-PHAs), such as cyanophenoxy and phenyl groups (Kim et al. 1995; Foster, 1996; Steinbüchel, 1996; Kim et al. 2000). In addition to variations in side chain composition, PHAs with elongation of monomer units in the chain backbone have also been reported, such as Poly(3HB–*co*-4HB), (Nakamura et al. 1992).

3. Poly(ethyelene glycol)s (PEGs)

Poly(ethylene glycol)s (PEGs) are oligomers or polymers of ethylene oxide (Figure 2). These synthetic polymers, produced through acidic or base catalysed polycondensation reactions, constitute a family of neutral, water-soluble polyethers with various molecular weights ranging from the oily, viscous liquid of PEG with a molecular weight of 106, also known as diethylene glycol (DEG, PEG-106), to the waxy, crystalline solid of PEG-4000. PEGs with molecular weights around 25,000 contain no hydroxyl end group and are referred to as Polyethylene oxide (PEO), (Harris, 1992).

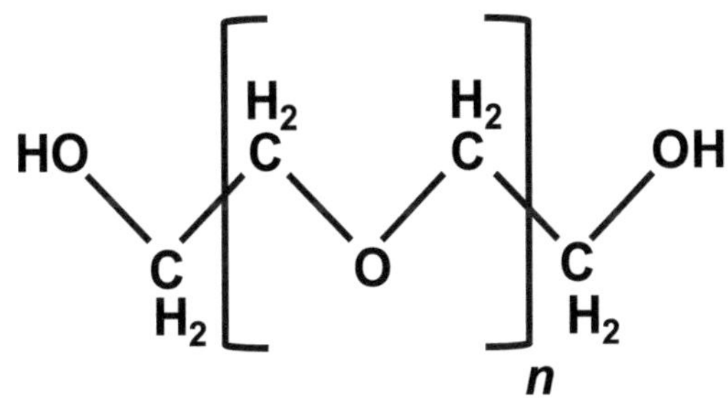

Fig. 2. Chemical formula for poly(ethylene glycol)s.

A variety of PEGs are employed as commodity chemicals in various industrial products ranging from cosmetics and lubricants to plasticisers and antifreeze agents. However some PEGs are biocompatible and utilised in the biomedical field, such as in the preservation of organs and tissues. Because of their hydrophilic nature, PEGs are also applied in the pharmaceutical industry to solubilise hydrophobic components and in protein purification. Protein-PEG conjugates exhibit a longer retention time in blood, this consequently supports their prolonged activity and, in the case of medication, longer dosing intervals. Classic examples include PEG-interferon α for the treatment of hepatitis-C and PEGylated granulocyte colony stimulating factor for neutropenia (Harris, 1992; Harris and Chess, 2003; Steward, 2005). In 1990, Krause and Bittner used PEGs with varying molecular weights to morphologically fuse severed axons in invertebrates and continuing research in this field has demonstrated that PEG can support nerve repair in mammals (Krause and Bittner, 1990; Borgens & Bohnert, 2001; Laverty et al. 2004).

4. PEGylation of PHAs

PEGylation is an established process for attaching or conjugating PEG chains to other molecules, including biomacromolecules such as peptides, proteins and antibody fragments for the production of biopharmaceuticals. PEGylation modifies the physiochemical properties including chain conformation, electrostatic binding and most noticeably, hydrophobicity. Conjugation of PHB with PEGs through chemical routes has been investigated by a number of authors.

Triblock copolymers of PHB-*co*-PEG-*co*-PHB have been synthesised by Kumagai et al. and Shuai et al. using ring-opening polymerisation of [R,S]-β-butyrolactone with PEG based macroinitiators. These triblocks possessed atatic PHB segments and had good solubility but lacked sufficient crystallinity as solids. However, there are issues with atatic PHB which can have severe implications for biological properties such as biodegradability and biocompatibility (Kumagai et al. 1993; Shuai et al. 2000). Chen used such triblocks to form biodegradable nanoparticles for drug release, which showed better release characteristics than triblocks of PHB with poly(ethylene oxide), (PEO-PHB-PEO), (Chen et al. 2006). Low molecular weight PEGylated PHB triblocks also demonstrated self-assembly and formed strong hydrogels (Kerh et al. 2010).

In contrast to studies using synthetic PHB, Marchessault and coworkers used microbial PHB which are 100% isotatic, highly crystalline and insoluble in water. Diblock copolymers of this bacterial PHB with monomethoxy-PEG (mPEG) were synthesised using the dehydrating agent dicyclocarbodi-imide (DCC) and (dimethyleamino)pyridine (DMAP) as a catalyst (Figure 3). However, yields and reaction times were low (Marchessault and Yu, 1999). Using the same PHB and mPEG-2000 components, but with bis(2-ethylehexanote) tin as a catalyst in a transesterification reaction resulted in yields of up to 77% within 60 minutes. This single-step synthesis of mPEGylated PHB proceeded with a concomitant depolymerisation to lower molecular weights, such that molecular weights of the resulting hybrids were reduced to approximately 2,300 to 7,300 (Ravanelle and Marchassault, 2002).

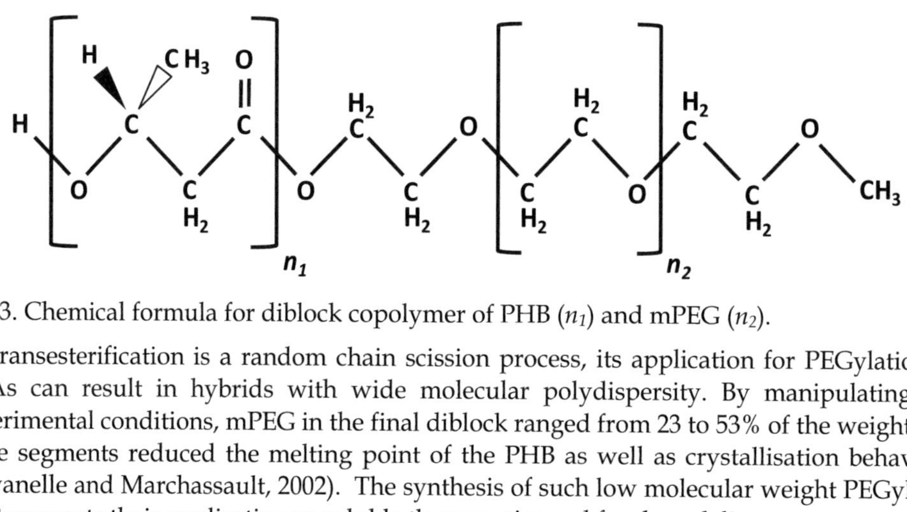

Fig. 3. Chemical formula for diblock copolymer of PHB (n_1) and mPEG (n_2).

As transesterification is a random chain scission process, its application for PEGylation of PHAs can result in hybrids with wide molecular polydispersity. By manipulating the experimental conditions, mPEG in the final diblock ranged from 23 to 53% of the weight and these segments reduced the melting point of the PHB as well as crystallisation behaviour (Ravanelle and Marchassault, 2002). The synthesis of such low molecular weight PEGylated PHB supports their application as soluble therapeutics and for drug delivery.

5. BioPEGylation of PHAs

5.1 Modulation of PHA composition using PEG

While copolymerisation of Poly(3-hydroxybutyrate) with Poly(4-hydroxybutyrate), P(3HB-*co*-4HB), results in copolymer with more favourable physiochemical and material properties than its P(3HB) counterpart, yields are low. PEG is known to associate with phospholipids' head groups in cell membranes, and this promotes an increase in membrane fluidity and consequently ions and small metabolites permeation (Yamazaki & Ito, 1990; Ingram & Buttke, 1984). In 1996, Gross and coworkers working on the theory that adding PEG-200 to the cultivation medium of the microorganism *Ralstonia eutropha* during the production of P(3HB-*co*-4HB) would promote the synthesis of PHAs with an enhanced composition of 4-

HBA units (Shi et al. 1996a). By extension of this reasoning, PEG modulated bioprocessing could also increase the diversity of monomeric units incorporated, potentially leading to the synthesis of new PHAs.

Shi et al. cultivated *R. eutropha* for the production of P(3HB-*co*-4HB) in Erlenmeyer flasks with PEG-200 loadings of 1 to 4% (w/v). When cultivated on 4-hydroxybutyrate as carbon source, *R. eutropha* synthesised a random copolymer with 3-HBA and 4-HBA monomeric units. However, addition of only 2% (w/v) PEG-200 to the bioprocessing significantly reduced the proportion of 3-HBA units from 34 to 11 mol %, with an increase in 4-HBA units from 66 to 86 mol %. Furthermore, this *scl*-PHA also contained approximately 3 mol % 3-hydroxyvalerate units, P(3HB-*co*-3HV-*co*-4HB), (Shi et al. 1996a). Steinbüchel and coworkers have compared the influence of PEG-200 to a variety of chemical inhibitors of fatty acid oxidation and the citric acid cycle to probe the influence of such polyethers on lipid and PHA synthesis in various *Rhodococcus* and *Nocardia* species (Alvarez et al. 1997). In this study they report that 0.2 to 5% (w/v) PEG-200 in the cultivation media caused decreases in *scl*-PHA contents in *R. ruber*, but under the conditions studied also stimulated greater incorporation of 3HV monomer units.

Foster and coworkers report that PHO produced by *Pseudomonas oleovorans* in the presence of 2% (w/v) PEG-106 had a significant change in the relative proportions of its C6, C8 and C10 monomeric hydroxyalkanoate components (Sanguanchaipaiwong et al. 2004; Foster et al. 2005). Bioprocessing using another strain of *P. oleovorans* in the presence of PEG-200 and PEG-400 for the production of a *fcl*-PHA with some degree of unsaturated bonds in the side chain, also saw a change in its composition, with the C8 component increasing from 47 to 56 mol % (Ashby et al. 2002). Results suggest that the presence of relatively low molecular weight PEGs in microbial growth media provided some measure of control over the PHA monomeric composition by promoting the main monomer in the carbon feed source. Therefore, *R. eutropha* cultivated with 4-hydroxybutyrate as carbon source incorporated more of these units in its *scl*-PHA, while different strains of *P. oleovorans* followed this trend for their *mcl*- and *fcl*-PHAs when cultivated with octanoic and oleic acids respectively (Ashby et al. 2002; Foster et al. 2005).

5.2 Modulation of PHA molecular weight using PEG

In addition to influencing PHA composition, the presence of PEGs in the bioprocessing systems also affected the molecular mass of the biopolymers (Figure 4). With the exception of *Azotobacter vinelandii* UWD, PEGs with molecular weights greater than approximately 2,000 had no apparent impact on the molecular weights of the *scl*-PHAs synthesised by a variety of microbial species. However, PEGs with molecular weights below 2,000 supported PHA chain termination and resulted in PHA molecular weights significantly below those for *scl*-PHAs synthesised in the absence of PEGs. The lower the PEG molecular weight the greater the PHA molecular weight reduction (Figure 4).

PHB synthesis by *R. eutropha* with PEG-350 and its methoxy end-capped derivatives in the cultivation media have shown that capping of one hydroxyl group on the PEG chain had no apparent affect on the ability of PEG-350 to interact with polymer biosynthesis and effect molecular weight change. However, when both hydroxyl groups on the PEG were capped molecular weight reduction did not occur (Shi et al. 1996b). These results suggest that PEGs interact with the biosynthetic system, presumably through the PHA synthase, to increase the rate of *scl*PHA chain termination relative to propagation.

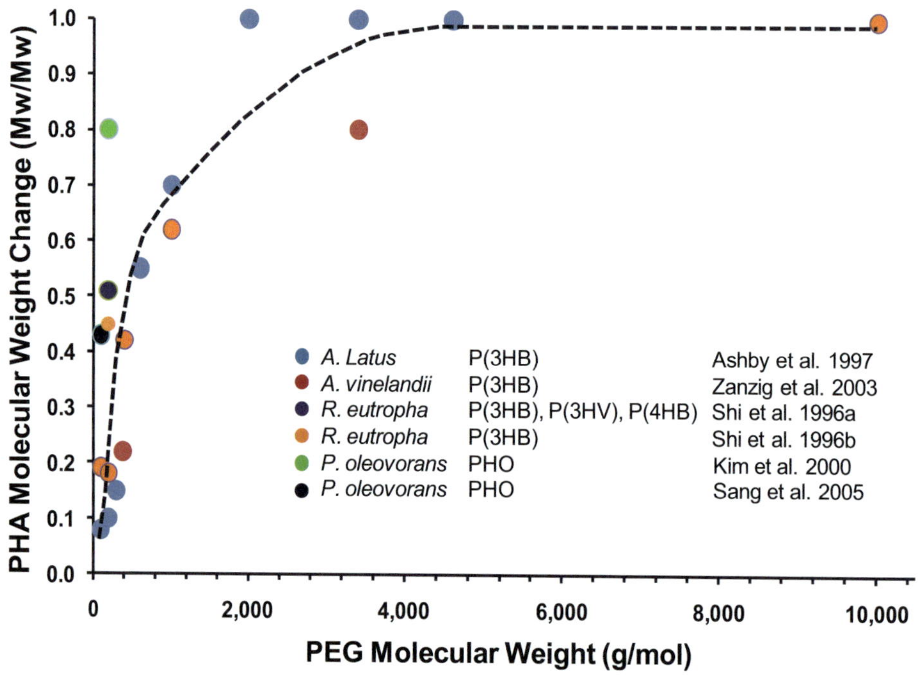

Fig. 4. Graph illustrating the relative change in molecular weight of PHAs due to presence of PEGs in the cultivation media.

Molecular weight is an important factor influencing the biodegradability and physico-mechanical properties of a biomaterial. Consequently, it has been suggested that PEGs may also offer some degree of PHA molecular weight control, a so-called 'tunable switch' (Ashby et al. 1999). The addition of 2% (w/v) PEG-200 to cultivations of *A. latus* and *R. eutropha* at different growth stages showed that the PHB synthesised had bimodal distributions of comparatively high and low molecular weights. In contrast, additions of PEG-200 at the beginning of the cultivations produced PHB with low to high ratios of 1:0.10 and 1:0.19 for *A. latus* and *R. eutropha* respectively; these ratios increased to 1:7.33 and 1:3.17 when the PEG was added during the established growth phases (Ashby et al. 1999).

The addition of comparatively low molecular weight PEGs to bioprocessing systems using P. *oleovorans* and *P. putida* induced manipulation of *mcl*-PHAs molecular weight. Additions of 2% (w/v) of PEG-106, PEG-200 or PEG-400 reduced the molecular weights of PHO synthesised by these species (Kim, 2000; Sanguanchaipaiwong et al. 2004). However, the impact appeared to be greater for *P. oleovorans* that *P. putida* with this *mcl*-PHA reduced to 40 and 80% respectively (Kim 2000).

5.3 Biosynthesis of PHA-PEG natural-synthetic hybrid copolymers

Modification of PHA molecular weight and composition during their biosynthesis is dependent upon the microbial species, the conditions of its cultivation and the molecular weight of PEG utilised, with PEGs possessing a relatively low degree of polymerisation influenced PHA compositions and reduced their molecular weights. Changes in molecular weight suggest that PEG molecules promote PHA chain scission. ^{1}H- and ^{13}C-NMR analysis of PHAs from PEG modulated cultivations support this, with results proving the termination of PHA chains by the PEG chains (Madden et al. 1994; Shi et al. 2000b; Sangaunchaipaiwong et al. 2004; Foster et al. 2005). Thus, PHA chains are 'end-capped' with PEGs forming a new class of natural-synthetic hybrid block copolymers. This biological end-capping is referred to as 'bioPEGylation' (Foster, 2007).

Foster and coworkers have quantitatively monitored the concentrations of PEG-106 in cultivation media during the synthesis of PHO by *P. oleovorans* and for PHB production in *A. latus*. PEG-106 was also shown to be readily associated with the cell membranes and penetrated into the cytoplasm (Sanguanchaipaiwong et al. 2004; Foster et al. 2005; Sanguanchaipaiwong, 2007). ^{13}C-labelled ethylene glycol (EG) was utilised by Shah *et al.* to probe the influence of these polyethers on the PHB synthetic pathway of *R. eutropha*. The results clearly showed that the EG units acted as PHB chain terminators (Shah et al. 2000a, 2000b).

	EG	DEG	Polyethylene Glycol (PEG)										*reference*
Species	62	106	200	300	350	400	600	1,000	2,000	3,400	4,600	10,000	
***scl* PHA**													
Comomonas testosteroni **	Y Y												Shah et al. 2000
Ralstonia eutropha *	Y Y												Madden et al. 1999; Shah et al. 2000
" " **			Y Y										Shi et al. 1996a
" " *		Y N	Y N		Y N	Y N		Y N				N N	" " 1996b
" " *			Y N										Ashby et al.1999
Alcaligenes latus *			Y N										" " "
" " *		Y Y	Y Y	Y Y			Y Y	Y Y	Y N		N N	N N	Ashby et al. 1997
" " *		Y Y				Y N			N -	N –	N –		Zanzig et al. 2003
Azotobacter vinelandii UWD *		Y Y				Y N			N -	N -	N –		" " "
Pseudomonas oleovorans * *B14682*			Y Y				Y Y						Ashby et al. 2002
***scl* PHA + *fcl* PHA**			Y Y			Y Y							
P. oleovorans B778 *†													Ashby et al. 2002
***mcl* PHA**			Y N			Y N							
P. putida KT2442 °			Y N			Y N							Kim 2000
P. oleovorans ATCC29347 °		Y Y	Y Y			Y N							" "
" " " °													Sanguanchaipaiwong et al. 2005, 2007
***fcl* PHA**			Y N			Y N							
" " *B14683* ‡													Ashby et al. 2002

Table 1. Summary of PHA bioPEGylation and/or molecular mass reduction due to the addition of various molecular weight PEGs to microbial cultivations for the production of PHAs. (Y = molecular mass reduction, Y= bioPEGylation, N = No effect, - not measured), *(3HB); **(3HB)(4HB)(3HV); †90 mol% (3HB)+10 mol% fclPHA; °fclPHA, ‡PHO

The influence of PEG molecular weights on PHA hybrid biosynthesis and molecular weight reduction for a number of species is summarised in Table 1. From Table 1 it can be clearly seen that bioPEGylation of PHA chains does not always occur and appears to be favoured in cultivation systems where the PEG molecular weight is 300 or below. However, this is not entirely surprising given the original premise that these polyethers would support small molecule penetration into the cell.

From the summary in Table 1, we can speculate that PEGs with molecular weights up to 300 facilitate PHA chain scission as well as the synthesis of bioPEGylated hybrids. PEG molecular weights between 300 and 2,000 may support fluidity of the cell membrane but due to their large size, they fail to enter the cell and do not act as chain terminating agents. Finally, PEGs with molecular masses exceeding 2,000 do not appear to influence PHA properties. While PEGs with molecular weights between 300 and 2,000 supported PHA composition and molecular mass control, these influences may be a function of the additives influence on cell viability.

3.4 Influence of PEGs on PHA bioprocessing parameters

Cell yields for *R. eutropha* for PHB production decreased by 30% as the concentration of PEG-200 in growth media increased from 1 to 4 % (w/v). In addition, the *scl*PHA yield also decreased by 50%. In companion experiments, the viability of this species was also shown to be reduced with the addition of 1 to 10% (w/v) of PEG-200 or PEG-400 (Shi et al. 1996b). Thus, higher concentrations of these comparatively low molecular weight PEGs in the microbial growth medium, the greater the reduction in cell viability. In contrast, increases in PEG molecular weight had comparatively less influence on cell viability. Thus, the addition of similar loadings of PEG-10,000 had no apparent effect on *R. eutropha* cell viability (Shi et al. 1996b). Studies reporting reductions in cell viability during PEG modulated cultivations for PHA production support the theory proposed here that PEGs of molecular weights between 300 and 2,000 influences PHA molecular mass and composition through modifications to cell viability in the bioprocessing system. In contrast, PEGs with molecular weights below approximately 300 modify PHA synthesis by acting as chain terminating agents forming natural-synthetic hybrids.

With the exception of PEG-10,000, the presence of PEGs in the cultivation media reduced the yield of *scl*-PHAs in *R. eutropha*. This was partly a consequence of their influence on cell mass (Shi et al. 1996b). Similar studies using *A. latus* have shown that PEGs with molecular weights between 106 and 600 had a greater impact on polymer synthesis compared to their higher molecular weight counterparts (1,000 - 10,000), with PHB productivity also being reduced depending on PEG concentration in the cultivation medium (Ashby et al. 1997). Thus, while 5% (w/v) PEG-106 in the growth media resulted in no PHA synthesis, 10% (w/v) of PEG-10,000 still showed some biopolymer synthesis, although the impact was greater than that observed when the studies were conducted with *R. eutropha* (Ashby et al. 1997). Similarly, additions to cultivations of *A. vinelandii* UWD of 2% (w/v) of various PEGs from PEG-106 to PEG-3400 also resulted in variable polymer yields (Zanzig et al. 2003). P(HB-*co*-HV) synthesis by *R. ruber* was influenced by the presence of 0.2-5% (w/v) PEG-200, with a reduction in *scl*-PHA yield of up to 50% (Alvarez et al. 1997).

Such variations in polymer yields are obviously due to the influence of PEGs on the bioprocessing systems. However, the utilisation of Erlenmeyer flasks in these studies limits their value. Foster and coworkers have cultivated *A. latus* for the production of PHB in the presence of 2% (w/v) PEG-106 using Braun 5 L bioreactors, which provides greater control over cultivation parameters. In these studies PEG was demonstrated to support an increase in cell mass (12%) as a consequence of the additional carbon, but through reduced microbial viability a reduction in cellular PHA synthesis was measured (9%), (Sanguanchaipaiwong, 2007). This increase in cell mass is consistent with the studies of Shah et al. who reported that EG was metabolised by *R. eutropha* for the production of PHB (Shah et al. 2000a)

Microbial PHA synthesis is species dependent; the diversity of *fcl*- and *mcl*-PHAs are significantly greater than their *scl*PHA companions. Kim investigated the influence of PEG-200 on the synthesis of *mcl*-PHAs by *P. oleovorans* and *P. putida* and reported that PHO yield was reduced with increasing PEG concentration, while at 8% PEG-200 loading no PHO synthesis was found (Kim, 2000). Using the same strain of *P. oleovorans* cultivated in the presence of PEG-106, Foster and coworkers reported an increase in cell concentration but a decrease in PHO yield (Sanguanchaipaiwong et al. 2004). Similarly, the addition of PEG-200 or PEG-400 to another strain of *P. oleovorans* (B-778) primarily producing PHB but mixed with 10 mol% of a *fcl*PHA, also showed an increase in cell mass but no change in polymer yield. In contrast, other strains of *P. oleovorans* producing PHB (B-14682) and a *fcl*PHA (B-14683), showed no apparent change in cell mass but did show a reduction in their polymer yields (Ashby et al. 2002). Thus, the influence of PEGs on PHA synthesis also has a degree of species dependence.

4. Properties and potential

4.1 Physiochemical and material properties

The majority of current studies in this emerging field have investigated the influence of PEGs on PHA synthesis, with a focus on their potential to control composition and molecular mass of these biopolyesters. Thus, PEG modulated bioprocessing of PHAs may offer a technique to control PHA physiochemical and material properties. In addition, characterisations of natural-synthetic hybrids of bioPEGylated PHAs suggest that these hybrids may also possess additional properties when compared to their PHA counterparts (Zanzig et al. 2003; Foster et al. 2005). Concurrent cultivations of *A. latus* from the same inoculums have shown that the PHB-*co*-PEG106 hybrid, while possessing a lower molecular weight compared to its PHB counterpart, also had significantly different physiochemical and material properties. In particular the elongation to break increased from 8.4 to 20.6 % (Sanguanchaipaiwong, 2007). In contrast, the change in elongation-to-break for bioPEGylated *mcl*-PHA was not as noticeable, 580 (±9.4) and 540 (±8.3) MPa for PHO and PHO-*co*-PEG106 respectively.

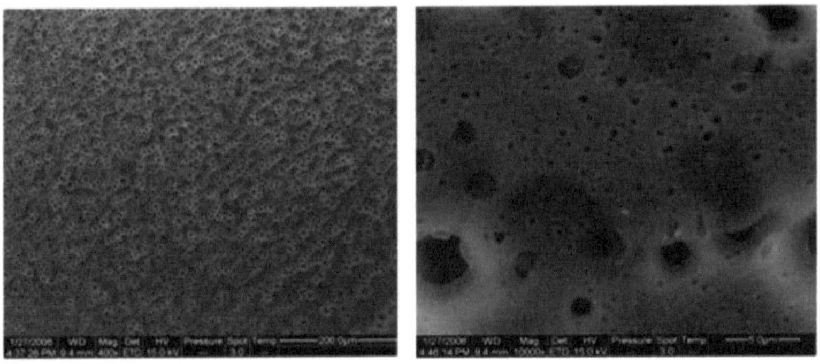

Fig. 5. SEMs of PHO-co-PEG106 films fabricated though solvent casting in the presence of humidity reveal the presence of pores not observed with PHO.

Small angle neutron scattering (SANS) was utilised to probe the chain conformations of protonated and deuterated *mcl*-PHA samples as well as their bioPEGylated hybrids. The comparatively small hydrophilic PEG group terminating much longer hydrophobic PHA chains exhibited no apparent influence on chain conformation when solvated in chloroform (Foster et al. 2008). However, with the addition of microquantities of water to the bioPEGylated PHA solutions, the hybrids formed stable microcrystalline suspensions. Consequently, when processed under humid airflow, bioPEGylated PHAs exhibited a degree of self-assembly resulting in the formation of disordered microporous films (Figure

5), (Foster et al. 2005; Sanguanchaipaiwong, 2007). Microporous and honeycomb films are of considerable interest based on their biotechnological and biomedical potential, such as immobilisation of biomolecules (Stenzel, 2002; Nishikawa et al. 1999).

4.2 Biological properties

Scholz and coworkers investigated the interactions of various bioPEGylated PHB samples with skin melanoma and human breast adenocarcinoma cells (Zanzig et al. 2003). In this study, cell adhesion was significantly affected by the presence of the PEG hydrophilic end groups, with reductions in cell attachments to hybrids ranging from 22 to 74%. The greatest reduction was found with PHB-*co*-PEG106, (Zanzig et al. 2003).

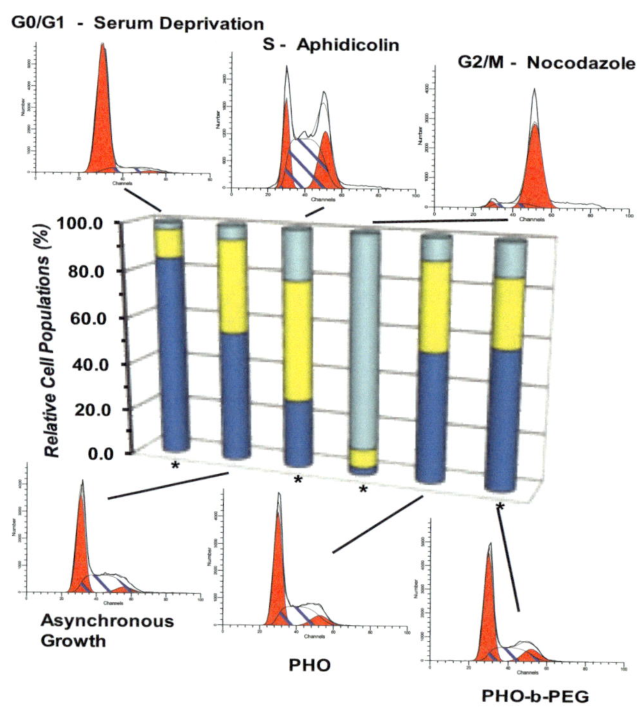

Fig. 6. Summary of relative cell growth populations determined from FACS sorting histograms, expressed as percentages of total populations. BioPEGylated PHO films showed statistically significant differences for cell populations in the G2/M ($p < 0.1$) and S ($p < 0.001$) phases, when compared to the same population cultivated in the absence of biomaterials. (* indicates statistically significant differences from asynchronous growth).

Figure 6 shows the changes in cell cycle when myoblastic satellite stem cells were cultivated in the presence of PHO and its bioPEGylated hybrid. The hybrid showed comparatively small but significant changes in the DNA population distributions for the stem cells compared to PHO, with significant differences between the two PHAs. Hence, the cycle of stem cell differentiation from G0-G1 through S to G2-M phase proceeded in the presence of PHO and its hybrid, but with significant differences in the respective populations of cells at these phases when compared to asynchronous growth, with the PHO-*co*-PEG106 showing the greatest deviation (Figure 6), (Marcal et al. 2008).

5. Conclusions and future trends

Conjugations of the *scl*-PHA, PHB with various PEGs, i.e, PEGylation, have been performed through chemical routes. The resulting diblocks and triblocks of this PHA have a low molecular weight and possess potential as pharmaceutical agents and polymers for injectable hydrogels or nanoparticles for drug delivery.

Strategic additions of PEGs to microbial cultivations for the production of *scl*- and *mcl*-PHAs not only permit a degree of control in PHA molecular mass and composition but may also provide routes for BioPEGylation. While bioPEGylated PHAs have a significantly reduced molecular weight compared to their PHA counterparts produced in the absence of PEG, they are considerable greater than the PEGylated counterparts and retain their material properties. Nevertheless, bioPEGylation permits significant changes in physiochemical and material properties, with subsequent implications for their material processing. Furthermore, the presence of hydrophilic end groups has significant implications for the biological properties of these hybrids, suggesting they may serve as novel biomaterials in their own right.

In general, PEGs with molecular weights below approximately 300-600 apparently increase membrane fluidity, supporting bioPEGylation. While PEGs with molecular masses from 600 to 2,000 facilitate PHA molecular mass and composition control, without bioPEGylation; PEGs above 2,000 appear to have little or no effect. For those PEGs exerting some degree of control over PHA synthesis, polymer yields are significantly reduced. Furthermore, this review suggests that, as with PHA biosynthesis, the influence of PEGs in cultivations systems may be strain specific.

Despite the molecular weight disparities between the hydrophobic PHA chain and it's comparatively tiny hydrophilic PEG terminal in bioPEGylated PHAs, there is sufficient evidence to suggest that the hybrid has subtly different physiochemical and material properties. Thus, work in this field is consistent with current trends to increase the diversification of PHAs and their conjugates (Hazer and Steinbüchel, 2007). Future work requires more detailed study of PHA-*co*-PEG hybrids as biomaterials, as well as investigations in the potential application of PEGs to bioprocessing technology for the production control of PHAs.

While PEGylation and bioPEGylation studies are still in their infancy, results to date suggest that these natural-synthetic hybrids have great potential as medical biomaterials.

6. References

Alvarez HM, Kalscheuer R, Steinbüchel A (1997) Accumulation of storage lipids in species of *Rhodococcus* and *Nocardia* and effect of inhibitors and polyethylene glycol. *Fett Lipid* 99:239-246.

Ashby RD, Shi F, Gross RA (1997) Use of poly(ethylene glycol) to control the end group structure and molecular weight of poly(3-hydroxybutyrate) formed by *Alcaligenes latus* DSM 1122. *Tetrahedron* 53:15209-15223.

Ashby RD, Shi F, Gross RA (1999) A tunable switch to regulate the synthesis of low and high molecular weight microbial polyesters. *Biotech Bioeng* 62:106-113.

Ashby RD, Solaiman DKY, Foglia TA (2002) Poly(ethylene-gycol)-mediated molar mass control of short-chain- and medium-chain-length poly(hydroxyalkanoates) from *Pseudomonas oleovorans*. *Appl Microbiol Biotechnol* 60:154-159.

Anderson AJ, Dawes EA (1990) Occurance, metabolism, metabolic role, and industrial uses of bacterial polyhydroxyalkanoates. *Microbiol Rev* 54:450-472.

Bondy PK, Blom WL, Whitner VS, Farrar BW (1949) Studies of the role of the liver in human carbohydrate metabolism by venous catheter technic. II. Patients with diabetic ketosis, before and after the administration of insulin. *J Clin Invest* 28:1126- 1125.

Borgens RB, Bohnert D. (2001) Rapid recovery from spinal cord injury after subcutaneously administered polyethylene glycol. *J Neurosci Res* 66(6):1179–1186.

Chen C, Yu CH, Cheng YC, Yu PH, Cheung MK. (2006) Biodegradable nanoparticles of amphiphilic triblock copolymers based on poly(3-hydroxybutyrate) and poly(ethylene glycol) as drug carriers. *Biomaterials* 27:4804-4814.

Chen G.-Q, Wu Q. (2005) The application of polyhydroxyalkanoates as tissue engineering materials. *Biomaterials* 26:6565-6578.

Cheng GX, Cai ZJ, Wang L. (2003) Biocompatibility and biodegradability of poly(hydroxybutyrate)/poly(ethylene glycol) films, *J Mater Sci-Mater M* 14,1073–8.

Doi Y, Microbial Polyesters. New York: VCH Publishers; 1990

Foster LJR, Tighe BJ (1995) Enzymatic assay of hydroxybutyric acid monomer formation in poly(β-hydroxybutyrate)degradation studies. *Biomaterials* 10:314-343.

Foster LJR, Fuller RC, Lenz RW (1996) Activities of extracellular and intracellular depolymerases of polyhydroxyalkanoates. In: Ottenbrite RM, Huang SJ, Park K. Hydrogels and biodegradable polymers for bioapplications. ACS books, Washington DC, pp 68-92.

Foster LJR (2000) The structural organisation of polyhydroxyalkanoate inclusion bodies. In: Scholz C, Gross RA. Polymers from renewable resources: biopolyesters and biocatalysis. ACS books, Washington DC, pp 42-66.

Foster LJR, Sanguanchaipaiwong V, Gabelish CL, Hook J, Stenzel M (2005) A natural-synthetic hybrid of polyhydroxyoctanoate-diethylene glycol: biosynthesis and properties. *Polymer* 46:6587-6594.

Foster LJR, Schwahn D, Pipich, V, Holden, PJ, Richter, D. (2008) SANS characterisation of polyhydroxyalkanoates and their bioPEGylated hybrids in solution. *Biomacromolecules* 9:314-320.

Hängii UJ, (1995) Requirements of bacterial polyesters as future substitutes for conventional plastics for consumer goods. *FEMS Microbiol Rev* 16:213-220.

Harris (1992) Introduction to biotechnical and biomedical applications of poly(ethylene glycol). In: Harris JM, (ed) Poly(ethylene glycol) chemistry: biotechnical and biomedical applications. Plenum press, new York, pp 1-14.

Harris JM, Chess RB (2003) Effect of PEGylation on Pharmaceuticals. *Nat Rev Drug Discuss* 3:214-221.

Hazer B, Steinbüchel A (2007) Increased diversification of polyhydroxyalkanoates by modification reactions for industrial and medical applications. *Appl. Microbiol Biotechnol* 74:1-12.

Ingram LO, Buttke TM (1984) Effects of alcohols on microorganisms. *Adv Micro Phys* 25:253-300.

Kerh LL, Jing-ling Z, Jun L. (2010) Elucidating rheological property enhancements in supramolecular hydrogels of short poly[(*R,S*)-3-hydroxybutyrate]-based amphiphilic triblock copolymer and α-cyclodextrin for injectable hydrogel applications. *Soft Matter* 6:2300-2311.

Krause TL, Bittner GD. (1990) Rapid morphological fusion of severed myelinated axons by polyethylene glycol. PNAS 87(4):1471–1475.

Kim DY, Kim YB, Rhee YH (2000) Evaluation of various carbon substrates for the biosynthesis of polyhydroxyalkanoates bearing functional groups by *Pseudomonas putida*. *Int J Bio Macromol* 28:23-29.

Kim O (2000) Biological effects of poly(ethylene glycol) on microbial poly(β-hydroxyalkanoates) produced by pseudomonads microorganisms. *J Polym Res* 7:91-96.

Kim YB, Lenz, RW, Fuller RC (1995) Poly-3-hydroxyalkanoates containing unsaturated repeating units produced by *Pseudomonas oleovorans*. *Macromolecules* 25:1852-1857.

Kumagai Y, Doi Y (1993) synthesis of a block copolymer of poly(3-hydroxybutyrate) and poly(ethyelene glycol) and its application to biodegradable polymer blends. *J Polym Environ* 1:81-87.

Laverty PH, Leskovar A, Breur GJ, Coates JR, Bergman RL, Widmer WR, Toombs JP, Shapiro S, Borgens RB. (2004) A preliminary study of intravenous surfactants in paraplegic dogs: Polymer therapy in canine clinical SCI. *J Neurotrauma* 21, 12, 1767-1777.

Lenz RW (1993) Biodegradable polymers. *Adv Polym Sci* 107:1-40.

Lenz RW, Marchessault RH (2005) Bacterial polyesters: biosynthesis, biodegradable plastics and biotechnology. *Biomacromolecules* 6:1-8.

Madden LA, Anderson AJ, Shah DT, Asrar J (1999) Chain termination in polyhydroxyalkanoate synthesis: involvement of exogenous hydroxyl-compounds as chain transfer agents. *Int J Biol Macromol* 25:43-53.

Marçal H, Wanandy NS, Sanguanchaipaiwong V, Woolnough, CE, Lauto A, Mahler SM, Foster LJR. (2008) BioPEGylation of polyhydroxyalkanoates: influence on properties and satellite stem cell cycle. *Biomacromolecules* 9(10):2719-2726.

Marchessault RH, Yu G.-E. (1999) *Polym. Prepr.* 40:527-528.

Martin DP, Peoples OP, Williams SF, Zhong LH. (1999) Nutritional and therapeutic uses of 3-hydroxyalkanoate oligomers, *US Patent Appl* 359086.

Nakamura S, Doi Y, Scandola M (1992) Microbial synthesis and characterisation of poly(3-hydroxybutyrtae-co-4-hydroxybutyrate). *Macromolecules* 25:4237-4241.

Nishikawa T, Nishida J, Ookura R, Mishimura S-I, Wada S, Karino T, Shimomura M (1999) Honeycomb-patterned thin films of amphiphillic polymers as cell culture substrates. *Mater Sci Eng* C8:495-500.

Ravanelle F, Marchessault RH (2002) One-step synthesis of amphiphillic diblock copolymers from bacterial poly([R]-3-hydroxybutyric acid). Biomacromolecules 3:1057-1064.

Reusch RN, Sparrow AW, Gardiner J (1992) Transport of poly-beta-hydroxybutyrate in human plasma. *Biochim Biophys Acta* 1123:33-40.

Sanguanchaipaiwong V, Gabelish CL, Hook J, Scholz C, Foster LJR (2004) Biosynthesis of natural-synthetic hybrid copolymers: polyhydroxyoctanoate-diethylene glycol. *Biomacromolecules* 5:643-649.

Sanguanchaipaiwong V (2007) Biosynthesis and characterisation of polyhydroxyalkanoate based natural-synthetic hybrid copolymers. Ph.D. thesis, University of New South Wales, Sydney, Australia.

Shah DT, Tran M, Berger PA, Aggarwal P, Asrar J, Madden LA, Anderson AJ (2000a) Synthesis and Properties of hydroxyl-terminated poly(hydroxyalkanoates). *Macromolecules* 33:2875-2880.

Shah DT, Berger PA, Tran M, Asrar J, Madden LA, Anderson AJ (2000b) Synthesis of poly(3-hydroxybutyrate) by *Ralstonia eutropha* in the presence of ^{13}C-labelled ethylene glycol. *Macromolecules* 33:6624-6626.

Shi F, Ashby R, Gross RA (1996a) Use of poly(ethylene glycol) to regulate poly(3-hydroxyybutyrate) molecular weight during *Alcaligenes eutrophus* cultivations. *Macromolecules* 29:7753-7758.

Shi F, Gross RA, Rutherford DR (1996b) Microbial polyester synthesis: effects of poly(ethylene glycol) on product composition, repeat unit sequence, and end group structure. *Macromolecules* 29:10-17.

Shuai XT, Jedlinski Z, Luo Q, Nozirov F (2000) Synthesis of novel block copolymers of poly(3-hydroxybutyric acid) with poly(ethyelene glycol) through anaionic polymerisation. *Chin J Polym Sci.* 18:19-23.

Steinbüchel A (1996) PHB and other polyhydroxyalkanoic acids. In: Rehm H-J, Reed G, Pühler A, Stadler P (ed) Biotechnology John Wiley & Sons, Weinheim, vol 6. pp 403-464.

Steinbüchel A, Füchtenbusch B (1998) Bacterial and other biological systems for polyester production. *TIBTECH* 16:419-427.

Steinbüchel A (2001) Perspectives for biotechnological production and utilisation of biopolymers: metabolic engineering of polyhydroxyalkanoate biosynthesis pathways as a successful example. *Macromol Biosci* 1:1-24.

Steward A (2005) Polymers under one umbrella. Chem Ind 18/04/05 18-20.

Stenzel M (2002) Formation of regular honeycomb-patterned porous films by self-organisation. *Aust J Chem* 555239-243.

Wu, Q.; Wang Y. & Chen, G-Q. (2009) Medical application of microbial biopolyesters polyhydroxyalkanoates. *Artificial Cells Blood Sub Biotech* 37, 1, 1-12.

Yamazaki M, Ito T (1990) Deformation and instability in membrane structure of phospholipid vesicles caused by osmophobic association - mechanical stress model for the mechanism of poly(ethylene glycol) induced membrane fusion. *Biochemistry* 29:1309-1314.

Yasin M, Tighe BJ (1993) Strategies for the design of biodegradable polymer systems - manipulation of polyhydroxybutyrate based materials. *Plast Rubb Compos Proc Appl* 19:15-27.

Zanzig J, Marimuthu B, Werka J, Scholz C (2003) Investigation of the impact of poly(ethylene glycol)-modulation of poly(β-hydroxybutryate) syntheses on cell interactions of the resulting polymers. *J Bioact Compat Polym* 18:339-354.

13

Lubrication and Adhesion by Charged Biopolymers for Biomedical Applications

Roberto Andresen Eguiluz, Rebecca M. Schur and Delphine Gourdon
Materials Science & Engineering, Cornell University
U.S.A

1. Introduction

Charged polymers (polyelectrolytes) are water-based lubricants that perform efficiently in biological systems where the charged hydrophilic surfaces provide an electrostatic "double-layer" repulsion, in addition to the "steric" repulsion of any protruding polyelectrolyte segments and the hydration layer of tightly bound water molecules, all leading to low friction (Urbakh et al., 2004). Polymer conformation and strength of attachment to surfaces also play essential roles in lubrication, and recent experiments (Raviv et al., 2003) indicate that brushes of polyelectrolytes grafted to surfaces rubbing across aqueous media result in superior lubrication (friction coefficient, μ <0.001) at low sliding velocities and pressures up to a few atmospheres. This suggests that some biolubricating systems, such as in the eyes, may be mediated by brushlike polyelectrolyte layers. However, other experiments (Lee et al., 2003; Benz et al., 2004) show that brushes may not function well at high pressures where wear (damage) and/or higher friction coefficients are usually measured, indicating that high-pressure, high-impact systems, such as in the hip or knee joints, are likely mediated by more complex mechanisms. In fact, no model system currently exhibits both low friction and good wear resistance at high pressure (and over a large-range of velocities); however, polyelectrolytes seem to be the best way to go. Most specifically, naturally occurring polyelectrolytes (charged biopolymers) have gained interest because of their (controlled) degradable capabilities, abundance in nature (low cost), and biocompatibility, essential requirements for biomedical applications.

In this chapter, we first review the challenges of biolubrication, including the need for controlling the properties of both the lubricant and the confining (shearing) surfaces. We then report the lubrication and adhesive properties of a variety of naturally occurring polyelectrolytes (from glycoproteins to polysaccharides) in particular, the potential use of (nano)thin films of those polymers (adsorbed from aqueous solutions, grafted, or free in solution) to improve the lubrication properties of tribological systems composed of molecularly smooth mica surfaces (a chemically inert inorganic aluminosilicate clay). We focus on the capacity of the biopolymers to carry high loads and pressures (up to hundreds of atmospheres) and their ability to resist wear when sheared over large distances (larger than the contact diameter)—two crucial requirements of most lubricating systems—in addition to exhibiting low friction forces and average friction coefficients, to ensure low energy dissipation.

2. Biolubrication challenges

Biotribology, namely the study of adhesion, friction/lubrication, and wear in a biological environment, relates to the relative motion of surfaces either manmade (and implanted), natural tissues or the combination of both. Finding the best compromise between tribological requirements (low friction coefficients, efficient load bearing, wear resistance, chemical stability) and biocompatibility (since the host response to a biomaterial is critical in determining the success of an implant) makes it a challenging yet exciting task.

The tribological behavior of any (not necessarily biological) system is complex because it is governed by interacting factors:

- surface properties (topography, crystallinity, chemistry),
- environmental "global" parameters (presence of lubricant, temperature, p*H*),
- external "local" parameters (applied load/pressure, displacement, shear).

Surfaces are defects of a material *per se*, having peculiar atomic organization and reactivity (surface energy) responsible for their interactions with other media. Many methods are currently employed to modify the surface properties of artificial or natural bodies. They can be functionalized with coatings or self-assembled layers, gradually modified (gradients) through e.g. ion implantation, surface chemical reactions, surface active bulk additives or etching (Ratner & Hoffman, 1996). A crucial parameter is the texture of surfaces, characterized by its roughness. It determines how an object genuinely interacts with its environment. Surface roughness not only distorts the adhesive forces, it also modifies the friction forces. The surface asperities limit the average distance of closest approach between the bodies, usually reducing their normal interaction (adhesion). However, the highest asperities will also deform elastically and/or plastically when pressed against other, modifying the original surface roughness and chemistry, which may generate wear debris upon shear, acting as a new compound, called the tribolayer (Schouwenaars et al., 2007).

In biotribology, and particularly in biolubrication, the absence of damage (wearless sliding) does not only translate into mechanical efficiency, it also prevents from further inflammation occasioned by wear debris. For example, premature wear of the ultra high molecular weight polyethylene (UHMWPE) acetabular cup of a total hip replacement will not only cause loosening of the implant leading to failure, but it will also produce toxic third body particles promoting serious inflammation. To prevent the formation of such tribolayer, postpone wear, reduce friction forces, and increase efficiency, lubricants are commonly introduced between the shearing surfaces. Those lubricants can be solid, oil-based or water-based. This review focuses exclusively on naturally occurring water-based lubricants as described in the section 3. Current approaches to assess biolubrication.

Together with the surface properties and the presence of a lubricant, environmental conditions such as p*H*, ionic strength, temperature, etc. will ultimately determine the strength of the interactions (adhesion and friction) between surfaces, in particular when the confined lubricant is a highly charged polyelectrolyte. Although those parameters are not the subject of the present study, they are mentioned later, when discussing the remarkable stability of polysaccharides (showing slow and controlled degradation) over a wide range of p*H* and temperatures.

Finally, external parameters such as applied load (pressure) and shear act as the local and transient *stimuli* controlling the instantaneous response of the tribological system. Among the harsh conditions encountered in the body are highly compressive loads (in the knee, hip or spine), large sheared distances (in the shoulder), a wide range of velocities (static, walking, running) and high fatigue strength cycles (in the heart and eyelids). In the next sections, we evaluate the capacity of several charged biopolymers (HA, lubricin, chitosan, and a polysaccharide from the algae) to carry high loads and pressures, to resist wear when sheared over large distances and over a wide range of velocities, in addition to exhibiting low friction forces. We also introduce thick (non-biological) polymer brush layers, as they were found to function well in tribological systems subjected to low pressures, and therefore currently thought to be the best biomimetic lubricants.

3. Current approaches to assess biolubrication

The most efficient, energetically speaking, tribosystems are those found in nature, e.g. mammals joints. Natural joints have been the center of interest of biomedical research with the hope of mimicking such low energy dissipation and wear resistance in engineered systems: physiological coefficients of kinetic friction μ~0.002-0.006, billions of cycles without wear (high fatigue limit), high-load carrying capacity and robustness, all hardly reproducible "macroscopic" natural lubrication achievements (Hills, 2000). Synovial fluid (SF), the natural thick and viscous lubricant of joints, has for long been believed to be responsible for such remarkable lubrication properties and is addressed in section 3.2. However, as mentioned earlier, some biolubricating systems (such as in the eyes where the pressure is low) may be mediated by brushlike polyelectrolyte layers. Because polymer brush layers are currently thought to be the best biomimetic lubricants (hence promising for future biomedical applications), we start our next section by reporting friction experiments performed with non-biological (charged and uncharged) polymer brush layers used to lubricate mica surfaces.

3.1 Polymer brushes

Polymer brushes (PB) have been proposed to be responsible for efficient lubrication in biological tribosystems subjected to low external pressures (Raviv et al., 2003; Gourdon et al., 2008) such as in the eye or in the respiratory system, mediating not only low friction, but also permeability and/or bacteria and dust adhesion. PB are in general, a dense array of polymer chains (neutral or charged) endgrafted to a surface. The vicinity between adjacent polymer chains forces the polymer to adopt more favorable conformations. When the polymer chains stretch and point out perpendicularly to the supporting surface, the array is named PB (Brittain & Minko, 2007).

To control the formation of dense brushes at the surface, specific conditions have to be fulfilled, i.e., a complex balance between molecular weight (M_W), surface density, quality of solvent, charge distribution, etc. Instead of brushes, the polymer chains may also adopt less desirable conformations, such as pancake or mushroom, see figure 1, leading to weaker stability and non functional properties of the coatings. Among the controlled properties achieved by brushes are wettability, adhesion and lubricity (Klein et al., 1993; Brittain & Minko, 2007). The ability to manipulate easily these properties makes brushes smart surfaces that have attracted attention for applications like biosensors, tissue engineering, drug delivery and lubricants (Dong et al., 2009).

An increasing number of groups have been clarifying the main parameters governing the tribological behavior of PB coated surfaces (Liberelle & Giasson, 2008; Zappone et al., 2008; Kampf et al., 2004; Hartung et al., 2008; Chen et al., 2009). It is well established that under low loads, almost any pair of polymer coated surfaces will slide with low friction coefficients, μ=0.01 (Raviv et al., 2003; Liberelle & Giasson, 2008). An extreme case has been found to be of the order of μ=0.001 (Raviv et al., 2003) when end-grafted charged polymers were sheared at low pressure (0.5MPa), and across very small distances (0.7μm). This low friction behavior was attributed to the strong resistance to mutual interpenetration of the pressed brushes together with the presence of hydrated ions surrounding the long polymer chains, see figure 2 a) and c). In contrast, neutral brushes exhibited high friction due to chain interpenetration (figure 2 b) and d)) and a significant increase in viscosity upon shear rate, leading to solid-like friction behavior (Neelov et al., 1998). Thus, the friction coefficient could be directly correlated to the degree of interpenetration between the polymer layers, governed itself by solvent ionic strength and polymer ionization degree.

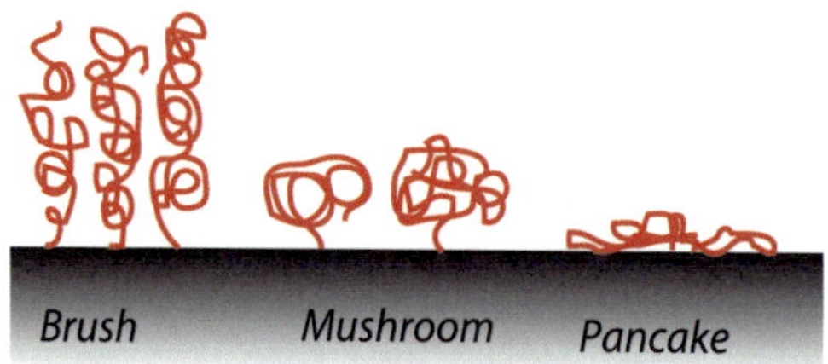

Fig. 1. Possible polymer conformations, depending on quality of solvent, polymer degree of ionization and grafting distance.

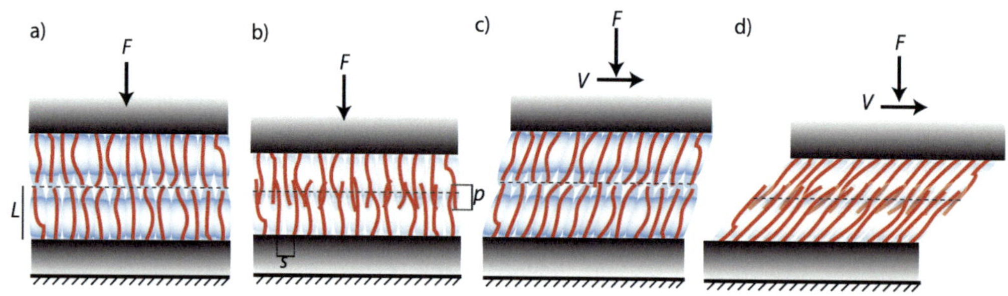

Fig. 2. a) Charged polymer brushes showing a strong resistance to mutual interpenetration, b) neutral brushes showing a weak resistance to mutual interpenetration. *L*, brush length; *s*, grafting distance; *p*, brush interpenetration; *F*, applied normal force, c) smooth sliding, consequence of (almost) no interpenetration, and d) dragging of brushes due to large interpenetration, yielding to high friction forces. *V*, sliding velocity.

PB appear to be efficient for modifying surface properties thus for controlling tribological behavior of shearing junctions, however, controlling the coverage density of brushes (and their attachment to surfaces) has been of great concern, especially to resist high pressures and large distances sheared. Highly packed synthetic PB have been used to successfully lubricate artificial joints at low pressures (Hartung et al., 2008). We next address another class of water-based lubricants, naturally occurring biological compounds, such as hyaluronan (the main component of synovial fluid), commonly used to heal degenerative joint diseases via direct injections at the site of inflammation.

3.2 Naturally occurring biological compounds

An intrinsic advantage of naturally occurring biological polymers is their similarity to macromolecules that biological systems are prepared to recognize, and handle metabolically. Thus, problems like toxicity or chronic inflammatory responses can be exluded (Yannas, 1996). In this section, we review the lubrication and adhesive properties of four naturally occurring charged polymers (hyaluronic acid, lubricin, chitosan, and an algae polysaccharide), and report the potential use of (nano)thin films of those biopolymers to improve the lubrication properties of tribological systems composed of molecularly smooth mica surfaces. We particularly focus on the capacity of the biopolymers to carry high pressures (up to hundreds of atmospheres), to resist wear when sheared over large distances, and to exhibit low friction forces, all required for efficient biolubrication.

3.2.1 Hyaluronan

Hyaluronan or hyaluronic acid (HA), one of the main glycoproteins present in synovial fluid and connective tissues, is found in many different concentrations, depending on the physiological location. A concentration of 3.5 $\frac{mg}{ml}$ is normally found in a regular adult knee (Tadmor et al., 2002). HA is a large, negatively charged and unbranched glycosaminoglycan (specialized glycoprotein with polysaccharide side-chains), having a high M_W, ranging from 6-7MDa for healthy to 3-5MDa for rheumatoidal fluid (Dahl et al., 1985), with a polymer length $\approx 2\mu m$ (Toole, 2004; Mow et al., 1984). It also possesses an hydrophilic face; due to the presence of carboxyl groups, figure 3 a). A number of specific functions like lubrication, macromolecular filtering, cartilage surfaces protection, water homeostasis, selective binding of proteins (known cell-surface receptors) have been attributed to HA (Laurent et al., 1996). Over the last decades, HA has been thought to be responsible for the ultra low friction and wear rates in articular joints, until Radin et al. questioned such claim (Radin et al., 1970). Despite still open questions, HA has been used for the treatment of degenerative joint diseases and total or partial joint replacements (Kirchner & Marshall, 2006), providing temporary pain relief for patients (HA intra-articular injections are recommended every six months).

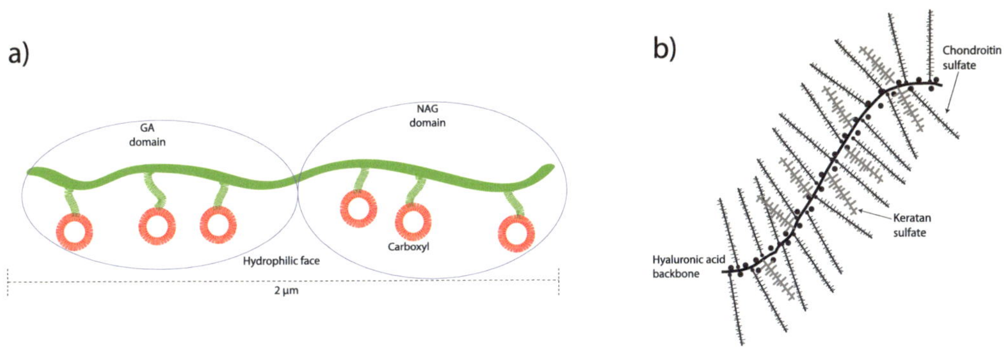

Fig. 3. a) Representation of a HA molecule. It is composed of a glucuronic acid (GA) domain and a *N*-acetylglucosamine (NAG) domain. b) Proteoglycan structures attached to a backbone of HA (adapted from (Neville et al., 2007)).

HA has been proved to be non-adhering to negatively charged surfaces, thus acting only as a viscous and bulky lubricant. In its native environment, the huge molecules can be locked mechanically by the pores of the cartilage, rather than bound, giving rise to a mechanical entrapment instead of physical or chemical binding. Experiments using bare and functionalized positively charged mica were performed using the SFA to measure the normal (adhesion) and friction forces of an HA solution of 3.4 $\frac{mg}{ml}$ (M_W=540kDA) (Tadmor et al., 2002).

The normal force profile of non adsorbing HA showed a short-range repulsive behavior. This was attributed to a decrease of the Debye length, consequence of the dissociated sodium cations from HA, and the squeezing of HA molecules out of the contact area. In the case of adsorbed polyelectrolytes to the surface (using Ca^{2+} as anchor), a steric, long-range repulsion was seen. At very short distances, the van der Waals and hydrophobic attraction forces increased (in comparison to the free HA solution), even though the system showed a more repulsive force. The weak binding of HA to the surface could be a consequence of the hydrated divalent cations that did not readily lose their water on binding to mica (Berg et al., 1993).
Friction experiments revealed an interesting behavior. While damage was always observed when shearing HA between bare mica surfaces, as indicated by a friction coefficient μ=0.25 (a typical value for wear), Ca^{2+} bonded HA to mica surfaces could resist damage upon initial shear at low pressures (0.5MPa), although the measured friction was relatively high. These results indicate that HA is capable to sustain lubrication as long as it is well confined between surfaces. Upon prolonged shear, the free HA molecules are expelled from the high pressure zones in the junction. This also suggests that HA works more like a visco-additive to enhance and maintain hydrodynamic lubrication in joints rather than like a boundary lubricant.
It has recently been proposed that HA is not the real responsible for the remarkable lubrication properties of synovial fluid, in spite of the high concentrations found in it (Neville et al., 2007). Hydrodynamic modes of lubrication are known to be enhanced by highly viscous lubricants, preventing contact between cartilage surfaces (e.g., during walking). The capability of HA to "wet" the highly hydrophobic cartilage surfaces (in order to promote hydrodynamic lubrication) has been hypothesized to be HA's main function (Hills, 2000). Other proteins, like lubricin, should then be responsible for the outstanding joint lubrication under extreme conditions, e.g., when no synovial fluid is present after prolonged periods of rest. More recently, HA has been proposed to be used together with bone grafts in order to accelerate the integration process of exograft bone implants, since HA has shown to facilitate cell migration and differentiation during tissue formation and repair (Aslan et al., 2006). This supports the idea that HA may function more as a carrier of wetting agents (to enhance binding), see figure 3 b), rather than as a boundary lubricant.

3.2.2 Lubricin: an example of mucins

Lubricin is another glycoprotein present in the synovial fluid, and is believed to lubricate and protect the cartilage surfaces in boundary lubrication, preventing cartilage-cartilage contact, and thus, wear (Schaefer et al., 2004). Like any other efficient boundary lubricant, lubricin is capable of binding to the cartilage surface. Apart from lubricating, it also protects the cartilage surfaces from excessive protein adsorption and cell overgrowth, which is one of the causes of joint failure diseases (Rhee et al., 2005). Lubricin is a structureless flexible high M_W molecule, of the order of 230kDa and has an extended length of 210nm. The molecular structure, represented in figure 4, is composed of a large heavily glycosylated central domain, carrying the negative charges, and two end-domains that are not glycosylated, carrying the positive charges (Zappone et al., 2007). These somatomedin (SMB), heparin and homepexin (PEX) like-domains are of special interest, since they are known to play specific roles in cell-cell and cell-extracellular matrix interactions (Jay, 2004).
In order to obtain a complete picture of the possible interactions of lubricin with different surfaces at physiological concentrations (250 $\frac{mg}{ml}$ (Jay, 2004)), a variety of substrates under controlled physical and chemical properties have been used. While negatively charged

lipids, sulphated proteoglycans and glycoproteins were mimicked by hydrophilic, negatively charged mica surfaces, basic amine groups of proteins were simulated by positively charged physisorbed poly-lysine to the mica surfaces. Chemisorbed hydrophobic monolayers of alkanethiol on gold were used to mimic collagen below the cartilage surface. The hydrophilic negatively charged mica, the physisorbed and chemisorbed hydrophilic, as well as the hydrophobic surfaces, showed similar normal force behaviors: a purely repulsive force, showing always a reversible compression-decompression cycle for different contact times and compressive forces, i.e., no hysteresis. This behavior was achieved with a saturated surface, corresponding to "brush" thicknesses L=70nm-100nm, i.e., approximately one half of the length of stretched lubricin (Zappone et al., 2007).

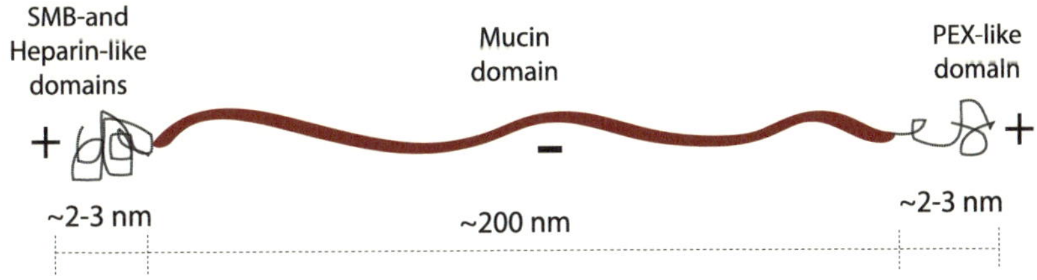

Fig. 4. Representation of lubricin. The end-domains of the molecule carry most of the positive charges and hydrophilic groups, while the central domain carries the negative charges, adapted from (Zappone et al., 2007).

The frictional properties of lubricin have been shown to be highly pressure-dependent. For contact pressures up to 0.4MPa, the friction coefficients were as low as μ=0.02-0.04, but increased abruptly by one order of magnitude (μ=0.2-0.6) at higher pressures (Zappone et al., 2007). This high friction eventually led to wear of the lubricin monolayer, but no damage of the underlying mica was reported. The hydrophobic surfaces showed a substantial different behavior in their tribological properties, exhibiting high friction coefficients even at low pressures, μ=0.3-0.5 (Zappone et al., 2007). Wear of the lubricin monolayer could also be seen, but when reaching much higher loads ($P\approx$1MPa) than those needed to cause wear in the hydrophilic samples ($P\approx$500kPa).

Since the semilogarithmic normalized force runs ($\frac{F}{R}$) versus surfaces separation (D) can be well described by the Alexandre-deGennes model, lubricin is believed to form a monolayer of brushes arranged in the various conformations shown in figure 5. This simple model can explain the low friction coefficients ($\mu\approx$0.03) found at low pressures, and the sudden change to higher friction coefficients ($\mu\approx$0.4) measured when increasing pressure. Similar behavior has been previously reported for charged polymer brushes (Raviv et al., 2003), and attributed to weak interpenetration of opposing brushes occurring at high pressures.

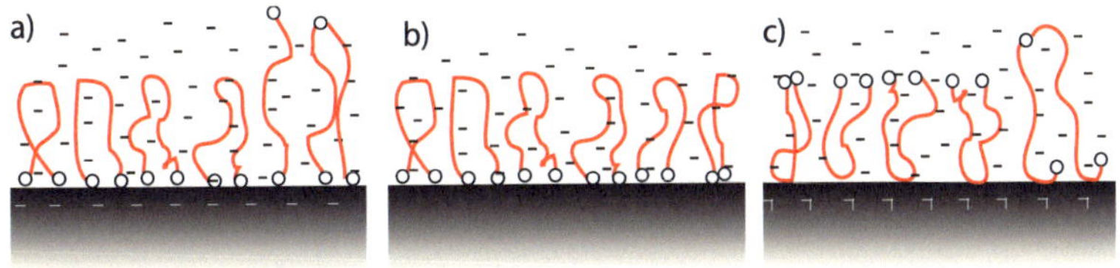

Fig. 5. Likely configurations of lubricin when physisorbed to a) negatively charged surfaces, b) hydrophobic surfaces and c) positively charged surfaces as proposed by Zappone et al. End-domains of lubricin carry most of the positive charge of the molecule.

These experiments suggest that the elongated lubricin molecules can attach to the mica surface by both its positively charged end-domains (leading to high repulsion as a consequence of entropic resistance to confinement). They also indicate that lubricin alone may not be responsible for the ultra low friction found in joints, although it is able to ensure good protection of the "contacting" surfaces against damage, even at high pressures.
The naturally occurring polyelectrolytes described so far (HA and lubricin) are both found in the human body. We now report the lubrication and adhesion properties of polyelctrolyte extracted from other natural sources (crustaceans or algae). Those have also been used successfully in specific biomedical applications, and they deserve our interest as complementary approaches to assess biolubrication.

3.2.3 Chitosan

Chitosan is a linear polysaccharide produced from chitin, the structural element in the exoskeleton of crustaceans and some unicellular organisms. It has gain considerable attention due to its biological properties, uses in food industries (water filtration), agricultural fields (biopesticides) and potential uses in medicine (antihemorrhagic agents). On controlling the M_W, pH, ionic strength and degree of deacetylation, one is able to drastically change its properties (Claesson & Ninham, 1992). Chitosan has a peculiar property; it is the only cationic polysaccharide occurring in nature in abundant quantities. Its cationic nature comes from the presence of amino groups at acidic to neutral pH, adopting an extended conformation at low ionic strengths, see figure 6, due to strong electrostatic segment-segment repulsion, in contrast to the large majority of polysaccharides that carry a negatively charge or are neutral. It is composed of a relatively large backbone, that is not very flexible due to the presence of bulky sugar rings. Solutions containing the approppiate concentration of chitosan are capable of forming very stable layers (with varying pH), as well as gels in the presence of cross-linking agents, such as carboxylic acid or polyphosphates. Being hypoallergenic and having antibacterial properties, chitosan has also been successfully used as an antihemorrhagic agent (Pusateri et al., 2003). Because of so many potential applications, Claesson & Ninham initiated the investigation of the molecular mechanisms of chitosan adsorption to surfaces.
Chitosan is a polyelectrolyte that fully dissociates at low pH, having a pK_a=6.5. Its interactions with surfaces are therefore highly pH dependent. At very acidic values, the system presented a steric repulsive force starting at 10nm. The smallest measured distance (hard wall), D, was 1.5nm, indicating that the layer on each surface was of the order of 0.7-0.8nm. In spite of pure repulsion during approach, a small adhesive force during receding was measured, suggesting the formation of weak polyelectrolyte bridges. Adsorbed chitosan dissolves poorly in free polyelectrolyte solution, once reached the surface saturation. Free cationic polymer molecules in solution could not contribute to the lubricating behavior, because the molecules were squeezed out when the surfaces approached, as indicated by changes in the refractive index. Changing the pH to less acidic environments, the approach/receding experiments showed similar behavior, decreasing only the Debye length and preserving a small 1 nm-thick polymer layer (Claesson & Ninham, 1992). When approaching to the pK_a value, pH=6.2, no double-layer force repulsion was present, leaving only the attractive van der Waals forces. Beyond this critical value (at higher pH values) the double layer repulsion reappeared and the layer thickness increased to 1.6nm (the force needed to reach the hard wall was also larger than at lower pH values). Chitosan has a strong positive charge at low pH, being therefore attracted to the negatively

charged mica surface, forcing the polysaccharide to adsorb in a flat conformation. The likely conformations of chitosan at various p*H* are shown in figure 6. Gelled (cross-linked) chitosan using sodium hexametaphosphate as linking agent showed some differences: hysteresis was observed between compression and decompression, increasing the repulsion force and range. Besides these differences, the behavior of the cross-linked chitosan was similar to the non cross-linked one.
Friction forces were measured for both free adsorbed and gelled chitosan layers. Friction forces, in the former case, could only be detected at distances, D=20nm and increased monotonically when decreasing D, being slightly above the noise level of the system (2μN). After gellation, friction was large even at large D, showing a sharp initial rise but final (steady state) friction could not be achieved because of the very small distances sheared. (Kampf et al., 2004). The friction forces between freely and gelled chitosan were also investigated as a function of velocity. Freely adsorbed chitosan showed little frictional force dependency on velocity, in contrast to what was seen in crosslinked chitosan, where a significant increase in the friction force (with increasing velocity) was observed. For a shearing velocity of 200 $\frac{nm}{s}$, an effective friction coefficient, μ_{ef} = 0.003 was measured for non cross-linked molecules, increasing up to μ_{ef} = 0.5 for the cross-linked gel (Kampf et al., 2004).

3.2.4 Algae polysaccharide

Algae polysaccharides have diverse biological functions and physicochemical properties and are increasingly being used in various fluid formulations. They dissolve easily in aqueous solutions, increasing their viscosity, and controlling their rheological properties that ultimately define the various applications of the biopolymers (Geresh & Malis, 1991). Biological properties of the polysaccharides include anti-inflammatory, anti-oxidant and anti-viral activity (Gourdon et al., 2008). They also own chemical and biological inertness (high stability) under extreme conditions of salinity, p*H*, temperature, light and dehydration (Geresh & Malis, 1991). Tribological studies at the macroscopic scale showed that their friction coefficients were lower than the ones found in HA, making them excellent candidates for biomedical applications. The naturally occurring polysaccharide studied is synthesized by the algae *Porphyridium* sp. and is composed mainly of three sugars: xylose (38%), galactose (24%) and glucose (22%), and glucuronic acid units (10%) (Percival & Foyle, 1979). It is an anionic high M_W (2.3MDa) molecule. Surprisingly, this polysaccharide is able to adsorb even to negatively charged surfaces, despite its anionic nature, showing a stable low friction coefficient, robustness (high load carrying capacity), good wear protection and a weak dependence of the friction force on sliding velocity (Gourdon et al., 2008).
Atomic force microscopy (AFM) imaging of mica surfaces in polysaccharide solution indicated that physisorbed small polysaccharide globular domains could built on the surface, increasing in number but neither in size nor height with increasing incubation time. These globules were measured to be 10nm in diameter and ≈1.1nm in height and appeared to remain highly mobile at the surface (easily dragged upon shearing with the AFM tip). Normal forces measurements carried out with the SFA (at room temperature) revealed that, in polysaccharide solution, the forces during approach were exponentially repulsive, due to the large-range double layer force. Below 10nm, an additional short-range repulsive hydration force could be seen (Gourdon et al., 2008). In contrast, adhesion was measured during receding. Such adhesion was found to increase with the maximum load applied during each compressive cycle, before reaching a maximum at $W \approx 6\frac{mJ}{m^2}$.

The friction behavior was recorded as a function of applied load/pressure, shearing velocity and shearing distance at different polysaccharide concentrations. At early stages of shear, the measured friction force was high and proportional to the contact area (indicative of adhesion-controlled friction). However, further shear drastically modified the frictional regime, showing an abrupt transition to low and stable (load-controlled) friction: average friction coefficients as low as $\mu \approx 0.015$, and no wear observed up to pressures of 10MPa and upon shearing over 3 decades of velocity. This tribological transition was attributed to the reorienting capability of the polysaccharide, similar to the one seen in nanoparticle systems when the particles minimize their energy via shear-induced alignment (Gourdon et al., 2004). Experiments performed over shearing distances greater than the contact area revealed that the polysaccharide could be carried out to new unsheared areas, capable of expanding the low friction and wear resistances domains. One of the most outstanding properties of the polysaccharide lubricating films is that only a subnanometric (0.5-1nm) monolayer was enough to provide such efficient lubrication and wear protection. Figure 7 shows the likely molecular mechanisms, involving confinement and shear-induced alignment of the polysaccharide molecules, allowing them to act as both good lubricants and adhesives. Such dual behavior can be explained by the highly flexible polysaccharide backbone, allowing for the rapid formation of hydrogen bonds between the biopolymer sugar moieties and the mica surfaces. Damage could only been observed when shearing over distances larger than twice the contact diameter, at pressures of 11MPa (higher than any pressure found in healthy human joints under normal daily activity).

The remarkable properties of the *Porphyridium* sp. polysaccharide should not be surprising, since it is found in the extracellular matrix of the algae cells, tightly bound to the outer membrane surface, and functions simultaneously as a lubricant and as a protecting film for the algae. In summary, the most notable properties exhibited by the naturally-occurring polyelectrolyte were:

- steady low friction
- stability at high pressures
- stability at high and low velocities
- wear protection
- stability at large shearing distances

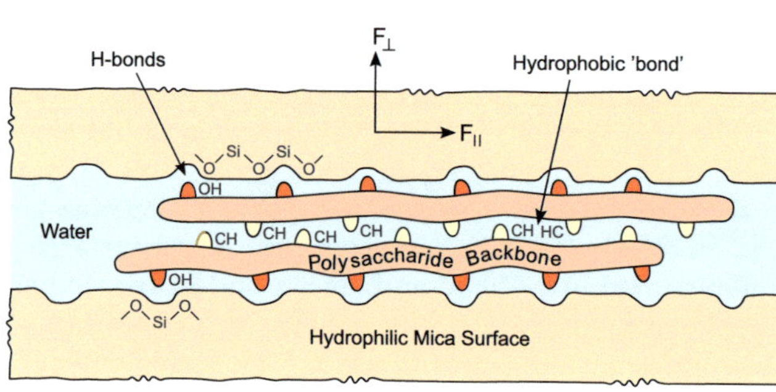

Fig. 7. Likely molecular mechanisms, involving confinement and shear-induced alignment of the polysaccharide molecules, allowing them to act as good lubricants and adhesives. Adapted form (Gourdon et al., 2008).

None of the biopolymers investigated so far has been able to combine *at once* these five characteristics. However, the algae polysaccharide is the one fulfilling most of the tribological requirements for efficient biolubrication. In such a context, polysaccharides seem to be the best way to go and further investigation (experimental and numerical) of this class of polyelectrolytes is needed to identify the crucial sequence(s) that determine the lubricating functionalities of the biopolymers.

4. Experimental methods

The surface forces apparatus (SFA) and the atomic force microscope (AFM) have proven to be ideal tools for combining nano-, micro-, and macroscale tribological experiments as well as for nanoscale imaging and characterization of surfaces in presence of lubricants. These two techniques are briefly described below.

4.1 Surface forces apparatus

The surface forces apparatus (SFA) was developed in the 70's and optimized in the 90's of last century to measure the normal and friction forces at the molecular and atomic scale, on a wide variety of systems, such as colloids, amorphous polymers, and metals. Details of the SFA can be found in (Israelachvili & McGuiggan, 1990). Two cylindrical silica lenses of radius R (R=1cm), wearing a thin sheet of back-silvered and atomically smooth mica, face each other and used as preliminary surfaces (these surfaces can be modified before being mounted into the SFA). The surfaces separation, D, is obtained by multiple beam interferometry, together with surface shape, contact radius, contact area and refractive index in both static and dynamic conditions. Normal and friction forces are measured by means of double cantilever springs while shearing is ensured by a piezoelectric bimorph slider allowing large displacements. A scheme of the SFA is shown in figure 8 a). Two enormous advantages of the direct visualization of the contact region with multiple beam interferometry using fringes of equal chromatic order (FECO) are: 1) the contact area is measured rather than calculated and 2) any damage of the surfaces can be easily detected, as well as undesirable particles/contaminants. A typical normal force $F(D)$ profile is presented in figure 8 b), for both, repulsive and adhesive behaviors.

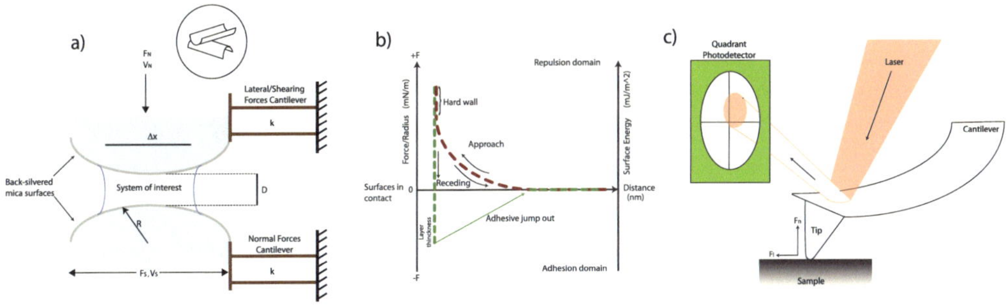

Fig. 8. Scheme of a) the SFA, where F_N is the normal applied load; V_N, approach speed of surfaces; F_S, the shear force; V_S, the shearing velocity; k, spring constant; Δx, shearing distance; R, curvature radius of the optical lenses, b) typical distance-force curves obtained with the SFA and the AFM. In sepia shown a purely repulsive behavior, during approach and receding of the surfaces. An induced adhesive behavior can be seen in the approach (sepia) combined with receding (green), extending the force to the negative (adhesion) domain, and c) the AFM, where F_n, is the normal force between the tip and sample, and F_l, is the lateral friction force, adapted from (Caprick & Salmeron, 1997).

4.2 Atomic force microscope

Another useful tool for tribological studies is the atomic force microscope (AFM), developed in 1986. The AFM can be used to measure normal and frictional forces as well as to image samples with atomic scale resolution, and has been implemented for applications ranging from semiconductors to biological samples. Details about the AFM can be found in (Carpick & Salmeron, 1997). In this instrument (scheme shown in figure 8 c)), a tip of radius *R* (*R*=30nm) mounted onto a compliant cantilever is brought into contact with the surface of a stationary sample. Adhesion and friction forces are measured by recording the normal and lateral deflections of the cantilever beam. For that, a laser beam is reflected off the back of the cantilever into a quadrant photodiode detector, which records changes in the vertical and horizontal deflection of the probe.

5. Conclusions and future work

Although thick polymer brush layers are currently thought to be the best biomimetic lubricants, we have found that subnanometer-thick monolayers of of an algae polysaccharide can exhibit stable low friction and good wear protection, up to 11MPa and over a 1000-fold range of sliding velocities. Such remarkable tribological behavior is attributed to the ability of the polysaccharide to adhere weakly to the surfaces (ensuring that the lubricant is not easily squeezed out of the shearing junction) and to be flexible and mobile enough to be shear-aligned and transferred into new unsheared areas. However, despite the superior lubrication demonstrated by the algae polysaccharide compared to HA, lubricin, and chitosan, we conclude that none of the systems investigated so far possesses the ideal combination of friction and adhesion properties when subjected to cyclic loading, high pressures, high impact, and large sliding distances, as found in natural tribological systems such as in the hip or knee joint. Indeed even the algae polysaccharide monolayer breaks down when sheared (at high pressures) over distances larger than twice the diameter of contact between surfaces.

Hence, the best candidate for use in water-based lubricant fluids and/or as potential additives to synovial fluid in joints and other biolubricating fluids still needs to be identified. By further investigating this class of polysaccharides but also other natural polysaccharides and synthetic peptides, we expect to be able to identify the crucial chemical groups and/or sequences that determine their different functionalities, with obvious potential for biomedical (medical synthesis) and/or simply biomimetic applications.

6. Acknowledgments

This work was partially supported by the Cornell Center for Materials Research with funding from the National Science Foundation under Award Number DMR-0520404.

7. References

Aslan, M., Simsek, G. & Dayi, E. (2006). The effect of hyaluronic acid-supplemented bone graft in bone healing: Experimental study in rabbits, *Journal of Biomaterials Applications* 20: 209–220.

Benz, M., Chen, N. & Israelachvili, J. (2003). Lubrication and wear properties of grafted polyelectrolytes, hyaluronan and hylan, measured in the surface forces apparatus, *Journal of Biomedical Materials Research A* 71: 6–15.

Berg, J., Claesson, P. & Neuman, R. (1993). Interactions between mica surfaces in sodium polyacrylate solutions containing calcium ions, *Journal of Coloid and Interface Science* 161: 182–193.

Brittain,W. & Minko, S. (2007). A structural definition of polymer brushes, *Journal of Polymer Science Part A: Polymer Chemistry* 45: 3505–3512.

Carpick, R. & Salmeron, M. (1997). Scratching the surface: fundamental investigations of tribology with atomic force microscopy, *Chemical Reviews* 97: 1163–1194.

Chen, M., Briscoe, W., Armes, S. & Klein, J. (2009). Lubrication at physiological pressures by polyzwitterionic brushes, *Science*, 323: 1698-1701.

Claesson, P. & Ninham, B. (1992). p*H*-dependent interactions between adsorbed chitosan layers, *Langmuir* 8: 1406–1412.

Dahl, L., Dahl, I., Engstrom-Laurent, A. & Granath, K. (1985). Concentration and molecular weight of sodium hyaluronate in synovial fluid from patients with rheumatoid arthritis and other arthropaties, *Annals of Rheumatic Diseases* 44: 817–822.

Dong, R., Lindau, M. & Ober, C. (2009). Dissociation behavior of weak polyelectrolyte brushes on a planar surface, *Langmuir* 25: 4774–4779.

Geresh, S. & Malis, S. (1991). The extracellular polysaccharide of the red microalgae: chemistry and rheology, *Bioresource Technology* 38: 195–201.

Gourdon, D., Qi, L., Oroudjev, E., Hansma, H. & Golan, Y. (2008). Adhesion and stable low friction by a subnanometer-thick monolayer of a natural polysaccharide, *Langmuir* 24: 1534–1540.

Gourdon, D., Yasa, M., Godfrey Alig, A., Youli, L., Safinya, C. & Israelachvili, J. (2004). Mechanical and structural properties of $BaCrO_4$ nanorod films under confinement and shear, *Advanced Functional Materials* 14: 238–242.

Hartung, W., Drobek, T., Lee, S., Zürcher, S. & Spencer, N. (2008). The influence of anchoring-group structure on the lubricating properties of brush-forming graft copolymers in an aqueous medium, *Tribology Letters* 31: 119–128.

Hills, B. (2000). Boundary lubrication in vivo, *Proceedings of the Institution of Mechanical Engineers, Part H: Journal of Engineering in Medicine* 214: 83–94.

Israelachvili, J. & McGuiggan, P. (1990). Adhesion and short-range forces between surfaces. Part I: New apparatus for surface forces measurements, *Materials Research Society* 5: 2223–2231.

Jay, G. (2004). Lubricin and surfacing of articular joints, *Opinion in Orthopaedics* 15: 355–359.

Kampf, N., Raviv, U. & Klein, J. (2004). Normal and shear forces between adsorbed and gelled layers of chitosan, a naturally occurring cationic polyelectrolyte, *Macromolecules* 37: 1134–1142.

Kirchner, M. & Marshall, D. (2006). A double-blind randomized controlled trail comparing alternate forms of high molecular weight hyaluronan for the treatment of osteoarthritis of the knee, *Osteoarthritis and Cartilage* 14: 154–162.

Klein, J., Kamiyama, Y., Yoshizawa, H., Israelachvili, J., Fredrickson, G., Pincus, P. & Fetters, L. (1993). Lubrication forces between surfaces bearing polymer brushes, *Macromolecules* 26: 5552–5560.

Laurent, T., Laurent, U. & Fraser, J. (1996). The structure and function of hyaluronan: an overview, *Immunology and Cell Biology* 74: A1–A7.

Lee, S., Müller, M., Ratoi-Salagean, M., Voros, J., Pasche, S., DePaul, S., Spikes, H., Textor, M. & Spencer, N. (2003). Boundary Lubrication properties of Poly(L-lysine)-g-poly(ethylene glycol) (PLL-g-PEG) as an additive in aqueous media, *Tribology Letters* 15: 231–239.

Liberelle, B. & Giasson, S. (2008). Friction and normal interaction forces between irreversible attached weakly charged polymer brushes, *Langmuir* 24: 1550–1559.

Mow, V., Holmes, M. & Lai, W. (1984). Fluid transport and mechanical properties of articular cartilage: a review, *Journal of Biomechanics* 17: 377–394.

Neelov, I., Borisov, O. & Binder, K. (1998). Stochastic dynamics simulation of grafted polymer brushes under shear deformation, *Macromolecular Theory and Simulations* 7: 141–156.

Neville, A., Morina, A., Liskiewicz, T. & Yan, Y. (2007). Synovial joint lubrication—does nature teach more effective engineering lubrication strategies?, *Proceedings of the Institution of Mechanical Engineers, Part C: Journal of Mechanical Engineering Science* 221: 1223–1230.

Percival, E. & Foyle, R. (1979). The extracellular polysaccharides of porphyridium cruentum and porphyridium aerugineum, *Carbohydrate Research* 72: 165–176.

Pusateri, A., McCarthy, S., Gregory, K., Harris, R., Cardenas, L., McManus, A. & Goodwin, C. (2003). Effect of a chitosan-based hemostatic dressing on blood loss and survival in a model of severe venous hemorrhage and hepatic injury in swine, *The Journal of Trauma: Injury, Infection, and Critical Care* 54: 177–182.

Radin, E., Swann, D. & Weissner, P. (1970). Separation of a hyaluronate-free lubricating fraction from synovial fluid, *Nature* 228: 377–378.

Ratner, B. & Hoffman, A. (1996). Thin films, grafts, and coatings, *in* B. Ratner, F. Shoen & J. Lemons (eds), *Biomaterials Science*, Academic Press, U.S.A., pp. 105–116.

Raviv, U., Giasson, S., Kampf, N., Gohy, J.-F., Jérôme, R. & Klein, J. (2003). Lubrication by charged polymers, *Nature* 425: 163–165.

Rhee, D. K., Marcelino, J., Baker, M., Gong, Y., Smits, P., Lefebvre, V., Jay, G., Stewart, M., Wang, H.and Warman, M. L. & Carpten, J. D. (2005). The secreted glycoprotein lubricin protects cartilage surfaces and inhibits synovial cell overgrowth, *Journal of Clinical Investigation* 115: 622–631.

Schaefer, D., Wendt, D., Moretti, M., Jakob, M., Jay, G., Herberer, M. & Martin, I. (2004). Lubricin reduces cartilage-cartilage integration, *Biorheology* 41: 503–508.

Schouwenaars, R., Jacobo, V. & Ortiz, A. (2007). Microstructural aspects of wear in soft tribological alloys, *Wear* 263: 727–735.

Tadamor, R., Chen, N. & Israelachvili, J. (2002). Thin film rheology and lubricity of hyaluronic acid solutions at a normal physiological concentration, *Journal of Biomedical Materials Research* 61: 514–523.

Toole, B. (2004). Hyaluronan: from extracellular glue to pericellular cue, *Nature Reviews Cancer* 4: 528–539.

Urbakh, M., Klafter, J., Gourdon, D. & Israelachvili, J. (2004). The nonlinear nature of friction, *Nature* 430: 525–528.

Yannas, I. (1996). Natural materials, *in* B. Ratner, F. Shoen & J. Lemons (eds), *Biomaterials Science*, Academic Press, U.S.A., pp. 84–94.

Zappone, B., Greene, G., Oroudjev, E., Jay, G. & Israelachvili, J. (2008). Molecular aspects of boundary lubrication by human lubricin: effect of disulfide bonds and enzymatic digestion, *Langmuir* 24: 1495–1508.

Zappone, B., Ruths, M., Greene, G., Jay, G. & Israelachvili, J. (2007). Adsorption, lubrication, and wear of lubricin on model surfaces: polymer brush-like behavior of a glycoprotein, *Biophysical Journal* 92: 1693–1708.

14

The Role of Biofilm Exopolysaccharides on Biocontrol of Plant Diseases

Wafaa M. Haggag
Plant Pathology Department, Agricultural Research and Biology Division
Centre of Excellence for Advanced Science, National Research Centre
Egypt

1. Introduction

A huge variety of biopolymers, such as polysaccharides, polyesters, and polyamides, are naturally produced by microorganisms. These range from viscous solutions to plastics and their physical properties are dependent on the composition and molecular weight of the polymer. Exopolymers have been associated with the initial adhesion of microbes which is the primary step for biofilm formation. Moreover, the polymeric matrix of biofilms has a considerable influence on some of the most important physical and physiological properties of biofilms. Many microorganisms in the natural environment exist in multicellular aggregates generally described as biofilms, associated with solid surfaces and in intimate contact with other microbial cells. Cells adhere to surfaces and each other through a complex matrix comprising a variety of extracellular polymeric substances (EPS) including exopolysaccharides, proteins and DNA.

2. Biofilm identification

A biofilm is an aggregate of microorganisms in which cells adhere to each other and/or to a surface. In fact, archaeal, bacterial, and eukaryotic microbes produce the biopolymers. Biofilms may form on living or non-living surfaces, and represent a prevalent mode of microbial life in natural and industrial (Hall-Stoodley *et al.*, 2004). The microbial cells growing in a biofilm are physiologically distinct from planktonic cells of the same organism, which, by contrast, are single-cells that may float or swim in a liquid medium. Biofilm cells respond to nutrient and waste product diffusion gradients, modulate their metabolism as a function of their position within the biofilm, contact adjacent cells, and engage in cell–cell communication. Microbes form a biofilm in response to many factors, which may include cellular recognition of specific or non-specific attachment sites on a surface, nutritional cues, or in some cases, by exposure of planktonic cells to sub-inhibitory concentrations of antibiotics (Karatan Watnick , 2009).

3. Biofilm formation

Formation of a biofilm begins with the attachment of free-floating microorganisms to a surface. These first colonists adhere to the surface initially through weak, reversible van der

Waals forces. The first colonists facilitate the arrival of other cells by providing more diverse adhesion sites and beginning to build the matrix that holds the biofilm together. Once colonization has begun, the biofilm grows through a combination of cell division and recruitment. The final stage of biofilm formation is known as development, and is the stage in which the biofilm is established and may only change in shape and size. The development of a biofilm may allow for the aggregate cell colony(ies) to be increasingly antibiotic resistant. For example, the bacteria colonize root elongation zones and root hairs, forming dense biofilms. Microscopy of rhizobial cells within curled root hairs reveals small biofilm-type aggregates that provide the inocula for root invasion; the rhizobial cells migrate down infection threads as biofilmlike filaments towards the root interior (Figure 1) (Gage 2004). *Agrobacterium tumefaciens* and rhizobia can form dense, structurally complex biofilms on root surfaces, extensively coating the epidermis and root hairs, and these bacteria also form elaborate biofilms on abiotic surfaces (Figure 1b) (Ramey *et al.*, 2004).

Microorganisms synthesize a wide spectrum of multifunctional polysaccharides including intracellular polysaccharides, structural polysaccharides and extracellular polysaccharides or exopolysaccharides (EPS). Production of exopolysaccharide is generally important in biofilm formation, and likewise can effect the interaction of microbes with roots and root appendages (Bianciotto *et al.*, 2004). Exopolysaccharides generally constitute of monosaccharides and some non-carbohydrate substituents (such as acetate, pyruvate, succinate, and phosphate). Owing to the wide diversity in composition, exopolysaccharides have found multifarious applications in various food and pharmaceutical industries. Recent findings suggest that multiple polysaccharides modulate the chemical and physical attributes of the *Pseudomonas fluoresces* and *P. aeruginosa* biofilm matrix on abiotic surfaces (Figure 2) (Friedman and Kolter 2004). Such complexity may explain variable observations regarding the requirement for specific exopolysaccharides in biofilm formation and root association.

4. Extracellular matrix

These adherent cells are frequently embedded within a self-produced matrix of extracellular polymeric substance (Figure 3). EPS are biopolymers of microbial origin in which biofilm microorganisms are embedded. Contrary to common belief, EPS are certainly more than only polysaccharides. An HPLC analysis showed similar pattern in the polysaccharide production by *P. polymyxa*. According to stander sugar, suggesting that the biopolymer is a homopolysaccharide, which is consistent of various sugars and sugar derivatives such as glucose, galactose, mannose and xylose (Haggag, Wafaa, 2007). They comprise, in addition, a wide variety of proteins, glycoproteins, , glycolipids, enzymes,, and in some cases, surprising amounts of extracellular DNA (e-DNA) (Figure 4 and 5). In environmental biofilms, polysaccharides are frequently only a minor component (Frølund *et al.*, 1961). Some EPS components deserve particular attention.

Alginate, a polyanion polysaccharide, is the best-investigated component of mucoid *Pseudomonas aeruginosa* biofilms (Figure 4). Alginate is themain Traditionally, the self-assembly processes of algal alginates were mainly used in biotechnology for encapsulation purposes but given the option of fine-tuning its material properties, bacterial alginates are more and more considered for the production of micro- or nanostructures suitable for medical applications. Representative of a family of polysaccharides that neither show branching nor repeating blocks or unit patterns and this property distinguishes it from to

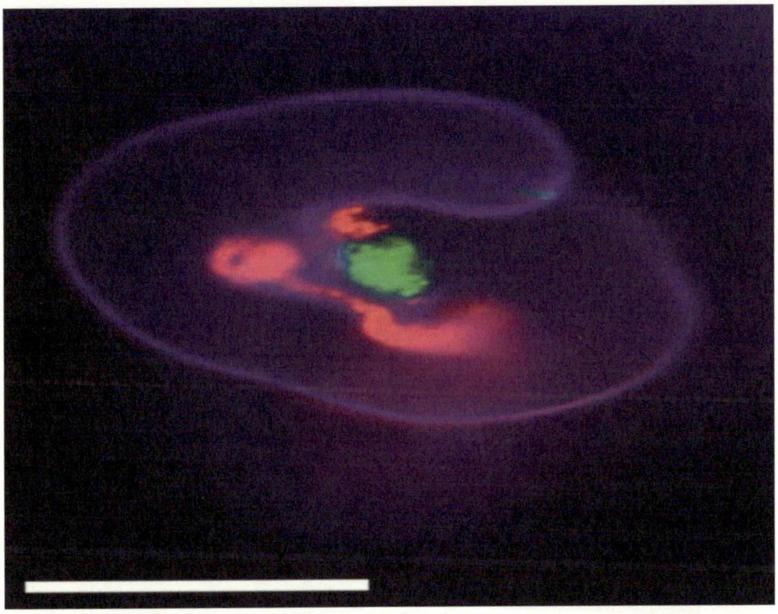

a

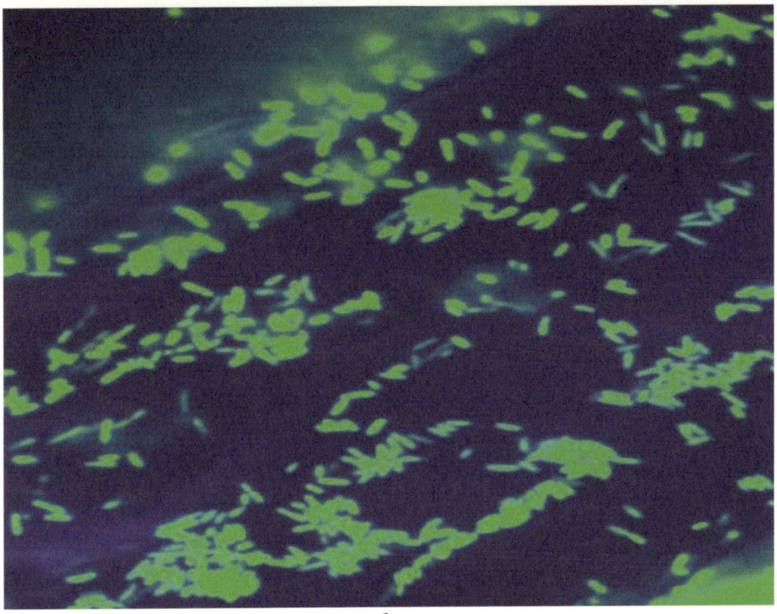

b

Fig. 1. Colonization of plants by the Rhizobiaceae. (a) Curled root hair of alfalfa with red (DsRed) and green (GFP)-expressing Sinorhizobium meliloti in a mixed microcolony occupying the interior bend of the curl. The DsRed-labeled cells have initiated an infection thread. (Image courtesy of (b) Epifluoresence micrograph of *Agrobacterium tumefaciens* C58 (Ptac-gfp) adhered to Arabidopsis thaliana Landsberg root segment. Overlay of gfp fluorescence and autofluorescence of plant tissue.

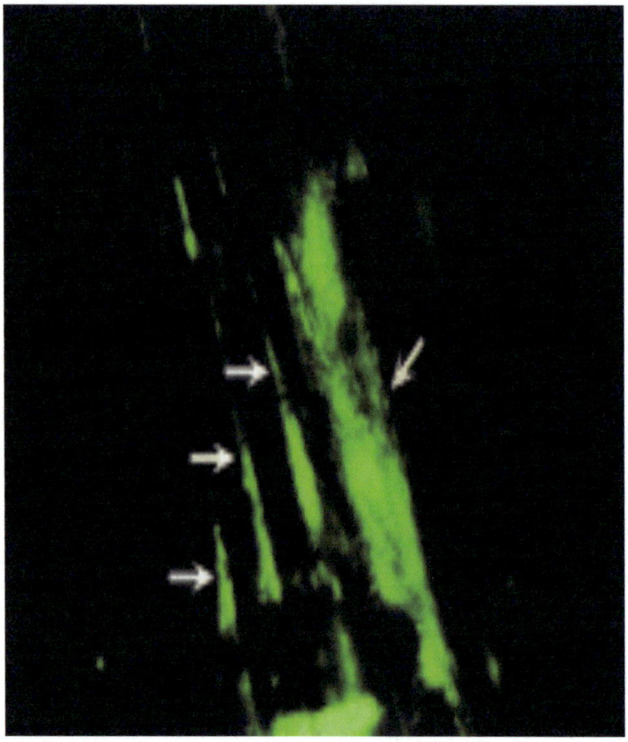

Fig. 2. Biofilm on a plant root. Biofilm of GFP-labeled *Pseudomonas fluorescens* WCS365 on the root of a tomato plant. A large microcolony of bacteria is apparent on the root surface and is indicated by the yellow arrow. The white arrows highlight three smaller colonies that have formed at plant root cell boundaries, which may be the site of release of root exudates used by bacteria as nutrient sources. The diffuse appearance of some bacterial cells in the large microcolonies suggests that these bacteria are covered by an EPS. EPS may play a role in formation of these microcolonies , suggesting that these communities have many of the characteristics of typical bacterial biofilms.

other polymers like xanthan or dextran. However, several recent reports have shown that other polysaccharides contribute to biofilms formed by nonmucoid *P. aeruginosa* strains, which are believed to be the first to colonize cystic fibrosis patients. A recent example is the expression of the *psl* operon, which is required in order to maintain the biofilm structure after attachment. Overproduction of the Psl polysaccharide led to enhanced cell surface and intercellular adhesion of *P. aeruginosa*, which translated into significant changes in the architecture of the biofilm (Ma, *et al.*, 2006). Some biofilms have been found to contain water channels that help distribute nutrients and signalling molecules. This matrix protects the cells within it and facilitates communication among them through biochemical signals. This matrix is strong enough that under certain conditions, biofilms can become fossilized. Microbe living in a biofilm usually have significantly different properties from free-floating microbe of the same species, as the dense and protected environment of the film allows them to cooperate and interact in various ways. One benefit of this environment is increased resistance to detergents and antibiotics, as the dense extracellular matrix and the outer layer of cells protect the interior of the community.

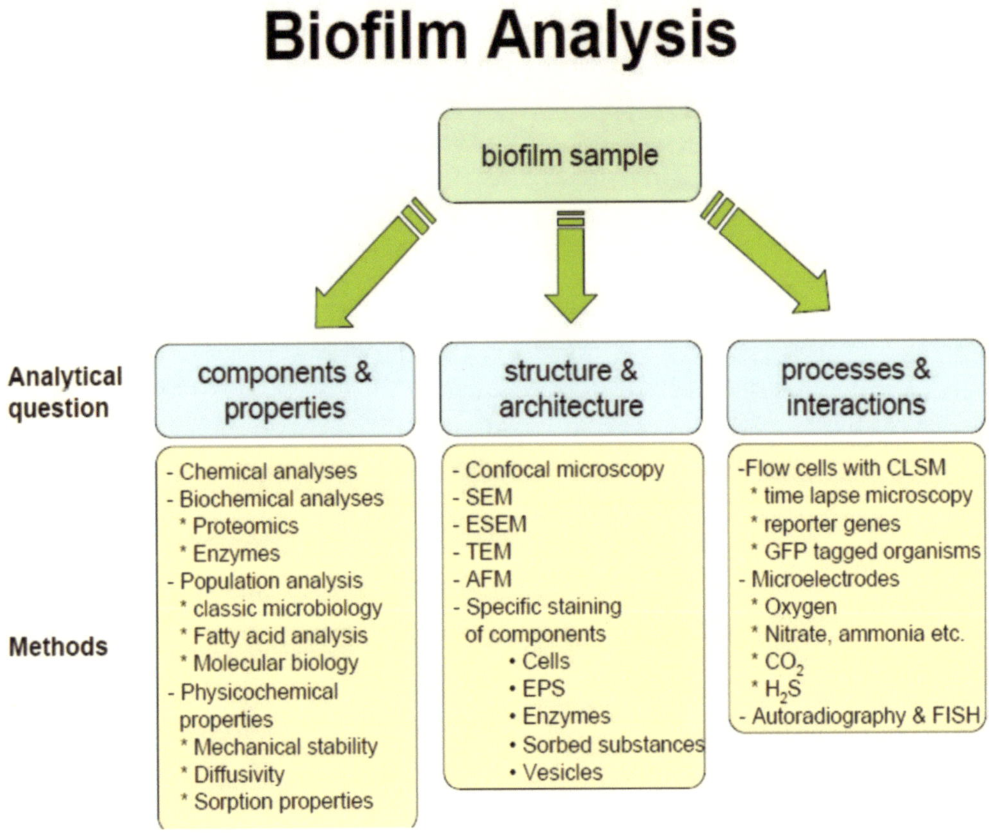

Fig. 3. Biofilm analysis

5. Metabolic engineering of microorganisms for oligosaccharide and polysaccharide production

Microorganisms naturally produce a wide variety of carbohydrate molecules, yet large-scale manufacturing requires production levels much higher than the natural capacities of these organisms. Metabolic engineering efforts generate microbial strains capable of meeting the industrial demand for high synthesis levels. Metabolically engineered strains have successfully produced many carbohydrate products, and many unexplored strategies, made available from recent progress in systems biology, can be used to engineer even better microbial catalysts.

For example, all species and strains of *Pseudomonas* are Gram-negative rods, and have historically been classified as strict aerobes. Exceptions to this classification have recently been discovered in *Pseudomonas* biofilms. A significant number can produce exopolysaccharides that are known as biofilms. Exopolysaccharide production contributes to surface-colonising biofilms which are difficult to remove from food preparation surfaces. *Pseudomonas* have the ability to metabolism a variety of diverse nutrients. Combined with the ability to form biofilms, they are thus able to survive in a variety of unexpected places (Hassett *et al.*, 2002).

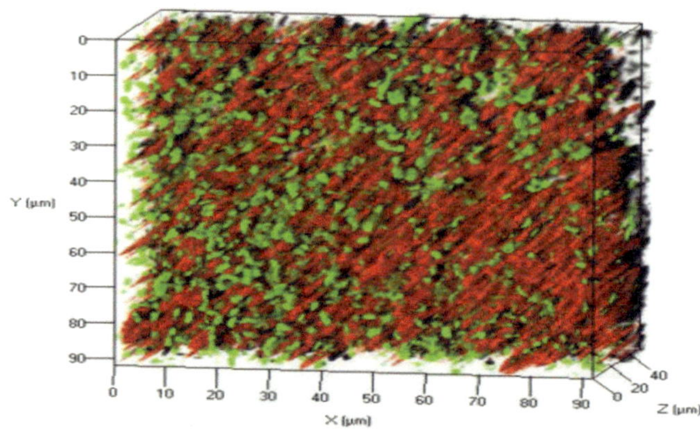

(a)

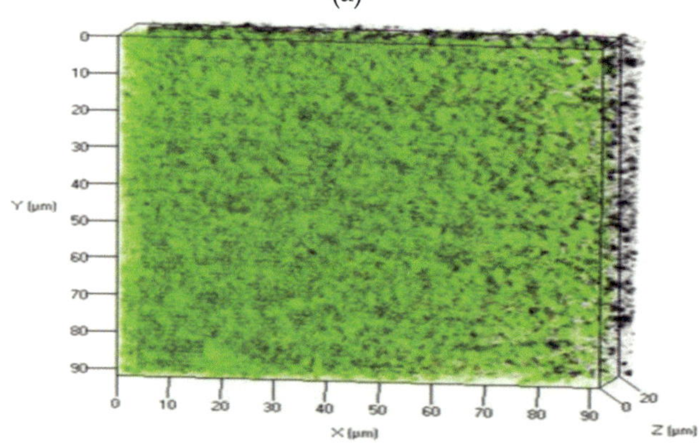

(b)

Fig. 4. Confocallaser scanning microscopy (CLSM) 3-D micrographsof a *P. aeruginosa* biofilm with lectin Concana valin A (red, alginate spec.) and SYTO 9 (green, cells), magnification1000 x

*P. aeruginosa*SG81 (mucoid): thick(40 !m), heterogeneous biofilm, lots of alginate between cells

b) P. aeruginosa SG81R1 (non mucoid): thin(20 !m), unstructured biofilm, cells densely packed

6. The role of biofilm on biocontrol of plant diseases

The ability of biocontrol agents to control plant disease is dependent on colonization of plant surfaces. Effective colonization of plant roots by biocontrol agents plays an important role in growth promotion, irrespective of the mechanism of action (production of metabolites, production of antibiotics against pathogens, nutrient uptake effects, or induced plant resistance) . It is now common knowledge that bacteria in natural environments persist by forming biofilms (Davey and O'Toole 2000). Biofilms are highly structured, surface-attached communities of cells encased in a self-produced extracellular matrix (Costerton, 1995). Biofilm formation is a dynamic process and different mechanisms are involved in their attachment and growth. Extracellular polymeric substances play an important role in the attachment and colonization of microorganisms to food-contact surfaces. Frequently, bacteria live in the environment as biofilms, which are highly structured, surface-attached communities of cells encased within a self-produced extracellular polymeric substance matrix (Branda *et al.* 2005 and Kolter & Greenberg 2006).

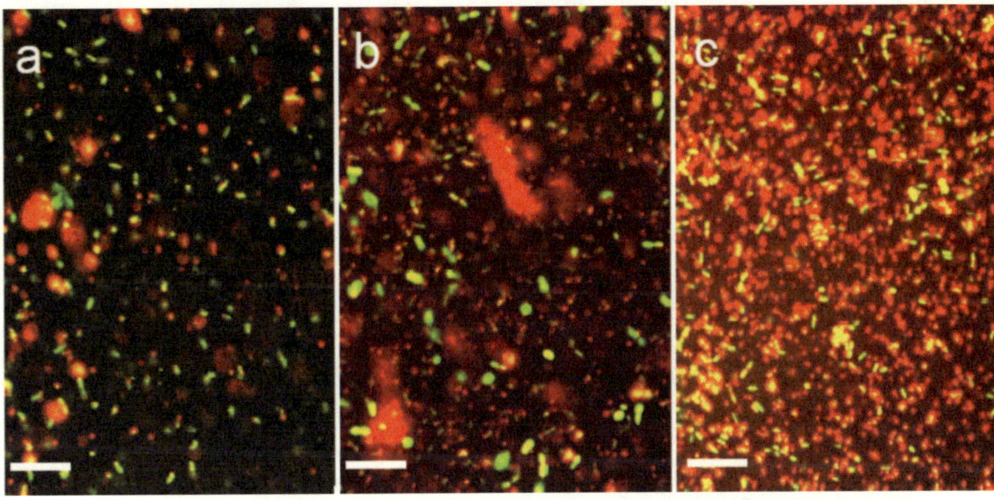

Fig. 5. edox-Enzymes demonstrated in the EPS matrix in distinct clusters Homogenous distribution in the matrix Spatia l and temporal heterogeneity
a b c Example2: Extracellular redox enzymes Visualized by CTC reduction(red)Cells: green(SYTO 9) Bar: 10 µm, magnification1000 x Visualization of enzyme activity in biofilms

Various techniques have been adopted for the proper study and understanding of biofilm attachment and control. This research describes work on understanding the role of biofilm formation (process by which microbes from very large aggregates on surfaces) in the biological control of plant diseases and how this knowledge has been applied to enhance commercial biocontrol agents. Bacterial biofilms established on plant roots could protect the colonization sites and act as a sink for the nutrients in the rhizosphere, hence reducing the availability of root exudate nutritional elements for pathogen stimulation or subsequent colonization on the root (Weller and Thomashow 1994).

Beneficial plant rhizobacteria (PR) are associated with the surfaces of plant roots and may increase plant yield by mechanisms that impart improved mineral nutrient uptake, disease suppression, or phytohormone production. One beneficial rhizobacterium is *Bacillus subtilis,* which is ubiquitous in soil, can promote plant growth, protect against fungal pathogen attack, and play a role in the degradation of organic polymers in the soil (Emmert and Handelsman, 1999). Recently, it has been reported that *B. subtilis* forms adhering biofilms on inert surfaces under the control of a variety of transcription factors (Hamon and Lazazzera, 2001; Stanley et al., 2003). Kinsinger *et al.* (2003) and Bais *et al.* (2004) used infection model and they demonstrated the biocontrol ability of a wild-type *B. subtilis* strain 6051 against *P. syringae.* Arabidopsis root surfaces treated with *B. subtilis* were analyzed with confocal scanning laser microscopy to reveal a three-dimensional *B. subtilis* biofilm.

Root-associated pseudomonads have been studied extensively, and many of these promote the growth of host plants or are used as biocontrol agents (Lugtenberg *et al.,* 2001). *Pseudomonas putida* can respond rapidly to the presence of root exudates in soils, converging at root colonization sites and establishing stable biofilms networks (Espinosa-Urgel *et al.,*

2002). Haggag, Wafaa and Salme Timmusk (2008) investigated the role of biofilm-forming *Paenibacillus polymyxa* strains in controlling crown root rot disease (*Aspergillus niger*) and highlighted importance of biofilms in biocontrol initiation. Both strains were able to suppress the pathogen but the superior biofilm former offers significantly better protection against crown rot. To make biocontrol effective and reproducible, successful colonization with the biocontrol agent must be ensured. Hence, we used SEM and solid surface assay to study colonization and biofilm formation ability of the strains (Figure 6). The strain B5 produces strikingly larger amounts of extracellular polymers (Figure 7).

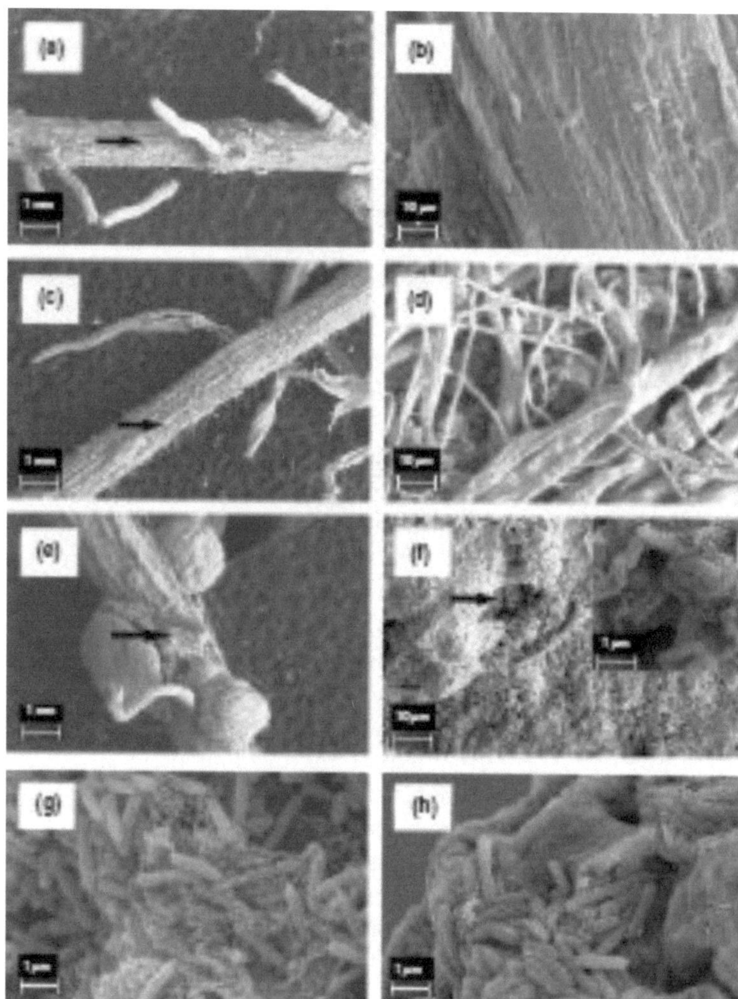

Fig. 6. Scanning electron micrographs of plant roots colonized by *Paenibacillus polymyxa* and Aspergillus niger (greenhouse experiment and gnotobiotic experiment). Peanut plant roots (a and b) in potting soil were infected with A. niger (c and d). *Paenibacillus polymyxa* B5 biofilm formation on peanut plant inhibits A. niger colonization (e and f). *Paenibacillus polymyxa* B5 (g) and B6 (h) biofilm formation on A. thaliana root. Note that *Paenibacillus polymyxa* B5 isolate produces remarkably more extracellular matrix.

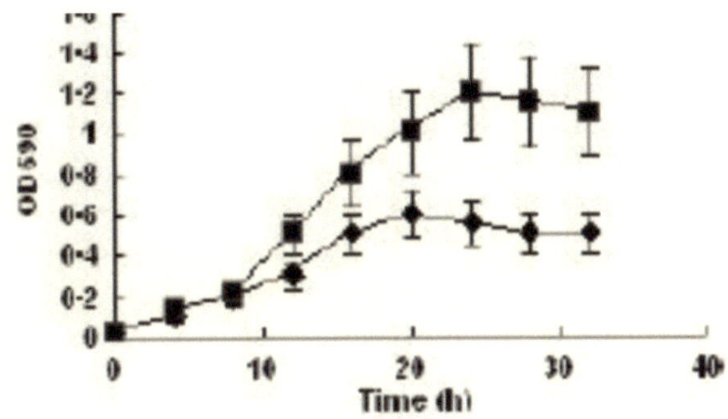

Fig. 7. Solid surface assay of *Paenibacillus polymyxa* B5 (r) and B6 (j) biofilm formation. The crystal violet assay was used to measure the solid surface biofilm formation at 30_C. Preparation and analysis as described in Materials and Methods.

Microorganisms often inhabit the leaf surface in organized structures termed biofilms. *Burkholderia* sp., FP62 is a biocontrol agent of *B. cinerea* in geranium and forms extensive biofilms in the phyllosphere. Scanning electron micrographs demonstrate extensive phyllosphere colonization (60-70% of the leaf surface). FP62 biofilms appeared to be many cells layers thick and enveloped in a polymer-like matrix. The biofilm phenotype of this strain is related to biocontrol. Isolation of transposon mutants that are deficient in biofilm formation in an *in vitro* biofilm assay, also lacked the capacity to control *B. cinerea* when applied to geranium leaves. The biofilm mutants are less efficient in phyllosphere colonization lacking many of the characteristics of wild-type biofilms. The biofilm mutation and biocontrol capacity could be restored through the addition of exogenous polymer to the biocontrol formulation of the mutants. The addition of polymers to the formulation of several other biocontrol agents also improved their biocontrol capacity suggesting that biofilms contribute to biocontrol efficacy and are an important aspect of phyllosphere competence.

7. Molecular genetics

The application of molecular technologies have initiated research on biofilms focusing on the characterization of biofilm-associated genes, their exploration as a reservoir of pathogenic microorganisms, their role in antimicrobial resistance and chronic diseases, evaluation of control strategies, and development of new detection methods. The genetic manipulation of microorganisms opens up an enormous potential for the biotechnological production of biopolymers with tailored properties suitable for high-value medical application such as tissue engineering and drug delivery. Understanding the molecular mechanisms of biofilm formation has facilitated the exploration of novel strategies to control bacterial biofilms.

Technological progress in microscopy, molecular genetics, genome analysis has significantly used for understanding of the structural and molecular aspects of biofilms, especially of extensively studied model organisms such as *Pseudomonas aeruginosa* . Biofilm development was divided into several key steps including attachment, microcolony formation, biofilm maturation and dispersion; and in each step bioagent may recruit different components and molecules including flagellae, type IV pili, DNA and exopolysaccharides .

Biofilms are also an ideal place for exchanging genetic material and maintaining a large and accessible gene pool. Horizontal gene transfer is facilitated, since the cells are maintained in close proximity to each other, are not fully immobilized, and can exchange genetic information. In 1999, Hausner and Wuertz reported significantly higher rates of conjugation in bacterial biofilms than in planktonic populations (Hausner and. Wuertz. 1999). High rates of conjugation in bacterial biofilms as determined by quantitative in-situ analysis. (Allesen-Holm *et al.*, 2006) found that in *P. aeruginosa* biofilms, the e-DNA is likely derived from whole genomic DNA. Surprisingly, e-DNA was organized in distinct patterns in the biofilms of this organism, forming grid-like structures, suggesting a structural role for e-DNA. Yang *et al.* (2007) also found that e-DNA was one of the major matrix components in *P. aeruginosa* biofilms, functioning as an intercellular connector; they supported the concept of the stabilizing role of e-DNA for the biofilm matrix. In *P. aeruginosa*, the release of e-DNA is under the control of quorum-sensing systems as well as iron regulation (Yang, *et al.*, 2007). Stanley *et al.* (2003) studied biofilms structured of *Bacillus subtilis* used DNA microarrays They identified 519 genes that were differentially expressed at one or more time points as cells transitioned to a biofilm. Approximately 6% of the genes of *B. subtilis* were differentially expressed at a time when 98% of the cells in the population were in a biofilm. These genes were involved in motility, phage-related functions, and metabolism. By comparing the genes differentially expressed during biofilm formation with those identified in other genomewide transcriptional-profiling studies, they were able to identify several transcription factors whose activities appeared to be altered during the transition from a planktonic state to a biofilm (Figure 8). Two of these transcription factors were Spo0A and sigma-H, which had previously been shown to affect biofilm formation by *B. subtilis*. A third signal that appeared to be affecting gene expression during biofilm formation was glucose depletion.

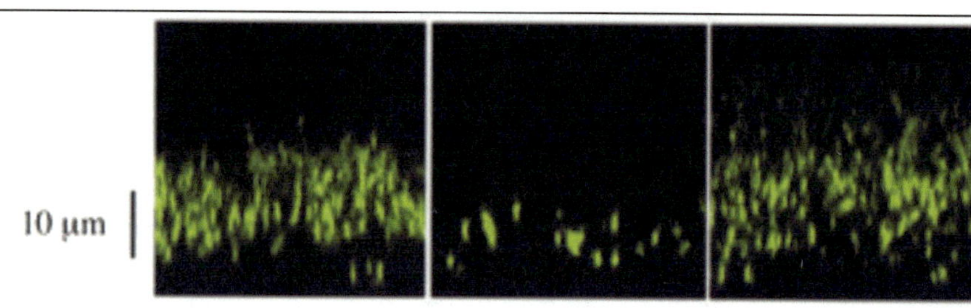

Strain	WT	WT	*ccpA*
[Glucose]	0.1%	1.0%	1.0%
Depth μm	13.4±1.2	7.15±0.69	15.9±1.2

Fig. 8. Effect of catabolite repression on the structure of *B. subtilis* biofilms. Shown are representative CSLM images of biofilms of *B. subtilis* expressing GFP in the *xz* plane. The depth of biofilm (± standard error of the mean), calculated from a minimum of three independent experiments , is indicated. WT, strain BAL218; *ccpA*, strain BAL795.

P. fluorescens '1100-6' was genetically tagged to allow exploration of its localization in grapevines (Eastwell *et al.*, 2006) . In growth chamber studies, *P. fluorescens* '1100-6' was found to survive in the rhizoplane of grapevines for 6 months. Microscopic examination showed that the genetically modified *P. fluorescens* '1100-6' predominantly occupied xylem and pith tissues, and occasionally persisted on external surfaces of grapevines (Figure 9).

The role of extracellular polymers in biofilm formation was studied using three mutants of *Sphingomonas paucimobilis* with increasing capabilities for exopolymer production. The physical, biochemical and physiological properties of three different layers of each biofilm were determined. The layers were detached by submitting the biofilm to increasing shear stress. The results revealed that the presence of exopolymers in the growth medium was essential for biofilm formation. The mutant producing the highest amount of exopolymer formed very thick biofilms, while the biofilms formed by the medium exopolymer producer were on average 8 times thinner. The lowest exopolymer producer did not form biofilm. In both types of biofilms, exopolymer density increased with depth, although this tendency was more significant in thinner biofilms. Cell distribution was also more heterogeneous in thinner biofilms, exhibiting a greater accumulation of cells in the inner layers. The thicker biofilms had very low activity in the inner layer. This was related to a high accumulation of proteins and DNA in this layer due to cell lysis and hydrolytic activity. Activity in the thin biofilm was constant throughout its depth, suggesting that there was no nutrient limitation. The production of exopolymers by each cell was constant throughout the depth of the biofilms, although it was greater in the case of the higher producer.

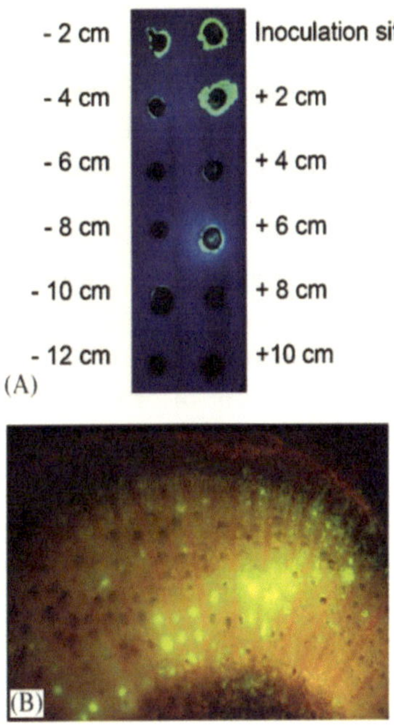

Fig. 9. Detection of the green-fluorescent-protein from the genetically tagged biocontrol agent *Pseudomonas fluorescens* isolate 1100-6 (*P. fluorescens* '1100-6-gfp') in grapevines 6 onths after stem inoculation using a bullet-piercing valve. (A) Fluorescence surrounding tissues sections after incubation on nutrient agar medium. Tissue sections were taken at 2 cm intervals acropetally (+) and basipetally from the inoculation site. Sections are observed under black light after growth on nutrient agar for 2 days. The *P. fluorescens* '1100-6-gfp' diffused out of the tissue sections and was evident by green fluorescent areas surrounding the sections. (B) A section of grapevine taken 2 cm above the point of injection and observed with an epifluorescent microscope. The green fluorescence in the xylem vessels is indicative of the presence of the gfp-gene product.

8. Conclusion

The use of microorganisms to control plant diseases offers an attractive alternative to the use of synthetic chemicals. In accordance, the diversity of microbial communities provides a rich source of potential biocontrol agents. The abundance of a beneficial strain of microorganism in the vicinity of plant roots may suppress plant pathogens without producing lasting effects on the rest of the soil microbial and plant communities. Biofilms serve as a new model system for the study of microbial development providing insights into microbial biology and ecology. One of the keys to studying complex biological systems is to develop accurate and realistic models of natural communities in the laboratory. Currently, many of these effects can only be speculated upon, although application of novel probes and improved analytical methods will gradually expand on our current, rather limited, and perhaps blinkered view of what biofilm structures really are and the extent to which they are determined by EPS. Their ability to interact with other polysaccharides and with other macromolecules and cells, as well as with ions and low-molecular-mass solutes, provide a multitude of microenvironments within any biofilm.

9. References

Allesen-Holm, M., Barken, K. B., Yang, L., Klausen, M., Webb, J. S., Kjelleberg, S., Molin, S., Givskov, M. and Tolker-Nielsen. T. (2006). A characterization of DNA release in *Pseudomonas aeruginosa* cultures and biofilms. Mol. Microbiol. 59:1114-1128.

Bais, H., Ray Fall and Jorge M. V. (2004). Biocontrol of *Bacillus subtilis* against Infection of Arabidopsis Roots by *Pseudomonas syringae* Is Facilitated by Biofilm Formation and Surfactin Production. Plant Physiology 134:307-319 .

Bianciotto, V., Andreotti, S., Balestrini, R., Bonfante, P. and Perotto S. (2004). Mucoid mutants of the biocontrol strain *Pseudomonas fluorescens* CHA0 show increased ability in biofilm formation. Current Opinion in Microbiology, 7:602–609.

Branda, S.S., Vik, S., Friedman, L. and Kolter, R. (2005). Biofilms: the matrix revisited. Trends Microbiol 13, 20–26.

Costerton, J. W. (1995). Overview of microbial biofilms. J. Ind. Microbiol. 15:137-140

Davey, M. E., and O'Toole, A. G. (2000). Microbial biofilms: from ecology to molecular genetics. Microbiol. Mol. Biol Rev. 64:847-867

Emmert, E.A.B. and Handelsman, J. (1999) Biocontrol of plant disease: a Gram-positive perspective. FEMS Microbiol Lett 171: 1-9

Eastwell, K.C., Sholberg, P.L. and Sayler R.J. (2006). Characterizing potential bacterial biocontrol agents for suppression of Rhizobium vitis, causal agent of crown gall disease in grapevines. Crop Protection. 25:1191-1200

Espinosa-Urgel, M., Kolter, R. and Ramos, J.L. (2002). Root colonization by *Pseudomonas putida*: love at first sight. Microbiology 148:341-343.

Friedman, L. and Kolter, R. (2004). Genes involved in matrix formation in *Pseudomonas aeruginosa* PA14 biofilms. Mol. Microbiol, 51:675-690.

Frølund, B., Palmgren, R. Keiding, K. and Nielsen, P. H. (1996). Extraction of extracellular polymers from activated sludge using a cation exchange resin. Water Res. 30:1749-1758.

Gage D.J. (2004). Infection and invasion of roots by symbiotic, nitrogen-fixing rhizobia during nodulation of temperate legumes. Microb Mol Biol Rev, 68:280-300.

Haggag, Wafaa, M. (2007). Colonization of exopolysaccharide-producing *Paenibacillus polymyxa* on peanut roots for enhancing resistance against crown rot disease . African Journal of Biotechnology, 6 (13): 1568-1577.

Haggag, Wafaa, M. and Salme Timmusk (2008). Colonization of peanut roots by biofilm-forming *Paenibacillus polymyxa* initiates biocontrol against crown rot disease. Journal of Appl Microbiol. 104 (4): 961-969.

Hall-Stoodley, L., Costerton, J.W. and Stoodley, P. (2004). Bacterial biofilms: from the natural environment to infectious diseases. Nature Reviews. Microbiology 2 (2): 95–108

Hassett, D., Cuppoletti, J., Trapnell , B., Lymar, S., Rowe, J., Yoon, S., Hilliard, G., Parvatiyar, K., Kamani, M., Wozniak, D., Hwang, S., McDermott, T. and Ochsner, U. (2002). "Anaerobic metabolism and quorum sensing by *Pseudomonas aeruginosa* biofilms in chronically infected cystic fibrosis airways: rethinking antibiotic treatment strategies and drug targets". Adv Drug Deliv Rev. 54 (11): 1425–43.

Hausner, M., and S. Wuertz. (1999). High rates of conjugation in bacterial biofilms as determined by quantitative in-situ analysis. Appl. Environ. Microbiol. 65:3710-3713.

Karatan, E.and Watnick, P. (2009). Signals, regulatory networks, and materials that build and break bacterial biofilms. Microbiology and Molecular Biology Reviews : MMBR 73 (2): 310–47

Kinsinger, R.F., Shirk, M.C. and Fall , R. (2003) Rapid surface motility and biofilm formation in *Bacillus subtilis* is dependent on extracellular surfactin and potassium ion. J Bacteriol 185: 5627-5631

Kolter, R. and Greenberg, E.P. (2006) Microbial sciences: the superficial life of microbes. Nature 441, 300–302.

Lugtenberg, B.J.J., Dekkers, L. and Bloemberg, G.V. (2001). Molecular determinants of rhizosphere colonization by Pseudomonas. Annu Rev Phytopathol 39:461-490.

Ma, L., K. D. Jackson, R. M. Landry, M. R. Parsek, and D. J. Wozniak. (2006). Analysis of *Pseudomonas aeruginosa* conditional Psl variant reveals roles for the Psl polysaccharide in adhesion and maintaining biofilm structure postattachment. J. Bacteriol. 188:8213-8221.

Ramey B.E., Matthysse A.G., Fuqua C. (2004). The FNR-type transcriptional regulator SinR controls maturation of *Agrobacterium tumefaciens* biofilms. Mol Microbiol. 52:1495-1511.

Stanley N.R., Britton RA, Grossman A.D, Lazazzera BA (2003) Identification of catabolite repression as a physiological regulator of biofilm formation by *Bacillus subtilis* by use of DNA microarrays. J Bacteriol 185: 1951-1957

Weller, D.M. and Thomashow, L.S. (1994). Current challenges in introducing beneficial microorganisms into the rhizosphere. In Molecular Ecology of Rhizosphere

Microorganisms ed. O'Gara, F., Dowling, D.N. and Boesten, B. pp. 1-18.NY: Academic Press.

Yang, L., K. B. Barken, M. E. Skindersoe, A. B. Christensen, M. Givskov, and T. Tolker-Nielsen. (2007). Effects of iron on DNA release and biofilm development by *Pseudomonas aeruginosa*. Microbiology 153:1318-1328.

15

Bacterial Type II PMIs: Exploitable Bifunctional Enzymes for Biotechnological Applications and the Rational Design of Antimicrobials

Sílvia A. Sousa, Christian G. Ramos, Joana Feliciano and Jorge H. Leitão
IBB – Institute for Biotechnology and Bioengineering, CEBQ, IST
Portugal

1. Introduction

Phosphomannose isomerases (PMIs, E.C. 5.3.1.8), first described by Slein (1950) and isolated from brewers' yeast by Gracy and Noltmann (1968), are metal-dependent aldose–ketose isomerases that catalyze the reversible isomerization of D-fructose-6-phosphate (F6P) into D-mannose-6-phosphate (M6P) in prokaryotic and eukaryotic cells (Fig. 1).

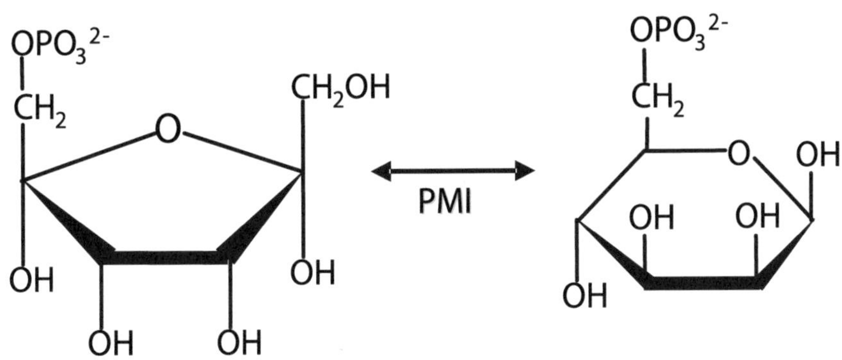

Fig. 1. The reversible conversion of fructose-6-phosphate (left) into mannose-6-phosphate (right), catalyzed by PMI enzymes.

The reaction catalyzed by PMIs is the first step of the mannose pathway leading to the generation of guanosine diphosphate (GDP)-D-mannose (Fig. 2). In this pathway, M6P is subsequently converted into mannose-1-phosphate (M1P) by the phosphomannomutase (PMM, E.C.5.4.2.8) enzymatic activity, followed by the conversion of M1P into GDP-D-mannose by GDP-D-mannose pyrophosphorylase (GMP, E.C.2.7.7.22). GDP-D-mannose is an important precursor of many mannosylated structures such as glycoproteins, nucleotide sugars, glycolipids, cell wall components found in fungi, and bacterial polysaccharides (Dunwell et al., 2000). As shown in Fig. 2, GDP-D-mannose is also the precursor of the activated sugar nucleotides GDP-L-fucose, GDP-D-rhamnose, GDP-colitose, and GDP-perosamine, that are required for the biosynthesis of several glycoconjugates, including lipopolysaccharide (LPS) O-antigens, exopolysaccharides (EPS) and glycoproteins (Richau et al., 2000b; Vinion-Dubiel and Goldberg, 2003, Fig. 2).

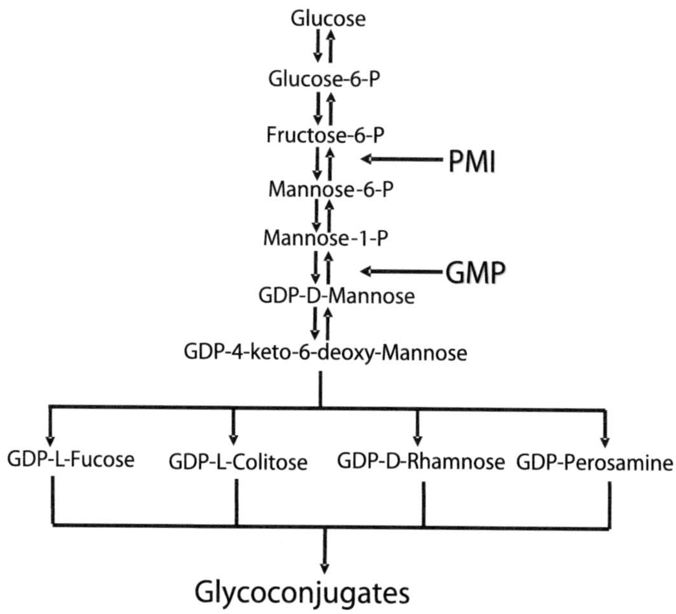

Fig. 2. Enzymatic conversions leading to GDP-D-mannose, the precursor for the synthesis of the sugar residues L-fucose, L-colitose, D-rhamnose, and D-perosamine, sugar residues commonly found in bacterial lipopolysaccharide O-antigens and other glycoconjugates.

2. The PMI family of proteins is divided in four classes

PMIs are a family of proteins that are members of the cupin superfamily of prokaryotic and eukaryotic proteins. Besides PMIs, the cupin superfamily includes several enzymes and proteins that bind sugars and other compounds (Dunwell et al., 2000). The term cupin derives from the latin word *cupa* (small barrel), and was given to this protein superfamily since their members possess a six-stranded β-barrel structural domain (the cupin domain) (Khuri et al., 2001).

The PMI protein family was divided into four distinct classes that are structurally unrelated, except for a small conserved amino acid sequence motif that belongs to the active site for the M6P to F6P isomerization reaction (Jensen & Reeves, 1998; Hansen et al., 2004). The type I class of PMIs is composed of monofunctional enzymes mainly found in eukaryotes and that only catalyze the F6P to M6P isomerization reaction. For example, the type I PMI isolated from *Saccharomyces cerevisiae* has been shown to be a zinc-dependent metalloenzyme with one metal ion per monomer (Gracy & Noltmann, 1968).

The type II class of proteins is only found in prokaryotes, and are bifunctional proteins with both PMI and GMP enzyme activities in separate catalytic domains of the protein (Jensen & Reeves, 1998). The GMP enzyme activity catalyses the reversible conversion of mannose-1-phosphate into GDP-D-mannose (Fig. 2).

The type III class of proteins comprises only a single protein from *Sinorhizobium meliloti* (Jensen & Reeves, 1998).

The type IV class includes atypical proteins from several aerobic crenarchaeota such as *Aeropyrum pernix, Thermoplasma acidophilum, Archaeoglobus fulgidus,* and *Pyrobaculum aerophilum*. This class of proteins is also bifunctional, with both phosphoglucose isomerase

(PGI, EC 5.3.1.9) and PMI enzyme activities (Hansen et al., 2004). PGI catalyses the reversible isomerization of glucose-6-phosphate (G6P) to F6P.

3. PMIs are critical for microbial survival and pathogenesis

PMIs have been reported as important enzymes for the survival and / or pathogenesis of several bacterial species (*Escherichia coli, Salmonella enterica, Mycobacterium smegmatis, Pseudomonas aeruginosa, Helicobacter pylori, Burkholderia cepacia complex*), yeasts (*Saccharomyces cerevisiae, Candida albicans, Cryptococcus neoformans,* and *Aspergillus nidulans*) and protozoan parasites (*Leishmania mexicana*) (Patterson et al., 2003; Shinabarger et al., 1991; Wu et al., 2002; Sousa et al., 2007a; Garami and Ilg, 2001; Payton et al., 1991; Smith et al., 1995; Wills et al., 2001; Smith and Payton, 1994; Jensen and Reeves, 2001). The requirement of a functional PMI activity for the survival and pathogenesis was recently demonstrated in the case of *Leishmania*. Species of the *Leishmania* genus synthesize large amounts of mannose-containing glycoconjugates, including the unusual glycoinositolphospholipids (GIPLs), the conserved protein-linked glycosylphosphatidylinositol (GPI) membrane anchors, glycoproteins with uncommon *N*-linked glycans, and phosphoglycan-modified molecules such as lipophosphoglycan (LPG), and proteophosphoglycans (PPGs) (Garami and Ilg, 2001). These glycoconjugates are responsible for the remarkable resistance of *Leishmania* parasites against the hostile habitats they find within their host organisms (Garami and Ilg, 2001). To gain insights on the role played by the PMI activity in the biosynthesis of these glycoconjugates, *L. mexicana* deletion mutants on the *lmexpmi* gene (encoding the protein with PMI activity) were generated (Garami and Ilg, 2001). As a consequence of the deletion of the *lmexpmi* gene, the mutants had a lower PMI activity. However, they were still able to grow in media deficient for mannose, but were unable to synthesize the phosphoglycan repeats [-6-Galβ1-4Manα1-PO_4^--], and the mannose-containing glycoinositolphospholipids. As a consequence, these mutants exhibited a reduced ability to express the glycosylphosphatidylinositol-anchored dominant surface glycoprotein leishmanolysin. When compared to the wild-type strain, the mutants were less virulent in mice and exhibited a lower ability to colonize macrophages *in vitro* (Garami and Ilg, 2001).

Mannose is also a key component of several cell-wall and intracellular molecules in mycobacteria, including mannolipids (phosphatidylinositol mannoside, lipomannan, lipoarabinomannan), the cytoplasmic 3-*O*-methylmannose polysaccharide and O-mannosylated glycoproteins. To understand the PMI role in the mannose metabolism in mycobacteria, Patterson et al. (2003) generated a *M. smegmatis* mutant auxotrophic for mannose by deleting the *manA* gene. The mutant cells were found to be shorter in lenght than the wild-type, and presented a mild hyperseptation phenotype. These changes caused an exponential loss of cell viability after 10 hours of growth in mannose-free medium, revealing the essentiality of the mannose metabolism for growth and viability of *M. smegmatis* (Patterson et al., 2003).

The PMI activity is also essential for the viability of the yeast *C. albicans*. In this pathogenic yeast, cell lysis was reported to occur in the absence of the *pmi* gene (Smith et al., 1995).

In the bacterium *H. pylori*, the LPS O-antigen contains fucosyloligosaccharides similar to the human Lewis X and Lewis Y antigens (Wu et al., 2002). These molecules contribute to the mimicry of the host and to the development of an autoimmune response, leading to the pathogen increased persistence in the host (Wu et al., 2002). The major pathway for biosynthesis of GDP-L-fucose, the precursor for L-fucose, starts from GDP-D-mannose (Fig.

2). Wu et al. (2002) have shown that the type II protein HP0043 was the point control for GDP-D-mannose biosynthesis in *H. pylori*.

Due to the important roles played by these enzymes for the survival and pathogenesis, PMIs have been considered as promising targets for the development of antibacterial, antifungal, and antiparasitic agents.

4. Type II phosphomannose isomerases

In contrast with the well-studied type I PMIs, type II PMI proteins remain poorly characterized. The bacterial type II PMIs that have been functionally characterized so far include the *E. coli* and *S. enterica* ManC (Jensen and Reeves, 2001), the *P. aeruginosa* AlgA (Shinabarger et al., 1991), the *Xanthomonas campestris* XanB (Köplin et al., 1992), the *Gluconacetobacter xylinum* AceF (Griffin et al., 1997), the *Sphingomonas chungbukensis* DJ77 PMI (Tran et al., 2009), the HP0043 protein from *H. pylori* (Wu et al., 2002), and the BceA and $BceA_J$ proteins from *B. cepacia* IST408 and *B. cenocepacia* J2315, respectively (Sousa et al., 2007a; Sousa et al., 2008). The sequence analysis of these type II PMIs revealed that they are composed of two domains, with four conserved motifs, as shown in Fig. 3: a GMP domain at the N-terminus and a PMI domain at the C-terminus. The GMP active site motif, that includes the conserved sequence FVEKP, is present in the N-terminus of the amino acid sequence and is essential for M1P binding, being the lysine (K) residue responsible for the binding of the phosphate moiety (May et al., 1994). The N-terminal region of the proteins also contain the highly conserved pyrophosphorylase signature sequence, GXGXR(L)-PK (where X represents any amino acid residue), similar to the activator-binding site of the bacterial XDP-sugar pyrophosphorylases (Jackson et al., 2004). In the C-terminal domain, two conserved motifs are also present, the zinc-binding motif QXH, and the putative PMI active site EN(Q/E)SX(Y/F)I (Jackson et al., 2004).

Sousa et al. (2008) have performed a prediction analysis of the secondary structure of the type II PMI $BceA_J$. These authors have found that, while the GMP domain of $BceA_J$ was putatively composed of α-helices interspaced by β-strands, the PMI domain of the protein was almost composed of β-strands. This predicted secondary structure is in good agreement with the occurrence of two distinct domains in type II PMIs (Sousa et al., 2008).

Both the PMI and GMP enzyme activities were detected in all the purified type II PMIs mentioned before. The requirement for divalent metal ions (*e.g.* Mg^{2+}, Ca^{2+}, Co^{2+}, Mn^{2+}, Ni^{2+}, or Zn^{2+}) for catalysis has also been demonstrated for those proteins. In fact, the type II PMIs extracted and purified from *X. campestris* and *B. cepacia* were totally inhibited by the metal chelator ethylenediaminetetraacetic acid (EDTA) (Sousa et al., 2007a; Papoutsopoulou & Kyriakidis, 1997).

In the type II PMI from *H. pylori*, a difference in pH dependence between the forward and reverse direction of PMI reaction was observed, the forward reaction exhibiting a pH optimum of 7, while the reverse reaction had an optimal pH range of 7-9 (Wu et al., 2002). Due to these results, it was suggested that this enzyme requires proper ionization of either M1P and/or GTP for binding and catalysis.

GDP-D-mannose was found to inhibit the PMI activity of the *P. aeruginosa* PlsB and the *H. pylori* HP0043 proteins, suggesting feedback regulation of the pathway in these bacterial species (Lee et al., 2008; Wu et al., 2002). This observation led to the suggestion that the mannose group in GDP-D-mannose might compete with M6P at the PMI active site to regulate the mannose utilization in these pathogens. However, no inhibition by GDP-D-mannose was observed for the *P. aeruginosa* AlgA protein (Shinabarger et al., 1991). In

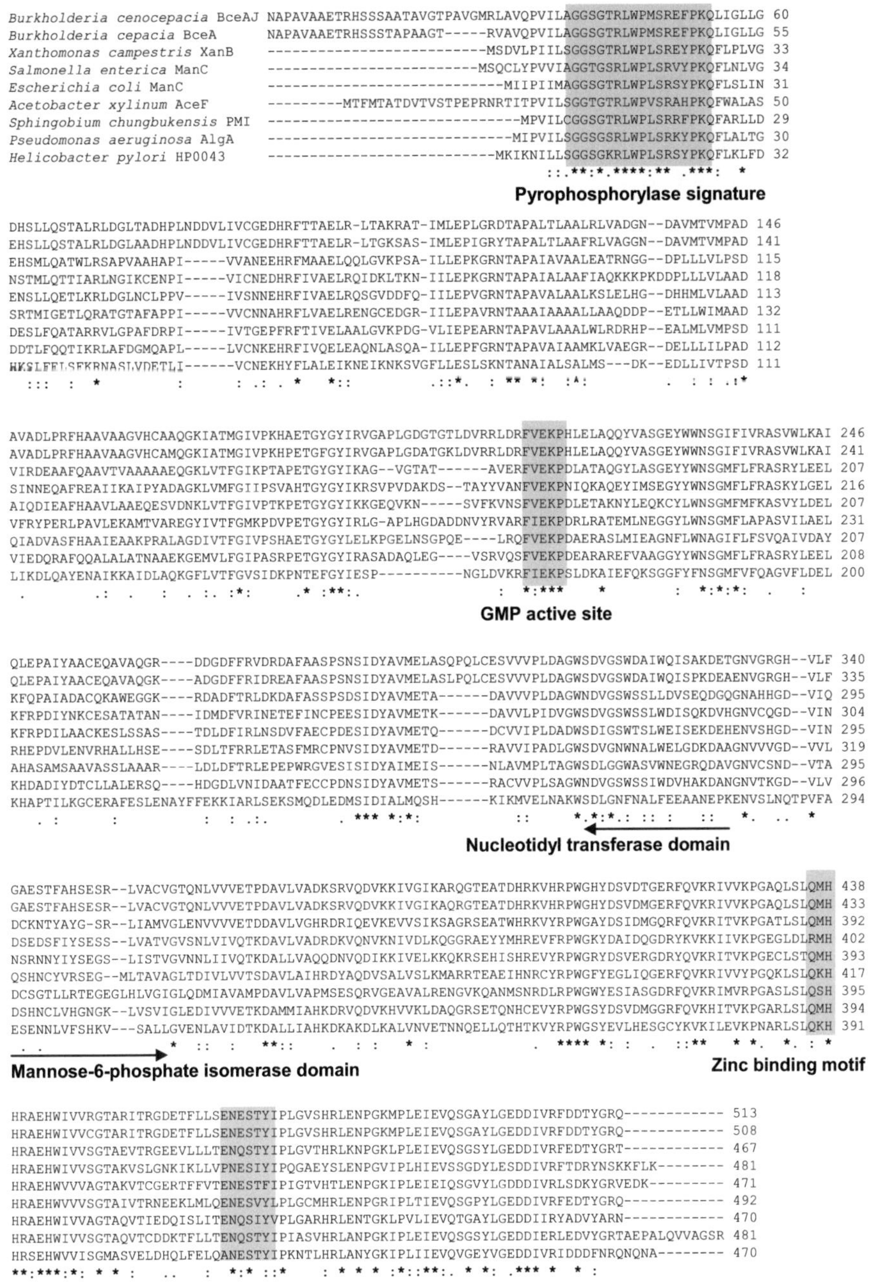

Fig. 3. Amino acid sequence alignment of experimentally caracterized type II PMIs from *B. cenocepacia*, *B. cepacia*, *X. campestris*, *S. enterica*, *E. coli*, *A. xylinum*, *S. chungbukensis*, *P. aeruginosa*, and *H. pylori*. The putative GMP and PMI domains are indicated by arrows. The conserved motifs are boxed in grey. Asterisks indicate the amino acid residues that are identical in all proteins. One or two dots indicate semi-conserved or conserved substitutions, respectively. Alignments were performed with ClustalW software.

addition, some reducing agents such as DTT and β-mercaptoethanol, have also been shown to strongly inhibit the PMI activity, suggesting the involvement of disulfide bonds in the formation of the active site and / or the substrate binding site of the enzyme (Shinabarger et al., 1991; Papoutsopoulou & Kyriakidis, 1997). The XanB protein was also inhibited by diethyl pyrocarbonate (DEPC), being the protection from inactivation obtained after addition of the substrate M6P, revealing that the amino acid residue affected is located at or near the active site (Papoutsopoulou & Kyriakidis, 1997). DEPC modifies lysyl (K), histidyl (H), cysteinyl (C), seryl (S) and tyrosyl (Y) residues of proteins. To identify the modified amino acid residues, different compounds were tested, such as PMSF for serine residues, N-acetylimidazole for tyrosine and lysine residues, hydrogen peroxide for cysteine and methionine (M) residues, and p-mercurybenzoate for cysteine residues (Papoutsopoulou & Kyriakidis, 1997). None of these compounds were able to inhibit the PMI activity. However, the treatment of the carbethoxy-PMI with hydroxylamine, that reverses histidine and serine modifications, restored the activity to about 82% of the initial activity, revealing that the modified amino acid was histidine (Papoutsopoulou & Kyriakidis, 1997). The PlsB protein was inhibited by 2,3-butanedione, suggesting the presence of a catalytic arginine (R) residue (Lee et al., 2008). Site-directed mutagenesis revealed that the residue R408 of PlsB was required for the PMI catalysis, but not for the GMP activity (Lee et al., 2008). Therefore, it was concluded that this residue must participate in the interconversion of sugar moieties by providing the binding of the sugar phosphate group or by forming the hydrogen bond of sugar hydroxyl group, stabilizing the binding between substrate and enzyme (Lee et al., 2008).

A summary of the kinetic parameters of the type II PMIs AlgA and PlsB from *P. aeruginosa* PAO1, AceF from *A. xylinum*, XanB from *X. campestris* ATCC13951, BceA from *B. cepacia* IST408, $BceA_J$ from *B. cenocepacia* J2315, HP0043 from *H. pylori* 26695 and PMI from *S. chungbukensis*, is presented in Table 1.

In most of these bacterial species, the type II PMIs are involved in the biosynthetic pathway of sugar precursors that are required for the production of polysaccharide-containing polymers, such as EPSs and LPSs. In bacteria, EPSs have been shown to provide protection against dehydration, engulfment by macrophages, bacteriophages, antibiotics and other toxic compounds like heavy metal ions (Ferreira et al., 2010). Due to their physico-chemical properties, like high viscosity and pseudoplasticity, the EPSs from some bacterial species are also used in many applications ranging from food processing to pharmaceutical production (Pettitt, 1979). We will focus on the EPS xanthan as an example of an EPS with several industrial applications, and on alginate and cepacian, both known to play a role on the pathogenesis of the producing organisms to the human host. In Fig. 4, the chemical structures of these three bacterial EPSs are shown, as well as photographs of Petri plates containing colonies of the respective EPS producing bacteria.

Type II PMIs XanB, AlgA, and BceA are involved in the biosynthesis of the sugar nucleotides necessary for polymerization of the EPS xanthan, alginate and cepacian, respectively. In the case of xanthan synthesis, the GDP-D-mannose is the activated sugar precursor for the mannose moiety found in the repeating unit (Fig. 5). However, in the case of the two other EPSs, the GDP-D-mannose is either directly used to form the repeating unit (cepacian), or further converted into GDP-D-rhamnose or to GDP-D-mannuronic acid, before being incorporated, respectively, into the repeating unit of cepacian or added into the growing poly-mannuronate polymer produced by *P. aeruginosa*.

Protein	Metals ions for PMI activity	V_{max} PMI activity (U/mg)	K_m PMI activity (μM)	Metals ions for GMP activity	V_{max} GMP activity (U/mg)	K_m GMP activity (μM)	Inhibitors	References
$BceA_J$	$Mg^{2+}>Ca^{2+}>Mn^{2+}>Co^{2+}>Ni^{2+}$	27.03	12390	$Mg^{2+}>Ca^{2+}$	4.56	24	ND	Sousa et al., 2008
BceA	$Ca^{2+}>Mn^{2+}>Mg^{2+}>Co^{2+}>Ni^{2+}$	[a]21.10	[a]9010	$Mn^{2+}>Ca^{2+}>Mg^{2+}>Ni^{2+}$	212.8	2940	EDTA	Sousa et al., 2007a
AlgA	$Co^{2+}>Ni^{2+}>Mn^{2+}>Mg^{2+}>Ca^{2+}>Zn^{2+}$	0.83	3030	$Mg^{2+}>Mn^{2+}$	5.17	14.2	DTT	Shinabarger et al., 1991
PlsB	$Co^{2+}>Mn^{2+}$	ND	1180	$Mg^{2+}>Co^{2+}>Mn^{2+}$	ND	110	GDP-D-mannose, 2,3-butanedione	Lee et al., 2008
HP0043	$Co^{2+}>Mg^{2+}>Mn^{2+}>Zn^{2+}$	ND	55.56	$Mg^{2+}>Mn^{2+}$	ND	101	GDP-D-mannose	Wu et al., 2002
XanB	$Co^{2+}>Zn^{2+}>Mn^{2+}>Ni^{2+}>Ca^{2+}$	33.5	2000	ND	ND	ND	EDTA, DTT, β-mercaptoethanol, DEPC	Papoutsopoulou and Kyriakidis, 1997
AceF	ND	[b]0.04	ND	ND	[b]0.01	ND	ND	Griffin et al., 1997
PMI	$Co^{2+}>Mg^{2+}>Ca^{2+}>Ni^{2+}>Zn^{2+}$	ND	ND	Mg^{2+}	ND	ND	GDP-D-mannose	Tran et al., 2009

[a]The PMI reaction was performed in presence of $MgCl_2$.
[b]The specific activity was determined in crude cell extracts of the *E. coli* CD1 (*pmi*-) and CWG152 (*gmp*-) strains.

Table 1. Comparison of the kinetic parameters of different bacterial type II PMIs. The PMI and GMP reactions were performed in the presence of the best metal ion activator. Abbreviations: ND, not determined.

Since its discovery in 1950, the xanthan EPS produced by the phytopathogen *X. campestris*, generated a great scientific and industrial interest, and was first approved as a food additive in 1969 by the Food and Drug Administration (FDA, USA) (Born et al., 2002). This high molecular weight acidic heteropolysaccharide consists of repeating units containing D-glucose, D-mannose and D-glucuronic acid in a molar ratio of 2:2:1, respectively (Fig. 4). The glucose residues are linked to form a β-1,4-D-glucan cellulosic backbone, with alternate glucose residues decorated with a short branch with a glucuronic acid residue sandwiched between two mannose residues (Born et al., 2002). The type II PMI XanB is one of the key enzymes in the pathway of xanthan biosynthesis, leading to the formation of the precursor GDP-D-mannose (Papoutsopoulou & Kyriakidis, 1997). Work performed by Papoutsopoulou & Kyriakidis (1997) revealed that the half life of this enzyme was 3 times higher in the presence of zinc ions, implying a structural role for this metal ion. This study also revealed that a histidyl residue at or near the PMI active site was essential for the isomerization reaction.

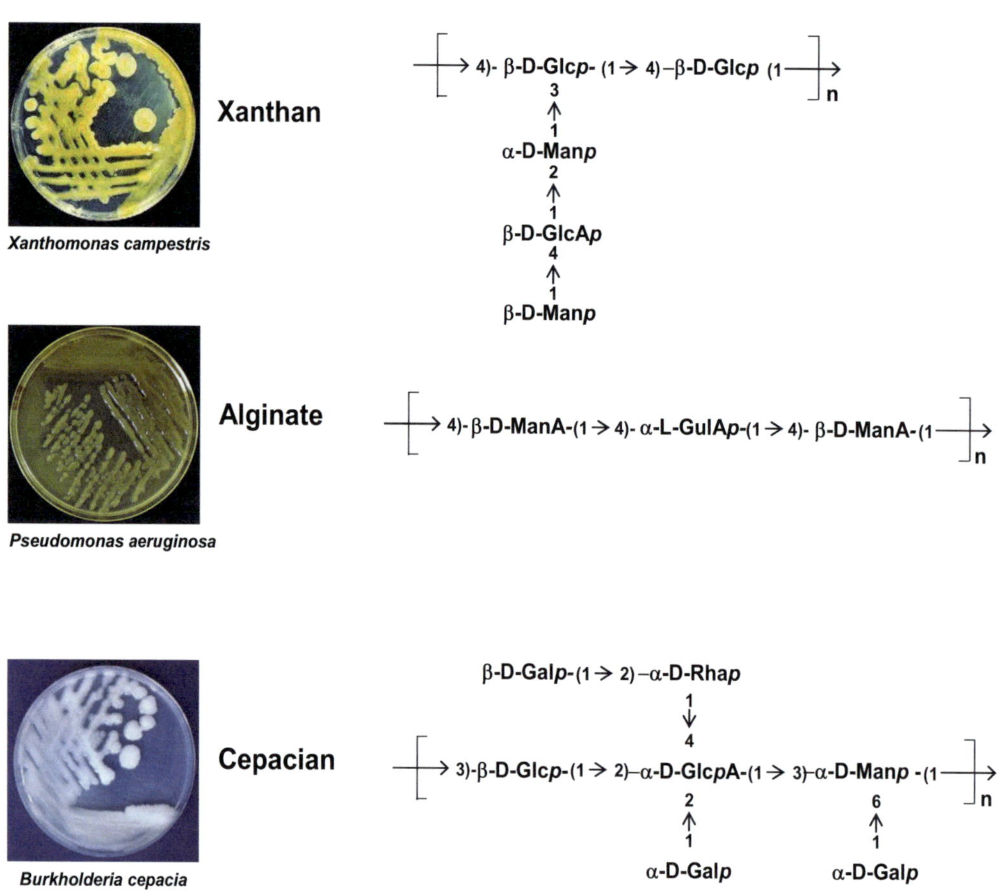

Fig. 4. Chemical structures of xanthan, alginate and cepacian (right) and photographs of Petri plates with isolated colonies of the respective producing bacterial strains, evidencing their heavy mucoid phenotype (left). Abbreviations: Glc, glucose; Man, mannose; GlcA, glucuronic acid; ManA, mannuronic acid; GulA, guluronic acid; Gal, galactose; Rha, rhamnose.

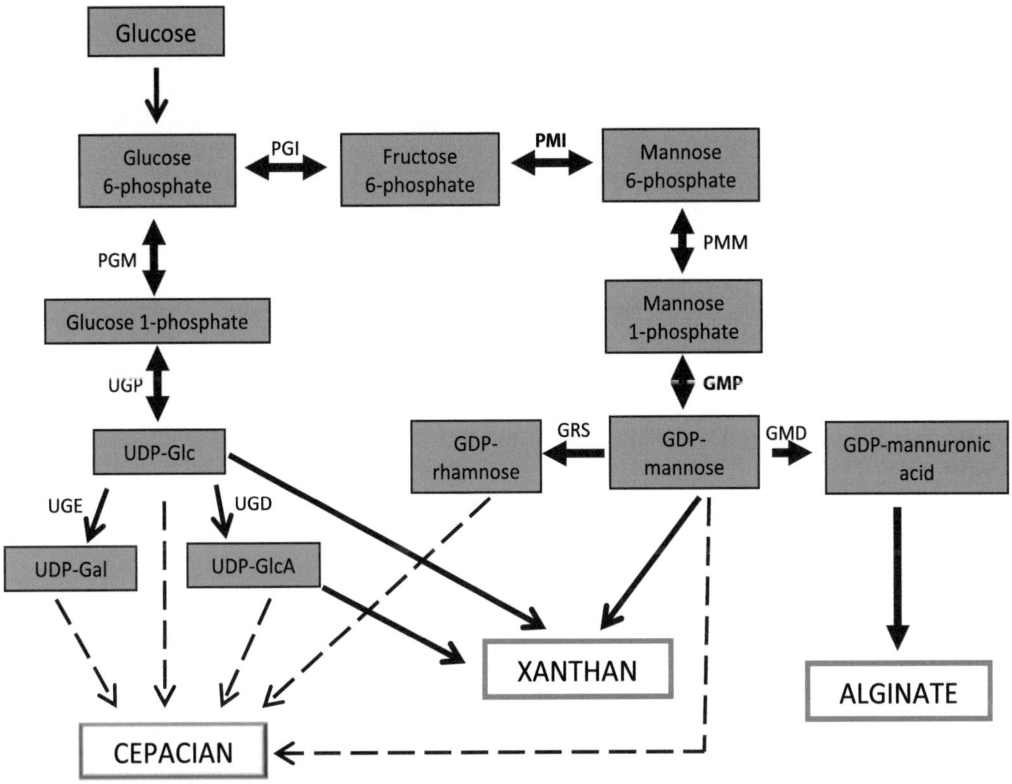

Fig. 5. Biosynthetic pathways leading to the activated sugar precursors necessary for assembly of the repeating units of cepacian, xanthan and alginate. The enzyme activity catalyzing each step is shown close to the arrow. Abbreviations: **PMI**, phosphomannose isomerase; **GMP**, GDP-D-mannose pyrophosphorylase; PGI, phosphoglucose isomerase; PGM, phosphoglucomutase; UGP, UDP-glucose pyrophosphorilase; UGE, UDP-galactose epimerase; UGD, UDP-glucose dehydrogenase; PMM, phosphomannomutase; GRS, GDP-rhamnose synthase; GMD, GDP-mannose dehydrogenase. The **PMI** and **GMP** activities of bifunctional type II PMIs are evidenced in bold.

In *P. aeruginosa* PAO1, three genes (*algA, wbpW, pslB*) encoding proteins homologous to type II PMIs have been described (Sá-Correia et al., 1987; Rocchetta et al., 1998; Jackson et al., 2004). These genes are located within three distinct polysaccharide biosynthesis gene clusters. WbpW participates in the production of the A-band LPS, while PslB is required for EPS production and biofilm formation (Rocchetta et al., 1998; Jackson et al., 2004). The occurrence in *P. aeruginosa* of multiple type II PMIs dedicated to the synthesis of distinct polysaccharide molecules is thought to enable the independent regulation of each specific pathway and thus to promote the adaptation of the bacterium to different environments. Nevertheless, AlgA is the best known characterized type II PMI from *P. aeruginosa*. This 56-kDa bifunctional enzyme catalyzes the first and third steps in the biosynthesis of alginate and was originally described by Sá-Correia et al. (1987). The *P. aeruginosa* alginate is a highly acetylated polymer, mostly composed of 1,4-linked β-D-mannuronic acid residues with few interspaced guluronic acid residues (Shinabarger et al., 1991; Fig. 4). This contrasts with the structure of the alginates produced by brown algae, which contain blocks of L-guluronic acid, responsible for the ability of alginate to originate gels in the presence of Ca^{2+} (Sabra et al., 2001). *P. aeruginosa* alginate has medical significance due to its production during the conversion of *P. aeruginosa* strains to a mucoid phenotype, in association with chronic infections in the lungs of cystic fibrosis patients (Shankar et al., 1995). This conversion is induced by several conditions, including nutrient starvation, use of energetically poor substrates, and presence of metabolic inhibitors (Leitão and Sá-Correia, 1997). The production of alginate by *P. aeruginosa* protects the bacterium from the host immune responses and against the action of antibiotics (Govan and Deretic, 1996). Leitão and Sá-Correia (1993) have shown that the alginate pathway from *P. aeruginosa* can be manipulated by increasing the amounts of the type II PMI AlgA. This result revealed that the modification and control of critical steps in complex microbial EPS pathways at the genetic level can allow the increase of the EPS production yield and the alteration of the rheological properties of aqueous solutions prepared with the biopolymer.

In another CF pathogen, the Bcc bacteria, multiple type II PMI encoding genes were also found (2 to 5). In particular, three type II PMI encoding genes were associated with polysaccharide biosynthetic clusters, the *bceA*, of the cepacian biosynthetic cluster, the *pmi* of the capsular polysaccharide genomic island, and a third *pmi* involved in LPS biosynthesis (Sousa et al., 2007a). Cepacian is the major exopolysaccharide produced by a large percentage of clinical isolates of Bcc (Cunha et al., 2004; Herasimenka et al., 2007; Richau et al., 2000a; Zlosnik et al., 2008; Ferreira et al., 2010). The production of cepacian has been correlated with higher persistence and virulence of the producing bacterium in animal models, inhibition of neutrophil chemotaxis and scavenging of reactive oxygen species *in vitro*, suggesting a role for cepacian in the protection of bacteria against the host immune response (Sousa et al., 2007b; Bylund et al., 2006). This EPS is composed of a branched heptasaccharide repeating-unit, with D-glucose, D-rhamnose, D-mannose, D-galactose and D-glucuronic acid, in the ratio 1:1:1:3:1, respectively (Cescutti et al., 2000; Fig. 4). The GDP-D-mannose formed by the type II PMI BceA is the activated sugar precursor for the incorporation of D-mannose in cepacian and is also the precursor for the synthesis of GDP-D-rhamnose, the activated sugar precursor of the D-rhamnose moiety of cepacian (Richau *et al.*, 2000b). The 55.3-kDa BceA protein was also shown to be required for thick biofilm formation and for the production of EPSs that lead to aqueous solutions with higher viscosity (Sousa et al., 2007a). However, the lack of a functional *bceA* gene did not affect the EPS production yield, suggesting that other *bceA* functional homologues may compensate the *bceA* mutation (Sousa et al., 2007a). The BceA protein from *B. cepacia* IST408 and the $BceA_J$ from *B. cenocepacia* J2315 exhibited no PMI and GMP activities in the presence of zinc, in spite of the existence, in their primary sequences, of a zinc-binding motif (Sousa et al., 2007a; Sousa et al., 2008; Fig. 3). Similarly, the PMI and GMP activities of PlsB could not be activated in presence of $ZnCl_2$ (Lee et al., 2008). Zn^{2+} was also reported to be the less effective activator of the type II PMI HP0043 from *H. pylori*, suggesting that the combination between the zinc metal ion and the PMI domain is loose and less specific than in the type I PMI proteins (Wu et al., 2002).

5. Biotechnological potential of PMIs

L-ribose has been proposed as a potential starting material for the synthesis of many L-nucleoside-based pharmaceutical compounds (*e.g.* the antiviral drug for the treatment of hepatitis B, Clevudine) (Okano, 2009). However, L-ribose is not an abundant sugar in nature, and therefore L-ribose has been produced mainly by chemical synthesis from L-arabinose, L-xylose, D-glucose, D-galactose, D-ribose, or D-mannono-1,4-lactone (Okano, 2009). Recently, the isomerization activity of the PMI enzyme from *B. subtilis* was reported to be specific for aldose substrates possessing hydroxyl groups oriented in the same direction at the C-2 and C-3 positions, such as the D and L forms of ribose, lyxose, talose, mannose, and allose (Yeom et al., 2009). This enzyme also exhibited the highest activity with L-ribulose, among all pentoses and hexoses tested (Yeom et al., 2009). Recently, experimental conditions have been reported as leading to the production of L-ribose at a final concentration of 213 g $liter^{-1}$ from 300 g $liter^{-1}$ of L-ribulose, using mannose-6-phosphate isomerase at 40°C for 3 h, with a conversion yield of 71% and a volumetric productivity of 71 g $liter^{-1}$ h^{-1} (Yeom et al., 2009). This was the highest volumetric productivity and product concentration reported until now for the biological manufacture of L-ribose. In addition, the downstream purification methodology was simpler than the previously used for the preparation of L-ribose, either by chemical methods or fermentation. The production of the expensive sugar L-ribose from the rare sugar L-ribulose by PMI may prove to be a valuable industrial process, because the L-ribulose sugar can be produced from the low-cost sugar L-arabinose, using the L-arabinose isomerase from *Geobacillus thermodenitrificans* (Yeom et al., 2008).

Antibiotic and herbicide resistance genes have been used for the selection of transgenic plants (Min et al., 2007). However, the use of these genetic markers has generated a widespread public concern. For this reason, alternative methods of selection have been developed, in particular the use of mannose as the carbon source in selective media, since plants have an inefficient ability to metabolize this sugar (Lee and Matheson, 1984). The *pmi* gene isolated from *E. coli* has been used as a selectable marker for the transformation of several plant species, including sugar beet, cassava, maize, wheat, *Arabidopsis*, pepper, sweet orange, pearl millet, tomato, papaya, onion, almond, Chinese cabbage, and cucumber (Miles and Guest, 1984; Min et al., 2007). The transgenic plants expressing the *E. coli pmi* gene and growing in mannose medium, are able to convert the M6P to F6P, thus providing a carbon and energy source to survive to the high selective pressure.

The study of the biological function of oligosaccharides from human milk such as lactose, D-galactose, N-acetylglucosamine, sialic acid, and L-fucose, has received an increasing interest in the last years. This interest derives from the properties of fucosylated oligosaccharides (*e.g.* the Lewis blood group antigen) in protecting infants against enteric pathogens (Boehm and Stahl, 2007). The availability of large amounts of fucosylated oligosaccharides will be useful for the development of therapeutic and protective strategies to prevent infections by pathogens, to improve the immune system response, and to reduce the inflammatory process (Newburg et al., 2004). Enzymatic fucosylation of oligosaccharides requires GDP-L-fucose as the donor of L-fucose. However, the high cost of GDP-L-fucose limits its application for large-scale production of these oligosaccharides. The conversion F6P to GDP-L-fucose requires the activity of five enzyme activities (mannose-6-phosphate isomerase ManA, phosphomannomutase ManB, mannose-1-phosphate guanyltransferase ManC, GDP-D-mannose-4,6-dehydratase Gmd, and GDP-L-fucose synthase WcaG), with GDP-D-mannose as the intermediate. Recently, a recombinant *E. coli* BL21star(DE3) strain overexpressing *gmd, wcaG, manB* and *manC* genes was developed to maximize the production of GDP-L-fucose (Lee et al., 2009). This advance might lead to the future availability of affordable fucosylated oligosaccharides.

6. PMIs are promising targets for the development of new antimicrobials

PMIs have been considered as suitable targets for the development of antibacterial, antiparasitic, and antifungal agents (Bhandari et al., 1998; Roux et al., 2004; Roux et al., 2007). Roux et al. (2004) reported the inhibition of a yeast type I PMI and a *P. aeruginosa* type II PMI by 5-phospho-D-arabinohydroxamate (5PAH). However, phosphorylated compounds have a limited therapeutical interest, not only because of their ionic character which do not allow them to freely cross the barrier that is the hydrophobic cell membrane, but also because of their high susceptibility to hydrolysis by endogenous phosphatase enzyme activity (Foret et al., 2009). Recently, the 6-deoxy-6-(dicarboxymethyl)-D-mannopyranose, a non-hydrolysable M6P analogue in which the phosphate group was replaced by a dicarboxymethyl group, was reported to be a strong inhibitor of the enzymatic activity of PMIs from *E. coli* and *S. cerevisiae* (Foret et al., 2009).

Sequence alignment studies have revealed that the type I PMIs from pathogenic microorganisms such as fungi and protozoa, exhibit a high level of amino acid sequence identity (>40%) with the humans type I PMIs, especially in the binding site region (Jensen and Reeves, 1998). This high degree of identity hampers the development of species-specific inhibitors against fungal or bacterial type I PMIs that do not inhibit the human type I PMIs. In most human tissues, the inhibition of the PMI activity will most probably do not impair the global metabolism, because the majority of the M6P that is utilized for glycoprotein synthesis is most likely not derived from F6P, but originates from efficient uptake of D-mannose through a specific exogenous mannose transporter, followed by its phosphorylation by hexokinases (Panneerselvam et al., 1997). However, in organs such as the liver and the intestine, where this pathway is less efficient, a deficiency in human PMI activity leads to the carbohydrate deficient glycoprotein syndrome type 1b (CDGS 1b), a severe metabolic disorder with hepatic and intestinal manifestations (Niehues et al., 1998). Nevertheless, this disease has been successfully treated by oral administration of D-mannose (Niehues et al., 1998). Therefore, a therapeutical strategy combining the enzyme inhibitor and D-mannose supplementation should alleviate the side-effects of PMI inhibition on humans.

In the pathogenic yeast *C. neoformans*, the pathway for exogenous mannose uptake was reported to be much less efficient than in humans, and the type I PMI from this pathogen was considered as an excellent therapeutic target (Wills et al., 2001). However, the efficiency of the pathway for M6P formation in other microorganisms remains poorly studied. The PMI inhibition therapy will also be possible for local treatments, such as against yeast infections (e.g. candidiasis) and cutaneous leishmaniasis, without affecting human metabolism.

No significant sequence identity has been found between Type I and Type II PMI enzymes, except for a very small conserved amino acid sequence motif in the active site of the protein (Jensen and Reeves, 1998). For this reason, specific inhibition of the type II PMI activity, while leaving the human type I protein unaffected, should be achieved more easily than for the type I PMI proteins from pathogens. This opens the door for the rational design of potent and highly species-specific inhibitors against the targeted PMIs from pathogens. However, for this purpose, structural information on type I and type II PMIs is necessary. High-resolution X-ray crystal structures have been reported for the type I PMIs of *Candida albicans* (Cleasby et al., 1996; PDB code 1pmi), *Salmonella typhimurium* (Gowda et al., 2008; PDB codes 2wfp, 3h1y, 3h1w, and 3h1m), *Helicobacter pylori* (PDB code 2qh5), *Bacillus subtilis*

(PDB code 1qwr) and the type IV PMIs from *A. fulgidus* (PDB code 1zx5) and *P. aerophilum* (Swan et al., 2004; PDB codes 1x9i, 1x9h, 1tzb and 1tzc). The analysis of the crystal structure of the type I PMI from *C. albicans* allowed the identification of the active site and the zinc metal cofactor binding site of the enzyme (Cleasby et al., 1996). Although the roles of the individual amino acid residues of the active site and the catalytic mechanisms are still poorly known, Gracy and Noltman (1968) proposed that the zinc ion coordinates with the carbonyl and hydroxyl oxygens on Cl and C2, activating the α-hydrogen to the carbonyl, which is abstracted by the nonprotonated nitrogen of an imidazole group to form the transient enediol intermediate. In contrast, the type II PMIs don´t bind Zn^{2+} specifically. This observation reveals that there are some differences between the different classes of PMIs. Contrasting with the wealth of structural information on types I and IV PMIs, no structural information on type II and III PMIs is available, difficulting the determination of the roles of individual amino acid residues of the active site and the catalytic mechanisms of these proteins. Therefore, knowledge on the three-dimension structure of these enzymes is necessary to allow the understanding of the mechanisms of the isomerization reaction and the role played by the metal ions in catalysis. In turn, this information will certainly give crucial information for the exploitation of these important enzymes as targets for the rational design of inhibitors, thus enabling us with new antimicrobials to fight infections caused by microorganisms resistant to the clinically relevant antibiotics available nowadays.

7. Acknowledgments

This work was partially funded by FEDER and Fundação para a Ciência e Tecnologia (FCT), Portugal, through contract PTDC/EBB-BIO/098352/2008, and a post-doctoral grant to Sílvia A. Sousa. Christian G. Ramos acknowledges a doctoral grant from Fundação Calouste Gulbenkian (FCG).

8. References

Bhandari A, Jones DG, Schullek JR, Vo K, Schunk CA, Tamanaha LL, Chen D, Yuan Z, Needels MC, Gallop MA (1998) Exploring structure-activity relationships around the phosphomannose isomerase inhibitor AF14049 via combinatorial synthesis. *Bioorg Med Chem Lett* 8:2303-2308.

Boehm G, Stahl B (2007) Oligosaccharides from milk. *J Nutr* 137:847S–849S.

Born K, Langendorff V, Boulenguer P (2002) Xanthan. In: *Biopolymers,* Steinbüchel A, Vandamme EJ, De Baets S (eds.), Weinheim: Weley-VCH, 5:259-291.

Bylund J, Burgess LA, Cescutti P, Ernst RK, Speert DP (2006) Exopolysaccharides from *Burkholderia cenocepacia* inhibit neutrophil chemotaxis and scavenge reactive oxygen species. *J Biol Chem* 281:2526–2532.

Cescutti P, Bosco M, Picotti F, Impallomeni G, Leitão JH, Richau JA, Sá-Correia I (2000) Structural study of the exopolysaccharide produced by a clinical isolate of *Burkholderia cepacia. Biochem Biophys Res Commun* 273:1088-1094.

Cleasby A, Wonacott A, Skarzynski T, Hubbard RE, Davies GJ, Proudfoot AE, Bernard AR, Payton MA, Wells TN (1996) The x-ray crystal structure of phosphomannose isomerase from *Candida albicans* at 1.7 angstrom resolution. *Nat Struct Biol* 3: 470-479.

Cunha MV, Sousa SA, Leitão JH, Moreira LM, Videira PA, Sá-Correia I (2004) Studies on the involvement of the exopolysaccharide produced by cystic fibrosis-associated isolates of the *Burkholderia cepacia* complex in biofilm formation and in persistence of respiratory infections. *J Clin Microbiol* 42:3052-3058.

Dunwell JM, Khuri S, Gane PJ (2000) Microbial relatives of the seed storage proteins of higher plants: conservation of structure and diversification of function during evolution of the cupin superfamily. *Microbiol Mol Biol Rev* 64: 153-179.

Ferreira AS, Leitão JH, Silva IN, Pinheiro PF, Sousa SA, Ramos CG, Moreira LM (2010) Distribution of cepacian biosynthesis genes among environmental and clinical *Burkholderia* strains and role of cepacian exopolysaccharide in resistance to stress conditions. *Appl Environ Microbiol* 76:441-50.

Foret J, de Courcy B, Gresh N, Piquemal JP, Salmon L (2009) Synthesis and evaluation of non-hydrolyzable D-mannose 6-phosphate surrogates reveal 6-deoxy-6-dicarboxymethyl-D-mannose as a new strong inhibitor of phosphomannose isomerases. *Bioorg Med Chem* 17:7100-7107.

Garami A, Ilg T (2001) The role of phosphomannose isomerase in *Leishmania mexicana* glycoconjugate synthesis and virulence. *J Biol Chem* 276:6566-6575.

Govan JR, Deretic V (1996) Microbial pathogenesis in cystic fibrosis: mucoid *Pseudomonas aeruginosa* and *Burkholderia cepacia*. *Microbiol Rev* 60:539-574.

Gowda G, Sagurthi SR, Savithrib HS, Murthya MRN (2008) Cloning, expression, purification, crystallization and preliminary X-ray crystallographic analysis of the mannose 6-phosphate isomerase from *Salmonella typhimurium*. *Acta Crystallogr Sect F Struct Biol Cryst Commun* 64:81-84.

Gracy RW, Noltmann EA (1968) Studies on phosphomannose isomerase. II. Characterization as a zinc metalloenzyme. *J Biol Chem* 243:4109-4116.

Griffin AM, Poelwijk ES, Morris VJ, Gasson MJ (1997) Cloning of the *aceF* gene encoding the phosphomannose isomerase and GDP-mannose pyrophosphorylase activities involved in acetan biosynthesis in *Acetobacter xylinum*. *FEMS Microbiol Lett* 154:389-396.

Hansen T, Wendorff D, Schönheit P (2004) Bifunctional phosphoglucose/phosphomannose isomerases from the archaea *Aeropyrum pernix* and *Thermoplasma acidophilum* constitute a novel enzyme family within the phosphoglucose isomerase superfamily. *J Biol Chem* 279:2262–2272.

Herasimenka Y, Cescutti P, Impallomeni G, Campana S, Taccetti G, Ravenni N, Zanetti F, Rizzo R (2007) Exopolysaccharides produced by clinical strains belonging to the *Burkholderia cepacia* complex. *J Cyst Fibros* 6:145-152.

Jackson KD, Starkey M, Kremer S, Parsek MR, Wozniak DJ (2004) Identification of psl, a locus encoding a potential exopolysaccharide that is essential for *Pseudomonas aeruginosa* PAO1 biofilm formation. *J Bacteriol* 186:4466 -4475.

Jensen SO, Reeves PR (2001) Molecular evolution of the GDP-mannose pathway genes (*manB* and *manC*) in *Salmonella enterica*. *Microbiology* 147:599-610.

Jensen SO, Reeves PR (1998) Domain organisation in phosphomannose isomerases (type I and II). *Biochim Biophys Acta* 1382: 5-7.

Khuri S, Bakker FT, Dunwell JM (2001) Phylogeny, function, and evolution of the cupins, a structurally conserved, functionally diverse superfamily of proteins. *Mol Biol Evol* 18:593-605.

Köplin R, Arnold W, Hötte B, Simon R, Wang G, Pühler A (1992) Genetics of xanthan production in *Xanthomonas campestris*: the *xanA* and *xanB* genes are involved in UDP-glucose and GDP-mannose biosynthesis. *J Bacteriol* 174:191-199.

Lee BT, Matheson NK (1984) Phosphomanno isomerase and phophogluco isomerase in seeds of *Cassia coluteoides* and some other legumes that synthesize galactomannan. *Phytochemistry* 23:983–987.

Lee HJ, Chang HY, Venkatesan N, Peng HL (2008) Identification of amino acid residues important for the phosphomannose isomerase activity of PslB in *Pseudomonas aeruginosa* PAO1. *FEBS Lett* 582:3479-3483.

Lee WH, Han NS, Park YC, Seo JH (2009) Modulation of guanosine 5'-diphosphate-D-mannose metabolism in recombinant *Escherichia coli* for production of guanosine 5'-diphosphate-L-fucose. *Bioresour Technol* 100:6143-6148.

Leitão JH, Sá-Correia I (1993) Manipulation of *Pseudomonas aeruginosa* alginate pathway by varying the level of biosynthetic enzymes and growth temperature. *J Appl Bacteriol* 74:452-459.

Leitão JH, Sá-Correia I (1997) Oxygen-dependent upregulation of transcription of alginate genes *algA*, *algC* and *algD* in *Pseudomonas aeruginosa*. *Res Microbiol* 148: 37–43.

May TB, Shinabarger D, Boyd A, Chakrabarty AM (1994) Identification of amino acid residues involved in the activity of phosphomannose isomerase-guanosine 5'-diphospho-D-mannose pyrophosphorylase. A bifunctional enzyme in the alginate biosynthetic pathway of *Pseudomonas aeruginosa*. *J Biol Chem* 269:4872-4877.

Miles JS, Guest JR (1984) Nucleotide sequence and transcriptional start point of the phosphomannose isomerase gene (manA) of *Escherichia coli*. *Gene* 32: 41–48.

Min BW, Cho YN, Song MJ, Noh TK, Kim BK, Chae WK, Park YS, Choi YD, Harn CH (2007) Successful genetic transformation of Chinese cabbage using phosphomannose isomerase as a selection marker. *Plant Cell Rep* 26:337–344.

Newburg DS, Ruiz-Palacios GM, Altaye M, Chaturvedi P, Meinzen-Derr J, Guerrero MD, Morrow AL (2004) Innate protection conferred by fucosylated oligosaccharides of human milk against diarrhea in breastfed infants. *Glycobiology* 14:253–263.

Niehues R, Hasilik M, Alton G, Körner C, Schiebe-Sukumar M, Koch HG, Zimmer KP, Wu R, Harms E, Reiter K, von Figura K, Freeze HH, Harms HK, Marquardt T (1998)

Carbohydrate-deficient glycoprotein syndrome type Ib. Phosphomannose isomerase deficiency and mannose therapy. *J Clin Invest* 101:1414-1420.

Okano K (2009) Synthesis and pharmaceutical application of L-ribose. *Tetrahedron* 65:1937–1949.

Panneerselvam K, Etchison JR, Freeze HH (1997) Human fibroblasts prefer mannose over glucose as a source of mannose for N-glycosylation. Evidence for the functional importance of transported mannose. *J Biol Chem* 272:23123 -23129.

Papoutsopoulou SV, Kyriakidis DA (1997) Phosphomannose isomerase of *Xanthomonas campestris*: a zinc activated enzyme. *Mol Cell Biochem* 177:183-191.

Patterson JH, Waller RF, Jeevarajah D, Billman-Jacobe H, McConville MJ (2003) Mannose metabolism is required for mycobacterial growth. *Biochem J* 372: 77-86.

Payton MA, Rheinnecker M, Klig LS, DeTiani M, Bowden E (1991) A novel *Saccharomyces cerevisiae* secretory mutant possesses a thermolabile phosphomannose isomerase. *J Bacteriol* 173:2006-2010.

Pettitt DJ (1979) In: *Polysaccharides in food*, Blanshard JMV, Mitchell JR (Eds.), pp. 263-280.

Richau JA, Leitão JH, Correia M, Lito L, Salgado MJ, Barreto C, Cescutti P, Sá-Correia I (2000a) Molecular typing and exopolysaccharide biosynthesis of *Burkholderia cepacia* isolates from a Portuguese cystic fibrosis center. *J Clin Microbiol* 38:1651-1655.

Richau JA, Leitão JH, Sá-Correia I (2000b) Enzymes leading to the nucleotide sugar precursors for exopolysaccharide synthesis in *Burkholderia cepacia*. *Biochem Biophys Res Commun* 276:71-76.

Rocchetta HL, Pacan JC, Lam JS (1998) Synthesis of the A-band polysaccharide sugar D-rhamnose requires Rmd and WbpW: identification of multiple AlgA homologues, WbpW and ORF488, in *Pseudomonas aeruginosa*. *Mol Microbiol* 29:1419-1434.

Roux C, Lee JH, Jeffrey CJ, Salmon L (2004) Inhibition of type I and type II phosphomannose isomerases by the reaction intermediate analogue 5-phospho-D-arabinonohydroxamic acid supports a catalytic role for the metal cofactor. *Biochem* 43: 2926-2934.

Roux C, Gresh N, Pereira LE, Piquemal J-P, Salmon L (2007) Binding of 5-phosphoarabinohydroxamate and 5-phospho-D-arabinonate to zinc phosphomannose isomerase from *Candida albicans* studied by polarizable molecular mechanisms and quantum mechanics. *J Comput Chem* 28:938 -957.

Sá-Correia I, Darzins A, Wang SK, Berry A, Chakrabarty AM (1987) Alginate biosynthetic enzymes in mucoid and nonmucoid *Pseudomonas aeruginosa*: overproduction of phosphomannose isomerase, phosphomannomutase, and GDP-mannose pyrophosphorylase by overexpression of the phosphomannose isomerase (*pmi*) gene. *J Bacteriol* 169:3224-3231.

Sabra W, Zeng AP, Deckwer WD (2001) Bacterial alginate: physiology, product quality and process aspects. *Appl Microbiol Biotechnol* 56:315-325.

Shankar S, Ye RW, Schlictman D, Chakrabarty AM (1995) Exopolysaccharide alginate synthesis in *Pseudomonas aeruginosa*: enzymology and regulation of gene expression. *Adv Enzymol Relat Areas Mol Biol* 70:221–255.

Shinabarger D, Berry A, May TB, Rothmel R, Fialho A, Chakrabarty AM (1991) Purification and characterization of phosphomannose isomerase - guanosine diphospho-D-mannose pyrophosphorylase. A bifunctional enzyme in the alginate biosynthetic pathway of *Pseudomonas aeruginosa*. *J Biol Chem* 266:2080-2088.

Slein MW (1950) Phosphomannose isomerase. *J Biol Chem* 186: 753-761.

Smith DJ, Payton MA (1994) Hyphal tip extension in *Aspergillus nidulans* requires the *manA* gene, which encodes phosphomannose isomerase. *Mol Cell Biol* 14:6030-6038.

Smith DJ, Proudfoot AE, Detiani M, Wells TN, Payton MA (1995) Cloning and heterologous expression of the *Candida albicans* gene *PMI 1* encoding phosphomannose isomerase. *Yeast* 11:301-310.

Sousa SA, Moreira LM, Wopperer J, Eberl L, Sá-Correia I, Leitão JH (2007a) The *Burkholderia cepacia bceA* gene encodes a protein with phosphomannose isomerase and GDP-D-mannose pyrophosphorylase activities. *Biochem Biophys Res Commun* 353:200-206.

Sousa SA, Ulrich M, Bragonzi A, Burke M, Worlitzsch D, Leitão JH, Meisner C, Eberl L, Sá-Correia I, Döring G (2007b) Virulence of *Burkholderia cepacia* complex strains in gp91$^{phox-/-}$ mice. *Cell Microbiol* 9:2817–2825.

Sousa SA, Moreira LM, Leitão JH (2008) Functional analysis of the *Burkholderia cenocepacia* J2315 BceAJ protein with phosphomannose isomerase and GDP-D-mannose pyrophosphorylase activities. *Appl Microbiol Biotechnol* 80:1015-1022.

Swan MK, Hansen T, Schönheit P, Davies C (2004) A novel phosphoglucose isomerase (PGI)/phosphomannose isomerase from the crenarchaeon *Pyrobaculum aerophilum* is a member of the PGI superfamily: structural evidence at 1.16-A resolution. *J Biol Chem* 279:39838-39845.

Tran ST, Le DT, Kim YC, Shin M, Choi JD (2009) Cloning and characterization of phosphomannose isomerase from *Sphingomonas chungbukensis* DJ77. *BMB Rep* 42:523-528.

Vinion-Dubiel AD, Goldberg JB (2003) Lipopolysaccharide of *Burkholderia cepacia* complex. *J Endotoxin Res* 9:201-213.

Wills EA, Roberts IS, Del Poeta M, Rivera J, Casadevall A, Cox GM, Perfect JR (2001) Identification and characterization of the *Cryptococcus neoformans* phosphomannose isomerase-encoding gene, *MAN1*, and its impact on pathogenicity. *Mol Microbiol* 40:610-620.

Wu B, Zhang Y, Zheng R, Guo C, Wang PG (2002) Bifunctional phosphomannose isomerase/GDP-D-mannose pyrophosphorylase is the point of control for GDP-D-mannose biosynthesis in *Helicobacter pylori*. *FEBS Lett* 519:87-92.

Yeom SJ, Ji JH, Yoon RY, Oh DK (2008) L-Ribulose production from L-arabinose by an L-arabinose isomerase mutant from *Geobacillus thermodenitrificans*. *Biotechnol. Lett* 30:1789–1793.

Yeom SJ, Ji JH, Kim NH, Park CS, Oh DK (2009) Substrate specificity of a mannose-6-phosphate isomerase from *Bacillus subtilis* and its application in the production of L-ribose. *Appl Environ Microbiol* 75:4705-4710.

Zlosnik JE, Hird TJ, Fraenkel MC, Moreira LM, Henry DA, Speert DP (2008) Differential mucoid exopolysaccharide production by members of the *Burkholderia cepacia* complex. *J Clin Microbiol* 46:1470-1473.

16

Thermal Degradation of Ligno-Cellulosic Fuels: Biopolymers Contribution

Valérie Leroy, Eric Leoni and Dominique Cancellieri
University of Corsica (SPE UMR CNRS 6134)
France

1. Introduction

Every year, thousands hectares of forest do burn in southern Europe. The Mediterranean area is especially affected during the dry season. Nevertheless, in spite of considerable efforts in fire research, our ability to predict the impact of a fire is still limited, and this is partly due to the great variability of fire behaviour in different plant communities (De Luis et al., 2005).
The combustion of forest fuels is partially governed by their thermal behaviour since this step produces a flammable gas mixture. Therefore, the analysis of the thermal degradation of lignocellulosic fuels is decisive for wildland fire modelling and fuel hazard studies (Dimitrakopoulos, 2001; Balbi et al., 2000; Stenseng et al., 2001). We propose in this work to focus on the thermal degradation of different forest fuels and their main components. Following a literature survey, we noticed that there is a lack in the description of the thermal degradation of forest fuels concerned by wildland fires (Grishin et al., 1983; Larini et al., 1998; Sero-Guillaume & Margerit, 2002; Linn & Cunningham, 2005). Even if these models are very different, it's well known that the energy emitted remains a crucial data. Classic approaches are based on the consideration of the low heat content value obtained by bomb calorimeter (Rothermel, 1983; Andrews, 1986; Nunez-Regueira et al., 2005). The experiments are led in constant volume what bring about a strong temperature raising, and with an excess of pure oxygen. These conditions are far from those met during a wildfire at atmospheric pressure in the air. DSC seems to be a convenient tool in order to follow the thermal degradation at the laboratory (Liodakis et al., 2002).
The degradation of forest fuels begins with the pyrolysis process from 373K to 773K (Simeoni et al., 2001; Shanmukharadhya & Sudhakar, 2007; Tonbul, 2008; Yuan & Liu, 2007). Non-combustible products, such as carbon dioxide, traces of organic compounds and water vapour, are emitted between 373K and 473K. Above 473K, the pyrolysis breaks down the fuels components into low molecular mass gases (volatiles), and carbonaceous char. Around 773K all the volatiles are gone; the remaining char is oxidized in a glowing combustion (Beall & Eickner, 1970).
Wood is a complex organic material, composed of cellulose (40 to 45% for coniferous trees and 38 to 50% for leafy trees), lignin (26 to 34% for coniferous trees and 23 to 30% for leafy trees), hemicellulose (7 to 15% for coniferous trees and 19 to 26% for leafy trees), extractives (<15%), ashes (< 1%) water and mineral matter (Orfao et al., 1999; Weiland et al., 1998). The chemical composition varies from species to species and within the same variety it varies with the botanical origin, age and location in the tree (trunk, branches, crown and roots). In

general, previous works refer to different commercial biopolymers (cellulose, xylan to represent hemicellulose...) whose structure is quite different from the native one (Koufopanos et al., 1989; Kohler et al., 2005; Alen et al., 1996; Cozzani et al., 1997). To have a realistic representation of biopolymer behaviour, it is essential to extract it directly from its source and to perform the analysis on the extracted matter.

Similarly to other authors (Koufopanos et al., 1989; Kohler et al., 2005; Miller & Bellan, 1997; Caballero et al., 1996), we made the assumption that the principal components of the fuel contributes to the thermal behaviour of the whole fuel itself and we showed in the present work that the enthalpy reaction of the thermal degradation follows this law.

The aim of this work was to quantify the participation of the principal components of the fuel to the reaction enthalpy of thermal degradation of the fuel. In a first time we have developed and used an analytical procedure in order to determine the content of each component in the forest fuels. In a second time, we have used DSC to record the enthalpy of the thermal degradation of the fuels and their components (previously extracted) under air sweeping. Then, calculated reaction enthalpy and experimental ones are compared.

2. Material and methods

2.1 Vegetal samples

We chose to study the thermal degradation of rockrose (*Cistus monspeliensis*: CM), heather (*Erica arborea*: EA), strawberry tree (*Arbutus unedo*: AU) and pine (*Pinus pinaster*: PP) which are representative species of the Corsican vegetation concerned by wildland fires. Naturally, the methodology developed hereafter is applicable to every lignocellulosic fuel.

Plant materials were collected from a natural Mediterranean ecosystem situated away from urban areas in order to prevent any pollution on the samples. Small particles (< 6mm) are considered in governing the dynamic of fire spread (Morvan & Dupuy, 2004; Burrows, 2001; Morvan, 2005). So we sampled the foliage and aerial parts of previously cited species. For each species, a bulk sample from six individual plants was collected in order to minimize interspecies differences. Current year, mature leaves were selected, excluding newly developed tissues at the top of the twigs. About 500g of each species were brought to the laboratory, washed with deionised water and oven-dried for 12 hours at 333K (Leoni et al., 2003). Dry samples were then grounded and sieved to pass through a 600μm mesh, then kept to the desiccator. The sieved powdery sample was stored in airtight plastic containers for future use. All the analysis (chemical and thermal) were performed on these powders. The moisture content coming from self-rehydration was about 4% four percent for all the samples before the analysis.

The elemental analysis was carried out at the SCA (Service Central d'Analyse) USR 59 CNRS, and the results are shown in Table 1.

	Carbon (%)	Hydrogen (%)	Oxygen (%)	N, mineral matter (%)
AU	48.24	6.15	40.33	5.28
EA	52.43	6.98	35.92	4.63
CM	46.58	6.22	37.68	9.52
PP	50.64	6.76	41.53	1.70

Table 1. Elemental composition of the different samples.

2.2 Extractions of biopolymers

Several analytical methods are available for the determination of lignocellulosic components. We used 2 kinds of techniques: one for the determination of the plant composition (quantitative analysis) and the other for the extraction of native constituents of the fuel (qualitative analysis).

Lignocellulosic materials were determined by different gravimetric methods, according to normalized or published methods. Figure 1 shows the experimental procedure performed on every fuel.

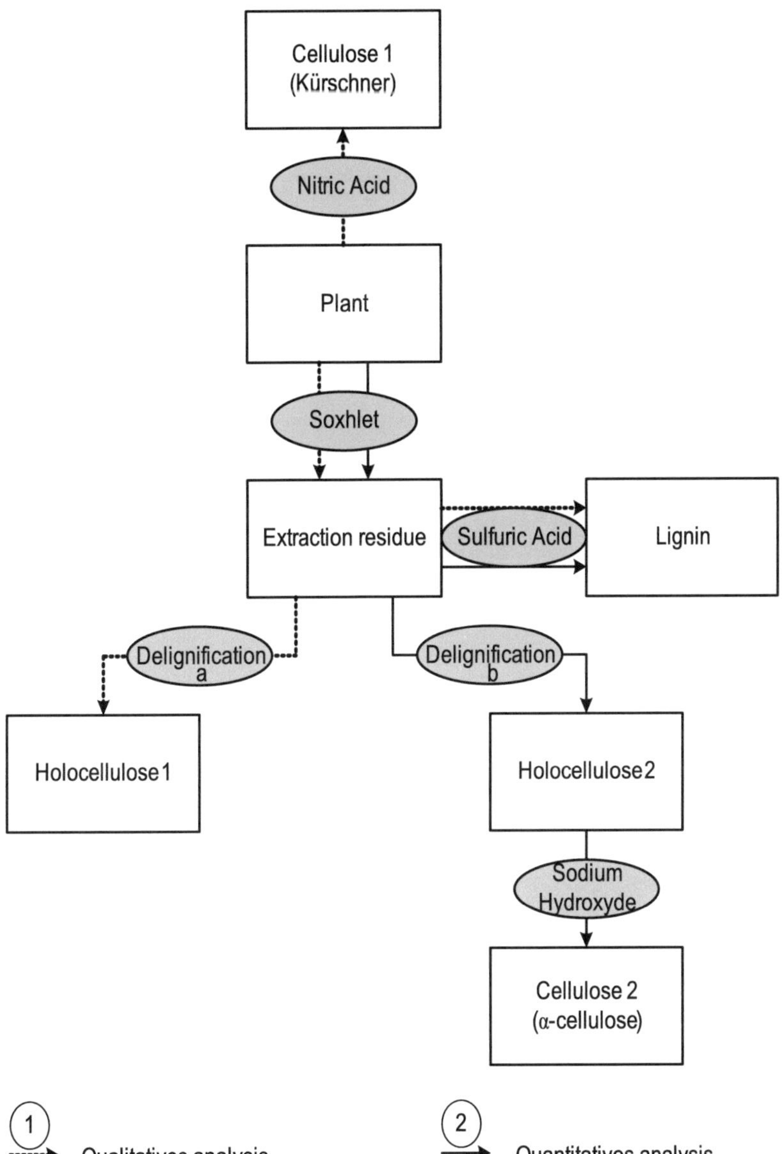

Fig. 1. Extractions diagram.

The extraction methods are summarized below. The extraction process was a multi-stage one excepted for cellulose 1. First of all, the extractives were determined by Soxhlet extractions with dichloromethane for six hours. Then, the sample was washed with distilled water and oven-dried (Ona et al., 1994). From the extraction residue, we determined the lignin, the holocelluloses 1, 2 and the cellulose 2. The lignin "Klason" content was obtained by gravimetric analysis after a sulphuric acid attack (Tappi, 1974). Holocellulose (i.e cellulose + hemicellulose) content was also obtained by gravimetric analysis after reaction of the extraction residue with sodium chlorite in acetate buffer; this step is called delignification. The difference between delignification "a" and "b" comes from the concentrations of reagent used and time of stoking (Wise et al., 1946; Kaloustian et al., 1996). The action of hydroxyl sodium on holocelluose 2 allowed obtaining cellulose 2. The cellulose 1 content was determined according to the Kürschner method (Peterssen, 1984; Kaloustian et al., 2003) directly on the powders by a gravimetric analysis after reaction with nitric acid.

As the chemical structure of hemicellulose and cellulose is very similar, there is no way to extract the native hemicellulose from the plant. That is why the proportions and the thermogram of this biopolymer were deducted from the difference between holocellulose and cellulose.

2.3 Thermal analysis procedure

There are only a few DSC studies in the literature concerning the thermal decomposition of lignocellulosic materials which is preferably studied by TGA (Bilbao et al., 1997; Branca & Di Blasi, 2003; Safi et al., 2004; Liodakis et al., 2005; Pappa et al., 2006; Liodakis & Kakardakis, 2008; Pappa et al. 2004). We recorded the emitted heat flow vs. temperature with a power compensated DSC (Perkin Elmer®. Pyris® 1). Thermal degradation was investigated in the range 400-900K under dry air (80% N_2 / 20% O_2) with a gas flow of 30 ml/min, measured at ambient temperature and atmospheric pressure.

We adapted the DSC for thermal degradation studies by adding an exhaust cover disposed on the measuring cell (degradation gases escape and pressure do not increase in the furnaces). The furnaces temperature was programmed for a linear heating from 400 to 900 K and we chose an intermediate heating rate β = 20K/min allowing very reproducible experiments. Open aluminium crucibles were used for both the sample and the reference (constituted of an empty crucible) (Tanaka, 1992). In every experiment, the powders were uniformly disposed on the bottom of the sample crucible. Samples consisted of about 5.0±0.1mg for all the experiments. The DSC calibration was performed at 20 K/min and verified every 5 runs using the melting point reference temperature and enthalpy reference of pure indium and zinc (Tmelt (In) = 429.8K. ΔHmelt(In) = 28.5J/g. Tmelt (Zn) = 692.8K. ΔHmelt(Zn) = 107.5J/g). The four species and their components were thermally analysed according to this procedure. Data of interest were the enthalpy values obtained by numeric integration of the DSC signal and peak top temperatures visualised on the thermogram. The uncertainty caused by weighting gave an error of 2-3% on the experimental values of enthalpy reactions. The pc-DSC allowed the measurement of very repeatable heat flows thanks to the micro furnaces. The energy measured in those experiments corresponds to the oxidation of vegetative fuels in an open atmosphere, of course this are not the same values obtained with calorimetric bombs, but the study with an adapted pc-DSC (with the exhaust cover) is more representative of forest fire conditions.

3. Chemical analysis results

All the results were expressed as a percentage on the dry matter (%DM).
The amount of Soxhlet extract (Extracted Matter: EM) was calculated according to equation (1).

$$Extracted\ Matter\ (\%DM) = \frac{final\ sample\ mass\ (g)}{initial\ sample\ mass\ (g)} \times 100 \quad (1)$$

Holocellulose content was calculated according to equation (2):

$$Holocellulose\ (\%DM) = \left[\frac{final\ sample\ mass\ (g)}{initial\ sample\ mass\ (g)} \times \frac{100 - EM}{100}\right] \times 100 \quad (2)$$

Where EM refers to the percentage of Extracted Matter.
Lignin content was obtained according to equation (3):

$$Lignin\ (\%DM) = \left[\frac{final\ sample\ mass\ (g)}{initial\ sample\ mass\ (g)} \times \frac{100 - EM}{100}\right] \times 100 \quad (3)$$

Finally, cellulose 1 was given by (4):

$$Cellulose\ 1\ (\%DM) = \frac{final\ sample\ mass\ (g)}{initial\ sample\ mass\ (g)} \times 100 \quad (4)$$

And cellulose 2 by (5):

$$Cellulose\ 2\ (\%DM) = \left[\frac{final\ sample\ mass\ (g)}{initial\ sample\ mass\ (g)} \times \frac{holocellulose}{100}\right] \times 100 \quad (5)$$

Table 2 presents the results obtained from the analytical process and the previous equations for the species studied herein. The presented results are averages, calculated from several attempts on the same sample.
The measured total does not reach 100%, because extractions are in series and there is a small amount of nitrogenous compounds, tannins and free sugars non extracted. For these fuels two dominants components were identified: lignin and cellulose.

	Cellulose (%)	Lignin (%)	Holo-cellulose (%)	Hemi-cellulose (%)	Extractive (%)	Total (%)
AU	38.0 ± 0.4	41.6 ± 0.4	43.2 ± 0.4	5.2 ± 0.1	12.9 ± 0.2	97.9 ± 1.0
EA	40.7± 0.4	39.7 ± 0.4	54.3 ± 0.5	13.6 ± 0.2	5.8 ± 0.1	99.8 ± 1.0
CM	39.4 ± 0.4	34.4 ± 0.3	52.0 ± 0.5	12.6 ± 0.2	9.2 ± 0.1	95.6 ± 1.0
PP	38.3 ± 0.4	38.9 ± 0.4	43.4 ± 0.4	5.1 ± 0.1	13.1 ± 0.2	95.2 ± 1.0

Table 2. Composition of the different samples

4. Thermal analysis results

The results from the thermal analysis of the four species and their components are shown in the present section; we chose to present only the curves of CM samples. This plant is known to have a high flammability in wildland fires.

4.1 Vegetation

Table 3 presents the reaction enthalpy and the peak top temperatures obtained from the DSC data on every fuel. Values of enthalpies are expressed for one gram of the fuel used for each experiment.

	$\Delta_r H°$ (J/g)	Peak 1 (K)	Peak 2 (K)
AU	-11410 ± 342	641	791
EA	-12540 ± 376	646	787
CM	-11070 ± 332	630	776
PP	-10885 ± 327	643	756

Table 3. Enthalpy value and peak top temperature of the different samples.

The enthalpy reaction (numeric integration of DSC signal) is varying between -10885J/g to -12540J/g for these different fuels. EA fuel was found to be the most energetic one and PP fuel was found to be the less energetic one. As the sample preparation and the experimental conditions were constant for all the fuels, the differences observed on the enthalpy values should come from the chemical composition of each fuel.

Figure 2 shows the thermal behavior of CM sample. For all the fuels the DSC curves exhibit a similar profile with two overlapped exothermic events. Two maxima were recorded around 630K and 775K; in a previous work we ascribed these thermal events (Cancellieri et al., 2005) corresponding to the oxidation of volatiles and chars. The behavior is similar and reproducible for all the species since we have triplicate experiments for all the species.

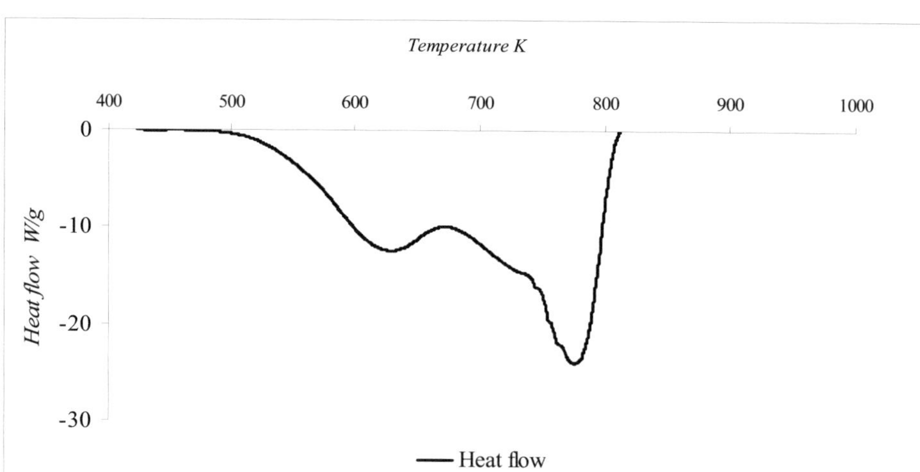

Fig. 2. DSC curve for CM at 20 K/min under air sweeping.

4.2 Extracted cellulose

Following the extraction, the cellulose from each fuel was thermally studied and table 4 presents the results obtained from the DSC data. Values of enthalpies are expressed for one gram of the cellulose used for each experiment.

It is important to note the strong resemblances of the thermal characteristics of the cellulose extracted from the various species. The values of enthalpy of the thermal degradation of cellulose are relatively constant for four plants (mean value: $\Delta rH° = -6892 \pm 213$ J/g) assuming that the cellulose extracted from these plants possesses a very close structure. Ours experiments show that cellulose is the lowest energetic component of the fuels.

	ΔH_{exp} (J/g)	T_{exo} 1 (K)	T_{exo} 2 (K)
AU	-6935 ± 208	623	766
EA	-7179 ± 215	627	803
CM	-6581 ± 197	623	778
PP	-6871 ± 206	628	799

Table 4. Enthalpy value and peak top temperature of extracted cellulose.

Figure 3 shows the thermal behavior of CM cellulose:

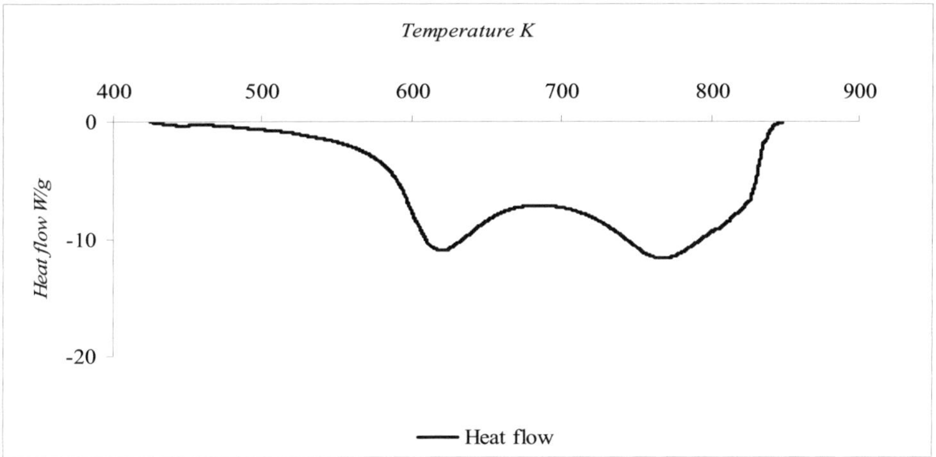

Fig. 3. DSC curve for CM cellulose at 20 K/min under air sweeping.

For all the fuels the DSC curves exhibit a similar profile with two overlapped exothermic events, the thermal behaviour of the cellulose extracted from the various species is similar to that of fuel. However, existing energetic interactions between the components are shown by comparing the curves of the fuel and the cellulose: the offset temperature is higher for the cellulose. In the modeling work we have neglected these interactions.

4.3 Extracted holocellulose

Following the extraction, the holocellulose from each fuel was thermally studied and table 5 presents the results obtained from the DSC data. Values of enthalpies are expressed for one gram of the holocellulose used for each experiment.

Values of enthalpies are globally higher for the holocellulose than the cellulose; this component can be considered as highly energetic in regard to the fuels. We noted a global trend similar to that of the plant, except the appearance of a supplementary phenomenon (peak 2') between both processes already shown on the fuels. This additional step is probably coming from the thermal degradation of hemicellulose mixed with cellulose in this component; unfortunately, we did not have the possibility to isolate hemicellulose from holocellulose. Figure 4 shows the thermal behavior of CM holocellulose obtained following the previously described method.

	$\Delta_rH°$ (J/g)	Peak 1 (K)	Peak 2' (K)	Peak 2 (K)
AU	-10565 ± 317	626	687	786
EA	-10891 ± 327	609	687	759
CM	-8974 ± 269	615	684	767
PP	-9789 ± 294	625	697	754

Table 5. Enthalpy value and peak top temperature of extracted holocellulose.

Figure 4 shows the thermal behavior of CM holocellulose obtained following the previously described method.

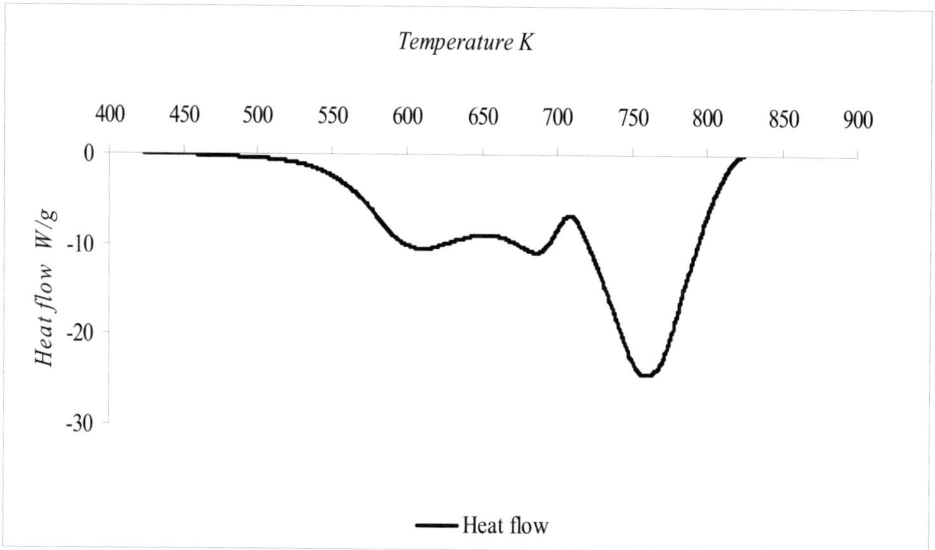

Fig. 4. DSC curve for CM holocellulose at 20 K/min under air sweeping.

This biopolymer showed a degradation process more complex than the other constituents. These observations are coherent as the holocellulose is a mixture of polysaccharides.

4.4 Extracted lignin

Following the extraction, the lignin from each fuel was thermally studied and table 6 presents the results obtained from the DSC data. Values of enthalpies are expressed for one gram of the lignin used for each experiment.

For all the species, we recorded a more marked difference between extracted lignin and the fuel. In regard with others biopolymers the value of reaction enthalpy is much higher and the major part of the exotherm is shifted toward high temperatures. Lignin is the highest energetic component of the fuel with a mean value of reaction enthalpy higher than the fuel itself.

	$\Delta_r H°$ (J/g)	Peak 1 (K)	Peak 2 (K)
AU	-16278 ± 488	653	846
EA	-16853 ± 505	668	831
CM	-15172 ± 455	662	815
PP	-17012 ± 510	682	799

Table 6. Enthalpy value and peak top temperature of extracted lignin.

As shown on the example of figure 5, DSC curves present an important amplitude for the peak 2 and a peak 1 almost flooded by the second. As for the cellulose, lignin curves shows existing energetic interactions between the components. Comparing the curves of the fuel and the cellulose we saw that the temperature of the end of reaction is higher for the lignin. In the modeling work we have neglected these interactions.

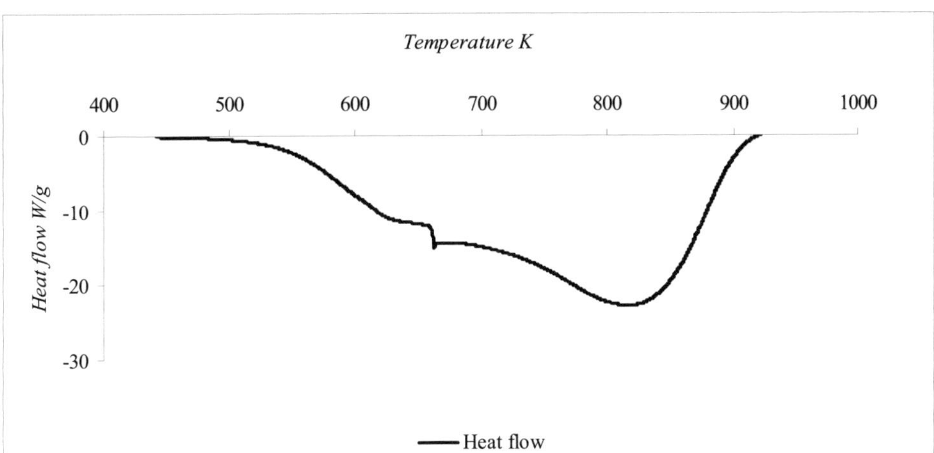

Fig. 5. DSC curve for CM lignin at 20 K/min under air sweeping.

All the thermograms also present an artefact owed to a condensation of products during the acidic extraction of lignin. This thermal behaviour agrees with previous studies (tsujiyama & Miyamori, 2000) indicating an only important peak for temperatures superior to 773K.

5. Modelling

Mean values of reaction enthalpies are presented in the following table 7 and were calculated from the data presented in tables 4, 5 and 6. The standard deviation was determined according to equation 6.

$$\sigma_x = \sqrt{\frac{\sum \left(\Delta_r H^\circ_x - \overline{\Delta_r H^\circ_x}\right)^2}{n}} \quad (6)$$

where $\overline{\Delta_r H^\circ}$ is average of $\Delta_r H^\circ$, x is a component and n is the number of values.

	Cellulose	Lignin	Hemicellulose	Extractives
$\overline{\Delta_r H^\circ}(J/g)$	- 6892	- 16329	- 15109	- 8176
σ	245,9	833,1	1318,7	2031,0
$\frac{\sigma}{\left\lvert\overline{\Delta_r H^\circ}\right\rvert}\times 100$	3,6 %	5,1 %	8,9 %	24,8 %

Table 7. Mean values of enthalpy reaction of the components in the 4 species.

The enthalpy variations of the various constituents are relatively constant and close to mean values. The relative error for every constituent is lower than 9%, except for the extractives (25%) which constitute a complex mixture of various chemicals. This result is however coherent, as far as the nature of compounds contained in the extracted matters varies according to the species (wax quantities, tannins, essential oil and colouring agents differ strongly according to the species).
According to assumption presented in the introduction section, in this work the thermal degradation of forest fuels is viewed as the thermal degradation of the components of the fuel as shown in figure 6.

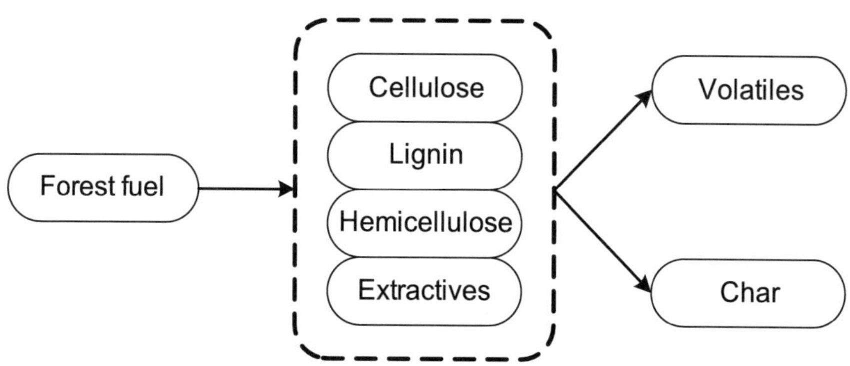

Fig. 6. Schematic thermal degradation of a forest fuel.

Compared to others studies, (Koufopanos et al., 1989; Kohler et al., 2005; Miller & Bellan, 1997; Caballero et al., 1996) we added the thermal behaviour of the extractives components which play an important role in wildland fires spread (Oasmaa et al., 2003a; Oasmaa et al., 2003b; Pappa et al., 2000).
The calculated reaction enthalpy can then be expressed as a combination of reaction enthalpy of each component of the fuel:

$$\Delta rH^{\circ}_{Mod} = a\,\Delta rH^{\circ}_{(cellulose)} + b\,\Delta rH^{\circ}_{(lignin)} + c\,\Delta rH^{\circ}_{(hemicellulose)} + d\,\Delta rH^{\circ}_{(extractives)} \quad (7)$$

where: a, b, c and d are the percentages of cellulose, lignin, hemicellulose and extractives respectively in the fuel.

This model is applicable and sturdy only if the enthalpy reaction of every constituent is constant for all the species. We thus made the following hypotheses of modelling which are valid in regard to the experimental results presented in table 7.

$$\Delta rH^{\circ}_{(cellulose)EA} \approx \Delta rH^{\circ}_{(cellulose)AU} \approx \Delta rH^{\circ}_{(cellulose)CM} \approx \Delta rH^{\circ}_{(cellulose)PP}. \quad (8)$$

$$\Delta rH^{\circ}_{(lignin)EA} \approx \Delta rH^{\circ}_{(lignin)AU} \approx \Delta rH^{\circ}_{(lignin)CM} \approx \Delta rH^{\circ}_{(lignin)PP}. \quad (9)$$

$$\Delta rH^{\circ}_{(hemi.)EA} \approx \Delta rH^{\circ}_{(hemi.)AU} \approx \Delta rH^{\circ}_{(hemi.)CM} \approx \Delta rH^{\circ}_{(hemi.)PP.} \quad (10)$$

However it was impossible to study the thermal behaviour of extractives (too unstable and giving very noisy DSC curves). Their enthalpy reaction was deducted by subtraction between the enthalpy reaction of the fuel and the enthalpy reaction of the residue from the solvent extraction. We thus made the following assumption:

$$\Delta rH^{\circ}_{(extra.)EA} \approx \Delta rH^{\circ}_{(extra.)AU} \approx \Delta rH^{\circ}_{(extra.)CM} \approx \Delta rH^{\circ}_{(extra.)PP.} \quad (11)$$

Equation 7 gave allowed the calculation of modelled enthalpies (ΔrH°_{Mod}) for the four species according to their chemical composition.

Figure 7 shows the experimental enthalpy (ΔrH°_{exp}) of thermal degradation of the fuels versus the results of calculated enthalpy. This figure clearly shows that according to its high enthalpy of reaction and its high content in the four species, lignin contributes strongly to the reaction enthalpy of the fuel. Compared to lignin, the low value of enthalpy of reaction of cellulose implies a weak contribution to the reaction enthalpy of the fuel for this component though cellulose content is as important as lignin. Our model takes into account the extractives percentage in the fuel; even if their content is low, these components are highly energetic and they should not be omitted in regard to the thermal behaviour of the fuels.

As shown in figure 7, the calculated enthalpy values get very close to the experimental values recorded from the thermal degradation of dry forest fuels.

The small difference between the model and experiments results indicates that energetic interactions between the various constituents of the plant can be neglected.

Regarding to the fuels: EA fuel was found to be the most energetic with the higher hemicellulose content and the lowest extractives content whereas PP fuel was found to be the less energetic fuel with the lowest hemicellulose content and the higher extractives content.

This approach is interesting since the reaction enthalpy of the thermal degradation can be calculated with the knowledge of the composition of the fuel. The reaction enthalpy of the thermal degradation is an important data in the field wildland fire modelling. In a future work we will study others fuels with a final objective to get an energetic profile of the forest fuels according to their chemical composition.

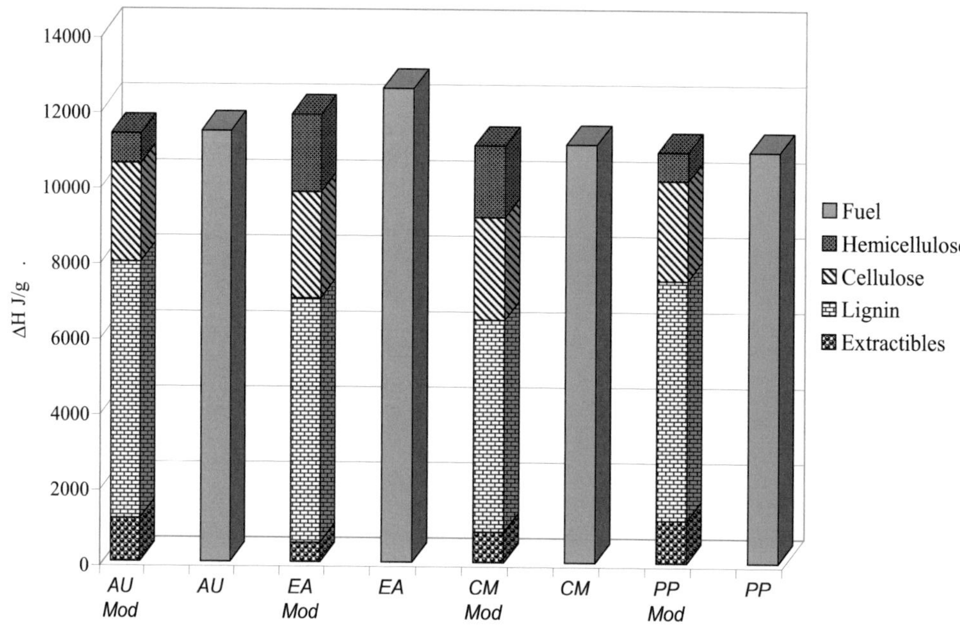

Fig. 7. Comparison between modelled and experimental enthalpy.

6. Conclusion

In wildland fire modelling and forest fuel hazard studies, the thermal degradation of the solid is a fundamental stage. Two ways are suitable: the first one considers the thermal degradation of the whole fuel giving a complex mixture of gas, tars and chars; the second one considers the thermal degradation as the sum of the contributions from the principal components of the fuel. Our aim was to verify the validity of the second approach. DSC analyses were performed in order to get the enthalpy reaction of the thermal degradation under air for four forest fuels and their components. With the assumption of constant enthalpies values for each component, we calculated the enthalpy reaction of the thermal degradation.

Calculated enthalpy was compared to the experimental data and we showed a good agreement. Thus, we can say that the reaction enthalpy of the fuels can be viewed as the sum of the reaction enthalpy of each component (taking into account their percentages in the fuel). The present study shows only a weak influence of interactions between the components in the fuel on the reaction enthalpy of thermal degradation.

This approach seems very interesting in order to get the enthalpy reaction of a forest fuel when its composition is available. This data is important for modelling purpose since the rate of consumption of fuels could be calculated from the energy emitted by gram of fuel when the power of the fire is available. Although this approach is attractive by its simplicity, the attempt to reproduce the complete thermal behaviour of a fuel turns out to be a delicate spot at the moment.

7. References

Alèn, R.; Kuoppala, E.; Oesch, P.J. (1996) Formation of the main degradation compound groups from wood and its components during pyrolysis. J. Anal. Appl. Pyrolysis, Vol.36, 137-148, 0165-2370.

Andrews, P.L. (1986) Behave: Fire behaviour prediction and fuel modelling system Burn subsystem. Part I. USDA Forest Service Research, Paper INT-194, 0363-6259.

Balbi, J.H.; Santoni, P.A.; Dupuy, J.L. (2000) Dynamic modelling of fire spread across a fuel bed. Int. J. Wildland Fire, Vol.9, 275-284, 1049-8001.

Beall, F.C.; Eickner, H.W. (1970) Thermal degradation of wood components: a review of literature, Forest Service Research, Paper FPL 130, 0363-6259.

Bilbao, R.; Mastral, J.F.; Aldea, M.E.; Ceamanos, J. (1997) The influence of the percentage of oxygen in the atmosphere on the thermal degradation of lignocellulosic materials, J. Anal. Appl. Pyr.,, Vol.42, 189-202,0165-2370.

Branca, C.; Di Blasi, C. (2003) Kinetics of the isothermal degradation of wood in the temperature range 528-708K, J. Anal. Appl. Pyr., Vol.67, 207-219, 0165-2370.

Burrows, N.D. (2001) Flame residence times and rates of weight loss of eucalypt forest fuel particles. Int. J. Wildland Fire, Vol.10, 137-143, 1049-8001.

Caballero, J.A.; Font, R.; Marcilla; A. (1996) Comparative study of the pyrolysis of almond shells and their fractions, holocellulose and lignin. Product yields and kinetics. Thermochim. Acta, Vol.276, 57-77, 0040-6031.

Cancellieri, D.; Leoni, E.; Rossi, J.L. (2005) Kinetics of the Thermal Degradation of Erica Arborea by DSC: Hybrid Kinetic Method. Thermochim. Acta, Vol.438, 41-50, 0040-6031.

Cozzani, V.; Lucchesi, A.; Stoppato, G. (1997) A new method to determine the composition of biomass by thermogravimetric analysis. The Cana. J. Chem. Eng., Vol.75, 127-133, 0008-4034.

De Luis, M.; Baeza, M.J.; Raventos, J. (2004) Fuel characteristics and fire behaviour in mature Mediterranean gorse shrublands. Int. J. Wildland Fire, Vol.13, 79-87, 1049-8001.

Dimitrakopoulos, A.P. (2001) Thermogravimetric analysis of Mediterranean plant species. J. Anal. Appl. Pyrolysis, Vol.60, 123-130, 0165-2370.

Grishin, A.M.; Gruzin, A.D.; Zverev, V.G. (1983) Matematical modelling of the spreading of hight level forest fires. Sov.phys.Dokl., Vol.28, 328-330, 0038-5689.

Kaloustian, J.; Pauli, A.M.; Pastor, J. (1996) Etude par analyse thermique de quelques végétaux méditerranéens, J. Therm. Anal. Cal.,, Vol.46, 1388-6150.

Kaloustian, J.; El-Moselhy, T.F.; Portugal, H. (2003) Chemical and thermal analysis of the biopolymers in thyme, Thermochim. Acta, Vol.401, 77-86, 0040-6031.

Kohler, S.; Mauviel, G.; Ferrer, M; Lédé, J. (2005) Flash pyrolysis of biomass model components and of their mixtures. 14th European Biomass Conference and Exhibition , Paris, Octobre 2005, Biomass for Energy, Industry and Climate Protection, Paris.

Koufopanos, C.A.; Maschio, G.; Lucchessi, A. (1989) Kinetics modelling of the pyrolysis of biomass and biomass components. The Can. J. Chem. Eng., Vol.67, 75-84, 0008-4034.

Larini, M.; Giroux, F.; Porterie, B.; Loraud, J.C. (1997) A multiphase formulation for fire propagation in heterogeneous various analytical methods. Int. J. Heat and Mass Transfer, Vol.41, 881-897, 0017-9310.

Leoni, E.; Cancellieri, D.; Balbi, N.; Tomi, P.; Bernardini, A.F.; Kaloustian, J.; Marcelli, T. (2003) Thermal degradation of Pinus pinaster Needles by DSC, Part 2: Kinetics of Exothermic phenomena. J. Fire Sciences, Vol.21, 117-130, 0734-9041.

Linn, R.R.; Cunningham, P. (2005) Numerical simulations of grass fires using a coupled atmosphere fire model: Basic fire behavior and dependence on wind speed. Journal of Geophysical Research, Vol.110, 1-19, 0148-0227.

Liodakis, S.; Bakirtzis, D.; Dimitrakopoulos, A.P. (2002) Ignition characteristics of forest species in relation to thermal analysis data. Thermochim. Acta, Vol.390, 83-91, 0040-6031.

Liodakis, S.; Katsigiannis, G.; Kakali, G. (2005) Ash properties of some dominant Greek forest species. Thermochim. Acta, Vol.437, 158-167, 0040-6031.

Liodakis, S.; Kakardakis, T. (2008) Measuring the relative particle foliar combustibility of WUI forest species located near Athens. J. Therm. Anal. Cal., Vol.93, 627-635, 1388-6150.

Miller, R.S.; Bellan, J. (1997) A generalized biomass pyrolysis model based on superimposed cellulose, hemicellulose and lignin kinetics. Combustion Science and Technology, Vol.126, 97-137, 0010-2202.

Morvan, D.; Dupuy, J.L. (2004) Modeling the propagation of a wildfire through a Mediterranean shrub using a multiphase formulation, Combustion and flame, Vol.138, 199-210, 0010-2180.

Morvan, D. (2005) Physique et modélisation du comportement des feux de forêt. 17ème Congrès Français de Mécanique, Plenary conference August 2005, Troyes, France.

Núnez-Regueira, L.; Rodríguez-Anón, J.A.; Proupín, J.; Mourino, B.; Artiaga-Diaz R. (2005) Energetic study of residual forest biomass using calorimetry and thermal analysis. J. Therm. Anal. Cal., Vol.80, 457-464, 1388-6150.

Oasmaa, A.; Kuoppala, E.; Gust, S.; Solantausta, Y. (2003) Fast Pyrolysis of Forestry Residue. 1. Effect of Extractives on Phase Separation of Pyrolysis Liquids. Energy & Fuels, Vol.17, 1-12, 0887-0624.

Oasmaa, A.; Kuoppala, E.; Solantausta, Y. (2003) Fast Pyrolysis of Forestry Residue. 2. Physicochemical Composition of Product Liquid. Energy & Fuels, Vol.17, 433-443, 0887-0624.

Ona, T.; Testsuya, S.; Masaru, S.; Kazumi, F. (1994) Small-scale method to determine the content of wood components from multiple eucalypt samples. Tappi journal, Vol.78, No 3, 121-126, 0734-1415.

Órfão, J.J.M.; Antunes, F.G.A.; Figueiredo, J.L. (1999) Pyrolysis kinetics of lignocellulosic materials – three independent reactions model. Fuel, Vol.78, 349-346, 0016-2361.

Pappa, A.; Tzamtzis, N.; Statheropoulos, M.; Fasseas, C. (2000) The pyrolytic behavior of Pinus halepensis needles observed by transmission light microscopy and stereoscopy. J. Anal. Appl. Pyr., 55, 195-202, 0165-2370.

Pappa, A.; Mikedi, K.; Tzamtzis, N. Statheropoulos, M. (2006) TG-MS analysis for studying the effects of fire retardants on the pyrolysis of pine-needles and their components. J. Therm. Anal. Cal., Vol.84, 655-661, 1388-6150.

Pappa, A.; Kyriakou, S.; Mikedi, K.; Tzamtzis, N.; Statheropoulos, M. (2004) Design considerations and an example of application of an in-house made TG-MS interface. J. Therm. Anal. Cal., Vol.78, 415-426, 1388-6150.

Petterssen, R.C.; (1984) Chemical composition of wood. Chapter 2. The Chemistry of Solid. R.R. Wood (Ed.), Advances in Chemistry Series 207, American Chemical Society, Washington, DC.

Rothermel, R.C. (1983) How to predict the spread and intensity of forest and range fires. USDA Forest Service Research, Paper INT-143, 0363-6259.

Safi, M.J.; Mishra, I.M.; Prasad, B. (2004) Global degradation kinetics of pine needles in air, Thermochim. Acta, Vol.412, 155-162, 0040-6031.

Sero-Guillaume, O.; Margerit, J. (2002) Modelling forest fires. Part I: a complete set of equations derived by extended irreversible thermodynamics, Int. J. Heat and Mass Transfert, Vol.45, 1705-1722, 0017-9310.

Shanmukharadhya, K. S.; Sudhakar K. G. (2007) Experimental investigations for the location of reaction zones in a bagasse fired furnace. J. Therm. Anal. Cal., Vol.90, 299-306, 1388-6150.

Simeoni, A.; Santoni, P.A.; Larini, M.; Balbi, J.H. (2001) Proposal for theoretical contribution for improvement of semi-physical forest fire spread models thanks to a multiphase approach: application to a fire spread model across a fuel bed. Combustion Science and Technology, Vol.162, 59-84, 0010-2202.

Stenseng, M.; Zolin, A.; Cenni, R.; Frandsen, F.; Jensen, A; Dam-Johansen, K. (2001) Thermal Analysis in Combustion Research. J. Therm. Anal. Cal., Vol.64, 1325-1354, 1388-6150.

Tanaka, S. (1992) Theory of power-compensated DSC, Thermochim. Acta, Vol.210, 67-76, 0040-6031.

Tappi,(1974) Acid insoluble lignin in wood and pulp. rapport T 222 05-74.

Tonbul, Y. (2008) Pyrolysis of pistachio shell as a biomass. J. Therm. Anal. Cal., Vol.91, 641-647, 1388-6150.

Tsujiyama, S.I.; Miyamori, A. (2000) Assignment of DSC thermograms of wood its components, Thermochim. Acta, Vol.351, 177-181, 0040-6031.

Weiland, J.J. ; Guyonnet, R.; Gibert, R. (1998) Analyse de la pyrolyse ménagée du bois par couplage TG-DSC-IRTF. J.Therm. Anal., Vol.51, 265-274, 1388-6150.

Wise, L.E.; Murphy, M.; D'Addieco, A.A. (1946) Chlorite holocellulose, its fractionation and bearning on sommative wood analysis and on studies on the hemicelluloses, Tappi Paper Trade Journal, Vol.122, 11-19, 0031-8655.

Yuan, H.R.; Liu R.H. (2007) Study on pyrolysis kinetics of walnut shell. J. Therm. Anal. Cal. , Vol.89, 983-986, 1388-6150.

17

Functional Properties of Some Non-conventional Treated Starches

Monica R. Nemţanu, Mirela Braşoveanu
National Institute for Laser, Plasma and Radiation Physics
Romania

1. Introduction

Starch is one of the most widespread and used biopolymers in different applications such as in the paper, textile, food, pharmaceutical and cosmetics industries (Richardson & Gorton, 2003). Even starch is useful in various industrial fields, its use as a native form has been restricted by some limitations (high viscosity, low solubility in cold water, paste instability, etc.) in specific applications due to its structure. Taking into account this aspect, starch is frequently submitted to the chemical, physical and enzymatic treatments or their combination in order to modify it and to obtain some functional properties suitable in specific industrial applications. The most used way to obtain modified starch is by chemical methods which are most of the time complex, expensive and time consuming.

Modern, non-conventional methods of starch modification are generally physical techniques (e.g., ionizing and non-ionizing radiation treatments, plasma treatment) which are fast, low cost and environmentally friendly because they do not use pollutant agents, do not allow the penetration of some toxic substances in the treated products and do not generate undesirable residual products, do not require catalysts and laborious preparation of sample.

There are reported several studies related to starch treatment with ionizing radiation, especially with gamma radiation (MacArthur & D'Appolonia, 1984; Sokhey et al., 1993; Kang et al., 1999; Wu et al., 2002; Ezekiel et al., 2007). Recently, information regarding starch treatment with electron beam (e-beam) in different irradiation dose range, either at low or relative high doses (0 - 25 kGy) (De Kerf et al., 2001; Pimpa et al., 2007) or at very high doses (> 50 kGy) (Kamal et al., 2007; Shishonok et al., 2007) were reported. The studies concerning the effects of ionizing radiation were performed on starches extracted from various vegetal sources such as corn (Kang et al., 1999; De Kerf et al., 2001; Adeil Pietranera & Narvaiz, 2001), wheat (MacArthur & D'Appolonia, 1984; Koksel et al., 1996), potato (Ezekiel et al., 2007; Shishonok et al., 2007), barley endosperms (Faust & Massey Jr., 1966), rice (Wu et al., 2002; Bao et al., 2005) or sago (Pimpa et al., 2007). The reported data showed that ionizing radiation treatment generates free radicals on starch molecules that can alter their size and structure (Raffi et al., 1980; Ciesla et al., 1991; Grant & D' Appolonia, 1991; Sabularse et al., 1991; Sokhey et. al. 1993). Therefore, the hydrolyze of the chemical bonds occurs by cleavage of starch macromolecule into lower fragments of dextrin, which can be either electrically charged or uncharged as free radicals (Kang et al., 1999). This type of treatment may change the functional and physicochemical properties of starch leading to the increase of solubility (Bao & Corke, 2002), the reduction of swelling power (MacArthur & D'Appolonia, 1984) as well as the reduction of relative viscosity of starch paste (Adeil Pietranera & Narvaiz, 2001) as a function of the irradiation dose. The use of radioisotopic sources (^{60}Co or ^{137}Cs) handling sources with continuous emission of radiation involves an additional risk of radiological accident compared to the usage of charged particle accelerators which send out radiation only during the technological process and, nowadays, are preferred for applications because they do not provide any risk of radioactive contamination. This is the reason explaining the late interest to study the influence of accelerated electrons on starch characteristics.

Also, investigations on starch modification by treatment with microwave or plasma were reported lately. Microwave treatment on starch were carried out especially on samples with high water content, over 20%, and less on samples with low (1 - 5%) or limited (7 - 15%) moisture content. The results showed that microwaves may determine the rearrangement of intermolecular structure indicating that the starch granule structure is disintegrated and, consequently, they lead to slight reduction of the water absorption ability, of solubility and swelling power as well as changes of the gelatinization temperature, syneresis or paste viscosity (Lewandowicz et al., 1997; Lewandowicz et al., 2000; Gonzalez & Perez, 2002; Gonzalez Parada & Perez Sira, 2003; Szepes et al., 2005; Luo et al., 2006; Lares & Perez, 2006) strongly correlated with experimental treatment parameters. Starch exposed to electrical discharges (plasma) may suffer either crosslinking phenomenon (Zou et al., 2004, Nemtanu & Minea, 2006) or depolymerization to dextrins with different molecular weights (Lii et al., 2002a; Lii et al., 2002b) depending on the experimental conditions (atmosphere, pressure, time of exposure, etc.).

The goal of our work is to present the experimental results concerning the comparison of physicochemical characteristics of starches from different botanical sources subjected to the accelerated electron beam treatment.

2. Experimental

Native starches from different botanical sources as cereals (corn, wheat and rice) and tubers (potato) were used in the experiments. The packed starch samples were irradiated with accelerated electron beam using a linear accelerator with medium energy of 6 MeV (INFLPR, Bucharest-Magurele, Romania), at ambient temperature and normal pressure. The irradiation doses between 10 and 50 kGy were checked using cellulose triacetate film.

pH was measured for 1% aqueous solutions of starch at 25°C with a Denver pH-conductivity meter (Denver Instruments, USA).

Starch was dispersed in water at temperature of 85°C by stirring for 30 min and the mixture was centrifuged (2000 x *g*, 15 min), and then the supernatant was collected. The supernatant was evaporated at 110°C and weighed. The solubility was the ratio in weight of the dried supernatant to the initial weight of the starch. The sediment was weighed and the swelling power (SP) was calculated according Wattanachant et al. (2002):

$$SP = \frac{w_s}{w_i(100 - s\%)} \cdot 100 \tag{1}$$

where: w_s - weight of sediment [g],
w_i - weight of initial sample [g],
s% - solubility of sample [%].

The paste clarity was determined according to method described by Craig et al. (1973) slightly modified. The clarity was measured on starch aqueous pastes of 1% concentration and expressed as a transmittance (T%) at 620 nm. The measurements were carried out against water blank in a Cary 100 Bio spectrophotometer (Varian, Inc., USA) and were repeated after 24, 48 and 72 hours for the same samples stored at room temperature (25 °C) to evaluate the stability of the clarity pastes.

The colorimetric characteristics of 1% starch aqueous pastes were measured with Cary Bio 100 spectrophotometer (Varian, Inc., USA) using transmittance in spectrum visible region (360 - 830 nm) with standard illuminant D65, and expressed as color attributes of CIELAB and CIELCH systems. The data have been analyzed using the Color Application of the Cary Win UV v. 3.10 software.

Gel consistency determination was carried out according Tang method cited by Wu et al. (2002). Gel consistency was measured for starch pastes of 5% in 0.2N KOH solution at 25°C as the length of the gel in a test tube held horizontally for 1 hour.

The rheological measurements were performed on 5% aqueous starch suspensions using a rotational HAAKE VT® 550 viscometer (ThermoHaake, Germany) with NV coaxial cylinder. The shear stress, τ and apparent viscosity, η_a of the samples were measured at different temperatures for different shear rates, $\dot{\gamma}$. The obtained data were analyzed with RheoWin v. 3.5. software.

The calorimetric investigation was realized on mixture of starch with 70-80% water using a dynamic scanning calorimeter (PerkinElmer, Inc., USA). The measurements were carried out at temperature within 20 - 90°C in nitrogen atmosphere. The samples were heated in sealed aluminum pans at a heating rate of 5 $K \cdot min^{-1}$.

The samples for GPC analysis were prepared according to the method described by Han & Lim (2004) slightly modified. Thus, starch (10 mg) was wetted with ethanol (20 µl), and then dispersed in 1M NaOH (500 µl). The mixture is then magnetically stirred (1000 rpm) for 10 minutes at 35°C. The starch solution obtained was diluted to 50mM NaOH with water and gently stirred again (150 rpm) for 30 minutes at room temperature, followed by centrifugation (3500 x *g*, 10 min). GPC analysis was performed using a Breeze system (Waters, USA) which consists in 1525 binary pump, autosampler 717+, 2414 differential refractive index detector, in-line degasser AF and temperature modul system. The calibration was carried out with pullulan standards for two Ultrahydragel columns (Waters, USA). The mobile phase used for GPC was 50mM NaOH and the flow rate was 0.5 ml/min. The obtained data were analyzed with Breeze v. 3.30 SPA software.

The experimental data statistic interpretation was performed using factor analysis of principal components with StatistiXL v.1.8 (2007) software in order to correlate the functional properties of studied starches.

3. Results and discussion

Free radicals are produced when polymers are exposed to ionizing radiation treatment (Woods & Pikaev, 1994). These active species are responsible for initiating chemical reactions leading to modifications like (1) formation of network in polymer chains called crosslinking, (2) molecular weight reduction by chain scission and (3) polymerizing a monomer and grafted onto the base polymer chain called grafting (Sarma, 2005). The effects of radiation depend on radiation type and irradiation dose as well as on chemical structure of polymer, thus the evaluation of physicochemical properties of starches treated with electron beam represents an important stage in order to elaborate such modification method.

3.1 pH evaluation

pH is an important property in starch industrial applications being used generally to indicate the acidic or alkaline properties of liquid media. A variation of pH value can significantly modify the evolution of other characteristics of starch after electron beam treatment.

Aqueous solutions of native studied starches had relative neutral pH that decreased by electron beam treatment with irradiation dose (Table 1) so that for a doses of 50 kGy, the solutions had an acidic character (pH = 4.7 - 4.8) for corn (Nemtanu et al., 2006a, Nemţanu, et al., 2006b), wheat and potato starches. In the case of rice starch, it was observed just a slight change of pH value as a consequence of e-beam treatment (Table 1).

Irradiation dose [kGy]	**Starch source**			
	Corn	*Wheat*	*Rice*	*Potato*
0	7.4 ± 0.2	7.7 ± 0.3	7.5 ± 0.1	6.7 ± 0.1
10	6.8 ± 0.3	5.6 ± 0.3	7.3 ± 0.1	6.2 ± 0.1
20	5.9 ± 0.1	5.3 ± 0.1	7.2 ± 0.1	5.8 ± 0.1
30	5.5 ± 0.2	5.1 ± 0.1	7.2 ± 0.1	5.4 ± 0.2
40	5.0 ± 0.3	5.0 ± 0.1	7.1 ± 0.2	5.1 ± 0.2
50	4.7 ± 0.2	4.7 ± 0.1	7.3 ± 0.3	4.8 ± 0.2

Table 1. pH values of the studied starch aqueous solutions

The decrease of solution pH value corresponds to the increase of its hydrogen ion concentration. The descending change of solution pH after e-beam treatment of starch - similar results on gamma-irradiated dry bean starch (Duarte & Rupnow, 1994) or e-beam treated sago starch (Pimpa et al., 2007) being previously reported by other research groups - indicates the formation of chemical groups with acidic character such as carboxyl, carbonyl or peroxide groups. This behavior is sustained also due to the fact that e-beam treatment was carried out in presence of oxygen determining thus the appearance of free radicals, compounds with carbonyl bonds (aldehydes/ketones), organic peroxides or other polysaccharide degradation products (Ershov, 1998) that can lead to the increase of starch acidity.

3.2 Solubility and swelling power

Solubility and swelling power are properties which provide the evidence of the magnitude of interaction among starch chains within the amorphous and crystalline domains, the extent of this interaction being influenced by amylose/amylopectin ratio, and by the characteristics of amylose and amylopectin in terms of molecular weight/distribution, degree and length of branching, and conformation (Hoover, 2001; Ratnayake et al., 2002).

The changes that appear in starch granule in an aqueous system depend on temperature and water availability from system. When starch molecules are heated in excess of water, the crystalline structure is disrupted and the water molecules become linked by hydrogen bonding to the exposed hydroxyl groups of amylose and amylopectin, which causes an increase in granule swelling and solubility (Singh et al., 2003).

The fact that the cereal starches (corn and wheat) exhibited generally lower solubility can be attributed to the more compact structure and different cristallinity than tuber starch (potato) according Collision (1968). The e-beam treatment led to the solubility value increase for all studied starches (Table 2). It was remarked the significant value of the sample solubility reached after treatment as the irradiation dose increase. This notable increase of solubility demonstrates clearly that the starch molecules suffer important changes as consequence of a degradation phenomenon induced by irradiation.

Irradiation dose [kGy]	**Starch source**			
	Corn	*Wheat*	*Rice*	*Potato*
0	78.73 ± 7.60	58.68 ± 2.47	58.80 ± 6.38	77.23 ± 11.53
10	86.99 ± 10.71	86.00 ± 3.58	88.87 ± 7.63	81.61 ± 10.57
20	90.36 ± 8.75	87.21 ± 5.09	92.66 ± 7.11	86.36 ± 8.98
30	90.85 ± 7.50	89.40 ± 4.19	88.91 ± 5.32	86.18 ± 12.58
40	90.97 ± 7.18	90.57 ± 3.87	88.33 ± 8.56	85.94 ± 13.06
50	91.29 ± 5.11	94.41 ± 3.05	87.54 ± 11.22	86.41 ± 13.22

Table 2. Solubility values [%] for studied starches

The improvement of starch solubility in water is an important request in the application where starch is involved, especially nowadays when modern technologies impose higher and higher quality standards and very simple and fast technological operations.

Swelling power is a measure of hydration ability of starch granule and it is determined by weighing the swelled starch and retained water. Such property is very important for certain starch application like those from food industry where the quality of starch-based products is strongly related to the capacity of starch granule to retain water and swell. A direct consequence of starch swelling is the increase of solubility, paste clarity and viscosity - properties of interest, especially in food industry.

The swelling power of native starch increases with the temperature increasing as a result of crystalline structure relaxation so that both amylose and amylopectin can form easily hydrogen bonds with water molecules.

Irradiation dose [kGy]	**Starch source**			
	Corn	*Wheat*	*Rice*	*Potato*
0	29.79 ± 2.75	21.51 ± 4.71	28.01 ± 2.58	22.51 ± 4.33
10	23.03 ± 1.49	11.12 ± 2.42	21.64 ± 1.85	10.65 ± 0.65
20	20.19 ± 1.19	11.29 ± 1.34	20.32 ± 1.37	13.36 ± 1.67
30	19.15 ± 1.42	12.27 ± 1.08	18.28 ± 1.19	6.73 ± 1.03
40	15.67 ± 0.75	11.65 ± 0.99	18.11 ± 0.63	7.42 ± 0.63
50	13.23 ± 0.89	11.26 ± 0.82	18.80 ± 1.88	7.26 ± 0.38

Table 3. Swelling power values [%] for studied starches

The e-beam treatment of starch caused the reduction of swelling power as the increase of the irradiation dose (Table 3). This behavior could be attributed to the fact that starch granules become sensitive being weaker and easier to break after e-beam treatment. In addition, a consequence of starch degradation induced by irradiation can be also the inhibition of granule ability to trap water and provoke the swelling explaining thus the swelling power reduction by irradiation.

3.3 Visual characteristics

Material or product quality is strongly related to its visual aspect (form, color, and transparency/opacity) especially as this visual aspect, by conferred information, influences consumer's perception and so the acceptance or rejection of that material or product.

3.3.1 Clarity and its stability

Clarity is one of the most important attributes in the starch paste properties, particularly in food industry where it influences directly brightness and opacity of products, an important factor to determine the clarity of starch paste being the physical arrangement of molecules which contribute to swelling ability of granules (Craig et al., 1973).

Light transmittance offers information regarding the starch paste behavior when light goes through it, thus the higher transmittance value the paste is more transparent.

Native corn starch paste had a low clarity ($T_{620\ nm} < 10\%$) (Nemţanu et al., 2006b, Nemţanu et al., 2007) which increased spectacularly ($T_{620\ nm} \approx 85\%$) after 24 hour storage (Fig. 1a). The sample clarity changed insignificantly during further storage time.

After electron beam treatment, the clarity of corn starch paste, immediately after preparation (Nemţanu et al., 2006b), increased as the increase of the irradiation dose getting to a value of approximately 30%(Nemţanu et al., 2007) (Fig. 1a). However, for doses of 40 and 50 kGy, the clarity was slightly lower than the samples treated with doses up to 30 kGy. Such behavior appeared due to the modifications induced by electron beam treatment in starch structure, if we take into account de study reported by Craig et al. (1973). During storage time, the samples treated with 10 and 20 kGy showed similar evolution like native starch, meaning an increase of the clarity, but not so spectacularly as control sample, after 24 hours, value kept even after 72 hours of storage. Therefore, the clarity of native and 10 - 20 kGy treated starch pastes showed instability in the first 24 hours followed by its stabilization unlike the samples treated with doses of 30 - 50 kGy which stabilized before 24 hour storage time.

The value of native wheat starch clarity was higher than of corn starch being approximately 19% (Fig. 1b), but it showed similar evolution during storage time. However, it was remarked the increase of clarity only up to ~60% after 24 hour storage time, namely lower than for corn starch value. The clarity of pastes prepared from wheat starch treated with electron beam at different doses varied insignificantly in comparison to native wheat starch immediately after preparation. The paste clarity of e-beam treated wheat starch increased during storage time in a similar way as paste of native wheat starch. Taking into account these observations, it can be considered the paste clarity of wheat starch treated with electron beam in range of 10 - 50 kGy was stable after 24 hour storage period.

The native rice starch had also an opaque paste ($T_{620\ nm} \approx 5\%$) (Nemtanu, 2005) and its clarity increased ($T_{620\ nm} \approx 50\%$) after storage of 24 hours (Fig. 1c) without significant change even after 72 hours.

The e-beam treatment led to an increase of the rice starch paste clarity, immediately after preparation, as the irradiation dose increase (Nemtanu et al., 2005), so that the clarity was ~60% for irradiation dose of 50 kGy (Fig. 1c). The samples treated with doses in the range of 10 - 30 kGy showed similarly evolution of clarity increase as control sample. As a result of these findings, it can be certainly considered the rice starch samples treated with doses of 40 - 50 kGy had stable clarity of their pastes during storage of 72 hours while the clarity of samples treated with 10 - 30 kGy has been stabilized after 24 hours from their preparation.

Native potato starch paste showed very good clarity with $T_{620\ nm} \approx 90\%$ which decreased nevertheless as the storage time (Fig. 1d). The clarity of potato starch paste was improved as a result of applying e-beam treatment on starch. The clarity decreased slowly after 24 hours and significantly after 72 hours for all samples. However, the clarity stabilization occurred after 24 hours for samples treated with doses of 40 - 50 kGy. Therefore, the potato starch samples treated with electron beam showed very clear pastes, the samples treated with doses of 40 - 50 kGy presenting even an obvious stability after 24 hour storage period.

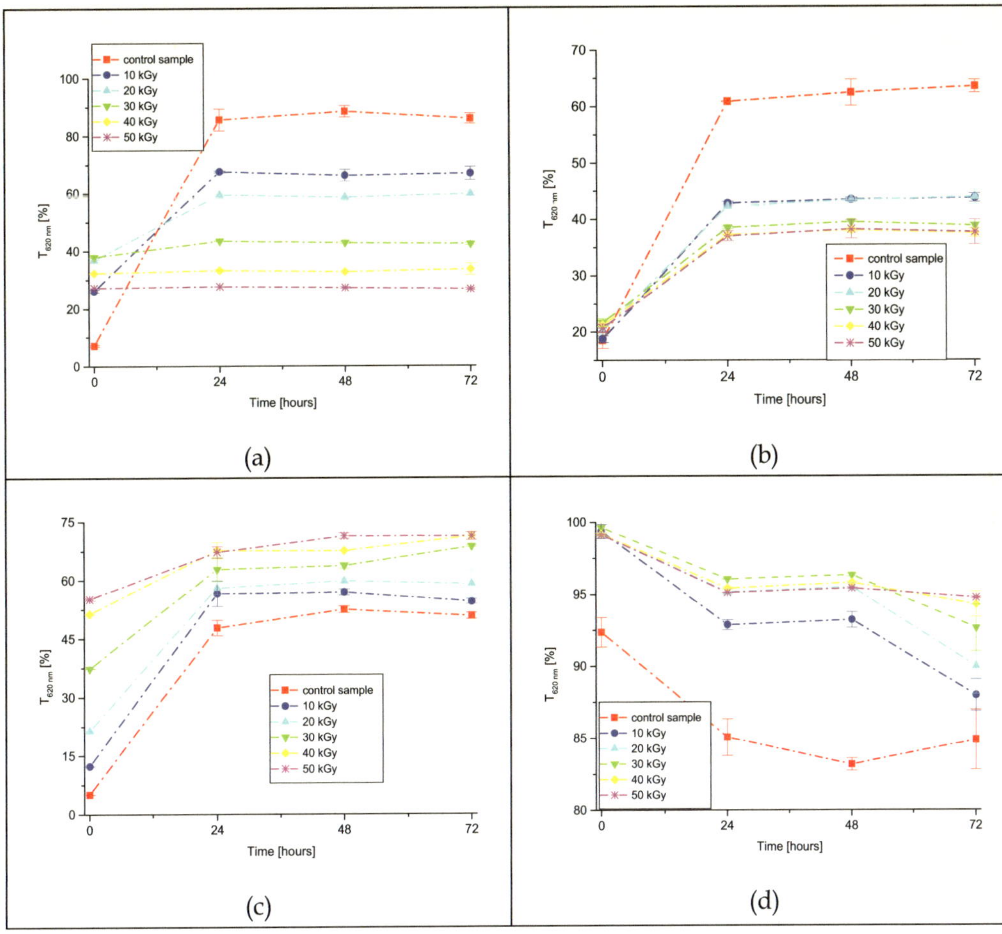

Fig. 1. Clarity of starch pastes: (a) corn, (b) wheat, (c) rice and (d) potato, before and after e-beam irradiation

As it was mentioned for solubility and swelling power, different values of clarity and its stability of cereal starches in comparison to tuber starch can be due to the physical arrangement of molecules within granules, which obstructs their ability to swell and disperse in water. In addition, the starch paste clarity during storage can be influenced by the interaction of several factors such as granule swelling, granule remnants, leached amylose and amylopectin, amylose and amylopectin chain length, intra- or inter-bonding, lipid and cross-linking (Jacobson et al., 1997).

The decrease of the transmittance value, meaning the increase of the starch paste turbidity indicates the re-association of initial broken bonds in orderly structure (retrogradation) and it could be noticed by liquid phase separation and sediment.

The paste behavior of e-beam treated starches studied using clarity indicates that this treatment can lead to the improvement of this property as a consequence of starch molecule degradation and formation of lower molecular fragments. According to Song et al. (2006), the increase in the transmittance value of starch pastes might be attributed to the introduction of carboxyl group, which retained the water molecules to form hydrogen bonds in the starch granules, thereby, increasing the clarity of starch pastes. Taking into account that, the formation of carboxyl groups after e-beam treatment of starch was suggested by the increased acidity of the irradiated starch pastes, we can consider this also as a factor which contributes to the increase of the clarity. Moreover, the evolution in time of irradiated starch paste clarity showed that the retrogradation phenomenon (formation of an orderly structure) was delayed by treatment of starch with accelerated electron beam.

3.3.2 Colorimetric parameters

Color is a fundamental property or, in other words, one of the vital components of visual quality of any material or product playing major role for the quality evaluation and consumer's decisions. Moreover, color variations confirm the changes that appear in structure or composition of a material or product and may be used as a quality index.
Color perception is three dimensional, meaning that three terms are needed to describe a color (Marcus, 1999): hue, lightness, and chroma (Fig. 2).

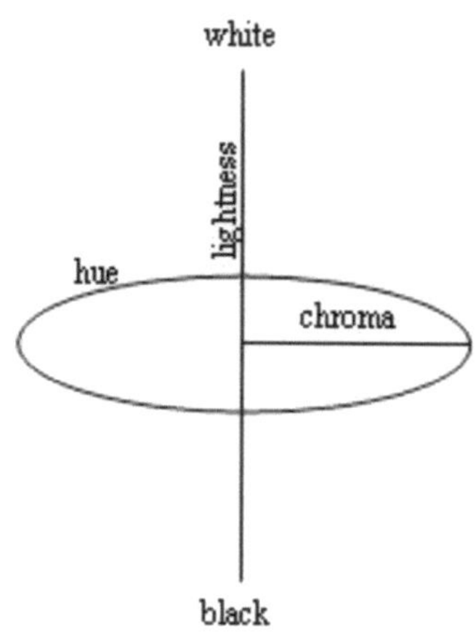

Fig. 2. The three terms describing a color

The International Commission on Illumination (CIE - Commission Internationale de l'Eclairage in French) (Ohta & Robertson, 2005), the organization responsible for international recommendations for photometry and colorimetry, recommended standard colorimetric systems including illuminants, colorimetric observer, light sources and the methodology used to derive values for describing color. The CIE Color Systems utilize three coordinates to locate a color within colorimetric space. Nowadays, the most used colorimetric spaces are CIE 1976 or CIELAB (L*a*b*) and CIELCH (L*C*h°). When color is expressed in CIELAB (Fig. 3), L^* defines lightness, a^* denotes the red/green value and b^* the yellow/blue value being presented in Cartesian coordinates of color. Thus, L^* has values from 0 (black) to 100 (white); $a^*>0 \rightarrow$ red color and $a^*<0 \rightarrow$ green color; $b^*>0 \rightarrow$ yellow color and $b^*<0 \rightarrow$ blue color.

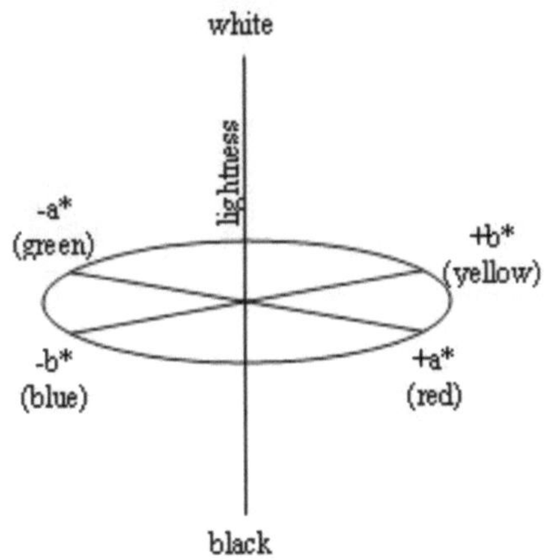

Fig. 3. CIELAB colorimetric system

CIELCH (D65/CIE 1964 10^0) is a polar representation of the CIELAB rectangular coordinate system and describes color in terms of lightness L^*, chroma (saturation) C^* and hue $h°$.

Investigated spectrocolorimetric parameters in our study showed different colorimetric behavior of cereal starches than tuber starch after application of electron beam treatment. In addition, the evolution of these parameters indicated clearly that changes occurred globally in the chemical structure of studied starches.

Therefore, it was noticed that yellow-blue and red-green coordinates modified after e-beam treatment of starches from perspective of L*a*b* color coordinates. The pastes of cereal starches exhibited the color movement in the $+a^*$ direction depicted a shift toward red while $+b^*$ movement indicated the shift toward yellow as the increase of the absorbed dose. But, the pastes of potato starch showed the color movement in the $-a^*$ direction toward green and $-b^*$ toward blue with irradiation dose increase.

Transposing the color in L*C*h° coordinates, it was observed that all native starch pastes showed very low values of chroma C^* (4 – 10) indicating the tendency of color to grey. The values of chroma $C^*$and hue $h°$for cereal starch pastes increased and decreased, respectively, exhibiting red-yellowish color after irradiation with electron beam (Fig. 4a-c). The pastes of the irradiated potato starch showed an opposite effect to that presented by cereal starches (Fig. 4d), meaning that the chroma decreased and the hue intensified indicating the movement of color to green.

Regarding the lightness, it intensified for all starches when the irradiation dose increased. It was remarked that the high lightness value ($L^* \approx 90\%$) of potato starch paste which tended to maximum by irradiation.

All these aspects related to the chromatic attributes of the investigated starch pastes confirmed the e-beam treatment influence on the starch macromolecular structure. The color change toward red-yellowish can be due to caramelization reaction of compounds with low molecular weight resulting from starch macromolecule degradation.

3.4 Gel consistency

The origin of rigid and weak gel properties is specific to molecular association of polymer chains (Dea, 1989). Cereal starches in their native form showed medium consistency (Nemţanu et al., 2006a; Nemţanu et al., 2006b; Yu & Wang, 2007) while the native form of potato starch was characterized by high consistency.

After treatment with accelerated electron beam, the gel consistency value increased with the increase of the irradiation dose indicating the reduction of this property (Table 4).

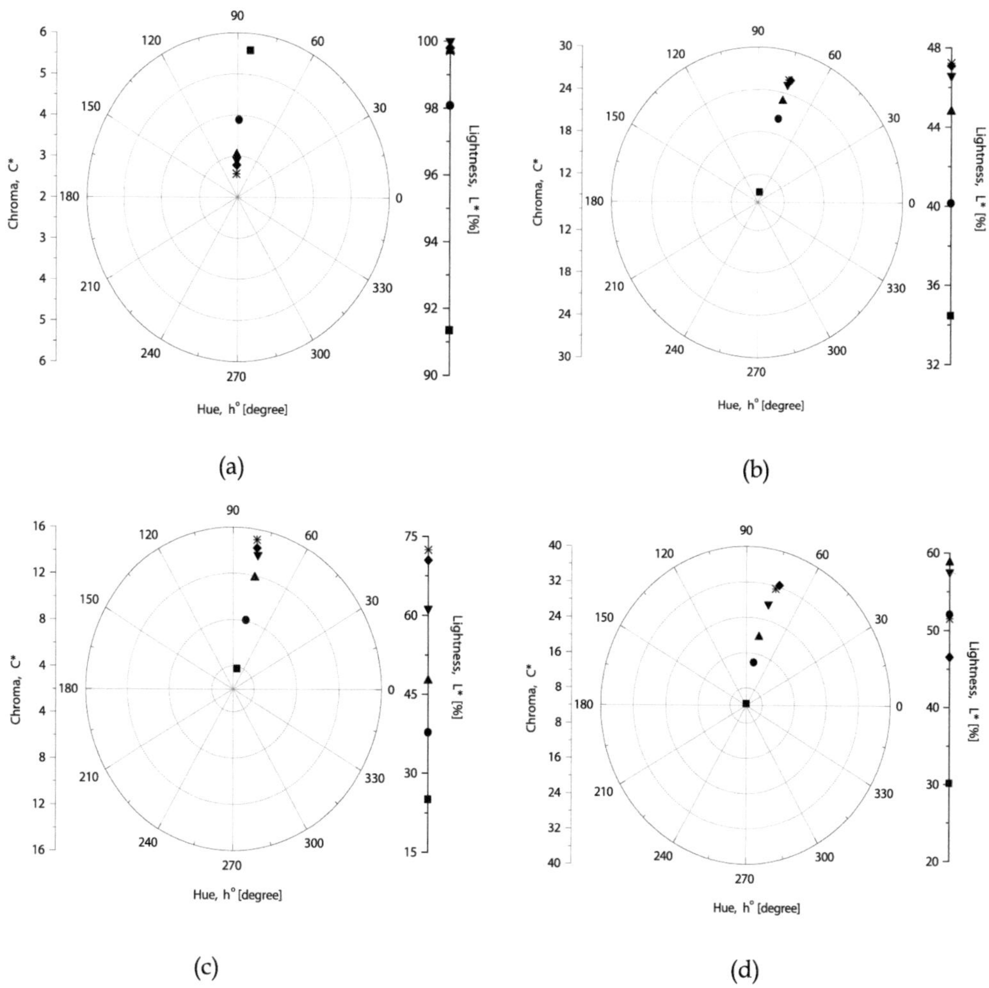

Fig. 4. CIEL*C*h parameters of starch pastes: (a) corn, (b) wheat, (c) rice and (d) potato, for ■ control sample and doses of • 10 kGy, ▲ 20 kGy, ▼30 kGy, ♦ 40 kGy, and * 50 kGy

Irradiation dose [kGy]	Starch source			
	Corn	*Wheat*	*Rice*	*Potato*
0	85 ± 2	88 ± 7	70 ± 7	45 ± 3
10	102 ± 9	108 ± 12	107 ± 10	83 ± 6
20	104 ± 10	112 ± 11	112 ± 9	87 ± 8
30	109 ± 4	116 ± 11	113 ± 9	89 ± 9
40	120 ± 7	128 ± 9	122 ± 8	95 ± 2
50	127 ± 10	132 ± 8	116 ± 12	101 ± 8

Table 4. Gel consistency values in mm, for studied starches

There are studies in literature reporting similar effect for gamma irradiated rice starch (Wu et al., 2002; Yu & Wang, 2007), and besides this, in their study, Yu & Wang (2007) found out that the increase of gel consistency is due to the reduction of the apparent amylose content.

3.5 Rheological properties

Rheological properties of a material are a consequence of the molecular interactions occurring within its molecular structure. During gelatinization, the starch granules swell to several times its initial size and, consequently, their constituents are leached out from inside granules, mainly amylose, leading to a three dimensional network (Steeneken, 1989; Tester & Morrison, 1990). These changes are responsible of rheological characteristics showed by starch suspensions during heating and shear rate. The rheological behavior of starch is led by several factors such as amylose content, granule size distribution, granule form or granule-granule interaction (Okechukwu & Rao, 1995; Kaur et al., 2002; Morikawa & Nishinari, 2002; Singh et al., 2003; Singh & Kaur, 2004). Joslyn (1970) claimed the viscosity is strongly related to molecular chain length and, on the other hand, to space arrangement of molecule determined by intermolecular bonds as well as hydrogen or van de Waals bonds.

The control samples of the studied starches showed non-Newtonian behavior for which the ratio of the shear stress and shear rate is not constant, the shear stress dependence of shear rate indicating the pseudoplastic character of the starch suspensions. Electron beam treatment of starches affected this kind of behavior and it was noticed the tendency toward Newtonian behavior as the irradiation dose increase.

Taking into consideration the experimental results which characterize gel consistency and behavior flow of starches, it was expected the potato starch to show high viscosity in comparison to cereal starches. The determination of the starch apparent viscosity at 25^0 C and 200 s^{-1} confirmed our expectations. Thus, native potato starch exhibited the highest apparent viscosity in the investigated starches while the cereal starches can be classified in the following order: rice starch > corn starch > wheat starch.

The apparent viscosity (25^0 C, $\dot{\gamma}$ =200 s^{-1}) of the starch suspensions presented descending evolution with dose increasing (Fig. 5) for all investigated starches. Thus, the viscosity of corn starch suspension decreased significantly from 294.24±27.95 mPa·s to 16.35±0.82 mPa·s after irradiation with 30 kGy (Fig. 5a). This significant decrease of viscosity can be found again for the other starches. For instance, the wheat starch suspension with initial viscosity value of 136.94±11.92 mPa·s reached to the viscosity of 19.07±1.02 mPa·s (Fig. 5b) after 20 kGy irradiation, and the rice starch suspension with higher initial value of viscosity (321.53±20.15 mPa·s) got to de 16.67±1.53 mPa·s after irradiation with 40 kGy (Fig. 5c). The potato starch viscosity having a very high value of 692.79±34.64 mPa·s, by irradiation with 30 kGy, decreased dramatically to 16.82±1.42 mPa·s (Fig. 5d), value comparable with that of corn starch irradiated with 30 kGy, so indicating high radiosensitivity of potato starch.

Temperature is one of the main interesting variables in technological processes which can affect rheological parameters. Studies of molecular thermodynamics showed that temperature has a direct effect on molecular motion, thus as the temperature increases, molecular motion takes place at a faster rate and consequently viscosity also depends on temperature (Figura & Teixeira, 2007). Viscosity decreases with increasing temperature, and fluidity - the inverse quantity of viscosity - increases with increasing temperature because the degree of molecular motion is different at different temperatures.

The rheological analysis of the investigated starch pastes, at different temperatures, showed that the apparent viscosity ($\dot{\gamma}$ =200 s^{-1}) decreased with increasing temperature, as we expected, for all studied suspensions, both control and e-beam irradiated samples. The obtained results indicated that the dependence of rheological characteristics on temperature is insignificantly influenced by e-beam treatment.

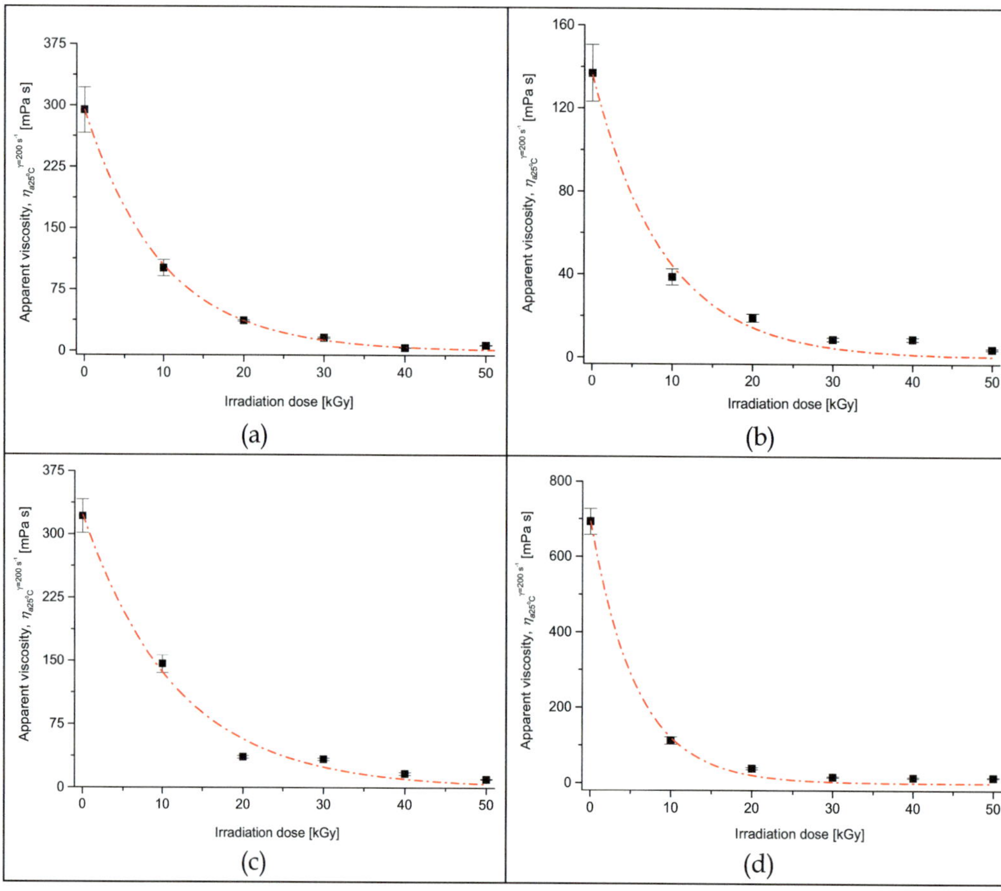

Fig. 5. Variation of apparent viscosity of starch suspensions: (a) corn, (b) wheat, (c) rice and (d) potato, with irradiation dose

3.6 Gelatinization

Gelatinization is one of the most important properties of starch when it is heated in excess of water. Gelatinization phenomenon of starch is practically a phase transition associated with transformation of granule crystalline phase in amorphous one, which causes irreversible changes of different functional properties such as granule swelling power, solubility, lose of optical birefringence. Gelatinization occurs initially in the amorphous regions, as opposed to the crystalline regions, of the granule, because hydrogen bonding is weakened in these areas (Singh et al., 2003). Thus, the differences which can appear in transition temperatures for different kinds of starches may be attributed to the differences in the cristallinity degree. High transition temperatures have been reported to result from a high degree of cristallinity, which provides structural stability and makes the granule more resistant towards gelatinization (Barichello et al., 1990). Both gelatinization and swelling are properties partially controlled by the molecular structure of amylopectin (unit chain length, extent of branching, molecular weight, and polydispersity), amylose to amylopectin ratio and granule architecture (crystalline to amorphous ratio) (Tester & Morrison, 1990).

Differential scanning calorimetry (DSC) is known as an extremely valuable tool to characterize and control the gelatinization phenomenon of starch (Stevens & Elton, 1971) because it provides a quantitative measurement of the enthalpy, ΔH, which means the energy required to take place the gelatinization process, and a determination of temperature range where gelatinization occurs as well.

DSC curves recorded for all starches showed an endotherm peak associated to gelatinization phenomenon and thus, based on it, the temperatures involved in this process and gelatinization enthalpy were determined. It was noticed that the temperatures and gelatinization enthalpy were different for each investigated starch and, as we mentioned previously, these differences are due to the different chemical composition, granule morphology and structure of starches.

It was expected that the possible radioinduced changes in starch structure and suggested by previous analyzed properties to lead also to changes of thermal characteristics of starches. The DSC studies reported (Ciesla & Eliasson, 2002; Ciesla & Eliasson, 2003) in the literature regarding the influence of gamma radiation on gelatinization properties of potato and wheat starches can be added here. The recorded DSC curves for samples treated with electron beam indicated that the height of the peak attributed to gelatinization decreased with dose increasing suggesting a decrease of the gelatinization enthalpy. Also, it was noticed the peak shift which indicates structural changes appeared within macromolecule.

Table 5 shows the temperatures and enthalpy characterizing the gelatinization process for all analyzed starches. We noticed that the e-beam treated starches presented a decrease of the peak gelatinization temperature (T_p) as irradiation dose increasing, excepting the potato starch where this temperature slightly increased. The onset gelatinization temperature (T_o) and conclusion temperature (T_c) of gelatinization process were affected also with the increase of the irradiation dose. Therefore, the onset temperature for cereal starches shifted to lower values simultaneously with the increase of temperature range of gelatinization. On the other hand, the onset temperature of potato starch increased while the temperature range narrowed as irradiation dose increasing.

The gelatinization enthalpy value decreased with the increase of the irradiation dose for all studied samples even the specific temperature range of gelatinization phenomenon modified according the starch source.

Most of the times, the enthalpy is directly associated to the granule cristallinity degree. Taking into account this aspect and the tabled values of gelatinization enthalpy, it might assume that electron beam treatment of analyzed starches led to the reduction of crystalline regions.

We reminded previously that the gelatinization begins in the amorphous regions and after that it affects crystalline regions. Therefore, one can consider the enthalpy of whole gelatinization process as a sum of the enthalpies attributed to amorphous region gelatinization, ΔH_{amorph}, and crystalline region, $\Delta H_{crystalin}$, respectively according the following relation (2):

$$\Delta H = \Delta H_{amorph} + \Delta H_{crystalin} \quad (2)$$

Consequently, in our opinion, the gelatinization enthalpy should be carefully used as a cristallinity index because it is practically the quantity required to the phenomena occurring during gelatinization process like intermolecular bond disruption, granule hydration and swelling, and crystallite melting.

Irradiation dose [kGy]	T_o [°C]	T_p [°C]	T_c [°C]	$T_c - T_o$ [°C]	ΔH [J/g]
			Corn starch		
0	64.1	69.1	72.3	8.2	10.2
10	63.7	66.4	72.6	8.9	7.4
20	63.4	66.9	71.7	8.3	7.8
30	61.0	66.9	69.2	9.2	6.1
40	61.1	65.8	70.2	9.1	6.2
50	60.9	66.5	70.7	9.8	5.1
			Wheat starch		
0	51.2	58.3	62.3	11.1	8.2
10	51.7	58.0	63.2	11.5	7.9
20	52.1	57.0	62.9	10.8	7.1
30	52.3	56.0	64.1	11.8	6.8
40	50.3	57.7	62.9	12.6	6.5
50	50.1	57.1	63.1	13.0	6.8
			Rice starch		
0	61.7	66.2	69.3	7.6	14.0
10	60.8	65.0	72.1	11.3	13.9
20	58.8	65.0	71.6	12.8	12.2
30	57.6	65.3	71.0	13.4	12.3
40	56.9	64.9	70.8	13.9	10.9
50	56.6	64.9	71.1	14.5	10.3
			Potato starch		
0	48.9	58.4	61.7	12.8	17.8
10	50.7	58.8	63.6	12.9	16.5
20	51.7	58.7	62.5	10.8	13.8
30	51.7	59.5	62.8	11.1	13.7
40	53.0	59.2	63.4	10.4	10.7
50	53.7	59.9	63.4	9.7	7.9

T_o – onset temperature; T_p – peak temperature; T_c – conclusion temperature;
T_c - T_o – temperature range of gelatinization; ΔH – enthalpy of gelatinization
Table 5. Starch thermal characteristics observed from DSC curves

3.7 Molecular weight

Molecular weight is a characteristic of polysaccharides which influences most of their physicochemical properties. Polysaccharides contain generally chains of different constituent monosaccharides conferring certain molecular weight distribution. This distribution is variable depending on different factors like the biosynthesis or extraction conditions used to isolate the polysaccharide.

Number average molecular weight (M_n), weight average molecular weight (M_w), z-average molecular weight (M_z) were determined for studied samples in order to reveal the influence of the e-beam treatment on molecular weights and distribution. Table 6 shows the molar mass distribution of the samples treated with electron beam in comparison with the control samples.

Irradiation dose [kGy]	M_n x 10^{-4} [g/mol]	M_w x 10^{-5} [g/mol]	M_z x 10^{-5} [g/mol]	*PI*
Corn starch				
0	5.52 ± 0.45	3.07 ± 0.22	8.36 ± 0.31	5.56 ± 0.23
10	4.19 ± 0.14	2.34 ± 0.02	6.43 ± 0.03	5.58 ± 0.12
20	3.62 ± 0.37	1.78 ± 0.08	5.43 ± 0.02	4.92 ± 0.27
30	3.42 ± 0.21	1.50 ± 0.03	4.75 ± 0.12	4.41 ± 0.37
40	2.87 ± 0.03	1.16 ± 0.02	3.82 ± 0.10	4.04 ± 0.28
50	2.84 ± 0.13	1.12 ± 0.05	3.48 ± 0.39	3.95 ± 0.34
Wheat starch				
0	6.46 ± 0.27	3.43 ± 0.01	7.67 ± 0.01	5.31 ± 0.11
10	3.53 ± 0.02	1.87 ± 0.01	5.87 ± 0.01	5.30 ± 0.04
20	2.58 ± 0.24	1.33 ± 0.01	4.62 ± 0.18	5.20 ± 0.53
30	2.46 ± 0.08	0.94 ± 0.05	3.13 ± 0.14	3.83 ± 0.09
40	2.13 ± 0.20	0.79 ± 0.02	2.37 ± 0.05	3.69 ± 0.34
50	2.31 ± 0.01	0.71 ± 0.07	1.79 ± 0.13	3.07 ± 0.27
Rice starch				
0	7.10 ± 0.03	3.80 ± 0.04	9.14 ± 0.02	5.36 ± 0.10
10	5.38 ± 0.15	2.62 ± 0.09	7.86 ± 0.20	4.86 ± 0.45
20	4.74 ± 0.09	2.24 ± 0.19	6.69 ± 0.33	4.72 ± 0.40
30	4.41 ± 0.43	2.04 ± 0.09	6.17 ± 0.02	4.64 ± 0.27
40	4.31 ± 0.24	1.96 ± 0.03	5.90 ± 0.08	4.55 ± 0.19
50	4.11 ± 0.02	1.86 ± 0.09	5.71 ± 0.25	4.53 ± 0.41
Potato starch				
0	8.85 ± 0.04	4.41 ± 0.04	9.81 ± 0.06	4.98 ± 0.20
10	6.72 ± 0.54	3.67 ± 0.23	9.38 ± 0.17	5.46 ± 0.45
20	5.57 ± 0.44	2.98 ± 0.09	8.50 ± 0.28	5.35 ± 0.54
30	4.16 ± 0.11	2.22 ± 0.01	7.60 ± 0.08	5.33 ± 0.13
40	3.38 ± 0.15	1.83 ± 0.10	6.57 ± 0.07	5.40 ± 0.39
50	3.20 ± 0.14	1.73 ± 0.05	6.01 ± 0.23	5.41 ± 0.50

Table 6. Molecular weight distributions of starch samples

It was noticed the decrease of the molecular weights as the irradiation increase for each studied starch, indicating the degradation phenomenon of the macromolecules. The decreasing evolution of the molecular weights with the irradiation dose suggests the break of polymeric chain and formation of the fragments with different molecular weights, which modify the mass molecular distribution of each sample. Even M_n and M_w decreased concomitantly, though their evolution was differently influenced by irradiation and it was reflected by polydispersity index as well. The polydispersity decreased by e-beam treatment of cereal starches showing the change of their molecular weight distribution so that the M_w decreased faster than M_n with the irradiation dose. For potato starch, it was observed a slight increase of the polydispersity after e-beam treatment showing that M_n decreased faster than M_w. The molecular fractions with low molecular weights appeared with a higher percentage than the fractions with high molecular weight as a consequence of the e-beam treatment.

3.8 Factor analysis

The factor analysis was used in order to find a correlation between the most important properties of the irradiated starches in respect of anticipation of the radio-induced change degree. This statistical method is generally used to analyze large numbers of dependent variables to detect certain aspects of the independent variables (factors) affecting those dependent variables - without directly analyzing the independent variables.

The statistic processing of our experimental data pursued the interdependence relationship of physicochemical properties of starches, among them and with the irradiation dose, being considered as system variables. This analysis assumes the volume reduction (condensation) of the initial variables and a new set of independent variables named factors, with a minimal loss of information. The variables are modeled as linear combinations, plus "error" terms.

In the first analysis stage, taking into account the experimental data introduced in the statistic analysis software, the correlation matrix of those 15 variables (Table 7) was calculated so that to highlight the relationship relevance between every two variables.

Thus, for each studied starch, it was globally noticed that, for instance, the apparent viscosity is strongly positive correlated both with number average molecular weight (r = 0.869) and mass average molecular weight (r = 0.857), less correlated with pH (r = 0.617) and gelatinization enthalpy (r = 0.684), at the same time showing strongly negative correlation with irradiation dose (r = - 0.801) and reduced one with gel consistency (r = -0.687). Also, properties like pH, swelling power, gelatinization enthalpy are positive correlated with number and mass average molecular weights, while they present negative correlation with the irradiation dose. Unlike these, solubility, paste clarity or other visual attributes present positive correlation with irradiation dose and negative correlation with number and mass average molecular weights. It was noticed as well the correlation among thermal properties and colorimetric aspects emphasized especially by strong positive correlation of gelatinization temperature and paste hue (r = 0.802), while it presents strong negative correlation with chroma (r = - 0.827) and reduced one with clarity (r = - 0.441) and lightness (r = - 0.432) of pastes. The gelatinization temperature showed also quite good positive correlation with polidispersity, despite of its quite weak correlation with number and mass average molecular weights.

In the second stage of statistic analysis, based on the variable correlation matrix, it was considered that every analyzed variable represents a factor so that the table of individual values of factors was further generated (Table 8). The factor individual value, known as *eigenvalue,* represents the specific value of variable variance explained by factor and it is given by saturations square sum from this factor.

It was observed that first 3-4 factorial axis bring a high percent of explicative factors of variance, ~90%, while the contribution of the other axis is insignificant so that their omission leads to minor loss of information. To reduce the number of factors, *Kaiser's criterion* was considered. According to this principle the contribution of factors is significant when the eigenvalue is higher than 1. Therefore, the total number of factors (independent artificial variables) was simplified to four, covering ~90% of variance with the following proportions: Factor 1 with ~47%, Factor 2 with ~26%, and Factor 3 and 4 with ~21%.

	η_a	CG	clarity	pH	s_{85}	P_u^{85}	L^*	C	h_o	T_p	ΔH	M_n	M_w	PI	Dose
η_a	1.000														
CG	-0.687	1.000													
clarity	-0.364	0.095	1.000												
pH	0.617	-0.200	0.156	1.000											
s_{85}	-0.489	0.132	0.295	-0.276	1.000										
P_u^{85}	0.858	-0.773	-0.003	0.706	-0.236	1.000									
L^*	-0.424	0.107	0.950	0.152	0.497	-0.010	1.000								
C	-0.401	-0.199	0.689	-0.363	0.231	-0.103	0.596	1.000							
h^o	-0.019	0.656	-0.159	0.398	-0.407	-0.197	-0.163	-0.659	1.000						
T_p	0.204	0.469	-0.441	0.292	-0.329	-0.120	-0.432	-0.827	0.802	1.000					
ΔH	0.684	-0.538	-0.424	0.515	-0.017	0.625	-0.347	-0.460	-0.102	0.164	1.000				
M_n	0.869	-0.584	-0.063	0.831	-0.538	0.876	-0.146	-0.226	0.096	0.112	0.623	1.000			
M_w	0.857	-0.453	-0.210	0.841	-0.619	0.776	-0.284	-0.427	0.285	0.310	0.638	0.965	1.000		
PI	0.336	0.219	-0.385	0.505	-0.563	0.064	-0.414	-0.712	0.759	0.710	0.268	0.383	0.599	1.000	
Dose	-0.801	0.605	0.448	-0.626	0.411	-0.715	0.438	0.440	-0.046	-0.153	-0.851	-0.818	-0.863	-0.528	1.000

Table 7. Correlation matrix of variables

Factor	*Eigenvalue*	% of variance	% cumulative
1	7.029	46.861	46.861
2	3.843	25.618	72.479
3	2.080	13.864	86.343
4	1.111	7.407	93.750
5	0.423	2.823	96.573
6	0.175	1.165	97.739
7	0.164	1.093	98.832
8	0.076	0.504	99.336
9	0.042	0.280	99.616
10	0.026	0.172	99.788
11	0.015	0.100	99.889
12	0.012	0.078	99.967
13	0.003	0.021	99.988
14	0.002	0.010	99.998
15	0.000	0.002	100.000

Table 8. Explained variance (eigenvalues)

Variable	Factor 1	Factor 2	Factor 3	Factor 4
η_a	0.900	0.279	0.111	0.068
CG	-0.456	-0.789	-0.323	-0.094
clarity	-0.462	0.439	-0.754	0.032
pH	0.739	0.126	-0.614	-0.195
s_{85}	-0.576	0.234	0.055	-0.768
P_u^{85}	0.732	0.603	-0.121	-0.054
L^*	-0.494	0.427	-0.713	-0.206
C	-0.616	0.653	-0.131	0.346
h^o	0.295	-0.812	-0.465	0.021
T_p	0.425	-0.803	-0.117	-0.153
ΔH	0.747	0.235	0.249	-0.489
M_n	0.881	0.350	-0.253	0.128
M_w	0.953	0.121	-0.230	0.114
PI	0.644	-0.609	-0.216	0.068
Dose	-0.914	-0.195	-0.141	0.085

Extraction method: principal component analysis, 4 extracted components
Table 9. Unrotated factor loadings

Next stage consisted in generation of matrix of factor loadings (Table 9) which indicate the correlation level existing between each variable and correspondent factor. For a better „outlook" of data and simplification of their interpretation, it was used the rotation method *varimax*, which minimize the number of variables with factor loadings. The matrix of factor loadings after rotation is showed in Table 10. The positive values of coefficients indicated an

increase of variable contribution simultaneously with the coefficient increase, while their negative values indicated the variable contribution decrease with the coefficient increase.
For instance, the coefficient positive values of apparent viscosity, swelling power, pH, gelatinization enthalpy and average molecular weights indicated the contribution increase of these variables to the increase of contribution decrease of these two variables to Factor 1 with the coefficient increase.

Variable	Factor 1	Factor 2	Factor 3	Factor 4
η_a	0.883	-0.018	0.305	0.177
CG	-0.705	-0.643	-0.177	-0.046
pH	0.810	-0.452	-0.341	-0.005
P_u^{85}	0.939	0.178	-0.064	0.022
ΔH	0.776	-0.055	0.385	-0.402
M_n	0.946	-0.098	-0.045	0.267
M_w	0.899	-0.302	0.074	0.287
Dose	-0.870	0.110	-0.362	-0.036
C	-0.247	0.754	-0.532	0.176
h^o	-0.060	-0.955	-0.043	0.214
T_p	0.026	-0.878	0.301	0.024
PI	0.294	-0.793	0.225	0.268
clarity	-0.086	0.197	-0.964	-0.006
L^*	-0.097	0.167	-0.937	-0.247
s_{85}	-0.306	0.255	-0.229	-0.877

Extraction method: principal component analysis
Rotation method: Varimax with Kaiser normalization
Rotation completed in 7 iterations
Table 10. Rotated factor loadings

The casewise factor scores (Table 11) of those four factors extracted previously were obtained in the last stage of the statistic analysis. The score matrix showed the contribution of each factor to the explanation of each sample behavior.
Graphical representation of samples and attributes as against principal components allowed the synthetic correlation among those and the analysis of the sample similitude. All previous remarks can be found again in the mentioned plot. To exemplify, there is displayed the plot for the first two factors (Fig. 6).
Sample distribution in all quadrants showed their differentiation as a whole. Nevertheless, the corn and wheat starch samples treated with irradiation doses over 20 kGy are well grouped and therefore similar, while the rice starch samples are rather isolated from the other samples and distributed each in only one quadrant, being similar only between them.

The potato starch samples kept the group identity even when they were in the same quadrants with other samples.

Sample	Factor 1	Factor 2	Factor 3	Factor 4
c0	1.709	-0.066	0.622	0.586
c10	0.656	0.355	-0.456	0.954
c20	0.143	0.630	-0.565	0.687
c30	-0.325	1.101	-0.652	1.233
c40	-0.751	1.816	-0.339	1.372
c50	-1.081	1.547	-0.581	1.610
w0	1.709	-0.066	0.622	0.586
w10	0.073	0.355	1.126	-1.033
w20	-0.360	0.547	1.017	-0.984
w30	-0.592	1.166	0.907	-1.093
w40	-0.869	0.815	0.913	-1.242
w50	-1.079	1.038	0.829	-1.264
r0	1.709	-0.066	0.622	0.586
r10	0.830	-0.185	-0.051	-1.312
r20	0.408	-0.139	-0.691	-1.354
r30	0.277	-0.366	-1.455	-1.251
r40	0.102	-0.579	-2.373	-0.984
r50	-0.208	-0.328	-2.458	-0.232
p0	1.709	-0.066	0.622	0.586
p10	0.170	-1.335	0.450	0.286
p20	-0.264	-1.422	0.340	0.339
p30	-0.924	-1.559	0.591	0.242
p40	-1.335	-1.434	0.523	0.666
p50	-1.706	-1.763	0.435	1.017

c – corn starch, w - wheat starch, r – rice starch, p – potato starch
10….50 – irradiation dose [kGy]
Table 11. Casewise factor scores

4. Conclusion

The treatment of starch with accelerated electron beam can produce several structural and functional changes. For the studied starch types, these modifications have generally similar tendency, but different quantitatively, depending on the specificity of each starch.

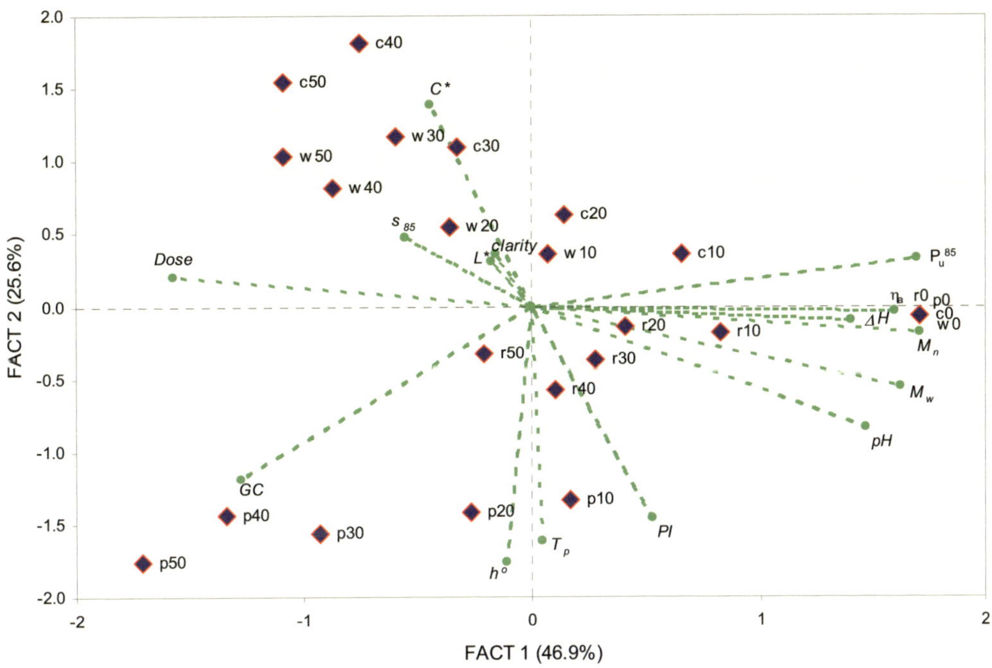

Fig. 6. Factor score plot for Factor 1 and 2

The acidity and solubility increased with the irradiation dose, while the swelling power and gel consistency decreased with the irradiation dose increase. The viscosity - the most important property of starch in applications - showed exponential decrease as irradiation dose increasing with a high radiosensitivity for tuber starch. The paste clarity and its stability improved at the same time with the delay of retrogradation phenomenon as the irradiation dose increase. The gelatinization temperature and enthalpy reduced in a dose-dependent manner. All these radioinduced changes depended on the vegetal source of starch and are globally confirmed also by the changes of the paste colorimetric characteristics.

Also, the radioinduced changes are strongly correlated with the structural organization of starch, especially amylopectin, the component with complex branched structure. Thus, for cereal starches with short chains, the molecular weight distribution was affected especially by formation of the fractions with higher molecular weight than the formation of the fractions with low molecular weight. For potato starch with long chains, the molecular weight distribution is just slightly changed by e-beam treatment, the scissions in fractions with high molecular weight being closer to that of the fractions with low molecular weight.

The statistic interpretation using principal component analysis of studied properties pointed out that the multitude of these properties depended on four independent variables which cover 94% of variance. The different evolution of starch properties could be explained mainly by the contribution of Factor 1 containing the apparent viscosity, gel consistency, pH, swelling power, gelatinization enthalpy, number and weight average molecular masses, and irradiation dose. The change of one of these variables led implicitly to similar changes of the others. The modification of one variable leads inherently to other modifications.

The behavior of the structural and physicochemical properties of treated starches highlights the effectiveness of the applied treatment in order to modify them. Both the choice of the starch type and the degree of the treatment depend on the desired characteristics for the proposed application, so that the final product can meet user's specific needs.

5. Acknowledgement

This work has been partially funded by project PN II 51-007/2007.

6. References

Adeil Pietranera, M.S & Narvaiz, P. (2001). Examination of some protective conditions on technological properties of irradiated food grade polysaccharides, *Radiat. Phys. Chem.*, Vol. 60, Issue 3, (February 2001) 195-201, ISSN 0969-806X

Bao, J. & Corke, H. (2002). Pasting properties of gamma-irradiated rice starches as affected by pH, *J. Agric. Food Chem.*, Vol. 50, Issue 2, (January 16, 2002) 336-341, ISSN 0021-8561

Bao, J.; Ao, Z. & Jane, J.-L. (2005). Characterization of physical properties of flour and starch obtained from gamma irradiated white rice, *Starch/Stärke*, Vol. 57, Issue 10, (October 2002) 480-487, ISSN 0038-9056

Barichello, T.; Yada, R.Y.; Coffin, R.H. & Stanley, D.W. (1990). Low temperature sweetening in susceptible and resistant potatoes: starch structure and composition, *J. Food Sci.*, Vol. 55, Issue 4, (July 1990) 1054-1059, ISSN 1750-3841

Ciesla, K.; Zoltowski, T. & Mogilevsky, L.Y. (1991). Detection of starch transformation under γ–irradiation by smal-angle X-ray scattering, *Starch/Die Starke*, Vol. 43, Issue 1, 11-12, ISSN 0038-9056

Ciesla K. & Eliasson, A.C. (2002). Influence of gamma radiation on potato starch gelatinization studied by differential scanning calorimetry, *Radiat. Phys. Chem.*, Vol. 64, Issue 2, (May 2002) 137-148, ISSN 0969-806X

Ciesla K. & Eliasson, A.C. (2003). DSC studies of gamma irradiation influence on gelatinization and amylose–lipid complex transition occurring in wheat starch, *Radiat. Phys. Chem.*, Vol. 68, Issue 5, (December 2003) 933-940, ISSN 0969-806X

Collision, R. (1968). Swelling and gelation of starch, In: *Starch and its derivatives*, J.A. Radley (Ed.), 4th ed., 168-193, Chapmann & Hall, London

Craig, S.A.S.; Maningat, C.C.; Seib, P.A. & Hoseney, R.C. (1989). Starch paste clarity, *Cereal Chem.*, Vol. 66, No. 3, (May - June 1989) 173-182, ISSN 0009-0352

Dea, I.C.M. (1989). Industrial polysaccharides, *Pure&Appl. Chem.*, Vol. 61, No. 7, 1315-1322, ISSN 0033-4545

De Kerf, M.; Mondelaers, W.; Lahorte, P.; Vervaet, C. & Remon, J.P. (2001). Characterisation and desintegration properties of irradiated starch, *Int. J. Pharm.*,Vol. 221, Issues 1-2, (June 2001) 69-76, ISSN 0378-5173

Duarte, P.R. & Rupnow, J.H. (1994). Gamma-irradiated dry bean (*Phaseoulus vulgaris*) starch: physicochemical properties, *J. Food Sci.*, Vol. 59, Issue 4, (July 1994) 839-843, ISSN 1750-3841

Ershov, B.G. (1998). Radiation-chemical degradation of cellulose and other polysaccharides, *Russ. Chem. Rev.*, Vol. 67, No. 4, 315-334, ISSN 0036-021X

Ezekiel, R.; Rana, G.; Singh, N. & Singh, S. (2007). Physicochemical, thermal and pasting properties of starch separated from γ -irradiated and stored potatoes, *Food Chem.*, Vol. 105, Issue 4, 1420-1429, ISSN 0308-8146

Faust, M. & Massey Jr., L.M. (1966). The effects of ionizing radiation on starch breakdown in barley endosperm, *Radiat. Res.*, Vol. 29, Issue 1, (September 1966) 33-38, ISSN 0033-7587

Figura, L.O. & Teixeira A.A. (2007). *Food Physics. Physical Properties - Measurement and applications*, Springer, ISBN 978-3-540-34191-8, Germany

Gonzalez, Z. & Perez, E. (2002). Evaluation of lentil starches modified by microwave irradiation and extrusion cooking, *Food Res. Int.*, Vol. 35, Issue 5, 415-420, ISSN 0963-9969

Gonzalez Parada, Z.M. & Perez Sira, E.E. (2003). Evaluacion fisicoquimica y funcional de almidones de yuca (*Manihot Esculenta* Crantz) pregelatinizados y calentados con microondas, *ACV (Acta Científica Venezolana)*, Vol. 54, No. 2, (April 2003) 127-137, ISSN 0001-5504

Grant, L.A. & D' Appolonia, B.L. (1991). Effect of low-level gamma radiation on water-soluble non-starchy polysaccharides isolated from hard red spring wheat flour and bran, *Cereal Chem.*, Vol. 68, No. 6, (November - December 1991) 651-652, ISSN 0009-0352

Han, J.A. & Lim, S.T. (2004). Structural changes in corn starches during alkaline dissolution by vortexing, *Carbohydr. Polym.*, Vol. 55, Issue 2, (January 2004) 193-199, ISSN 0144-8617

Hoover, R. (2001). Composition, molecular structure, and physicochemical properties of tuber and root starches: a review, *Carbohydr. Polym.*, Vol. 45, Issue 3, (July 2001) 253-267, ISSN 0144-8617

Jacobson, M.R.; Obanni, M. & BeMiller, J.N. (1997). Retrogradation of starches from different botanical sources, *Cereal Chem.*, Vol. 74, No. 5, (September - October 1997) 511-518, ISSN 0009-0352

Joslyn, M.A. (1970). Viscosity, consistency, and texture. Conductivity measurements and gas analysis, In: *Methods in Food Analysis*, 2nd ed., Academic Press, New York, (cited in Adeil Pientranera, M.S. & Narvaiz, P. (2001). Examination of some protective conditions on technological properties of irradiated food grade polysaccharides, *Radiat. Phys. Chem.*, 60, Issue 3, (February 2001) 195-201, ISSN 0969-806X)

Kamal, H.; Sabry, G.M.; Lotfy, S.; Abdallah, N.M.; Ulanski, P.; Rosiak, J. & Hegazy, E.-S.A. (2007). Controlling of degradation effects in radiation processing of starch, *J. Macromol. Sci. Part A*, Vol. 44, Issue 8, 865-875, ISSN 1060-1325

Kang, J.; Byun, M.W.; Yook, H.S.; Bae, C.H.; Lee, H.S., Kwon, J.H. & Chung, C.K. (1999). Production of modified starches by gamma irradiation, *Radiat. Phys. Chem.*, Vol. 54, Issue 4, (April 1999) 425-430, ISSN 0969-806X

Koksel, H.; Celik, S. & Tuncer, T. (1996). Effects of gamma irradiation on durum wheats and spaghetti quality, Cereal Chem., Vol. 73, No. 4, (July - August 1996) 507-509, ISSN 0009-0352

Kaur, L.; Singh, N. & Sodhi, N.S. (2002). Some properties of potatoes and their starches II. Morphological, thermal and rheological properties of starches, *Food Chem.*, Vol. 79, Issue 2, (November 2002) 183-192, ISSN 0308-8146

Lares, M. & Perez, E. (2006). Determination of the mineral fraction and rheological properties of microwaves modified starch from *Canna edulis*, *Plant Food Hum. Nutr.* Vol. 61, No. 3, (September 2006) 109-113, ISSN 0921-9668

Lewandowicz, G.; Fornal, J. & Walkowski, A. (1997). Effect of microwave radiation on physico-chemical properties and structure of potato and tapioca starches, *Carbohydr. Polym.*, Vol. 34, Issue 4, (December 1997) 213-230, ISSN 0144-8617

Lewandowicz, G.; Jankowski, T. & Fornal, J. (2000). Effect of microwave radiation on physico-chemical properties and structure of cereal starches, *Carbohydr. Polym., Vol.* 42, Issue 2, (June 2000) 193-199, ISSN 0144-8617

Lii, C.-y.; Liao, C.-d.; Stobinski, L. & Tomasik, P. (2002a). Effects of hydrogen, oxygen, and ammonia low-pressure glow plasma on granular starches, *Carbohydr. Polym.*, Vol. 49, Issue 4, (September 2002) 449-456, ISSN 0144-8617

Lii, C.-y.; Liao, C.-d.; Stobinski, L. & Tomasik, P. (2002b). Exposure of granular starches to low-pressure glow ethylene plasma, *Eur. Polym. J.*, Vol. 38, Issue 8, (August 2002) 1601-1606, ISSN 0014-3057

Luo, Z.; He, X.; Fu, X.; Luo, F. & Gao, Q. (2006). Effect of microwave radiation on the physicochemical properties of normal maize, waxy maize and amylomaize V starches, *Starch/Stärke*, Vol. 58, No. 9, (September 2006) 468-474, ISSN 0038-9056

MacArthur, L.A. & D'Appolonia, B.L. (1984). Gamma radiation of wheat. II. Effects of low dosage radiations on starch properties, *Cereal Chem.*, Vol. 61, No. 4, (July - August 1984) 321-326, ISSN 0009-0352

Marcus, R.T. (1999). Colorimetry, In: *Measurement, instrumentation, and sensors handbook*, J.G. Webster (Ed.), 58.1-58.22, CRC Press, Boca Raton, ISBN 3-540-64830-5, Florida, USA

Morikawa, K. & Nishinari, K. (2002). Effects of granule size and size distribution on rheological behaviour of chemically modified potato starch, *J. Food Sci.*, Vol. 67, Issue 4, (May 2002) 1388-1392, ISSN 1750-3841

Nemtanu, M.R.; Brasoveanu, M. & Minea, R. (2005). Studies on modified starches by irradiation, *Proceedings of Euro Food Chem XIII*, vol. 1, pp. 97-101, ISBN 3-936028-31-1, Hamburg, Germany, September 21 - 23, 2005, Gesellschaft Deutscher Chemiker e. V., Frankfurt am Main

Nemtanu, M.R.; Brasoveanu M. & Minea, R. (2006a). Studies on the rheological and functional properties of corn starch treated with accelerated electron beams and electrical discharges, *Proceedings of the 4th International Symposium on Food Rheology and Structure (ISFRS 2006)*, pp. 637-638, ISBN 3-905609-25-8, Zurich, Switzerland, February 19 - 23, 2006, Laboratory of Food Process Engineering, Institute of Food Science and Nutrition, ETH Zurich

Nemţanu, M.R.; Minea, R. & Mitru, E. (2006 b). Application of accelerated electron beams for modifying corn starch, *Elektrotechnica & Elektronica*, 5-6, 224-226, ISSN 0861-4717

Nemtanu, M.R. & Minea, R. (2006). Functional properties of corn starch treated with corona electrical discharges, *Macromol. Symp.*, Vol. 245-246, Issue 1, (December 2006) 525-528, ISSN 1521-3900

Nemţanu, M.R.; Minea, R.; Kahraman, K.; Koksel, H.; Ng, P.K.W.; Popescu, M.I. & Mitru, E. (2007). Electron beam technology for modifying the functional properties of maize starch, *Nucl. Instrum. Meth. Phys. Res. A*, Vol. 580, Issue 1, (September 2007) 795-798, ISSN 0168-9002

Ohta, N. & Robertson, A.R. (2005). *Colorimetry. Fundamentals and applications*, John Wiley & Sons, Ltd, ISBN 0-470-09472-9, USA

Okechukwu, P.E. & Rao, M.A. (1995). Influence of granule size on viscosity of cornstarch suspension, *J. Texture Stud.*, Vol. 26, Issue 5, (December 1995) 501-516, ISSN 1745-4603

Pimpa, B.; Muhammad, S.K.S.; Hassan, M.A.; Ghazali, Z.; Hashim, K. & Kanjanasopa, D. (2007). Effect of electron beam irradiation on physicochemical properties of sago starch, *Songklanakarin J. Sci. Technol.*, Vol. 29, No. 3, (May - June 2007) 759-768, ISSN 0125-3395

Raffi, J.; Michel, J.P. & Saint-Lebe, L. (1980). Theoretical study of the radiopolymerization of starch, *Starch/Die Starke*, Vol. 32, Issue 7, 227-229, ISSN 0038-9056

Ratnayake, W.S.; Hoover, R. & Warkentin, T. (2002). Pea starch: composition, structure and properties - a review, *Starch/Stärke*, Vol. 54, Issue 6, (June 2002) 217-234, ISSN 0038-9056

Richardson, S. & Gorton, L. (2003). Characterisation of the substituent distribution in starch and cellulose derivatives, *Anal. Chim. Acta*, Vol. 497, Issues 1-2, (November 2003) 27-65, ISSN 0003-2670

Sabularse, V.C.; Liuzzo, J.A.; Rao, R.M. & Grodner, R.M. (1991). Cooking quality of brown rice as influenced by gamma-irradiation, variety and storage, *J. Food Sci.*, Vol. 56, Issue 1, (January 1991) 96-98, ISSN 1750-3841

Sarma, S.K.S.S. (2005). Electron beam technology in industrial radiation processing, *IANCAS Bulletin* Vol. IV, No. 2, (April 2005) 128-134

Shishonok, M.V.; Litvyak, V.V.; Murshko, E.A.; Grinyuk, E.V.; Sal'nikov, L.I.; Roginets, L.P. & Krul', L.P. (2007). Structure and properties of electron beam irradiated potato starch, *High Energy Chem.*, Vol. 41, No. 6, (November 2007) 425-429, ISSN 0018-1439

Singh, N.; Singh, J.; Kaur, L.; Sodhi, N.S. & Gill, B.S. (2003). Morphological, thermal and rheological properties of starches from different botanical sources, *Food Chem.*, Vol. 81, Issue 2, (May 2003) 219-231, ISSN 0308-8146

Singh, N. & Kaur, L. (2004). Morphological, thermal and rheological properties of potato starch fractions varying in granule size, *J. Sci. Food Agr.*, Vol. 84, Issue 10, (August 2004) 1241-1252, ISSN 1097-0010

Sokhey, S.; Hanna, M.A.; Wooton, M. & Gallant D.J. (1993). Properties of irradiated starches, *Food Struct.*, Vol. 12, No. 4, 397-410, ISSN 1046-705X

Song, X.Y.; Chen, Q.H.; Ruan, H.; He, G.Q. & Xu, Q. (2006). Synthesis and paste properties of octenyl succinic anhydride modified early *Indica* rice starch, *J. Zhejiang Univ. SCIENCE B*, Vol. 7, No. 10, (October 2006) 800-805, ISSN 1673-1581

Steeneken, P.A.M. (1989). Rheological properties of aqueous suspensions of swollen starch granules, *Carbohydr. Polym.*, Vol. 11, Issue 1, 23-42, ISSN 0144-8617

Stevens, D.J. & Elton, G.A.H. (1971). Thermal properties of starch/water system. I. Measurement of heat of gelatinization by differential scanning calorimetry, *Starch/Die Starke*, Vol. 23, Issue 1, 8-11, ISSN 0038-9056

Szepes, A.; Hasznos-Nezdei, M.; Kovacs, J.; Funke, Z.; Ulrich, J. & Szabo-Revesz, P. (2005). Microwave processing of natural biopolymers—studies on the properties of different starches, *Int. J. Pharm.*, Vol. 302, Issues 1-2, (September 2005) 166-171, ISSN 0378-5173

Tester, R.F. & Morrison, W.R. (1990). Swelling and gelatinization of cereal starches. II. Waxy Rice Starches, *Cereal Chem.*, Vol. 67, No. 6, (November - December 1990) 558-563, ISSN 0009-0352

Wattanachant, S.; Muhammad, S.K.S.; Mat Hashim, D. & Rahman, R.A. (2002). Suitability of sago starch as a base for dual-modification. *Songklanakarin J. Sci.Technol.*, Vol. 24, No. 3, (Jul.-Sep. 2002) 431-438, ISSN 0125-3395

Woods, R.J. & Pikaev, A.K. (1994). *Applied Radiation Chemistry: Radiation Processing*, John Wiley & Sons, Inc., ISBN 978-0-471-54452-3, New York

Wu, D.; Shu, Q.; Wang, Z. & Xia, Y. (2002). Effect of gamma irradiation on starch viscosity and physicochemical properties of different rice, *Radiat. Phys. Chem.*, Vol. 65, Issue 1, (August 2002) 79-86, ISSN 0969-806X

Yu, Y. & Wang, J. (2007). Effect of γ-ray irradiation on starch granule structure and physicochemical properties of rice, *Food Res. Int.*, Vol. 40, Issue 2, (March 2007) 297-303, ISSN 0963-9969

Zou, J.J.; Liu, C.J. & Eliasson, B. (2004). Modification of starch by glow discharge plasma, *Carbohydr. Polym.*, Vol. 55, Issue 1, (January 2004) 23-26, ISSN 0144-8617

18

Bacterial Cellulose-Based Biomimetic Composites

Thi Thi Nge[1,2], Junji Sugiyama[2] and Vincent Bulone[3]
[1]*Present Address: Biomass Technology Research Centre, National Institute of Advanced Industrial Science and Technology, 3-11-32, Kagamiyama, Higashi-Hiroshima, Hiroshima 739-0046;*
[2]*Research Institute for Sustainable Humanosphere, Kyoto University, Uji, Kyoto 611-0011;*
[3]*School of Biotechnology, Royal Institute of Technology, SE-10691 Stockholm*
[1,2]*Japan*
[3]*Sweden*

1. Introduction

As the primary component of plant cell walls, cellulose is a major constituent of plant biomass and represents the most abundant biopolymer on earth. It is also a representative of microbial extracellular polymers synthesized by some bacteria. The cellulose synthesized by the *Gram-negative* bacterium *Gluconacetobacter xylinus* (previously named *Acetobacter xylinum*) has been extensively studied since the pioneering work of Hestrin and Schramm (1954). A single *G. xylinus* cell may polymerize up to 200,000 glucose molecules as a linear β-1-4 glucan chain that are excreted extracellularly as subelementary fibrils. The latter are assembled into microfibrils that aggregate to form ribbons with a hierarchical structure (Hestrin & Schramm, 1954). The matrix of interwoven ribbons constructs a pellicle which has a dense surface side and a gelatinous layer on the opposite side that remains in contact with the liquid medium (Klemm et al., 2001; Nge & Sugiyama, 2007).

Typically, networks of well-separated nano- and microfibrils of bacterial cellulose create extensive surface area and hold a large proportion of water while maintaining a high degree of structural coherence. The water content of never-dried bacterial cellulose pellicles is about 99% (w/w). A high density of inter- and intra-fibrillar hydrogen bonds offers a great deal of mechanical strength. The elastic modulus of dried bacterial cellulose is known to be around 15-30 GPa. Besides being chemically identical to plant cellulose, bacterial cellulose is produced in a virtually pure form free from hemicelluloses, pectins and lignin, which are present in plant cellulosic matrices. Moreover, the in vivo biocompatibility evaluation of bacterial cellulose in rats has demonstrated that it is well integrated into the host tissues and does not elicit any chronic inflammatory reaction, making it a potentially interesting scaffolding material for tissue engineering (Helenius et al., 2006). The unique physical and mechanical properties of bacterial cellulose as well as its purity can be exploited for multiple applications that range from high quality audio membranes, electronic paper, and fuel cell to biomedical materials (Klemm et al., 2001; Czaja et al., 2006).

In the biomedical area, bacterial cellulose can be used for wound healing applications (Czaja et al., 2006), micro vessel endoprothesis (Klemm et al., 2001), scaffolds for tissue engineered cartilage (Svensson et al., 2005) and tissue engineered blood vessels (Bäckdahl et al., 2006). Some of the materials based on bacterial cellulose, such as new skin substitutes and wound dressing materials, are now commercially available (Czaja et al., 2006). Other biomedical applications such as the use of bacterial cellulose as a regenerative aid to correct skeletal defects are under investigation.

Aging is accompanied by the progressive deterioration of the skeletal system, such as in osteoporosis, and by increasing dental problems. There is an increasing demand to design and fabricate high-performance biocompatible materials that mimic the unique quality of natural bones. Bone tissues composite materials composed of specialized cells, an organic matrix rich in collagen fibers and an inorganic mineral phase consisting essentially of calcium and phosphate. The organic fraction synthesized by osteoblasts mediates the formation of apatite (calcium-phosphate minerals) crystals into distinct microstructures that accommodate the mechanical forces encountered in bone tissues (Gilmcher, 1998).

Since the mineral phase of bone is mainly composed of apatite, cements based on synthetic hydroxyapatite (HA; $Ca_{10}(PO_4)_6(OH)_2$) and other calcium phosphate salts are among the most investigated materials for dental and orthopaedic applications in reconstructive surgery. Because of their biocompatibility, moldability, bone bonding ability, osteoconductivity, non-toxic and noninflammatory effects (de Groot, 1983), these bioceramics are widely employed as bone substitute materials. For instance outstanding results were reported for the treatment of periodontal osseous defects and alveolar ridge augmentation (Murugan & Ramakrishna, 2004). However, these materials have some disadvantages, such as brittleness, rapid resorption, migration of cement particles from the implant site and considerably inferior mechanical strength compared to natural bone, which limit their application as load bearing structures at the site of skeletal defects.

Fabrication of matrix-assisted bioceramic composites by in situ precipitation or bone-like apatite coating of biocompatible and/or biodegradable polymers by biomimetic approaches has become an attractive alternative method to improve the mechanical stability and strength of the desired implant. The polymer matrices also serve as binders to prevent migration of the cement particles, in addition to enhance mechanical integrity owing to their toughness and flexibility. Biomimetic approaches involve soaking polymer matrices in a simulated body fluid (SBF) whose ion concentration is equal to that of human blood plasma at physiological pH and temperature. Alternatively, the polymer matrices are soaked at 37°C in calcium and phosphate solutions prepared with a Ca/P ratio of 1.67, similar to the Ca/P ratio of synthetic HA. The concept of apatite formation includes (i) the heterogeneous nucleation of apatite crystals in the presence of surface functional groups of polymer matrices and (ii) the increased supersaturation of the surrounding fluid to accelerate the nucleation process and growth. Several studies have reported the use of various kinds of organic polymers, e.g. silk fibre (Takeuchi et al., 2003), gelatin (Bigi et al., 2002), phosphorylated chitin fibres (Yokogawa et al., 1997) and polylactic acid (Maeda et al., 2002).

In the present study, the biocompatible polymer bacterial cellulose (BC) with its ultra-fine network nature, abundant surface hydroxyl groups and natural mechanical properties is used to investigate the possibility of fabricating BC-based materials that mimic the properties of collagen fibrils in bone. It is widely accepted that surface chemical structures exert a significant influence on the formation of apatite layers on materials surfaces in physiological conditions. They promote cells attachment, proliferation, growth and differentiation when the materials are exposed to biological environments. Surface modification of BC and engineering of the culture medium of *G. xylinus* were performed to render the BC materials biomimetic. Surface modification was performed by TEMPO (2,2,6,6 – tetramethylpyperidine-1-oxyl)-mediated oxidation to introduce carboxyl functional groups, which are known to be effective inducers of apatite nucleation. TEMPO, a water-soluble and stable nitroxyl radical, is well-known for its catalytic and selective oxidation of primary hydroxyl groups of polysaccharides under aqueous conditions (Saito & Isoagi, 2004; Saito & Isogai, 2005; Montanari et al., 2005). Culture medium modification was performed by the addition of an amino sugar (*N*-acetylglucosamine) in the culture medium normally used for BC production. This amino sugar moiety is the monomer unit of chitin, the second most abundant natural biopolymer, which is commonly found in shells of marine crustaceans and cell walls of fungi (Roberts, 1992). It also shares the structural feature of glycosaminoglycan (GAGs) which are unbranched heteropolysaccharides consisting of the repeating unit [uronic acid-amino sugar]. GAGs are extracellular matrix (ECM) constituents of skeletal tissues, in addition to collagen fibrils and apatite minerals. GAGs are considered to play an important role in stimulating chondrogenesis by modulating chondrocyte morphology, differentiation, and function (Suh & Matthew, 2000; Di Martino et al., 2005).

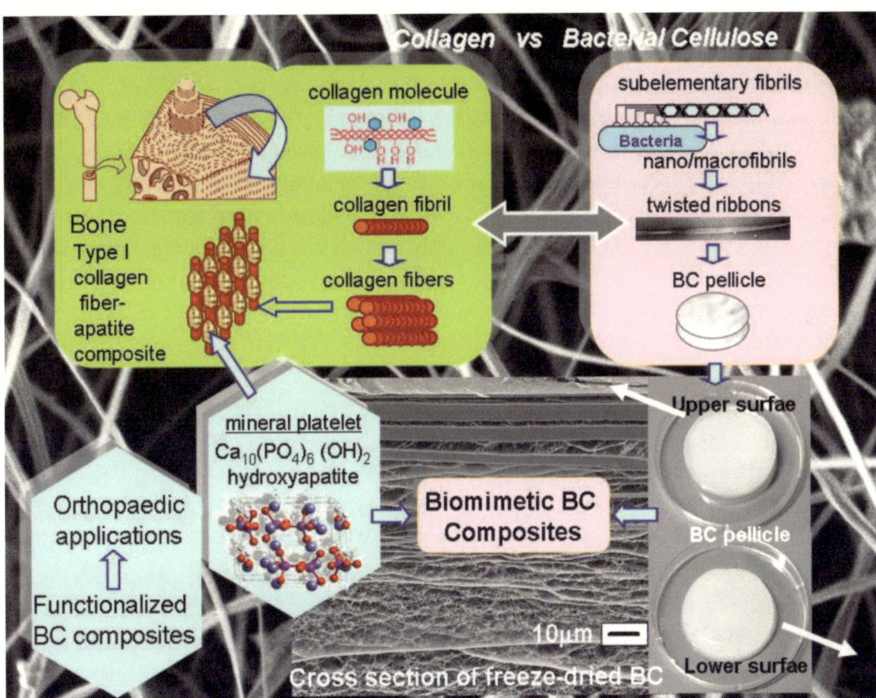

Fig. 1. Schematic representation of bacterial cellulose-based biomimetic composites with potential osteological applications

Considering the function of GAGs in ECM of skeletal tissues, our strategy was to attempt incorporating an amino sugar in bacterial cellulose to prepare a biomimetic ECM. The overall objective is to fabricate a new generation of BC-based biomimetic composites with potential osteological applications and to study the morphological and structural properties of the modified BC. The workflow comprises the following three main parts (see also Figure 1):

1. production of bacterial cellulose from standard culture medium as well as modified culture medium (addition of *N*-acetylglucosamine), and surface modification by TEMPO-mediated oxidation;
2. apatite formation by soaking the surface modified bacterial cellulose in a simulated body fluid;
3. apatite formation by alternatively soaking the surface modified bacterial cellulose and the bacterial cellulose containing *N*-acetylglucosamine in calcium and phosphate solutions.

2. Preparation of bacterial cellulose

2.1 Microbial synthesis of bacterial cellulose

The bacterial strain *Acetobater aceti* (AJ12368), a cellulose producing species closely related to *G. xylinus*, and the Schramm-Hestrin (SH) medium (Hestrin & Schramm, 1954) were used for bacterial cellulose (BC) production. The culture medium consists of 2% glucose, 0.5% peptone, 0.5% yeast extract, 0.27% sodium hydrogen phosphate (Na_2HPO_4) and 0.115% citric acid monohydrate. The pH of the medium was adjusted to 5 with 1N HCl and autoclaved at 120°C for 20 min. The cells used as the inoculum were statically pre-cultured in test tubes at 27°C for two weeks. The thick gelatinous membrane (BC pellicle) was squeezed aseptically to remove cells embedded inside the pellicle. A 25 mL cell suspension (~1.3×10^7 cells/mL) was then transferred to 500 mL fresh medium and successively distributed (9 mL/well) into 6-well culture plates (Iwaki-Asahi Techno Glass, Japan). The BC pellicles (~35 mm Φ) recovered after static culture at 27°C for 9 days were purified in two steps. First, the pellicles were placed in boiling distilled water for 1h, followed by an incubation in 0.1 M NaOH at 80°C for 2h. The pellicles were then washed with distilled water until neutral pH was reached. Some of the purified pellicles were used for TEMPO-oxidation and some were subjected to freeze-drying after having been flash-frozen in liquid nitrogen. Hereafter, the freeze-dried BC is designated as "native" BC.

2.1.2 Surface modified bacterial cellulose (BC-TEMPO and BC-TEMPO-Ca)

Surface modification of the never-dried BC was performed by TEMPO-mediated oxidation to introduce carboxyl functional groups at C(6) primary hydroxyl groups of the cellulose chains (Nge & Sugiyama, 2007). The never-dried BC pellicles (dry weight of 0.15 ± 0.003 g) were suspended in distilled water (80 mL) containing TEMPO (0.0075 g) and sodium bromide (0.075 g) under continuous stirring. An appropriate amount of sodium hypochlorite solution corresponding to 2.42 mmole/g cellulose was added slowly. The reaction was carried out for 2h at pH 10.5 and 20°C by addition of 0.5 M NaOH. The oxidation was quenched by adding 2 mL ethanol followed by the addition of 0.1 N HCl to reach neutral pH. Subsequently, the oxidized pellicles were washed with distilled water several times to remove residual reagents and subjected to freeze-drying as mentioned above. Hereafter, the freeze-dried TEMPO-oxidized bacterial cellulose pellicles are designated as BC-TEMPO. The carboxylate content of BC-TEMPO measured by conductimetric titration (Nge & Sugiyama, 2007) was 0.25 mmol/g cellulose. The oxidation conditions used for BC-TEMPO in this study were not the most optimal to limit the level of oxidation.

Some of the TEMPO-oxidized pellicles recovered after washing with distilled water were immersed in 0.1M $CaCl_2$ with stirring at room temperature for 6 h to exchange the sodium counterion of the carboxylate groups to calcium. The ion-exchanged TEMPO-oxidized BC pellicles were subjected to freeze-drying after washing with distilled water several times. Hereafter, these bacterial cellulose pellicles are designated as BC-TEMPO-Ca. The amount of calcium ions in BC-TEMPO-Ca measured by X-ray fluorescence analysis (MESA-500, Horiba Co., Japan) was 0.22 mmol/g cellulose. With a carboxylate content of 0.25 mmol/g cellulose and an ion exchange ratio of approximately 1:1, the formation of BC-$COOCa^+$ was predominant over $(BC\text{-}COO)_2Ca$ (Saito & Isogai, 2005).

2.2 Bacterial cellulose prepared from modified culture medium (BC-GlcNAc)

The amino sugar *N*-acetylglucosamine (GlcNAc) was added in the culture medium as a source of sugar together with the glucose normally used for BC production. The objective was to determine the optimal medium composition that allows the incorporation of a maximum number of GlcNAc residues in BC. The different compositions tested are presented in Table 1 and Figure 2.

	Medium ID	Glucose (mg/100mL)	GlcNAc (mg/100 mL)
SH medium	2000	20	0
Modified SH medium	1604	16	4
Modified SH medium	1010	10	10
Modified SH medium	0416	4	16
Modified SH medium	0020	0	20

* A medium ID was given for the different sugar compositions. The BC produced from the different media are named after the corresponding medium ID

Table 1. Composition of glucose and *N*-acetylglucosamine at a fixed carbon source of 2%

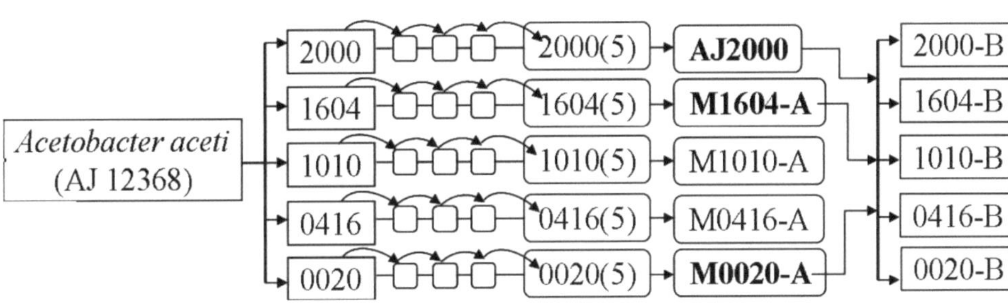

M: modified medium containing GlcNAc (starter culture obtained by repeated transfer (5 times) of inoculum at 3 days of growth)
A: GlcNAc-containing medium/GlcNAc-containing BC (inoculation using the same sugar contents as for the starter culture)
B: GlcNAc-containing BC with different sugar content (inoculation from AJ 2000, M1604 A, and M0020A as starter cultures)

Fig. 2. Experimental setup for *N*-acetylglucosamine (GlcNAc) incorporation in BC; the first 2 digits represent the glucose concentration and the later 2 represent the GlcNAc concentration (mg/100mL).

As shown in Figure 2, adaptation of *Acetobacter* cells to the presence of the amino sugar was performed by 5 repeated transfers of 3-day-old inoculum into the medium containing different GlcNAc concentrations. The inoculum culture time of 3 days was selected based on cell density estimated by measuring the optical density at 600 nm. The volume ratio of inoculum to medium was 1:10 for all compositions. The microbial production of cellulose was performed by static culture fermentation as described in section 2.1.1.
The effect of sugar composition on the rate of cellulose production was studied. The cellulose dried mass produced in the medium devoid of GlcNAc (AJ2000) increased over 14 days and reached 6.41 mg/mL, with an exponential mass production between days 3 to 7 followed by a slower production rate until day 14 (Figure 3a). In the presence of GlcNAc, the lag phase was shorter than with the AJ2000 medium as judged by a higher mass production at days 3 and 4. This lag phase was followed by an increasing mass phase until day 7 and levelling-off. However, the masses produced over 14 days in the media containing the sugars mixtures were lower than in the absence of GlcNAc, with 4.74, 3.33, and 1.29 mg cellulose/mL for M1604-1604A, M1604-1010B and M1604-0416B, respectively. The cellulose production in the medium containing GlcNAc only (M1604-0020B) was low (0.19 mg/ mL) throughout the culture period, most likely reflecting a low cell growth. At a fixed total carbon source of 2%, the decrease in cellulose mass with increasing GlcNAc concentrations reflects the limitation of cellulose formation although GlcNAc enhances the initial production rate. There was a greater consumption of glucose rather than GlcNAc by the cells. The mass therefore decreased with decreasing glucose concentrations at a fixed total carbon source. Shirai et al. also reported the accumulation of GlcNAc during incubation in mixed sugars media. For their analyses, the authors determined the time course of sugar consumption by using Schales' modified procedure for total sugar and a glucose oxidase/peroxidase system for glucose (Shirai, et al., 1994).

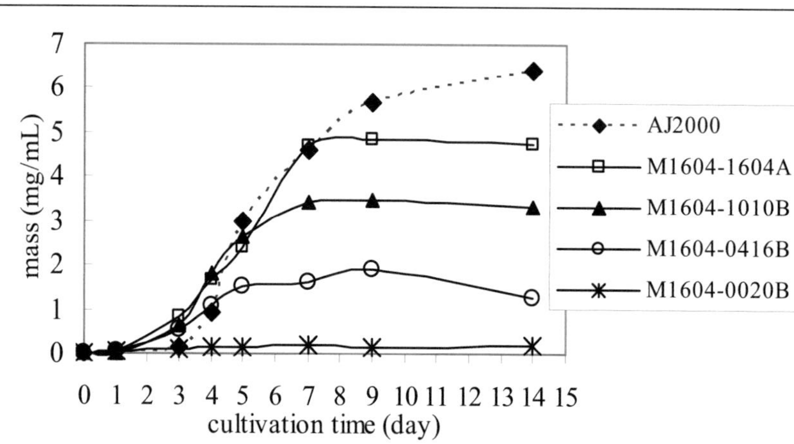

Fig. 3a. Cellulose production in the presence of glucose, mixed sugars, or GlcNAc as a function of cultivation time

The pH (Figure 3b) of the mixed sugar medium decreased rapidly from its initial value of 5.0 to the lowest value of 3.53-3.98 within 4 days, whereas a gradual decrease in pH was observed in the medium containing glucose only (AJ2000), with the lowest pH of 3.5 at day 7. The phases were followed by an increase of the pH value of from 4.9 to 5.3 in the media

AJ2000 to M1604-0416B (Figure 3b). The decrease in pH was due to the conversion of glucose to keto/gluconic acids by a membrane-bound *Acetobacter* dehydrogenase (Klemm et al., 2001). Klemm et al. reported that the gluconic and 5-keto gluconic acids were detected from the second day and third day of cultivation in the culture broth of *G. xylinus* AX 5 by means of HPLC. The acid may affect cellulose production by first lowering the environmental pH and then possibly the intracellular pH, resulting in the stabilization and/or activation of key enzyme(s) for cellulose biosynthesis. On the other hand, a sharp decrease in pH may limit not only cellulose formation, but also lower the medium pH to suboptimal levels for cell viability and cellulose synthesis. The highest mass gain obtained until day 7 (Figure 3a) during the gradual decrease of pH in AJ2000 is consistent with the above assumption. The pH of the medium containing GlcNAc only (M1604-0020B) showed a different trend: the pH remained nearly constant after it had dropped to 3.98 (Figure 3b).

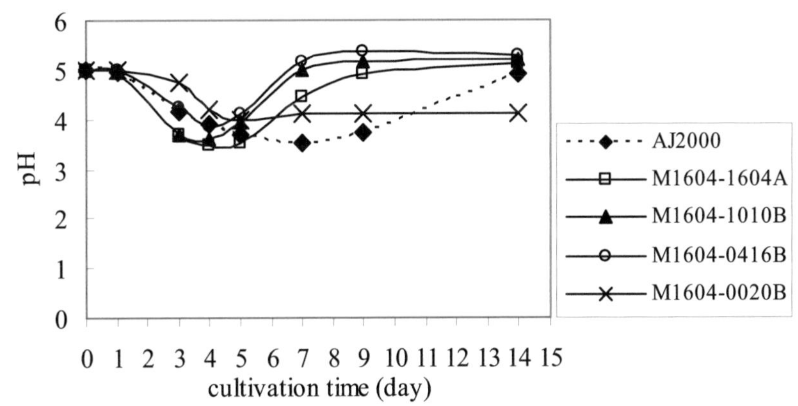

Fig. 3b. pH of the media containing glucose, mixed sugars, or GlcNAc only as a function of cultivation time

Based on the results shown in Figure 3, the cultivation time in further experiments was fixed to 7-9 days for the media containing GlcNAc. Figure 4 shows the effect of the type of starter culture on cellulose dry mass as a function of GlcNAc content in the medium.

In general, the mass decreased with increasing GlcNAc concentrations in all 3 types of starter cultures. Cellulose produced in AJ2000 and M1604A showed the similar trend, whereas a slightly lower mass profile was observed for cellulose produced in M0020A. Hence, cell adaptation was not critical in starter cultures containing small amounts of GlcNAc (M1604A).

2.2.1 Determination of GlcNAc incorporation using a radiolabelling method

A previous study has reported the determination of GlcNAc incorporated in bacterial cellulose by using GC-MS (Lee et al., 2001). Here, the amount of incorporated GlcNAc was examined by a labelling method based on the use of radioactive GlcNAc (1.85 MBq *N*-acetyl-D-[1-^{14}C] glucosamine; Moravek Biochemicals, Inc., California, U.S.A.) and Liquid Scintillation Counting (LSC, Packard Tricarb, Perkim Elmer Inc.). Small-scale labelling was carried out in 1 mL culture media containing different amounts of non-radioactive GlcNAc and inoculated using starter cultures in AJ2000, M1604A, M1010A, M0416A, and M0020A.

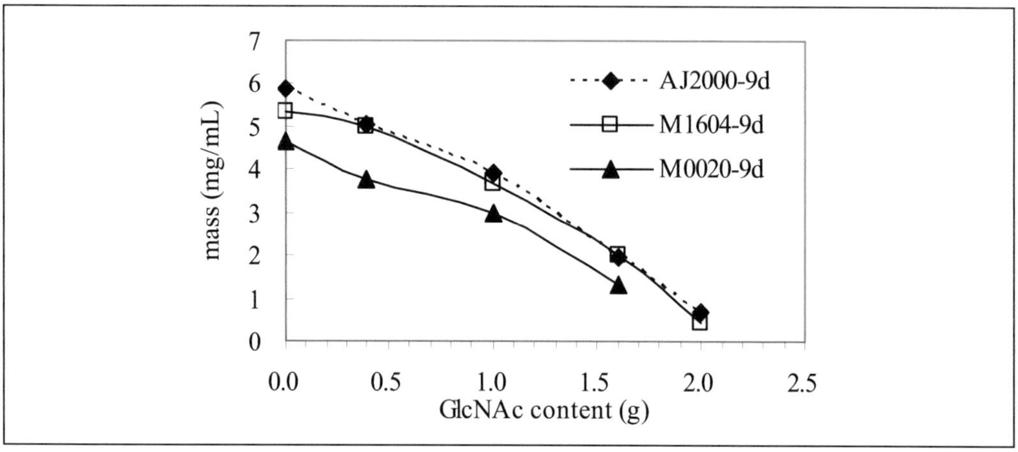

Fig. 4. Cellulose production from different starter cultures as a function of GlcNAc content in the culture medium

The molar concentration of the added radioactive GlcNAc was negligible compared to the concentration of non-radioactive sugar in the culture medium. The static cultures were incubated for 7 days in the same conditions as for bulk production (see section 2.2). The small pellicle formed was taken out at days 2, 3, 5, and 7 followed by purification (washing with distilled water, boiling with 0.1 M NaOH at 80°C for 90 min, and washing with distilled water until neutral pH was reached). After addition of scintillation cocktail (Ultima Gold, Packard) in the vials containing the purified BC pellicles, the radioactivity corresponding to the incorporated GlcNAc was measured by liquid scintillation. The total amount of incorporated GlcNAc was then calculated from the levels of radioactivity measured. Parallel runs were also performed in the absence of radioactive GlcNAc to determine the cellulose dry mass.

The mole% of incorporated GlcNAc as a function of cultivation time from different starter cultures is shown in Figure 5a. A detectable level of GlcNAc incorporated was observed at day 2. The radioactivity measured increased until day 3 in M0416-0416A, M0020-0416B and M0020-1010B; and day 5 in M1010-1010A, M1604-1604A, M0020-1604B, and AJ2000-1604B. They were fairly stable after 5 days. It was concluded that the incorporation of GlcNAc occurred essentially in early incubation times, within 5 days. This finding was in agreement with the higher cellulose mass observed in the media containing mixed sugars compared to the medium containing glucose only (Figure 3a). The level of GlcNAc incorporated after 7 days was in the range 0.6-0.36 mole%, depending on the amount of non-radioactive GlcNAc present in the culture medium and the type of starter culture used.

Generally, the incorporation of GlcNAc decreased with increasing GlcNAc concentrations (decreasing glucose content) in the culture medium. The presence of high GlcNAc concentrations with concomitant decrease in glucose content may limit the activity of cellulose biosynthesis (Figure 3a) since a negligible cellulose mass was produced in the medium containing GlcNAc only. The time course of cellulose production in the medium containing a high concentration of non-radioactive GlcNAc (Glucose : GlcNAc - 0416) showed a similar trend regardless of the type of starter culture used. The effect of the type of starter culture was observed in the glucose:GlcNAc - 1010 and 1604 media. The starter cultures containing the same sugar content (e.g. inoculation of M1604 with an inoculum

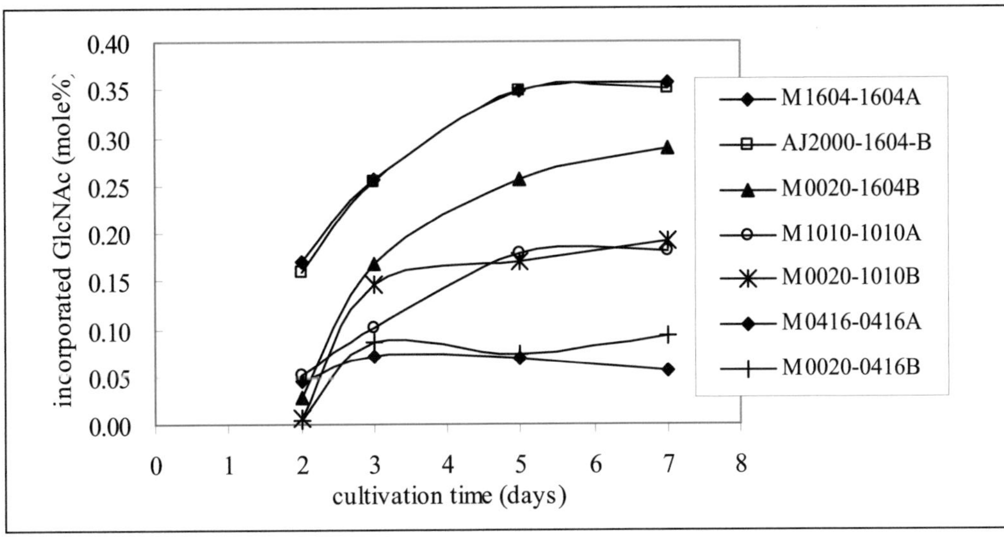

Fig. 5a. Amount of incorporated GlcNAc in bacterial cellulose as a function of cultivation time

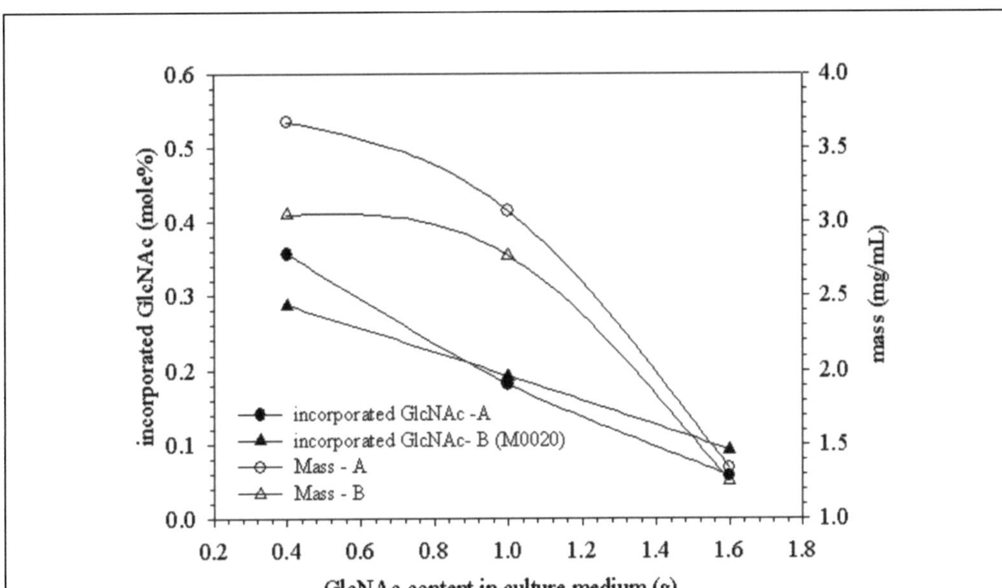

Fig. 5b. Cellulose dry mass and amount of incorporated GlcNAc as a function of non-radioactive GlcNAc content in culture medium (A denotes M1604-1604A, M1010-1010A, and M0416-0416A; B denotes M0020-1604B, M0020-1010B, and M0020-0416B).

prepared in the 1604 medium) led to higher amounts of GlcNAc incorporation than inoculums from the medium containing GlcNAc only (i.e. inoculation of the 1604 medium with a M0020 inoculum). This effect was more significant for the 1604 medium than for the 1010 medium, where a higher incorporation rate was observed throughout the cultivation time. Moreover, the M1604-1604A cultures showed a similar trend as sample AJ 2000-1604B

(direct inoculation from a starter culture containing glucose only). This observation confirmed that no adaptation period in the presence of GlcNAc was necessary to obtain an increased GlcNAc incorporation into cellulose. The 1604 medium is the best suited for high GlcNAc cellulose incorporation and cellulose production under the conditions studied (Figure 5b) although the maximum incorporation obtained (0.36 mole%) is lower compared to other earlier studies (Shirai et.al, 1994; Lee et. al., 2001) which were based on the use of different *Acetobacter* strains and methods.
As only small amounts of GlcNAc can be incorporated during BC biosynthesis, we switched our strategy to the use of chitosan (deacetylated form of chitin) instead of GlcNAc and fabricated bacterial cellulose/chitosan porous scaffolds by freeze-drying TEMPO-oxidized BC microfibrils in suspension in a chitosan solution (Nge et al., 2010). By using this approach, one can control the ratio of BC and chitosan to obtain the desired properties.

2.2.2 Analysis of the morphology of BC-GlcNAc by electron microscopy

Microscopic observations of the synthesized BC-GlcNAc composite was performed by transmission electron microscopy (TEM; JEOL JEM 2000-EXII microscope) and field emission scanning electron microscopy (FE-SEM; JEOL JSM-6700F microscope).
In order to observe the nascent BC microfibrils after ribbon assembly and before the formation of interwoven network mats, on-grid sample preparation was carried out. Drops (10 µl) of cultivation medium inoculated with different starter cultures were placed on carbon-coated copper grids (200-mesh) and incubated at 27 °C for 2 days followed by a gentle washing with distilled water. Negative staining was performed with 2% uranyl acetate. The TEM micrographs of nascent BC microfibrils synthesized in media containing different concentrations of GlcNAc or glucose only are shown in Figure 6.
The network assembly had already formed in M1604-1604A, whereas ribbons were observed in AJ2000 and M1010-1010A. The formation of the network in M1604-1604A even at a 10-µl medium scale was an indication of the faster production rate during early incubation times, in agreement with the mass production profile shown in Figure 3a. A small increase in the lateral dimension of fibril aggregates (~80 - 125 nm) was observed in M1010 compared to AJ2000 (~60 - 105 nm), as judged from the flatten parts of the ribbons. Individual finer cellulose fibrils with lateral dimensions of 18-23 nm (arrow in Figure 6b) were observed in the M1604 medium. Because of network assembly in M1604, the lateral dimensions of the fibril aggregates were found to be in a wider range comprising 18-23 nm, 46-60 nm, 70-90 nm, and ~110 nm bundles.

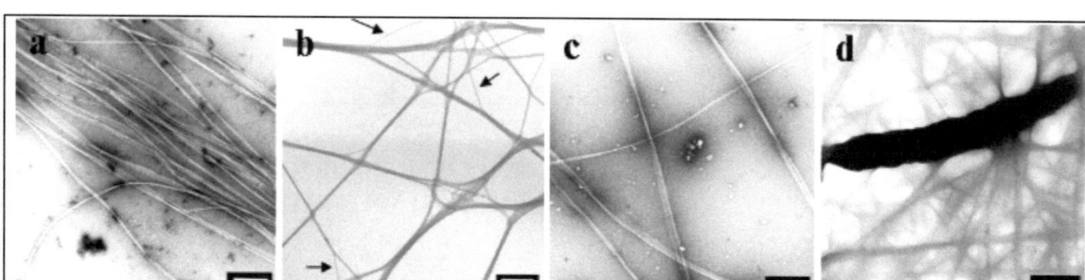

Fig. 6. TEM micrographs of bacterial cellulose ribbons from 2-day cultures of (a) AJ2000, (b) M1604-1604A, (c) M1010-1010A, and (d) an *Acetobacter* cell immobilized within a fibril network (7 day culture of M1010), scale bars = 500 nm

The presence of water soluble polysaccharides in the culture medium can affect the structure and formation of BC (Uhlin et al., 1995). The polymers can interfere with the assembly of microfibrils and bundles during extrusion of cellulose as elementary fibrils. In general, the sorption of these polysaccharides results in less ordered BC, as demonstrated by electron microscopy, X-ray diffraction, FT-IR, and solid-state NMR spectroscopy. Addition of the water soluble amino sugar GlcNAc in the culture medium in our study not only limits the cell activity but also affects the lateral dimensions of fibril aggregates to some extent. The reason of the latter effect is not clear. Either the interaction occurs after the glucan chain has been extruded in the extracellular medium or GlcNAc is first taken up in the intracellular compartment and incorporated in the glucan chains before their extrusion. According to the assumption of Ogawa and Tokura (1992), an acetamido-group-transferring enzyme is present in the bacterium and the UPD-GlcNAc synthesized in the cytoplasm as a precursor of peptidoglycan or lipopolysaccharides becomes a substrate of cellulose synthase.

The BC-GlcNAc pellicles obtained in a 6-well culture plate after 9 days of culture are shown in Figure 7. Because oxygen is required for the aerobic growth of *Acetobacter* cells and for cellulose production, the formation of cellulose occurs only at the upper film/air interface (aerobic zone). It is therefore assumed that the cellulose produced is gradually pushed down, while the new cellulose layers are constantly being built on the top of the mature cellulose (Scharamm & Hestrin, 1954). A surface layer built at the interface between the culture medium and air shows a dense pellicle surface (Figure 8a), especially in AJ2000. Because of the immobilization of the bacteria between the newly formed layer of cellulose fibril network and the gradually sinking older layers during cellulose production, the zone comprising the viable cells that are able to produce cellulose is also gradually drawn into the anaerobic part of the pellicle. The bacteria present in deeper zones are inactive for cellulose production. Therefore, the lower layer that remains in contact with the liquid medium is gelatinous and consists of a loose network. The middle compartment is a stack of several layers and the network density is intermediate to that of the upper and lower surface layers.

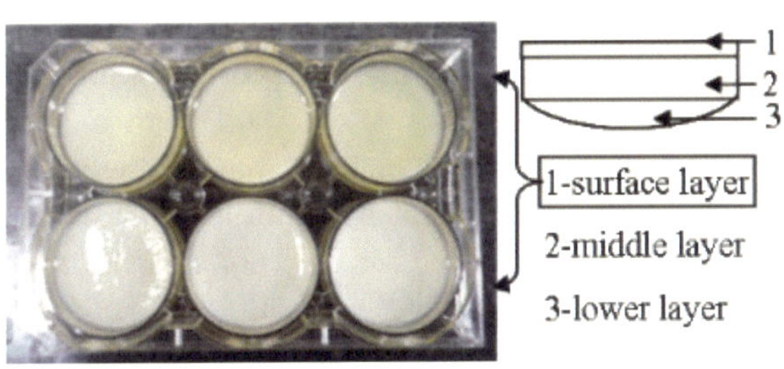

Fig. 7. Typical appearance of bacterial cellulose pellicles after 9 days of culture and schematic drawing of a pellicle layer

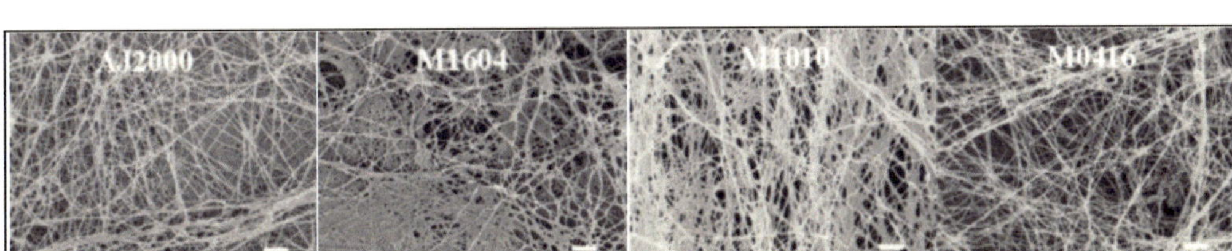

Fig. 8a. FE-SEM images of the surface layer of BC-GlcNAc; scale bar – 1 μm

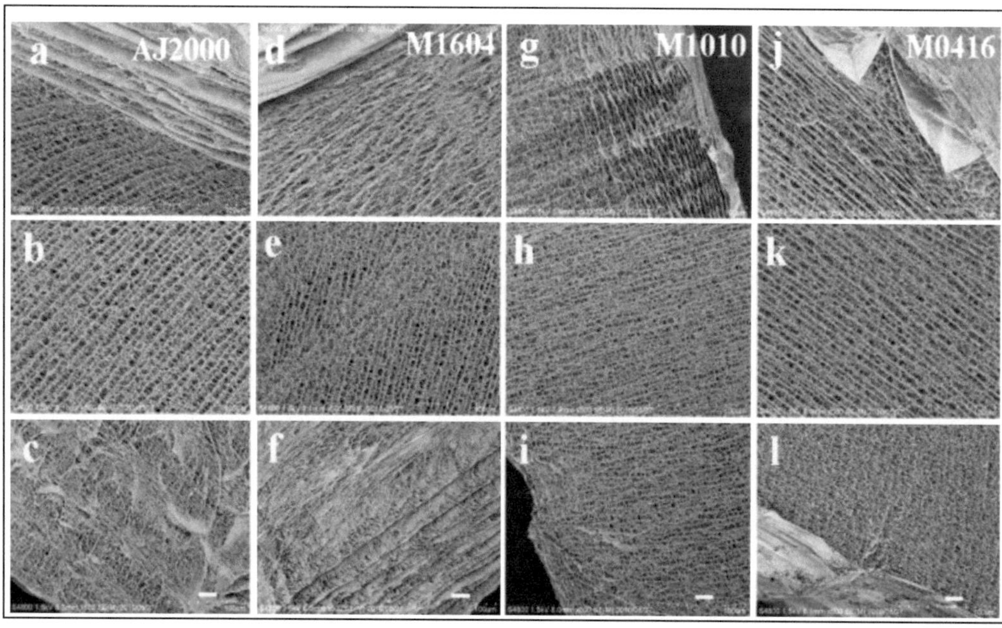

Fig. 8b. FE-SEM images of cross sectional surfaces of freeze-dried BC-GlcNAc pellicles; the first row (a,d,g,j) represents the upper part (pellicle-air interface during cultivation), the second row (b,e,h,k) the middle part, and the third row (c,f,i,l) the lower part (pellicle-culture medium interface) of the pellicles; scale bar - 20 μm

The ultrafine network structure of BC can be seen by FE-SEM in Figure 8b, which presents the cross sectional surface of the freeze-dried BC-GlcNAc pellicles. The physical three-dimensional microfibril network exhibits a hierarchical structure (Figure 8b).

The dense surface layer normally formed in native BC was not seen in M1010 and M0416. A thin surface layer is present instead, while the M1604 layer is similar to that of native BC (AJ2000). The network architecture of the frontier area between the surface layer and middle compartment is also varying between samples obtained from media containing different GlcNAc concentrations. An array of specific network patterns was observed in M1010 and M0416 compared with M1604. The texture of the middle compartment reflects the highly swollen network nature of the BC pellicle in its cultivation state. The never-dried BC pellicle is known to contain ca. 99% of water. This high water holding ability or high hydrophilicity is explained by the presence of pores forming ultrafine network structures, in addition to the occurrence of abundant hydroxyl groups in the cellulose molecular structure. The pore network structure was retained by during freeze-drying. The hollow space or pore size as well as the stacking lamellar pattern are different from each other. Judging from the FE-SEM images, the decreasing order of pore size is AJ2000 ≥ M0416 > M1604 > M1010, with sizes of 4-8 μm, 4-7 μm, 3-7 μm, 2-5 μm, respectively. The thickness of the lower layer also varied from sample to sample, with a few layers of fibril mat in AJ2000 and M1604 and only a thin layer in M1010 and M0416.

Variations in the formation of the dense surface layer and the loose lower layer as well as variations of their thickness reflects the activity of viable cells, which in turn reflects the dry mass of the BC pellicles produced from the respective culture media.

3. Formation of apatite on bacterial cellulose using a biomimetic approach

3.1 Soaking in a simulated body fluid (SBF)

In their study of bioactive glass-ceramic, Kokubo et al. (1990) developed a simulated body fluid (SBF), in which ion concentrations were nearly equal to those of human blood plasma at physiological pH and temperature.

Order	Reagent	Amount	
		SBF	1.5 SBF
		1000 mL	1000 mL
0	ultra-pure water	750 mL	750 mL
1	NaCl	7.996 g	11.994 g
2	$NaHCO_3$	0.350 g	0.525 g
3	KCl	0.224 g	0.336 g
4	$K_2HPO_4.3H_2O$	0.228 g	0.342 g
5	$MgCl_2.6H_2O$	0.305 g	0.458 g
6	1 M HCl	40 mL	60 mL
7	$CaCl_2$	0.278 g	0.417 g
8	Na_2SO_4	0.071 g	0.107 g
9	$(CH_2OH)_3CNH_2$	6.057 g	9.086 g
10	1 MHCl	Appropriate amount for adjusting pH	

Table 2. Reagents and corresponding concentrations used for the preparation of SBF and 1.5 SBF (Kokubo et. al., 1990)

In our study, an SBF containing 1.5 time the ion concentrations of SBF (Table 2; 1.5 SBF) was prepared by dissolving sequentially reagent grade sodium chloride (NaCl), sodium hydrogen carbonate ($NaHCO_3$), potassium chloride (KCl), dipotassium hydrogen phosphate ($K_2HPO_4.3H_2O$), magnesium chloride hexahydrate ($MgCl_2.6H_2O$), calcium chloride ($CaCl_2$), and sodium sulphate (Na_2SO_4) (Wako Pure Chemical Ltd, Osaka Japan) in distilled water. The pH was adjusted to pH 7.4 ± 0.01 with tris-hydroxymethyl aminomethane ($NH_2C(CH_2OH)_3$) and 1 M HCl while maintaining the solution temperature at 36.5°C. The stock solution was filtered through a 0.2 µm cellulose acetate filter and stored at 4°C until used.

Freeze-dried BC samples, namely native-BC, BC-TEMPO, and BC-TEMPO-Ca, with a size of 10×10×5 mm and known weight were soaked in 10 mL of 1.5 SBF for three weeks at 37°C. The SBF solution was renewed every 7 days. Samples were taken out at 2, 7, 14, and 21 days, washed thoroughly with distilled water and freeze-dried. The freeze-dried samples were weighed to determine the extent of mineral deposition. Four to five pieces were used for each BC sample for each designated soaking time.

3.2 Soaking in calcium and phosphate solutions

For alternate soaking processes, 0.05 M $CaCl_2$ (dissolving $CaCl_2$ in 50 mM Tris-HCl buffer at pH 7.4) and 0.03 M NaH_2PO_4 were prepared to obtain a Ca/P ratio of 1.67. Samples of native-BC, BC-TEMPO, BC-TEMPO-Ca and BC-GlcNAc of the same size as above were alternatively soaked in the calcium and phosphate solutions at 37°C for 30 min with reciprocal shaking (50 strokes/min). The samples were rinsed thoroughly with distilled water between each soaking cycle. A total of 5 soaking cycles were performed. Samples were recovered at each soaking cycle and washed thoroughly with distilled water prior to freeze-drying and weighing.

4. Morphological and structural analyses

A field emission scanning electron microscope (FE-SEM, S 4100B, Hitachi Co., Japan and FE-SEM, JEOL, JSM-6700F, Japan) operating at an accelerating voltage of 1.5 kV was used to observe the ultrafine BC networks and the deposited minerals on the BC nano/microfibril surfaces. The BC samples were sputter coated with Pt-Pd and Pt for the FE-SEM-Hitachi and FE-SEM-JEOL microscopes, respectively. Formation of the calcium phosphate phases was determined by Fourier transformed infrared spectroscopy (Spectrum 2000 FT-IR, PerkinElmer) equipped with an attenuated total reflection (ATR) accessory and a single reflection diamond crystal ATR top-plate. The calcium phosphate (Ca/P) ratio of the deposited apatite was measured by inductively coupled plasma atomic emission spectroscopy (ICP-AES, Model SPS7800, Seiko Instrument, Japan). A known amount of apatite-deposited BC samples were soaked in 2% HNO_3 for 1 day to dissolve the calcium-phosphate mineral, followed by filtration through a 0.2 µm cellulose acetate filter and dilution to a concentration of about 5 ppm (detectable range of ICP-AES). The determination of the Ca/P ratios of all samples was based on calibration using synthetic hydroxyapatite (Wako Pure Chemical Ltd., Japan) as a standard.

4.1 BC samples soaked in a simulated body fluid (SBF)

Native-BC, BC-TEMPO and BC-TEMPO-Ca samples were used to investigate the effect of surface functional groups on apatite formation by soaking in 1.5 SBF. The SEM images of the apatite-deposited BC surfaces after 21 days of soaking in 1.5 SBF are shown in Figure 9. Calcium-phosphate minerals appear as hemispherical globules on the BC microfibril surfaces. These morphologies are similar to those reported in other studies using other substrates than BC. It can be seen that the globule sizes are clearly different between native-BC and BC-TEMPO-Ca (about 3 µm and <1 µm, respectively). The variation in size of the apatite globules under the same soaking condition reveals the importance of the microfibril surface structure in the nucleation process of apatite at initial soaking time (Sato et al., 2001; Uchida et al., 2003).

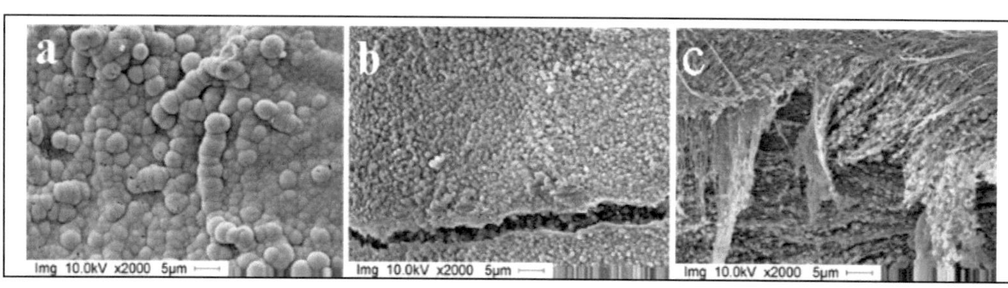

Fig. 9. SEM images of apatite-deposited BC surfaces: (a) Native-BC, (b and c) BC-TEMPO-Ca after 21 days of soaking in 1.5 SBF

The globule size in the BC-TEMPO sample is about 1.5 µm. The small apatite globules firmly adhere on the microfibril surface and form a uniform calcium-phosphate layer (compare Figures 9b and c). The presence of carboxylate groups on the BC microfibril surface (BC-TEMPO) and further ion-exchange treatment with calcium (BC-TEMPO-Ca) significantly affects the rate of apatite nucleation. A large number of nuclei grew competitively on BC-TEMPO-Ca leading to a smaller average size than for the BC-TEMPO and native-BC samples. In other words, multiplication of mineralization or formation of more crystals may occur in BC-TEMPO-Ca, whereas a progressive mineralization by increase in size of the crystals or crystal growth may occur in native-BC, as in bone matrix mineralization and calcification of enamel, respectively (Rey et al., 1991a).

It was found that the apatite globules were composed of a cluster of thin-filmed flakes, which can be seen by TEM (Figure 10a). The specimen was prepared by a slight scratching of the BC-TEMPO surface with a razor blade and dispersion in 10% ethanol. A drop of the sample suspension was loaded on a 200-mesh carbon-coated copper grid. Microdiffraction was performed using a low dose electron probe on a ~100 nm diameter area, with a camera length of 15 cm and 2.8 s exposure. The micrograph and diffraction diagram were recorded on MEM films (Nge & Sugiyama, 2007). Depending on the particular orientation, the crystals exhibit an electron-dense needle-like appearance. The electron diffraction pattern (Figure 10b) taken from crystal clusters on microfibril surfaces indicates the (002) and (211) planes of the apatite crystal (Kim et al, 1996) together with the (004), (200), and (110) planes of cellulose microfibrils (Koyama et al., 1997).

By examining the weight gain at each soaking time interval, it was found that the relative amount of calcium-phosphate phase deposited on the BC microfibril surfaces increased with the soaking time, with a continuous consumption of Ca and P ions from the surrounding fluid. The BC-TEMPO-Ca samples, which exhibited the smallest apatite globules, showed the highest deposit weight, whereas the larger globule size in Native-BC was characterized by the lowest deposited weight after 21 days. The Ca/P molar ratio analysed by ICP-AES was 1.52 ± 0.09, 1.50 ± 0.08, and 1.45 ± 0.04 for Native-BC, BC-TEMPO, and BC-TEMPO-Ca, respectively, after 21 days soaking in 1.5 SBF, while it was of only 1.37 ± 0.08, 1.39 ± 0.02 and 1.31 ± 0.05 for a soaking time of 2 days (Nge & Sugiyama, 2007). Although the Ca/P ratio increased with the soaking time, the values for all the BC samples are lower than the theoretical value of 1.67 for hydroxyapatite ($Ca_{10}(PO_4)_6(OH)_2$).

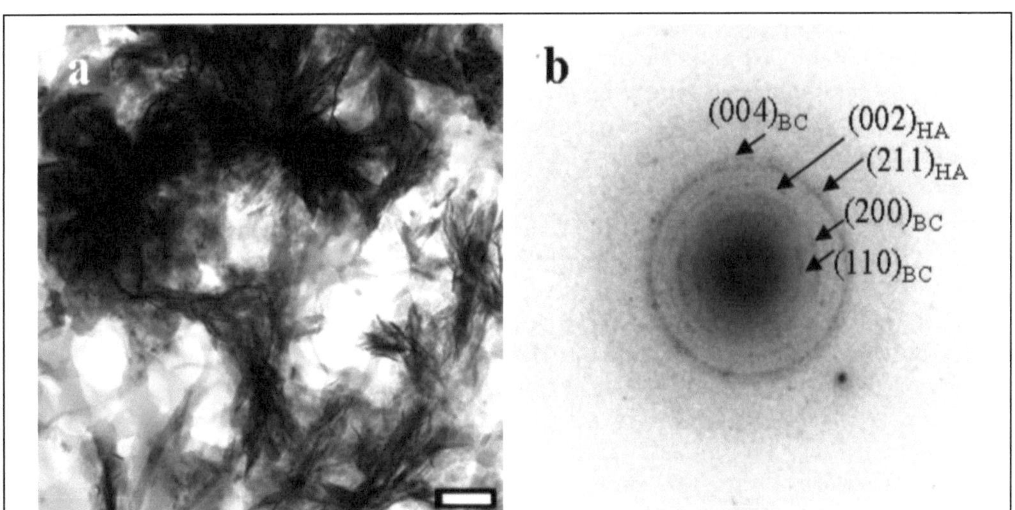

Fig. 10. (a) TEM micrograph of a cluster of apatite crystals deposited on a BC-TEMPO microfibril surface during 21 days of soaking in 1.5 SBF, and (b) Electron diffraction pattern of a cluster of individual crystals from a. scale bar – 500 nm

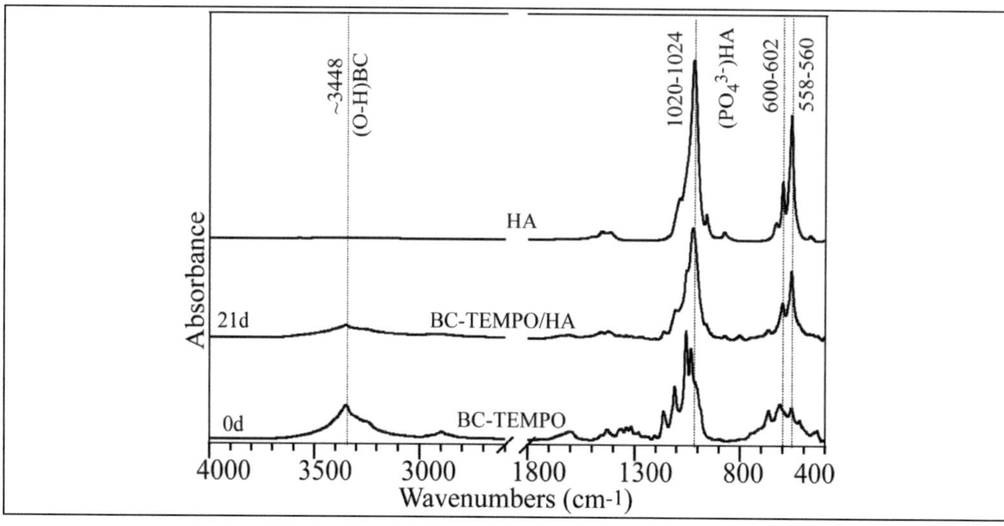

Fig. 11. Typical FT-IR spectra of BC-TEMPO, synthetic hydroxyapatite (HA) and apatite-deposited BC-TEMPO (BC-TEMPO/HA) after 21 days of soaking in 1.5 SBF

The calcium-phosphate phases thus formed are calcium deficient compared to hydroxyapatite. The lower Ca/P ratio was due to the substitution of the trivalent PO_4^{3-} ions by divalent ions such as acid phosphate groups (HPO_4^{2-}) and carbonate groups (CO_3^{2-}). As a result, vacancies at Ca^{2+} cationic sites in the apatite structure occurred during the formation of the calcium-phosphate phase (Elliott et al., 1985; Ray et al., 1989). The detailed information relative to the calcium-phosphate phases obtained with soaking times of 2 d, 7 d, 14 d, and 21 d was revealed by the ATR-FTIR spectral features, as described in our recent paper (Nge & Sugiyama, 2007). The FTIR spectral feature corresponding to the 2-21 days of soaking time indicated that the crystalline phase nucleated on the BC microfibril surfaces was calcium-deficient carbonated apatite resulting from the initial formation of octacalcium phosphate (OCP) or OCP-like calcium phosphate phases. The formation of calcium-deficient carbonated apatite can be explained by the presence of a PO_4^{3-} absorption peak at 1020-1024 cm^{-1} as shown in Figure 11. This peak arises from crystalline imperfections of nonstoichiometric apatite containing HPO_4^{2-} and/or CO_3^{2-} groups (Rey et al.1991b). The stoichiometric hydroxyapatite usually displays this characteristic peak at ~1030 cm^{-1}. In addition, the peaks at 600-602 cm^{-1} and 558-560 cm^{-1} observed in the BC-TEMPO/HA samples are also characteristic of the phosphate absorption band.

4.2 BC samples soaked in calcium and phosphate solutions

Native-BC, BC-TEMPO, BC-TEMPO-Ca, and BC-GlcNAc samples were used in alternate soaking processes. The biomimetic deposition of a calcium-phosphate coating on several BC samples was performed by alternative soaking of 1 cm^2 freeze-dried BC samples in 0.05 M $CaCl_2$ and 0.03 M Na_2HPO_4 at 37 °C, with 50 strokes/min and 30 min/cycle. A total of 5 cycles were performed for all BC samples. The weight-gain of calcium-phosphate deposited on the BC samples at each soaking cycle, the Ca/P ratio and morphology of the Ca-P crystals were determined.

The alternate soaking method showed a similar result as described in section 4.1., where the newly formed calcium-phosphate globules varied in size between the different BC samples.

FE-SEM images of apatite-deposited BC surfaces after 5 soaking cycles are shown in Figure 12. The size of the newly formed apatite globules on native-BC (4-6 µm) was much larger than for BC-TEMPO (1.5- 2 µm). In both cases, the globules were larger than those formed in 1.5 SBF, with respectively ~ 3 µm and ~1.5 µm for the native-BC and BC-TEMPO soaked in 1.5 SBF. The globules on BC-TEMPO-Ca, however, have a similar size with both soaking methods. The larger sizes formed using the alternate soaking method may be due to a relatively higher ion concentration in the calcium and phosphate soaking solutions than in 1.5 SBF, as well as to a more rapid soaking cycle (30 min interval as opposed to 21 days for the 1.5 SBF soaking method). As the lower layer of the BC pellicle (Figure 7) consists in general of a looser network than the surface layer, the apatite crystals formed on the microfibril surfaces in the lower layer exhibit a slightly different morphology. The apatite globules in the BC-TEMPO sample exhibit a more defined shape with a honeycomb-like three-dimensional structure composed of thin plate-like crystals (Figure 12e). Instead, thin plate-like crystals spread upward in all direction in the BC-TEMPO-Ca samples (Figure 12f). It is noteworthy that the surface functional groups or surface structure plays an important role in the initial step of apatite nucleation and governs the following mineralization process by either multiplication or progressive mineralization.

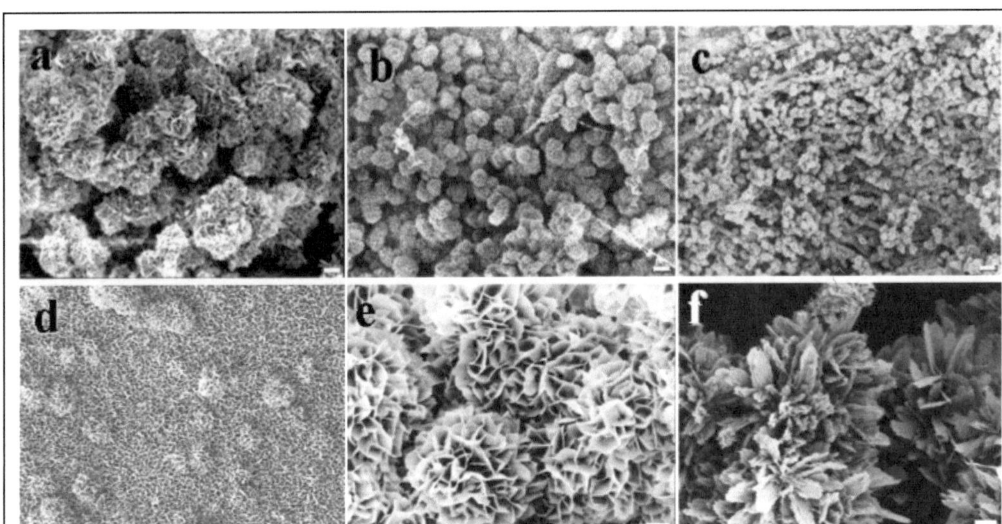

Fig. 12. FE-SEM images of apatite-deposited BC surfaces after 5 cycles of alternate soaking in calcium and phosphate solutions. (a) Native-BC, (b) BC-TEMPO, (c) BC-TEMPO-Ca, (d) BC-GlcNAc (M1604-1604A), (e) magnified view of apatite globules formed on the lower surface of the BC-TEMPO sample and (f) on the lower surface of the BC-TEMPO-Ca sample. Scale bars, 2 µm for a-d, 500 nm for e and f.

The apatite-deposited BC-GlcNAc (M1604-1604A) surface in Figure 12d, however, showed a different morphology compared with native-BC and BC-TEMPO. The crystal growth seemed to spread along the surface, with the initial formation of a monolayer followed by the formation of apatite globules of about 1-1.5 µm on the existing monolayer. The effect of the proportion of incorporated GlcNAc and type of starter culture used for the inoculation on apatite formation was also observed by FE-SEM (Figure 13).

In general, the Ca-P mineralization may take place by spreading as a monolayer along the BC surface layer, followed by further crystal growth upon the existing layer. Sample M1604-1604A, which exhibited the highest GlcNAc incorporation among all BC-GlcNAc samples, showed a distinct morphology. The globules were smaller in the BC samples which contained lower amounts of GlcNAc, such as M1010-1010A and M0416-0416A (Figure 13).

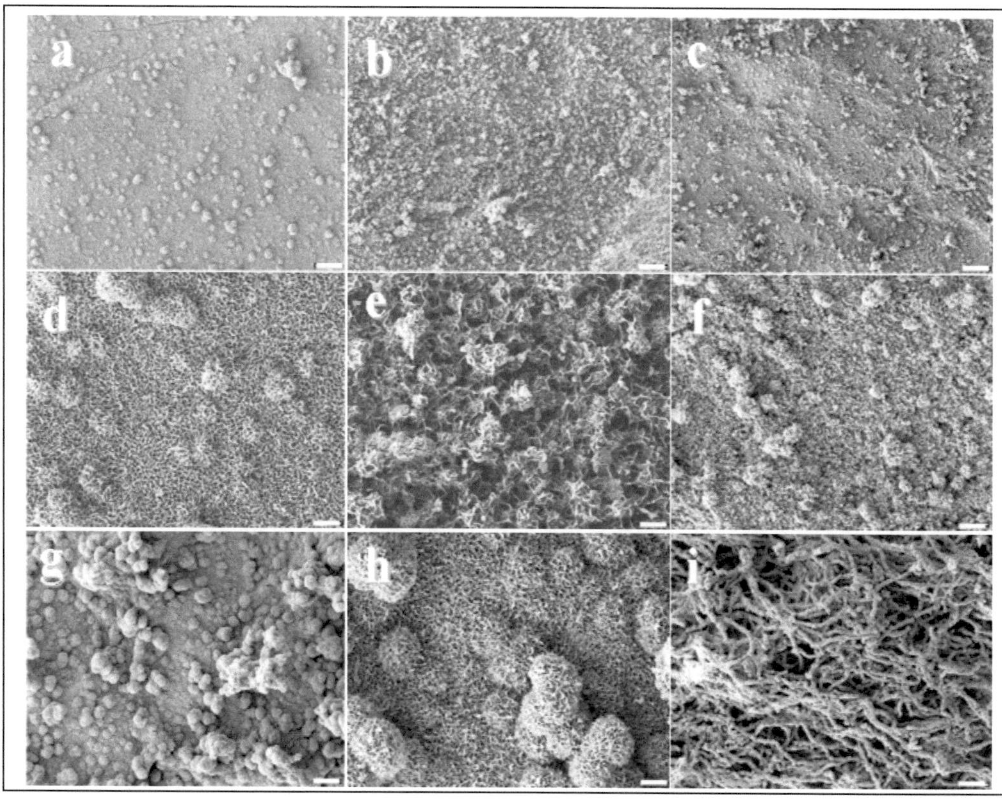

Fig. 13. FE-SEM images of apatite-deposited BC-GlcNAc surfaces after 5 cycles of alternate soaking in calcium and phosphate solutions.
(a,d) M1604-1604A, (b,e) M1010-1010A, (c,f) M0416-0416A, (g,h) surface layer of M0020-1010B, (i) lower layer of M0020-1010B.
d,e,f , and h are high magnification of a,b,c, and g, respectively.
Scale bars: 10 μm for a-c, g, i and 2 μm for d-f, h.

The type of starter culture used for inoculation also affects the formation of apatite. Comparison of Figures 13 b and g (low magnification) or Figures 13 e and h (high magnification) reveals a significant difference in globule size for the BC-GlcNAc 1010 samples. The BC-GlcNAc-M1010-1010A sample in Figures 13 b and e corresponds to an inoculum from starter culture M1010 (Glucose : GlcNAc 10 :10), while the BC-GlcNAc-M0020-1010B sample in Figures 13 g and h was inoculated with starter culture M0020 containing GlcNAc only. The further growth of apatite crystals in M0020-1010B (Figure 13 g, h) led to the formation of globules of about 1-2 μm, whereas globules are likely to grow further in M1010-1010A under the same conditions of soaking. Although the amounts of incorporated GlcNAc are nearly the same in both samples, the type of starter culture used for inoculation may affect the microfibril network structure, which in turn governs the mineral nucleation and further crystal growth. Figure 13i shows the typical morphology of apatite formed on the lower side of the BC microfibril surface in all BC-GlcNAc samples. It was clear that the mineral nucleation took place around the surface of each microfibril/bundle of BC-GlcNAc. The apatite-coated microfibril diameter (width) was found to be about 1 μm. Based on the observed lateral dimensions of fibril aggregates or ribbons of about 100-110 nm (Figure 6) and 1 μm in width after 5 cycles (~ 5 h) of apatite coating, the mineral deposition rate can be estimated to be 3 nm per min. This interesting observation indicates that the surface properties of BC can be tailored by modifying the BC during biosynthesis.

The weight of calcium-phosphate mineral deposited as a function of the number of soaking cycles is shown in Figure 14. The highest amounts of deposit were observed in the native-BC, BC-TEMPO and BC-TEMPO-Ca samples throughout the soaking cycles. Compared to the BC-GlcNAc samples, the surface modified BC-TEMPO and BC-TEMPO-Ca were characterized by the highest amount of deposit. The presence of carboxylate groups affects not only the crystal nucleation process but also the amount of mineral deposited.

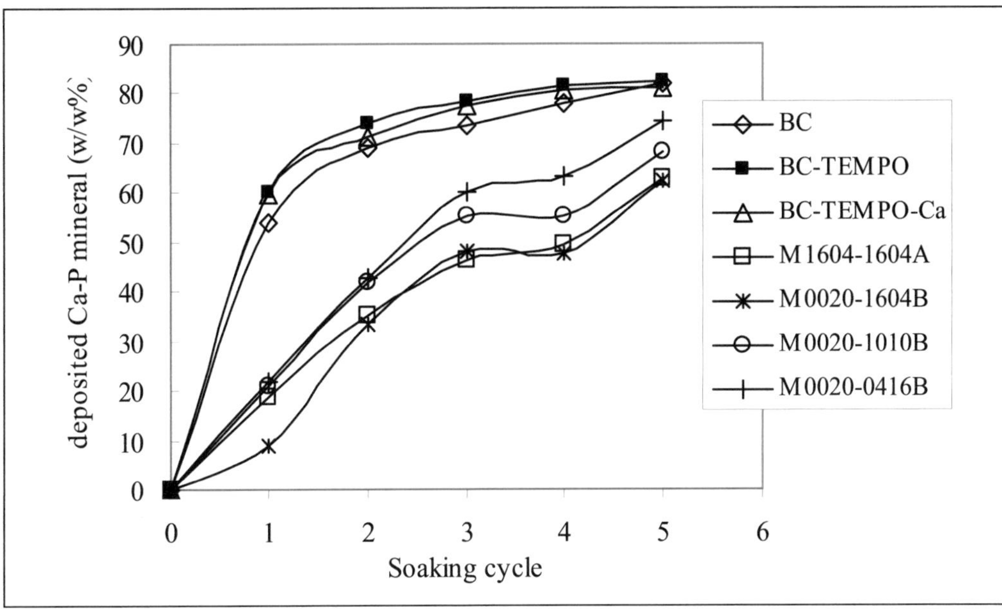

Fig. 14. Relative weight of Ca-P mineral deposit on different BC-GlcNAc as a function of the number of soaking cycles

The BC-GlcNAc samples prepared from the inoculum containing GlcNAc only (M0020) exhibit larger deposits than the samples inoculated with cultures containing various proportions of glucose and GlcNAc. The samples containing lower amounts of incorporated GlcNAc, like M0020-0416B and M0020-1010B, present higher amounts of deposits than M0020-1604B (which exhibits higher incorporated GlcNAc). It can be concluded that the presence of GlcNAc limits the rate of Ca-P nucleation to some extent, although the detailed mechanism remains to be elucidated. The ultrafine network nature of BC (as seen in Figure 8) favours the infiltration of Ca and P ions in the microfibrilar network during the first

soaking cycle and facilitates Ca-P nucleation at the microfibril surface. The presence of incorporated GlcNAc somehow hinders mineral nucleation; as a consequence the overall rate of mineral nucleation on BC microfibril surfaces is slow and the mineral deposit observed at the first soaking cycle is very low. After a gradual increase of the mineral deposition during the first three cycle, the deposition became was fairly stable between the third and fourth cycles in all BC-GlcNAc samples. The entire BC surface was covered by a Ca-P deposit at this stage, which is in agreement with the observation of Ca-P layers in all BC-GlcNAc samples. The increase of Ca-P deposition between the fourth and fifth cycles reflects the further growth of apatite globules visible in FE-SEM images (Figure 13).
The Ca/P molar ratios of all the BC samples analyzed by inductively coupled plasma atomic emission spectroscopy (ICP-AES) are presented in Table 3. Judging from the Ca/P molar ratio, the octacalcium phosphate phase (OCP) is likely to form after the 3rd cycle. Ca/P ratios of all BC samples after the 5th cycle were lower than the theoretical value of hydroxyapatite (1.67). Thus, the Ca-P phase formed on all BC surfaces was a calcium-deficient hydroxyapatite. The BC-GlcNAc samples that were produced from the 0416 medium (Glucose:GlcNAc – 0.4:1.6) inoculated with any type of starter culture (M1604, M0020, M0416) present a Ca/P ratio (1.45-1.51) higher than that of other BC-GlcNAc samples. The highest Ca/P ratio of this sample correlated with its highest deposit weight, as shown in Figure 14.

	1st cycle	3rd cycle	5th cycle
Native-BC	1.25 ± 0.002	1.29 ± 0.003	1.41 ± 0.03
BC-TEMPO	1.23 ± 0.001	1.31 ± 0.002	1.46 ± 0.01
BC-TEMPO-Ca	1.23 ± 0.001	1.32 ± 0.003	1.39 ± 0.01
BC-GlcNAc			
M1604-1604A	NA	NA	1.39 ± 0.01
M1604-1010B	NA	NA	1.46 ± 0.03
M1604-0416B	NA	NA	1.46 ± 0.01
M0020-1604B	NA	NA	1.42 ± 0.03
M0020-1010B	NA	NA	1.41 ± 0.04
M0020-0416B	NA	NA	1.45 ± 0.07
M1010-1010A	NA	NA	1.38 ± 0.04
M0416-0416A	NA	NA	1.51 ± 0.03

Table 3. Ca/P molar ratios of different BC samples after 5 cycles of alternate soaking in Ca and P solutions, as measured by ICP-AES

The surface of the BC microfibril network may be considered as a matrix of chemical groups. Therefore, every monomer unit at the surface is a potential site for interaction with calcium and/or phosphate ions. The accessibility of these units at the surface to the ions in 1.5 SBF or calcium and phosphate solutions depends on the orientation of the polymer chains making up the matrix as well as the pore size and pore distribution within the network. Subtle morphological changes at the surfaces of several BC-GlcNAc during apatite formation as well as varying weight deposits and Ca/P ratios are observed accordingly.

5. Conclusion

Microbial cellulose has proven to be a remarkably versatile biomaterial and can be used in a wide variety of fields, to produce for instance paper products, electronics, acoustics, and biomedical devices. Various biodegradable and biocompatible polymeric materials have recently been investigated to fabricate inorganic-organic hybrid composites by mimicking the mineralization system of natural bone, with some successful outcomes. However, the search for an ideal biomaterial with properties and functionalities similar to natural bone is a continuing process because no single material can satisfy all the requirements for creating optimal scaffolding properties, such as strength, toughness, osteoconductivity, osteoinductivity, controlled degradation, inflammatory response, and deformability. In this study, the ultrafine 3-D BC network structure with its native unique properties is exploited for the synthesis of materials analogous to natural bone. Our study showed that the formation of apatite is dependent on the presence and type of surface functional groups in the microfibrillar BC network.

Degradation of BC has not been fully evaluated in in vitro and in vivo settings. Other cellulose-based materials have however shown limited degradation. Although the complete degradability of materials for tissue engineering applications is very attractive, it is difficult to practically optimize and synchronize the degradation time and mechanical properties of the materials. Modification of BC by incorporation of lysozyme (an enzyme with anti-bacterial action that is found in body fluids, saliva, sweat and tears) susceptible sugars such as analogues of N-acetylglucosamine (GlcNAc) was performed during microbial synthesis. In addition, GlcNAc shares the structure of some repeated disaccharide units of glycosaminoglycans, which are essential components of extracellular matrices. It is expected that the incorporation of GlcNAc will make BC more degradable and more relevant for end use, for instance in the biomedical area. Subtle changes were observed in the formation of apatite deposits on various BC-GlcNAc surfaces. However, the GlcNAc content of the BC-GlcNAc produced was low (0.36 mole%) compared to other studies (4-18 mole%). The bacterial strain used in our study is different from those reported by others. It is possible that the type of cellulose producing strain influences the incorporation of GlcNAc in BC. This observation suggests that it is worth pursuing this type of investigations to tailor the surface properties of BC to meet the main criteria of mineralized collagen composites such as natural bone and teeth. Such investigations should be complemented by in vitro and in vivo degradation studies.

Since BC is an environmentally friendly biopolymer, its use for materials fabrication for a broad range of applications can be envisaged as an alternative to forest resources. A limitation however is the large-scale production of BC-based composites.

6. Acknowledgments

This research was supported by the Japan Society for the Promotion of Science (JSPS) (Grant-in-Aid for Scientific Research number 16004160) and the RISH-Mission project (2006-2007 fiscal year), the Center for Exploratory Research on Humanosphere and the Research Institute for Sustainable Humanosphere at Kyoto University. Prof. Yamanaka from Shinshu University is thanked for the generous gift of the *Acetobacter* strain used in this study. TTN thanks Prof. Isogai and Dr. Saito (University of Tokyo) for their guidance to conduct conductimetric titration, and Dr. Hattori from RISH (Kyoto University) for his guidance to conduct the radio-labelling experiments.

7. References

Bäckdahl, H. ; Helenius, G. ; Bodin, A. ; Nannmark, U. ; Johansson, B.R. ; Risberg, B. ; Gatenholm, P. (2006). Mechanical properties of bacterial cellulose and interactions with smooth muscle cells. *Biomaterials*, 27 :2141-2149.

Bigi, A. ; Boanini, E. ; Panzavolta, S. ; Roveri, N. ; Rubini, K. (2002). Bonelike apatite growth on hydroxyapatite-gelatin sponges from simulated body fluid. *J. Biomed. Mater. Res.*, 59 : 709-714.

Czaja, W. ; Krystynowicz, A. ; Bielecki, S. ; Brown, R.M. Jr. (2006). Microbial cellulose- The natural power to heal wounds. *Biomaterials*, 27 : 145-151.

de Groot, K. (1983). Bioceramics of calcium phosphate, In : Ceramic of calcium phosphate : Preparation and properties, de Groot, K., (Ed.), 100-114, CRC, Boca Raton, FL.

Di Martino, A. ; Sittinger, M. ; Risbud, M.V. (2005). Chitosan : a versatile polymer for orthopaedic tissue-engineering (review). *Biomaterials*, 26 : 5983-5990.

Elliot, J.C.; Holcomb, D.W.; Young, R.A. (1985) Infrared determination of the degree of substitution of hydroxyl by carbonate ions in human dental enamel. *Calcif. Tissue Int.*, 37: 372-375.

Gilmcher, M.J. (1998). The nature of the mineral phase in bone : biological and clinical applicaitons, In : *Metabolic bone disease and related disorders*, Avioli, L.V. ; Krane, S.M., (Ed.), 23-50, Academic Press.

Helenius, G. ; Bäckdahl, H. ; Bodin, A. ; Nannmark, U. ; Gatenholm, P. ; Risberg, B. (2006). In vivo biocompatibility of bacterial cellulose. *J. Biomed. Mater. Res.*, 76A : 431-438.

Hestrin, S. & Schramm, M. (1954). Synthesis of cellulose by *Acetobacter xylinum*, Part 2 : Preparation of freeze dried cells capable of polymerizing glucose to cellulose. *Biochem J.*, 58 : 345-352.

Kim, H.M.; Rey, C.; Glimcher, M.J. (1996) X-ray diffraction, electron microscopy, and Fourier transform infrared spectroscopy of apatite crystal isolated from chicken and bovine calcified cartilage. *Calcif. Tissue Int.*, 59: 58-63.

Klemm, D.; Schumann, D.; Udhardt, U.; Marsch, S. (2001). Bacterial synthesized cellulose-artificial blood vessels for microsurgery. *Prog. Polym. Sci.*, 26: 1561-1603.

Kokubo, T. ; Kushitani, H. ; Sakka, S. ; Kitsugi, T. ; Yamamuro, T. (1990). Solutions able to reproduce in vivo surface-structure changes in bioactive glass-ceramic A-W. *J. Biomed. Mater. Res.*, 24 : 721-734.

Koyama, M.; Helbert, W.; Imai, T.; Sugiyam, J.; Henrissat, B. (1997) Parallel –up structure evidences of the molecular directionality during biosynthesis of bacterial cellulose. *Proc. Natl. Acad. Sci. USA.*, 94: 9091-9095.

Lee, J.W. ; Deng, F. ; Yeomans, W.G. ; Allen, A.L. ; Gross, R.A. ; Kaplan, D.L. (2001). Direct incorporation of glucosamine and *N*-acetylglucosamine into exopolymers by *Gluconacetobacter xylinus* (*Acetobacter xylinum*) ATCC 10245 : Production of chitosan-cellulose and chitin-cellulose exopolymers. *Appl. Environ. Microbiol.*, 67(9) : 3970-3975.

Maeda, H. ; Kasuga, T. ; Nogami, M. (2002). Biomimetic apatite formation on poly(lactic acid) composites containing calcium carbonates. *J. Mater. Res.*, 17 : 727-730.

Montanari, S. ; Roumani, M. ; Heux, L. ; Vignon, M.R. (2005) Topochemistry of carboxylated cellulose nanocrystals resulting from TEMPO-mediated oxidation. *Macromolecules*, 38 : 1665-1671.

Murugan, R. ; Ramakrishna, S. (2004). Bioresorbable composite bone paste using polysaccharide based nano hydroxyapatite. *Biomaterials*, 25(17) : 3829-3835.

Nge, T.T. & Sugiyama, J. (2007). Surface functional groups dependent apatite formation on bacterial cellulose microfibrils network in a simulated body fluid. *J. Biomed. Mater. Res.*, 81A : 124-134.

Nge, T.T. ; Nogi, M. ; Yano, H. ; Sugiyama, J. (2010). Microstructure and mechanical properties of bacterial cellulose/chitosan porous scaffold. *Cellulose*, 17: 349-363.

Ogawa, R. & Tokura, S. (1992). Preparation of bacterial cellulose containing *N*-acetylglucosamine residues. *Carbohydr. Polym.*, 19 : 171-178.

Rey, C.; Collins, B.; Goehl, T.; Dickson, I.R.; Glimcher, M.J. (1989) The carbonate environment in bone mineral: A resolution-enhanced fourier transform infrared spectroscopy study. *Calcif. Tissue Int.*, 45: 157-164.

Rey, C.; Renugopalakrishnan, V.; Shimizu, M.; Collin, B.; Glimcher, M.J. (1991) A resolution-enhanced Fourier transform infrared spectroscopic study of the environment of the CO_3^{2-} ion in the mineral phase of enamel during its formation and maturation. *Calcif. Tissue Int.*, 49: 259-268. (a)

Rey, C.; Shimizu, M.; Collins, B.; Glimcher, M.J. (1991). Resolution-enhanced Fourier-transform infrared spectroscopy study of the environment of phosphate ion in the early deposits of a solid phase of calcium phosphate in bone and enamel and their evolution with age, Part 2: Investigations in υ_3 PO_4 domain. *Calcif. Tissue Int.*, 49: 383-388. (b)

Saito, T. & Isogai, A. (2004). TEMPO-mediated oxidation of native cellulose : The effect of oxidation conditions on chemical and crystal structures of the water-insoluble fractions. *Biomacromolecules*, 5 : 1983-1989.

Saito, T. & Isogai, A. (2005). Ion-exchange behavior of carboxylate groups in fibrous cellulose oxidized by TEMPO-mediated system. *Carbohydr. Polym.*, 61 : 183-190.

Sato, K.; Kogure, T.; Kumagai, Y.; Tanaka, J. (2001). Crystal orientation of hydroxyapatite induced by ordered carboxyl groups. *J. Colloid Interface Sci.*, 240: 133-138.

Schramm, M. & Hestrin, H. (1954). Factors affecting production of cellulose at the air/liquid interface of a culture of Acetobacter xylinum. *J. Gen. Microbiol.*, 11: 123-129.

Shirai, A. ; Takahashi , M. ; Kaneko, H. ; Nishimura, S.I. ; Ogawa, M. ; Nishi, N. ; Tokura, S. (1994). Biosynthesis of a novel polysaccharide by *Acetobacter xylinum*. *Int. J. Biol. Macromol.*, 16(6) : 297-300.

Suh, J.F.K. & Matthew, H.W.T. (2000). Application of chitosan-based polysaccharide biomaterials in cartilage tissue engineering : review. *Biomaterials*, 21 : 2589-2598.

Svensson, A. ; Nicklasson, E. ; Harrah, T. ; Panilaitis, B. ; Kaplan, D.L. ; Brittberg, M. ; Gatenholm, P. (2005). Bacterial cellulose as potential scaffold for tissue engineering of cartilage. *Biomaterials*, 26 : 419-431.

Takeuchi, A. ; Ohtsuki, C. ; Miyazaki, T. ; Tanaka, H. ; Yamazaki, M. ; Tanihara, M. (2003). Deposition of bone-like apatite on silk fiber in a solution that mimics extracellular fluid. *J. Biomed. Mater. Res.*, 65A :283-289.

Uchida, M.; Kim, H.M.; Kokubo, T.; Fujibayashi, S.; Nakamura, T. (2003). Structural dependence of apatite formation on titania gels in a simulated body fluid. *J. Biomed. Mater. Res.*, 64A: 164-170.

Uhlin, K.I.; Atalla, R.H.; Thompson, N.S. (1995). Influence of hemicelluloses on the aggregation patterns of bacterial cellulose. *Cellulose*, 2:129-144.

Yokogawa, Y. ; Pazreyes, J. ; Mucalo, M.R. ; Toriyama, M. ; Kawamoto, Y. ; Suzuki, T. ; Nishizawa, K. ; Nagata, F. ; Kamayama, T. (1997). Growth of calcium phosphate on phosphorylated chitin fibers. *J. Mater. Sci. Mater. Med.*, 8(7): 407-412.

19

Precise Depolymerization of Poly(3-hydroxybutyrate) by Pyrolysis

Haruo Nishida[1], Hidayah Ariffin[2], Yoshihito Shirai[1] and Mohd Ali Hassan[2]

[1]Kyushu Institute of Technology
[2]Universiti Putra Malaysia
[1]Japan
[2]Malaysia

1. Introduction

Poly(3-hydroxybutyrate) (PHB) is a well known microbial and biodegradable polymer, which is accumulated and stored by prokaryotic microorganisms at levels up to 90% of their cellular dry weight (Steinbüchel & Valentin, 1995). PHB has been attracting much interest from researchers not only as an environmentally compatible thermoplastic, but also as a polymeric material obtainable from renewable resources and having a high melting temperature of around 180 °C (Marchessault et al., 1981).

A major problem of PHB when used as a thermoplastic is its thermal instability during melt-processing. Therefore, intense interest has been shown in the thermal degradation of PHB and other related poly(hydroxyalkanoate)s (PHAs). Recently, it has been demonstrated that PHB is a chemically recyclable material with end products such as crotonic acid, linear oligomers having a crotonate end group (Morikawa & Marchessault, 1981), and a cyclic trimer (Melchiors et al., 1996). Melchiors et al. found that cyclic oligomers were obtained via back-biting reactions in a toluene solution with catalysts such as dibutyltin dimethoxide. However, the present review concentrates upon the thermal degradation of PHB in melt.

If plastic materials originating from renewable resources can be efficiently recycled through precise control of their thermal degradation, an ideal recycling system could be constructed for plastic products, in which the resources and production energy of the materials are minimized.

2. Accumulative results on thermal degradation behavior

The thermal degradation behavior of PHB, including other PHAs, has been discussed in many reports with the main studies listed in Table 1.

2.1 Analytical procedures

Several thermoanalytical procedures have been used to investigate the thermal degradation behavior of PHB, including thermogravimetry (TG) for the analysis of weight loss behavior (Kopinke et al., 1999; Galego & Rozsa, 1999; Li et al., 2001; He et al., 2001; Lee et al., 2001; Aoyagi et al., 2002; Carrasco et al., 2006; Kim et al., 2006; Kawalec et al., 2007; Liu et al., 2009;

Ariffin et al., 2008, 2010), differential scanning calorimetry (DSC) for monitoring the heat of reaction (Kopinke et al., 1996), fast atom bombardment mass spectrometry (FAB-MS) (Ballistreri et al., 1989), electrospray ionization mass spectrometry (ESI-MS) (Kawalec et al., 2007), pyrolysis-mass spectrometry (Py-MS) (Abate, 1994; Kopinke et al., 1996), pyrolysis-gas chromatography (Py-GC) (Lehrle, 1994), pyrolysis-GC/mass spectrometry (Py-GC/MS) (Kopinke et al., 1996, 1997; Aoyagi et al., 2002; Ariffin et al., 2008, 2010), TG/Fourier transform infrared spectroscopy (TG/FTIR) (Li et al., 2003), NMR (Melchiors et al., 1996; Kopinke et al., 1996; Ariffin et al., 2009), and pyrolysis-GC/FTIR (Li et al., 2003; Gonzalez et al., 2005), for the analysis of volatile products, and size exclusion chromatography (SEC) (Grassie et al., 1984; Kunioka & Doi, 1990; Nguyen et al., 2002; Kim et al., 2006) for the analysis of changes in molecular weight of residual PHB.

Year	First author	Analytrical procedure	Heating method	Kinetic parameters		
				E_a (kJ mol^{-1})	A (s^{-1})	n / random L
1981	Morikawa, H.	GC, MS, NMR	multiple isothermal heating	-	-	-
1984	Grassie, N.	TG, DSC, TVA, GPC	multiple isothermal heating	247	-	random degradation
1989	Ballistreri, A.	TG, EI-MS, CI-MS, FAB-MS	single constant heating rate	-	-	-
1990	Kunioka, M.	GPC, NMR	multiple isothermal heating	212	-	random degradation
1993	Gogolewski, S.	Injection moulding SEC, DSC	heating at 135 - 160°C	-	-	-
1994	Abate, R.	TG, Py-MS, NMR	single constant heating rate	-	-	-
1994	Hoffman, A.	Mechanical strength, GPC	multiple isothermal heating	-	-	-
1994	Lehrle, R. S.	Py-GC with microthermocouple -controlled filament	isothermal heating at 350°C	-	-	$n = 1$
1995	Lehrle, R.	Py-GC with microthermocouple -controlled filament, MS	isothermal heating at 350°C methanolysis at 84°C	-	-	random degradation + chain-end reactions
1996	Kopinke, F. D.	TG, DSC, Py-MS, Py-GC/MS	single constant heating rate	DSC: 235 Py-GC/MS: 380	1 x 10^{19}	$n = 1, 1.5$
1996	Melchiors, M.	GPC, NMR	multiple isothermal heating, in refluxing solvent	DG^0_{depoly}= -70 kJ mol^{-1} DH^0_{depoly}= -43 kJ mol^{-1}		
1997	Kopinke, F. D.	TG, DSC, Py-MS, Py-GC/MS	single constant heating rate	-	-	-
1999	Galego, N.	TG, FTIR	single constant heating rate	PHB: 296 PHBV: 233-311	PHB: 6.36 x 10^{22} PHBV: 10^{22} - 10^{29}	$n = 1$

Year	First author	Analytrical procedure	Heating method	Kinetic parameters		
				E_a (kJ mol^{-1})	A (s^{-1})	n / random L
2001	Li, S. D.	TG	multiple constant heating rates	-	-	-
2001	He, J. D.	TG	multiple constant heating rates	-	-	-
2001	Lee, M. Y.	TG/DTA	single constant heating rate	PHB: 296 PHA: 253-378	-	β-elimination
2002	Aoyagi, Y.	TG, Py-GC/MS	multiple constant heating rates	111	-	β-elimination
2002	Nguyen, S.	NMR、GPC、FAB$^+$ -MS	Isothermal heating monitoring MW changes	-	-	random degradation
2003	Li, S. D.	TG, TG-FTIR, Py-GC/MS	multiple constant heating rates (Coats-Redfern integral method)	298-307	4.81 x 10^{22} - 4.43 x 10^{25}	n=0.6-0.8
2005	Gonzalez, A.	Py-GC/FTIR, Hi-Res TGA	multiple isothermal heating	-	-	-
2006	Kim, K.-J.	TG, GPC, NMR, Py-GC/MS	multiple constant heating rates isothermal heating	GPC: 182±23 178±10 (Ca$^+$) TG: 140±4 119±2 (Ca+)	-	random chain scission via β-elimination reaction
2006	Carrasco, F.	TG	single constant heating rate	PHB: 304.1 PHBV: 325.4-367.4	PHB: 2.50 x 10^{27} PHBV: 10^{28} - 10^{31}	nth-order reaction
2007	Kawalec, M.	TG, GPC, NMR, MS	plural isothermal heating	-	-	random chain scission E1cB mechanism
2008	Kim, K.-J.	TG, GPC, NMR, Py-GC/MS	multiple constant heating rates isothermal heating	-	-	random chain scission
2008	Ariffin, H.	TG, FTIR, FAB-MS, Py-GC/MS	multiple constant heating rates molecular orbital calculation	117±3 (crude) 135±4 (purified)	3.3 x 10^8 - 1.18 x 10^9 1.9 x 10^{10} - 3.5 x 10^{10}	n = 0 unzipping β-elimination
2009	Ariffin, H.	TG, FTIR, FAB-MS, Py-GC/MS	multiple constant heating rates	-	-	unzipping β-elimination
2009	Liu, Q. S.	TG	single constant heating rate	PHBV: 340 - 550	-	random chain scission
2010	Ariffin, H.	TG, NMR, Py-GC/MS	multiple constant heating rates	122±2 (MgO) 119±3 ($Mg(OH)_2$)	2.9 x 10^{10} 1.2 x 10^{10}	n = 0 - 1 unzipping β-elimination

Table 1. List of studies on thermal degradation behaviors of PHB and PHAs.

2.2 Evaluation of degradation kinetics

2.2.1 Activation energy, E_a

Previous reports have shown that the activation energy (E_a) value of the degradation falls within a wide range of 110-550 kJ mol^{-1}. This range of values can be classified into three categories: 1) 235-550 kJ mol^{-1} calculated with single constant heating rate methods based on weight loss (Kopinke et al., 1996; Galego & Rozsa, 1999; Li et al., 2001; Carrasco et al., 2006; Liu et al., 2009), 2) 111-140 kJ mol^{-1} calculated with multiple constant heating rate methods based on weight loss (Aoyagi et al., 2002; Kim et al., 2006; Ariffin et al., 2008, 2010), and 3) 178-247 kJ mol^{-1} calculated with non-auto-catalytic random degradation analysis based on changes in molecular weight (Grassie et al., 1984; Kunioka & Doi, 1990; Kim et al., 2006).

2.2.2 Single and multiple constant heating rate methods

One reason for the higher values obtained using the single constant heating rate method (e. g., Eq. 1, Kopinke et al., 1996) is probably due to the different reactions proceeding in a mixed and complex manner. If this is true, it is difficult to evaluate the correct E_a value of each elementary reaction, because the E_a value is calculated as an apparent value from the slope of the total plot made up of multiple reactions.

$$\begin{aligned} -dw/dt &= k \cdot \exp(-E_a/RT) \cdot (1-w)^n \\ \ln(-dw/dt) - n \cdot \ln(1-w) &= \ln k - E_a/RT \end{aligned} \tag{1}$$

Thus, this method sometimes produces a meaningless value. On the other hand, if the multiple constant heating rate method is used, which is explained in section 3.1, and if the calculated E_a value is constant in a range of residual weight (w), it can be assumed that a constant reaction state is maintained within the range of w, giving assurance that the obtained E_a value is significant for the reaction. However, if the plot of the calculated E_a value against w shows a slope in the range of w, it indicates that multiple reactions are proceeding in a mixed manner. By using the multiple constant heating rate method, Aoyagi et al. (Aoyagi et al., 2002), Kim et al. (Kim et al., 2006), and our group (Ariffin et al., 2008, 2010) reported the E_a value of PHB thermal degradation as being in a relatively narrow range of 111-140 kJ mol^{-1}. It was proposed that this small variation in E_a value was due to the influence of residual metal ions such as Ca^{2+}, Mg^{2+} and Zn^{2+}.

2.2.3 Estimation from molecular weight change

In order to avoid the influence of some complex factors on pyrolysis at high temperatures, such as the overlapping of multiple reactions and the occurrence of secondary reactions, the E_a value of β-elimination has also been calculated from molecular weight changes in a lower temperature range of 170-200 °C. Interestingly, in spite of the same expected degradation mechanism for β-elimination, obtained E_a values were in a medium range of 178-247 kJ mol^{-1}. Kim et al. (Kim et al., 2006) suggested that a discrepancy in the obtained E_a values, i.e. 140 and 182 kJ mol^{-1} from the weight loss and the molecular weight change, respectively, was caused in part by the auto-acceleration effect of low molecular weight compounds.

2.3 Thermal degradation mechanisms

In Scheme 1, previously reported mechanisms of the thermal degradation of PHB and PHAs are illustrated. Even though many approaches have been taken on the subject of PHB thermal degradation, a clear explanation for the wide variation in E_a value has yet to be provided. Despite this uncertainty, the random β-elimination scission has been widely held as the exclusive degradation mechanism of PHB based on typical structures of pyrolysis products, i.e. crotonic acid and oligomers with a crotonate end-group (Morikawa & Marchessault, 1981; Ballistreri et al., 1989). Recently, an E1cB mechanism proceeding via α-deprotonation by a carboxylate anion to produce the same products was proposed (Kawalec et al., 2007).

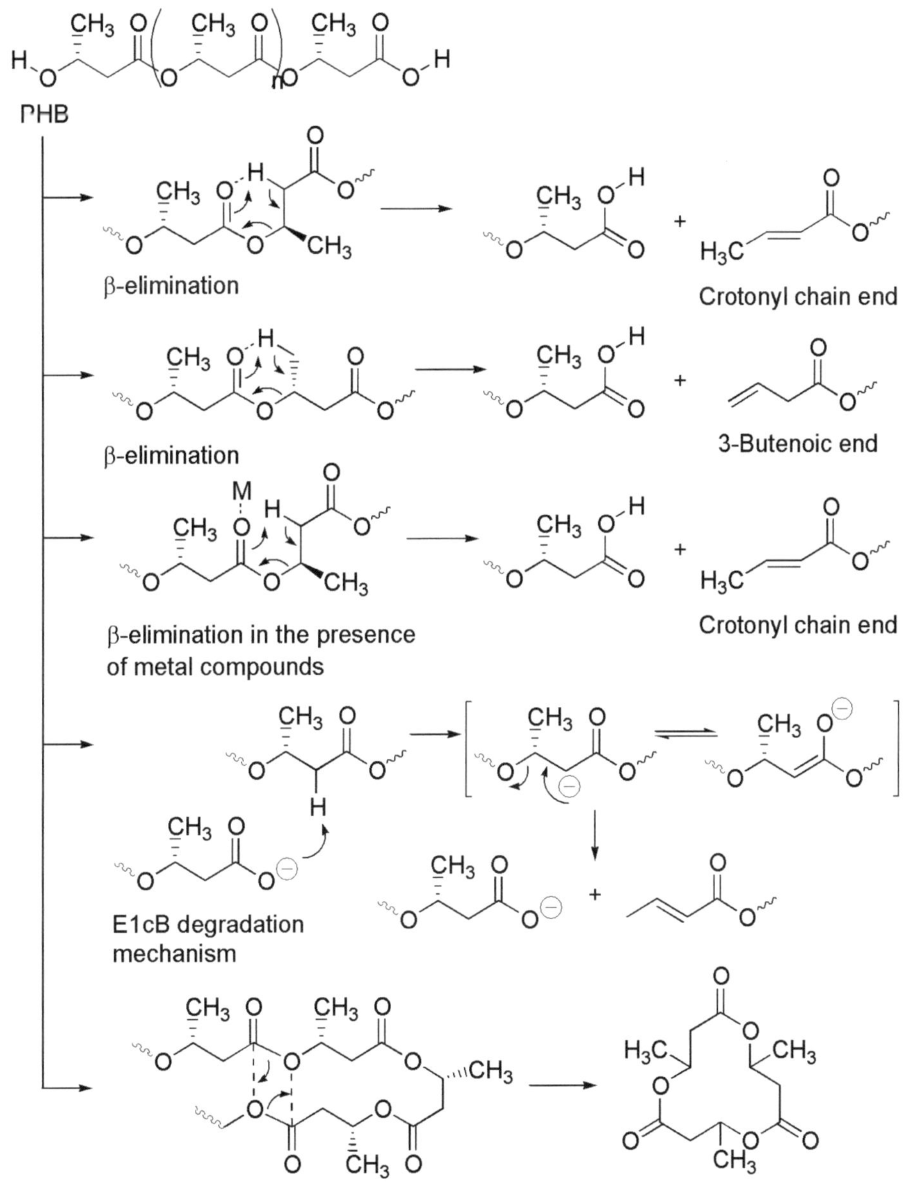

Scheme 1. Reported thermal degradation mechanisms of PHAs.

2.4 Suggestions of other degradation mechanisms

There have been suggestions that some other reactions may be occurring in the course of PHB thermal degradation. Lehrle et al. reported that the analytical results of partial pyrolysis of PHB showed some deviations from theoretical predictions of random scission statistics (Lehrle et al., 1994, 1995). These deviations were probably due to reactions induced by the end groups. Kopinke et al. found that the isothermal degradation curve at 270 °C did not agree with the model curves of random degradation reactions, and that non-isothermal thermogravimetric (TG) and differential TG (DTG) curves of PHB seemed to better fit a first-order degradation reaction curve (Kopinke et al., 1996). However, they also commented that the high E_a value of 380 kJ mol^{-1} obtained by using a single heating rate method was not chemically meaningful and may have been due to an auto-catalytic reaction. Finally, they suggested that the observed degradation behavior was characteristic of a chain reaction with a short kinetic chain length. Galego and Rozsa (Galego & Rozsa, 1999) came to similar conclusions to those of Kopinke et al. (Kopinke et al., 1996). Nguyen et al. (Nguyen et al., 2002) also suggested the occurrence of auto-acceleration in the PHB thermal degradation at 190 °C based on the changes in molecular weight.

3. Precise depolymerization of PHB

Based on abovementioned results, three features are noteworthy in relation to PHB thermal degradation, i.e. 1) during the pyrolysis of PHB, other kinds of degradation mechanisms in addition to β-elimination proceed at the same time, making the kinetics analysis complicated; 2) obtained plausible E_a values may incorporate some parameters of other reactions; and 3) the thermal degradation of PHB may be auto-accelerated.

3.1 Kinetic analysis by a multiple constant heating rate method

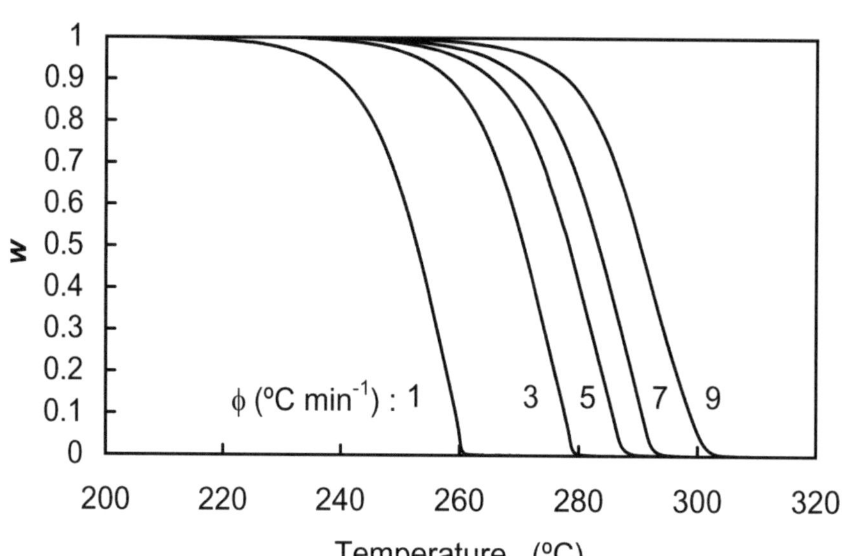

Fig. 1. TG curves of PHB degradation at multiple heating rates (φ=1, 3, 5, 7 and 9 °C min^{-1}) in a constant N_2 flow (100 mL min^{-1}).

In order to analyze the thermal degradation kinetics by the multiple constant heating rate method (Nishida et al., 2000), thermograms at heating rates varying from 1 to 9 °C min^{-1} are illustrated in Fig. 1, in which each thermogram showed smooth and one-step decomposition from the beginning to almost completion. The TG traces are shifted up the temperature scale with increase in the rate. If the pyrolysis of PHB proceeds by the same degradation process with fixed apparent parameters, E_a and A, over the examined temperature range, the thermogravimetric curves can be superposed by shifting them along the abscissa as shown by Eq. 2 (Qzawa, 1965).

$$1/T_1 = 1/T_2 + (R/bE_a)(\log\phi_2 - \log\phi_1) \quad (2)$$

where T, R and ϕ are temperature in degrees Kelvin, the molar gas constant and heating rate, respectively, and b is a constant value for Doyle's approximation (Doyle, 1961, 1962) as follows:

$$\log p(y) \cong -a - bEa/RT \quad (3)$$

where $y = Ea/RT$. If $y > 20$ in the function $p(y)$, values of constants a and b are 2.315 and 0.4567, respectively.
Where the thermogravimetric curves are given as plots of fractional weight ratio, w, versus the right-hand side in Eq. 2, superposition produces a single master curve. This master curve means that the thermogravimetric curves of PHB pyrolysis can be analyzed as a single degradation process with fixed apparent parameters, E_a and A (Flynn & Wall, 1966a).

3.1.1 Activation energy, E_a

The logarithms of the heating rates, $\log\phi$, are plotted against $1/T$ based on Eq. 2. Straight lines for each fraction w are drawn by the method of least squares. From the slopes of these lines, tentative E_a values are calculated using the tentative b value, 0.4567. With an average tentative E_a value, the function y is calculated in the corresponding temperature range. For this range of function y, more accurate values of constants, a and b, were recalculated. The recalculated values of a and b are then used to re-determine a more accurate E_a value. The E_a value can also be determined from the slope of the plots of $\ln(T^2/\phi)$ versus $1/T$ using the same fractional w based on Eq. 4.

$$E_a = R\,\{\ln(T_1^2/\phi_1) - \ln(T_2^2/\phi_2)\}/(1/T_1 - 1/T_2) \quad (4)$$

Eq. 4 was proposed by Reich (Reich, 1964). The method is based on taking the exponential integral in $p(y)$ as approximately equal to the sum of the first two terms of the asymptotic expansion (Doyle, 1961). These two differently derived E_a values are compared to each other to select a plausible E_a value.

3.1.2 Order of reaction, n, and pre-exponential factor, A

The general rate equation of decomposition can be written as follows:

$$-dw/dt = k_d g(w) \quad (5)$$

where k_d is the apparent specific rate constant of the decomposition reaction. The specific form of $g(w)$ depends on the type of kinetic process but usually takes the form w^n, where n is the order of reaction. The temperature dependence of the rate constant is obtained by the application of the Arrhenius equation. Considering the relationship $dT/dt = \phi$, Eq. 5 becomes as below (Flynn & Wall, 1966b).

$$-dw/dT = (A/\phi)\exp(-Ea/RT)g(w) \tag{6}$$

In the differential method, to evaluate the reaction order n for pyrolysis, $-dw/dT$ is plotted against $(1 - w)$ based on Eq. 6. Experimental plots are then compared with the plots for the model reactions, *i. e.*, zero, half, 1st, and 2nd order reactions, and random degradation (Qzawa, 1965).

Although the differential value, $-dw/dT$, is effective for determining n, as shown above, it is ineffective in the lower range of $(1 - w)$ during the beginning period. To examine the behavior of the reaction in this early period for variation, especially, in comparison with a random degradation process, the integration method is examined. For some typical kinetic processes, the integral, $-\int dw/g(w)$ has been given by Simha and Wall (Simha and Wall, 1952) and Ozawa (Ozawa, 1965), *e. g.*, $-\int dw/g(w) = 1-w$, $2(1-w^{1/2})$, $-\ln w$, $(1/w) -1$, and $-\ln\{1-(1-w)^{1/2}\}$ for zero, half, 1st, 2nd order reactions and random degradation. The integral is expressed as follows:

$$-\int dw/g(w) = (AE_a/\phi R)p(y) = A\theta \tag{7}$$

where $\theta = E_a/\phi R p(y)$ was demonstrated as the "reduced time" (Ozawa, 1965). In the integral method, the residual weight value w is plotted against $A\theta$ with model reaction plots.

Consequently, the reaction-order: n and the pre-exponential factor: A, can be determined properly by comparing with the plots for n^{th}-order model reactions.

3.1.3 Random degradation: *L* parameters

When a random degradation process is suggested to proceed, it is analyzed on the basis of the linear relation: $\ln[-\ln\{1-(1-w)^{1/2}\}]$ vs. $1/T$ for TG data (Nishida et al., 2000). However, the exact linear relation is only applicable to the random degradation (L = 2), where L is the least number of repeating units of oligomer not volatilized.

The general relation between L and w for all random reactions was derived as Eq. 8 (Nishida et al., 2002).

$$\ln\left\{1-(1-w)^{\frac{1}{2}}\right\} = -\frac{L}{2}A\theta + \ln\left\{e^{\frac{L}{2}A\theta} - \left(e^{LA\theta} - Le^{A\theta} + L - 1\right)^{\frac{1}{2}}\right\} \tag{8}$$

When L=2,

$$\ln\left\{1-(1-w)^{\frac{1}{2}}\right\} = -A\theta \tag{9}$$

$$\ln\left[-\ln\left\{1-(1-w)^{\frac{1}{2}}\right\}\right] = \ln A\theta = \ln\frac{AE_a}{\varphi R} - a' - b'\frac{E_a}{RT} \quad \propto \frac{1}{T} \tag{10}$$

where a' and b' are constants.

The nth-order model-reactions' plots converge and indicate a linear relation in the beginning period, after which the reaction plots separate as the reaction progresses. On the other hand, all random degradation model reactions can be plotted linearly and in parallel over the whole period. The linear parts in the beginning period of the group of nth-order model reaction curves and the random degradation model reaction curves do not converge. Slopes of the beginning plots are obviously different between the nth-order and random degradation model reactions. Therefore, by analyzing the plots, the random degradation and the nth-order reactions can be distinguished even in the beginning period.

3.2 Products during PHB thermal depolymerization

In the dynamic degradation process of PHB in Py-GC/MS (EI), the total ion current (TIC) chromatograms of the evolved gas show two or three peaks attributable to crotonic acid monomer, related dimer and trimer.

Isothermal degradation of PHB in a glass tube oven was conducted at 260 °C *in vacuo* to determine the components of distilled pyrolyzates. The pyrolyzates were collected and analyzed by ^{1}H NMR and ^{1}H-^{1}H COSY measurements with the major component: *trans*-crotonic acid, making up 67.7 % of the total PHB pyrolyzates. Other components were *cis*-crotonic acid and its derivatives (3.1 %) and oligomers having crotonyl end groups 29.2 %.

In order to examine the selective transformation of PHB during the catalytic thermal degradation, compositions of pyrolyzates generated in various temperature ranges were analyzed from the peak area in Py-GC/MS chromatograms and integrated with the fractional weight loss values from the TG data (Ariffin et al., 2010). The products from PHB/Mg compounds showed maximum values for monomeric acid composition at a temperature range of 230–240 °C, which was lowered by 40-50 °C in comparison with the non-catalytic pyrolysis. Especially, products from PHB-H/$Mg(OH)_2$ gave the highest selectivity for monomeric acids of ~100 %. These results indicate that $Mg(OH)_2$ is the most effective catalyst for the selective transformation of PHB. As a general tendency, it is noted that the lower the temperature, the higher the monomers' composition, indicating that the catalytic reaction is more favorable for monomer formation at lower temperatures.

3.3 Kinetic analysis of PHB thermal degradation

In order to analyze the thermal degradation behavior of PHB precisely, kinetic analysis of PHB degradation is carried out by using the improved random degradation analytical method (Nishida et al., 2002) and integral method (Ozawa, 1965; Nishida et al., 2000). The result of the improved random degradation analysis is illustrated in Fig. 2, in which the kinetics parameters: E_a = 120 kJ mol^{-1} and A = 1.0 × 10^9 s^{-1} were employed for the model reactions: zero (n=0), half (n=0.5), 1st (n=1), 2nd-order (n=2), and random degradations (random L=2-6).

The degradation of PHB (1 °C min^{-1}) is initially faithful to a simulation plot of random degradation (L=2). The L value was actually measured to be around 2 (crotonic acid) or 3 (dimer) by Py-GC/MS analysis. Thereafter, the observed curve gradually diverges from the random (L=2) simulation with an associated rise in L value (L~6) (Fig. 2).

Plots in Fig. 3 show the integral analysis of the observed data (1 °C min^{-1}) and the model reaction plots with parameter values: E_a=120 kJ mol^{-1} and A=1.0 ×10^9 s^{-1}. Interestingly, the observed plots gradually decreased in weight similar to the random degradation process in the initial period, and thereafter the weight loss curve showed a linear relationship between w and $A\theta$ in parallel to the 0th-order simulation plot.

According to Flynn and Wall (Flynn & Wall, 1966a), if the total material may be volatilized by two alternative degradation paths, for which the respective Arrhenius parameters are E_{a1}

and A_1 for path 1, and E_{a2} and A_2 for path 2, and if the two paths are expressible by a combined function $f(w)$, then the integral function will be given by Eq. 11.

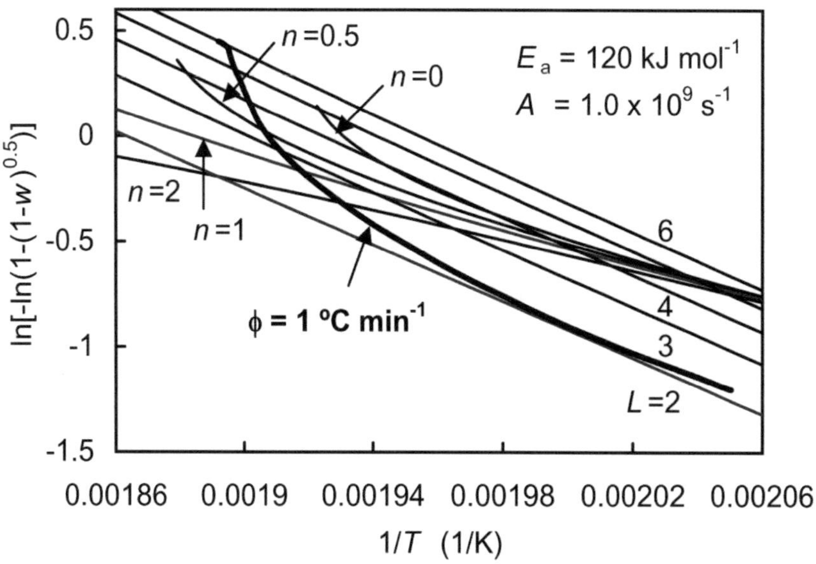

Fig. 2. Plots of ln[-ln{1-(1-w)$^{0.5}$}] vs. $1/T$ for thermogravimetric data of PHB at a heating rate of 1 °C min^{-1} (E_a=120 kJ mol^{-1}, A=1.0×10^8 s^{-1}), and for model reactions. Model reactions: zero (n=0), half (n=0.5), 1st (n=1), 2nd-order (n=2), and random degradations (random L=2-6).

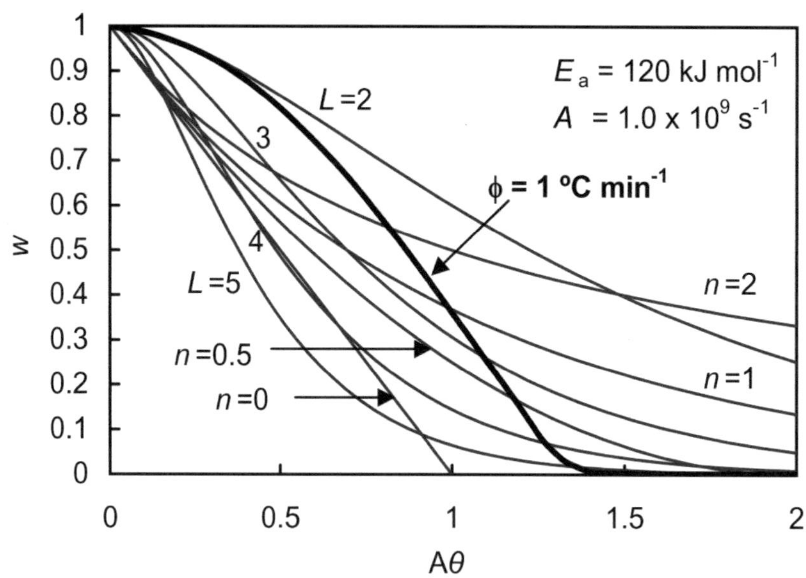

Fig. 3. Plots of w vs. $A\theta$ for thermogravimetric data of PHB at a heating rate of 9 °C min^{-1} (E_a=120 kJ mol^{-1}, A=1.0×10^9 s^{-1}), and for model reactions. Model reactions: zero (n=0), half (n=0.5), 1st (n=1), 2nd-order (n=2), and random degradations (Random L=2-5).

$$
\begin{aligned}
-\int\frac{dw}{f(w)} &= \frac{A_1}{\phi}\int\exp(-Ea_1/RT)+\frac{A_2}{\phi}\int\exp(-Ea_2/RT) \\
&= \frac{A_1Ea_1}{\phi R}p(x_1)+\frac{A_2Ea_2}{\phi R}p(x_2)=A_1\theta_1+A_2\theta_2
\end{aligned}
\tag{11}
$$

where x_1, x_2, θ_1, θ_2, and $p(x)$ are E_{a1}/RT, E_{a2}/RT, $(E_{a1}/\phi R)p(x_1)$, $(E_{a2}/\phi R)p(x_2)$, and function $p(x)$ (Nishida et al., 2003). If each path occurs within an isolated temperature range, the two paths may appear as consecutive reactions. Eq. 11 suggests that the linear relation parallel to the 0th-order simulation plot, which appeared as one of the consecutive processes, should be regarded as a 0th-order reaction. Thus, the time lag, $A_1\theta_1$, between the linear part of the observed plot and the 0th-order simulation plot in Fig. 3 can be regarded as a retardation time or induction period needed to shift from the initial random degradation (L=2) process as shown in Fig. 2 into the 0th-order weight loss process in the middle stage.

It is considered that the 0th-order weight loss process may be due to an unzipping degradation from chain ends. Kopinke et al. (Kopinke et al., 1996) also suggested the presence of a chain reaction with a short kinetic chain length during the PHB degradation. Thus, the above kinetically analyzed results based on the weight loss demonstrate that the PHB degradation proceeds by at least two-steps: the initial random degradation and a subsequent 0th-order weight loss process.

3.4 Theoretical calculation of activation energy of β-elimination

Kopinke et al. (Kopinke et al., 1996) discussed in their report that the decomposition rate of PHB is very sensitive to the reactivity of the β-C–H bonds. They speculated that, if carboxyl groups at chain ends of fractured molecules accelerate the scission of neighboring ester groups faster than the original ester groups, the decomposition results would be equivalent to a chain reaction. Nguyen et al. (Nguyen et al., 2002) also suggested that the conjugation of an unsaturated group with an ester group generates an inductive effect, increasing the scission rate of the neighboring ester linkages.

To confirm these suggested acceleration reactions, the activation energy, E_a, value of the β-elimination via a pseudo-six-membered transition structure was estimated by well-known molecular orbital calculation methods (Table 2) (Ariffin et al., 2008). To calculate the E_a value, various model compounds are employed. The equilibrium geometries of these model compounds are calculated by the Hartree-Fock/3-21G* method and the reaction coordinate systems of the β-elimination were derived from the energy profile obtained by using the PM3 Hamiltonian. Estimated results indicated that the lowest E_a value (253.91 kJ mol^{-1}) was associated with the reaction of an internal ester group of a trimer model molecule, supporting the observed internal random scission as the initial reaction. The reaction at the chain-end neighboring crotonyl group produced a slightly high E_a value of 260.46 kJ mol^{-1}, but this is considerably lower than the 365.22 kJ mol^{-1} at carboxyl chain-ends. The reaction at the chain-end must be accelerated by the chain-end's higher mobility when compared to that at internal chain units. Moreover, it is clearly indicated that for the cases where R_2 = H in Table 2, the crotonyl group has a fundamental acceleration effect for β-elimination on the neighboring ester group.

Thus, this acceleration effect at the crotonyl chain-end induces the generation of a crotonic acid and a new crotonyl chain-end, repeating the same reaction in the same manner as the unzipping reaction.

Model reactions	E_a (kJ mol^{-1})
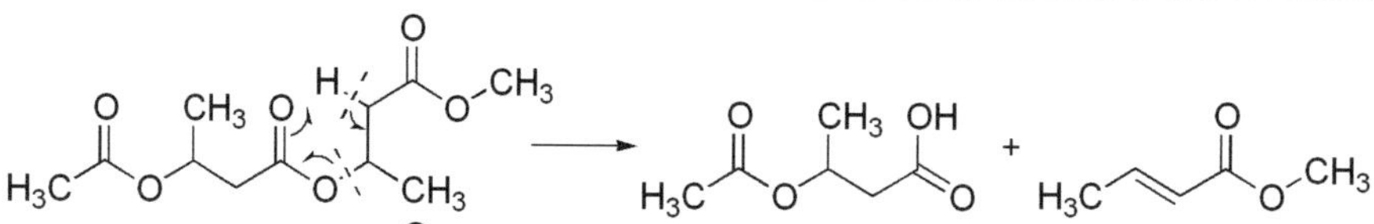	253.91
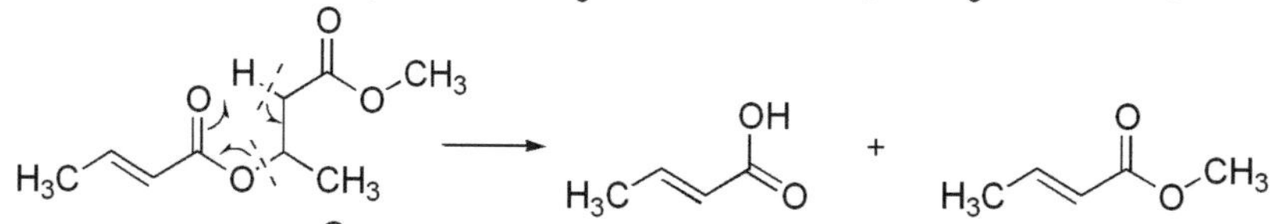	260.46
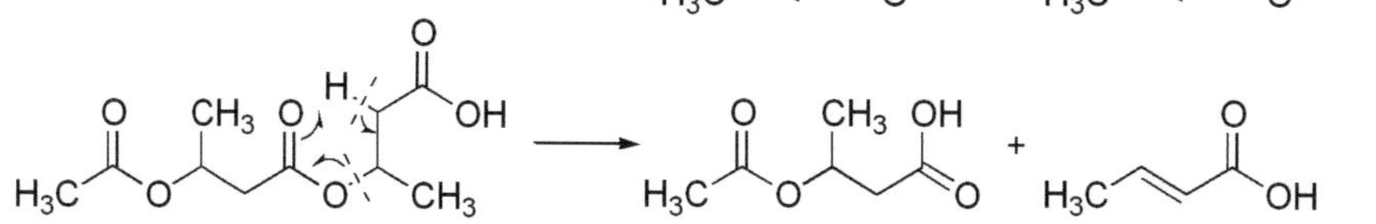	365.22
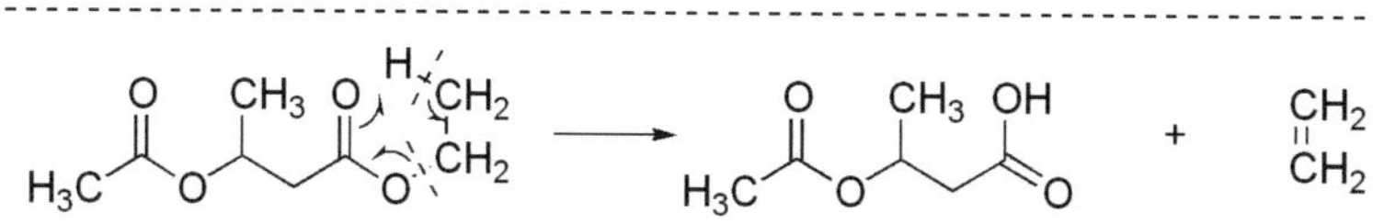	339.04
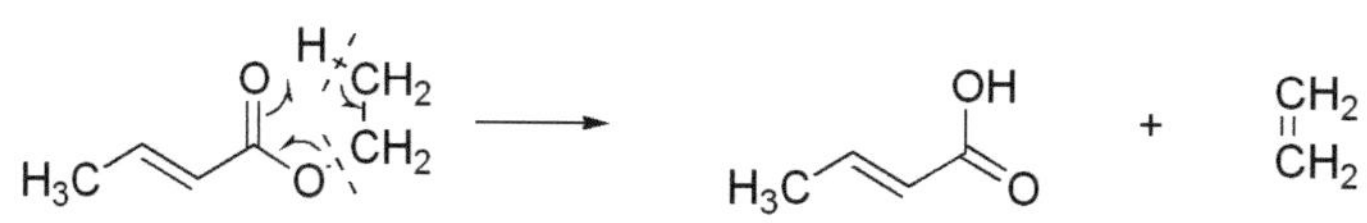	287.40

Table 2. Activation energy values of β-elimination calculated by the semi-empirical molecular orbital method PM3.

3.5 Effects of alkali earth compounds as depolymerization catalysts

In order to transform PHB into a specific monomer, precise control of the thermal degradation is crucial. In spite of the wealth of previous studies, there are few reports thus far which focus on the selective depolymerization of PHB into the monomer. In cases of PHB pyrolysis without a catalyst, net yield of all butenoic acids (*trans/cis*-crotonyl acids (CAs) + 3-butenoic (BA)) as main pyrolysis products ranged from 39.5 wt.-% using thermal volatilization analysis (Grassie et al., 1984) to 87 total ion count % (TIC-%) using Py-GC/MS analysis (Gonzalez et al., 2005). Recently, it was reported that ^{1}H NMR analysis of PHB pyrolyzates at 260 °C revealed the ratio *trans*-CA:*cis*-CA:oligomers = 67.7:3.1:29.2 (monomeric unit-%) (Ariffin et al., 2008). Kopinke et al. (Kopinke et al., 1996) also reported on the production of a large amount of oligomers during the PHB pyrolysis, with the recovered monomer fraction being composed of *trans*-CA:*cis*-CA:BA = 100:10:2 (GC-area-%). When pyrolysis catalysts were used in PHB pyrolysis, Kim et al. (Kim et al., 2006, 2008) reported the enhancement of PHB depolymerization by $CaCl_2$ or $MgCl_2$. This enhancement was explained as being due to the Lewis acidity of Ca^{2+} and Mg^{2+} facilitating the formation of a double bond by the elimination of β-hydrogen. Kawalec et al. (Kawalec et al., 2007) also reported on the enhancement of the thermal degradation of PHB by Na^+, K^+, and Bu_4N^+ counter cations at chain-ends, proposing an E1cB mechanism. However, there was no discussion on the selectivity of pyrolyzates.

Completely selective transformation of poly[(*R*)-3-hydroxybutyric acid] (PHB) into *trans*-crotonic acid was achieved by thermal degradation using Mg compounds: MgO and

$Mg(OH)_2$ as catalysts (Ariffin et al., 2010). Through catalytic action, not only were the temperature and E_a value of degradation lowered by 40-50 °C and 11-14 kJ mol^{-1}, respectively, but also significant changes in the selectivity of pyrolyzates were observed. Notably, $Mg(OH)_2$ and MgO showed nearly complete selectivity (~100%) to *trans*-crotonic acid.

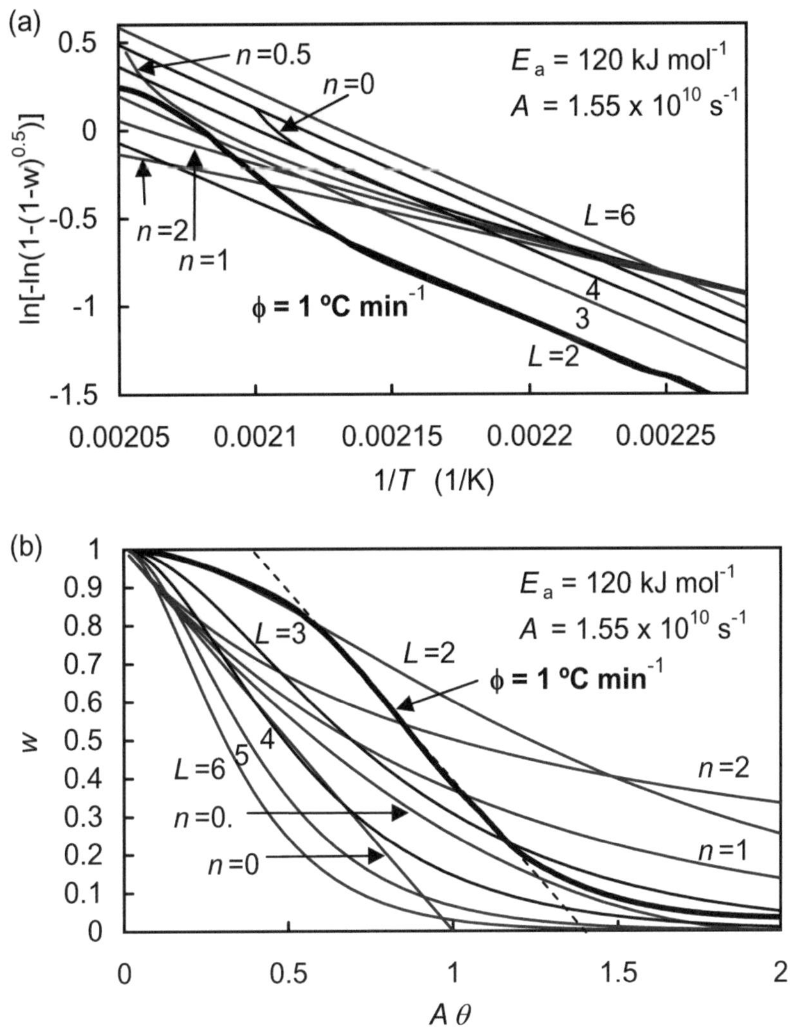

Fig. 4. Plots of (a) $\ln[-\ln\{1-(1-w)^{0.5}\}]$ vs. $1/T$ and (b) w vs. $A\theta$ for thermogravimetric data of PHB/MgO (100:5 wt/wt) at a heating rate of 1 °C min^{-1} (E_a=120 kJ mol^{-1}, A=1.55×10^{10} s^{-1}), and for model reactions. Model reactions: zero (n=0), half (n=0.5), 1st (n=1), 2nd-order (n=2), and random degradations (random L=2-6).

In Fig. 4, the random degradation and the integral analyses of PHB/MgO (100:5 wt/wt) are illustrated. The pyrolysis of PHB-H/MgO was shown to start by a random degradation (L=2), and gradually shift into an nth-order weight loss behavior. PHB/$Mg(OH)_2$ also showed the same kinetics. These results show that the thermal degradation mechanisms of all the samples are similar to each other, *i.e.* random β-elimination and/or E1cB mechanisms as the initial processes followed by the unzipping β-elimination accelerated by the chain-end crotonate group as the main process.

3.6 Degradation mechanisms

Under the pyrolysis conditions without any other reactant from the outside, the auto-catalytic or auto-accelerated random degradation may be induced by a large number of carboxyl compounds, which self-generate as the reaction progresses. The carboxyl compounds randomly attack ester groups on polymer chains to induce the trans-esterification, resulting in the reproduction of carboxyl groups and a remarkable decrease in molecular weight as demonstrated by Nguyen et al. (Nguyen et al., 2002).

In order to induce the auto-accelerated random degradation, carboxylic groups have to be self-proliferated in bulk. The unzipping reaction occurring at the end of molecules, which was suggested to be the 0th-order reaction by the kinetic analysis and the theoretical calculation in the above sections, is just a kinetically favored scission of polymer chains that repeatedly generate a large number of crotonic acids. This is an ideal self-proliferation reaction of carboxyl compounds and is the key factor in the auto-accelerated random degradation.

According to predictions based on the above measured and analyzed results, an expected overall thermal degradation pathway of PHB is illustrated in Scheme 2.

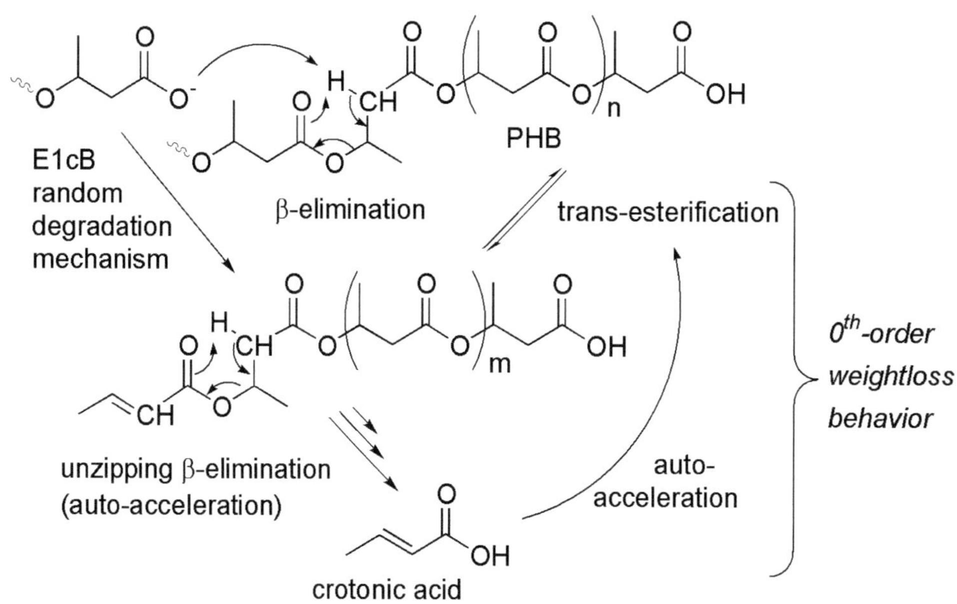

Scheme 2. Expected overall thermal degradation pathway of PHB.

The Mg compounds as catalysts may accelerate both the β-elimination reactions in the beginning and main stages and/or the E1cB mechanism by carboxyl anions as Lewis acids and counter cations, respectively. Thereby the initial random degradation was promoted as shown by the increases in *A* values, consequently increasing the crotonate chain-end groups, resulting in the acceleration of the main unzipping β-elimination (*cis*-elimination) as shown in Scheme 3.

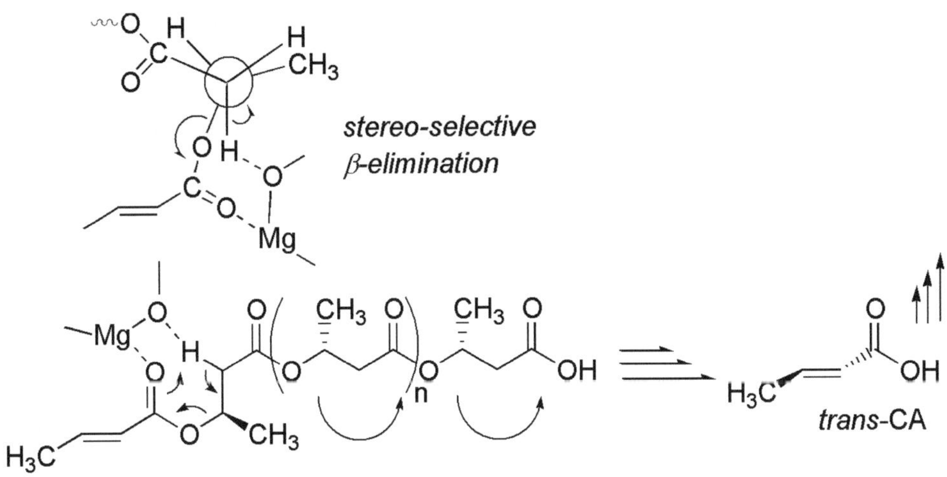

Scheme 3. *Trans*-form preferential depolymerization of PHB.

4. Conclusions

The thermal degradation of PHB was investigated by various kinetic analysis methods in detail to clarify its complex degradation behavior, resulting in a finding of mixed mechanisms comprising at least a thermal random degradation with subsequent auto-accelerated trans-esterification, and a kinetically favored chain reaction from crotonate chain ends.

From the kinetic analysis of changes in molecular weight, it was found that a random degradation proceeding in the initial period was followed by an auto-accelerated reaction in the middle period. From the kinetic analysis of weight loss behavior, it was proposed that there were some kinetically favored scissions occurring at the chain ends, where the degradation proceeded by a 0th-order weight loss process in the middle stage. The observed 0th-order weight loss process was assumed to be an unzipping reaction occurring at ester groups neighboring the crotonate end groups.

The completely selective transformation of PHB into *trans*-crotonic acid was achieved by thermal degradation using catalysts: MgO and $Mg(OH)_2$. Through catalytic action, not only were the temperature and E_a value of degradation lowered, but also significant changes in the selectivity of pyrolyzates were observed. Notably, $Mg(OH)_2$ showed nearly complete selectivity (~100%) to *trans*-crotonic acid. It was suggested that the Mg catalysts promote the totality of the β-elimination reactions by acting throughout the beginning and main processes, resulting in a lowering in the degradation temperature and the completely selective transformation of PHB.

By the above approaches, some complicated problems on the thermal degradation mechanisms of PHB have been solved, and the controlled depolymerization of PHB into *trans*-crotonic acid, which is a versatile monomer for many applications, has been achieved.

5. References

Abate, R.; Ballistreri, A.; Montaudo, G. & Impallomeni, G. (1994). Thermal degradation of microbial poly(4-hydroxybutyrate). *Macromolecules*, 27, 2, 332-336, ISSN: 0024-9297.

Abe, H. (2006). Thermal Degradation of Environmentally Degradable Poly(hydroxyalkanoic acid)s. *Macromol. Biosci.*, 6, 7, 469–486, ISSN: 1616-5187.

Aoyagi, Y.; Yamashita, K. & Doi, Y. (2002). Thernal degradation of poly[(R)-3-hydroxybutyrate], poly[ε-caprolactone], and poly[(S)-lactide]. *Polym. Degrad. Stab.*, 76, 1, 53-59, ISSN: 0141-3910.

Ariffin, H.; Nishida, H.; Shirai, Y. & Hassan, M. A. (2008). Determination of multiple thermal degradation mechanisms of poly(3-hydroxybutyrate). *Polym. Degrad. Stab.*, 93, 8, 1433-1439, ISSN: 0141-3910.

Ariffin, H.; Nishida, H.; Shirai, Y. & Hassan, M. A. (2009). Anhydride production as an additional mechanism of poly(3-hydroxybutyrate) pyrolysis. *J. Appl. Polym. Sci.*, 111, 1, 323–328, ISSN: 0021-8995.

Ariffin, H.; Nishida, H.; Shirai, Y. & Hassan, M. A. (2010). Highly selective transformation of poly[(*R*)-3-hydroxybutyric acid] into *trans*-crotonic acid by catalytic thermal degradation. *Polym. Degrad. Stab.*, in press, ISSN: 0141-3910.

Ballistreri, A.; Garozzo, D.; Giuffrida, M,; Impallomeni, G. & Montaudo, G. (1989). Analytical degradation: An approach to the structural analysis of microbial polyesters by different methods. *J. Anal. Appl. Pyrolysis*, 16, 3, 239-253, ISSN: 0165-2370.

Carrasco, F.; Dionisi, D.; Martinelli, A. & Majone, M. (2006). Thermal stability of polyhydroxyalkanoates. *J. Appl. Polym. Sci.*, 100, 3, 2111–2121, ISSN: 0021-8995.

Doyle, C. D. (1961). Kinetics analysis of thermogravimetric data. J. Appl. Polym. Sci., 5, 15, 285-292, ISSN: 0021-8995.

Doyle, C. D. (1962). Estimating isothermal life from thermogravimetric data. J. Appl. Polym. Sci., 6, 24, 639-642, ISSN: 0021-8995.

Flynn, J. H. & Wall, L. A. (1966a). General treatment of the thermogravimetry of polymers. *J. Res. Nat. Bur. Stand.*, 70A, 6, 487-523, ISSN: 0160-1741.

Flynn, J. H. & Wall, L. A. (1966b). A quick, direct method for the determination of activation energy from thermogravimetric data. *J. Polym. Sci. Part C: Polym. Lett.*, 4, 5, 323-328, ISSN: 0449-2986.

Galego, N. & Rozsa, C. (1999). Thermal decomposition of some poly(β-hydroxyalcanoates). *Polym. Int.*, 48, 12, 1202-1204, ISSN: 0959-8103.

Gogolewski, S.; Jovanovic, M.; Perren, S. M.; Dillon, J. G. & Hughes, M. K. (1993). The effect of melt-processing on the degradation of selected polyhydroxyacids: polylactides, polyhydroxybutyrate, and polyhydroxybutyrate-co-valerates. *Polym. Degrad. Stab.*, 40, 3, 313-322, ISSN: 0141-3910.

Gonzalez, A.; Irusta, L.; Fernandez-Berridi, M. J.; Iriarte, M. & Iruin, J. J. (2005). Application of pyrolysis/gas chromatography/Fourier transform infrared spectroscopy and TGA techniques in the study of thermal degradation of poly (3-hydroxybutyrate). *Polym. Degrad. Stab.*, 87, 2, 347-354, ISSN: 0141-3910.

Grassie, N.; Murray, E. J. & Holmes, P. A. (1984a). The thermal degradation of poly(-(D)-β-hydroxybutyric acid): Part 1- Identification and quantitative analysis of products. *Polym. Degrad. Stab.*, 6, 1, 47-61, ISSN: 0141-3910.

Grassie, N.; Murray, E. J. & Holmes, P. A. (1984b). The thermal degradation of poly(-(D)-β-hydroxybutyric acid): Part 2-Changes in molecular weight. *Polym. Degrad. Stab.*, 6, 2, 95-103, ISSN: 0141-3910.

He, J. D.; Cheung, M. K.; Yu, P. H. & Chen, G. Q. (2001). Thermal analyses of poly(3-hydroxybutyrate), poly(3-hydroxybutyrate-co-3-hydroxyvalerate), and poly(3-hydroxybutyrate-co-3-hydroxyhexanoate). *J. Appl. Polym. Sci.*, 82, 1, 90-98, ISSN: 0021-8995.

Hoffmann, A.; Kreuzberger, S. & Hinrichsen, G. (1994). Influence of thermal degradation on tensile strength and Young's modulus of poly(hydroxybutyrate). *Polym. Bull.*, 33, 3, 355-359, ISSN: 0170-0839.

Kawalec, M.; Adamus, G.; Kurcok, P.; Kowalczuk, M.; Foltran, I.; Focarete, M. L. & Scandola, M. (2007). Carboxylate-induced degradation of poly(3-hydroxybutyrate)s. *Biomacromolecules*, 8, 4, 1053-1058, ISSN: 1525-7797.

Kim, K. J.; Doi, Y. & Abe, H. (2006). Effects of residual metal compounds and chain-end structure on thermal degradation of poly(3-hydroxybutyric acid). *Polym. Degrad. Stab.*, 91, 4, 769-777, ISSN: 0141-3910.

Kim, K. J.; Doi, Y. & Abe, H. (2008). Effect of metal compounds on thermal degradation behavior of aliphatic poly(hydroxyalkanoic acid)s. *Polym. Degrad. Stab.*, 93, 4, 776-85, ISSN: 0141-3910.

Kopinke, F. D.; Remmler, M. & Mackenzie, K. (1996). Thermal decomposition of biodegradable polyesters-I: Poly(β-hydroxybutyric acid). *Polym. Degrad. Stab.*, 52, 1, 25-38, ISSN: 0141-3910.

Kopinke, F. D. & Mackenzie, K. (1997). Mechanism aspects of the thermal degradation of poly(lactic acid) and poly(β-hydroxybutyric acid). *J. Anal. Appl. Pyrolysis*, 40/41, 43-53, ISSN: 0165-2370.

Kunioka, M. & Doi, Y. (1990). Thermal degradation of microbial copolyesters: poly(3-hydroxybutyrate-*co*-3-hydroxyvalerate) and poly(3-hydroxybutyrate-*co*-4-hydroxybutyrate). *Macromolecules*, 23, 7, 1933-1936, ISSN: 0024-9297.

Lee, M. Y.; Lee, T. S. & Park, W. H. (2001). Effects of side chains on the thermal degradation of poly(3-hydroxyalkanoates). *Macromol. Chem. Phys.*, 202, 7, 1257-1261, ISSN: 1022-1352.

Lehrle, R. S. & Williams, R. J. (1994). Thermal degradation of bacterial poly(β-hydroxybutyrate): Mechanism from the dependence of pyrolysis yields on sample thickness. *Macromolecules*, 27, 14, 3782-3789, ISSN: 0024-9297.

Lehrle, R.; Williams, R.; French, C. & Hammond, T. (1995). Thermolysis and methanolysis of poly(β-hydroxybutyrate): Random scission assessed by statistical analysis of molecular weight distributions. *Macromolecules*, 28, 13, 4408-4414, ISSN: 0024-9297.

Li, S. D.; Yu, P. H. & Cheung, M. K. (2001). Thermogravimetric analysis of poly(3-hydroxybutyrate) and poly(3-hydroxybutyrate-co-3-hydroxyvalerate). *J. Appl. Polym. Sci.*, 80, 12, 2237–2244, ISSN: 0021-8995.

Li, S. D.; He, J. D.; Yu, P. H & Cheung, M. K. (2003). Thermal degradation of poly(3-hydroxybutyrate) and poly(3-hydroxybutyrate-co-3-hydroxyvalerate) as studied by TG, TG-FTIR, and Py-GC/MS. *J. Appl. Polym. Sci.*, 89, 6, 1530-1536, ISSN: 0021-8995.

Liu, Q.-S.; Zhu, M.-F.; Wu, W.-H. & Qin, Z.-Y. (2009). Reducing the formation of six-membered ring ester during thermal degradation of biodegradable PHBV to enhance its thermal stability. *Polym. Degrad. Stab.*, 94, 1, 18–24, ISSN: 0141-3910.

Marchessault, R. H.; Coulombe, S.; Morikawa, H.; Okamura, K. & Revol, J. F. (1981). Solid state properties of poly-β-hydroxybutyrate and of its oligomers. *Can. J. Chem.*, 59, 1, 38-44, ISSN: 1480-3291.

Melchiors, M.; Keul, H. & Höcker, H. (1996). Depolymerization of poly[(R)-3-hydroxybutyrate] to cyclic oligomers and polymerization of the cyclic trimer: An example of thermodynamic recycling. *Macromolecules*, 29, 20, 6442-6451, ISSN: 0024-9297.

Morikawa, H. & Marchessault, R. H. (1981). Pyrolysis of bacterial polyalkanoates. *Can. J. Chem.*, 59, 15, 2306-2313, ISSN: 1480-3291.

Nguyen, S.; Yu, G. & Marchessalt, R. H. (2002). Thermal degradation of poly(3-hydroxyalkanoates): Preparation of well-defined oligomers. *Biomacromolecules*, 3, 1, 219-224, ISSN: 1525-7797.

Nishida, H.; Tokiwa, Y.; Yamashita, M. & Endo, T. (2000). Thermal decomposition of poly(1,4-dioxane-2-one). *Polym. Degrad. Stab.*, 70, 3, 485-496, ISSN: 0141-3910.

Nishida, H.; Yamashita, M. & Endo, T. (2002). Analysis of the initial process in pyrolysis of poly(p-dioxanone). *Polym. Degrad. Stab.*, 78, 1, 129-135, ISSN: 0141-3910.

Nishida, H.; Mori, T.; Hoshihara, S.; Fan, Y.; Shirai, Y. & Endo T. (2003). Effects of tin on poly(L-lactic acid) pyrolysis. *Polym. Degrad. Stab.*, 81, 3, 515-523, ISSN: 0141-3910.

Ozawa, T. (1965). A new method of analyzing thermogravimetric data. *Bull. Chem. Soc. Japan*, 38, 11, 1881-1886, ISSN: 0009-2673.

Reich, L. (1964). A rapid estimation of activation energy from thermogravimetric traces. *J. Polym. Sci. Part C: Polym. Lett.*, 2, 6, 621-623, ISSN: 0449-2986.

Simha, R. & Wall, L. A. (1952). Kinetics of chain depolymerization. *J. Phys. Chem.*, 56, 6, 707-715, ISSN : 0022-3654.

Steinbüchel, A. & Valentin, H. E. (1995). Diversity of bacterial polyhydroxyalkanoic acids. *FEMS Microbiol. Lett.*, 128, 3, 219-228, ISSN: 0378-1097.

20

Biotechnological Production and Application of Hyaluronan

Chiara Schiraldi, Annalisa La Gatta and Mario De Rosa
Department of Experimental Medicine, Faculty of Medicine, Second University of Naples, Via L. De Crecchio 7, 80138 Naples, Italy

1. Introduction

Glycosaminoglycans (GAGs) are linear polysaccharides formed from repetitions of a disaccharide unit composed of one aminosugar and one uronic acid residue. Among these, hyaluronic acid (HA) ([D-glucuronic acid (1-β-3) N-acetyl-D-glucosamine (1-β-4)]$_n$) (figure 1), that differs from the other for not presenting sulphate groups, is a biopolymer of broad scientific interest and largely applied in different biomedical fields.

This macromolecule is most frequently referred to as hyaluronan, because of the many different forms the molecule can assume in physiological conditions (i.e. the acid form, HA, and the salts, such as sodium hyaluronate) (Balazs & Gibbs, 1970).

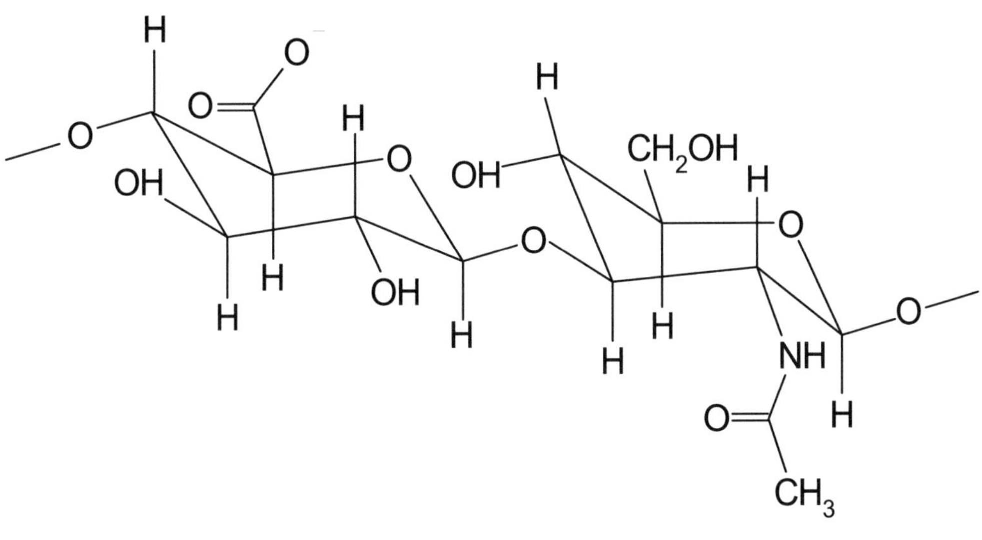

Fig. 1. Disaccharide repeating unit ([D-glucuronic acid (1-β-3) N-acetyl-D-glucosamine (1-β-4)]$_n$) of hyaluronic acid.

This biopolymer is widespread in nature, having been identified in vertebrate soft tissues (e.g. joints, synovial fluid, skin, vitreous humour of the eye, umbilical cords, roster combs) (Balazs et al., 1993), in algae (De Angelis, 1999), in molluscs (Volpi & Maccari, 2003), and also in cultured eukaryotic cell lines, and certain prokaryotes, where it occurs as a mucoid capsule surrounding the cell (O'Regan et al., 1994). HA present in all vertebrates is a main component of the extracellular matrix: it is the major constituent in the vitreous of human eye (0.1 mg/mL wet weight), and in the synovial joint fluid (3-4 mg/mL wet weight). However, the largest amount of HA (7-8 g of hyaluronate per average adult human, or approximately 50% of the total in the body) resides in the skin, where it is present in both the dermis and the epidermis (0.5 mg/g wet tissue). The embryo is covered by a thick HA coating during certain stages of development, it is also abundant in the umbilical cord (4 mg/mL) (Toole, 1997; Marcellin et al., 2009). Interestingly rooster combs, a specialized piece of skin, contain even higher amounts of HA (up to 7.5 mg/mL), and in fact they are a preferred source for HA industrial extraction.

In vertebrates, HA has a wide variety of functions: in the skin it maintains tissue hydration (Bettelheim & Popdimirova, 1992); in the cartilage it fastens proteoglycans to regulate water and ion content, sustaining tissue physical properties and cell-substrate interactions. The biological effects associated with HA-receptor binding, furthermore, induce rate changes in cell proliferation, cell migration, and angiogenesis (Goldberg & Toole, 1987; Alho & Underhill, 1989). Moreover, overproduction of HA is observed in diseases associated with inflammation, fibroses and cancer. Recently, direct evidences demonstrate the involvement of HA in cancer metastasis (Stern, 2005; Heldin, 2003).

In the present chapter, production processes, chemico-physical properties, and established and foreseen applications of hyaluronan and derivatives will be analysed and critically presented.

2. Hyaluronan production

Currently there are two competing methods for industrial HA production that are extraction from animal sources, such as bovine eyes and rooster combs, and microbial production through the use of large scale fermenters. Both will be discussed in the following sections, in addition the opportunity of using novel genetically engineered microbial factories and a chemo-enzymatic synthesis approach will be reported from the very recent literature.

2.1 Traditional extraction processes

The traditional method for HA production is based on solvent extraction from animal tissue extracts, eventually using cetylpiridinium chloride (CPC) precipitation.

One of the first paper presented by Swann (1968), reported the following procedure: (1) mechanical slicing of the rooster combs to obtain small pieces, (2) washing with ethanol (4 L ethanol to 1 Kg comb), this operation could be repeated until the solvent would not appear cloudy; (3) extracting the minced combs with a water/chloroform mixture (2.5 Kg combs: 10 L water: 0.5 L chloroform), while stirring to allow combs to swell; (4) filtering the solids from the broth and adding NaCl, successively carrying an additional chloroform extractions; (5) accomplishing protease (pronase) digestion, followed by chloroform extraction and centrifugation.

In alternative methods (Swann, 1968.) the crude extracts were purified by epichlorohydrin triethanolamine- (ECTEOLA-) chromatography and by fractionation with CPC. In addition, repeated ethanol precipitation (1:3 water/ethanol ratio), before and after CPC (1%) HA precipitation, were reported (Prescott, 2003). In all the cases the product is then filtered through sterilizing filters, followed by solvent precipitation, finally the HA is formulated

into medical devices and pharmaceutical products. HA purified by these procedures was recovered with a yield greater than 90% with respect to the uronic acid evaluated in the starting material.
However the collection of rooster combs and the extraction and purification procedures of HA from these tissues are time-consuming and labour intensive, making hyaluronan production very costly (O' Regan et al., 1994). In fact in animal tissues hyaluronan is complexed with proteoglycans and often contaminated with HA degrading enzymes, making the isolation of high purity and high molecular sized polysaccharide very difficult. Moreover the use of animal-derived biomolecules for biopharmaceutical applications is facing growing opposition because of the risk of cross-species viral and other adventitious agent contaminations. Hence, since '80 microbial production is gradually replacing extraction from animal tissues in HA industrial manufacturing.

2.2 Biotechnological production of HA

Bacteria known to be capable of the synthesis of HA are Streptococci of groups A and C, gram-positive bacteria such as *Streptococcus equi*, an equine pathogen, *Streptococcus equisimilis*, that is infective for different animals, *Streptococcus pyogenes*, a human pathogen and *Streptococcus uberis*, a bovine pathogen. These β-hemolytic bacteria, able to digest blood based agar medium, also present a slimy translucent layer surrounding bacterial colonies that can be attributed to HA synthesis (figure 2).

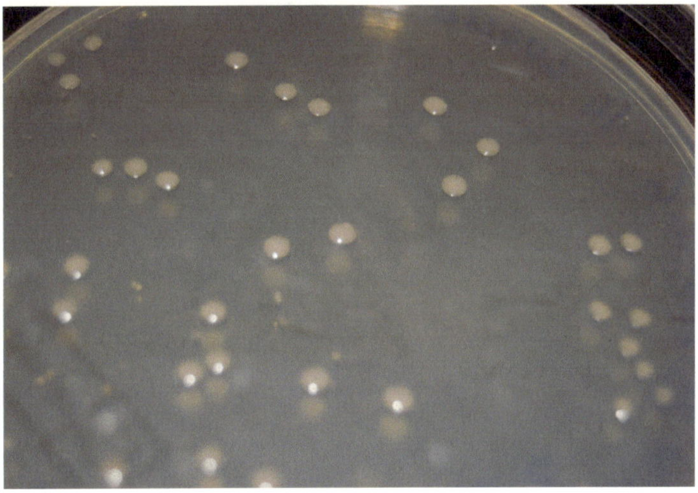

Fig. 2. Slimy colonies of *Streptococcus zooepidemicus* HA producing cells.

A gram-negative pathogenic bacteria, *Pasteurella multocida* was also found to produce HA in its capsule (De Angelis et al., 1998). The HA capsule is a virulence factor in both Streptococcus and Pasteurella, probably providing bacteria a stealth function that result in the failure of the immune system to recognise the HA capsule as a foreign entity (Schmidt et al., 1996). The capsule may also protect the bacteria against reactive oxides released by leukocytes, attempting to prevent infection. Finally, it helps the migration through epithelial layers, exploiting CD44 mediated tissue response (Cywes & Wessels, 2001). Thus the HA capsule contributes in large part to the pathogenicity of these microorganisms.

Both *Streptococcus zooepidemicus* and *Pasteurella multocida* produce HA starting from activated substrates (nucleotidic sugars) through specific membrane bound glycosyltransferases, so-called HA synthases (HASs). The latter have been exploited in few recent studies for the chemoenzymatic synthesis of HA, also reporting the possibility to obtain biopolymers of defined molecular weight using *P. multocida* HAS (De Angelis et al., 2003). Nevertheless the established industrial production process today is based on fermentation of mutagenized streptococcal cells.

It has been estimated that hyaluronan synthesis in bacterial fermentation accounts for 5–10% of the carbon metabolised. The D-glucuronic acid and the N-acetyl-glucosamine moieties of HA are derived from glucose-6-phosfate and fructose-6-phosfate, respectively, as demonstrated for *S. zooepidemicus* through ^{13}C NMR studies (Matsubara et al., 1991). The proposed biosynthetic pathway for HA was well described in the recent literature and for readers convenience is schematically depicted in figure 3.

Overall, the synthesis of one mole of HA disaccharide consumes five moles of nucleosides triphosphates (3 as ATP and 2 as UTP), two moles of glucose and one mole of acetyl coenzyme A (Acetyl-CoA) and generates two moles of reducing equivalents (NADH) and, therefore, it is expected that the flux through the HA pathway is intimately related to the cellular needs of other pathways, (i.e. glycolysis and cell growth); it is also expected to be

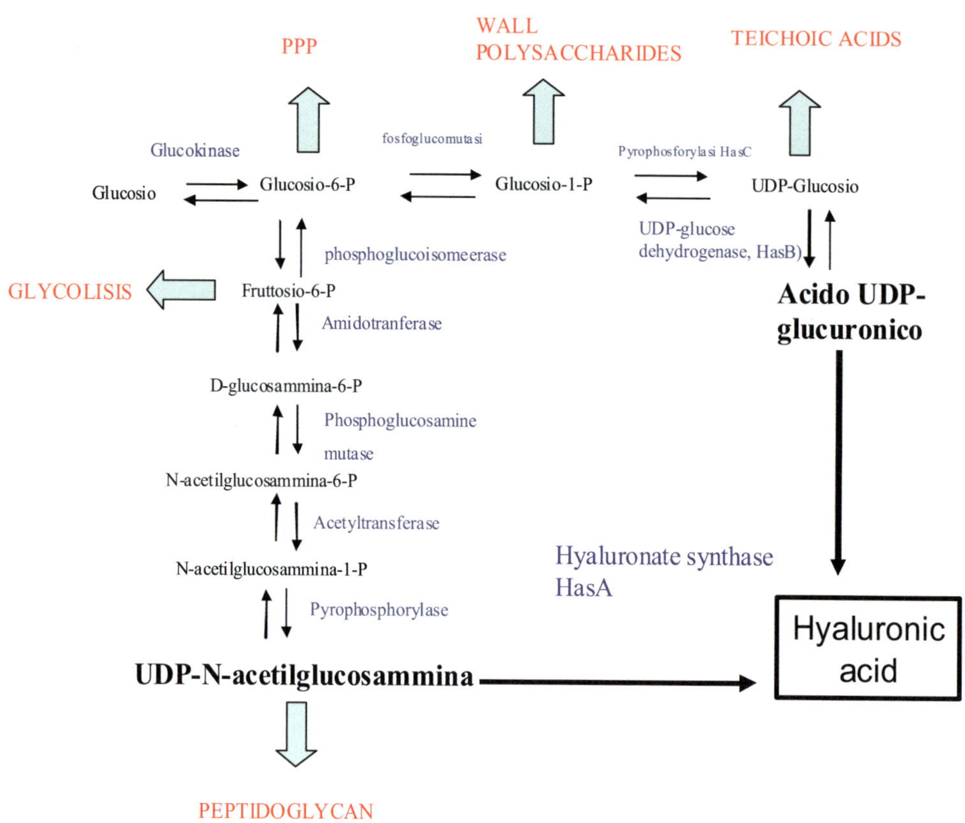

Fig. 3. Biosynthetic pathway responsible for hyaluronic acid production in streptococci: as shown few intermediates are also required for cell wall synthesis.

dependent on the energetic state of the cell. In fact, ATP levels and reducing equivalents in the cell play a key role in biosynthesis, being these substances consumed and generated in the production of hyaluronate (Chong & Nielsen, 2003). Besides furnishing precursors for HA synthesis, the two pathways showed in figure 3 also supply the structural constituents of the bacterial cell wall, specifically peptidoglycan, teichoic acids and antigenic wall polysaccharides: these three major wall components accounts for 20% (w/w) of the cell dry weight and represent a significant drain on the precursors pool used to synthesize HA.

2.2.1 Enzymes involved in HA synthesis

HA is polymerised on the cytoplasmatic side of the plasma membrane as a free linear polymer, differently from the other glycosaminoglycans which are synthesized by resident Golgi enzymes and covalently attached to core proteins. HAS (Has A) is the only protein required for HA synthesis and it functions as a monomer stabilised by phospholipids (De Angelis & Weigel, 1994; Kumari & Weigel, 1997). After the discovery in 1993 of the first gene encoding a hyaluronan synthase from Group A *Streptococcus*, many others similar hyaluronan synthase genes were identified in other bacteria and in a wide range of eukaryotes (De Angelis 1999; Itano & Kimata, 2002). The molecular masses of the streptococcal (49 KDa) or eukaryotic (65 KDa) HASs are relatively small in view of the multiple functions mediated by these enzymes in order to synthesize HA (table 1). HAS binds UDP-glucuronic acid (UDP-GlcUA) and UDP-N-acetylglucosamine (UDP-GlcNAc) in the presence of $MgCl_2$ and catalyzes two distinct intracellular glycosyltransferase reactions. HAS also binds and translocates the growing HA chain through the cell membrane. The pendulum hypothesis was proposed by De Angelis and Weigel (1994) to explain how these functions coordinates to synthesized and transfer a growing HA chain (www.glycoforum.gr.jp/science/hyaluronan). In HA-producing streptococci two unique genes encoding for HA-synthase (*has* A) and UDP-glucose dehydrogenase (*has* B) were found on the so-called *has* operon. The latter was firstly sequenced in *S. pyogenes* presenting also a third gene, *has* C, encoding for the UDP-glucose phosphorylase (Crater & van De rijn, 1995). More recently, Blank and co-workers (2008) found other two genes on the operon of *S. zooepidemicus*, GlmU and pgi, the former is responsible for the two final steps of UDP N-acetyl glucosamine biosynthesis, while the latter is an additional phosphoglucoisomerase involved in the Embden-Mayerhof Parnas pathway.

Multiple functions of Class I HA synthases
Addition of GlcNAc to the growing HA-GLcUA-UDP chain (left)
UDP-GlcNAc acceptor binding
HA-GlcUA-UDP donor binding
HA-GlcUA-UDP: UDP-GlcNAc, β-1-3 (HA-GlcUA) transferase
HA traslocation through the membrane
Addition of GlcUA to the growing HA-GlcNAc-UDP chain (right)
UDP-GlcUA acceptor binding
HA-GlcNAc-UDP donor binding
HA-GlcNAc-UDP: UDP-GlcUA, β-1-4 (HA-GlcNAc) transferase
HA traslocation through the membrane

Table 1. Multiple functions of Class I HA synthases.

Based on differences in protein structure and mechanism of action, the known HASs have been classified into two classes (De Angelis, 1999). Class I members include HASs from *Streptococcus*, mammals, and other eukaryotes, whereas the bacterial HAS from *Pasteurella multocida* is the only class II member (table 2). The major mechanistic difference is that the two classes of synthases extend hyaluronan at opposite ends of the polysaccharide. The Class II pmHAS has a two-domain modular structure, with two transferase activities, that alternatively bind and release hyaluronan chains to add new sugars to the non-reducing end by typical glycosyltransferase activity. The Class I enzymes are the first glycosyltransferases that has been unanimously demonstrated to function at the reducing end of a growing glycosaminoglycan chain (Tlapak-Simmons et al., 2005).

	Class I	**Class II**
Members	HASs of *Streptococcus* ssp., mammalian, avian and amphibian	HAS of *Pasteurella multocida*
Size (amino acids)	417-588	972
Membrane attachment domain	6-8 membrane-associated domains	C-terminal membrane anchor
HA chain growth	At reducing end	At non reducing end
Primer oligosaccharide	No evidence for extension	HA extension

Table 2. Classes of hyaluronan synthases.

The hyaluronan polymerization rates for the streptococcal hyaluronan synthases in isolated membranes were estimated to be ~1200-2400 sugars/min: at this elongation rate one active hyaluronan synthase molecule would take about 8-16 minutes to synthesize a single hyaluronan chain with a mass of 2 MDa. The rate of hyaluronan chain elongation in live cells has not been determined, but is likely to be faster than what has been measured *in vitro* as the elongation rate increases with substrate concentration until when too high concentration determines the release of the HA chain from the cell. Very little is known about the enzyme properties that control hyaluronan chain length and how different hyaluronan synthases make hyaluronan products of different size distributions. However, it has been demonstrated that specific hyaluronan synthase mutations can create variants that produce HA of altered size.

2.2.2 Streptococcal fermentation

Streptococci are non-sporulating and non-motile bacteria that at the optical microscopy appear as small spherical or ovoid cells that usually grow as pairs or chains surrounded by an extensive extracellular capsule: typically, the hyaluronan capsule is one to three times the diameter of the cell body (figure 2).

HA has been produced commercially since the early 1980s through fermentation of group C streptococci, in particular *Streptococcus equi* subs. *equi* and subs. *zooepidemicus* (Yamada and Kawasaki, 2005). Given the high viscosity of HA solutions, it is not practical to ferment HA beyond 5–7 g/L of product: usually the yield of polysaccharide on consumed carbon source is around 0.05-0.1 g/g and the molecular weight of the polysaccharides is averagely 1-2 MDa, being the maximum molar mass reported up to date 4 MDa (Rangaswamy and Jain, 2008). In table 3 the most important fermentation processes described in literature articles are briefly depicted.

Microorganism	Fermentation mode	Main nutrients	Oxygenation parameters	Biomass and HA yield HA molecular weight	References
S. equi subsp *zooepidemicus* (ATCC 35246)	Batch 2L	Mussel processing wastewater 50 g/L and tuna peptone 8 g/L	500 rpm 0 vvm	X: 3.67 g/L; [HA]: 2.46 g/L MW: 2.5MDa	Vazquez et al., 2010
S. equi subsp *zooepidemicus* (ATCC 35246)	Batch 2.5 L	Maltose 20 g/L, CDM	10 Hz 1.3 vvm	X: 2 g/L; [HA]: 2.14 g/L M_W: 2.1 MDa	Chong & Nielsen., 2003
S. equi subsp *zooepidemicus* (ATCC 35246)	Batch 2L	Glucose 60 g/L, CDM	600 rpm 1 vvm	X: 3.5 g/L; [HA]: 4.2 g/L MW: 3.2	Armstrong & Johns, 1997
Streptococcus sp. ID9102 (KCTC 1139BP)	Batch 75L	4% glucose, 0.75% YE, 1% casein peptone, Gln+Glu+ oxalic acid	400rpm 0.5 vvm	X: 3 OD_{600}; [HA]: 6.94 g/L MW: 5.9 MDa	Im et al, 2009
S. zooepidemicus (ATCC 39920)	Batch 3L	Glucose 20 g/L, YE 10 g/L, + acetoin and acetate	300 rpm 1 vvm	X: 2.43 g/L; [HA]: 2.15 g/L MW: n.d.	Wu et al., 2009
S. equi subsp *zooepidemicus* (ATCC 39920)	Batch 10 L	Sucrose 50 g/L, 10 g/L of casein hydrolysate	400 rpm 2 vvm	X: 6.5 g/L; [HA]: 5.1 g/L MW: 3.9 MDa	Rangaswamy & Jain 2008
S. zooepidemicus G1 (mutant of ATCC 39920)	Batch + pulse 5L	40 g/L glucose, 20 of polypeptone, 10 of YE	n.d. 10-80% DO	X: n.d.; [HA]: max 3.5 g/L MW: max 2.19	Duan et al, 2008
S. zooepidemicus WSH 24	Fed-batch 7L	Sucrose 70 g/L, 25 of YE	200 rpm 0.5 vvm	X: 16.3 g/L; [HA]: 6.6 g/L MW: n.d.	Liu et al., 2008
Streptococcus	continuous	Chemically defined medium (CDM)	High dilution rate	25% higher than batch cultures	Blank et al. 2008

Table 3. Overview of the different fermentation conditions reported in literature for HA production in Streptococci fermentations, with specific reference to medium components and aeration strategies. X:Biomass CDM: chemically defined medium; n.d.: not determined or not descripted; NTG: N-methyl-N'-nitro-N-nitrosoguanidin; *phbCAB* genes: polyhydroxybutyrate synthesis genes; YE: yeast extract.

The HA production from streptococci may be influenced from genetic factors and bioprocess parameters. First it must be considered that these microorganisms can produce hyaluronidases (HAase), extracellular enzymes that hydrolyze the external polysaccharide, leading to the decrement of both concentration and molecular weight of the product.
Consolidated strain improvement procedures have been implemented (i.e. chemical mutagenesis with N-methyl-N'-nitro-N-nitrosoguanidin), followed by a serial selection scheme, to obtain colonies lacking HAase and β-haemolytic activity, among those also selecting fast growing and overproducing HA cells (Kim et al., 1996).
Culture conditions affecting hyaluronan production are various, like medium composition, pH and dissolved oxygen concentration and geometry and speed velocity of the stirrer.
Because streptococci have specific nutritional requirements, being auxotrophic for some aminoacids and vitamins, medium formulations typically include yeast or animal extract, or casein hydrolysate as well as divalent metal ion (Mg^{2+} and Mn^{2+}), to permit polysaccharide synthesis. Occasionally calf blood or serum, as growth factors, and sometimes lysozyme,

have been added to the medium to stimulate HA production (Chong at al., 2005). However, increasing restriction of regulatory agency in Europe (EMEA) and United States (FDA), and the specific concern strictly related to every compound coming from animal sources nowadays prevent exploitation of such components in production processes. A few chemically defined media have been formulated for microbial metabolism studies but because of low yield they proved not suitable for industrial use (Chong at al., 2005).

The carbohydrate (glucose or sucrose) concentration during the fermentation process is proportionally correlated to intracellular substrates levels of HAS, whose enzymatic activity and conversion rate depends on cytoplasmatic activated sugar levels. Therefore, differently from reported fermentation strategies (to avoid growth inhibition due to substrate accumulation and overflow metabolism), sugar should be maintained high during HA production process. It has been proposed that HAS activity mechanism consists in a single protein synthesizing a single HA chain during its lifespan.

In traditional batch processes, that are widely applied for HA production, optimal sugar concentration at the inoculum is about 60-70g/L; batch fermentation procedures implemented by pulsed carbohydrate concentrated feed, have been reported successful in increasing yield, two spike feed additions (from 20 to 50 g/L of sugar) after 8-10 hour of fermentation are generally used. Continuous fermentation strategies have been exploited (table 3) only at lab scale but they are not yet industrially applied, probably due to the instability of HA-producing phenotype of streptococcal strains.

Cooney et al. (1999) aimed to increase the ATP yield of catabolism, unfortunately, by using a glucose limitation strategy the average molecular weight of HA produced resulted lower presumably due to inadequate supply of UDP-sugars to the HA synthase during its half-life. However it may be feasible to obtain a higher ATP yield using a different sugar source such as maltose, that is slowly metabolized by streptococcal cells. In fact, Chong and Nielsen (2003) succeeded, in batch fermentation, to increase the yield of HA produced on carbon source consumed from 0.088 g/g using glucose to 0.1 g/g employing maltose as carbon source. At a molecular level analysis, it was found an up-regulation of cytosolic NADH oxidases (NOX) gene.

Aeration is another key parameter during hyaluronate production process. The biopolymer can be produced in both anaerobic and aerobic fermentation, however the latter favours a higher yield and molecular weight of hyaluronan (table 3); in particular Armstrong and Johns (1997) observed a 20% increase in HA yield when *S. zooepidemicus* was grown under aerobic conditions. First attempts of producing HA using streptococci include anaerobic fermentations but the product had a low molecular weight (e.g BrackeJW & Thacker K, 1985; Park et al., 1996). Successively aerobic conditions (0.5-2 volume of gas per volume of culture per minute (vvm)) proved to favour HA production; this phenomenon can be explained by the following: (1) oxygen may stimulate HA synthesis as the aggregation of streptococcal cells mediated by their HA capsule shielded them from oxygen metabolites (Cleary & Larkin, 1979; Chong & Nielsen., 2003); (2) in presence of oxygen the energetic yield on glucose increase, due to the presence of the NOX that catalyzes the following reaction: 1 O_2+ 2 NADH→ 1 H_2O+ 2 NAD^+, contributing in such way to the energetic flux of bacterial metabolism; (3) dissolved oxygen in the medium can redirect part of carbon flux to acetate production in place of lactic acid ($Y_{ATP/glucose}$ is 3 mol/mol with acetate production against 2 mol/mol for homolactic metabolism). The extra ATP concurrently generated during the formation of acetate by acetate kinase facilitates the attainment of the higher growth yields and also the increase of hyaluronan titer.

The effect of stirring is unclear: the need for "vigorous" mixing is described, probably to enhance oxygen transfer, yet the polymer chain is reported susceptible to mechanical stress (Chong et al., 2005).

2.2.3 Genetics tools to improve HA production

Metabolic engineering of streptococci has been improved immensely over the last decade, thanks to genomic sequence now available for a number of relevant streptococci, including *S. pyogenes* and *S. equi* (Yamada & Kawasaki, 2005). Nevertheless there are few examples reported in literature of recombinant DNA techniques resulting in strain improvement towards better HA production, probably because of the number of genes involved in HA production pathway and its regulation.

Chong and Nielsen (2003) tried to maximize HA production overexpressing the endogenous *nox* gene in a *S. zooepidemicus* strain: in shaking flask experiments lactic acid and ethanol production decreased to advantage the catabolic pathway towards acetate, with the consequent increase of ATP yield; however no increase on HA yield was observed.

Krahulec and Krahulcova (2006) succeeded to increase final sodium hyaluronate concentration in the medium of about 29% using streptococci where β-glucuronidase gene was deleted; again this result was obtained in laboratory scale experiments.

Hyaluronan production in heterologous host may be an alternative way to overcome issues associated with streptococcal HA production.

De Angelis et al. (1998) were able to express *hasA* gene of *P. multocida* in an *E. coli* strain and confer to the host the capability to produce hyaluronan capsule *in vivo*.

A new system for HA synthesis was reported (Yamada & Kawasaki, 2005): *Chlorovirus* (virus of single-celled green algae, *Chlorella*) PBCV-1 was found to produce fibrous material on the cell wall of the host, that was shown to be HA. Experimentally, approximately 0.5-1 g/L of hyaluronan was recovered from a culture of *Chlorella* cells infected with *Chlorovirus*.

Recently, HA produced by using a genetically modified *Bacillus subtilis* strain has been developed by Novozymes (Widner et al., 2005). The advantages of employing *B. subtilis* to produce HA are various: first of all this bacterium is a generally recognized as a safe strain and the produced HA is free of exotoxins and endotoxins; moreover it is easy to grow in industrial fermenters; furthermore its genome has been sequenced and genetic modifications can be easily achieved; besides as HA producing streptococci, *B. subtilis* is a gram-positive microorganism that has the potential to biosynthesize HA, possessing all enzymatic activity necessary except HAS.

In particular, Widner and co-workers (2005) overexpressed in a *Bacillus subtilis* strain the *hasA* gene from *Streptococcus equisimilis*, which encodes the enzyme hyaluronan synthase along with the endogenous *tuaD* gene encodes for UDP-Glc dehydrogenase resulting in the production of HA in the 1 MDa range in 3L fermentation experiments.

Successively, also Chien and Lee (2007) succeeded in producing hyaluronan from a *B. subtlis* strain. The recombinant *B. subtilis* strain developed contained *VHb* (*Vitreoscilla* haemoglobin) gene, *S. zooepidemicus hasA*, and endogenous *tauD* genes in the expression cassette, by cultivation of these recombinant strains in 250 mL shaked flasks (30 h) they obtained about 1.8 g/L of HA.

2.3 Recovery and purification of HA from fermentation broth

In all the fermentation processes reported HA is released in the medium during fermentation mostly in the late deceleration-stationary phase of the growth curve.

Purification is then obtained directly from fermentation broth after cell removal. The separation of streptococcal cells is quite tricky. It has to be considered that in high yield fermentation medium viscosity (dynamic viscosity) increase overtime reaching 2000-3000 Pa · s. This creates a very strong buoyance force that preclude successful centrifugation unless using diluted broth (i.e.5/10 fold). Recently we studied the influence of earth aided filtration on biomass separation and HA recovery from fermentation broth, also evaluating the effect on average molecular weight of the biopolymer during these first step of downstream processing (Schiraldi et al., 2009). However to accomplish separation and recovery, repeated precipitation, ultrafiltration, CPC precipitation have been reported so far. In all cases specific attention on endotoxin removal should be carefully planned when a pharmaceutical grade product is needed.

For instance an efficient process was recently reported by Rangaswamy and Jain (2008). In this paper the fermentation broth of *Streptococcus zooepidemicus* cultivated in a 10 L reactor, was treated following a novel downstream process. Cell removal was obtained after dilution in pyrogen free water (1:1, v/v), with high speed centrifugation (17686 *g*), the supernatant was then precipitated with 2-propanol, resuspended in 3% w/v sodium acetate, and treated on silica gel and carbon prior to diafiltration and microfiltration. This process is schematically represented in the flowchart in figure 4, and permitted to recovered HA with specification meeting the Europenan Pharmacopaeia standards (2003) with a satisfying yield of 65%.

In recent years, studies aimed at accomplishing accurate and complete characterization of hyaluronan chains have remarkably intensified. In fact, because of the well-established dependence of HA biological activity on its molecular weight, basic research is interested in well-characterized HA fragments (covering a wide range of chain lengths and with low polydispersity) that could be used in experimental models to unravel the correlation. SEC systems coupled with a multi-angle light scattering detector and a refractometer (SEC-MALS-RI) are commonly used for the analysis of hyaluronan and, generally, of biopolymers for which molecular weight standards are difficult to obtain (Jing et al, 2006). Likewise a complete characterization of HA fragments generated during enzymatic hydrolysis was obtained by our group using a Viscotek instrument equipped with triple detector (La Gatta et al., 2010).

Hyaluronan obtained by both animal cell extraction and biotechnological processes is at the basis of many applications that will be presented in the following paragraph.

3. Hyaluronan applications

3.1 The properties of HA exploited in the biomedical applications.

HA finds a broad range of biomedical applications due to a unique combination of properties such as (1) high hygroscopicity; (2) viscoelastic nature; (3) magnificent biocompatibility; (4) non immunogenicity; (5) capacity to degrade in safe products.

1. The great capacity of the polymer to retain water is related to its hydrophilic chemical nature. Due to the presence of carboxylic groups on the chains, it behaves as a polyelectrolyte at physiological pH (HA pK_a = 2.9); in the presence of water, HA molecules can expand in volume up to 1000 times and form loose hydrated matrices. (Lapcik & Lapcik, 1998; Brown & Jones, 2005).
2. The viscoelastic nature refers to the rheological behaviour of HA aqueous solutions that exhibit the elasticity of a gel combined with the viscosity of a fluid. Undergoing

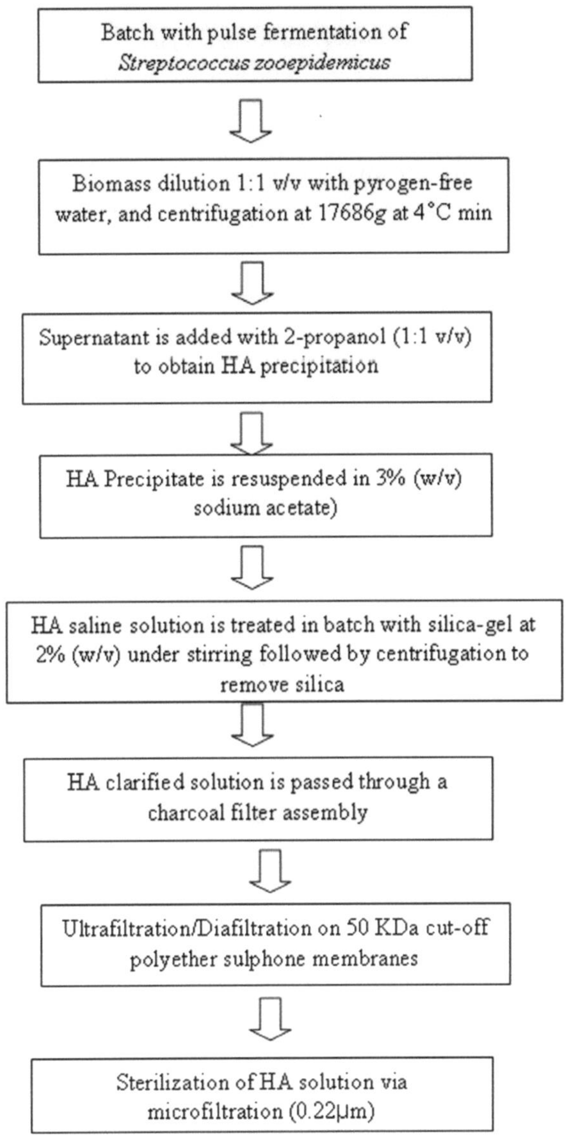

Fig. 4. Overview of hyaluronic acid biotechnological production process from *Streptococcus zooepidemicus* fermentation to recently proposed downstream procedure as described by Rangaswamy and Jain (2005).

rheological measurements, HA solutions behave as pseudo-plastic materials decreasing their viscosity at the increasing of the shear rate. Many of the HA medical uses are based on this shear thinning behavior. Rheological properties are functions of the polymer molecular weight (i.e. chain length), concentration and environmental conditions, like pH (Lapcik & Lapcik, 1998).

3. Being a natural component of many human tissues, HA is highly biocompatible, a property that is essential for the application in biomedicine.

4. HA molecules present the same structure in all species and all tissues and therefore they never "alert" the immune system (Matarasso, 2004).
5. HA is *in vivo* mainly degraded through hydrolysis catalyzed by hyaluronidases giving safe products. It has been estimated that the half-life of HA in the skin is about 24h, in the eye 24-36h, in the cartilage 1-3 weeks and 70 days in the vitreous humor (Laurent & Reed, 1991; Stern et al., 2007; Murray et al., 2005).

Because of the above highlighted properties, the development and commercialization of HA based products are in continuous intensification (Widner et al., 2005). HA is principally used in the treatment of osteoarthritis, in cosmetics, in ophthalmology, in aesthetic medicine, in surgery and wound healing, in topical drug delivery, and in tissue engineering. (Brown & Jones, 2005; Girish & Kemparaju, 2007)

3.2 HA in commercial formulations: linear, derivatized and crosslinked forms.

In some of the aforementioned fields of application, HA is used in its natural occurring linear form. However, for many purposes, it requires chemical modifications. In particular, it is usually subjected to derivatization processes (modification of the linear chain) or crosslinking processes (formation of covalent bonds between HA chains resulting in three-dimensional HA networks).

Delivered modifications allow to overcome the high rate of HA *in vivo turn over* that is required in specific applications. For instance, if linear HA is used for intra-dermal injections, it would be too rapidly degraded to provide its advantageous effects over a significant period of time. Modified HA, on the contrary, being less susceptible to chemical and enzymatic hydrolysis, shows a prolonged *in vivo* persistence thus performing better (Brown and Jones, 2005). Modification processes, especially crosslinking ones, also enhance specific mechanical properties of the material (Brown & Jones, 2005).

A schematic representation of linear, derivatized and crosslinked HA is shown in Figure 5.

In several commercially available formulations, HA (linear or chemically modified) is also found in combination with other polymers (chondroitin sulphate, carboxy methyl cellulose etc.).

HA chemical modifications are generally performed involving the hydroxyl or the carboxyl groups of the polymer.

Strategies for HA derivatization include esterification and sulphation processes. Sulphation is performed at the hydroxyl groups of the HA chains, giving products that exhibit an heparin-like activity correlated to the sulphation degree (Magnani et al., 1996). Esterification processes involve the carboxylate moieties of the polymer, that are converted in ester groups, thus causing a decrease in the total polymer charge contemporary increasing hydrophobicity (Vindigni et al., 2009). As a consequence, polymer solubility in water is reduced depending on the degree of modification thus making HA more stable in physiological environment. Among the derivatized HA based products, benzyl esters of HA are the most diffuse on the market.

In the last decade many strategies have been developed for the production of crosslinked HA, some of them are commonly employed in marketed formulations. These strategies include bis-carbodiimide crosslinking (Sadozai et al., 2005), polyvalent hydrazide crosslinking mediated by carbodiimide (i.e. EDC: 1-ethyl-(3,3-dimethylaminopropyl)carbodiimide) and co-activators (i.e. N-hydroxysulfosucinimide -sulfo-NHS- or 1-hydroxybenzotriazole –HOBt-) (Bulpitt & Aeschlimann, 1999; Prestwich et

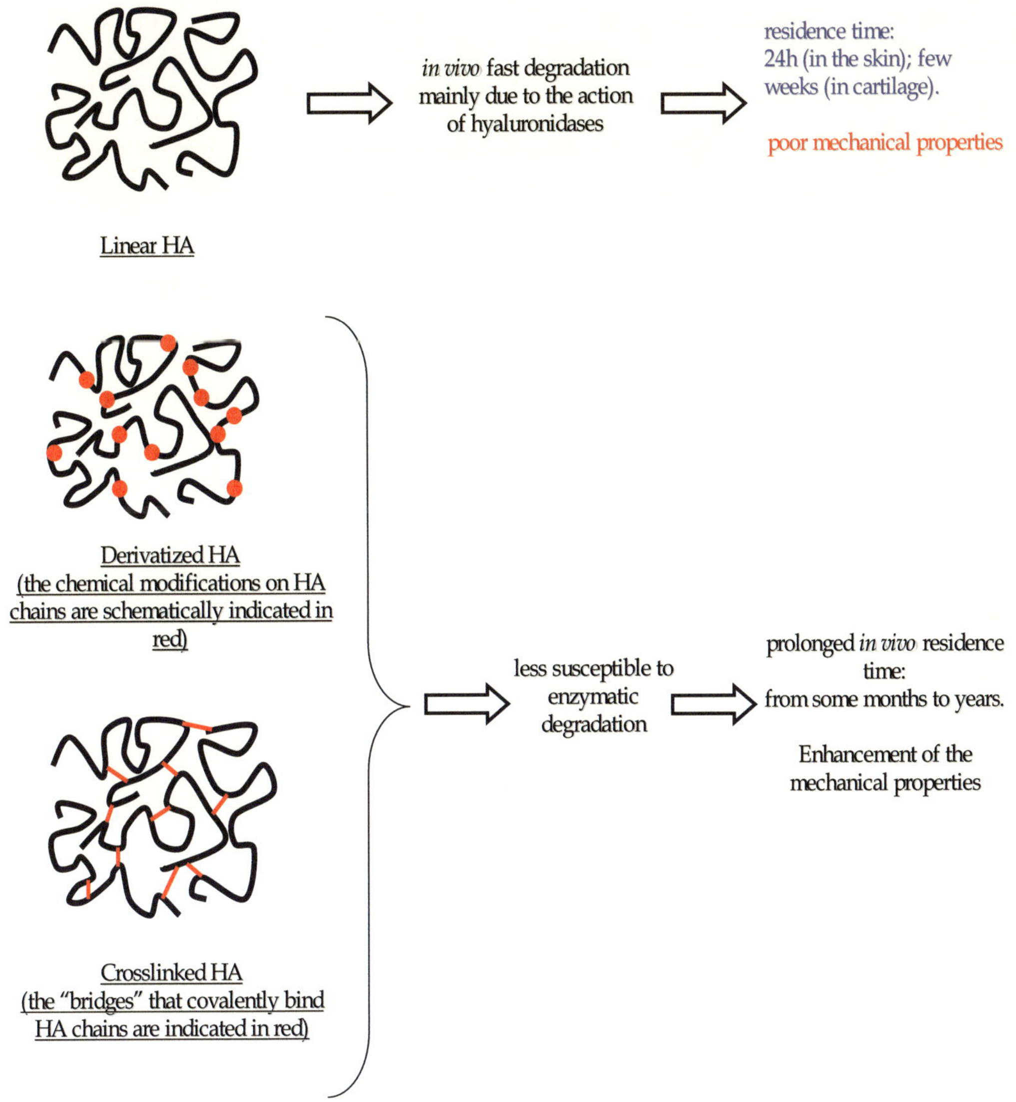

Fig. 5. Cartoon of the HA forms used in commercial formulations: linear, derivatized and crosslinked. Some applications require the use of chemically modified HA in order to enhance HA stability and tune the material mechanical properties.

al., 1998), disulfide crosslinking (Shu et al., 2003), auto-crosslinking mediated by carbodiimide and a co-activator or by 2-chloro-1-methylpyridinium iodide (CMPI) (Radice et al., 2002; Young et al., 2004), photocrosslinking (Park et al., 2003; Leach et al., 2003). All aforementioned processes involve the carboxyl groups of the HA chains. Other strategies involving the HA hydroxyl groups include divinyl sulfone crosslinking (Larsen et al., 1993, Ibrahim et al., 2010) and di-epoxide crosslinking (Agerup, 1998; Segura et al., 2005). The reported crosslinking processes are schematized in table 4.

HA group involved	Crosslinking agent	Product	Reference
Carboxyl	bis-carbodiimide	HA crosslinked via N-glucuronil urea or O-glucuronil isourea groups.	Sadozai et al., 2005
	polyvalent hydrazide coupled with carbodiimide and co-activator	HA crosslinked via hydrazide bonds	Bulpitt et al., 1999; Prestwich et al., 1998
	Carbodiimide + co-activators (sulfo NHS/HOBt) or CMPI	HA auto-crosslinked via ester bonds	Radice et al., 2002; Young et al., 2004
	Ditiobishydrazide coupled with carbodiimide	HA disulfide-crosslinked via air oxidation	Shu et al., 2003
	Metacrylating agent coupled with carbodiimide	HA photocrosslinked after exposition to light	Leach et al., 2003; Park et al., 2003
Hydroxyl	di-epoxide	HA crosslinked via ether bonds	Agerup, 1998; Segura et al., 2005
	divinyl sulfone	HA crosslinked via ether bonds	Balazs et al., 1986 Larsen et al., 1993, Ibrahim et al., 2010

Table 4. The table resumes the main strategies for HA crosslinking presented to date in patent reports and scientific literature

3.3 Linear HA applications.

Linear HA finds application mainly in cosmetics, in ophthalmology and in wound healing.
In **cosmetics** it is used as a moisturising component due to its hydrophilic nature. Almost all the main cosmetic brands present a line of hyaluronan based creams.

Wrinkles appear due to the depletion of HA in the skin with aging. The use of beauty products like creams containing HA helps to hydrate the skin and restore elasticity thus reducing the wrinkles depth. In fact, when HA solutions are applied on skin surface, they are supposed to form a light coating which absorbs moisture from the air thus hydrating the skin (and filling the wrinkles). Additionally, HA is supposed to stimulate epidermal cells migration. Besides, this HA coating should allow biologically active substances contained in the cosmetics to persist on site and eventually to penetrate more easily into epidermis. Cosmetic HA formulations proved capable to protect human skin from UV irradiation (Trommer et al., 2003).

Linear HA received significant attention also in the **topical delivery of drugs**. *Solaraze* (Pharma Derm, US), for example, is a formulation consisting of 3% w/w diclofenac in 2.5% w/w HA gel. It is indicated for the local treatment of actinic keratoses (Brown et al., 2001; Wolf et al., 2001). In fact, HA proved to significantly enhance the penetration of diclofenac through the stratum corneum (which normally acts as a barrier to the permeation of molecules into deeper skin layers) and the retention and localization of the drug in the epidermis with respect to the control or other carriers (Brown et al., 2001).

Preparations based on linear HA are used to favour the healing in the general **treatment of skin irritations and injuries**. *Jaloplast Cream* (Fidia, Italy), for instance, is a preparation containing HA as the main component (0.2% w/w sodium hyaluronate) intended for coating of acute and chronic wounds (abrasions, areas of skin grafts, post-surgical incisions, first and second degree burns, metabolic and vascular ulcers and pressure sores). The product permitted faster cicatrisation and recovery of the lesions (Lopex et al., 2005).

Plastic surgery bio-revitalization is a treatment of intradermal delivery of HA intended to counteract and prevent the skin's aging process. It is based on the use of HA to correct and smoothen facial lines and wrinkles. The HA is injected in the skin in small doses to help, restore and preserve its elasticity and healthy appearance. It is applied to frown lines, crow's feet and marionette lines. The result is smoother, more compact and more luminous skin.

In **ophthalmologic surgery**, linear HA physiological solutions are used to protect the delicate eye tissues and to provide space during surgical manipulations (Brown & Jones, 2005; Arshinoff et al., 2002; Neumayer et al., 2008). Viscoelasticity is the main HA feature responsible for this application. When stationary (static), the high viscosity of the HA solution allows to manipulate ophthalmologic tissues and to maintain the surgical space. The low viscosity of the solution at high shear rates permits easy injection and removal by pushing and sucking it through a cannula. The elasticity of the solution protects ocular cells from surgical instruments and implants.

One of the most utilized products belonging to this category is *Healon*, by Abbott Medical Optics Inc. (AMO) (USA). It is a viscoelastic physiological solution of highly purified, high molecular weight fraction of sodium hyaluronate 1% (w/w, pH 7.0-7.5) indicated for use as a surgical aid in cataract extraction, Intra Ocular Lens (IOL) implantation, corneal transplant, glaucoma filtration and retinal attachment surgery (Arshinoff et al., 2002, Oshika et al., 2004).

Viscoat (Intraocular Viscoelastic Injection) by Cilco (USA) is another product indicated as a surgical aid in anterior segment procedures including cataract extraction and IOL implantation (table 5). *Viscoat* has been formulated as a combination of sodium hyaluronate (medium molecular weight fraction), 30mg/mL, and sodium chondroitin sulphate, 40mg/mL, in a physiological buffer because the cornea contains the greatest concentration of chondroitin sulphate, respect to the vitreous and the aqueous humor where HA is prevalent (Rainer et al., 2005).

In **ophthalmology**, linear HA is also used as the active ingredient of many eyewash formulations. *Hyalistil* by Sifi (Italy) is, for instance, a 0.2% w/w hyaluronate solution indicated for the stabilization of the tear film and for the hydration and the lubrication of the cornea. It is useful in increasing the comfort during contact lenses application. Once more HA hygroscopicity and viscoelasticity are the basis for this application. *Blink Contacts* by AMO (USA) are eye drops for contact lenses users containing HA 0.15% w/w indicated for prolonging the comfort of the device.

In **urology,** intravesical instillation of linear HA has been recently used as effective alternative treatment of interstitial cystitis, recurrent urinary tract infections, and hemorrhagic cystitis. In fact HA is a protective barrier of the urothelium. A damaged glycosaminoglycan layer may increase the possibility of bacterial adherence and infection. This damage is proposed to be a causative factor in the development of the pathologies listed above, and hemorrhagic cystitis due to posthematopoietic stem cell transplantation.

Product	Source	HA form	Co-formulation with other active compounds	Crosslinking Agent	Manufacturer	Indications
Jaloplast	n.r.	linear	NO	-	Fidia Advanced Biopolymers s.r.l; Italy.	treatment of skin injuries
Healon	rooster combs	linear	NO	-	Abbott Medical Optics Inc.; USA	ophtalmology
Hyalistil	n.r.	linear	NO	-	Sifi ; Italy	eye wash
Solaraze	n.r.	linear	contains diclofenac	-	Pharma Derm; USA	topical drug release
Viscoat	biotechnological production	linear	combined with chondroitin sulphate	-	Cilco; USA	ophtalmology
Synovial	biotechnological production	linear	NO	-	IBSA, Switzerland	osteoarthritis
Hylaform	rooster combs	crosslinked	NO	DVS	Genzyme Corp.; USA	dermal filler
Synvisc	rooster combs	crosslinked	NO	DVS	Biomatrix; USA	osteoarthritis
Restylane	biotechnological production	crosslinked	NO	BDDE	Q-Med ; Sweden	dermal filler
Amalian	biotechnological production	crosslinked	NO	n.r.	S&V technologies AG; Germany	dermal filler
Viscofill	biotechnological production	crosslinked	NO	n.r.	IBSA Pharmaceuticals srl; Italy	dermal filler
Incert	n.r.	crosslinked	NO	biscarbodiimide	Anika Therapeutics, Inc.; USA	anti-adhesion barrier
ACP gel	n.r.	auto-crosslinked	NO	EDC-coactivator/CMPI	Fidia Advanced Biopolymers s.r.l.; Italy	dermal filler and anti-adhesion barrier
Hyaff	n.r.	modified by esterification	in some formulations it is combined with autologous chondrocytes	-	Fidia Advanced Biopolymers s.r.l.; Italy	tissue engineering
Seprafilm	n.r.	derivatized and partially crosslinked	combined with carboxymethyl cellulose	EDC	Genzyme Biosurgery ; USA	anti-adhesion barrier

Table 5. Summary of major marketed products based on linear, crosslinked and derivatized hyaluronic acid

However, the available clinical data regarding the effectiveness of HA as a potential treatment of patients with interstitial cystitis, recurrent urinary tract infections, and hemorrhagic cystitis are up to now limited.

3.4 Crosslinked HA applications.

Crosslinked HA derivatives find application especially in aesthetic medicine, in the treatment of osteoarthritis and in tissue engineering.

The use of crosslinked HA in **aesthetic medicine** considerably increased in the last decade (Lupo, 2006; Andre, 2004). In fact, HA based dermal fillers have become the most successful response to the current massive demand for non-surgical soft tissue augmentation. Intra-dermal injections of HA fillers are performed to fill wrinkles and to augment the volume of soft tissues such as lips and breast (Brown & Jones, 2005). According to the American Society of Aesthetic Plastic Surgery, more than 85% of all dermal filler procedures performed in 2008 occurred with HA based products (Beasley et al., 2009) Because of the great clinical and commercial impact, almost each company producing medical devices for aesthetic medicine has launched an HA based dermal filler.

HA fillers are generally made of micrometric differently crosslinked HA particles suspended in physiological solution. Often, they also contain linear un-crosslinked HA to facilitate the injectability (Allemann & Baumann, 2008; Beasley et al., 2009). They differ for HA concentration, the crosslinking agent used, the crosslinking degree, the particle size, the swelling capacity, the amount of soluble HA present in the formulation and the elastic modulus (Allemann & Baumann, 2008; Beasley et al., 2009). These properties strictly affect their final clinical performance.

Among the commercially available products, *Restylane* (Q-med, Uppsala, Sweden) and *Hylaform* (Genzyme Corp., Boston MA) exhibit the longest clinical history. *Restylane* is made of HA (biotechnological product) micrometric particles crosslinked with BDDE (Matarasso, 2004; Beasley et al., 2009; Manna et al., 1999) at a final HA concentration equal to 20mg/mL (table 5). *Hylaform*, also known as *Hylan B* gel, consists in HA of animal origin crosslinked with divinyl sulfone (Matarasso, 2004; Beasley et al., 2009; Manna et al., 1999). Micrometric HA particles of *Hylaform* are suspended in physiological solution at a concentration of 5.5 mg/mL (Matarasso, 2004). In the table 5, the *Amalian* and the *Viscofill* products, more recently appeared on the market, are also indicated.

Crosslinked and also linear HA based products are used in the treatment of osteoarthritis. HA is a physiological component of the synovial fluid and its concentration is reduced in osteoarthritic joints (Mathieu et al., 2009). Intra-articular injections of crosslinked and linear HA were found to have therapeutic effects on osteoarthritic pathologies. Several studies have been performed to investigate such effects revealing that HA is able to suppress cartilage degeneration, to protect the soft tissue surfaces of joints, to normalize the rheological properties of the synovial fluid and to reduce pain perception (Altman, 2000; Uthman et al. 2003; Girish & Kemparaju, 2007). FDA approved *Synvisc* (Biomatrix), as a medical device since 1997, this product is made of Hylan GF-20, a DVS cross-linked HA derivative (Conrozier & Chevalier, 2008). *Hyalgan* (Fidia), *Orthovisc* (Anika) and *Synovial* (IBSA) are examples of linear HA based commercial products widely used in the osteoarthritis treatment. A survey by Frost and Sullivan (2007), reported the global market for HA in the treatment of osteoarthritis worthed $940 million, pointing out that the major markets were United States, Japan and Europe, the latter accounting for $121.2 million in

2006. The Frost and Sullivan analysis also permitted a projection of continuous market growth in Europe till 2013 of 2.1% per year, leading to a final hypothetical value of $139.7 million.

Crosslinked HA was also proposed for using in the **prevention of post surgical adhesions.** For instance *ACP* (AutoCrosslinked Polymer) by Fidia (Italy) is an autocrosslinked HA derivative (in which intra-and inter-molecular ester bonds are formed involving hydroxyl and carboxyl groups of HA chains) that was found to be effective in reducing adhesions after abdominal surgery in animal models and in the clinical practice (Belluco et al., 2001). *Seprafilm*, manufactured by Genzyme Biosurgery (USA), is an adhesion barrier (membrane) made of hyaluronan and carboxymethylcellulose (CMC) chemically modified with EDC. Presumably, such product is partially derivatized, partially crosslinked (Young et al., 2004). It is indicated for use in patients undergoing abdominal or pelvical laparotomy to reduce the incidence, the extent and the severity of postoperative adhesions (Chuang et al., 2008). A similar application is proposed for Incert by Anika Therapeutics, Inc. (Woburn, MA) (Haney & Doty, 1998) (table 5).

Crosslinked HA is diffusely proposed for **tissue engineering** applications, though to our knowledge no product is present at the moment on the market. However, great part of the scientific research in polymeric biomaterials is currently focused on the development of novel constructs including HA as the main component of the scaffold. This topic will be extensively discussed in a following paragraph.

3.5 Derivatized HA applications.

Esters of HA are the most utilized derivatized HA based products. They find applications especially in **tissue engineering**. One of the most endowed materials is represented by *HYAFF* (Fidia Advanced Biopolymers, Italy), a benzyl ester of hyaluronan. In particular, *HYAFF-11* (a completely esterified hyaluronan derivative) is used in many medical applications for tissue repair, controlled drug release, nerve regeneration, wound dressing. It proved effective as a scaffold for skin and cartilage regeneration (Caravaggi et al., 2003; Grigolo et al., 2002; Tonello et al., 2003). It is available in several forms: films, gauzes, sponges, tubes and microsphere. *Laserskin* and *Hyalograft C Autograft* are examples of Hyaff-based commercialized materials. *Laserskin* consists in sheets of HYAFF-11 in which micro-perforations with diameter of 40-500μm were made (Price et al., 2007). It was successfully applied in the treatment of burns and skin lesions (Lobmann et al., 2003). *Hyalograft C Autograft* is a commercial 3D HYAFF-11 scaffold enriched with autologous chondrocytes successfully applied for the treatment of cartilage defects since 1999 (Vindigni et al., 2009).

4. Novel hyaluronan based scaffolds for tissue engineering applications

Because of its role in the extracellular matrix, hyaluronan is addressed as the more suitable among natural polymers for the development of novel functional constructs in **Tissue Engineering** (TE) applications. These TE constructs are generally made of scaffolds combined with appropriate cell lines and/or bioactive substances. As known, the role of the scaffold is essentially to provide an appropriate physical and mechanical support and to act as an artificial extracellular matrix able to properly interact with the cells guiding their proliferation and leading to tissue formation. It can be reasonably argued that the scaffold-cell interaction is the basis of TE successful outcome.

Since the surface chemistry and the 3-D structures of the scaffolds are key parameters affecting the scaffold-cell interaction, researchers are exploring a large number of chemical compositions and architectures.

Considering that ECM is made of polysaccharides and proteins, several formulations in which HA is combined with collagen, gelatin, chondroitin sulphate have been investigated. For example, a bi-layer micro-porous membrane made of gelatin, chondroitin-6-sulphate and HA crosslinked via 1-ethyl-3(3-dimethylaminopropyl)carbodiimide (EDC) has been produced and evaluated by Wang and co-workers in 2007. They demonstrated that keratinocytes and dermal fibroblasts were well attached on the bi-layer membrane. Collagen II/hyaluronan/chondroitin-6-sulfate tri-copolymer was investigated as scaffold for nucleus pulpous tissue engineering by Huang and co-workers (2010).

HA combinations with synthetic degradable polymers have also been applied. Nesti and co-workers (2008) combined HA with poly(L-lactic acid); the resulting scaffold was then successfully combined with human mesenchymal stem cells proved a promising material for intervertebral disc regeneration (Nesti et al., 2008).

In order to optimize the 3D architecture of the scaffold, a wide variety of methods have been used. The main target is to obtain a porous structure, with interconnected pores (to facilitate the transport of nutrients and oxygen inside the scaffold and the removal of waste products of cellular metabolism) contemporary mimicking the natural ECM structure. Conventional techniques for the production of micro-porous scaffold include solvent casting, fiber bonding, phase separation, porogen leaching and gas foaming.

Recently, the importance of nanostructured matrices that can mimic the nanofibrous structure of the natural ECM has been recognized. The most promising technique up-rising is electrospinning, by which HA nanofibers have been obtained.

Kim and co-workers (2008) fabricated a nanofibrous and macroporous scaffold of HA added with different amount of collagen by combining the electrospinning process with a salt leaching technique and using EDC for the crosslinking of the electrospun polymeric fibers. They aimed by this technical approach to achieve colonization of the scaffold core by cells. They found bovine chondrocytes satisfyingly adhering on the surface of the scaffold with improvement at increasing collagen content in the matrix.

Xu and co-workers (2009) used electrospinning to obtain HA/gelatin nanofibrous scaffolds stabilized by crosslinking through EDC/NHS. They found an increased resistance to degradation with the increase in gelatin amount also proving biocompatibility contemporary to deficient mouse embryonic fibroblasts adherence.

Besides the mimicking of 3D extracellular matrix, the reproduction of the surface characteristics of the ECM is also desirable since it is known that **surface chemistry** is also responsible for the regulation of cellular behaviour. One of the mechanisms of cell adhesion to the ECM is based on the interaction of trans-membrane proteins, particularly integrins, with ECM proteins such as fibronectin, osteopontin, vitronectina, collagen, laminina. In particular, integrins recognize a preserved sequence of three amino acids Arg-Gly-Asp, also called RGD, that is present in many ECM proteins. On this basis, efforts to increase the biological activity of the scaffold surface through the introduction of "adhesive" signals have been made. The most common chemical approaches are surface coating with ECM proteins and surface functionalization by the chemical anchoring of the RGD sequence.

Hyaluronan-alginate-chitosan based scaffold was proposed for cartilage regeneration: the scaffold surface was modified with an RGD-containing protein. Cytocompatibility studies demonstrated that the addition of the RGD-containing protein enhanced the cellular adhesion and proliferation. *In vitro* and *in vivo* studies demonstrated the suitability of the polymeric material for the proposed application (Hsu et al., 2004). Finally HA hydrogels with RGD peptides were proposed for brain tissue engineering (Cui et al., 2006).

5. Conclusions

Hyaluronan is a strategic biopolymer of primary scientific interest also because of the multiplicity of applications in cosmetic and biomedical fields. For this reason research is continuously growing in many interdisciplinary fields attempting on one side to the improvement of biotechnological production processes on and another side to the development of new hyaluronan formulations/HA-based new materials. Research is promoted by the commercial demand for satisfying improvements in any established application or foreseen novel uses. Scientific discussion is still open from a metabolic engineering side and also on the development of new biotranformation processes, aiming to the production of biopolymer of specific molecular weight. This particular aspect is strictly related to biological function as many literature reports point out. Despite it was firstly isolated eighty years ago, we are far apart from a comprehensive knowledge of hyaluronan related chemico-physical and biological phenomena and strong scientific effort is still needed to completely exploit its potentiality.

6. Acknowledgement

We gratefully aknowledge Dr Iolanda Marzaioli, and Dr Sara Vinciguerra that contributed through their PhD thesis to part of the literature review.

7. References

Agerup B, 1998. Polysaccharide gel composition. USP 5827937.

Alho AM and Underhill CB, 1989. The hyaluronate receptor is referentially expressed on proliferating epithelial cells. J Cell Biol 108: 1557–1565.

Allemann IB, Baumann L, 2008. Hyaluronic acid gel (Juvederm) preparations in the treatment of facial wrinkles and folds. Clin Interv Aging 3(4): 629-634.

Altman RD, 2000. Intra-articular sodium hyaluronate in the osteoarthritis of the knee. Semin arthritis rheum 30: 11-18.

Andre P, 2004. Hyaluronic acid and its use as a "rejuvenation" agent in cosmetic dermatology. Semin Cutan Med Surg 23: 218-222.

Armstrong DC and Johns MR, 1997. Effect of culture conditions on molecular weight of hyaluronic acid produced by *Streptococcus zooepidemicus*. Appl Env Microbiol 63(7): 2759-2764.

Arshinoff SA, Albiani DA, Taylor-Laporte J, 2002. Intraocular pressure after bilateral cataract surgery using Healon, Healon 5, and Healon GV. J Cataract Refract Surg 28: 617-625.

Balazs EA and Gibbs DA, 1970. The rheological properties and biological function of hyaluronic acid. Chem Mol Biol Intercell Matrix 3: 1241-5.

Balazs EA, Leshchiner E, Larsen NE and Band P, 1993. Applications of hyaluronan and its derivates. In: Gebelein CG (ed) Biotechnological polymers. Technomic, Lancaster. 41-65.

Beasley KL, Weiss MA, Weiss MD, 2009. Hyaluronic acid fillers: a comprehensive review. Facial Plast Surg 25: 86-94.

Belluco C, Meggiolaro F, Pressato D, Pavesio A, Bigon E, Donà M, Forlin M, Nitti D, Lise M, 2001. Prevention of postsurgical adhesions with an autocrosslinked hyaluronan derivative gel. J Surg Res 100: 217-221.

Bettelheim FA and Popdimirova N, 1992. Hyaluronic acid—synergetic glycosaminoglycan. Curr Eye Res 11: 411–419.

Blank LM, McLaughlin RL, Nielsen LK, 2008. Stable production of hyaluronic acid in Streptococcus zooepidemicus chemostats operated at high dilution rate. Biotechnol Bioeng 20; 90(6): 685-93.

Bracke JW and Thacker K, 1985. Hyaluronic acid from bacterial culture. USP 4517295.

Brown MB, Hanpanitcharoen M, Martin GP, 2001. An in vitro investigation into the effect of glycosaminoglycans on the skin partitioning and deposition of NSAIDs. Int J Pharm 225: 113–121.

Brown MB, Jones SA, 2005. Hyaluronic acid: a unique topical vehicle for the localized delivery of drugs to the skin. J Eur Acad Dermatol Venereol 19(3): 308-18.

Bulpitt P, Aeschlimann D, 1999. New strategy for chemical modification of hyaluronic acid: preparation of functionalized derivatives and their use in the formation of novel biocompatible hydrogels. J Biomed Mater Res 47: 152-169.

Caravaggi C, De Giglio R, Pritelli C, Sommaria M, Dalla Noce S, Faglia E, Mantero M, Clerici G, Fratino P, Dalla Paola L, Mariani G, Mingardi R, Morabito A, 2003. Hyaff 11-based autologous dermal and epidermal grafts in the treatment of noninfected diabetic plantar and dorsal foot ulcers:a prospective, multi center, controller, randomized clinical trial. Diabets Care 26(10): 2853-2859.

Chien LJ and Lee CK, 2007. Enhanced Hyaluronic acid production in *Bacillus subtilis* by coexpressing bacterial hemoglobin. Biotechnol Prog 23(5): 1017-22.

Chong BF and Nielsen LK, 2003. Aerobic cultivation of *Streptococcus zooepidemicus* and the role of NADH oxidase. Biochem Eng J 16: 153–162.

Chong BF and Nielsen LK, 2003. Amplifying the cellular reduction potential of *Streptococcus zooepidemicus* J Biotech 100: 33-41.

Chong BF, Blank LM, McLaughlin R, Nielsen L, 2005. Microbial hyaluronic acid production. Appl Microbiol Biotechnol 66: 341-351.

Chuang YC, Fan CN, Cho FN, Kan YY, Chang YH, Kang HY, 2008. A novel technique to apply a Seprafilm (hyaluronate–carboxymethylcellulose) barrier following laparoscopic surgeries. Fertil Steril 90: 1959–1963.

Cleary PP and Larkin A, 1979. Hyaluronic acid capsule: strategy for oxygen resistance in group A streptococci. J Bacteriol 140(3): 1090-1097.

Cooney MJ, Goh LT, Lee PL and Johns M R, 1999. Structured Model-Based Analysis and Control of the Hyaluronic Acid Fermentation by *Streptococcus zooepidemicus*: Physiological Implications of Glucose and Complex-Nitrogen-Limited Growth. Biotechnol Prog 15(5): 898-910.

Cornozier T, Chevalier X, 2008. Long term experience with hylan GF-20 in the treatment of knee osteoarthritis. Expert Opin Pharmacoter 9(10): 1797-1804.

Crater DL and Van De Rijn I, 1995. Hyaluronic acid synthesis operon (has) expression in group A streptococci. J Biol Chem 270: 18452-18458.

Cui FZ, Tian WM, Hou SP, Xu QY, Lee IS, 2006. Hyaluronic acid hydrogel immobilized with RGD peptides for brain tissue engineering. J Mater Sci Mater Med 17(12): 1393-1401.

Cywes C and Wessels MR, 2001. Group A *Streptococcus* tissue invasion by CD44-mediated cell signalling. Nature 6(414): 648-52.

De Angelis PL and Weigel PH, 1994. Immunochemical confirmation of the primary structure of streptococcal hyaluronan synthase and synthesis of high molecular weight product by the recombinant enzyme. Biochemistry 9(33): 9033-9039.

De Angelis PL, Oatman LC, Gay DF, 2003. Rapid Chemoenzymatic Synthesis of Monodisperse Hyaluronan Oligosaccharides with Immobilized Enzyme Reactors J Biol Chem 278: 35199-35203.

De Angelis PL, 1999. Hyaluronan synthases: fascinating glycosyltransferases from vertebrates, bacterial pathogens, and algal viruses. Cell Mol Life Sci 56: 670-682.

De Angelis PL, Jing W, Drake RR and Achyuthan AM, 1998. Identification and Molecular Cloning of a Unique Hyaluronan Synthase from *Pasteurella multocida*. J Biol Chem 273: 8454-8458.

Duan XJ, Yang L, Zhang X, Tan WS, 2008. Effect of oxygen and shear stress on molecular weight of hyaluronic acid. J Microbiol Biotechnol 18(4): 718-724.

Frost and Sullivan, 2007. European osteoarthritis market. www.frost.com

Girish KS, Kemparaju K, 2007. The magic glue hyaluronan and its eraser hyaluronidases: a biological review. Life Sciences 80: 1921-1943.

Goldberg RL and Toole BP, 1987. Hyaluronate inhibition of cell proliferation. Arthritis Rheum 30: 769–778.

Grigolo B, Lisignoli G, Piacentini A, 2002. Evidence for redifferentiation of human chondrocytes grown on a hyaluronan-based biomaterial (Hyaff 11): molecular, immunohistochemical and ultrastructural analysis. Biomaterials 23: 1187-1195.

Haney AF, Doty E, 1998. A barrier composed of chemically crosslinked hyaluronic acid (Incert) reduces postoperative adhesion formation. Fertil Steril 70: 145–51.

Heldin P, 2003. Importance of hyaluronan biosynthesis and degradation in cell differentiation and tumour formation. Braz J Med Biol Res 36(8): 967-73.

Hsu SH, Whu SW, Hsieh SC, Tsai CL, Chen DC, Tan TS, 2004. Evaluation of chitosan-alginate-hyaluronate complexes modified by an RGD-containing protein as tissue-engineering scaffolds for cartilage regeneration. Artif Organs 28(8): 693-703.

Huang B, Li CQ, Zhou Y, Luo G, Zhang CZ, 2010. Collagen II/hyaluronan/chondroitin-6-sulfate tri-copolymer scaffold for nucleus pulposus tissue engineering. J Biomed Mater Res B Appl Biomater 92(2): 322-331.

Ibrahim S, Kang QK, Ramamurthi A, 2010. The impact of hyaluronic acid oligomer content on physical, mechanical, and biologic properties of divinyl sulfone-crosslinked hyaluronic acid hydrogels. J Biomed Mater Res 94A: 355–370.

Im JH, Song JM, Kang JH, Kang DJ, 2009. Optimization of medium components for high-molecular-weight hyaluronic acid production by Streptococcus sp. ID9102 via a statistical approach. J Ind Microbiol Biotechnol 36(11): 1337-44.

Itano N and Kimata K, 2002. Mammalian hyaluronan synthases. IUBMB Life. 54: 195-199.

Jing W, Haller FM, Almond A, De Angelis PL, 2006. Defined megadalton hyaluronan polymer standards. Anal Biochem 355: 183-188.

Kim JH, Yoo SJ, Oh DK, Kweon YG, Park DW, Lee CH and Gil GH, 1996. Selection of a *Streptococcus equi* mutant and optimization of culture conditions for the production of molecular weight hyaluronic acid. Enz Microbial Tech 19: 440–445.

Kim TG, Chung HJ, Park TG, 2008. Macroporous and nanofibrous hyaluronic acid/collagen hybrid scaffold fabricated by concurrent electrospinning and deposition/leaching of salt particles. Acta Biomater 4: 1611-1619.

Krahulec J and Krahulcová J, 2006. Increase in hyaluronic acid production by *Streptococcus equi* subs. *zooepidemicus* strain deficient in β-glucuronidase in laboratory conditions. Appl Microbiol Biotechnol 71: 415–422.

Kumari K and Weigel PH, 1997. Molecular cloning, expression, and characterization of the authentic hyaluronan synthase from group C *Streptococcus equisimilis*. J Biol Chem 19(51): 32539-32546.

La Gatta A, De Rosa M, Marzaioli I, Busico T, Schiraldi C, 2010. A complete hyaluronan hydrodynamic characterization using a size exclusion chromatography-triple detector array system during *in vitro* enzymatic degradation. Anal Biochem 404: 21-29.

Lapcik L., Lapcik L, 1998. Hyaluronan: preparation, structure, properties and applications. Chem Rev 98(8): 2663-2684.

Larsen NE, Pollak CT, Reiner K, Leshchiner E, Balazs EA, 1993. Hylan gel biomaterial: dermal and immunologic compatibility. J Biomed Mater Res 27(9):1129-34.

Laurent UBG, Reed RK, 1991. Turnover of hyaluronan in the tissues. Adv Drug Deliv Rev 7: 237-256.

Leach JB, Bivens KA, Patrick CW, Schmidt CE, 2003. Photocrosslinked hyaluronic acid hydrogels: natural, biodegradable tissue engineering scaffolds. Biotechnol Bioeng 82(5): 578-89.

Liu L, Wang M, Du G and Chen J, 2008. Enhanced hyaluronic acid production of *Streptococcus zooepidemicus* by an intermittent alkaline-stress strategy. Appl Microbiol 46(3): 383-388.

Lobmann R, Pittasch D, Muhlen I, Lehnert H, 2003. Autologous human keratinocytes cultured on membranes composed of benzyl ester of hyaluronic acid for grafting in nonhealing diabetic foot lesions A pilot study. J Diab Compl 17: 199–204.

Lopex RJ, Gomez ST, Palmero GA, Martinez BM, Bueno MAM, 2005. Hyaluronic acid: a new trend to cure skin injuries an observational study. Rev Enferm 28(6): 53-57.

Lupo MP, 2006 Hyaluronic acid fillers in facial rejuvenation. Semin Cutan Med Surg 25: 122-126.

Magnani A, Albanese A, Lamponi S, Barbucci R, 1996. Blood-interaction performance of differently sulphated hyaluronic acids. Thromb Res 81(3): 383-395.

Manna F, Dentini M, De Pità O, Mortilla E, Maras B, 1999. Comparative chemical evaluation of two commercially available derivatives of hyaluronic acid (Hylaform® from rooster combs and Restylane ® from streptococcus) used for soft tissue augmentation. J Eur Acad Dermatol Venereol 13: 183-192.

Marcellin E, Chen W, Nielsen LK, 2009. Microbial hyaluronic caid biosynthesis in Microbial production of biopolymers and polymers precursors application and perspectives. Caisters academic press, 7: 163-179.

Matarasso SL, 2004. Understanding and using hyaluronan. Aesthetic Surg J 24: 361-364

Mathieu P, Conrozier T, Vignon E, Rozand Y, Rinaudo M, 2009. Rheologic behavior of osteoarthritic synovial fluid after addition of hyaluronic acid. Clin Orthop Relat Res 467: 3002-3009.

Matsubara C, Kajiwara M, Akasaka H and Haze S, 1991. Carbon-13 nuclear magnetic resonance studies on the biosynthesis of hyaluronic acid. Chem Pharm Bull 39: 2446–2448.

Murray CA, Zloty D, Warshawski L, 2005. The evolution of soft tissue fillers in clinical practice. Dermatol Clin 23: 343-363.

Nesti LJ, Li WJ, Shanti RM, Jiang YJ, Jackson W, Freedman BA, kuklo TR, Giuliani JR, Tuan RS, 2008. Intervertebral disk tissue engineering using a novel hyaluronic acid-nanofibrous scaffold (HANFS) amalgam. J Tiss Eng Part A 14(9): 1527-1537.

Neumayer T, Prinz, A, Findl O, 2008. Effect of a new cohesive ophthalmic viscosurgical device on corneal protection and intraocular pressure in small-incision cataract surgery. J Cataract Refract Surg 34: 1362–1366.

O'Regan M, Martini I, Crescenzi F, De Luca C, Lansing M, 1994. Molecular mechanisms and genetics of hyaluronan biosynthesis. Int J Biol Macromol 16(6): 283-6.

Oshika T, Eguchi S, Oki K, Yaguchi S, Bissen-Miyajima H, Ota I, Sugita G, Miyata K, 2004. Clinical comparison of Healon5 and Healon in phacoemulsification and intraocular lens implantation. Randomized multicenter study. J Cataract Refract Surg 30: 357-362.

Park MG, Jang JD and Kang WK, 1996. *Streptococcus zooepidemicus* medium and process for preparing hyaluronic acid. USP 5,496,726.

Park YD, Tirelli N, Hubbel JA, 2003. Photopolymerized hyaluronic acid-based hydrogels and interpenetrating networks. Biomaterials 24: 893-900.

Prescott AL, 2003. Method for purifying high molecular weight hyaluronic acid. USP 6660853.

Prestwich GD, Marecak DM, Marecek JF, Vercruysse KP, Ziebell MR, 1998. Controlled chemical modification of hyaluronic acid: synthesis, applications, and biodegradation of hydrazide derivatives. J Control Rel 53: 93–103.

Price RD, Berry MG, Navsaria HA, 2007. Hyaluronic acid: the scientific and clinical evidence. J Plast reconstr Aesthet Surg 60: 1110-1119.

Radice M, Pastorello A, Pavesio A, Callegaro R, 2002. Injectable hyaluronic acid derivative with pharmaceuticals/cells. USP 00768110.

Rainer G, Menapace R, Schmid KE, Sacu S, Kiss B, Heinze G, Findl O, 2005. Natural course of intraocular pressure after cataract surgery with sodium chondroitin sulfate 4%–sodium hyaluronate 3% (Viscoat). Ophthalmology 112: 1714–1718.

Rangaswamy V and Jain D, 2008. An efficient process for production and purification of hyaluronic acid from streptococcus equi subsp. Zooepidemicus. Biotechnol Lett 30: 493-496.

Sadozai KK, Gooding TB, Bui K, Sherwood CH, 2005. Crosslinked hyaluronic acid composition for tissue augmentation. USP 0136122 A1.

Schiraldi C, Andreozzi L, Marzaioli I, Vinciguerra S, D'Avino A, Volpe F, Panariello A, De Rosa M, 2010. Hyaluronic acid degradation during initial steps of downstream processing. Biocatal Biotransform 28(1): 83-89.

Schmidt KH, Gunther E and Courtney HS, 1996. Expression of both M protein and hyaluronic acid capsule by group A streptococcal strains results in a high virulence for chicken embryos. Med Microbiol Immunol 184(4): 169-73.

Segura T, Anderson BC, Chung PH, Webber RE, Shull KR, Shea LD, 2005. Crosslinked hyaluronic acid hydrogels: a strategy to functionalize and pattern. Biomaterials 26: 359-371.

Shu XZ, Liu Y, Palumbo F, Prestwich GD, 2003. Disulfide-crosslinked hyaluronan-gelatin hydrogel films: a covalent mimic of the extracellular matrix for in vitro cell growth. Biomaterials 24: 3825-3834.

Stern R, 2005. Hyaluronan metabolism: a major paradox in cancer biology. Pathol Biol 53(7): 372-382.

Stern R, Kogan G, Jedrzejas M, Soltes L, 2007. The many ways to cleave hyaluronan. Biotechnol Adv 25: 537-557.

Swann DA, 1968. Studies on hyaluronic acid: I. The preparation and properties of rooster comb hyaluronic acid. Bioch Bioph Acta (BBA) - General Subjects 156(1): 17-30

Tlapak-Simmons VL, Baron CA, Gotschall R, Haque D, Canfield WM and Weigel PH, 2005. Hyaluronan biosynthesis by class I streptococcal hyaluronan synthases occurs at the reducing end. J Biol Chem 280(13): 13012-13018.

Tonello C, Zavan B, Cortivo R, Brun P, Panfilo S, Abatangelo G, 2003. In vitro reconstruction of human dermal equivalent enriched with endothelial cells. Biomaterials 24(7): 1205-1211.

Toole BP, 1997. Hyaluronan in morphogenesis. J Intern Med 242(1): 35-40.

Trommer H, Wartewig S, Bottcher R, Poppl A, Hoentsch J, Ozegowski JH, Neubert RHH, 2003. The effects of hyaluronan and its fragments on lipid models exposed to UV irradiation. Int J Pharm 254: 223-234.

Uthman I, Raynauld JP, Haraoui B, 2003. Intra-articular therapy of osteoarthritis. J Postgrad Med 79: 449-453.

Vazquez JA, Montemayor MI, Fraguas J, Murado MA, 2010. Hyaluronic acid production by Streptococcus zooepidemicus in marine by-products media from mussel processing wastewaters and tuna peptone viscera. Microb Cell Fact 9(1): 46.

Vindigni V, Cortivo R, Iacobellis L, Abatangelo G, Zavan B, 2009. Hyaluronan benzyl ester as a scaffold for tissue engineering. Int J Mol Sci 10: 2972-2985.

Volpi N and Maccari F, 2003. Purification and characterization of hyaluronic acid from the mollusc bivalve Mytilus galloprovincialis. Biochimie 85: 619-625.

Wang TW, Sun JS, Wu HC, Huang YC, Lin FH, 2007. Evaluation and biological characterization of bilayer gelatin/chondroitin-6-sulphate/hyaluronic acid membrane. J Biomed Mater Res Part B: Appl Biomater 82B: 390-399.

Widner B, Behr R, von Dollen S, Tang M, Heu T, Sloma A, Sternberg D, De Angelis PL, Weigel PH, Brown S, 2005. Hyaluronic acid production in *Bacillus Subtilis*. Appl Env MicroBiol 71: 3747-3752.

Wolf JE, Taylor JR, Tschen E, Kang S, 2001. Topical 3.0% diclofenac in 2.5% hjyaluronan gel in the treatment of actinic keratoses. Int J Dermatol 40: 709-713.

Wu TF, Huang WC, Chen YC, Tsay YG, Chang CS, 2009. Proteomic investigation of the impact of oxygen on the protein profiles of hyaluronic acid-producing *Streptococcus zooepidemicus* Proteomics. 9(19): 4507-4518.

www.glycoforum.gr.jp/science/hyaluronan.

Xu S, Li J, He A, Liu W, Jiang X, Zheng J, Han CC, Hsiao BS, Chu B, Fang D, 2009. Chemical crosslinking and biophysical properties of electrospun hyaluronic acid based ultra-thin fibrous membranes. Polymer 50: 3762-3769.

Yamada T and Kawasaki T, 2005. Microbial synthesis of hyaluronan and chitin: New approaches. J Biosci Bioeng 99(6): 521-528.

Young JJ, Cheng KM, Tsou TL, Liu HW, Wang HJ, 2004. Preparation of cross-linked hyaluronic acid film using 2-chloro-1-methylpyridinium iodide or water-soluble 1-ethyl-(3,3-dimethylaminopropyl)carbodiimide. J Biomater Sci polymer Edn 15(6): 767-780.

21

Biopolymers by *Azotobacter vinelandii*

Adriana Navarro da Silva and Crispin Humberto Garcia-Cruz
Department of Food Engineering and Technology, IBILCE/UNESP
Brazil

1. Introduction

Alginate is a polysaccharide typically extracted from cell walls of brown algae (*Phaeophyta*), where is as a structural component of cell walls and intracellular spaces. The Giant kelp *Macrocystis pyrifera*, which grows abundantly on the shores of North America and South America, New Zealand, Australia and Africa, is the main world alginate source supply. There are also used varieties of *Laminaria, Ecklonia* and *Aschophyllum nodosum*. These different brown algae species produce alginates with different proportions of polyguluronic acid structure, resulting in different properties and functions of these compounds (Glicksman, 1987; Garcia-Cruz et al., 2008). The alginic acid structure consists of acid β-D-mannuronic (M) linear chains connected by links type (1 $\rightarrow$ 4) and its epimer of acid, α-L-guluronic (G) in various proportions. These residues are arranged in mannuronic (M) or guluronic (G) blocks, connected so that the sequence of these residues in the molecule is cycled. The molecule of this polymer is composed of homo-blocks, M and G, and block heteropolymer MG (Smidsrød, 1970). (Fig. 1).

Currently, production is concentrated mainly in the brown algae cultivation, however several bacteria belonging to the *Pseudomonas* and *Azotobacter* genus produce alginate and monomer blocks structure is similar in alginate produced by marine algae and synthesized by *A. vinelandii*. In contrast, the alginate produced by *Pseudomonas* does not have a block G. Because the property of alginic acid being insoluble in water at room temperature, the sodium, calcium and potassium salts of the acid, soluble in water, are preferred to be employed in the food industry. The compound most widely used is sodium alginate, which becomes insoluble by the divalent cations addition, usually calcium, resulting in gels or films, and the gel strength depends on the divalent cation nature. The alginate gels are capable of forming micro-beds and incorporate enzymes or whole living cells and this application has generated much interest in the food industry, biotechnology and biomedical sector (Smidsrød & Skjåk-bræk, 1990; Sabra et al., 2001). In the food industry, alginates are widely used as additives with the ability to increase viscosity, stabilize, emulsify and gelling aqueous solutions. The main application is in the production of ice cream, which is used to prevent crystallization and shrinkage, resulting in a homogeneous product. Another application is in salad dressings, where the sodium alginate or ester of propylene glycol alginate (PGA) are used as stabilizers to prevent phase separation, with the same purpose as it is applied to stabilize the mayonnaise emulsion phase water/oil. Because of its low calorie (1.4 Kcal / g), these polymers can be used as "body agents" in the formulation of low calorie products such as mayonnaise and pasta. The bacterium *Azotobacter vinelandii* as well as

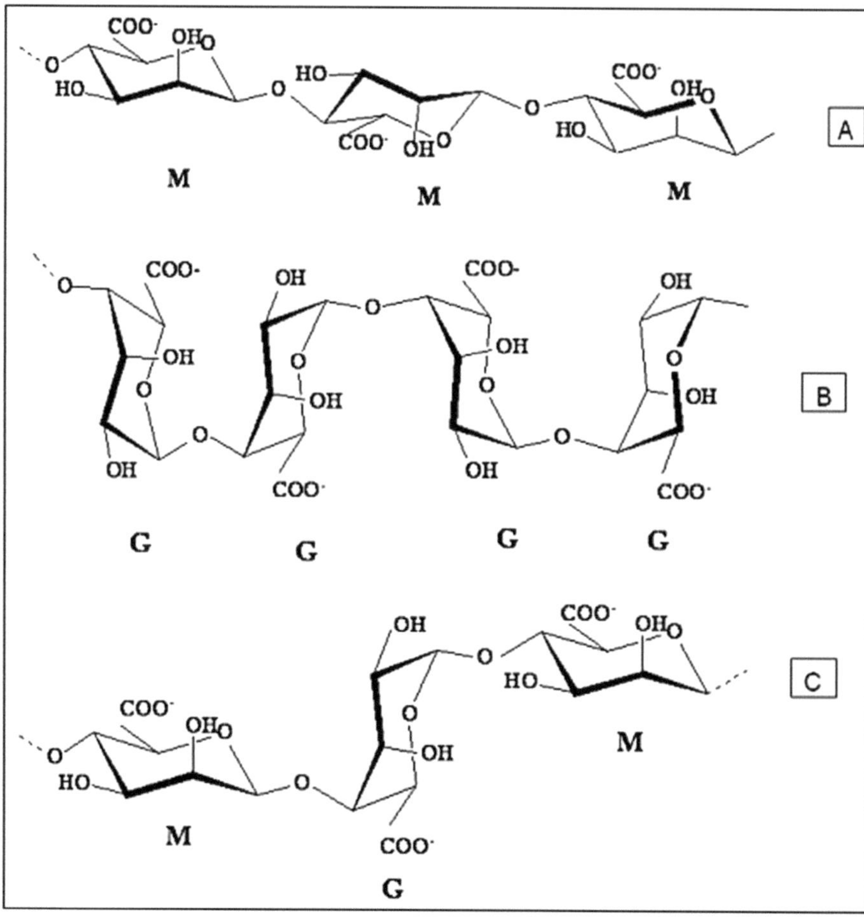

Fig. 1. Block structure homopolymeric M- and G-, and MG- block heteropolymeric, which constitute the molecule of alginate. At the top of the figure (A) has a sequence MM-; the center (B) and a GG- sequence at the bottom of the figure (C) a sequence MGM-. Adapted from Smidsrød (1970).

producing the alginate has another important feature: by limiting nutrients such as phosphorus, oxygen and the presence of an excess carbon source, produces polyhydroxyalkanoate (PHAs), intracellular polymers belonging to the polyesters family. The PHAs can be synthesized by many bacteria in bioreactors from sugars under stress. These polymers can represent up to 80% of total dry mass of the cell and are 100% biodegradable and biocompatible with the animal tissue. Conventional plastics of petrochemical origin, take decades to decompose in nature and also produce toxins during the degradation process. Therefore, there is a special interest in the plastics production from materials that can be easily eliminated from our environment (Suriyamongkol et al., 2007; Franchetti & Marconato, 2006). The PHAs are also known as bioplastics, they have thermoplastic properties and performance characteristics similar to those of conventional plastics, however, bioplastics are easily degraded by the microorganisms action in the environment. PHAs examples: polyhydroxybutyrate (PHB), poly-β-hydroxyvalerate (PHV) and polyhydroxybutyrate-co-valerate (PHB-V) (Franchetti & Marconato, 2006). (Fig. 2).

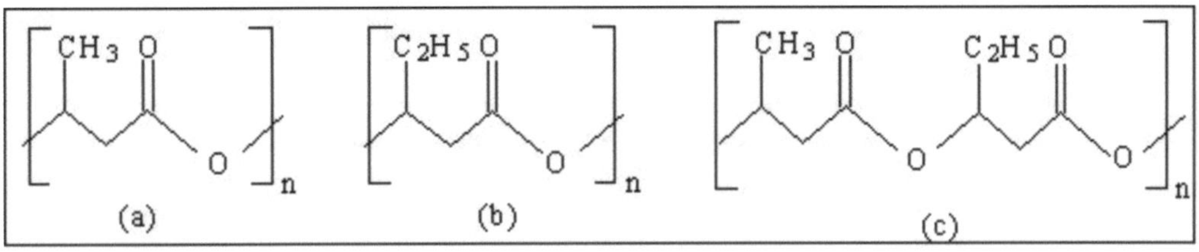

Fig. 2. Chemical structure of polyhydroxyalkanoates (PHAs): a) polyhydroxybutyrate, b) polyhydroxyvalerate, and c) polyhydroxybutyrate-co-valerate. Adapted from Franchetti & Marconato (2006).

Some possible PHAs applications include: biodegradable carriers that demonstrate the ability to deliver drugs for a given time within the individual's body, surgical needles, suture materials, bone tissue replacement, etc. The advantage of using biodegradable plastics is that it does not require surgical removal. Many microorganisms are producers of these PHAs and the *Azotobacter vinelandii* is a bacterium that can accumulate large PHB amounts with the advantage of use during its growth a wide variety of not necessarily refined sugars, like those found in cane sugar molasses, beet molasses and corn syrup.
Given the global alginate and polyhydroxybutyrate importance, this work shows the study of the both compounds production by the bacterium *Azotobacter vinelandii* in submerged fermentation using different parameters (pH, incubation temperature, incubation time, salt concentrations and also different molasses cane sugar concentrations as carbon source) and the production evaluation by methodology response surface used as a statistical tool.

2. Bacterium *Azotobacter vinelandii*

The *Azotobacteriaceae* family comprises of the *Azotobacter* genus, which are Gram-negative eubacteria that possess a cell wall complex consists of an outer membrane and an inner peptidoglycan layer containing muramic acid and murein. These bacteria reproduce by binary fission, live in soil and in fresh waters, and large ovoid cells have diameters of 1.5 to 2.0 micrometers. It can also exhibit characteristics such as the pleomorphism, varying their morphology from rod-shaped to cocci-shaped cells (Allman et al., 1990; García et al., 2002).
Bacteria of the *Azotobacter* genus are chemo-organotrophs, use sugars, alcohols and inorganic salts to grow. When free, setting an average of 10 mg of nitrogen per gram of carbohydrate (glucose) consumed and this activity requires molybdenum which can be partially replaced by vanadium. The optimum pH for growth when they fix nitrogen is 7.0 to 7.5. The *Azotobacter vinelandii* is polyploid, whose reproduction depends on the culture medium and cultivation conditions used, as well as the growth phase in which the organism (Allman et al., 1990). The biological functions that are attributed to the PHB present in *A. vinelandii* are different, being the main material to constitute a reserve of carbon and energy for the bacterial cell, which can be used during periods of nutrient limitation in the middle. These polymers store large amounts of carbon that remain condensed in the form of insoluble intracellular granules without affecting the osmotic pressure inside the cell (Page & Knosp, 1989).
Another function that is attributed to PHB in *A. vinelandii* is related to biological nitrogen fixation, specifically with the nitrogenase protection, ensuring the respiratory function of the bacterium in the absence of an exogenous carbon source by depriving the cell of a carbon and energy source rapidly oxidized. Thus, it allows maintaining an adequate respiratory

avoiding decrease the concentration of oxygen, thus contributing to the protection of this enzyme (Page et al., 1992; Almeida et al., 2004).

The *Azotobacter vinelandii* advantage is the PHB production during its growth through the use of a wide variety of refined sugars do not necessarily like those found in molasses cane sugar, beet sugar and corn syrup, for example. Another advantage is the easy attainment of this bacterium, since *A. vinelandii* is found in soils and in freshwater (Page et al., 1992).

3. Alginate

3.1 Alginate from seaweed

Brown algae (*Phaeophyceae*) containing alginate are in a diverse plants family which grow in rocky beaches or areas of the ocean with clear, rocky bottom. Some species are found at high tide, others occur in a belt along the beach, the depth shallower than 38 meters (the limit for sunlight penetration). Only a few brown algae species are used as commercial alginates source. The Giant kelp *Macrocystis pyrifera,* which grows abundantly on the shores of North America and South America, New Zealand, Australia and Africa, is the main alginate source supply of world. Are also used *Laminaria, Ecklonia* and *Aschophyllum nodosum* varieties which are also taken by some producers. These different brown algae species produce alginates with different ratios of polyguluronic acid structure, thus resulting in different properties and functions of these compounds (Glicksman, 1987).

Formaldehyde is usually added to the brown seaweed, post harvest, to prevent microbial growth during its storage and to fix the polyphenols in the algae before the alginate extraction. Knowing that formaldehyde is toxic, allergenic and possibly carcinogenic, the salt addition in *Ascophyllum nodosum* (*Phaeophyceae*) was tested as an alternative by Moen and colleagues (1999). The algae were harvested on the coast of Norway, where the salinity is about 30 ‰ from late autumn to early spring, and stored at 22 + / - 2 ° C under certain conditions. If the algae samples are stored without treatment loses quality well as the material for alginate production by up to 14 days. Treatments with formaldehyde (2% weight / weight) or 20% NaCl preserve algae for up to 46 days, but the technique is less aggressive with NaCl, it is still inconvenient and costly for the alginate industry, it is necessary controlling the temperature, pH, oxygen, and salt (Moen et al., 1999).

3.2 Bacterial alginate

The alginic acid structure consists of linear chains of residues β-D-mannuronic acid (M) joined by links type (1 → 4) and of its epimer, the α-L-guluronic acid (G) in various proportions. These residues are arranged in the form of blocks mannuronic acids (M) or guluronic (G), connected so that the sequence of these residues in the molecule is alternating. The polymer molecule is composed of homopolymer blocks, M-and G-and MG-block heteropolymer (Figure 1).

The physicochemical alginates properties depend on the molecular weight, the monomers M: G proportion along the chain and also the acetylation degree. The alginates are produced by bacteria and brown seaweed, and mannuronate residues of bacterial alginate O-acetylated in positions 2 and / or O-3.

Although seaweeds are usually the commercial alginates source, those of bacterial origin have been suggested as possible substitutes for the alginates from algae. For this purpose, studies initially focused on the opportunistic pathogenic bacterium *Pseudomonas aeruginosa* and then, three species non-pathogenic strains of *Pseudomonas*, including *P. mendocina*, *P.*

putida and *P. fluorescens*, and yet, the soil bacterium *Azotobacter vinelandii* (Brivonese & Sutherland, 1989). Most knowledge about the alginate biosynthesis comes from studies of *Pseudomonas aeruginosa*, mainly because of the medical relevance of this bacterium as an important opportunistic pathogen microorganism to humans, in patients suffering from cystic fibrosis (Alkawash et al., 2006; Govan & Harris, 1986; May & Chakrabarty, 1994). Herein, the alginates have an important role as a virulence factor. The reason for this seems to be alginate biofilm formation, which facilitates colonization of the lung (Gacesa & Russell, 1990).

A. vinelandii and *P. aeruginosa* produce alginate as an extracellular polysaccharide in vegetative cells, whereas the alginate production by *A. vinelandii* is involved in a differentiation process called "cyst" (Sadoff, 1975). This cyst is formed by the intracellular accumulation of polyhydroxybutyrate delimited within the cytoplasm by a membrane lipoprotein double wall when the bacterial cell is in an environment where, in general, there are large amounts of carbon source and nitrogen, phosphorus or oxygen limitation. When there is carbon source exhaustion, these cysts oxidize quickly (through the activation of the PHB depolymerases enzymes) and are used as energy sources.

The potential of alginate produced by bacteria, such as industrial polymers, is still a controversial subject. However, the possibility of using raw materials free of seasonal and geographical variations, and also selected strains under carefully controlled operating conditions, so as to meet specific applications in biotechnology and biomedicine may be sufficient to compensate for the relatively low production, and acetylation relatively high, the bacterial polymers (Clementi et al., 1999).

3.3 Bacterial alginate applications

The alginate gels are capable of forming micro-beds and incorporate whole living cells or enzymes, and this application has aroused interest in the food industry, biotechnology and biomedical sector (Smidsrød & Skjåk-bræk, 1990; Sabra et al., 2001).

The alginates are widely employed in the fruit analogues area or type imitation products. In 1946, Peschardt patented a process for making artificial cherries using a colored and flavored alginate solution with sugar, which was added in the form of drops in coagulant solution made with calcium salts soluble. A film of insoluble calcium alginate is formed immediately around the surface of the droplets. After dipping them in calcium salts solution, the ions penetrated by diffusion inside to gelation. Various textures types could be obtained through proper control of the type of calcium salt, concentration, time and temperature, etc. These cherries produced were not affected by heat and can be perfectly used in bakery products. Furthermore, had the advantage of providing uniformity in size, weight and quality. Similar products were developed using cherry puree in a system of alginate gelation, sold under the name "cherry shaped balls." Other type imitation products commercially found are: peppers imitation for stuffing olives, onion rings, caviar, meat and fish imitation, etc. (apud Glicksman, 1987).

3.4 Bacterial alginate biosynthesis

The alginate has an essential biological role in *Azotobacter vinelandii*, that when the vegetative cell becomes mature cysts layers of exine and intine present 32 and 13% of the dry weight of alginate, respectively. The mutant varieties of bacteria do not produce alginate and are unable to form mature cysts. The alginate extracellular accumulation acts as a barrier to oxygen diffusion or heavy metals or as protection against other environmental insults (Segura et al., 2003).

In bacterium *A. vinelandii* the alginate is synthesized from fructose-6-P, which is converted by the phospho-mannose isomerase (PMI) in mannose-6-P and this, in turn, becomes mannose-1-P by the enzyme phospho-mano-mutase (PMM) action. The next step is the mannose-1-P activation by GDP-mannose-pyrophosphorylase (GPMP) resulting in GDP-mannose formation, which is oxidized to GDP-mannuronic acid by the enzyme GDP-mannose dehydrogenase (GMD). The GDP-mannuronic acid is the substrate that is polymerized in the inner membrane to form polymannuronic acid. In periplasm some mannuronic residues of the polymannuronic acid are acetylated by acetylase. The polymer is secreted outside the cell where some manurônic residues not acetylated are epimerized to guluronic residues by multiple extracellular epimerases resulting in alginate. (Fig. 3).

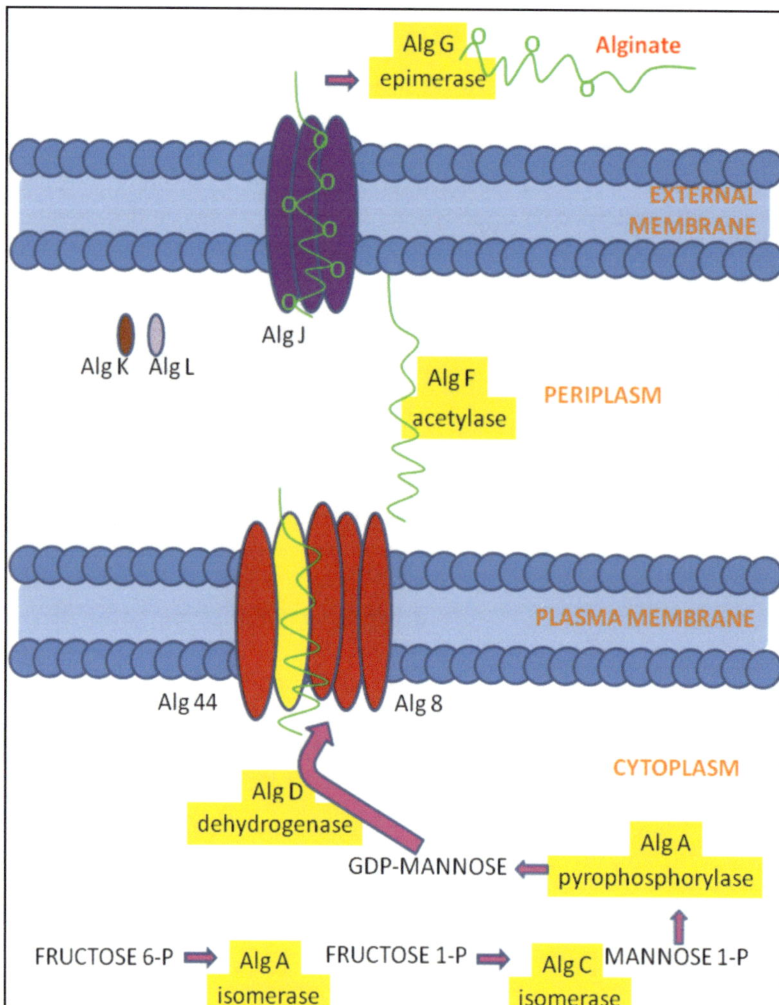

Fig. 3. Alginate biosynthesis by *Azotobacter vinelandii*. Adapted from Trujillo-Roldan et al., (2003).

The genes that encode enzymes involved in alginate synthesis, modification and excretion, were identified in *A. vinelandii*. All structural genes are clustered on the chromosome and are transcribed from multiple promoters, except algC gene that encodes the enzyme phospho-mano-mutase. The algD gene, which encodes GDP- mannose dehydrogenase (GMD) is transcribed from three promoters. Genes alg8, alg44, algK algJ which are located below the algD are organized into a transcription unit and their products participate in the alginate polymerization and excretion, the product of the alg8 is a glycosyl transferase, which has been proposed as responsible for polymerase activity, the alg44 encodes another inner membrane protein, which is responsible for part of the complex polymerization and also transports the polymer to the periplasm, the algJ encodes a protein of the outer membrane and present activity in the ion channel that is essential for the alginate excretion and algK gene product is a periplasmic protein that can participate in the AlgJ incorporation in the outer membrane. The algG gene encodes the epimerase and algL the enzyme alginase. Genes algX, algV, algI and algF are responsible for acetylation of mannuronic residues in the periplasm and as the algA product form bifunctional enzyme that catalyzes the first and third step of the biosynthesis (Gaona et al., 2004; Steigedal et al., 2008).

3.5 Bacterial alginate production

Horan and colleagues (1983) noted that when *Azotobacter vinelandii* mutant was grown in continuous culture, the polysaccharide amount produced depended on the dissolved oxygen concentration and also the carbon source. Changes in specific activity of key enzymes in the alginate biosynthesis (phosphomannose isomerase and GDP-mannose pyrophosphorylase), measured in cells extracts grown within a range of values of dissolved oxygen concentration were reflected by changes observed in the alginate production, whereas that the activity of GDP-mannose dehydrogenase remained unchanged. A similar correlation between the specific activities of these enzymes and the rate of alginate production was observed in this study during the transition of sorbitol to sucrose as sole carbon source, but the activity of GDP-mannose dehydrogenase also increased with increased alginate production. After prolonged continuous cultivation in sucrose the mutant lost the ability to produce alginate. The key enzymes in the alginate biosynthesis was not detected in extracts of this strain does not produce alginate, which had also lost the ability to form cysts. These results support the view that the alginate formation is controlled by the key enzymes induction in the alginate biosynthesis and the alginate has an important role in cyst formation by bacteria (Horan et al., 1983).
Some factors that influence the alginic acid production by *Azotobacter vinelandii* were investigated in batch culture by Brivonese & Sutherland (1989). We observed that the highest alginate yields (6.0 to 7.5 mg mL^{-1} supernatant) occurred in rich medium in nitrogen and phosphate, and glucose as carbon source, by aerating the medium with agitation at 280 rpm. The oxygen importance was evident when, at 120 rpm, the alginate production decreased to 1.4 mg mL^{-1}. A 120 rpm intracellular polyhydroxybutyric acid (a polymer whose intracellular accumulation is associated with the limitation of oxygen) accumulation was 40% versus 30% at 280 rpm. The inorganic phosphate presence was considered important for the growth, because in medium with low salts concentration resulted in decreasing production, which was not improved by the nitrogen sources addition like nitrate and glutamate. The glucose replacement by sucrose also reduced the alginate production.

In a similar study, Savalgi & Savalgi (1992) found that the largest alginic acid amounts produced by *Azotobacter vinelandii* NCIB 9068 (5.5 - 6.2 mg mL^{-1} supernatant) in flasks, were obtained when growth was conducted in a medium rich in nitrogen-limited phosphate, and sucrose as carbon source under shaking at 240 rpm. Also observed reduction in the alginate production of to 1.6 mg mL^{-1}, with the decrease in agitation (140 rpm). In contrast to Brivonese & Sutherland (1989), the sucrose replacement by glucose reduced the growth and the alginate production.
The authors Peña and colleagues (1997) had a higher alginate concentration under low aeration conditions (conventional Erlenmeyer flasks). This phenomenon may be related to oxygen sensitivity by some enzymes involved in the alginate biosynthesis by *A. vinelandii*. The results are consistent with those observed by Chen and colleagues (1985), which indicated that the alginate concentration was increased to 170 rpm (for values between 110 and 200 rpm) for *A. vinelandii* culture conducted in shake flasks. An increase in stirring speed (ie greater aeration) was unfavorable for the alginate production.
The culture broth viscosity was greater at lower aeration, suggesting an important effect of oxygen on the molecular characteristics (ie molecular weight) polymer. It is known that the alginate solutions viscosity is strongly influenced by the molecular weight distribution, but not both, content and distribution of the two monomers of the alginate molecule (Martinsen et al., 1991). However, no previous data published on the culture broth viscosity under different growing conditions. Peña and colleagues (1997) observed that the alginate acetylation degree produced by *A. vinelandii* was independent of the aeration conditions.
Clementi et al (1995) observed that the rate of alginate synthesis of drops rapidly to zero as a result of a drastic drop in pH in the *A. vinelandii* culture. However, the study by Peña and colleagues (1997), the alginate formation rate was not affected by pH drop experienced in the stationary phase of growth.

4. Bioplastics or biodegradable plastics

Biodegradable plastics are degraded by microorganisms when discarded in soil and landfills. This degradation results primarily from the microorganisms action such as fungi, bacteria and algae, generating CO_2, CH_4, cellular components and other products, as stipulated in "American Standard for Testing and Methods" (ASTM D-833). Thereby, biodegradable plastics are materials which degrade into carbon dioxide, water and biomass, as a result of the living organisms or enzymes action (Franchetti & Marconato, 2006; Suriyamongkol et al., 2007; Van-Thuoc et al., 2008).
The intracellular PHAs are polymers that can be synthesized by many bacteria in bioreactors from sugars in growth conditions characterized by carbon source excessive and other nutrients limitation such as nitrogen and phosphorus (Franchetti & Marconato, 2006; Albuquerque et al., 2007; Suzuki et al., 2008). Although many microorganisms have the ability to accumulate this polymer, a limited number is considered good candidates for industrial PHB production. Among them, *Cupriavidus necator* (formerly *Ralstonia eutropha*), *Alcaligenes latus*, *Azotobacter vinelandii* and recombinant *Escherichia coli* have demonstrated the highest polyester accumulations (about 70-90% of dry cell mass). The microorganism *Cupriavidus necator* requires limiting nutrients during fermentation to then produce PHB. This microorganism has accumulated more PHB during the stationary phase of growth. The other three bacteria do not require nutrient limitation for initiating the PHB synthesis despite the nutrient limitation in the culture medium promote greater accumulation of this

polymer within the cells. In some of these cases, these complex nitrogen sources such as yeast extract or fish peptone may be used to enhance cell growth and, in turn, the volumetric productivity (Quillaguamán et al., 2008).

Previous studies showed that enterobacteria did not accumulate PHAs naturally, requiring the cloned genes introduction. However, Lugg and colleagues (2008) studied enterobacteria of natural occurrence, *Serratia* sp, and observed that it was able to accumulate PHB (about 50% of cell dry weight) in medium containing excess carbon source and a deficiency in nitrogen.

The synthesis and incorporation of different monomers depend on the provision of a suitable substrate that can be converted into the desired hydroxyacyl-CoA through metabolic reactions in the bacterial cell. Furthermore, it is necessary that the bacterial cell contains an enzyme called PHA synthase capable of incorporating the hydroxyacyl-CoA synthesized a polymer chain.

4.1 Polyhydroxyalkanoate

The inclusions presence in the cytoplasm, like lipid inclusions, which were soluble in chloroform were first observed in *Azotobacter chroococcum* by 1900. The chemical inclusions composition found in similar bacterium *Bacillus megaterium* was identified in 1926 as the acid-poly 3-hydroxybutyrate (P(3HB)), by Lemoigne. In the late '50s, studies of the *Bacillus* genus suggested that the P(3HB) had the intracellular function of carbon reserves and energy for this bacterium. In 1974, Wallen and Rohwedder identified other polyhydroxyalkanoate beyond the P(3HB), extracted from activated sludge.. An interesting work was also done by Witholt and coworkers in 1983, when cultured *Pseudomonas oleovorans* on n-octane. An elementary analysis of the polymer showed that intracellular bacteria mainly accumulated 3-hydroxyoctanoate and small amounts of 3-hydroxyhexanoate (apud Sudesh et al., 2000, p.: 1504).

The copolymer P(3HB-co-3HV) was produced commercially by Zeneca Bioproducts, England, from glucose and propionic acid, using a mutant strain of *Ralstonia eutropha*. The polyester obtained from this process was put on the market, with the name Biopol (Byrom, 1992). The Biopol is used since 1990 in Germany in the shampoo bottles production for the cosmetics industry Wella (Braunegg et al., 1998). Other products such as disposable razors, also made by Biopol, were tested in Japan. Cups made of pure PHA or internally coated with a Biopol film, as sealant, have been used. The PHA blends with conventional plastics such as polypropylene, were also tested and currently Monsanto (USA) has the right to patent this product (Byrom, 1992).

PHB has properties similar to polypropylene with three unique features: thermoplastic processability, 100% water resistance and 100% biodegradability (Hrabak, 1992). Boom, Giacino and Selker (1994) established that PHB is an aliphatic homopolymer with melting point of 179 ° C and highly crystalline (80%). The molecular weight of PHB decreases by about half the original value when the latter is maintained at 190 ° C for 1 hour. The PHAs have physical properties ranging from brittle and thermally unstable until soft and tough, depending on their composition. The PHB physical properties, for example, crystallization and high tensile strength depends on the molecular weight, which is influenced by the species of microorganism used, growth conditions and purity of the samples (Punrattanasin, 2001; Heo et al., 2008).

4.2 Polyhydroxyalkanoate applications

An international standard ISO 10993-3 (1982) lays down specific requirements for biocompatibility, including tests based on the nature of contact and residence time of the implanted biomaterial. The standard requires that all materials that remain in contact with mucous membranes, bone or dental tissues, in which the contact exceeds 30 days, and all implantable devices in the contact exceeds 24 hours, should be tested for genotoxicity. Based on the ISO standard 10993-3 (1982) and ASTM (1987) recommendations, many test methods are needed to determine the genotoxic and mutagenic activities of the implant. These include tests for toxicity, mutagenesis, chromosomal aberrations, etc (Ali et al., 2008).

Although the P(3HB) degradation product, D(-)-3-hydroxybutyric acid, is an intermediate metabolite of many organisms (Lafferty et al., 1988; Lee, 1996), Ali and colleagues (2008) performed the mutagenicity study of polyhydroxybutyrate test using the *Salmonella*/microsome and found that the PHB is not genotoxic and does not alter the expression of proto-oncogenes and anti-apoptotic gene considered in the study. Thus, it is plausible that there animal biocompatibility with the P(3HB) and can therefore be deployed in their tissues without any toxicity.

Some possible PHAs applications include: biodegradable carriers that have the function of drug release for a given time within the individual's body; surgical needles; sutures; bone tissue replacement; etc. The biodegradable plastics advantage is not requiring surgical removal (Punrattanasin, 2001).

Another advantage of using PHB in drug delivery systems is that the drug concentration in patient blood and / or tissue can be maintained at a desired level for a long time. The main requirement for the drugs manufacture with a release system and prolonged action is the availability of a suitable capsule material, which should be absolutely harmless to the body and possess the physical and mechanical properties and biomedical necessary, including their degradation in the middle biological. Other materials are widely used in controlled drug, such as gelatin, proteins, etc. The PHAs are available in chemically pure form and their degradation rate is low when compared with the compounds mentioned above, indeed it is very important in prolonged treatments. By varying the PHAs chemical structure or their mixture with different materials, you can control the porosity and degradation rate of polymer matrix and, consequently, the rate of drug release. Currently, biodegradable carriers are used to carry anti-depressants, contraceptives, antineoplastic and antiinflammatory (Shishatskaya et al., 2008; Suzuki et al., 2008).

The PHAs are biodegradable in soil and can be used as porters to release pesticides, herbicides or fertilizers for a given time in the soil, such as containers and packaging, as biodegradable matrix for drug delivery in veterinary medicine, etc.

Possible PHAs applications include plastic films, bags, containers, razors, household items and a variety of products found in everyday life.

4.3 Polyhydroxybutyrate biosynthesis

The PHB production in the bacterium *A. vinelandii* involves three enzymes. At the beginning process is the condensation of two acetyl-CoA molecules by the enzyme β-ketothiolase to generate acetoacetyl-CoA, which is reduced by acetoacetyl-CoA reductase using NADPH, producing D(-)-β-hydroxybutyryl-CoA, which is finally polymerized by PHB synthase resulting in PHB (Segura et al., 2003).

In general, bacteria producing PHB present the accumulation of this polymer in response to interference in its growth, mainly due to lack of nutrients such as nitrogen, phosphorus, magnesium and oxygen, and the presence of an excess carbon source (Burns et al., 2007).

The control point is the major activity of the enzyme β-ketothiolase. This enzyme is activated when the acetyl-CoA concentration is high, a situation that results from the accumulation of NADH or NADPH in response to low oxygen concentration in the medium. With these metabolites, occurs enzymes inhibition (citrate synthase and isocitrate dehydrogenase) of the Krebs cycle by reducing the carbon flow into the cycle and leading to an increased of acetyl-CoA concentration, a fact that stimulates the β-ketothiolase activity. These conditions also favor the enzyme activity (acetoacetyl-CoA reductase) that catalyzes the second step of the process (Page et al., 1992).

Many researchers have identified the structural genes responsible for PHB synthesis in *A. vinelandii*. It was observed that the phbA gene encoding the β-ketothiolase and phbC and phbB genes codify the second and third enzyme, respectively, the step of PHB biosynthesis (Segura et al., 2003). (Fig. 4).

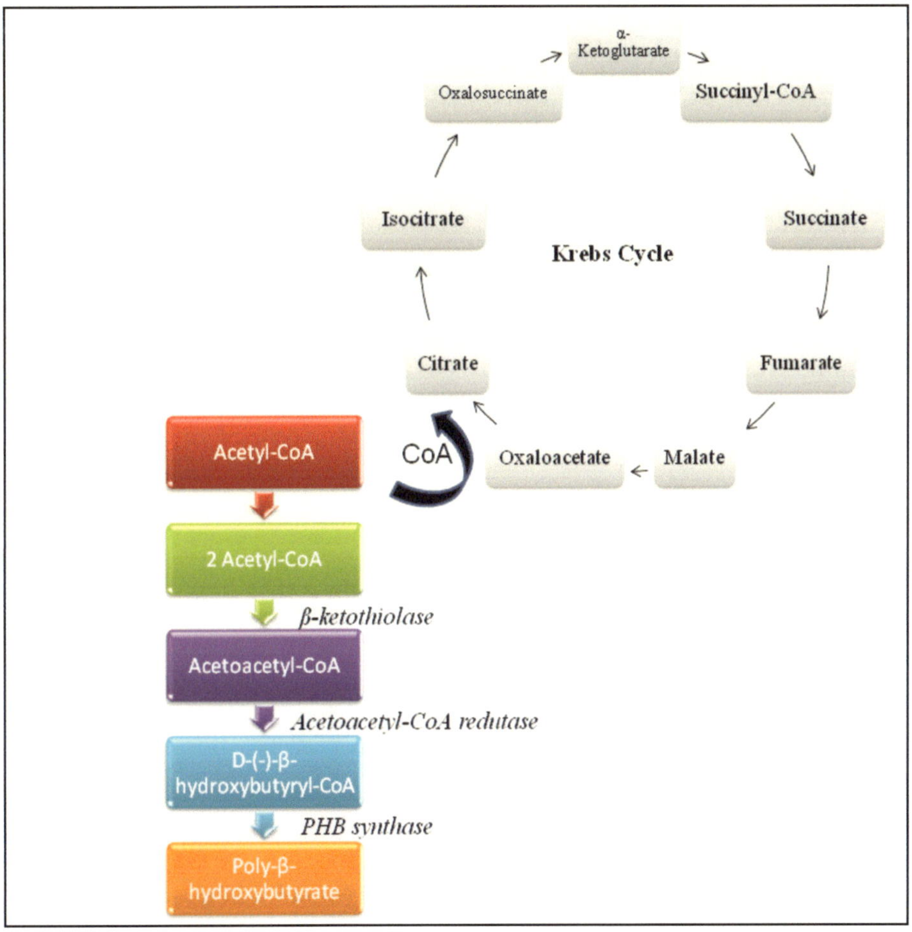

Fig. 4. Polyhydroxybutyrate biosynthesis and Krebs Cycle by *Azotobacter vinelandii*. Adapted from Segura et al., (2003).

4.4 Polyhydroxyalkanoate production

The PHA production using bacteria involves the following steps: maintaining the bacterial culture in appropriate medium, pre-fermentation and/or fermentation. During the pre-fermentation and fermentation, the microorganism is cultivated in medium containing mineral salts and one or more carbon sources. The carbon sources can be refined sugar like sucrose and glucose or more complex sources such as beet molasses, sugar cane molasses, whey and others agro-industrial residues (Page & Knosp, 1989).

5. Purpose of the study

The alginate is a polysaccharide extracted from cell walls of brown seaweed used in food, pharmaceutical and biotechnology industries. The production is concentrated in the brown seaweed cultivation, but several bacteria, *Pseudomonas* and *Azotobacter* genus, produce alginate. The chemical alginate structure by algae produced is similar to those synthesized by *A. vinelandii*. Thus there is the possibility of using raw materials free of seasonal and geographical variations, and also selected strains under carefully controlled operating conditions, so as to meet specific applications in biotechnology and biomedicine. This bacterium also produces intracellular polymers such as polyhydroxybutyrate (PHB), known as bioplastic. The PHB has properties similar to polypropylene with three unique features: thermoplastic processability, 100% water resistance and 100% biodegradability. The *Azotobacter vinelandii* advantage is the PHB production during growth using a wide sugars variety like those found in cane sugar molasses, beet molasses and corn syrup, for example. Another advantage is the easy attainment of this bacterium, because *A. vinelandii* is found in soil and also in freshwater

6. Materials and methods

6.1 Microorganism

The microorganism used was the *Azotobacter vinelandii* CCT 2841 obtained from the Research and Technology André Tosello Foundation, Campinas - SP, Brazil.

6.2 Culture medium maintenance

Azotobacter vinelandii CCT 2841 was maintained in agar Yeast Medium (YM) whose formula is given below (in grams per liter): yeast extract 3.0, malt extract 3.0, peptone 5.0, dextrose 10.0, agar 20.0 and 1.0 L distilled water.

6.3 Medium fermentation

For the polyhydroxybutyrate and alginate production was used minimal or basal medium adding cane sugar molasses as carbon source.

6.4 Minimal or basal medium

The minimal medium used was described by Garcia and colleagues (2002), composed for the following components (in grams per liter): KH_2PO_4 0.16, K_2HPO_4 0.64; $MgSO_4.7H_2O$ 0.4, NaCl 0.2, $CaSO_4.2H_2O$ 0.05; $FeSO_4.7H_2O$ 0.0025 and $Na_2MoO_4.2H_2O$ 0.001. The pH was adjusted according to the experimental design and sterilized in autoclave at 121 °C for 20 minutes. The minimal medium was sterilized separately the cane sugar molasses solutions at 121 °C for 20 minutes. Cane sugar molasses was diluted to 30 Brix with distilled water and subsequently clarified. After cool and sit for 24 h was used a refractometer to adjust the desired soluble solids concentration with sterile distilled water. Next, the flasks containing solution of cane sugar molasses in desired concentrations were sterilized in autoclave at 121 °C for 20 minutes. The basal or minimal medium was added to the fermentation stage.

7. Fermentation conditions

7.1 Pre-inoculum preparation

Azotobacter vinelandii CCT 2841 from the stock culture was transferred for tubes containing Plate Count Agar (PCA) inclined, which were incubated in an oven at 30 °C for 24 hours.

7.2 Pre-fermentation

With the pre-inoculum previously obtained, was realized the bacterial cells suspension by adding 5.0 mL of nutrient broth contained in the flask (total of 50.0 mL nutrient broth), and then moved the suspension to the same flask. The vials were incubated in an orbital shaker rotating at 30 °C for 24 h and 225 rpm (Page & Knosp, 1989; Page et al., 1992).

7.3 Polyhydroxybutyrate and alginate production

In a bucket spectrophotometer (Cintra 5 UV-VIS Doublebar ''), was added 3.0 mL of minimal or basal medium and culture medium drops obtained in the pre-fermentation to reach an optical density of 0.9 at 620 nm wavelength. For all experiments, the *Azotobacter vinelandii* CCT 2841 inoculum was standardized in 0.74 mg dry weight/mL, corresponding to 0.9 absorbance of the suspension at 620 nm. Soon after, the flasks were incubated in an orbital shaker rotating at 120 rpm, temperature and incubation time predetermined by the experimental design.

7.4 Fermentation parameters optimization

For the PHB (Y1 = polyhydroxybutyrate) and alginate (Y2 = alginate) optimization production were carried out two experimental design to determine the best production area of both compounds. Thus, in the first experiment was performed a statistical fractional factorial design 2^{6-2} and the independent variables were: X1 = soluble solids concentration (%), X2 = pH, X3 = incubation temperature (°C), X4 = ammonium acetate (mmol L^{-1}), X5 = ammonium citrate and iron (III) (mol L^{-1}) and X6 = incubation time (h), resulting in 16 experiments plus 2 replicates at the central point (Table 1).

Variables	Levels		
	-1	0	+1
X1 - Soluble solids concentration (%)	1.0	3.0	5.0
X2 - pH	6.0	7.0	8.0
X3 - Incubation temperature (°C)	25.0	32.5	40.0
X4 - Ammonium acetate (mmol L^{-1})	15.0	37.5	60.0
X5 - Ammonium citrate and iron (III) (µmol L^{-1})	30.0	60.0	90.0
X6 - Incubation time (h)	48.0	72.0	96.0

Table 1. Independent variables in the first experimental design 2^{6-2}.

With the prospect approaching the optimum production region was conducted a second experimental design for sugar cane molasses by means independent variables: X1 = carbon source; X3 = incubation temperature and X6 = incubation time, resulting in a full factorial statistical design 3^{3-0} (Table 2).

Variables	Levels		
	-1	0	+1
X1 - Soluble solids concentration (%)	5.0	15.0	25.0
X2 - Incubation temperature (°C)	40.0	50.0	60.0
X3 - Incubation time (h)	12.0	30.0	48.0

Table 2. Independent variables in the second experimental design 3^{3-0} for sugar cane molasses.

The second experiment for sugar cane molasses was made with the variables fixed: initial pH = 7.0, ammonium acetate = 60.0 mmol L^{-1}, ammonium citrate and iron (III) = 90.0 mmol L^{-1}.

7.5 Standard curve for biomass determination

To construct the calibration curve of cell concentration was used *Azotobacter vinelandii* CCT 2841 suspension obtained after the pre-fermentation. In addition, absorbance readings were made at 620 nm (spectrophotometer Cintra 5 UV-VIS "DoubleBeam") with samples of the original cell suspension and the diluted corresponding to cell dry weight of each sample for built up a calibration curve (relating the absorbance and biomass in mg mL^{-1}).

7.6 Biomass determination

The cell mass was determined by measuring in a spectrophotometer at 620 nm a cell suspension in distilled water, separating it by centrifugation at 7077g, 15 minutes at 4 °C. Previously, was obtained a correlation plot of absorbance versus cell concentration. All tests were performed in triplicate.

7.7 Polyhydroxybutyrate and alginate extraction

The cells were separated from the fermentation medium by centrifugation at 7077g, for 15 minutes at 4 °C. After this procedure, the PHB intracellular was extracted of the cell mass with chloroform and the solution was filtered with paper filter. Next, was added three times the absolute ethanol volume for PHB precipitation. The supernatant obtained on centrifugation was used to obtain the alginate. Thus, added three times the absolute ethanol volume for alginate precipitation. Both were dried in vacuum oven at 45 °C to constant weight.

7.8 Polyhydroxybutyrate and alginate purification

The purification was performed by dissolving PHB in sodium hydroxide 1 N and precipitated again with absolute ethanol (Lin & Sadoff, 1968; Pouton & Akhtar, 1996). The alginate dried was dissolved in distilled water and precipitated again with absolute ethanol. Both procedures were repeated three times.

7.9 Polyhydroxybutyrate purity determination

The PHB purity determination was performed the Law & Slepecky (1961) method based in hot reaction, the polymer obtained (PHB) with concentrated sulfuric acid of high purity analytical results in crotonic acid, whose absorbance is measured in 235 nm.

8. Production flowchart

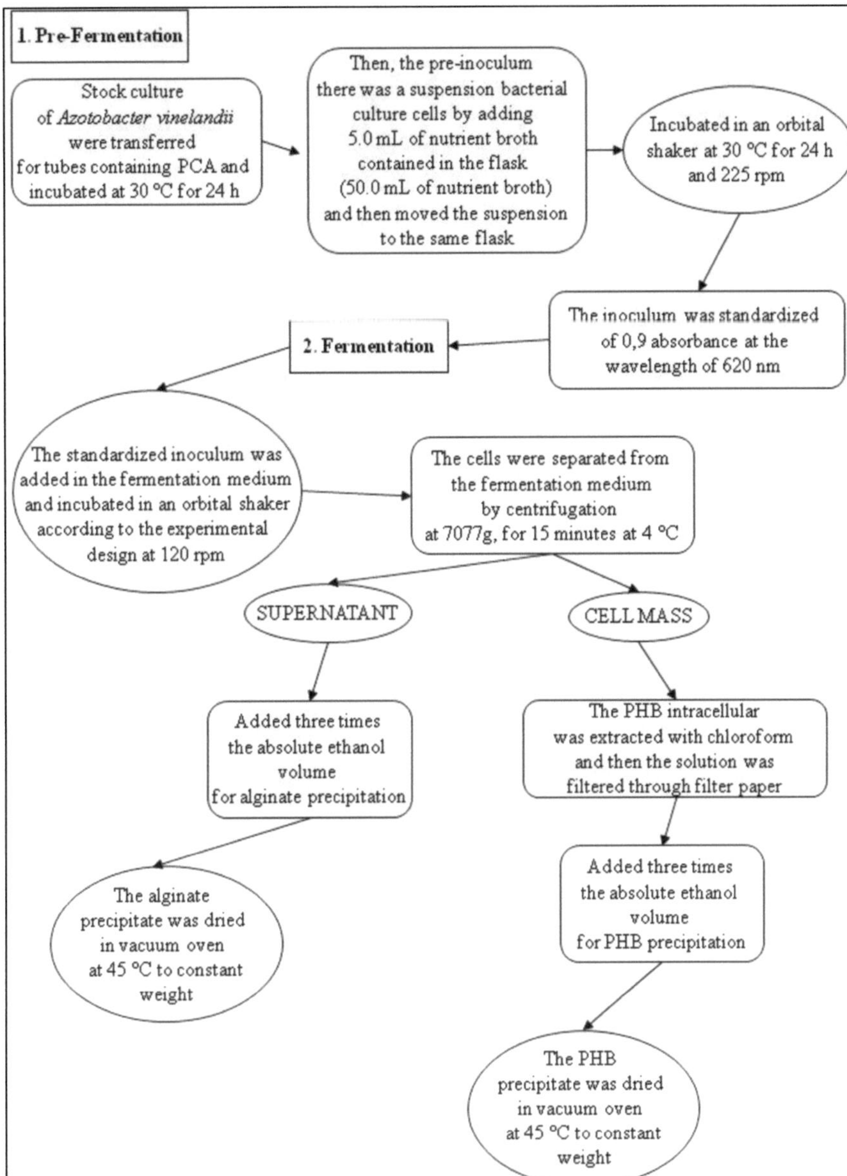

9. Results and discussion

9.1 First experimental design

With the help of the response surface methodology can be observed that the biomass (5 mg mL^{-1}) was favored by higher soluble solids concentrations. The lower incubation temperature (24 °C) and the longest incubation time favored cell growth, this is due to the need for more time to adapt to this fermentation medium, since the molasses is a medium nutritionally complex. (Fig. 5).

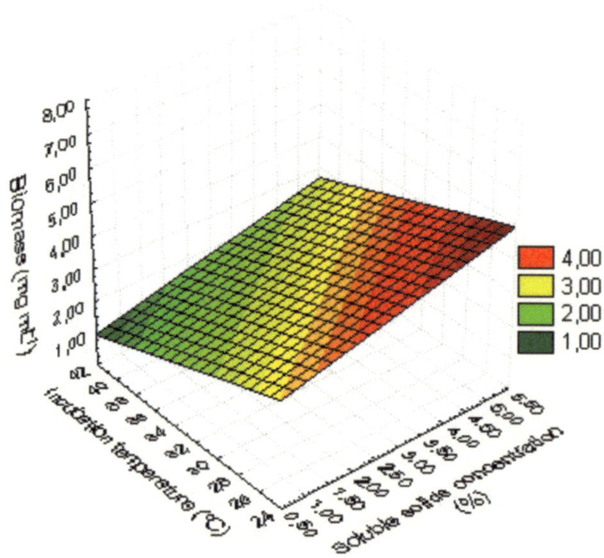

(a)

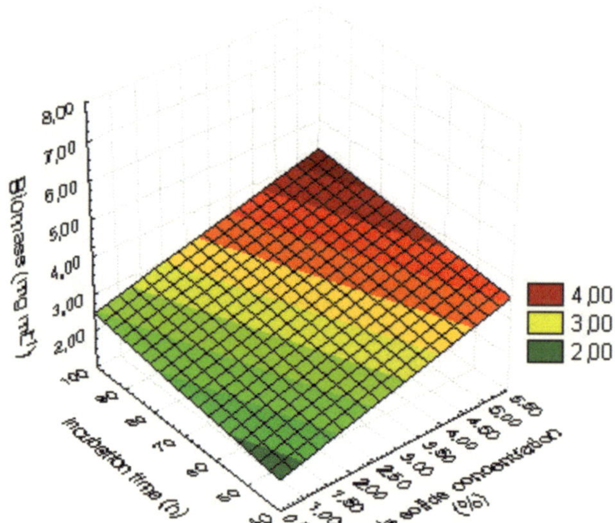

(b)

Fig. 5. (a) Surface response for biomass (mg mL^{-1}) around the optimal values of soluble solids concentration (%) and incubation temperature (ºC) (equation 6.17 + 0.34 * x – 0.13 * y), (b) biomass response areas (mg mL^{-1}) around the optimal values of soluble solids concentration (%) and incubation time (h) (equation 1.23 + 0.34 * x + 0.011 * y).

The temperature had a positive effect on the alginate production, with a higher exopolysaccharide synthesis by bacteria in response to high temperature (40 - 42 °C), functioning as a defense to the environment with a high temperature. The productivity was

also higher with soluble solids concentration increasing. Already, the shorter the incubation time (40 - 50 h) and pH (5.8 - 7.4) increased the alginate production. (Fig. 6).
The maximum alginate yield obtained using sugar cane molasses as carbon source was between 200 - 280 mg g cell^{-1} h^{-1}.

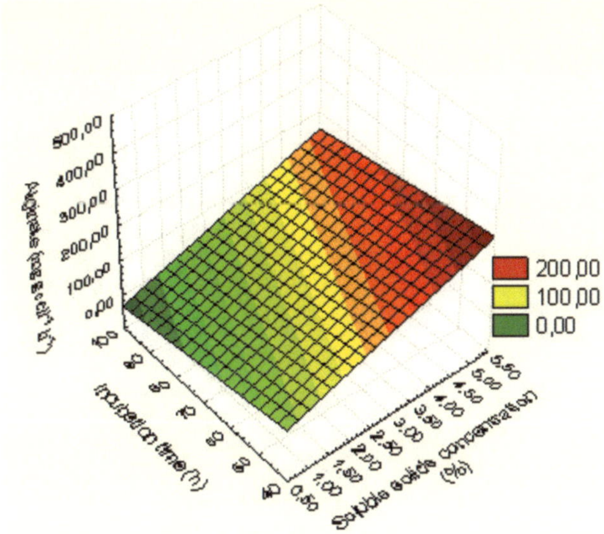

(a)

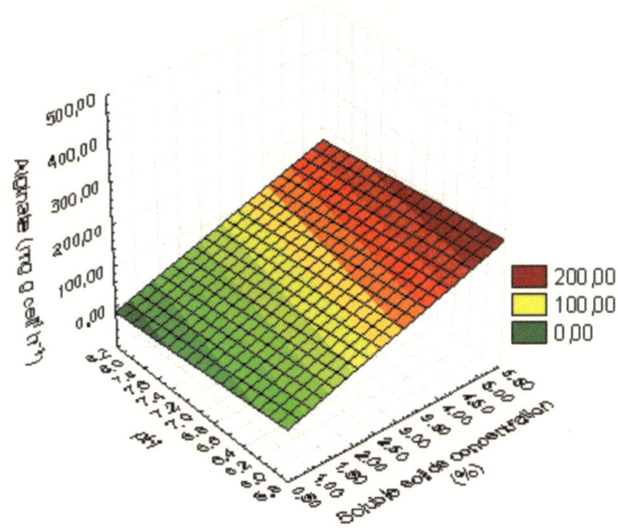

(b)

Fig. 6. Response surfaces for: (a) alginate productivity (mg g cell^{-1} h^{-1}) around the optimal values of soluble solids concentration (%) and incubation time (h) (equation: 88.73 + 36.68 * x - 1.39 * y), (b) alginate productivity (mg g cell^{-1} h^{-1}) around the optimal values of soluble solids concentration (%) and pH (equation: 102.06 + 36.68 * x - 16.26 * y).

The PHB production was maximum (PHB = 27 mg g cell^{-1} h^{-1}) in the shortest incubation time (40 h) and higher incubation temperature (42 °C). Since the sugar cane molasses concentrations tested were not significant for PHB production as shown in Figure 7.
The presented purity PHB extracted between 93.0 and 95.0%.

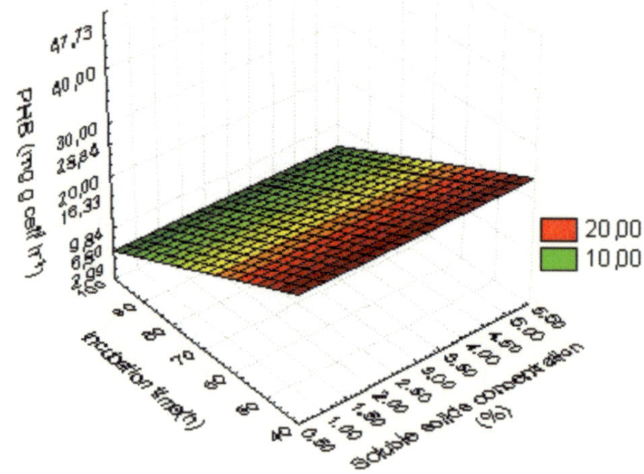

(a)

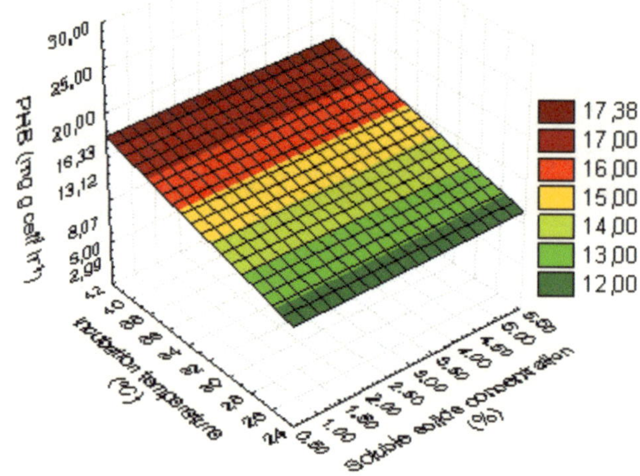

(b)

Fig. 7. Response surfaces for: (a) PHB yield (mg g cell^{-1} h^{-1}) around the optimal values of soluble solids concentration (%) and incubation time (h) (equation: 36.43 -0.0053 * x – 0.31 * y), (b) PHB yield (mg g cell^{-1} h^{-1}) around the optimal values of soluble solids concentration (%) and incubation temperature (°C) (equation 3.57 -0.0053 * x + 0.33 * y).

9.2 Second experimental design

With the second experimental design (3^{3-0}) using the soluble solids concentration, temperature and incubation time was possible to prove that the biomass increase (2.8 - 3.6 mg mL-1) was accompanied by the soluble solids concentration (12 - 26%). Noted in the first experimental design that bacterial growth was less pronounced in the incubation temperature tested (38 - 40 °C) and the longest incubation time measured (45 - 50 h). (Fig. 8).

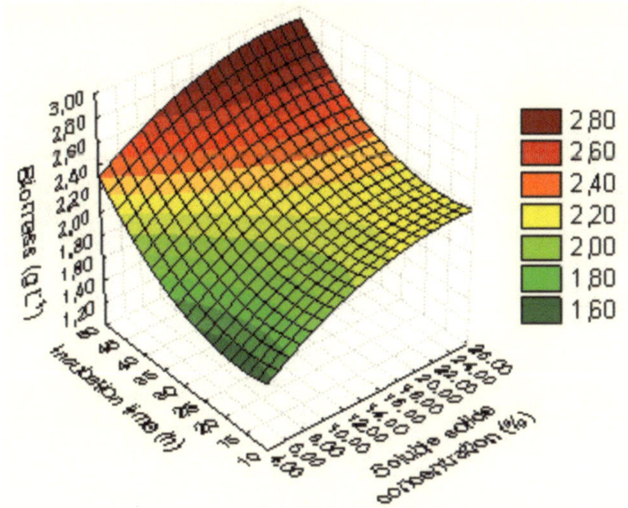

(a)

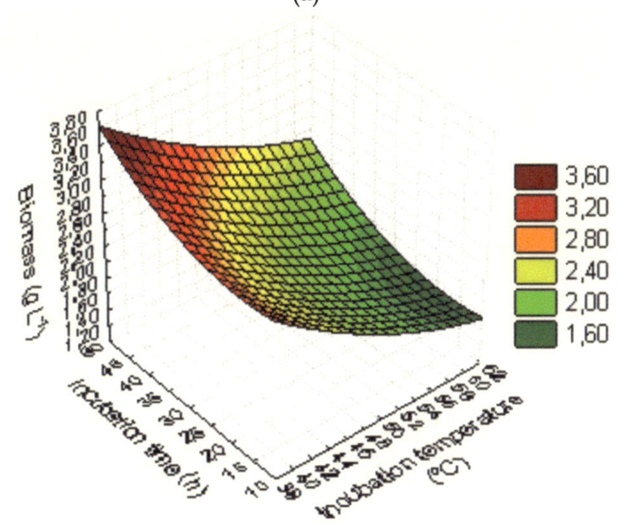

(b)

Fig. 8. (a) Surface response for biomass (mg mL^{-1}) around the optimal values of soluble solids concentration (%) and incubation time (h) (equation $7.32 + 0.072 * x - 0.0015 * x^2 - 0.025 * y + 0.00072 * y^2 - 5.83$), (b) biomass response areas (mg mL^{-1}) around the optimal values of incubation temperature (°C) and incubation time (h) (equation: $7.32 - 0.18 * x + 0.0012 * x^2 - 0.025 * y + 0.00072 * y^2 +0.75$).

During the study of alginate productivity (maximum 200 - 250 mg g cell^{-1} h^{-1}) can be observed that the optimal soluble solids concentration was between 4.0 and 8.0%, a fact consistent with the behavior of first experiment for molasses, in which the optimal concentration was 4.5 - 5.5%. The ideal temperature is between 58 - 62 °C, at high incubation temperatures and shorter incubation (10 - 40 h) had the highest alginate synthesis, as shown in Figure 9.

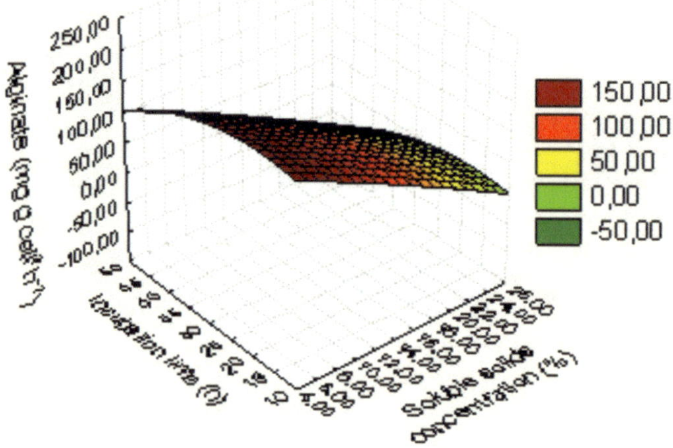

(a)

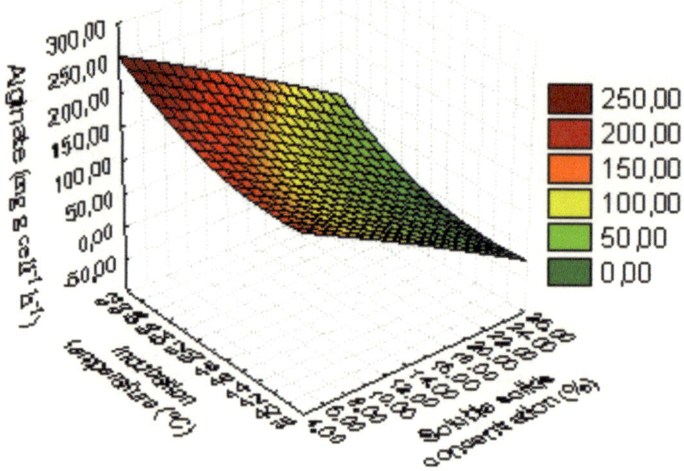

(b)

Fig. 9. Response surfaces for: (a) alginate productivity (mg g cell^{-1} h^{-1}) around the optimal values of soluble solids concentration (%) and incubation time (h) (equation: $544.83 - 7.40 * x - 0.028 * x^2 + 3.01 * y - 0.083 * y^2 - 354.12$), (b) alginate productivity (mg g cell^{-1} h^{-1}) around the optimal values of soluble solids concentration (%) and incubation temperature (°C) (equation: $544.83 - 7.40 * x - 0.028 * x^2 -18.01 * y + 0.22 * y^2 + 15.33$).

The Figure 10 shows an increase in the PHB yield from 27 to 100 mg g $cell^{-1}$ h^{-1} between the first and second experimental results for the molasses and it was noticeable that the soluble solids concentration was ideal in the range of 12 - 26 %, mostly incubation temperature (60 - 62 °C) and shorter incubation (10 - 15 h).

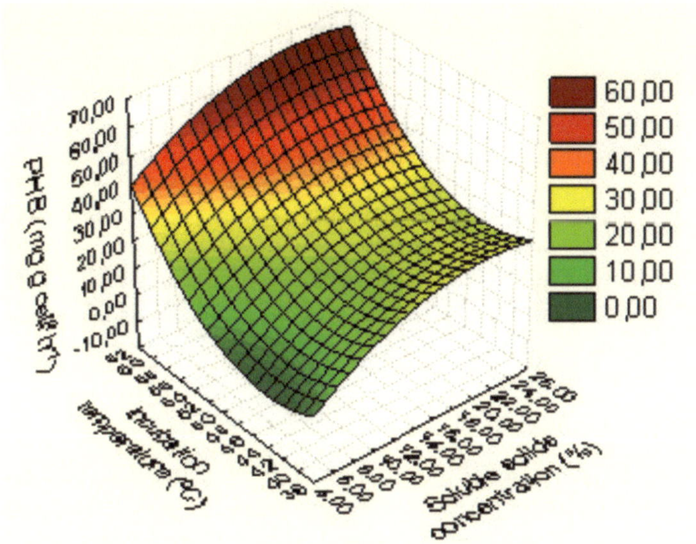

(a)

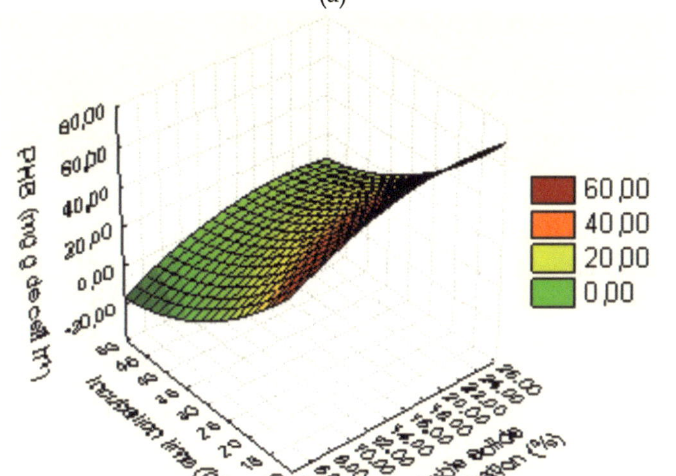

(b)

Fig. 10. Response surfaces for: (a) PHB yield (mg g $cell^{-1}$ h^{-1}) around the optimal values of soluble solids concentration (%) and incubation temperature (C) (equation: $317.71 + 3.52 * x - 0.082 * x^2 - 11.50 * y + 0.13 * y^2 - 77.04$), (b) PHB yield (mg g $cell^{-1}$ h^{-1}) around the optimal values of soluble solids concentration (%) and incubation time (h) (equation: $317.71 + 3.52 * x - 0.082 * x^2 - 3.52 * y + 0.032 * y^2 - 250.19$).

The presented purity PHB extracted between 93.4 and 95.1%.

10. Conclusion

The highest PHB yield (100 mg g cell^{-1} h^{-1}) using sugar cane molasses occurred in the incubation time of 10 h, 60.0 °C and the soluble solids concentrations between 14.0 - 25.0%. To alginate yield was observed that, using molasses, yield was greater (250 mg g cell^{-1} h^{-1}) also in the incubation time of 10 h, temperature of 60.0 °C and the soluble solids concentration between 4.0 - 6.0%. The PHB purity was between 93.0 to 97.5%. Thus, Cane sugar molasses was very promising for the alginate and PHB production.

11. Acknowledgments

We would like to thank PROPG/UNESP - Profa. Dra. Marilza Vieira Cunha Rudge and FUNDUNESP (Foundation for the Development of the State University of Sao Paulo) for the financial support for this chapter.

12. References

Albuquerque, M. G. E.; Eiroa, M.; Torres, C., Nunes, B. R. & Reis, M. A. M. (2007). Strategies for the development of a side stream process for polyhydroxyalkanoate (PHA) production from sugar cane molasses. *Journal of Biotechnology*, Vol. 130, No.4, July 2007, 411-421.

Ali, A. Q.; Kannan, T. P.; Ahmad, A. & Samsudin, A. R. (2008). *In vitro* genotoxicity tests for polyhydroxybutyrate - A synthetic biomaterial. *Toxicology in Vitro*, Vol. 22, No.1, August 2008, 57-67.

Allman, R.; Hann, A. C.; Phillips, A. P.; Martin, K. L. & Lloyde, D. (1990). Growth of *Azotobacter vinelandii* with correlation of Coulter size flow cytometric parameters, and ultrastructure. *Cytometry*, Vol.11, No.7, June 1990, 822-831.

Almeida, A.; Ruiz, J. A.; López, N. I. & Pettinari, M. J. (2004). Bioplásticos - una alternativa ecológica. *Revista Química Viva*, Vol.3, No.3, September 2004, 122 -133.

Alkawash, M. A.; Soothill, J. S. & Schiller, N. L. (2006). Alginate lyase enhances antibiotic killing of mucoid *Pseudomonas aeruginosa* in biofilms. *APMIS: Acta Pathologica, Microbiologica et Immunologica Scandinavica*, Vol.114, No.2, February 2006, 131-138.

Booma, M.; Selke, S. E. & Giacin, J. R. (1994). Degradable Plastics. *Journal of Elastomers and Plastics*, Vol.26, No.2, April 1994, 104-142.

Braunegg, G.; Lefebvre, G. & Genser, K. F. (1998). Polyhydroxyalkanoates, biopolyesters from renewable resources - Physiological and engineering aspects. *Journal of Biotechnology*, Vol.65, No.2-3, October 1998, 127–161.

Brivonese, A. C. & Sutherland, I. W. (1989). Polymer production by a mucoid strain of *Azotobacter vinelandii* in batch culture. *Applied Microbiology and Biotechnology*, Vol.30, No.1, January 1989, 97-102.

Burns, K. L.; Oldham, C. D.; Thompson, J. R.; Lubarsky, M. & May, S. W. (2007). Analysis of the *in vitro* biocatalytic productionof poly-(β)-hydroxybutyric acid. *Enzyme and Microbial Technology,* Vol.41, No.5, October 2007, 591–599.

Byrom, D. (1992). Production of poly-β-hydroxybutyrate - poly-β-hydroxyvalerate copolymers. *FEMS Microbiology Reviews,* Vol.103, No.2-4, December 1992, 247-250.

Chen, W.; Chen, J.; Chang, S. & Su, C. (1985). Bacterial alginate produced by a mutant of *Azotobacter vinelandii. Applied and Environmental Microbiology,* Vol.49, No.3, March 1985, 543-546.

Clementi, F.; Fantozzi, P.; Mancini, F. & Moresi, M. (1995). Optimal conditions for alginate production by *Azotobacter vinelandii. Enzyme and Microbial Technology,* Vol.17, No.11, November 1995, 983-988.

Clementi, F.; Crudele, M. A.; Parente, E.; Mancini, M. & Moresi, M. (1999). Production and characterization of alginate from *Azotobacter vinelandii. Journal of the Science of Food and Agriculture,* Vol.79, No.4, April 1999, 602-610.

Franchetti, S. M. M. & Marconato, J. C. (2006). Biodegradable polymers - a partial way for decreasing the amount of plastic waste. *Química Nova,* Vol.29, No.4, July/August 2006, 811-816.

Gacesa, P. & Russell, N. J. (1990). The structure and property of alginate, In: *Pseudomonas infection and alginates*. London: Chapman & Hall, p. 29-49.

Gaona, G.; Nunez, C.; Goldberg, J. B.; Linford, A. S.; Najera, R.; Castaneda, M.; Guzman, J.; Espin, G. & Soberon-Chavez, G. (2004). Characterization of the *Azotobacter vinelandii* algC gene involved in alginate and lipopolysaccharide production. *FEMS Microbiology Letters,* Vol.238, No.1, July 2004, 199-206.

García, M. C. V., López, M. J., Elorrieta, M. A., Suárez, F. & Moreno, J. (2002). Physiology of exopolysaccharide production by *Azotobacter vinelandii* from 4-hydroxybenzoic acid. *Journal of Industrial Microbiology and Biotechnol*ogy, Vol.29, No.3, September 2002, 129-133.

Garcia-Cruz, C.H.; Foggetti, U. & Silva, A.N. (2008). Bacterial alginate: technological aspects, characteristics and production. *Química Nova,* Vol.31, No.7, September 2008, 1800-1806.

Glicksman, M. (1987). Utilization of seaweed hydrocolloids in the food industry. *Hydrobiologia,*Vol.151-152, No.1, September 1987, 31-47.

Govan, J. R. W. & Harris, G. S. (1986). *Pseudomonas aeruginosa* and cystic fibrosis - unusual bacterial adaptation and pathogenesis. *Microbiology Science,* Vol.3, No.10, October 1986, 302-308.

Heo, K.; Yoon, J.; Jin, K. S.; Jin, s.; Sato, H.; Ozaki, Y.; Satkowski, M. M.; Noda, I. & Ree, M. (2008). Structural evolution in microbial polyesters. *Journal of Physical Chemistry B,* Vol.112, No.15, January 2008, 4571-4582.

Horan, N. J.; Jarman, T. R. & Dawes, E. A. (1983). Studies of some enzymes of alginic acid biosintesis in *Azotobacter vinelandii* grown in continuous culture. *Journal of General Microbiology,* Vol.129, No.10, February 1983, 2985-2990.

Hrabak, O. (1992). Industrial production of poly-b-hydroxybutyrate. *FEMS Microbiology Reviews*, Vol.103, No.12, 1992, 251-256.

Lafferty, R. M.; Korsatko, B. & Korsatko, W. (1988). Microbial production of poly-β-hydroxybutyric acid. In: *Biotechnology*. 1ª. ed. New York: VCH Publishers, p. 135-176.

Law, J. H. & Slepecky, R. A. (1961). Assay of poly-β-hydroxybutyric acid. *Journal of Bacteriology*, Vol.82, December 1961, 33-36.

Lee, S. Y. (1996). Bacterial polyhydroxyalkanoates. *Biotechnology and Bioengineering*, Vol.49, No.1, July 1996, 1-14.

Lin, L. P. & Sadoff, H. L. (1968). Encystment and polymer production by *Azotobacter vinelandii* in the presence of β-hydroxybutyrate. *Journal of Bacteriology*, Vol.95, No.6, June 1968, 2336-2343.

Lugg, H.; Sammons, r. L.; Marquis, P. M.; Hewitt, c. J.; Yong, P.; Paterson-Beedle, M.; Redwood, M. D.; Stamboulis, a.; Kashani, M.; Jenkins, M. & Macaskie, L. E. (2008). Polyhydroxybutyrate accumulation by a *Serratia* sp. *Biotechnology Letters*, Vol.30, No.3, March 2008, 481-491.

Martinsen, A.; Skjåk-Bræk, G.; Smidsrød, O.; Zanetti, F. & Paoletti, S. (1991). Comparison of different methods for determination of molecular weight and molecular weight distributions of alginates. *Carbohydrate Polymers*, Vol.15, No.2, 1991, 171-193.

May, T. B. & Chakrabarty, A. M. (1994). *Pseudomonas aeruginosa*: genes and enzymes of alginate synthesis. *Trends Microbiology*, Vol.2, No.5, May 1994, 151-157.

Moen, E.; Larsen, B.; Ostgaard, K. & Jensen, A. (1999). Alginate stability during high salt preservation of *Ascophyllum nodosum*. *Journal of Applied Phycology*, Vol.11, No.1, April 1999, 21-25.

Page, W. J. & Knosp, O. (1989). Hyperproduction of poly-β-hydroxybutyrate during exponential growth of *Azotobacter vinelandii* UWD. *Applied and Environmental Microbiology*, Vol.55, No.6, June 1989, 1334-1339.

Page, W. J.; Manchak, J. & Rudy, B. (1992). Formation of poly-(hydroxybutyrate-co-hydroxyvalerate) by *Azotobacter vinelandii* UWD. *Applied and Environmental Microbiology*, Vol.58, No.9, September 1992, 2866-2873.

Peña, C.; Campos, N. & Galindo, E. (1997). Changes in alginate molecular mass distributions, broth viscosity and morphology of *Azotobacter vinelandii* cultured in shake flasks. *Applied Microbiology and Biotechnology*, Vol.48, No.4, October 1997, 510-515.

Pouton, C. W. & Akhatar, S. (1996). Biosynthetic polyhydroxyalkanoates and their potential in drug delivery. *Advanced Drug Delivery Reviews*, Vol.18, No.2, January 1996, 133-162.

Punrattanasin, W. The utilization of activated sludge polyhydroxyalkanoates for the production of biodegradable plastics. 2001. 131 f. Doctor of Philosophy in Environmental Science and Engineering - Faculty of the Virginia Polytechnic Institute and State University, Blacksburg, EUA, 2001.

Quillaguamán, J.; Van-Thuoc, D.; Guzmán, H.; Guzmán, D.; Martín, J.; Everest, A. & Hatti-Kaul, R. (2008). Poly(3-hydroxybutyrate) production by *Halomonas boliviensis* in fed-batch culture. *Applied Microbiology and Biotechnology*, Vol.78, No.2, February 2008, 227-232.

Sabra, W., Zeng, A. P. & Deckwer, W. D. (2001). Bacterial Alginate - physiology, product quality and process aspects. *Applied Microbiology and Biotechnology*, Vol.56, No.3-4, August 2001, 315-325.

Sadoff, H. L. (1975). Encystment and germination in *Azotobacter vinelandii*. *Bacteriological Reviews*, Vol.39, No.4, December 1975, 516–539.

Savalgi, V. & Savalgi, V. (1992). Alginate production by *Azotobacter vinelandii* in batch culture. *Journal General Applied Microbiology*, Vol.38, No.6, September 1992, 641-645.

Segura, D.; Cruz, T. & Espin, G. (2003). Encystment and alkylresorcinol production by *Azotobacter vinelandii* strains impaired in poly-beta-hydroxybutyrate synthesis. *Archives of Microbiology*, Vol.179, No.6, June 2003, 437-443.

Shishatskaya, E. I.; Voinova, O. N.; Goreva, A. V.; Mogilnaya, O. A. & Volova, T. G. (2008). Biocompatibility of polyhydroxybutyrate microspheres - in vitro and in vivo evaluation. *Journal of Materials Science - Materials in Medicine*, Vol.19, No.6, June 2008, 2493-2502.

Smidsrød, O. (1970). Solution Properties of Alginate. *Carbohydrate Research*, Vol.13, No.3, June 1970, 359-372.

Smidsrød, O. & Skjåk-bræk, G. (1990). Alginate as immobilization matrix for cells. *Trends Biotechnology*, Vol.8, No.3, March 1990, 71-78.

Steigedal, M.; Sletta, H.; Moreno, S.; Maerk, M.; Christensen, B. E.; Bjerkan, T.; Ellingsen, T. E.; Espìn, G.; Ertesvåg, H. & Valla, S. (2008). The *Azotobacte vinelandii* AlgE mannuronan C-5-epimerase family is essential for the in vivo control of alginate monomer composition and for functional cyst formation. *Environmental Microbiology*, Vol.10, No.7, July 2008, 1760 -1770.

Sudesh, K.; Abe, H. & Doi, Y. (2000). Synthesis, structure and properties of polyhydroxyalkanoates - biological polyesters. *Progress in Polymer Science*, Vol.25, No.10, December 2000, 1503-1555.

Suriyamongkol, P., Weselake, R., Narine, S., Moloney, M. & Shah, S. (2007). Biotechnological approaches for the production of polyhydroxyalkanoates in microorganisms and plants - A review. *Biotechnology Advances*, Vol.25, No.2, March/April 2007, 148-175.

Suzuki, D. v.; Carter, J. m.; Rodrigues, M. F. A.; Silva, E. S. & Maiorano, a. E. (2008). Purification of polyhydroxybutyrate produced by *Burkholderia cepacia* IPT64 through a chemical and enzymatic route. *World Journal of Microbiology & Biotechnology*, Vol.24, No.6, June 2008, 771-775.

Trujillo-Roldan, M. A.; Moreno, S.; Segura, D.; Galindo, E. & Espin, G. (2003). Alginate production by an *Azotobacter vinelandii* mutant unable to produce alginate lyase. *Applied Microbiology and Biotechnology,* Vol.60, No.6, February 2008, 733-737.

Van-Thuoc, D.; Quillaguamán, J.; Mamo, G. & Mattiasson, B. (2008). Utilization of agricultural residues for poly (3-hydroxybutyrate) production by Halomonas boliviensis LC1. *Journal of Applied Microbiology,* Vol.104, No.2, February 2008, 420-428.

22

Biopolymer Surfactant Interactions

Lisa Sreejith, S.M.Nair and Jinu George
Soft Materials Research Laboratory, Department of Chemistry,
National Institute of Technology Calicut, Kerala
India-673601

1. Introduction

Biopolymers are polymers that are biodegradable. The input materials for the production of these polymers may be either renewable (based on agricultural plant or animal products) or synthetic. There are four main types of biopolymer based respectively on: **1. Starch 2. Sugar 3.Cellulose and 4.Synthetic materials.** Two main strategies may be followed in synthesizing a polymer. One is to build up the polymer structure from a monomer by a process of chemical polymerization. The alternative is to take a naturally occurring polymer and chemically modify it to give it the desired properties. A disadvantage of chemical modification is however that the biodegradability of the polymer may be adversely affected. Therefore it is often necessary to seek a compromise between the desired material properties and biodegradability. Current and future developments in biodegradable polymers and renewable input materials focus relate mainly to the scaling-up of production and improvement of product properties. Larger scale production will increase availability and reduce prices.

Surfactants are wetting agents that lower the surface tension of a liquid, allowing easier spreading, and lower the interfacial tension between two liquids. Surfactant solutions have a general tendency to solubilize a certain amount of additives, which can be correlated with their structural organization and mutual interactions.

Polymer-surfactant interactions [1-8] have been extensively investigated by researchers due to its manifold applications in the fields of the food industry, pharmaceutical industry and analytical biochemistry. The interaction of protein with other ingredients, mostly cationic and anionic surfactants, is of particular interest because they are used co-operatively in formulated complexes. The mechanisms for the protein-surfactant interactions are polyelectrolyte absorption [9], hydrophobic [10] and ionic interactions [11], depending on the substrate and type of proteins involved. The effect of adding a hydrophilic counterion, NaCl to aqueous CTAT solution was measured by dynamic light scattering. The obtained results indicate that the micelles grow rapidly upon salt addition and eventually achieve a constant size under static conditions [12]. The results were explained in terms of a competition between micellar growth induced by salt addition and changes in micellar flexibility caused by ionic screening effects. Adding hydrophobic non-polar additives to cationic surfactant solutions is also found to increase the micellar growth [13, 14].

Several research papers on the micellar solution of cationic surfactant and salt systems are available [15-31]. Many cationic micellar solution were prepared from mixing long-chain cationic surfactants such as hexadecyltrimethylammonium (CTA^+) and hexadecylpyridinium (CPy^+) with organic counterions such as, salicylate, Sal^- ,tosylate T−)), and (HNC−) [32-40]. For the above systems, the growth of the cationic micelle was mainly due to the strong binding of hydrophobic counterions. As the hydrophobic character of the cationic surfactant increases the cmc, degree of counterion dissociation (α) and Gibbs free energy of micellization, ΔGm, decreases while the aggregation number, N_{agg} increases. The cmc, α, ΔG_m and N_{agg} values of cationic surfactants CTAC, CTAB and cetyltrimethylammonium tosylate, CTAT at 30°C were compared [41]. It is noted that cmc, α and ΔG_m were lowest while N_{agg} is highest to CTAC and CTAB. This correlation between cmc, α, ΔG_m, N_{ag} and type of counterion indicate that the micellization process is more spontaneous in the case of CTAT compared to CTAC and CTAB. From an application point of view, the higher binding of T^- to CTA^+ explains the higher adsorption of CTAT onto silica [42]. It was concluded that on the silica surface, increasing the binding degree increases the cooperativity of adsorption below the bulk cmc, corresponding to a stabilization of interfacial self assembly. An increase in the surface excess concentration due to a decrease in the curvature of the adsorbed aggregates is noticed. The effect of adding a hydrophilic counterion NaCl to CTAT aqueous solution was measured by dynamic light scattering. The obtained results indicate that the micelles grow rapidly upon salt addition but eventually achieve a constant size under static conditions [43]. The results are explained in terms of a competition between micellar growth induced by salt addition and changes in micellar flexibility caused by ionic screening effects. Adding hydrophobic non polar additives to cationic surfactant solutions is also found to increase the micellar growth [44, 45].

There are many evidences for an entangled micellar phase in several aqueous cationic detergent systems in the presence of added salt [46-49]. The presences of neutral salts [50-53] showed a marked shift in phase transitions. It has been suggested that the protein unfolds in the cooperative binding region [54] and sodium chloride is reported to act as a denaturing agent in many protein systems [55-57]. According to Curtis et al. [58] in classical salting-out behavior, the protein-salt preferential interaction is unfavorable as addition of salt raises the surface free energy of the protein, thereby protein-protein attraction increases leading to a reduction in solubility. Systematic investigations of solubilisation of gelatin in AOT-isooctane-water microemulsions in high concentrations was found to be analogous to solubilization of an ionic polymer in a surfactant solution by Koteshwar et al. [59] where surfactant polymer interactions play a dominant role.

There would be definite interactions between the negatively charged surfactant and the biopolymer, since gelatin is amphoteric due to the presence of both carboxylic and amino groups. The organogel thus formed could well be described as a surfactant-polymer complex [60-62]. It is noted that, on increasing the surfactant concentration, this "complex" would redissolve in excess surfactant solution, and the rigidity of the system would be affected. This could lead to the formation of a gel with a flexible network and the interesting physical properties could be exploited for biotechnological applications. The viscous organogels formed from the solutions containing AOT, isooctane, water, and high amounts of gelatin (up to 6.8%) at high temperature (>40°C) were reported [63] to actually "melt" on cooling, and at low temperature (<10 °C), they are free-flowing liquids of moderate viscosity. On rewarming, these solutions progressively become viscous with increasing temperature. The phenomenon is completely reversible, and again on cooling the viscous gels becomes free-flowing liquids. The phenomenon is observed not only in the case of a surfactant with a negatively charged headgroup such as AOT but also in a surfactant such as CTAB with a positively charged headgroup. Such behavior is uncommon and very interesting from both the biophysical and biotechnological points of view. In solutions of nonionic surfactants, such as Tween 80-Tween 60 mixtures, solubilization of gelatin was too low to form an organogel.

Cationic CTAB forms worm-like micelles with Cl^- and Br^- ions due to counterion binding and charge screening but rod-like micelles with salicylate$^-$ ions are formed as a result of charge attraction, counterion binding, and hydrophobic interactions [64]. In dilute solutions, the length of rod-like CTAB: NaSal micelles can exceed 1000 A° [65] while that of worm-like CTAB: KBr micelles are around 500 A° [66]. These rod-like surfactant micelles display interesting rheological properties, which result from their polymer-like structure. In many applications, such as shampoos [67] and drag-reducing agents [68], rod-like micelles are used as viscosity modifiers. Over the same period of time, extensive studies have been conducted on interactions between water-soluble polymers and surfactants.

Schubert et al. [69] examined the microstructure and rheological properties of mixed CTAT and SDBS micelles. The effects of surfactant ratio on the addition of the sodium chloride and sodium tosylate were noted. Small angle neutron scattering studies revealed that the addition of sodium chloride induces electrostatic screening between micelles and reduces or eliminates intramicellar repulsions. Sodium tosylate had a similar effect at low concentrations; however, further addition resulted in the formation of branched micelles. The formation of branch points was the result of interactions between the surfactant and the added hydrotrope, and was not observed with the addition of sodium chloride. Varade et. al.[70] reported that the addition of three anionic hydrotropes affected the cmc, viscosity and CP temperature of the cationic CTAB, the anionic SDS and the nonionic polyoxyethylene t-octylphenol (Triton X-102). All the hydrotropes increased the CP of Triton X-102, while the sodium chloride decreased the CP. The cmc of CTAB decreased and the viscosity of concentrated surfactant solutions rose sharply with the addition of any of the hydrotropes. The decrease in cmc suggests an increase in micelle stability and a strong increase in solution viscosity is indicative of micelle elongation.

However, until now, stabilizing effect of salts in a gelatin-micellar media has not been reported. Physico chemical or biological treatments [71, 72] trigger protein modification i.e., modification in conformation/structure and there by functional properties of protein, leading to partial hydrolysis or aggregation of gelatin. The gelatin conformation is retained in many formulations to improve the functional properties in combination. The present work explores the effect of NaCl on the conformational properties and stability of gels formed by polyelectrolyte-surfactant–water interfaces stabilized by gelatin–surfactant complexes. This study is attempted to unravel the effect of brine on the gelatin-surfactant (CTAB) system by viscosity, conductivity, Circular Dichroism (CD) spectroscopy, Differential scanning calorimetry (DSC) and by Fourier Transform Infrared Spectroscopic (FTIR) measurements at 35°C.

2. Experimental procedure

i. **Materials and Methods:** An alkali processed deionized bone gelatin (Loba Chemie, India), Cetyl trimethyl ammonium bromide (CTAB) and sodium chloride (NaCl) were used as supplied.
ii. **Sample Preparation:** Gelatin solutions were prepared by weighing the required amount of gelatin flakes and soaking in hot water (~40°C) with stirring. Deionized distilled

water from Millipore (Milli-Q; 18 MΩ) was used for all the experiments. The concentrations quoted here are in weight percentage of gelatin. The surfactant solution was prepared by weighing out the requisite amount each time and dissolving it in a freshly prepared gelatin solution. The sample solutions after proper mixing were left for equilibration for 24 hrs.

Viscosity and CD measurements were done for the pure aqueous solution of gelatin and for the samples were done at 35± 2°C. Flow characteristic were carried out by Capillary Viscosity method and were measured using an Ostwald viscometer thermostated at 35 ± 2°C. Under Newtonian condition density correction is not made since these are found negligible. The solutions were kept at the experimental temperature for at least 30 min to attain thermal equilibrium. At the ambient temperature, gelatin is characterized macroscopically as a random coil in dilute solutions with a length of ~20 Å.

3. Influence of salt on cationic surfactant – biopolymer interactions in queous media

Changes in the bulk viscosity are caused by structural changes in the gelatin/surfactant associates, for example, by formation of intermolecular cross-bonds. Figure 1 shows the variation of relative viscosity, η_r of CTAB in presence of sodium chloride in 1.5% gelatin. A sharp increase in relative viscosity as a function of [CTAB + NaCl], expressed in terms of mole fraction of the salt, is observed. The abrupt increase in viscosity can be the onset of an association of gelatin with CTAB. The micelles at this stage act as transient cross-links i.e., inter polymer association are possible. According to Mukerjee [73], an additive, which is surface active, will mainly be solubulised near the micellar head group region, and facilitate the sphere→ higher order transition with a concomitant increase in η_r and the packing parameter, Rp.

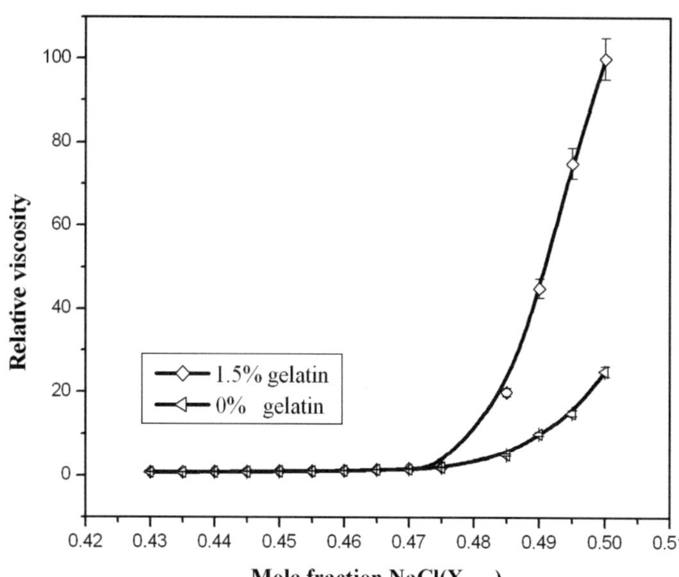

Fig. 1. Variation of relative viscosity, ηr, of [CTAB + NaCl] in 1.5% gelatins at 35 °C.

Gelatin is a polypeptide with 12% negatively charged amino acid residue is a random coil in dilute solutions with a persistence length of ~ 20A^0 and R_h ~ 220 A^0. DLS studies of CTAB - Gelatin interaction [74, 75] showed that beyond critical aggregation concentration gelatin - CTAB complexes were observed to grow significantly and were in equilibrium with CTAB micelles. The micellar shapes were found to be oblate ellipsoidal for CTAB micelles which were explained through the necklace -bead model through co-operative bonding [76]. Hydrophobic interaction may be operating between the hydrophobic part of CTAB and hydrophobic sites on the gelatin chain. Beyond CMC, the co-operative binding prevailed and gelatin - CTAB complexes were found to co-exist with free CTAB micelles. Increase in size of the hydrodynamic radius of Gelatin-CTAB complex is due to electrostatic repulsion between the positive sites of the chain and bulky cationic head groups. Hydrophobic interaction may be operating between the tail part of CTAB and the hydrophobic sites on the Gelatin chain. An abrupt increase in, ηr, can be due to the incorporation of halide ion leading to the screening of the electrostatic interaction of surfactant head groups. Decreased head group repulsion favors closer packing of the surfactant monomers and hence induces sphere → rod transition in micelles [76, 77]. At higher [CTAB+ NaCl], further increase in size is observed due to formation of polymer -micelle complex. Further, the interfacial portioning of additive is important for viscosity rise (micellar growth) while core solubilization (swollen micelle) enhances the sphericity of the micelle and works oppositely. The high viscosities observed in these systems are interpreted in terms of a micelle sphere-to-rod transition, which occurs over a certain range of mconcentration of either surfactant or added salt.

To gain information on the secondary structure present in gelatin in the system, CD measurements were performed at various concentrations of CTAB in the presence of NaCl (Fig. 2).

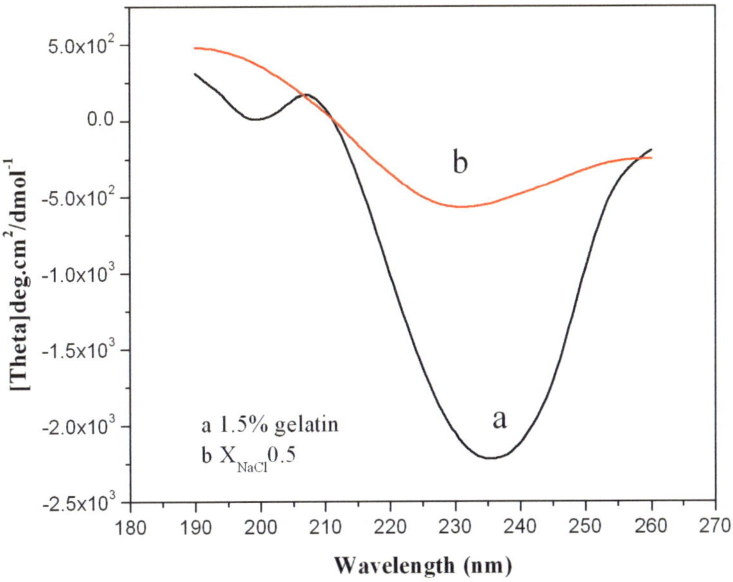

Fig. 2. Circular dichroism spectra on the effect of (CTAB+ NaCl) on gelatin in the wavelength region260 - 190 nm. (a) Gelatin alone and (b) gelatin+0.5M CTAB+0.5M NaCl.

The random coil structure found in gelatin gives rise to a characteristic negative peak at around 235 nm in the CD spectrum, which decreases in ellipticity upon addition of [CTAB +NaCl]. The general shape and peaks of the spectrum does not show much change in surfactant /gelatin/salt mixture of varying concentrations except for their magnitude. Ellipticity was maximum for the highest concentration and negative peaks with lower intensity were observed in comparison to gelatin alone. This implies that the gelatin has no conformational change at 35°C. The lower ellipticity indicates [78] the alteration or disturbance in the order or periodicity in the structure of surfactant i.e., aggregation or disruption as predicted from other measurements. The increase in molar ellipticity at 235 nm for the system is indicative of aggregated structures of surfactant/gelatin in the presence of salt and the effect is very much pronounced at higher concentrations of salt. This shows that the gelatin facilitates the structural transitions in CTAB solution in presence of salt without having any structural changes to it.
The utility of FTIR spectroscopy in studies of micellar growth induced by electrolyte has been increasingly demonstrated in recent years [79, 80]. In order to verify the expected trend of CTAB to alter the ordered structure of gelatin into a random structure especially at low concentrations Fourier transform infrared spectroscopy (FTIR) spectra was carried out for [CTAB + NaCl] in1.5٪ gelatin, which depict certain characteristic features (Fig.3).

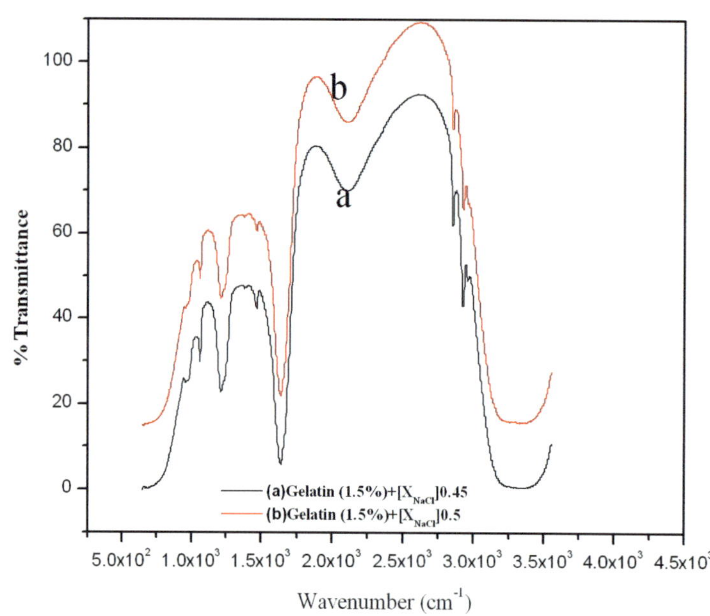

Fig. 3. FTIR spectra of 0.1M and 0.5 M [CTAB + NaCl] in 1.5% gelatin at 35 °C.

The transmittance bands at 3368, 2923 and 2853 cm^{-1} represents the -OH, $-CH_2$ and $-CH_3$ aliphatic groups whereas bands at 1466 and 1635 cm^{-1} represent the NH-group bending vibration and vibrations of -OH group in the primary alcoholic group, respectively. The amino group has a characteristic absorption band in the region of 3400–3500 cm^{-1}, which is masked by the broad absorption band from the -OH group. The shoulders at 1635 cm^{-1} represent the C=O groups and suggests gelatin as a partially deacetylated product. The conformation of protein in the interface changes slowly to allow the maximum number of hydrophobic segments to contact the surface. Above the saturated concentration, the further rearrangement may allow more binding of CTAB to hydrophobic patches on the protein surface. In the process to become hydrophilic, the electrostatic interactions of anionic amino-acid residue to cationic surfactant might have taken place. The sphere-to-rod shaped transition is accompanied by partial ordering of methylene chain.

DSC traces were recorded in a high sensitivity calorimeter, in the temperature range -80 to 80° C, with a heating rate of 5° C / minute. Fig. 4 shows the DSC heating curve for [CTAB+NaCl] in presence of gelatin. The two main transitions associated to the main peak can be explained in terms of different structures in different populations, and the different surfactant packing of CTAB, which determines the vesicle curvature. This study once again strengthens our observation that higher order structural transitions leading to gel formation has happened without undergoing any conformation changes to gelatin.

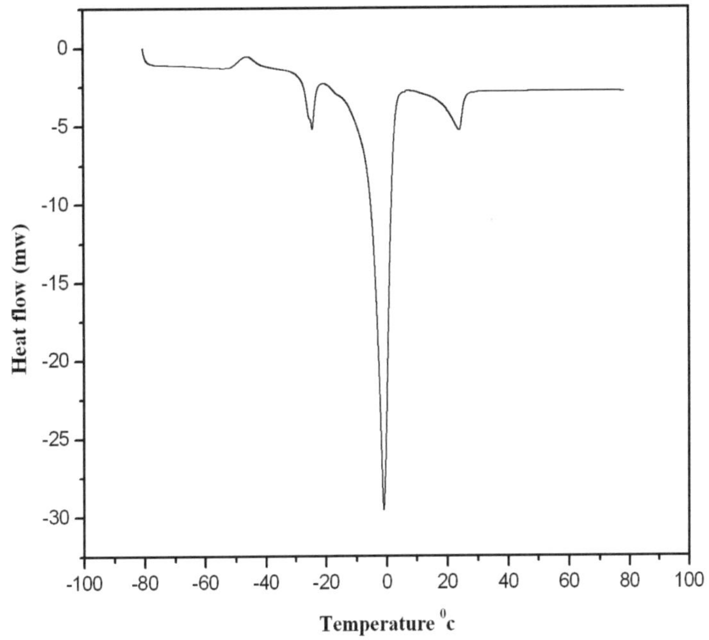

Fig. 4. DSC curve for X_{salt} =0.1M in the gelatin/CTAB/NaCl system.

The large endothermic peak, situated around 0°c is due to the melting of ice, together with another downward peak at a high temperature (about 25°c) which reflects the evaporation of the non-crystallisable water which might be an indication of the possible chemical cross-linking [81, 82].

4. Conclusion

The effect of sodium chloride on micellar property of CTAB in biopolymer gelatin were systematically studied. It was found that, micellisation and transition is favoured by increase in concentration of sodium chloride, however, without affecting the conformation of gelatin. The main findings from the present investigation refer to the stabilizing role of salt in presence of a biopolymer, gelatin, in micellar media. Increase in viscosity and gel formed is further explored by other techniques, such as CD spectroscopy, DSC and FTIR measurements that suggest a sphere - to - higher order micellar transition in the system. These results emphasize new possibilities offered by such systems in obtaining organized assemblies with novel architectures, for investigating the fundamental functional attributes in colloid studies and pharmaceutical studies. Hence, we strongly feel that this work will contribute to extend the field of utilization of gelatin, showing its true potential for specific utilizations in pharmaceuticals.

5. Acknowledgements

The authors gratefully acknowledge M/s. Sciyo publishers for their efforts to publish the work in their new book project titled "Biopolymers". We place on record our sincere thanks to Dr. E. Gopinathan, Dr. M. S. Sunitha, Mr. Gloy Augustine for careful reading of the manuscript. Financial support from Kerala State Council for Science,Technology and Environment under grant No. **(T)024/SRS/2009/CSTE** is gratefully acknowledged.

6. References

[1] Nerina A. Camino, Oscar E. Pérez, Ana M.R., Pilosof, Food Hydrocolloids, 23, (2009) 1089.
[2] Griffiths P.C, Roe J. A., Bales B. L., Pitt A. R., Howe A. M. Langmuir 16 (2000) 8248.
[3] Mitra, Debolina, Bhattacharya, Subhas C.; Moulik, Satya P. J. Phy. Chem. B, 112 (2008) 6609.
[4] Sonesson A.W., Blom H., Hassler K., Elofsson U.M., Callisen T.H., Widengren. J., Brismar H., J Colloid Interface Sci. 317 (2008) 449.
[5] Blanco. E , Ruso J.M., Prieto G., Sarmiento F, J. Colloid Interface Sci. 316 (2007) 37.
[6] Ruiz C.C, Hierrezuelo J.M, Aguiar J, Peula-García J.M, Biomacromolecules, 8 (2007) 2497.
[7] Gull. N., Kumar. S, Ahmmed B, Khan R.H, Kabir-Ud-Din, Colloids Surf., B, 51 (2006) 10.
[8] Santiago P.S, Moreira L.M., de Almeida, Tabak M Biochim. Biophys Acta., 506 (2007) 1770.
[9] Richert L., Boulmedais F., Lavalle P.,. Mutterer J, Ferreux E., Decher G., Schaaf P., Voegel, Biomacromolecules, 5 (2004) 284.
[10] Wolde P. R., J. Phys.: Condens. Matter, 14 (2002) 9445.
[11] Kato K., Sano S ., Ikada Y., : Colloids Surf..B, 4 (1995) 221.
[12] Torres M, González J, Rojas M, Müller A, Sáez A, Löf D, et al. J Colloid Interface Sci. 307 (2007) 221.
[13] Desai A, Varade D, Mataa J, Aswal V, Bahadur P. Colloids Surf A 259 (2005) 111.
[14] Thakur R, Dar A, Rather G. J Mol Liq 136 (2007) 83.
[15] Umme Salma Siddiqui, Goutam Ghosh, Kabir-ud-Din, Langmuir 22 (2006) 9874.
[16] Salabat, Alireza, Alinoori, Maryam Calphad: Thermochemistry, 32 (2008) 611.
[17] Zhao, Zhongkui Bi, Chenguang Qiao, Weihong Li, Zongshi Cheng, Lubo Colloids Surf.A, 294(2007) 191.
[18] Fisicaro E, Ghiozzi A, Pelizzetti E, Viscardi G, Quagliotto P. J Colloid Interface Sci 184 (1996) 147.
[19] Matsuda, Tsuyoshi, Annaka, Masahiko Langmuir 24 (2008) 5707.
[20] Ono Y, Kawasaki H, Annaka M, Maeda H. J, Colloid Interface Sci 287 (2005) 685.
[21] Gamboa C, Ríos H, Sepúlveda L. J Phys Chem. 93 (1989) 5540.

[22] Raghavan S, Kaler E. Langmuir 17 (2001) 300.
[23] Sakaiguchi Y, Shikata T, Urakami H, Tamura A, Hirata H. Colloid Poly Sci, 265 (1987) 750.
[24] Shikata T, Imai S. J Colloid Interface Sci 244 (2001) 399.
[25] Shikata T, Imai S, Morishima Y. Langmuir 14 (1998) 2020.
[26] Jitendra Mataa, Dharmesh Varadea, Prashant Bahadur, Thermochimica Acta 428 (2005) 147.
[27] Shikata T, Imai S. J Phys Chem B 103 (1999) 8694.
[28] Kuntz, D.M. , Walker, L.M. , J. Phy. Chem. B, 111(2007) 6417.
[29] Marı´ a L. Olivares, Marta B. Peirotti, Julio A. Deiber, Food Hydrocolloids 20 (2006) 1039.
[30] Akiada H, Ishıhara M, Nishi M, Higake M, Ishimaru S, Nishida J. Colloid Polym Sci 281 (2003) 993.
[31] Almgern M, Swarup S. J Phys Chem 87 (1983) 876.
[32] Acharya D, Kunieda H. Adv Colloid Interface Sci 123 (2006) 401.
[33] Walker L. Curr Opin Colloid Interface Sci 6 (2001) 451.
[34] Feitosa E, Brazolin M, Naal R, Lama M, Lopes J, Loh W, J Colloid Interface Sci 299 (2006) 883.
[35] Umlong I, Ismail K. J Colloid Interface Sci 291 (2005) 529.
[36] Varade D, Joshi T, Aswal V, Goyal P, Hassan P, Bahadur P. Colloids Surf A 295 (2005) 95.
[37] Ranganathan R, Okano L, , Quina F. J Colloid Interface Sci. 214 (2005) 238.
[38] Jiang N, Li P,Wang Y,Wang J, Yan H, Thomas R. J Colloid Interface Sci. 286 (2005) 755.
[39] Aswal V, Goyal P. Chem Phys Lett 364 (2002) 44.
[40] Horbaschek K, Hoffmann H, Thunig C, J Colloid Interface Sci. 206 (1998) 439.
[41] Mata J, Varade D, Bahadur P. Thermochimica Acta 428 (2005) 147.
[42] Velegol S, Tilton R. J Colloid Interface Sci. 249 (2002) 282.
[43] Torres M, González J, Rojas M, Müller A, Sáez A, Löf D, et al. J Colloid Interface Sci 307 (2007) 221.
[44] Desai A, Varade D, Mataa J, Aswal V, Bahadur P. Colloids Surf A 259 (2005) 111.
[45] Thakur R, Dar A, Rather G. J Mol Liq 136 (2007) 83.
[46] Sionkowska A., Wisniewski M., Skopinska J., Kennedy C. J., Wess T, Biomaterials 25 (2004) 795.
[47] Roderich Bott, Thomas Wolff, Karl Zierold, Langmuir 18 (2004) 2012 .
[48] Nilsson, P, Johan Unga, Hansson, P., J. Phy. Chem. B 111(2007) 10959.
[49] Toyoko Imae, Shoichi Ikeda, J. Phys. Chem., 90 (1986) 5216.
[50] Ting Lu, Jianbin Huang, Zihui Li, Shikai Jia, Honglan Fu, J. Phy. Chem. B, 112 (2008) 2909.
[51] Saxena A., Antony T., Bohidar. H.B, J. Phys. Chem. B, 102 (1998) 5063.
[52] Hone J. H. E., Howe A. M., J. Colloid Interface Sci., 251 (2002) 193.
[53] Norrman. J, Lynch. I., Piculell.L, J. Phy. Chem. B, 111 (2007) 8402.
[54] Wachem P. B., Beugelling T., Feijen J., Bantjes A., Detmers J. P., Van Aken W.G Biomaterials, 6 (1985) 403.
[55] Yakhno T., J. Colloid Interface Sci., 318 (2008) 225.
[56] Annunziata, L. Paduano, J.G. Albright, J. Phys Chem B. 111(2007) 10591.
[57] Hao, Li-Sheng, Nan, Yan-Qing Colloids Surf. A 325, (2008) 186.

[58] Curtis R.A, Ulrich J, Montaser A., Prausnitz J.M., Blanch H.W., Biotechnol Bioeng. 79 (2002) 367.
[59] Nitin W. Fadnavis, Kinnera Koteshwar, Biotechnol. Prog. 15 (1999) 98104.
[60] Lopez. F., Venditti. F., Ambrosone. L., Colafemmina. G., Ceglie. A., Palazzo. G, Langmuir 20 (2004) 9449.
[61] Wesley, R. D.; Cosgrove, T.; Thompson, L.; Armes, S. P.; Baines, F. L. Langmuir 18 (2002) 5704.
[62] Rao. A., Kim. J., Thomas. R. R., Langmuir 21 (2005) 617.
[63] Mya, K. Y.; Sirivat, A.; Jamieson, A. M. Macromolecules 34 (2001) 5260.
[64] Santipharp Panmai , Robert K. Prud'homme, Dennis G. Peiffer, J. Colloid Interface Sci. 257 (2003) 154.
[65] Goddard E.D., J. Am. Oil Chem. Soc. 71 (1994) 1.
[66] Lin M.Y., Hanley H.J.M., Sinha S.K., Straty G.C., Peiffer D.G., Kim M.W., Phys. Rev. E 53 (1996) 4302.
[67] Schramm, L.L., Stasiuk, E.N., Marangoni, D.G., Surfactants and their applications, Annual Reports on the Progress of Chemistry (2003) 3.
[68] Rose G.D., Teot A.S., in: Structure and Flow in Surfactant Solutions, Herb C.A., Prud'homme R.K. (Eds.); ACS Symposium Series 578, American Chemical Society, Washington, DC, 1994.
[69] Schubert BA, Kaler EW, Wagner NJ, Langmuir 19 (2003) 4079.
[70] Varade D, Bahadur P. J Surfactants Deterg. 7 (2004) 257.
[71] Hong Shen, Xixue Hu, Fei Yang, Jianzhong Bei, Shenguo Wang, Biomaterials, 28 (2007) 4219.
[72] Toledano O., Magdassi S., :Emulsification and Foaming Properties of Hydrophobically Modified Gelatin J. of Colloid Interface Sci., 200, 235 (1998).
[73] P. Mukerjee, J. Phys. Chem., 76 (1972) 565.
[74] Saxena A., Antony T.,.Bohidar H.B, J. Phys. Chem. B, 102 (1998) 5063.
[75] Romsted L. S. , Zhang J., Cuccovia I. M., M. J. Politi, H. Langmuir, 19 (2003) 9179.
[76] Vasilescu M., Angelesu D., Almgren M., Valstar A. Langmuir, 8 (1999) 2635.
[77] Pezron I. Djabourov. M., Leblond, J.Polymer. 32 (1991) 3201.
[78] Bhat M, Gaikar V.G., Langmuir, 15 (1999) 4740.
[79] Swati De, Agnishwar, S.Das, J. Colloid Interface Sci., 93, (2005) 562.
[80] Holler F., Callis J. B, J. Phys. Chem., 93 (1989) 2053.
[81] Umemura J., Cameron D. G., Mantsch H. H., J. Phys. Chem., 84 (1980) 2272.
[82] Geetha B., Mandal A. B., Ramasami T., Macromolecules 26 (1993) 4083.

23

Properties and Function of Pyomelanin

Charles E. Turick, Anna S. Knox, James M. Becnel,
Amy A. Ekechukwu and Charles E. Milliken
Savannah River National Laboratory, Aiken, SC 29808
USA

1. Introduction

Melanin pigments are the most common pigments produced in nature and these complex biopolymers are found in species of all biological kingdoms. There are several categories of melanins which include eumelanins, pheomelanins and allomelanins. Eumelanins and pheomelanins are produced from oxidation of tyrosine or phenylalanine to o-dihydroxyphenylalanine (DOPA) and dopaquinone. Pheomelanin results from cysteinylation of DOPA. Allomelanins include a heterogeneous group of polymers that include pyomelanin. Melanin biochemistry and synthesis has been reviewed previously (Plonka and Grabacka 2006). This chapter will focus on the properties and function of pyomelanin and their potential utility in biotechnology. Pyomelanin originates from the catabolism of tyrosine or phenylalanine (Lehninger, 1975) (Fig. 1). Complete breakdown of tyrosine to acetoacetate and fumarate requires the enzymes 4-hydroxyphenylpyruvic acid dioxygenase (4-HPPD) and homogentisic acid oxidase (HGA-oxidase). In the absence of HGA-oxidase, or if homogentisic acid (HGA) production exceeds that of HGA-oxidase activity, HGA is over-produced and excreted from the cell (Yabuuchi and Ohyama 1972; Ruzafa et al. 1994; Katob et al. 1995). Autooxidation and selfpolymerization of HGA then results in pyomelanin. In addition, deletion of the gene that encodes for HGA-oxidase results in hyper production of pyomelanin while deletion of the gene that encodes for 4-HPPD results in the inability to produce pyomelanin (Coon et al. 1994; Ruzafa et al. 1995).
In humans with loss-of-function mutations in HGA-oxidase, pyomelanin (also known as alkapton or ochronotic pigment) forms in the urine due to the spontaneous auto oxidation of excess HGA (Beltrán-Valero de Bernabé, et al. 1999). This condition is known as alkaptonuria in humans and can result in arthritis in adults. Pyomelanin production in microorganisms often is associated with numerous survival advantages and was first characterized in bacteria among numerous species of the genus *Pseudomonas* (Yabuuchi & Ohyama 1972). Since then several fungi and a number of bacteria, especially in the γ Proteobacteria have been shown to produce pyomelanin.

2. Microbial exploitation of pyomelanin

Melanins are structurally similar to humic acids and thus have similar redox cycling properties (Menter & Willis, 1997; Stott & Martin, 1990). Although melanin pigments are thought to primarily function as a protection mechanism against ultraviolet (UV) radiation,

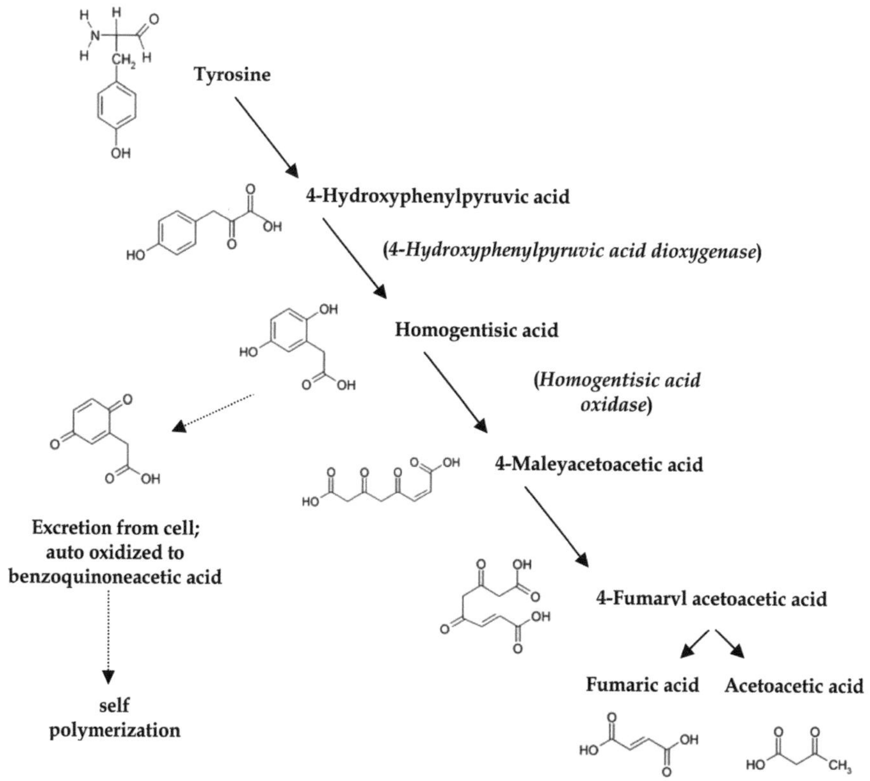

Fig. 1. Tyrosine degradation pathway showing tyrosine catabolism and pyomelanin production. Key enzymes in pyomelanin synthesis are in parentheses.

there is evidence that its useful properties to bacterial cells extend beyond this function. For instance, eumelanin has been reported to have cation exchange and metal sequestration properties (McLean et al., 1998; White, 1958). Cell cultures of the fungus *Aureobasidium pullulans* that produce eumelanin also show increased adsorption to numerous metals and tributyltin chloride relative to non-melanized cells (Gadd et al., 1990; Gadd and Mowll, 1985; Gadd et al., 1987). Uranium accumulation has also been reported in the eumelanin containing lichen *Trapelia involuta* when grown on uranium minerals (McLean et al., 1998). Hence, melanogenic organisms can be used in bioremediation and therefore have potential in biotechnology.

The properties of pyomelanin appear to be exploited by microorganisms for a survival advantage. Pyomelanin is formed abiotically outside the cell when excreted homogentisic acid autooxidizes to form benzoquinoneacetic acid. This is followed by self assembly into pyomelanin polymers. While pyomelanin at this point is a soluble exocellular pigment, it can also sorb to the cellular surface and thereby pigment the cell (Turick et al, 2003). The chemical structure of these polymers provides for a multifaceted biological utility such as; protection from light and oxidative stress, energy transduction and chemical reduction of soluble and solid phase metal oxides. For instance, pigmentation correlated with homogentisic acid production in *Legionella pneumophila* confers protection from light in the spectrum of 290-780 nm with a maximum at 635 nm (Steinert et al. 1995; Steinhert et al. 2001). Pyomelanin, like other melanin pigments is very reactive in the presence of reactive oxygen species. Oxidative stress from peroxide is decreased in *Pseudomonas aeruginosa* as a result of pyomelanin production, a trait common in clinical isolates of *P. aeruginosa* from lungs of chronically infected patients (Rodríguez-Rojas et al. 2009). Similarly, pyomelanin produced by *Burkholderia cepacia* has been shown to scavenge superoxide radicals and may play a significant role in the survival during phagocytosis by pulmoary macrophages, an important pathogenic mechanism of persistant pulmonary infections (Zughaier et al. 1999). Iron chelation and reduction has been demonstrated with pyomelanin by *Shewanella algae* BrY (Turick et al. 2008 b) and has been linked to iron acquisition by *L. pneumophilia* (Chatfield & Cianciotto, 2007). For iron assimilation to occur, Fe(III) must be reduced to Fe(II). The redox cycling properties of pyomelanin enable this process to occur.

When the redox cycling properties of pyomelanin are coupled to energy conservation by dissimilatory metal reducing bacteria, the rate of metal oxide reduction increases along with growth rates (Turick, et al. 2002; Turick, et al. 2003; Turick, et al. 2009). Similarly, dissimilatory metal reducing bacteria also are capable of using the pyomelanin they produce as a terminal electron acceptor (Turick, et al. 2002). This appears to only be the case of facultative anaerobes that experience daily fluctuations in oxygen levels. Pyomelanin production occurs during oxygen stress in the environment and may be a strategy to allow for increased electron transfer to inorganic terminal electron acceptors like Fe(III) oxides (Turick, et al. 2003; Turick, et al. 2009). Environmental concentrations of the precursors tyrosine or phenylalanine are likely sufficient in most locations to provide for enough pyomelanin production to significantly increase metal reduction rates (Turick, et al. 2009). However, pyomelanin production from cryptic growth also occurs in the absence of supplemental precursors in the laboratory (Turick, et al. 2009). Consequently, tyrosine or phenylalanine may not be required to be present in the environment for pyomelanin production to proceed, especially in biofilms where cell density is high.

The rate of exocellular electron transfer by dissimilatory metal reducing bacteria is enhanced by only small quantities of pyomelanin (femtograms per cell) due to the redox cycling nature of this polymer (Turick, et al. 2002; Turick, et al. 2009). Surface sorption of pyomelanin to the bacterial cell allows for its repeated use as an electron conduit (Turick, et al. 2003; Turick, et al. 2009). There is evidence of an optimum concentration on the cell surface whereupon; at higher concentrations electron transfer is slowed, possibly due to impedance from the polymeric structure (Turick, et al. 2009). Pyomelanin sorption onto bacterial surfaces may also have an indirect effect of increased electron transfer by trapping other soluble electron shuttles. The soluble electron shuttle riboflavin (Marsili et al. 2008) can be entrapped by pyomelanin polymers on the cell surface, which may result in the recycling of this electron shuttle (Turick, et al. 2009).

Given the nutrient and terminal electron acceptor limitations of biofilms, pyomelanin production would be considered a significant advantage for sustained metabolic activity in these environments. *B. cepacea* populations demonstrate an increase in the pyomelanin phenotype when grown in biofilms but not under pelagic growth. This biased generation of mutations is linked to mutagenic repair of double stranded DNA breaks caused by oxidative stress (Boles et al. 2004; Boles & Singh, 2008).

An understanding of the role of pyomelanin in microbial physiology and its genetic control will provide a better understanding of survival strategies utilized by specific pathogens linked to such diseases as cholera, cystic fibrosis and Legionnaires' disease. The multifaceted nature of pyomelanin provides opportunities for its exploitation for environmental and industrial biotechnology. This chapter will address research into the possible structure and related properties of pyomelanin and its application in environmental and industrial biotechnology.

3. Material and methods

3.1 Pyomelanin production and isolation

Bacterial pyomelanin was produced from regent grade tyrosine (2g/l) in bacterial cultures of *Shewanella algae* BrY, concentrated through dialysis and stored as a dry (65°C) powder, as previously described (Turick, et al. 2002). HGA melanin (artificial pyomelanin) was produced by auto oxidation of reagent grade homogentisic acid in 1N NaOH, followed by concentration as above.

3.2 In-situ production of pyomelanin

Tyrosine (0.2 - 2 g) in 50 ml deionized water was sterilized and added to constructed soil columns in a uranium contaminated riparian zone and monitored over time for pore water metals at 10, 30, and 50 cm below surface using installed lysimeters (Turick, et al. 2008 a).

3.3 Isolation of pyomelanin-producing bacteria

Methods followed those reported previously (Turick, et al. 2008 a). In general most probable number assays incorporating lactate basal salts medium (Turick, et al. 2002) with and without 1g/l tyrosine were inoculated with fresh soil from the test plot (above) and incubated at 28°C for one month. Cultures from pigmented tubes were isolated on lactate basal salts agar with 2 g/l tyrosine for isolation of pigment producing bacteria. Pigments were characterized chemically, physically and through enzyme inhibition assays.

3.4 Pyomelanin sorption studies

Bacterial cultures grown in tryptic soy broth at 28°C were concentrated and washed through centrifugation 3 times and added as a cell suspension to test tubes with/without 2g/l sterile bacterial pyomelanin. Following incubation for 5 days at 28°C, cells were concentrated via centrifugation and washed 3 times as above, followed by analysis for pyomelanin as previously reported (Turick, et al. 2002).

3.5 Electrochemistry

Electrochemical measurements incorporating cyclic voltammetry were performed in the three-electrode geometry with a carbon paste working electrode, a Pt counter electrode (Bioanalytical Systems,BAS,West Lafayette, IN) and an Ag/AgCl, 3M NaCl reference (BAS), all immersed in the electrochemical cell. Potentials measured relative to Ag/AgCl, 3M NaCl. Potential sweeps originated in the positive direction for all cyclic voltammograms in this study, using a model 100B/A potentiostat (BAS) or a 2 channel Versastat potentiostat (Princeton Instruments, Trenton, NJ USA). Between each set of measurements using different cultures, the working electrodes were rinsed in sterile 50 mM phosphate buffered saline. For each culture analyzed, a CV of the washed bare electrode in PBS was recorded, which served as a no-cell control for the next series of voltammetry studies. Sweep rates from 100 to 1700mVs^{-1} were conducted as above to determine surface or bulk phase electrochemical behavior of pyomelanin. These rates were analyzed and no significant peak distortions were detected at these rates. Studies incorporating dried (65°C) pore water from in-situ pyomelanin production experiments involved analysis in nonaqueous liquid as previously described (Nurmi & Tratnyek. 2002) using a Pt working electrode (BAS). Between each set of measurements the working electrodes were rinsed in deionized water, followed by a methanol rinse, and sonication for 10 min in deionized water. Carbon paste electrode composites were constructed by mixing 10% (wt/wt) of pyomelanin, or anthraquinone disulfonate with carbon paste and inserting into electrode housings. Chronoamperometry of carbon paste and pyomelanin/carbon paste composite electrodes were conducted with a UV light source and poised potential of the electrodes at -700mV.

3.6 Fe(III) oxide reduction

Hydrous ferric oxide (50mM) was used to measure Fe(III) reduction kinetics in 1 g fresh soil in the presence of 10 ml carbonate buffer with H_2 as the electron donor, according to a previously described methodology (Turick, et al., 2002; 2003). Fe(II) was measured spectrophotometrically using the ferrozine assay as described elsewhere (Turick, et al., 2002). For analysis of bacterial cultures with/without sorbed pyomelanin, washed cells were added to anaerobic carbonate buffer containing hydrous ferric oxide, following the procedures above.

4. Structure and properties

In order for the properties of pyomelanin to be optimized for biotechnology, a better understanding of its structure is beneficial. Structural characterization of melanins has been elusive due to their size and complexity. Pyomelanin ranges from 10-14 kDa (Turick, et al. 2002) and is smaller than other melanin pigments. FTIR analyses of pyomelanin produced from the auto oxidation of HGA (HGA-melanin) (David, et al. 1996 and Turick, et al. 2003) reveal absorbance peaks consistent with the OH stretch of polymeric structures, aliphatic C-H bonds, aromatic C=C bonds conjugated with C=O and or COO^- groups as well as phenolic OH groups. HGA-melanin provides a good starting point in beginning to determine the structure and function of microbial pyomelanin. The relative purity of HGA melanin provides for a consistent structure because microbial pigments often contain metabolic residues such as proteins, amino acids and carbohydrates (David, et al. 1996).

Based on FTIR data and other previous results, (David et al) there are several possible polymeric structures resulting from HGA auto oxidation and polymerization (Fig.2). Combining FTIR data and computational chemistry analysis allows us to predict some functional aspects of HGA-melanin and then correlate the predictions to experimental results.

The electron transfer properties of L-DOPA melanin pigments have been demonstrated through the coupling of hydroxybenzene depigmenting-compound oxidation to ferric cyanide reduction (Menter & Willis, 1997). The electron transfer properties of melanin polymers, including pyomelanin, also constitute a mechanism of electron transfer to solid electron acceptors like metal oxides (Ellis & Griffiths, 1974, Turick et al. 2002; 2003; 2009) and electrodes (Turick, et al. 2009). Based on this information, the structure that most likely will provide for optimum electron transfer capacity is structure B in Figure 2. This quality is due largely to the presence of quinones. Figure 3 illustrates the HGA melanin electron density iso-surface (at 0.2 $e/Å^3$) colored by susceptibility to radical attack (red = high, blue = low).

Going by these, the CH_2COOH group is probably not attacked, and that still leaves the ring and the OH groups. An expected increase in electron transfer potential exists in structure B in Figure 2, compared to the other structures in Figures 2&3. Based on the computational analyses, an alternating structure would possess minimal electron transfer properties due to the absence of quinones and hence is not a likely structure of HGA melanin or microbial pyomelanin.

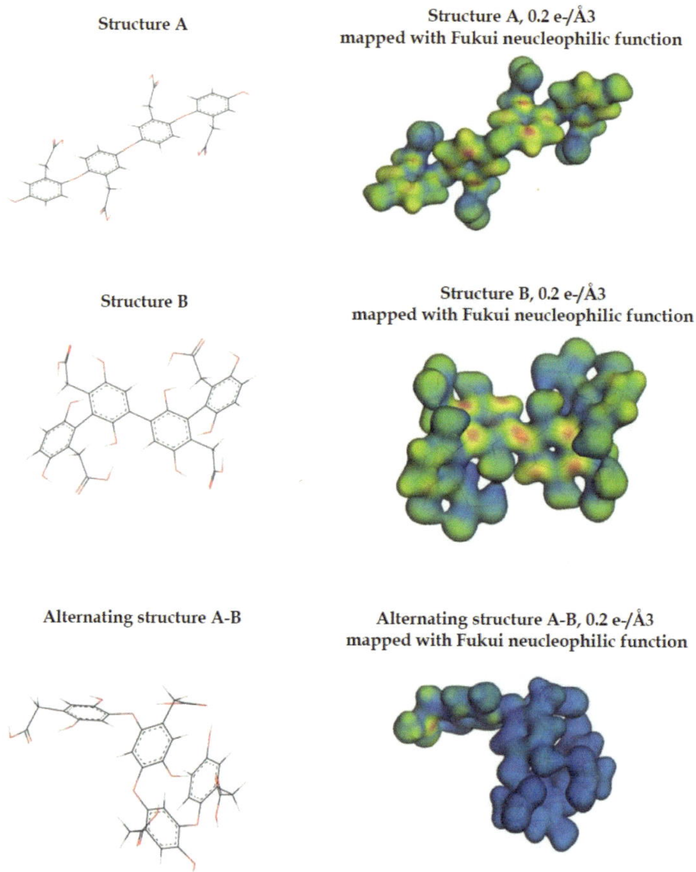

Fig. 2. Possible linear polymer structure of HGA-melanin. Line structures (left) are derived from David et al. (1996) and their corresponding simulated 3D structure mapped with Fukui neucleophilic function (right). The colorations are electrophilic Fukui Function values, for radical attack describes the relative ease of transferring a small amount of charge from the highest occupied molecular orbital to the lowest unoccupied molecular orbital.

Computations for HGA-melanin have shown the increased availability of conducting electrons in the quinone forms of the pyomelanin. Two possible structures presented in David, *et. al.* (1996) and Figure 2 are analyzed along with a quinone-analog structure. In order to confirm our computational analyses of the electrochemical properties of pyomelanin in response to radical attack we monitored the current response of a pyomelanin/carbon paste electrode compared to carbon paste alone in the presence of UV light. L-DOPA melanin interacts with UV light by offering protection from harmful radical formation. Microbial pyomelanin also displayed an electrochemical response to UV light when combined with a carbon paste electrode at a 1:10 wt/wt ratio of carbon paste (Fig. 4). These results demonstrate that pyomelanin reacts with UV light in a similar fashion as L-DOPA melanin. These data also offer further evidence that pyomelanin is composed of multiple quinone moieties.

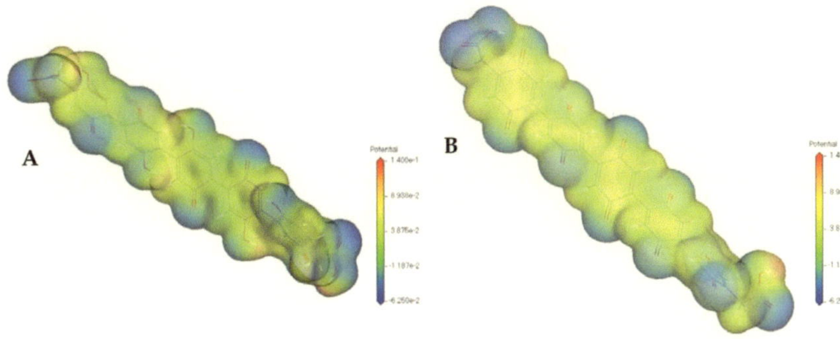

Fig. 3. HCA melanin electron density iso-surface (at 0.2 e/Å^3) colored by susceptibility to radical attack (red = high, blue = low). Structures are partially oxidized (semiquinone) (A) and fully oxidized (quinone) (B). The colorations are electrophilic Fukui Function values, for radical attack describes the relative ease of transferring a small amount of charge from the highest occupied molecular orbital to the lowest unoccupied molecular orbital.

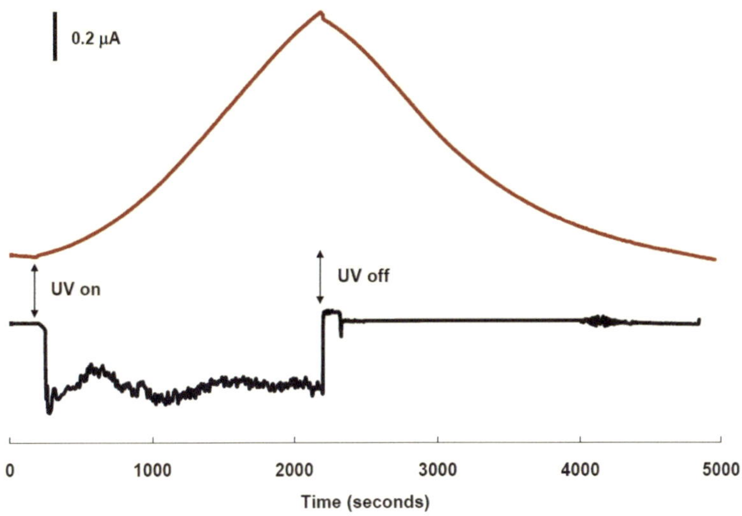

Fig. 4. Pyomelanin Response to UV Light (365 nm, 9100 mW/cm^2). Pyomelanin also displayed an increased current in response to UV light when combined with a carbon paste electrode at a 1:10 wt/wt ratio of carbon paste. Electrodes were poised at –700 mV (vs Ag/AgCl) throughout the study.

By examining a short, five-ringed segment of the proposed structure, QM calculations show that as the structure becomes more quinone-like, there is a higher density of electrons near the Fermi level, indicating more electrons are available for conduction. In Figure 5, Structure A has no quinone-like character, Structure B hydro-quinone groups that could become oxidized to quinones, and Structure C is Structure B oxidized to a full quinone state; all three structures have a net charge of 0. The figure shows the increase in the density of conducting electrons as we move to the quinone-like structure.

Figure 6 shows a structure of a small, 5kDa pyomelanin segment (of the quinone structure depicted in Figure 5C. after simulated annealing with the COMPASS force field. This simple simulation illustrates that the quinone-analog of the pyomelanin structure has a tendency to form a spiraled structure. To verify the presence of quinones, studies incorporating cyclic voltammetry were conducted to determine if the electrochemical behavior of pyomelanin was similar to a 2-step oxidation followed by a 2 step reduction of the model quinone anthraquinone-2,6-disulfonate (AQDS). A 2-step oxidation followed by a 2 step reduction was determined for AQDS as well as pyomelanin (Fig. 7) and thus providing strong evidence of the presence of quinones in the pyomelanin structure, as well as corroborating previous findings employing wet chemistry techniques (Turick, et al. 2003).

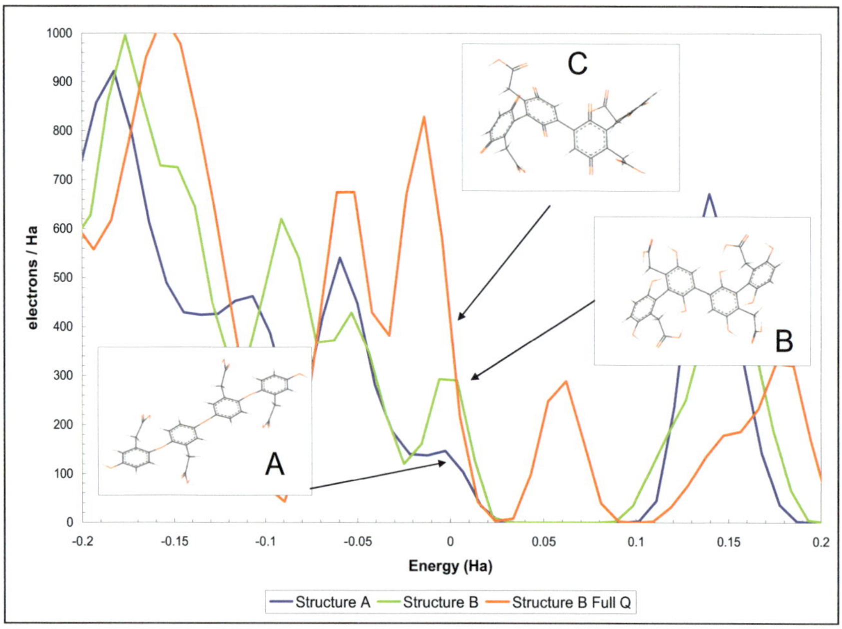

Fig. 5. Superimposed density of states for three postulated structures of HGA-melanin from Figure A. A significant difference is demonstrated here between the quinone state (C) and the hydro-quinone state (B).

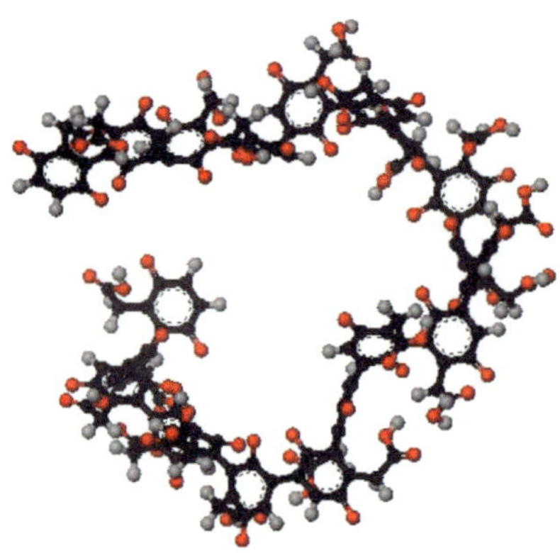

Fig. 6. Structure of a small, 5kDa pyomelanin segment of the quinone structure depicted in Fig.5-C after simulated annealing with the COMPASS force field.

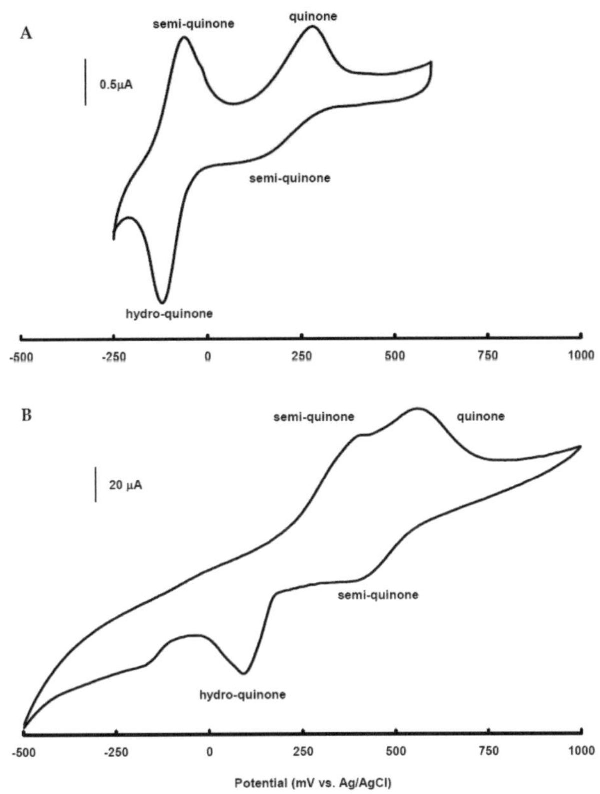

Fig. 7. Cyclic voltammetry of AQDS and pyomelanin. AQDS (A) and pyomelanin (B) were mixed individually in a 1:10 ratio (wt/wt) with carbon paste and analyzed in 50 mM phosphate buffered saline at pH 2. Scans began at -500 mV and moved in the positive direction at 25 mV/s. A 2-step oxidation (upper peaks) and 2-step reduction (downward peaks) are indicative of quinones.

5. Biotechnological applications

5.1 Biogeochemistry: metal transformation and redox manipulation

The metal binding and electron shuttling properties of pyomelanin offer a means of altering the biogeochemistry of specific environmental sites provided the conditions are appropriate for pyomelanin production as well as conducive to electron transfer and metal retention in the soil. Supplementation of soil with the chemical precursors tyrosine and/or phenylalanine was successful in resulting in pyomelanin production (Turick, et al. 2008 b). Due to the long-term affects of pyomelanin on the environment, a one-time application of these carbon sources may prove cost effective in many cases. Environmental conditions that may limit the production of pyomelanin include Fe(II) availability. The presence of Fe(II) in laboratory studies demonstrated decreased rates and degrees of pyomelanin production as a function of Fe(II) present (Turick, et al. 2008 b). While 4-HPPD is required for pyomelanin production and Fe(II) is required for this enzyme, increased Fe(II) concentrations decrease 4-HPPD activity (Lindstedt, et al. 1977). Similarly, we have found the rate of pyomelanin production to also follow a similar trend relative to Fe(II) in laboratory studies (Fig. 8)

These results suggest that pyomelanin production will be low in soils with higher concentrations of bioavailable iron. It is interesting to note that along with evidence that pyomelanin is used in Fe(II) acquisition, Fe(II) concentrations in the environment may limit the activity of 4-HPPD and as a result minimize pyomelanin production when Fe(II) is in abundance.

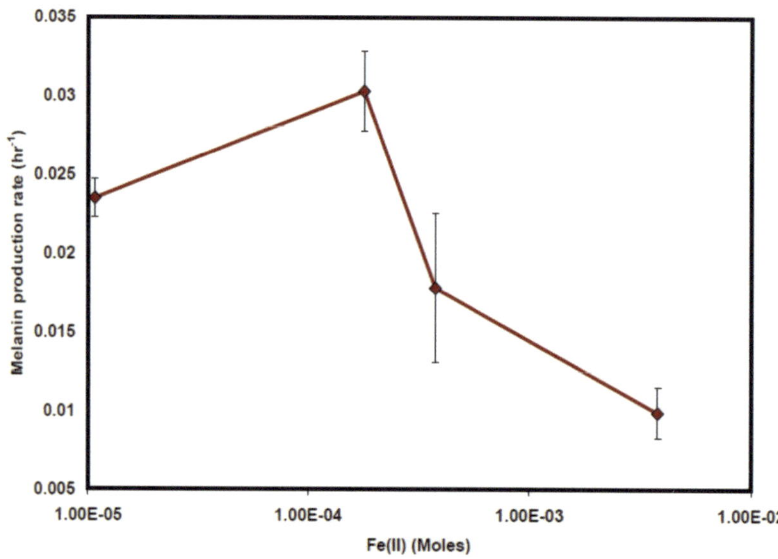

Fig. 8. Pyomelanin production rates by *S. algae* BrY in relation to Fe(II) concentrations. Fe(II) concentrations drastically inhibit the rate of pyomelanin production above 0.5 mM Fe(II).

The presence of naturally occurring bacteria capable of pyomelanin production eliminated the need to augment the soil with pure cultures of allochthonous pyomelanin producers (Turick, et al. 2008 b). A pyomelanin producing strain (Fig. 9) previously isolated from uranium contaminated soil (Turick, et al. 2008 b) has since been tentatively identified as *Bacillus mycoides* through fatty acid methyl ester analysis. Little is known however about the abundance of pyomelanin production in the environment. Since bacteria need only small quantities of this polymer to enhance their electron shuttling capacity, the low detection limit required to identify environmentally relevant quantities marginalizes a direct analytical chemical approach. However, given that numerous strains of soil bacteria are known to produce pyomelanin and if the rate of mutation of the HGA-oxidase gene is similar in bacteria to that in humans (1:100,000 - 250,000) (Zatkova, et al. 2000), pyomelanin-producing bacteria may be ubiquitous in the environment.

If present in a particular environment, pyomelanin producers may be too low in number to contribute to significant electron transfer and hence metal transformation. However, sufficient supplementation of tyrosine (1-10 mM) to the soil resulted in significant production of pyomelanin production. The result was a dramatic increase of the redox properties of the soil as determined through Fe(III) oxide reduction assays and electrochemistry (Fig. 10). With H_2 as the electron donor, Fe(III) reduction, as measured as Fe(II) evolution was greater in soils with pyomelanin compared to equal molar concentrations of the carbon and energy source lactate. Similarly, the redox activity as measured by cyclic voltammetry was greater in pyomelanin containing soils. Cyclic voltammograms revealed two distinct oxidation (upward) peaks and reduction (downward) peaks with pyomelanin-containing soils whereas only one smaller redox couple was detected in controls. Peaks less than -750 mV were likely due to solvent breakdown.

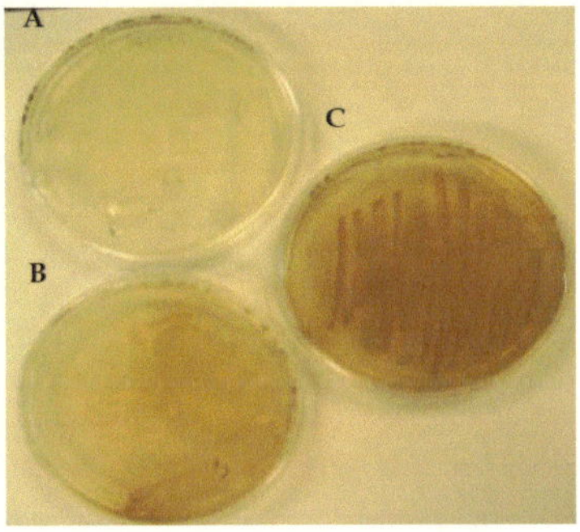

Fig. 9. Growth of melanin producing isolate from uranium contaminated soils. Results after 1 week growth on lactate basal salts medium. without tyrosine (A) and with tyrosine (B). Redish-brown coloration is a result of partial catabolism of tyrosine. More intense pigmentation was evident after 2 weeks growth on the same medium (C). Redish brown pigment was identified as pyomelanin.

Laboratory studies suggest that pyomelanin reduction may occur by other soil bacteria that are not pyomelanin producers. The sorptive capacity of pyomelanin to bacterial cells was not limited to pyomelanin producers (Table 1) and all bacterial strains tested under anaerobic conditions (50%H_2/50%N_2), in the presence of pyomelanin, demonstrated increased Fe(III) oxide reduction with H_2 as electron donor.

	Fe(III) oxide (mM) reduced in 72h		
Species	**Control**	With pyomelanin	Pyomelanin sorbed (μg/l)/10^8 cells/ml
B. cereus	0.02 (0.005)#	0.25 (0.02)	1.8 (0.7)
V. angullarium	0.02 (0.004)	0.12 (0.02)	ND*
P. putida	0.02 (0.001)	0.30 (0.03)	0.7 (0.4)
M. luteus	0.01 (0.005)	0.09 (0.01)	0.4 (0.2)
E. coli K-12	0.02 (0.009)	0.09 (0.01)	1.8 (0.2)
P. aeruginosa	0.01 (0.011)	0.04 (0.02)	0.2 (0.05)

standard deviation in parenthesis; * ND= not determined.

Table 1. Pyomelanin sorption to bacteria and resultant increased metal reduction capacity

The bacteria evaluated in Table 1 do not couple energy conservation and growth to metal reduction and hence do not typically play a significant role in metal transformation in the environment. However a small degree of metal reduction does occur by these bacteria. Metal reduction in this case is often viewed as a sink for excess reducing power of microorganisms or part of assimilatory Fe(III) reduction (Ehrlich 2002, & Schröder et al, 2003). However the proportion of bacteria capable of metal reduction, but not coupled to growth, would be expected to be far greater than dissimilatory metal reducers in

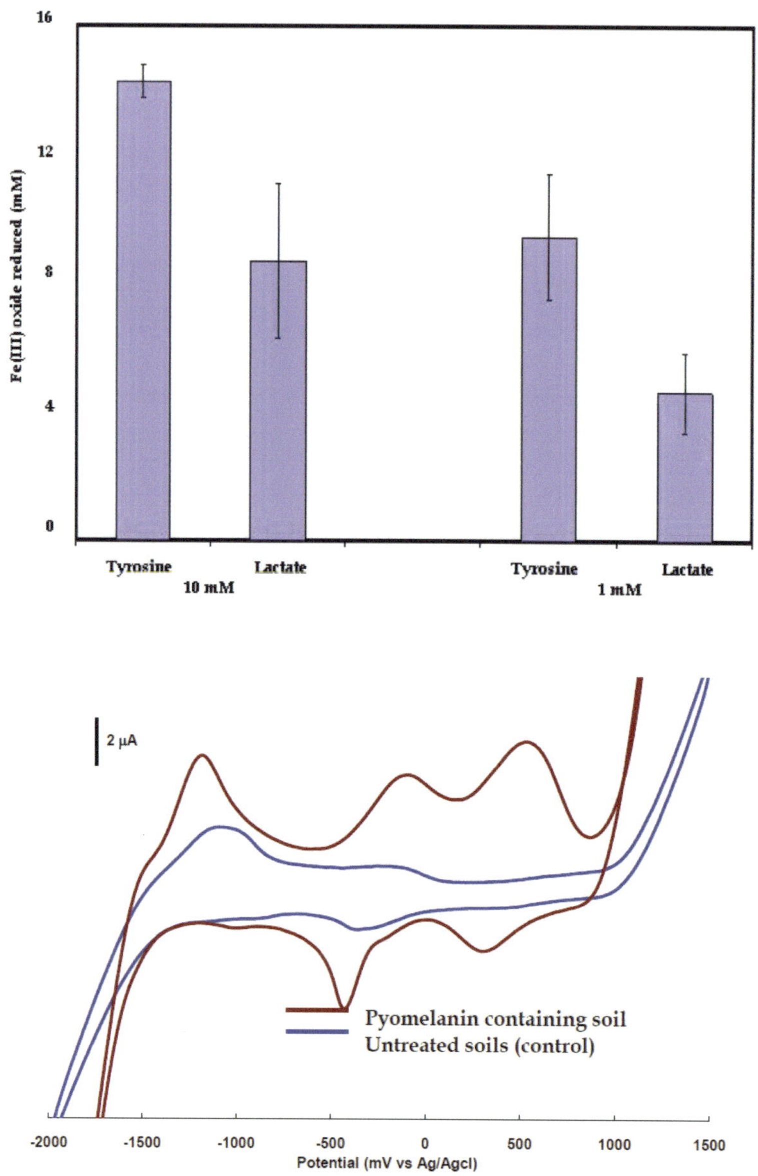

Fig. 10. Increased electron transfer capacity of soil with pyomelanin produced in-situ. Soils treated with tyrosine which resulted in pyomelanin production had an increased capacity for reduction of supplemental Fe(III) oxide under anaerobic conditions compared to equal molar concentrations of lactate after exposure to hydrogen (electron donor) for one week (top). Cyclic voltammograms (bottom) of dried (65°C) soils in DMSO also demonstrated increased redox activity of pyomelanin containing soils (brown) compared to controls (blue). Voltammograms were obtained with a Pt working electrode and Pt wire as counter electrode and the reference electrode was Ag/AgCl. Scan rate was 100 mV/s beginning at -2V and proceeding in the positive direction.

many subsurface environments. Indeed, dissimilatory metal reducing bacteria may be in very low numbers or even absent in some environments (Lehman, et al. 2001). Fe(III) reduction was also reported by the yeast *Cryptococcus neoformans* (Nyhus, et al., 1997) and is facilitated by extracellular melanin for Fe(III) reduction but does not appear to be coupled to energy conservation or growth, instead the resultant Fe(II) is assimilated. A similar mechanism was reported for *L. pneuophilla* where pyomelanin produced by this organism associates with the cell surface and reduces Fe(III), resulting in Fe(II) assimilation (Cianciotto, et al 2002). The action of various complexing agents, including quinones, to facilitate Fe(III) reduction for assimilation by bacteria are linked to dehydrogenase activity (Schröder et al, 2003).

Previous work has also demonstrated the ability of bacteria to reduce a variety of quinone-containing compounds, ultimately increasing the rates of azo dye reduction (Rau, et al. 2002). The quinone-reducing bacteria in the study of Rau et al. represented considerable diversity and included lactic acid bacteria, high and low GC Gram positive bacteria, as well as bacteria from the α, β, and γ-Proteobacteria, including *Escherichia.coli* K12. Therefore one should expect qunione reduction by soil bacteria to be common. Hence, the role of quinone-containing compounds as electron shuttles for non-dissimilatory metal reducing bacteria offers another possibility to control redox conditions and metal transformations in the environment. Based on these studies, pyomelanin production in-situ has the potential to accelerate metal reduction by a broad range of soil bacteria even though this process is not linked to energy conservation and growth.

5.2 Biogeochemistry: metal immobilization

The biotechnological use of pyomelanin is being explored for accelerated in-situ metal contaminant immobilization. The stimulation of pyomelanin production in-situ allows for inorganic contaminant immobilization due to metal binding and mineral sorption properties of pyomelanin. The properties of metal binding and surface sorption of pyomelanin were applied to field studies for uranium immobilization through the formation of a ternary structure with soil components (Turick, et al. 2008 b). Our previous work focused on the mechanisms of uranium complexation with pyomelanin and subsequent pyomelanin sorption to clay and mineral particles. This approach, which includes the addition of tyrosine to surface soils, results in production of pyomelanin pigments followed by a decrease in pore-water uranium concentrations. Through a one-time supplementation of tyrosine, uranium is immobilized for at least 13 months relative to controls. Follow-on studies were conducted to determine how in-situ pyomelanin production affected sorption phenomena in the soil.

Uranium immobilization was quantified by determining the partition coefficient of various treatments at different depths. The metal partition coefficient (K_d; also known as the sorption distribution coefficient) is the ratio of sorbed metal concentration to the dissolved metal concentration. Uranium K_d values were highest with the tyrosine treated soils (Table 2). The resulting conversion of tyrosine to pyomelanin played a significant role in retarding uranium transport from the soil. While lactate addition also increased K_d values, they were not as high as the tyrosine treatments. Lactate is a carbon and energy source for bacteria and hence likely contributed to microbial activities that influenced, to some degree, the biogeochemical behavior of uranium. This effect was not as long lasting as that of the tyrosine amended soils.

Depth (cm)	Control	10 mM Lactate	10 mM Tyrosine
	After 1 month		
10	9,954	28,198	47,569
30	6,359	17,154	36,744
50	4,394	15,133	26,592
	After 13 months		
10	4,151	15,789	16,503
30	8,592	5,022	13,187
50	3,514	3,629	8,810

*= ml/g

Table 2. K_d* values for uranium resulting from various soil treatments

The one-time, in-situ stimulation of pyomelanin production resulted in a considerable impact on uranium mobility over one year, down to at least 50 cm. We also examined other metals to get a better idea of the effect of this metal immobilization approach. Figure 11 illustrates the affect of pyomelanin production on molybdenum, lead, bismuth and arsenic one month and 13 months after treatment ensued. Pore-water metal concentrations in general were higher for the controls compared to the pyomelanin containing soil.

The mechanism of pyomelanin sorption to soil and its related effects on the sorptive capacity of metal ions was also examined one month after pyomelanin stimulation began. XRF data of control soils demonstrated that uranium, calcium, nickel and zinc displayed a strong correlation to iron and manganese oxides. This correlation did not exist to the same degree in the presence of pyomelanin (Fig. 12). A possible explanation for this phenomenon is that pyomelanin sorbed easily to clays at the subsurface pH of 3.9 - 5.2. Previously we demonstrated that pyomelanin sorbes readily to illite and goethite at pH 4. The greater abundance of clay than iron or manganese minerals in this soil likely contributed to the difference in metal sorption indicated in our XRF data. Interestingly, changes in metal sorption did not necessarily increase metal immobilization in all cases. Nickel and zinc in particular did not demonstrate any changes in metal concentration in the pore water.

This work demonstrated the biotechnological use of bacterial biopolymers in accelerated in-situ metal contaminant immobilization. The stimulation of pyomelanin production in-situ allowed for inorganic contaminant immobilization due to metal binding and mineral sorption properties of pyomelanin. Similarly, the electron transfer capacity of pyomelanin accelerates soluble and solid phase metal reduction. The ability for pyomelanin to be sorbed onto bacterial surfaces may also accelerate electron transfer between a significant portion of the soil microbial population to contaminating metal oxides.

5.3 Microbial production of electric current

The ability of some bacteria to access solid mineral oxides as terminal electron acceptors has elicited interest in understanding mechanisms of bacterial electron transfer to solid terminal electron acceptors, including electrodes. Bacterial production of electric current results when an anode is used as a sink for electrons produced through microbial physiological activities. This phenomenon is being exploited for electricity production through microbial fuel cells (Drapcho, et al. 2008) sensors, and new methods for studying microbial physiology. Because pyomelanin increases the rate of electron transfer by bacteria that transfer

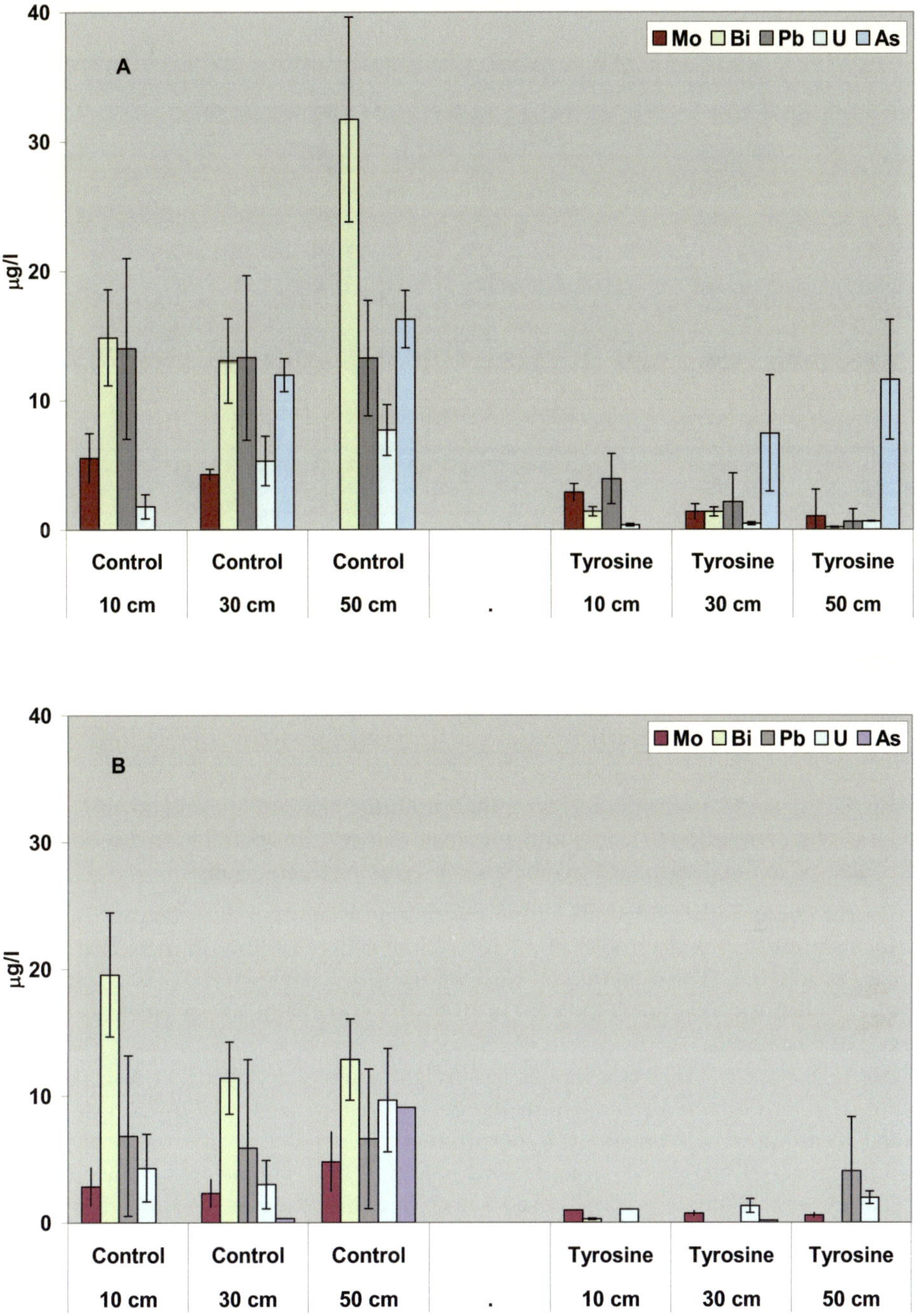

Fig. 11. Affect of pyomelanin production on molybdenum, gold, lead, bismuth and arsenic one month (A) and 13 months (B) after treatment ensued. Pore-water metal concentrations in general were higher for the untreated controls compared to the tyrosine treatments that resulted in pyomelanin production in soil.

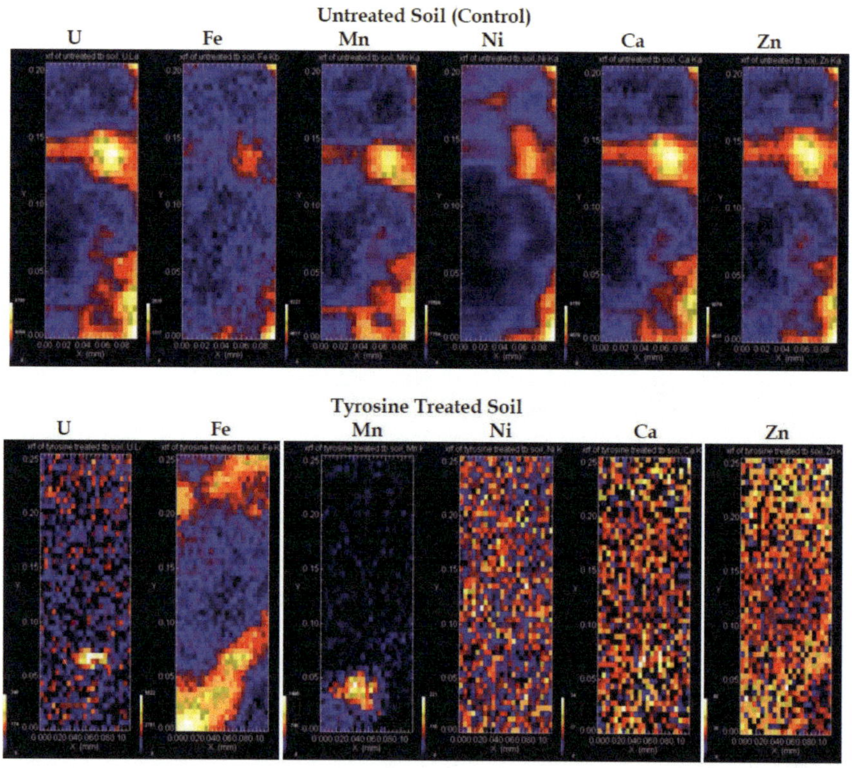

Fig. 12. XRF data of control soils demonstrate that uranium, calcium, nickel and zinc displayed a strong correlation to iron and manganese oxides. This correlation did not exist to the same degree in the presence of pyomelanin in tyrosine treated soils.

electrons to insoluble Fe(III) minerals, these microorganisms also demonstrate increased rates of electron transfer to electrodes in microbial fuel cells, resulting in increased current for electricity production. The dynamics of electron transfer from bacterial cells as examined through electrochemistry demonstrates the utility of pyomelanin as an electron conduit from the cell to the anode.

The bacteria *S. oneidensis* MR-1 is a model dissimilatory metal reducer that also produces pyomelanin. Through directed mutagenesis the gene that encodes for HGA oxidase (*hmgA*) was deleted resulting in a pyomelanin hyperproducer (*ΔhmgA*) and in another strain, the gene encoding for 4-HPPD (*melA*) was deleted resulting in a pyomelanin deficient strain (*ΔmelA*) (Turick, et al. 2009) (Fig. 13). These strains were employed in the following studies to understand the roles of pyomelanin in electron transfer to anodes. The sorptive nature of pyomelanin offers an advantage as a potential electron shuttle that is associated to the bacteria surface and the anode. To confirm this, electrochemistry studies employing cyclic voltammetry were conducted at various sweep rates (Fig. 14) with a pyomelanin hyperproducing strain *S. oneidensis ΔhmgA*. When the peak current is plotted against the scan rates, if the data demonstrate a high linear correlation, this indicates surface associated electron transfer, while a high correlation to the log of the scan rates indicates electron transfer in the bulk phase or a soluble electron shuttle. The data from Fig. 14 are indicative of surface associated electron transfer. The redox couple labeled as Peak 2 (ox. and red.) in

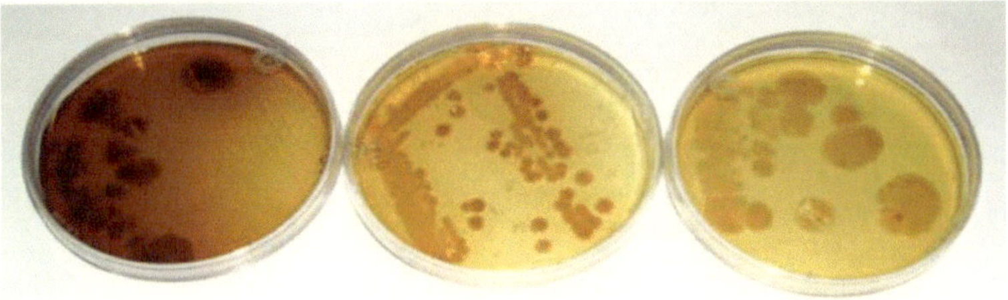

Fig. 13. Strains of the model dissimilatory metal reducing bacteria *S. oneidensis* used in these studies. Through directed mutagenesis, a pyomelanin hyper producing mutant (*ΔhmgA*) and a pyomelanin deficient mutant (*ΔmelA*) were generated and used in this study to determine the role of pyomelanin on electrogenic biofilms of the wild type strain *Shewanella oneidensis* MR-1.

Fig. 14 were defined previously as pyomelanin.(Turick et al 2009,). These data coincide with previous work (Turick, et al. 2003) demonstrating that pyomelanin is associated with the bacterial surface. Riboflavin is a soluble electron shuttle produced by *S. oneidensis* (Marsilli, et al. 2009) and is also associated with and entrapped by pyomelanin (Turick, et al. 2009). Its electrochemical signature is described in Figure 4 as Peak 1.

The three strains of *S. oneidensis* (above) were grown in TSB in the presence of sterile patterned electrodes poised at 400 mV (vs Ag/AgCl) to determine the affect of pyomelanin production on electron transfer to an anode. The patterned electrodes used had a very hydrophobic working electrode (diameter, 2mm). Cells were grown at 28°C with enough stirring to keep the cells in suspension but to also maintain microaerobic conditions. Cell densities were similar in all cultures analyzed throughout the study. Current was monitored over time and stabilized after 72 hrs. Electrodes were examined microscopically at the termination of the experiment and no biofilms were detected on the hydrophobic working electrode. This indicated that electrons were passed to the electrodes by the suspension culture and not by biofilms growing on the electrodes. Current production was relative to pyomelanin produced by these cultures (Turick, et al. 2009) (Fig. 15), demonstrating the utility of pyomelanin as an electron conduit.

In order to study current production from biofilms, cultures were grown as above but on identical graphite rods as working electrodes to provide a suitable surface for biofilm development. The working electrodes were immersed 5 cm into the medium to provide sufficient surface area to the culture. These studies demonstrated similar results with respect to increased current production by the pyomelanin hyper producer when compared to the pyomelanin deficient strain. During establishment of biofilms, *ΔhmgA* demonstrated increased current production at the termination of aeration (Fig. 16-A). Pyomelanin production occurs during oxygen stress and during this time the *ΔhmgA* culture broth began to turn redish brown (indicative of pyomelanin) compared to the *ΔmelA* culture. TSB has sufficient tyrosine and phenylalanine to support pyomelanin production (McCuen, 1988). Following establishment of a biofilm on the electrode and current production the electrodes were rinsed gently in sterile phosphate buffered saline (pH 7) and then placed in clean phosphate buffered saline for cyclic voltammetry analysis. These analyses demonstrated enhanced current production by the pyomelanin hyper producer with a current response similar to that of pyomelanin (Figure 16-B).

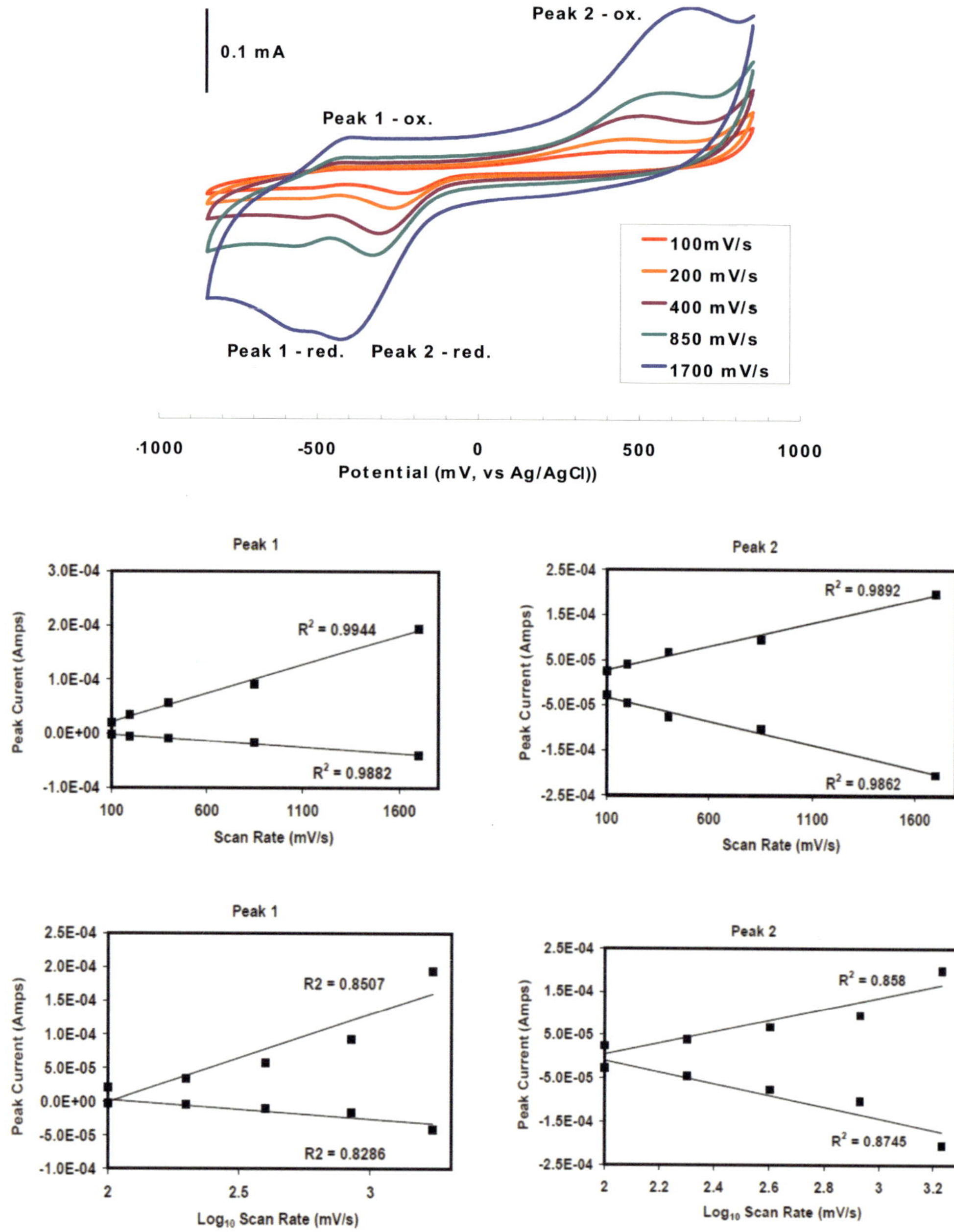

Fig. 14. Determination of surface associated, bacterial electron transfer by pyomelanin. Cyclic voltammetry conducted at various scan rates (top graph) was used to determine if electron transfer to the electrode by the pyomelanin over producer occurred at the cell surface or in the bulk phase. In these scans, oxidation peaks are upward and reduction peaks are downward. Peak currents plotted against scan rates demonstrated a higher linear correlation as compared to a logarithmic correlation, which would demonstrate surface associated electron transfer.

Directly following the previous study, the graphite rod working electrodes with biofilms were transferred to TSB (as above) and incubated for 4 days to allow further biofilm growth followed by a change to fresh TSB. The fresh static, microaerobic medium [dissolved oxygen =1.7(±0.09) $nMmL^{-1}$] demonstrated increasing current production from the *ΔhmgA* culture relative to the pyomelanin deficient strain *ΔmelA* (Fig. 16C). Cyclic voltammetry performed on the rinsed working electrodes demonstrated significant differences between the two strains where *ΔhmgA* revealed a greater current density and charge capacity as well as several redox couples, indicative of pyomelanin.

Upon termination of this study, biofilm thickness on the electrodes (determined with confocal microscopy) was similar for both cultures at the bottom of the 5 cm electrode but thicker at the top (air/medium interface) (Fig. 17). Current production in microaerobic environments exceeds that of strict anaerobic conditions and could be a result of production of electron shuttles under microaerobic conditions (Mohan, et al. 2008). This could explain why the biofilm of the pyomelanin hyper producer was thicker at the medium surface but the pyomelanin deficient strain had consistent biofilm thickness along the working electrode.

The only difference between these two strains was the ability to make pyomelanin and these results demonstrate the contribution of pyomelanin to electron transfer to solid terminal electron acceptors. *S. oneidensis* has numerous mechanisms for electron transfer to solid terminal electron acceptors and is evident with current production from *ΔmelA*. While it is conceivable that *ΔmelA* utilized other electron shuttleing mechanisms to a greater extent in the absence of pyomelanin production, the overall differences in current production between these two strains demonstrated the utility of pyomelanin as an electron conduit for bacterial growth.

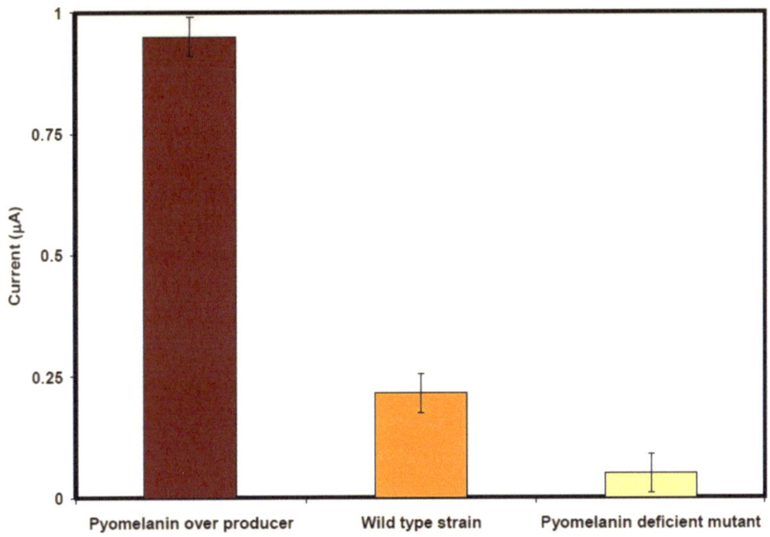

Fig. 15. Role of pyomelanin production on electric current production. Using a patterned composite electrode with hydrophobic graphite as working electrode stirred pelagic bacterial cultures demonstrated increased electron transfer, measured as electric current production as a function of pyomelanin production.

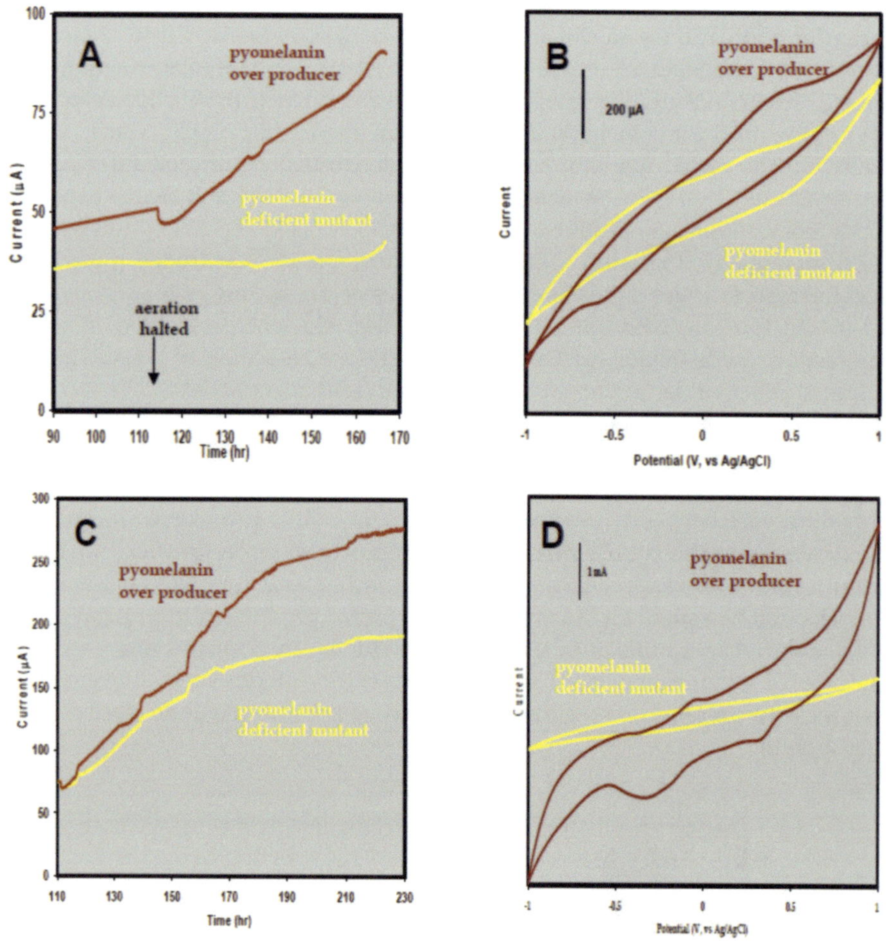

Fig. 16. Affects of pyomelanin production on current generation by bacteria. Current production from initiation of bacterial biofilms on a graphite rod electrode over time (A) increased by the pyomelanin over producer after aeration stopped, likely due to production of pyomelanin under low dissolved oxygen conditions. Cyclic voltammetry (B) demonstrated enhanced current production by the pyomelanin over producer with a current response similar to that of pyomelanin. Current production from established bacterial biofilms on a graphite rod electrode over time (C) increased by the pyomelanin over producer after dissolved oxygen levels dropped. Cyclic voltammetry (D) demonstrated an increase in redox couples and increased current density by the pyomelanin over producer. The increased current density is indicative of higher current production in the biofilm.

6. Conclusions and outlook

Biotechnological application of bacterial polymers, including pyomelanin have been discussed (David, et al, 1996; Weiner, 1997). The multifaceted nature of melanin, including pyomelanin offers numerous opportunities for applications, such as reporter genes, cosmetics, dyes, colorings and sunscreens (Weiner, 1997). In this report, we demonstrated further, the biotechnological utility of pyomelanin. Through the exploitation of its electron transfer properties metal reduction in relation to bioremediation can be enhanced in-situ. Hyper production of pyomelanin may lead to surface sorption by a wide range of microorganisms through interspecies sorption of the soluble polymers, thereby altering redox activity in soil microbial communities. The mineral sorption and metal chelation capacity of pyomelanin has been shown here to enhance contaminant immobilization in soils. As an electron conduit, pyomelanin also has the potential to increase the current response from biofilms for electricity production in microbial fuel cells.

Future applications of pyomelanin could include its use as an extracellular electronic material. A further understanding of the structure of this polymer and its metal chelation behavior will provide valuable insights into how to modify it chemically, potentially for enhanced electrochemical activity. Research into the biochemistry and microbial ecology of pyomelanin production and utilization will allow us to understand its role in the survival of microorganisms of clinical and environmental significance.

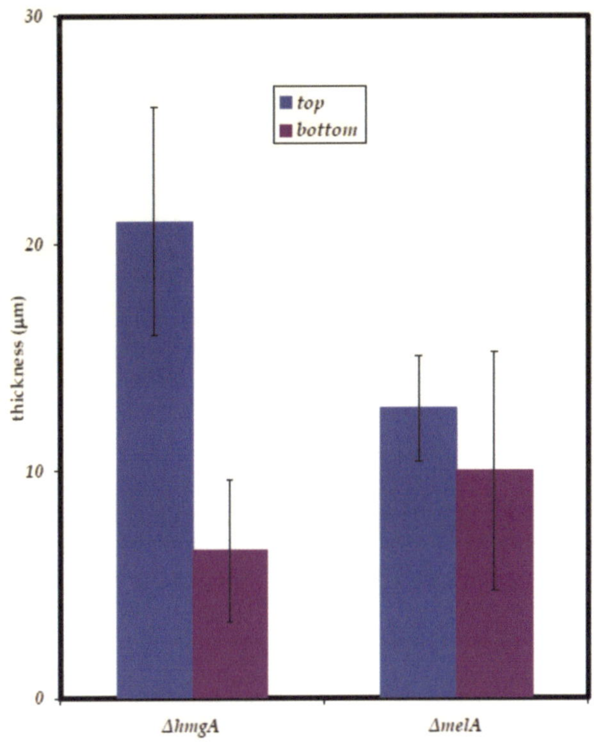

Fig. 17. Biofilm thickness on graphite electrodes. Biofilm thickness on the electrodes (determined with confocal microscopy) was similar for both cultures at the bottom of the 5 cm electrode but thicker at the top (air/medium interface).

7. Acknowledgements

This research was supported in part through funding by AES Hatch grant 389: the Department of Energy (DOE) Office of Science NABIR/ERSP and Genomics: GTL programs, the Savannah River National Laboratory Directed Research and Development Program, and Soil and Groundwater Closure Projects of the Savannah River Site funded through DOE-Environmental Management Program. This document was prepared in conjunction with work accomplished at SRNL under Contract No. DE-AC09-08SR22470 with the US Department of Energy. We would like to thank Dr. M. Duff for XRF analyses and T. Poppy, A. Maloney, W. Jones and Y.G. Kritzas for able technical assistance. We would also like to thank Dr. D.A. Lowy for assistance and valuable discussions in electrochemistry.

8. References

Beltrán-Valero de Bernabé, D., F.J. Jimenez, R. Aquaron & S. Rodríguez de Córdoba. 1999. Analysis of alkaptonuria (AKU) mutations and polymorphisms reveals that the CCC sequence motif is a mutual hot spot in the homogentisate 1,2 dioxygenase gene (HGO). Am. J. Hum. Gene. 64: 1316-1322.

Boles, B.R., M. Thoendel, & P.K. Singh. 2004. Self-generated diversity produces "insurance effects" in biofilm communities. Proc. Nat. Acad. Sci. 101:16630-16635.

Boles, B.R., & P.K. Singh. 2008. Endogenous oxidative stress produces diversity and adaptability in biofilm communities. Proc. Nat. Acad. Sci. 105:12503-12508.

Chatfield, C.H., & N.P. Cianciotto, 2007. The secreted pyomelanin pigment of *Legionella pneumophila* confers ferric reductase activity. Infect. Immun. 75:4062-4070.

Coon S.L., S. Kotob, B.V. Jarvis, S. Wang, W.C. Fuqua, and R.M.Weiner. 1994. Homogentisic acid is the product of MelA, which mediated melanogenesis in the marine bacteriaum *Shewanella colwelliana* D. Appl. Env. Microbiol.60:3006-3010.

Drapcho, C.M, N.M. Nhuan & T.H. Walker. 2008. Microbial Fuel Cells. in *Biofuells Engineering Process Technology*. pp 303-327. (Ed) L.S. Hager. McGraw-Hill Co. New York, NY. USA.

Ellis, D.H. & D.A. Griffiths. 1974. The location and analysis of melanins in cell walls of some soil fungi. Can. J. Microbiol. 20:1379-1386.

Ehrlich, H.L. 2002. *In Geomicrobiology*. Marcel Decker Inc. N.Y., N.Y., pp 378-380.

Gadd, G.M., D.J. Gray, & P.J. Newby. 1990. Role of melanin in fungal biosorption of tributyltin chloride. Applied Microbiology and Biotechnology 34, 116-121.

Gadd, G.M. & J.L. Mowll. 1985. Copper uptake by yeast-like cells, hyphae, and chlamydospores of *Aureobasidium pullulans*. Experimental Mycol. 9, 230-240.

Gadd, G.M., C. White, & J.L. Mowll. 1987. Heavy metal uptake by intact cells and protoplasts of *Aureobasidium pullulans*.FEMS Microbiol. Lett. 45:261-267.

Kotob, S.I., S.L. Coon, E.J. Quintero, & R.M. Weiner. 1995. Homogentisic acid is the primary precursor of melanin synthesis in *Vibrio cholera*, a *Hyphomonas* strain, and *Shewanella colwelliana*. Appl. Environ. Microbiol. 61:1620-1622.

Lehman, R.M., F.F.Roberto, D. Earley, D.F. Bruhn, S.E. Brink, S.P. O'Connell, M.E. Delwiche, & F.S. Colwell. 2001. Attached and unattached bacterial communities in a 120-meter corehole in an acidic, crystalline rock aquifer Appl. Environ. Microbiol. 67:2095-2106.

Lehninger AL (1975) Oxidative degradation of amino acids. In: *Biochemistry*. pp.568-570. Worth Publishers. N.Y. NY,

Lindstedt, S., B. Odelhög, & M. Rundgren. 1977. Purification and some properties of 4-hydroxyphenylpyruvate dioxygenase from *Pseudomonas* sp. P.J. 874. Biochem. 16:3369-3377.

Marsili, E., D.B. Baron, I.D. Shikhare, D. Coursolle, J.A. Gralnick & D.R. Bond. 2008. *Shewanella* secretes flavins that mediate extracellular electron transfer. Proc. Nat. Acad. Sci. 105:3968–3973.

McCuen, P.J. 1988. Culture media components In: *Manual of BBL Products and Laboratory Procedures.* Power, D.A. (ed). Beckton Dickinson Systems, Cockeysville, MD. pp. 293-294.

McLean, J., O.W. Purvis, B.J. Williamson, & E.H. Bailey. 1998. Role for lichen melanins in uranium remediation. Nature 391, 649-650.

Menter, J.M. & I. Willis. 1997. Electron transfer and photoprotective properties of melanins in solution. Pigment Cell Res. 10:214-217.

Mohan, S.V., G. Mohanakrishna, & P.N. Sarma. 2008. Effect of anodic metabolic function on bioelectricity generation and substrate degradation in single chambered microbial fuel cell. Environ. Sci. Technol. 42:8088-8094.

Nurmi, J.T. & P.G. Tratnyek. 2002. Electrochemical properties of
natural organic matter (NOM), fractions of NOM, and model biogeochemical electron shuttles. Environ. Sci. Technol. 36:617-624.

Nyhus, K.J., Wilborn, A.T. & Jacobson E.S. 1997. Ferric iron reduction by *Cryptococcus neoformans.* Infection and Immunity. 65:434-438.

Plonka, P.M. & M. Grabacka. 2006. Melanin sythesis in microorganisms - biotechnical and medical aspects. ACTA Biochimica Polonica. 53:429-443.

Rau,J., H-J. Knackmuss & A. Srolz. 2002. Effects of different quinoide redox mediators on the anaerobic reduction of azo dyes by bacteria Environ. Sci. Technol. 36:1497-1504.

Rodríguez-Rojas, A., A. Mena, S. Martín, N. Borrell, A. Oliver & J. Blázquez. 2009. Inactivation of the *hmgA* gene of *Pseudomonas aeruginosa* leads to pyomelanin hyperproduction, stress resistance and increased persistence in chronic lung infection. Microbiol. 155:1050-1057.

Ruzafa, C., F. Solano, & A. Sanchez-Amat. 1994. The protein encoded by the *Shewanella colwelliana melA* gene is *p*-hydroxyphenylpyruvate dioxygenase. FEMS Microbiol. Lett. 124:179-184.

Schröder, I., E. Johnson & S. deVries. 2003. Microbial ferric iron reductases. *FEMS Microbiol. Rev.* 27:427-447.

Scott, D.E. & J.P. Martin, in, *Humic Substances in Soil and Crop Sciences: Selected Readings,* P MacCarthy, C.E. Clapp, R.L. Malcolm, and P.R. Bloom, Eds. (Soil Science Society of America, Inc., Madison WI, USA,1990), chap. 3.

Steinert, M., M. Flügel, M. Schuppler, J.H. Helgig, A. Supriyono, P. Proksch & P.C. Lück. 2001. The Lyl protein is essential for p-hydroxyphenylpyruvate dioxygenase activity in *Legionella pneumophilia.* FEMS Microbiol. Lett. 203:41-47.

Steinert, M., H. Engelhard, M. Flügel, E. Wintermeyer & J. Hacker. 1995. The Lyl protein protects *Legionella pneumophilia* from light but does not influence its intracellular survival in *Hartmannella vermiformis.* Appl. Environ. Microbiol. 61:2428-2430.

Turick, C.E., L.S. Tisa, & F. Caccavo, Jr. 2002. Melanin production and use as a soluble electron shuttle for Fe(III) oxide reduction and as a terminal electron acceptor by *Shewanella algae* BrY. *Appl. Environ. Microbiol.* 68:2436–2444.

Turick, C.E., F. Caccavo, Jr., & L.S. Tisa. 2003. Electron transfer to Shewanella algae BrY to HFO is mediated by cell-associated melanin. FEMS Microbiol. Let. 220:99-104

Turick, C.E., A.S. Knox, C. L. Leverette & Y. G. Kritzas. 2008 a. In-situ uranium immobilization by microbial metabolites. Invited paper, J.Environ. Rad. 99:890-899.

Turick, C.E., F. Caccavo, Jr. & L.S. Tisa. 2008 b.Pyomelanin is Produced by *Shewanella algae* BrY and Affected by Exogenous Iron. Can. J. Microbiol. 54:334-339.

Turick, C. E. A. Beliaev, A.A. Ekechukwu, T. Poppy, A. Maloney, & D. A. Lowy. 2009. The role of 4-hydroxyphenylpyruvate dioxygenase in enhancement of solid-phase electron transfer by *Shewanella oneidensis* MR-1. FEMS Microbiol. Ecol. 68:223-235.

White, L.P. 1958. Melanin: A naturally occurring cation exchange material. Nature. 182:1427-1428.

Yabuuchi, E. & Omyama, A. 1972. Characterization of "pyomelanin"-producing strains of *Pseudomonas aeruginosa.* Internat. J. Syst. Bacteriol. 22:53:64.

Zatkova, A., D. Beltrán-Valero de Bernabé, H. Polakova, M. Zvarík, E. Ferákova, V. Bošák, V. Ferák, L. Kádasi & Rodríguez de Córdoba. 2000. High frequency of alkaptonuria in Slovakia: Evidence for the appearance of multiple mutations in HGO involving different mutational hot spots. Am. J. Hum. Genet. 67:1333-1339.

Zughaier, S.M., H.C. Ryley, & S.K. Jackson. 1999. A melanin pigment purified from an epidemic strain of *Burkholderia cepacia* attenuates monocyte respiratory burst activity by scavenging superoxide anion. Infect. Immun. 67:908-913.

24

Microbial Biopolymerization Production from Palm Oil Mill Effluent (POME)

Zaini Ujang, Salmiati and Mohd Razman Salim
Institute of Environmental and Water Resource Management (IPASA), Universiti Teknologi Malaysia, 81310 Johor Malaysia

1. Introduction

Malaysia is one of the world leaders in the production and export of crude palm oil. In Malaysia, the oil palm industry has contributed vastly towards the country's economic well being. During the economic crisis in the late 1990s, the industry has helped to cushion the impact of the economic downturn through its export-oriented activities, which provided the much needed foreign exchange for the country. Crude palm oil (CPO) production has increased from only 1.3 million tonnes in 1975, to 4.1 million tonnes in 1985 and 7.8 million tonnes in 1995 to 17.56 million tonnes in 2009 (Malaysian Palm Oil Board-MPOB, 2010).

The Malaysian palm oil industrial complex refers to the various direct linkages, processing chains and products created as a consequence of the cultivation of oil palm and the production of the main product (palm oil) and secondary products (palm kernel oil and cake). The oil palm products are employed in numerous food and non-food applications (Butler, 2006). In 2009, there were 418 crude palm oil mills, 59 refineries, 57 downstream industries and 18 oleochemical plants. The Malaysian Palm Oil Board's long-term programme is to establish a biodiesel plant that will produce methyl ester (biodiesel), which can be used to replace petroleum diesel. There were 28 biodiesel plants with a production of 2.7 million tonnes per year methyl ester, respectively (MPOB, 2010). Another potential revenue generator is to convert the large quantity of biomass (13.2 million tonnes dry weight) into added value products (Ming & Chandramohan, 2002). However, this important economic activity generates an enormous amount of liquid effluent from the milling processes (Ahmad et al., 2005). Palm oil mills with the wet milling process accounted for the major production of wastes (Kittikun et al., 2000). Hence, the increase in number of mills will generate more environmental problem.

In general, the palm oil milling process can be categorized into a dry and a wet (standard) process. The wet process of palm oil milling is the most common and typical way of extracting palm oil, especially in Malaysia. According to the industrial standard, the milling process produces wastewater in the range 0.44-1.18 m^3/tonne fresh fruit bunches (FFB) with the average figure of 0.87 m^3/tonne FFB. It is estimated that for each tonne of CPO that is produced, 5–7.5 tonnes of water are required, and more than 50% of this water ends up as palm oil mill effluent (POME) (Ahmad et al., 2003). It has been reported that for every tonne of the CPO produced, about 3.5m^3 of POME is generated, which indicates that with some 500 palm oil mills, more than 17.5 million tonnes of CPO is produced annually. It is estimated that about 55 million m^3 of POME is generated from the palm oil industry. POME is an oily wastewater generated by palm oil processing mills and consists of various suspended components. On average, for each tonne of FFB processed, a standard palm oil mill generated about 1 tonne of liquid waste with biochemical oxygen demand (BOD) 27 kg, chemical oxygen demand (COD) 62 kg, suspended solids (SS) 35 kg and oil and grease (O&G) 6 kg (Salmiati, 2008).

Despite the fact that the palm oil industry is one of the causes of the environmental pollution, not enough has been done on its improvement. The technology applied in almost all oil palm factories is based on methods developed since the 1970s and 1980s. The major steps in the oil palm processing are as follows:

i. Threshing - Removal of fruits from the bunches. The FFB consists of the fruits that are attached onto the spikelets growing on a main stem. The fruit-laden spikelets are cut from the bunch stem using axe for manual threshing before separating the fruits from spikelets.

ii. Sterilisation - Loose fruits are sterilised in batches using high temperature wet-heat treatment. This is carried out in autoclave by steam application at 120-140 °C at 3-3.5 bar, for 75 min. Sterilisation prevents fatty acid formations and assists in fruits stripping, as well as preparing the fruit fiber for the next processing steps. Besides, sterilisation fruits stripping can be carried out by cooking in hot water. Cooking breaks down oil-splitting enzyme and stops hydrolysis and autooxidation. The fruits stem is weakened allowing easier removed of fruits from the bunches. Cooking also solidifies protein that dispersed oil-bearing cell microscopically. It weakens the pulp structure, softening it and making it easier to detach the fibrous material and its content during the digestion process. The high heat is enough to partially disrupt the oil-containing cells in the mesocarp and permits oil to be released more readily.

iii. Crushing process - the palm fruits will be passed through shredder and pressing machine to separate oil from fibre and seeds.

iv. Digestion of the fruit - This process releases palm oil in the fruit through cracks the oil-bearing cells. The digestion consists of a steam-heated cylindrical vessel with central rotating shaft that is filled with several beater arms. The fruit is pounded by the rotary beater arms at high temperature, which reduces the oil viscosity. This destroys the exocarp fruits or the outer covering and completes the disruption of the oil cell already begun in the sterilization process. The digestor must be filled to ensure the maximum of storage and the effect of the agitation.

v. Extracting the palm oil – There are two distinct methods of extracting oil from the digested material. One system uses mechanical presses and is called the "dry" method. The other called the "wet" method uses hot water to leach out the oil.

vi. Kernel recovery - The residue from the press consists of a mixture of fibres and palm nuts which are then sorted. The sorted fibres are covered and allowed to heat by own internal exothermic reactions for about two or three days. The fibres are then pressed in spindle press to recover second grade (technical) oil that is used normally in soap-making. The nuts are usually dried and sold to other operators who process them into palm kernel oil.

vii. Refining process - Refining converts crude palm oil (CPO) into refined oil. The refined oil is then processed to segregate fat and obtained refined palm oil. The refined palm olein obtained through fractionation process is used in related industries.

viii. Oil Storage - Palm oil is stored in large steel tanks at 31 to 40°C to keep it in liquid form during bulk transport. The tank headspace is often flushed with CO_2 to prevent oxidation. Higher temperatures are used during filling and draining tanks. Maximum storage time is about 6 months at 31°C.

Governments have made various initiatives to reduce the undesired environmental pollution particularly among major corporations. An efficient treatment system is highly desirable in all palm oil mills in order to control the discharge of effluent to water bodies. The final effluent should meet the standards set by the authority, in the case of Malaysia, Department of the Environment. Alternative treatment with value added package will ensure that the treatment of the effluent will be more efficient, innovative and attractive for the mill's owners. The formulation of the CPO mill effluent discharge standards by phases has brought about a catalytic impact in the development of effluent treatment technology in the form of innovative or newly created technology (Yeoh, 2004; Yusoff, 2006, Ahmad et al., 2008). Some of the POME treatment options are described in Table 1.

Objectives		**Options technology**	**Impacts**
a) Treatment	Primary treatment	• Waste stabilization ponds (anaerobic, aerobic, facultative)	• odour problems • large footprint • Insufficient effluent quality • colour • low operation and maintenance cost
	Secondary treatment	• Enhanced waste stabilization ponds (aerated lagoon) • Up-flow anaerobic sludge bioreactor (UASB) • Sequencing batch reactor (SBR) • Extended aeration activated sludge	• large variation in effluentquality • insufficient effluent quality • high operation and maintenance cost • high power consumption • insufficient for colour removal
	Advanced treatment	• Membrane bioreactor (MBR) • Membrane filtration • Adding chemical and biochemical produts	• high operation and maintenance cost • need pre-treatment • small area required • good quality for reuse • advanced operation and maintenance
b) Resource recovery	Reuse	• Fertilizer • Animal feed	• Low digestibility of protein • Slow down their composition • High lignin content
	Recovery	• Volatile fatty acid • Polyhydroxyalkanoates • Biogas • Methane	• Produce value-added product • Reduce cost treatment that needs further treatment

Table 1. Options for POME treatment, reuse and recovery

Cleaner production (CP) inventiveness was introduced since 1990s. A study to develop a sustainable CP policy and framework was conducted and implemented to all indigenous industries including palm oil (Ujang, 2008). The cleaner production has been implemented in the crude palm oil industry through the support of government and international donor agencies. The result is an increasing number of mills that implement cleaner technology strategies and approaches. Although there are obvious environmental and often also economic benefits in implementing cleaner production strategies, cleaner production can and often does entail investments.

Currently, recovery of renewable organic-based product is a new approach in managing POME. The technology is aimed to recover by-products such as volatile fatty acid, biogas and polyhydroxyalkanoates to promote sustainability of the palm oil industry. In addition, it is envisaged that POME can be sustainably reused as a fermentation substrate in production of various metabolites through biotechnological advances. In addition, POME consists of high organic acids and is suitable to be used as a carbon source (Alias & Tan, 2005: Md Din et al., 2006a).

2. Sources of POME

The two main wastes resulting from palm oil production in a mill are the solid and liquid wastes. Solid wastes typically consist of palm kernel shells (PKS), mesocarp fruit fibres (MF) and empty fruit bunches (EFB). The liquid waste generated from the extraction of palm oil of wet process comes mainly from oil room after separator or decanter. This liquid waste combined with the wastes from steriliser condensate and cooling water is called palm oil mill effluent (POME). Figure 1 shows the different point sources of waste in palm oil milling.

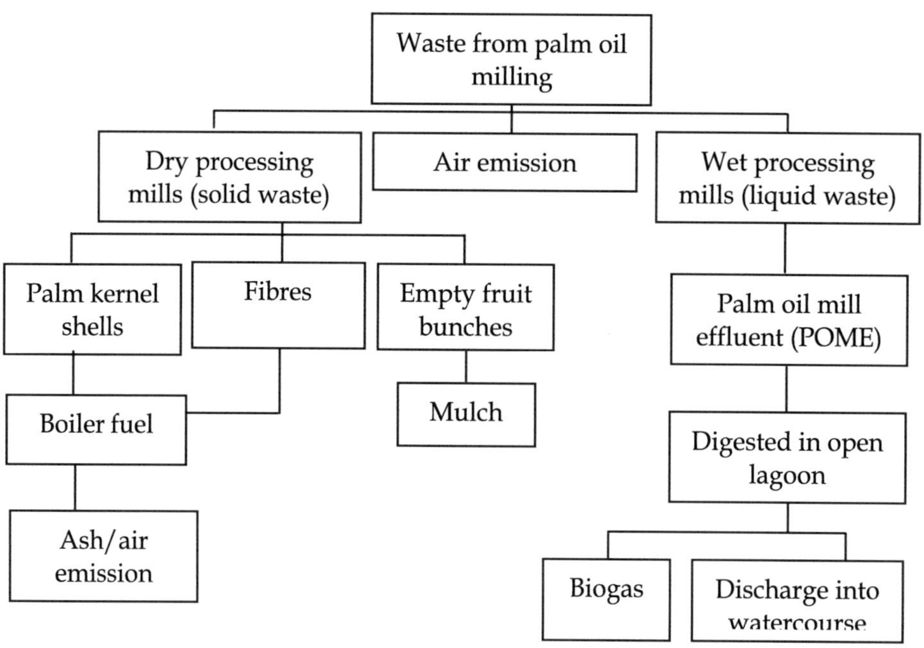

Fig. 1. Sources of waste from palm oil milling

Large quantities of water are used during the crude oil extraction process. Up to about 1.5 cubic meters of water are characteristically used to process one tonne of FFB. From this quantity, about 50% of the water results in the POME, the other 50% being lost as steam, mainly through sterilizer exhaust, piping leakages, as well as wash waters (Anonymous, 1999).

POME comprises a combination of the wastewaters which are principally generated and discharged from the following major processing operations as follows (DOE, 1999; Salmiati, 2008):

- Sterilization of FFB - sterilizer condensate is about 36% of total POME or about 0.9 tonnes POME for each produced tonnes of palm crude palm oil;
- Clarification of the extracted CPO - clarification wastewater is about 60% of total POME (approximately 1.5 tonnes of sludge obtained per tonnes of produced crude palm oil); and
- Hydrocyclone separation of cracked mixture of kernel and shell-hydrocyclone wastewater is about 4% of total POME.

The ratio from the mixture of sterilizer, condensate and separator sludge wastewater is 9:15:1 respectively (Wu et al., 2010). There are other minor sources of relatively clean wastewater that may be included in the combined with POME, which is sent to the wastewater streams. These include turbine cooling water and steam condensates, boiler blow-downs, overflows from the vacuum dryers and some floor washings. The volume of the combined POME discharged depends to a large extent on the milling operations.

3. Characteristic of POME

Palm oil mill effluent (POME) is mainly generated from sterilisation, hydrocyclone and clarification processes in which large amounts of steam and/or hot water are used (Ma, 1999). Distinctive quality characteristics of the individual wastewater streams from the three principal sources of generation are presented in Table 2. POME, when fresh is a thick brownish in colour colloidal slurry of water, oil and fine cellulosic fruit residues. POME is generated from mill operation at a temperature of between 80°C and 90°C and it is slightly acidic with a pH between 4 to 5 (Md Din et al., 2006a). The characteristics of a usual raw combined POME are presented in Table 3 which attests that POME has a very high Biochemical Oxygen Demand (BOD) and Chemical Oxygen Demand (COD), which is 100 times more than the municipal sewage. POME is a non-toxic waste, as no chemical is added during the oil extraction process, but will pose environmental issues due to large oxygen depleting capability in aquatic system due to organic and nutrient contents.

The high organic matter is due to the presence of different sugars such as arabinose, xylose, glucose, galactose and manose at the concentrations of 6.43, 0.44, 0.22, 0.15 and 0.10% dry weight, respectively (Agamuthu & Tan, 1985). The oil residue was in the range of 1-2% which depends very much on the quality of raw material (palm fruits), process control and machine efficiency. The suspended solids in the POME are mainly oil-bearing cellulosic materials from the fruits (Ma, 1999). Since the POME is non-toxic as no chemical is added in the oil extraction process, it is a good source of nutrients for microorganisms.

However, using BOD as a characteristic of POME was conventional or old method. Recently, Damayanti et al. (2010) used respirometric test to estimate model parameters for activated sludge modelling of POME. The study found the heterotrophic yield coefficient and some of the COD fractionations of POME are described in Table 4. These coefficients could be serving as basis for design and optimization of a POME treatment process.

Parameters	Sterilizer condensate	Oil clarification wastewater	Hydrocyclone wastewater
pH	5.0	4.5	-
Oil and Grease	4,000	7,000	300
$BOD_{3\text{-day}}$; 30°C	23,000	29,000	5,000
COD	47,000	64,000	15,000
Suspended solid	5,000	23,000	7,000
Dissolved solids	34,000	22,000	100
Ammonical nitrogen	20	40	-
Total nitrogen	500	1,200	100

All units are in mg//L except for pH
Source: DOE (1999)
Table 2. Characteristics of individual wastewater streams

General Parameters	Value	Metals & other Constituents	
		Element	Value
pH	4.2	Phosphorus	180
Oil & Grease	6000	Potassium	2270
BOD_3 @ 30°C	25000	Magnesium	615
COD	50000	Calcium	440
Total Solids (TS)	40500	Boron	7.6
Suspended Solids (SS)	18000	Iron	47
Total Volatile Solids (TVS)	34000	Manganese	2.0
Ammoniacal Nitrogen (AN)	35	Copper	0.9
Total Nitrogen (TN)	750	Zinc	2.3.

Note: All parameter's unit in mg/l except pH
Source: DOE (1999)
Table 3. Characteristics of combined POME

Model coefficient	Values
Total COD (mg/L)	45,000
S_s (mg/L)	50
S_I (mg/L)	16,600
X_s (mg/L)	25,550
X_I (mg/L)	2,800
Y_H (g COD oxidized)$^{-1}$	0.44
Cell COD	14,100
μ_A (/day)	0.76
μ_H (/day)	0.78
K_s (mgCOD/L)	100
b_H (/day)	0.33

Source: Damayanti et al. (2010)
Table 4. Estimated model parameters and state variable of POME

4. The science of biopolymer

The usage of conventional plastics has resulted in environmental degradation (Serafim et al., 2006). The production of biodegradable polymers is seen to be a viable alternative with the increasing environmental pressure to replace conventional plastics. The problem that is faced by industry is the high production costs of biopolymers. Biodegradable polymers derived from polyhydroxyalkanoates (PHAs) are considered to be good candidates for biodegradable plastics due to their large range application and capability of being produced from renewable resources (Dionisi et al., 2005; Bengtsson et al., 2008; Albuquerque et al., 2007). These materials have attracted interest because of their potential use as biodegradable alternatives to petroleum-based synthetic plastics such as polypropylene and polyethylene. Polyhydroxyalkanoates (PHAs) are mainly produced by microbial fermentation processes, and a major challenge is to reduce their production costs (Verlinden et al., 2007). A feasibility study using fermentative volatile fatty acids (VFAs) as carbon source to synthesise PHA by activated sludge was carried out to simultaneously reduce the production cost of PHAs and disposal amount of organic wastes (Salmiati et al., 2010).

Several efforts have been investigated to produce PHAs by microbial fermentation on organic wastes (palm oil mill effluent (POME), olive oil and kitchen wastes). The production of biodegradable polymers from oil palm industry can be seen as beneficial to the environment as well as contributing to sustainable development (Yu, 2003; Salmiati et al, 2007). Until recently, the remaining 90% (empty fruit bunches, fibres, fronds, trunks, kernels, POME) was discharge as waste, and either burned in the open air or left to settle in waste ponds (MPOC, 2007, Salmiati et al., 2010). By utilizing the POME and empty palm oil fibre bunch (EPFB) as carbon source and support matrix, the disposal of POME that needed further treatment could be reduced.

Although POME consists of high organic acids and is suitable to be used as a carbon source, POME is usually present in a complex form that cannot directly be utilised by PHA-producing bacterial species for PHA synthesis. Typically, raw POME is difficult to degrade because it contains significant amounts of oil (tryacylglycerols) and degradative products such as diacylglycerols and monoacylglycerols and fatty acids . (Alias & Tan, 2005; Salmiati et al, 2007). The fatty acids composition (C_{12} - C_{20}) of each of this fraction are different from one another and contribute to a high value of pollution load in POME. Therefore, anaerobic treatment has been proposed to reduce the POME characteristics. It is one of the naturally occurring processes involving decomposition and decay, in which complex organic matters are broken down into their chemical constituents. Hydrolysis and acidogenesis are the first step to convert the wastes to short-chain VFAs (i.e acetic, butyric and propionic acids). After that, the VFAs will be utilised by PHA-producers for PHA production (Bengtsson et al., 2009; Salmiati et al., 2010).

Considering the interest of using fermented substrate as carbon source for production of PHA, therefore, this study is aimed at understanding the mechanism of VFA influences on the yield of polymer production, and to optimise the PHAs production using fermented POME as carbon sources using mixed cultures. An improved system containing two separate bioreactors to satisfy the different physiologies and metabolic activities of the two types of microbes; one for acidogenesis of POME and a second for mixed microbial culture of PHA-producing strains was investigated.

5. Biopolymer production using POME

Ideal life cycle of eco-friendly exposure for PHA bioplastic made from renewable resources likes POME is a closed-loop process (as depicted in Figure 2). The production of bioplastic will subsequently serve as feed to a microbial fermentation process (at the end of cycle), to promote the environmental-friendly effect. Ideally, this process occurs aerobically (in natural and tropical conditions), yielding water (H_2O) and CO_2 in the same proportions that were originally used in photosynthesis. The harmless end products could also be generated from microbial fermentation to produce biofuel energy.

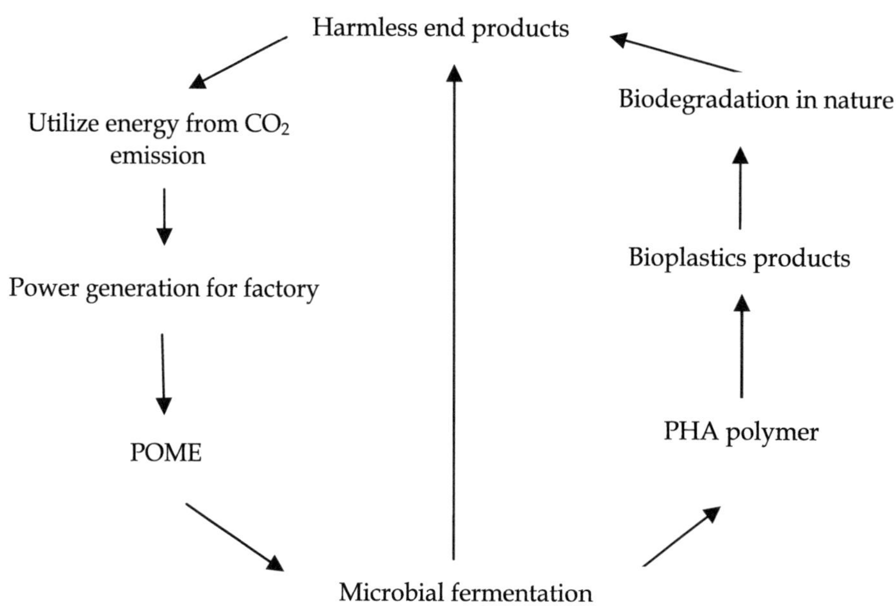

Fig. 2. Proposed cycle loop of regenerating waste from POME to biodegradable plastics, end-up with preventing pollution load to environment

The price of PHAs is mainly dependent on substrates cost, accounting for about 40% of the total production cost (Md Din et al., 2006a). In the last decade, a variety of low cost carbon substrates (e.g., starch, tapioca hydrolysate, whey and molasses) have been tested for PHA production to reduce the production cost. POME can be considered as an alternative, no cost reusable substrate for PHA production. The production of VFA and PHA using POME as substrates are listed in Table 5. According to Hassan et al., (1997a), with a content of 50% PHA in the dried cells and 2% dissolved in the chloroform, the calculated minimum cost for obtaining PHA from POME is below USD2/kg. By increasing the PHA content in the cell from 50 to 80%, the unit cost of PHA could be slightly reduced; whereas an increase in the amount of PHA dissolved in chloroform from 2% to 5% would result in a remarkable reduction of the PHA cost to less than US1/kg.

Nevertheless, POME is usually presented in complicated forms that cannot be directly reused by PHA-producing species such as *Ralstonia eutropha*, a representative bacterium for PHA synthesis (Salmiati et al., 2007). It was proposed that an anaerobic treatment of POME could be coupled with PHA production using heterotrophic bacteria to reduce PHA

Product	Microorganism, Fermentation medium and Condition based on POME	Maximum production	Reference
VFA	Mixed cultures, POME + palm oil sludge. 30°C, pH was controlled at 7, SBR, 24 h	7.8 g/L	Hassan et al. (1996)
VFA	Mixed cultures, POME + palm oil sludge in the ratio 1:1. 300 rpm, 30°C, pH was controlled at 7, stirred tank bioreactor fermentation. 84h	10 - 14 g/L	Yee et al. (2003)
VFA	Mixed cultures, POME + palm oil sludge, pH was controlled at 7. SBR. 96h	10.27 g/L	Cheong et al. (2004)
PHA	*Rhodobacter sphaeroides* IFO 12203, Synthetic waste based on organic acids profiles obtained during POME treatment. 30°C, pH was controlled at 7, photobioreactor fermentation, ≈ 200h	≈ 4g/L	Hassan et al. (1996)
PHA	*Rhodobacter sphaeroides* IFO 12203, Anaerobically digested, 30°C, pH was controlled at 7, photobioreactor fermenetation.	>2g/L	Hassan et al. (1997a)
PHA	*Alcaligenes eutrophus* H16(ATCC 17699), Standard medium with feeding of acetic acid obtained from anaerobically digested POME. 400 rpm with aeration rate of 0.75L/min. 30°C, pH was controlled at 7, stirred tank bioreactor fermentation, 17h	1.8 g/L	Hassan et al. (1997b)
PHA	*Ralstonia eutropha* ATCC 17699, Concentrated organic acids from anaerobically digested POME (100g/L of total acids with acetic:propionic = 3:1), 400 rpm with aeration rate of 0.75L/min. 30°C, pH was controlled at 7, bioreactor fermentation, ≈65h	≈ 6.25g/L	Hassan et al. (2002)
PHA	Mixed cultures, High concentration of POME with 490 COD/N ratio (g COD/ g N) and 160 COD/P ratio (g COD/ g P), 1000 rpm with aeration rate of 1.5L/min, 30°C. pH was controlled at 7. SBR	24.24 g/L	Md Din et al. (2006b)

Table 5. Various products in bioprocesses using POME as substrates

production costs (Salmiati, 2008). According to Hassan et al. (1996), it was critical to maintain the pH at 7 in the anaerobic treatment of POME by sludge in the first stage of the process, in order for only acetic and propionic acid to be produced and not formic acid and biogas. With increasing concentrations of formic acid (for a pH maintained below 4), the PHA yield and content in *Rhodobacter sphaeroides* IFO 12203 dropped from 0.50 g/g and 67% to 0.21 g/g and 18%, respectively. Hassan et al. (1997b) later found that the presence of sludge in the anaerobically treated POME inhibited PHA accumulation by *R. sphaeroides* IFO 12203. This was attributed to the PHA being produced in a POME without sludge as opposed to a treated POME with sludge. A low concentration of ammonium would accelerate the PHA production in a synthetic waste with an organic acid profile, which was observed during POME treatment. However, Hassan et al. (1997b) found that addition of ammonium and phosphate to anaerobically treated POME was required to maintain the cell activity and production of PHA since neither ammonium nor phosphate was present in the anaerobically treated POME. In total, the organic acid concentrations obtained from anaerobically treated POME were too low (Hassan et al., 1996; Salmiati et al., 2010) for it to be reused as raw material in the production of PHA on an industrial scale. The underlying reason was that this would require a production reactor with a much larger size than that of a reactor for normal bioplastic production.

The organic acids in the anaerobically digested POME could be concentrated by evaporation for use as substrates in the fed-batch non-sterile PHA fermentation system using *R. eutropha* ATCC 17699. Although the proposed overall zero emission system appeared to be practical, major drawbacks were found, including the rather low yield and productivity of PHA by *R. eutropha* when the concentrated organic acids from POME were used as compared to synthetic organic acids. This could be due to the high presence of ammonium (1.5 g/L) or other compounds in the anaerobically digested POME concentrate (Hassan et al., 2002).

Md Din et al. (2006b) proposed the suitability of using mixed cultures to produce PHA in POME since most prokaryotes are capable of PHA production. The study noted that by using mixed cultures and POME, different types of PHA-constituents could be obtained. The harvesting of these PHA-constituents was more reliable for use as biodegradable plastics material as opposed to a single PHA-constituent. A type of mixed culture was maintained in a SBR, and a high concentration of POME proposed for this system in order to generate autotrophic rather than heterotrophic bacteria in the production of PHA. However, the average PHA production by using POME could only reach 44% of the CDW, indicating that an optimisation of the PHA sludge content must be carried out by varying the oxygen rate, feeding regime or transient conditions.

Acidogenic fermentation POME and conversion of the produced VFAs to PHA by mixed microbial cultures has also been studied (Salmiati et al., 2007). VFA characterisation has been conducted using fermented POME in anaerobically digestion. Earlier studies have shown that the VFAs (acetic, butyric, propionic, etc) have been used as carbon source by bacteria, as individual or mixed of the VFAs (Dionisi et al., 2005; Salmiati, 2008).

6. Reactor design

In order to investigate the PHA production in POME and mixed culture, two bioreactors have been designed and used to meet metabolic activities in one cycle. The anaerobic part was designed in a cylindrical shape with a working volume of 19 litres, interior diameter 19.5 cm and height 90 cm. The sampling ports were designed along 10 cm height intervals from the bottom. A perforated piping system was used at the bottom of the reactor to ensure homogenous distribution of flow into the reactor and no recirculation of effluent was practiced. Support materials used in the fixed bed, which are: oil palm fibres; spherical in shape with approximately 1 cm diameter (total specific surface area was $0.15 cm^2 m^{-3}$). The acidogenesis process was conducted for 45 days for the inoculation. The reactor was operated at room temperature (28 ± 2°C) and at varied SRT/HRT. At each HRT, the reactor was operated for six weeks to reach steady state conditions. Steady state conditions were established when the variation in the product concentration were constant (effluent VFAs and COD concentrations). Typically, the reactor was operated without pH control and aeration nor stirring to avoid the acetogenesis process that transforms VFA to other forms.

Acidic slurry produced in anaerobic reactor was then pumped to the aerobic reactor as substrate feeding for the microorganisms that produce PHA. This supply can be varied depending on each SRT/HRT.

The aerobic part was fabricated in two double-jacketed laboratory-scale reactor with six litres effective volume. Figure 3 depicts the schematic drawing of reactor set-up. The operation of aerobic reactor was based on the sequencing batch reactor (SBR) system under feast-famine regime which was conducted on samples taken from fermented POME from anaerobic reactor. The dissolved oxygen (DO) concentration was measured as percentage of the saturation concentration (100% = 9.1 mgL^{-1}). DO concentration and pH were measured continuously. In order to control the oxygen concentration properly, the gas concentrations was controlled using gas flow meter.

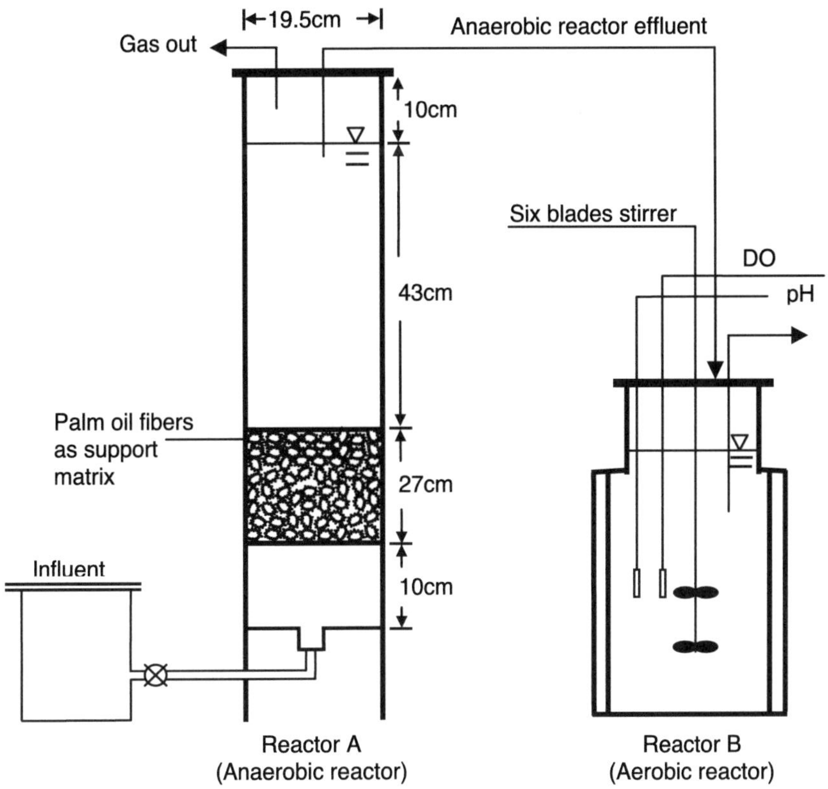

Fig. 3. Anaerobic and aerobic reactors used for the PHA production POME

Raw POME and waste sludge from a third sludge pond was collected from a local palm oil mill wastewater treatment plant. A fresh activated sludge taken from an aeration tank of a local municipal wastewater treatment was used as seed sludge for both of reactor. Waste sludge, POME and sewage sludge were introduced as inoculate to acclimatise the autotrophic and/or heterotrophic bacteria in each reactor. The ratio of the inoculums is 1:2:1. For the anaerobic reactor, the substrates were fed into the reactor at the bottom and the culture medium contained supernatant discharge circulated at the rate of about 40 mL/min through a granulated sludge bed. A perforated piping system was used at the bottom of the reactor to ensure homogenous distribution of flow into the reactor and no recirculation of effluent was practiced. For the aerobic reactor, at least more than 50% of the working volume must be designed to be discharged as supernatant. The operating principles of a batch activated sludge system are characterised in just three discrete periods: fill, react and draw (discharging). After two to three days, the aerobic system was continuously operated by supplying nutrient adaptation for several weeks, in order to reach a steady-state condition. At steady-state conditions, some parameters has been extensively monitored (e.g. pH, DO, and samples was collected every hours for COD, TOC, NH_4^+-N, P, VFAs CDW and ash constituent analysis). In order to maximise the growth rate and fast substrate uptake rate and storage polymer formation, the system was operated in continuous reaction period, which means no settling or allowing the idle phase (HRT=SRT). The length of HRT and SRT in each treatment depends on the microorganisms' reaction after feeding. The cycle of operation system depends on substrate concentrations. The detailed features of the aerobic are described in Table 6.

Reactor	Anaerobic reactor	Aerobic reactor
Working volume	19-L (laboratory scale)	6-L (laboratory scale)
Influent feed	Raw POME	VFA (fermented POME)
Temperature	28 ± 2°C	28 ± 2°C
pH	4.0 - 10.0	7 ± 0.1
Aeration and DO	No aeration	1 - 1.7 $Lmin^{-1}$
HRT /SRT	7-8 hours/ 5-10 days	1 hour/ 4-8 hours
Stirrer	-	300 - 400 rpm

Table 6. Reactor description and condition of each experiments

The acidic supernatant is pumped into the reactor and mixed with biomass that settled during the previous cycle until the time for filling is reached. A mineral solution with composition was added for the growth phase only (Md Din et al., 2006b; Salmiati, 2008). In general, the overall operation period of fermented POME cultivation is shown in Table 7. The filling phase can be mixed in either aerated (oxygen as electron donor) or microaerophilic-aerobic (controlling the oxygen level) conditions. In order to control the oxygen concentration properly, the gas concentrations was controlled using gas flow meter. The reaction phase can also be mixed in similar condition with the common environment needed by bacteria to live. This reactor was equipped with peristaltic pumps for influent feeding and effluent withdrawal, and air compressor was employed for aeration. The procedures of reactor operation, such as feeding, aerating and withdrawing were controlled automatically by timers and temperature was maintained at 28 ± 2°C in a temperature controlled-room. The dissolved oxygen (DO) concentration was measured as percentage of the saturation concentration (100% = 9.1 mgL^{-1}). DO concentration and pH were measured continuously.

Samples were taken from the reactor with a 60 mL syringe (Syphon, United Kingdom). The syringe was always rinsed with the content of the reactor before sampling. Part of the sample was stored in the refrigerator for analysis. The remaining supernatant was centrifuged at 10,000 rpm for 10 minutes. The centrifugation for separating the debris and supernatant was performed using Sorval RC-5B (Hermmicks, Germany) for 15 minutes at 2000 rpm and 4°C. The supernatant was then filtered by using PVDF-syringe filter. Samples for analysis of NH_4-N, PO_4-P, TOC and COD and VFA were immediately centrifuged and filtered using 0.45 µm filters to separate the bacterial cells from the liquid, 0.2 µm cone-syringe filters were used for soluble analysis. The supernatant was stored in the refrigerator at -4°C (for PHA analysis) and at -5°C (for VFA, MLSS, MLVSS, CDW, NH_4^+, NO_3^-, PO_4^{2-} and COD). All analytical measurements performed in this study were conducted according to *Standard Methods for the Examination of Water and Wastewater* (APHA, 2000).

Experiment (s)	Operating time (min)			
	Aerobic mineral feeding	**Aerobic feeding**	**Aerobic reactor**	**Draw/discharge**
Growth	0 – 60	0 - 60	60 -330	330 - 340
DO (pretreated POME)	No fill	-	0 - 600	-
Mic ae (pretreated POME)	No fill	-	0 - 500	-

Table 7. Operating phase with POME as substrate

The composition and concentration of lipids in POME measured using gas chromatography according to the type of carbon chain. The quantification of TSS performed using volatile suspended solid (VSS) and ash technique according to the Dutch Standard (NNI, NEN) and APHA (2000). VFA (acetic acid [HAc], propionic acid [HPr], *iso*-butyric acid, butyric acid [HBu], *iso*-valeric acid, valeric acid [HVa] and caproic acid) were quantified with a Varian 3400 gas chromatograph after filtration through filters with pore size 1.6 µm (Munktell MGA). The injection volume was 2 µL and the gas chromatograph was equipped with a Chromosorb 101 (80/100 mesh) column (length: 2.5 m, diameter: 2.3 mm) and a flame ionization detector. Nitrogen gas saturated with formic acid was used as carrier gas (30 mL/min). The temperatures of the injector and detector were 240 and 250 °C, respectively. The column temperature was 170 °C for the initial 2 min and was then increased at 10 °C/min to 200 °C and retained for 3 min. PHA determination was conducted according to the Comeou et al. (1988) and PHA concentration was determined according to the method of Braunegg et al. (1998) using benzoic acid as internal standard. CDW was determined as described previously by Wong (2001), and synthesis of PHA was analysis according to the method of Yu & Chen (2006).

7. COD Removal and VFA production from fermented POME

Organic wastes are usually complex in nature, and cannot be directly utilized by PHA-producing microbes for PHA synthesis. POME consists of high organic acids; therefore it is suitable to be used as a carbon source. Typically, raw POME is difficult to degrade because it contains significant amounts of oil (tryacylglycerols) and degradative products such as diacylglycerols and monoacylglycerols and fatty acids (Alias & Tan, 2005; Salmiati et al, 2007). The fatty acids composition (C_{12} - C_{20}) of each of this fraction are different from one another and contribute to a high value of pollution load in POME. The typical characteristics of raw POME are given in Tables 8 and 9 showing the composition of lipids in POME as used in this study. POME is usually present in a complex form that cannot directly be utilized by PHA-producing bacterial species for PHA synthesis. Therefore, anaerobic treatment has been proposed to reduce their POME characteristic. It is one of the naturally occurring processes involving decomposition and decay, in which complex organic matter is broken down into its chemical constituents (Tay et al., 1996). Hydrolysis and acidogenesis are the first step to convert the wastes to short-chain VFAs (i.e acetic, butyric and propionic acids). After that, the VFAs will be utilized by PHA-producers for PHA production (Lee & Yu, 1997; Yu, 2003).

Parameters	**Units**	**Results**
pH	-	4.8 ± 0.21
Total suspended solids	mg/L	35,000 ± 200
Turbidity	NTU	21,000 ± 300
COD	mg/L	65,000 ± 800
BOD_5	mg/L	27,000 ± 800
O&G	mg/L	8,000 ± 300
TOC	mg/L	12,300 ± 570
Phosphorus	mg/L	142 ± 19
Ammonia Nitrogen	mg/L	62 ± 10

Note: values represent means of triplicate determination

Table 7. Characteristics of raw POME obtained from a local palm oil mill factory used in the study

Lipids	**Concentrations (%)**
Tryglycerides	81.5
Diglycerides	7.0
Monogliycerides	0.5
Free fatty acids	11.0

Table 8. The composition and concentration of lipids in POME used in this study

The fermentation reactor started to produce acids immediately after inoculation. The fermenter biomass consisted of a large number of small free-living bacteria and some small aggregates. The COD removal achieved was as high as 80% in approximate SRT of 6 days, as shown in Figure 4. The COD removal actually occurred during the fermentation process where a high amount of COD removal was used in the generating and synthesizing new bacterial cells for the anabolism route (de la Rubia et al., 2006). This process begins with the hydrolysis of complex organic compounds in the initial POME to more soluble intermediates. Through the process of acidogenesis, these intermediates are broken down primarily into VFAs and other monomer species.

However, the differences in SRT influenced the VFA results; the concentration of the VFA was increased significantly for the SRT of 6 days and the plateau was reached after about 10 days of the fermentation. Then the production rate decreased after 11 days. As shown in Figure 4 where at SRT of 11 days, acidogenic conditions prevail while at SRT ≥ 11 days, methanogenic conditions prevail. It shows that the SRT strongly affects the type and rate of bioconversion process under anaerobic conditions (Miron et al., 2000).

Figure 5 shows the relative distribution of individual acids within VFAs at the end of fermentation. With regard to POME with pretreatment using oil palm fibre, four organic acids were detected as the major fermentative products. Among them, acetic acid was the main product in the whole range of initial concentrations (50-57%), with lower relative amount of butyric acid (25-33%), propionic acid (4-6%) and valeric acid (1-3%). Isobutyric and lactic acids were in negligible quantities or not produced at all.

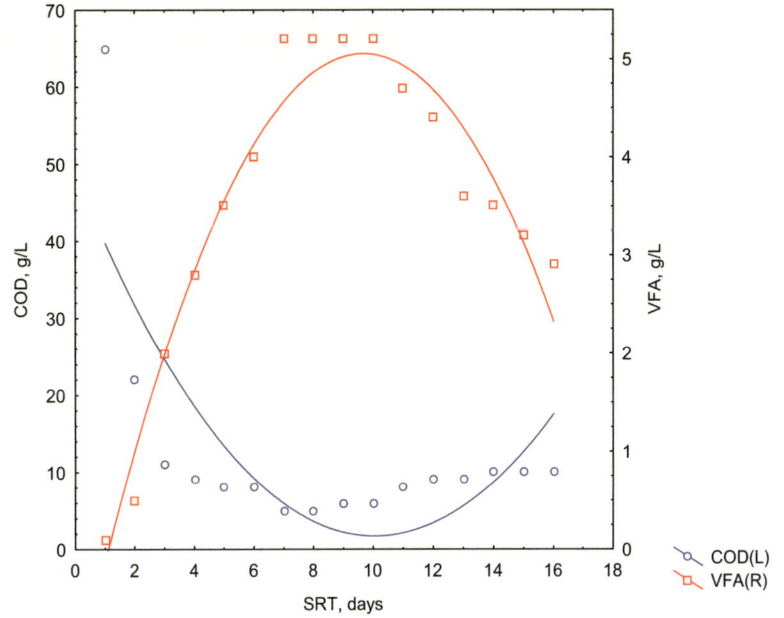

Fig. 4. Profile of VFA and COD concentration using anaerobic process at room temperature

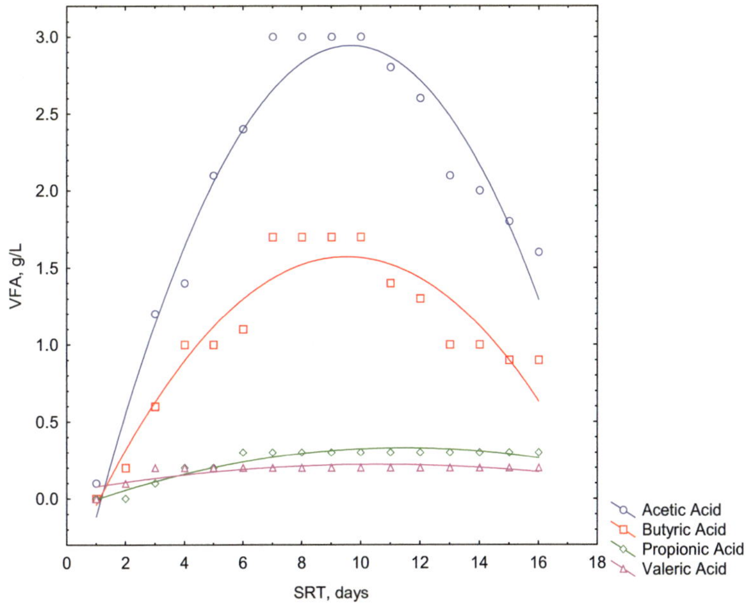

Fig. 5. The time courses of acidification of POME under anaerobic conditions

The VFAs in the anaerobic reactor are shown in Figure 5. The predominant VFA products in the effluent of this anaerobic reactor are acetic and butyric acids and their concentrations reached 3.0 g/L and 1.7 g/L starting after six days of the fermentation process and decreased after 12 days. About 0.3 g/L of propionic acid and 0.2 g/L of valeric acid were accumulated in anaerobic digestion and these are kept at a constant level for most of the time. The concentrations of propionic and valeric acids were relatively low in the effluent. The result indicated that even-numbered carbon fatty acids degraded more easily than odd-numbered carbon fatty acids. In addition, the degradation of the even-numbered fatty acids was not a rate-limiting step (Shin et al., 2001).

The different VFA distribution can have a great influence on the composition of the polymer produced in the subsequent steps, namely the percentage of HV monomers within the copolymer P(HB-HV). Indeed, VFAs containing an even number of carbon atoms (i.e. acetic and butyric acids) mostly lead to formation of HB monomers, whereas VFAs containing an odd number of carbon atoms (i.e. propionic and valeric acids) mostly lead to the formation of HV monomers (Yu, 2003; Dionisi et al., 2005). As an increase of HV content within the copolymer generally improves its thermal and mechanical properties, the negative effect of centrifugation on propionic and valeric acids production has to be considered along with the important beneficial effects discussed above (Dionisi et al., 2005).

VFA characterisation has been conducted using fermented POME in anaerobic reactor. Table 10 shows the composition of VFAs found in this study. Fermented POME composition constitute up to 50% of acetic acid, 30% of butyric acid, and 10% of propionic and lactic acids. Earlier studies have shown that the VFAs (acetic, butyric, propionic, etc) have been used as carbon source by bacteria, as individual or mixed or of the VFAs (Yu, 2003; Dionisi et al., 2005; Bengtsson et al., 2008).

Systematic name	Trivial name	Compound	Values
Ethanoic Acid	Acetic acid	CH_3COOH	57
Butanoic Acid	Butyric acid	C_3H_7COOH	33
Propanoic Acid	Propionic acid	C_2H_5COOH	6
Pentanoic Acid	Valeric acid	C_4H_9COOH	4

Table 10. VFAs composition (%) of fermented POME in Anaerobic reactor

For the cultivated fermented POME composition, an overall effect on growth and accumulation factor has been proposed, as shown in Figure 6. The concentration of PHA biomass component was determined as PHB and the value obtained (gPHB/gVSS) from the system could reach up to 4.0 x 10^{-2} g/gVSS. Under this situation, the cell growth (active biomass) increased with the initial feed of VFAs and reached steady after five hours, ranging from 2 to 5 g/L. Immediately, the PHB content increased to the maximum levels up to 4.20 x 10^{-2} g/g VSS, before sharply decreasing to 2.0 x 10^{-2} g/g VSS.

The low concentration of PHB during this preliminary experiment may be affected by low concentration of VFAs in the medium. However, the type of polymer was homopolymer (PHB). This suggests that the mixture of acetic acid and butyric acid is a good carbon source for PHA synthesis, but unlike propionic acid, valeric acid cannot be used for synthesis of hydroxyvalerate (HV) monomer, even though both acids have three carbons.

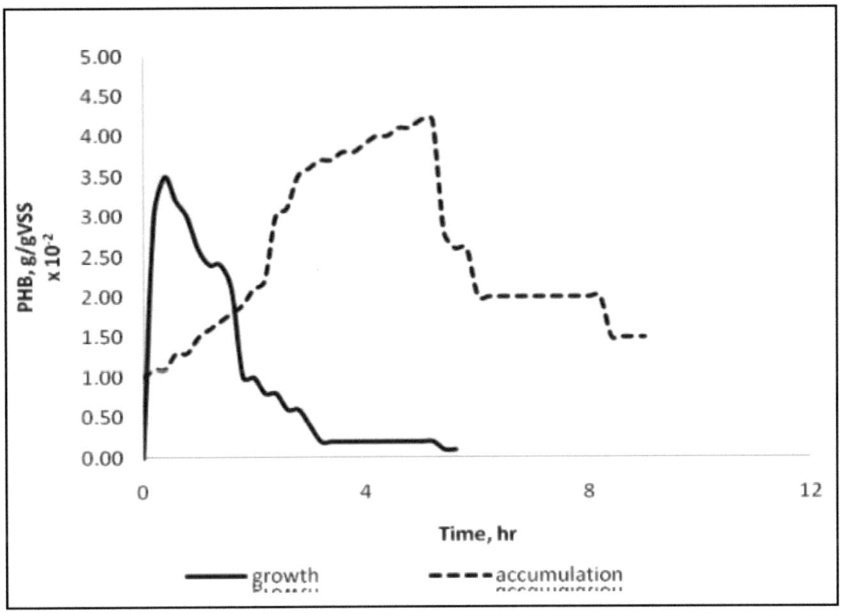

Fig. 6. PHB production during pre-determined growth and accumulation conditions

8. Substrates concentration

With current biological PHA production practice, the feedstock is estimated at 30 to 50% of the total production costs (Braunegg et al., 2002). These estimates vary somewhat with the microbial species utilized, carbon source, PHA yield and PHA production capacity (Gerngross & Slater, 2000). The quantity and form of the carbon substrate dictates the polymeric structure and yield of the PHA (Coats, 2005).

In this study, for 50% feeding (second phase of feeding), the biomass content is obviously higher than for 25% feeding (first phase) as shown in Table 11. The concentration of biomass produced increased directly with substrate volumes. In addition, the amount of polymer produced per substrate consumed increased concurrently with VFA production rate.

Parameters	Anaerobic (Stage A)		Aerobic (Stage B)	
	Influent	Effluent	25% substrate	50% substrate
COD(g/L)	65	8	25	32
COD removal (%)	-	80	45	60
VFA (gCOD/L)	NA	10.1	4.2	12.2
Cell dried weight (CDW) (%)	-	-	13	30
PHA (% CDW)	-	-	12	40

Table 10. The concentration of parameters for the two step process

The concentration of a substrate supplied affects the amount of polymer produced (Serafim et al., 2006). In other words, the amount of PHA in the biomass will be accumulated and increased directly in proportion to the amount of the initial carbon source for about 40% from cell dried weight as in Table 11. Besides, the volume of substrates influenced the time of accumulation. The length of the feast-famine period depends on the amount /volume of substrate. If the substrate fed was in smaller volume, then the usage of air for oxidizing and for substrate storage into microorganism becomes shorter as shown in Table 11.

9. Molecular weight of PHA

The original molecular size of PHA is usually very large (1,000-2,000 kDa) compared to those of synthetic polyesters (100-200 kDa). Reduction of the molecular weight to some extent does not affect the mechanical properties of the biopolymers. The bioplastics, however, do become brittle or less ductile when the molecular weight is greatly reduced (e.g., <100-200 kDa) (Yu & Chen, 2006). PHA recovered by NPCM or dispersion digestion can change the molecular weight (M_w) of PHA. PHA samples were measured to evaluate the effect of M_w on this digestion. Cells were digested with different ratios of $CHCl_3$: $NaOCl_2$ for 3 hours at 35ºC. Table 11 shows the M_w of PHA recovered. The number of weight (M_n) and average molecular weight (M_w) of intact PHA recovered by chloroform extraction were 9 x10^5 and 22 x10^5 respectively. The M_w of PHA recovery at the 0.5 ratio was 18 x10^5. At both ratios 1 and 2 the weights (20 x10^5) were similar. This suggests that there is a negligible degradation of PHA when sodium hypochlorite is used as extraction. The digestion at low temperature and low concentration of sodium hypochlorite is found to give negligible degradation of PHA (Choi & Lee, 1999; Salmiati, 2008).

Treatment	Number average molecular weight (Mn) (10^5)	Weighted average molecular weight (Mw) (10^5)	Polydispersity Index (PI)
Intact cell by chloroform	9.0	22	2.5
0.5	7.0	18	2.6
1	7.1	20	2.8
2	6.9	20	2.9

Table 11. Molecular weight of PHA recovered by $CHCl_3$ and $NaOCl_2$ for 3 hours at 35°C

The polydispersity index (PI) (weighted average molecular weight/number average molecular weight), reflects the structural degradation of PHA. Table 11 is the PI of the PHA recovered by chloroform and sodium hypochlorite extraction. As the sodium hypochlorite concentration increases, the PI increases significantly for mixed culture biomass. There has been a number of reports by Choi & Lee (1999); Wong (2001) and Hu (2004) in which the authors stated that bacteria PHA granules were in a mobile amorphous state, which is consistent with this result. On the basis of these findings, the patterns of the PI that occurred as the hypochlorite ratios increased were analyzed. The PHA in the bacteria was protected from hypochlorite digestion by its crystalline morphology. When PHA was recovered by using the dispersion technique, however, the crystalline morphology could not protect the molecule because the PHB is soluble in chloroform.

10. Conclusions

Palm oil mill effluent (POME) is to be managed under cleaner production framework, no longer end-of-pipe-engineering solutions. Thus, on top of pollution control, it must be create high value added to the investment. Therefore, this study has successfully shown the possibility of PHA production from POME in a two-stage-process consisting of anaerobic acidogenic fermentation and step aerobics processes. The reactor consisted of two components, anaerobic and aerobic systems. The experiment has demonstrated that POME is a potential feedstock substrate for the production of PHA from high storage capacity mixed cultures under microaerophilic-aerobic and feast–famine conditions. The cycle of microaerophilic-aerobic condition influences the PHA production.

11. Acknowledgment

The authors are pleased to acknowledge the Ministry of Science, Technology and Innovation, (ScienceFund-79004) for funding this research and Bukit Besar Palm Oil Mill, Kulai for providing samples of POME.

12. References

Agamuthu, P. & Tan, E.L. (1985). Digestion of dried palm oil mill effluent (POME) by *Cellulomonas spp*. *Microbios Letter*, 30, 109 -113.

Ahmad, A.L.; Ismail, S. & Bhatia, S. (2003). Water recycling from palm oil mill effluent (POME) using membrane technology. *Desalination*, 157, 87-95.

Ahmad, A.L.; Ismail, S. & Bhatia, S. (2005). Membrane treatment for palm oil mill effluent: effect of transmembrane pressure and crossflow velocity. *Desalination*, 179, 245-255.

Ahmad M.Z.; Salmiati. &Ujang Z. (2008) A novel approach for palm oil mill effluent (POME) treatment and recovery using hybrid membrane bioreactor. *Asiawater 2008*. Malaysia Water Association. Kuala Lumpur, April 2008.

Albuquerque, M.G.E.; Eiroa, M.; Torres, C.; Nunes, B.R. & Reis, M.A.M. (2007). Strategies for the development of a side stream process for polyhydroxyalkanoates (PHA) production from sugar cane molasses. *Journal of Biotechnology*, 130, 411-421.

Alias, Z & Tan, I.K.P. (2005). Isolation of palm oil-utilising polyhydroxyalkanoate (PHA)-producing bacteria by an enrichment technique. *Bioresource Technology*, 96, 1229–1234.

American Public Health Association, APHA (2000). *Standard Methods for the Examination of Water and Wastewater*. 20th ed. Washington, DC APHA.

Bengtsson, S.; Werker, A.; Christensson, M. & Welander, T. (2008). Production of polyhydroxyalkanoates by activated sludge treating a paper mill wastewater. *Bioresource Technology*, 99, 3, 509-616.

Bengtsson, S.; Hallquist.; Werker, A. & Welander, T. (2009). Acidogenic fermentation of industrial wastewater: Effects of chemostat retention time and pH on volatile fatty acids production. *Biochemical Engineering Journal*, 40, 492-499.

Braunegg, G.; Bona, R.; Schellauf, F. & Wallner, E. (2002). Polyhydroxyalkanoates (PHAs): Sustainable biopolyester production. *Polimery*, 47, 7-8, 479-484.

Butler, R.A. (2006). Why is palm oil replacing tropical rainforests? Why are biofuels fuelling deforestation? Mongabay, April 2, 2006. Available at: http://news.mongbay.com.

Cheong, W.C.; Hassan, M.A.; Abdul Aziz, S.; Abdul Karim, M.I.; Sabaratnam, V. & Shirai, Y. (2004). Treatment of palm oil mill effluent (POME) coupled with biohydrogen production. *Asia Water 2004*. The Mines Kuala Lumpur.

Choi, J.I. & Lee, S.Y. (1997). Process analysis and economic evaluation for poly (3-hydroxybutyrate) production by fermentation. *Bioprocess Engineering*, 17, 335-342.

Comeau, Y.; Hall, K.J. and Oldham, W.K. (1988). Determination of poly-betahydroxybutyrate and poly-beta-hydroxyvalerate in activated-sludge by gas-liquid-chromatography. *Appl. Environ. Microbiol.*, 54 , 9, 2325- 2327.

Coats. E.R. (2005). Sustainable production of biodegradable thermolastics through wastewater treatments, and new theory on biological phosphorus removal, PhD. Dissertation, Washington State University.

Chua, A.S.M.; Takabatake, H.; Satoh, H. & Mino, T. (2003) Production of polyhydroxyalkanoates (PHA) by activated sludge treating municipal wastewater: effect of pH, sludge retention time (SRT), and acetate concentration in influent. *Water Research,* 37, 3602-3611.

Damayanti, A.; Ujang, Z; Salim, M.R.; Olsson, G. & Sulaiman. A.Z . (2010). Respirometric analysis of activated sludge models from palm oil mill effluent. *Bioresource Technology,* 101, 144-149.

de la Rubia, M.A.; Perez, M.; Romero, L.I. & Sales, D. (2006). Effect of solid retention time (SRT) on pilot scale anaerobic thermophilic sludge digestion. *Process Biochemistry,* 14, 79-86.

Department of Environment (1999). *Industrial processes & the environmental - Crude Palm Oil Industry*. Handbook No. 3. Ministry of Sciences, Technology and the Environment, Malaysia.

Dionisi, D.; Caruccia, G.; Petrangeli Papinia, M.; Riccardi, C.; Majone, M. & Carrasco, F. (2005). Olive oil mill effluents as feedstock for production of biodegradable polymers. *Water Research,* 39, 2076-2084.

Grengross, T.U. & Slater, S.C. (2000). How green are green plastics? *Science Am.*, 8, 37-41.

Gurmit, S.; Lim Kim, H.; Teo, L. & David Lee, K. (1999). Oil palm and the environment al *Malaysian perspective,* vol. 1. Malaysian Oil Palm Growers Council, 1 , 253.

Hassan, M.A.; Shirai, Y.; Kusubayashi, N.; Abdul Karim, M.I.; Nakanishi, K. & Hashimoto, K., (1996). Effect of organic acid profiles during anaerobic treatment of palm oil mill effluent on the production of polyhydroxyalkanoates by *Rhodobacter sphaeroides. Journal Fermentation Bioengineering*, 82, 151-156.

Hassan, M.A.; Shirai, N.; Kusubayashi, N.; Abdul Karim, M.I.; Nakanishi, K. & Hashimoto, K. (1997a). The production of polyhydroxyalkanoates from palm oil mill effluent by *Rhodobacter spheroides. Journal Fermentation Bioengineering*, 83, 485-488.

Hassan, M.A.; Shirai, N.; Umeki, H.; Abdul Karim, M.I.; Nakanishi, K. & Hashimoto, K. (1997b). Acetic acid separation from anaerobically treated palm oil mill effluent for the production of polyhydroxyalkanoate by *Alcaligenes eutrophus. Bioscience Biotechnology Biochemical*, 61, 1465 -1468.

Hassan, M.A.; Nawata, O.; Shirai, Y.; Nor Aini, A.R.; Yee, P.L.; Arif, A. & Abdul Karim, M.I. (2002). A proposal for zero emission from palm oil industry incorporating the production of polyhydroxyalkanoates from palm oil mill effluent. *Journal Chem. Eng. of Japan*, 35, 1, 9 - 14.

Hu, W. (2004) Synthesis of polyhydroxyalkanoates (PHA) from excess activated sludge, PhD. Thesis, The Hongkong Polytechnic University, Hongkong.

Kittikun, A. H.; Prasertsan, P.; Srisuwan, G. & Krause, A. (2000). Environmental management for palm oil mill. *Material Flow Analysis of Integrated Bio-Systems.* March - October 2000. University of Tokyo, 1-12.

Lee, S.Y & Yu, J. (1997). Production of biodegradable thermoplastics from municipal sludge by a two-stage bioprocess. *Resources, Conservation and Recycling,* 19, 151-164

Ma, A.N. (1999). The planters, Kuala Lumpur Innovations in Management of Palm Oil Mill Effluent. *Palm Oil Research Institute of Malaysia (PORIM).*

Malaysian Palm Oil Board -MPOB. (2010). *Overview of the Malaysian oil palm industry 2009.* Ministry of Plantation Industries and Commodities Malaysia.

Malaysian Palm Oil Council-MPOC. (2007). Malaysia palm oil Industry calls for closer ties with EU on sustainability. *Press Release in Brussels,* Belgium (9 March 2007).

Md Din, M. F.; Ujang, Z.; Salmiati & Van Loosdrecht M. C. M. (2006a). Storage of polyhydroxyalkanoates (PHA) in fed-batch mixed cultures. *4th Seminar on Water Management* (JSPS-VCC), July 11-13, Johor, Malaysia.

Md Din, M.F.; Ujang, Z.; van Loosdrecht, M.C.M.; Salmiati & Ahmad, M.A. (2006b). Polyhydroxyalkanoates (PHAs) production from aerobic-mixed cultures. *Malaysian Journal of Civil Engineering,* 18, 2, 109-128.

Ming, K.K. & Chandramohan, D. (2002) *Malaysian palm oil Industry at crossroads and its future direction,* Malaysian Palm Oil Board, Kuala Lumpur.

Miron, Y.; Zeeman, G.; van Lier, J. & Lettinga, G. (2000). The role of sludge retention time in the hydrolysis and acidification of lipids, carbohydrates and proteins during digestion of primary sludge in CSTR systems. *Water Research,* 34, 5, 1705-1713.

Salmiati; Ujang, Z.; Salim, M.R.; Md Din, M.F. & Ahmad, M.A. (2007). Intracellular biopolymer productions using mixed microbial cultures from fermented POME. *Wat. Sci. Technol.,* 56,8, 179-185.

Salmiati (2008). Intracellular Biopolymer Production from Fermented Palm Oil Mill Effluent Using mixed Microbial Cultures, PhD. Thesis, Universiti Teknologi Malaysia, Malaysia.

Salmiati; Salim, M.R.; Ujang, Z. & G. Olsson (2010). Process development to produce PHAs from fermented palm oil mill effluent using anaerobic-aerobic sequencing batch reactor. *Proc. 1st IWA Malaysia Young Water Professionals Conference* (IWAYP2010), Kuala Lumpur, Malaysia. 2-4 March 2010.

Serafim, L. S.; Lemos, T.; Rosetti, S.; Levantesi, C.; Tandoi, V. & Reis, M. A. (2006). Microbial community analysis with a high PHA storage capacity. *Wat.Sci. Technol.,* 54, 1, 183-188.

Shin, P.K.; Kim, M.H. & Kim, J.H. (2001). Biodegradability of degradable plastic exposed to anaerobic digested sludge and simulated landfill conditions. *J. Environ. Polym.* 5, 1, 33-39.

Tay, J.H.; Show, K.Y. & Jeyaseelan, S.(1996). Effects of media characteristics on performance of upflow anaerobic packed-bed reactors. *J. Environ. Eng.* 122, 469–476.

Ujang, Z. (2008). Industrial waste management using cleaner production approach ©2007/39/81,UTMCopyright(Malaysia)

Verlinden, R.A.J.; Hill, D.J.; Kenward, M.A.; Williams, C.D. & Radecka, I. (2007). Bacterial synthesis of biodegradable polyhydroxyalkanoates. *Journal of Applied Microbiology,* 102, 1437–1449.

Wong, A.L. (2001). *Biopolymers Production with Carbon Sources from the Wastes of a Beer Brewery Industry*, PhD. Thesis, Hong Kong Polytechnic University

Wu, T.Y.; Mohammad, A.W.; Md Jahim, J. & Anuar, N. (2010). Pollution control technologies for treatment of palm oil mill effluent (POME) through end-of-pipe process. *Journal of Environmental Management,* 91. 1467-1490.

Yee, P.L.; Hassan, M.A.; Shirai, Y.; Wakisaka, M. & Karim, M.I.A . (2003). Continuous production of organic acids from palm oil mill effluent with sludge recycle by the freezing-thawing method. *J Chem Eng Jpn.*, 36, 707-710 .

Yeoh, B.G. (2004). A technical and economic analysis of heat and power generation from biomethanation of palm oil mill effluent. *Electricity Supply Industry in Transition: Issues and Prospect for Asia,* 20-63.

Yusoff, S. (2006). Renewable energy from palm oil-innovation on effective utilization of waste. *Journal of Cleaner Productio,* 14, 87-93.

Yu, J. (2003). Production biodegradable thermoplatic materials from organic wastes. *World Patent: WO03/062434 A1.*

Yu, J. & Chen L.X. L. (2006). Cost-effective recovery and purification of polyhydroxyalkanoates by selective dissolution of cell mass. *Biotechnol. Prog.*, 22, 547-553.

25

Calculation of Relaxation Spectra from Stress Relaxation Measurements

Vassilis Kontogiorgos
University of Huddersfield
United Kingdom

1. Introduction

Application of stress on materials increases the energy of the system. After removal of stress, macromolecules comprising the material shift towards equilibrium to minimize the total energy of the system. This process occurs through molecular rearrangements or "relaxation" during which macromolecules attain conformations of a lower energetic state. The time, however, that is required for these rearrangements can be short or long depending on the interactions between the macromolecular species that consist the material. When rearrangements occur faster than the time of observation (experimental timescale) then molecular motion is observed (flow) and the material is regarded as *viscous*. In this type of materials all the energy that is added in the system is dissipated as heat during flow. Conversely, when molecular rearrangements are very slow compared to the time scale of observation molecular motion cannot be observed and the material is regarded as *elastic*. In these types of materials the deformation energy is used to bring the molecules back to their original equilibrium positions. Rheological behavior of real materials, however, lies in between these two classic extremes. Some of the energy is stored in the material during deformation and is used towards return to equilibrium and some is dissipated as heat. These materials are called *viscoelastic* and the time necessary for the molecules to return to equilibrium (relax) is comparable with the time scale of observation (Ferry, 1980; Tschoegl, 1989). The importance of calculation of relaxation spectrum is that it gives us insights to the dynamics of macromolecular chains, which is connected to the molecular structure (molecular weight distribution, branching, network formation) (Malkin, 2006). For sufficiently small deformations the viscoelastic behavior can be described with separable ordinary linear differential equations and thus is termed *linear viscoelastic behavior*. The behavior of a viscoelastic material in the linear regime can be modeled with the aid of mechanical analogues, that is, a combination of dashpots and springs in series or parallel (Maxwell or Voigt elements, respectively), depending on the excitation that is applied on the system (stress or strain, respectively). The overall aim of modeling viscoelastic behavior is to be able to formulate it mathematically and calculate the relaxation spectrum of macromolecules so as to control, predict and eventually engineer their mechanical responses.

However, the relaxation spectrum cannot be measured directly but can be calculated from experimental data using the aforementioned mathematical formulations. An experimental method to test biopolymers that leads to the relaxation spectrum of material is the stress relaxation experiment. Stress relaxation can be modeled with the Maxwell element and adequate representation of stress relaxation behavior of the material may require models with large or infinite number of elements. In that case the stress relaxation function is given by:

$$\sigma(t) = \sigma_e + \int_0^{+\infty} \sigma(\tau)\exp(-\frac{t}{\tau})d\tau \tag{1}$$

where $\sigma(t)$ is the gradual relaxation of stress to the equilibrium stress (σ_e, complete material relaxation means that $\sigma_e = 0$), with $\sigma(\tau)$ being the distribution function of the elements with relaxation time, τ. Therefore, the relaxation function $\sigma(\tau)$ that corresponds to the relaxation spectrum of the material should be calculated from measurements of $\sigma(t)$. The integral (1) can be written in terms of the generic form of the first-kind Fredholm equation:

$$g(s) = \int_a^b K(s,t)f(t)dt,\ a \le s \le b \tag{2}$$

where, $K(s,t)$ is the kernel exp(- t/s) that describes the system, $g(s)$ is the measured signal, and $f(t)$ is the unknown integral solution. The objective in this type of analysis is to determine the spectral function $f(t)$ that represents the relaxation spectrum $\sigma(\tau)$ of the material. Numerical treatment of this procedure is not a straightforward task since the Fredholm integral equation is a classical example of an ill-posed problem that requires a special mathematical approach (Groetsch, 1984). In well-posed problems the solution exists, is unique and depends continuously on the data whereas ill-posed problems fail to comply with the these conditions (Hadamard, 1923). This means that arbitrarily small perturbations of the signal g result in large perturbations of the solution f or, in other words, the solution is extremely sensitive to experimental noise of the measurement (Hansen, 1992b). To overcome such a problem, regularization methods are employed, which allow incorporation of additional information about the sought solution. With the aid of numerical algorithms a regularized solution $\mathbf{x}_\lambda$ to a least squares problem is possible to be calculated as it will be described in a subsequent section (Groetsch, 1984; Provencher, 1982a; Provencher, 1982b; Tikhonov and Arsenin, 1977).

Ill-posed problems in mechanical spectroscopy and numerical methods to solve them have been discussed in the literature (Brabec et al., 1997; Friedrich et al., 1996; Honerkamp, 1989). Furthermore, these techniques are used in various fields of science such as in the measurement of emulsion droplet size distribution (Hollingsworth and Johns, 2003), NMR spectral analysis (van Beek et al., 2003), image deblurring (Dahl et al., 2009) or tomography (Correia et al., 2009) just to name a few. Stress relaxation experimentation is a fundamental and direct approach in the elucidation of molecular dynamics of relaxation processes, yet application of regularization methods to such measurements on biopolymer systems are not well documented in literature. The outcome of such an effort should be an improved understanding of the molecular processes involved in the relaxation of three-dimensional biopolymer structures. In this chapter, we will describe the methodology and the difficulties encountered in the collection of rheological data from biopolymer samples in the stress relaxation measurement. Subsequently, mathematical aspects that are used in Hansen's algorithms and the procedure for the calculation of the relaxation spectrum will be outlined. Finally, the application of these algorithms in a case study using hydrated gluten networks will be described.

2. Data collection

As it has been stressed in the previous section, the relaxation spectrum of the material cannot be measured but should be calculated from experimental data. Stress relaxation measurement in the linear viscoelastic region (LVR) of the material is an experimental technique that can be used to probe molecular relaxations. Stress relaxation experimental mode is usually available in all modern rheometers; however, appropriate experimental setup and procedures must be followed so as to collect meaningful data.

Biopolymers are usually dispersed in water and left to hydrate until the desired macrostructure is formed. It must be emphasized that stress relaxation experiment should be used with solid or semi-solid samples because liquid-like viscoelastic samples relax practically instantaneously and therefore there is not enough time for the instrument to record the decay of storage modulus. Before starting the measurement an appropriate geometry should be selected. The presence of water may result in slippage particularly when the sample is left to relax for several hours. Slippage contributes to noise generation and the resulting storage modulus data may not give accurate representation of sample's relaxation behavior. To alleviate this problem a serrated plate - serrated plate geometry (i.e., both surfaces in contact with the sample should be serrated) is recommended for solid-like aqueous biomaterials. It must be noted that noise generation is an inherent characteristic of any experimental technique something that the algorithms that are described below take into account. However, is important to minimize perturbations in order to identify a spectrum that is as close as possible to the "true" relaxation spectrum of the material. Furthermore, depending the temperature and time of experimentation, a solvent trap or silicon oil to avoid water evaporation and drying at the borders of the geometry maybe necessary. This is particularly important when the measurement is conducted for several hours at temperatures higher than 30 °C where evaporation of water results in drying thus modifying the mechanical response of the sample.

The first step for data collection after sample preparation should be the identification of the LVR of the material at the specified temperature of interest. This is performed with a stress-sweep (amplitude sweep) measurement usually at 1 or 6.28 rad/sec angular frequency of oscillation (0.16 or 1 Hz, respectively). Temperature as well as other factors (i.e., ionic strength, sugars, pH etc) influences biopolymer-biopolymer and water-biopolymer interactions thus modifying the mechanical properties of the material. Therefore, a stress-sweep measurement should be always conducted when the experimental conditions change. This is particularly important for biopolymer gels because network formation usually proceeds by physical interaction of the chains (hydrogen or hydrophobic interactions) that are greatly affected by changes in the surrounding environment where the biopolymer chains are dispersed (Kasapis et al., 2009). Before initiation of the stress relaxation experiment a time-sweep step must be added in the experimental protocol in order to identify the time that the material should rest at the geometry before initiation of the measurement. This experiment can be also performed at 1 or 6.28 rad/sec at a strain that has been determined in the previous experimental step. This ensures that a reproducible initial state of the material will be achieved and therefore reproducible data will be collected for the subsequent numerical analysis. Conclusion of the resting time is usually identified when storage modulus (G′) reaches a pseudo-equilibrium plateau at semi-logarithmic scale, which indicates that most of the measurable stresses introduced during loading of the material have been decayed. Figure 1 demonstrates the effect of resting time on the reproducibility of the stress relaxation experiment with data taken from the literature for highly hydrated gluten networks (Kontogiorgos et al., 2009). The curves superimpose closely in semi-logarithmic plots for all different relaxation periods (20 sec to 10 h) as a result of adequate resting time at the geometry of the instrument prior to measurement. Finally, stress relaxation experiment must be conducted on the material using an instantaneous strain within the LVR and left to relax for the appropriate period of time. The relaxation part of the experiment can be crucial for the subsequent calculation of the spectrum. Regularization algorithms usually require long baselines and optimally collected data points in order to calculate highly resolved spectra (Istratov and Vyvenko, 1999). Because the objective is to resolve exponential components with close time constants in the experimentally measured decay it is very important to record the transient until it decays completely or in other words until the specimen relaxes to equilibrium. Since the ratio of amplitudes of two exponentials with close decay rates $\exp(-\lambda_1 t)$ and $\exp(-\lambda_2 t)$ increases with the time as $\exp[(\lambda_2-\lambda_1)t]$, then these exponentials can be distinguished if the decay is monitored for a sufficiently long time (Istratov and Vyvenko, 1999). This condition, however, presents difficulties in the collection of rheological data in biopolymer samples. Often the transient cannot be left to decay completely owning to the prohibitive laboratory time required to observe such molecular equilibrium as well as instrumental limitations. In addition, biopolymer ageing during experimentation, possible slippage, and other time dependent effects (gelation, aggregation of the chains) contribute to experimental artifacts that may not correspond to true molecular motion of the biopolymer under investigation.

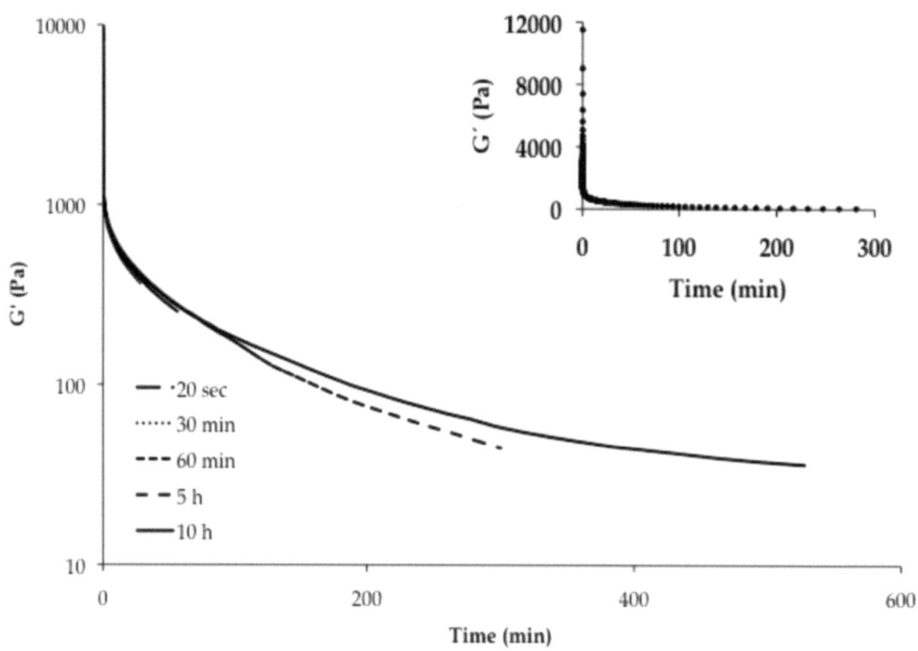

Fig. 1. Stress relaxation curves for hydrated gluten networks for different relaxation times. A pseudo-equilibrium storage modulus is attained after 10 h of relaxation. Inset shows a plot of data in linear mode for 5 h of relaxation (data from (Kontogiorgos et al., 2009)).

Therefore, experimentation with various relaxation times should be preferably conducted until a practically well resolved and stable spectrum is achieved (Kontogiorgos et al., 2009). This is illustrated in Figure 1 where hydrated gluten samples left to relax for different periods of time. Storage modulus decays one logarithmic cycle instantaneously followed by a slower exponential loss of 1.5 orders of magnitude. After 10 h of observation, the material loses more than 99% of its initial modulus value resulting in long baselines, which could be considered as a "fully" decayed transient. However, several trials should be performed so as to find the relaxation time that provides a well-resolved spectrum in a reasonable experimental timeframe. In the case of hydrated gluten 30 min relaxation time was found to result in spectra with practically same resolution with the samples that were left to relax for 10 h (Kontogiorgos et al., 2009). A final note on data collection concerns the scale that storage modulus should be recorded. It is important to collect data in semi-logarithmic scale otherwise storage modulus appears to decay faster than it actually does (Figure 1, inset). The inset that corresponds to the sample that was left to relax for 5 h shows that if data are plotted in linear mode then storage modulus reaches an apparent equilibrium in less than 100 min. However, plotting data in logarithmic mode reveals that relaxation time for 100 min is far from equilibrium storage modulus.

After collection of raw data, calculation of the relaxation spectrum is possible with a regularization method. In a previous work (Kontogiorgos et al., 2009) we have used and compared three different regularization algorithms (Hansen, 2002; Marino, 2007; Wendlandt, 2005) that employ different approaches in their structure. Hansen's algorithms were the fastest requiring only a few seconds to analyze large datasets. For that reason, in the present work we will describe theoretical aspects pertaining to Hansen's algorithms (Hansen, 1994; Hansen, 2002). These employ the L-curve criterion for calculation of the optimum regularization parameter (Hansen, 1992a) and Tikhonov regularization for the computation of the spectrum using MATLAB as analytical software.

3. Theoretical aspects and calculation of spectrum

Calculation of relaxation spectrum involves solving systems of linear equations utilizing concepts from numerical analysis and linear algebra. The basic theoretical aspects will be outlined below to help the reader to grasp the mathematical processes that are involved in the calculations using MATLAB with Hansen's algorithms. Equation 2 is a continuous function and should be formulated to a system of linear equations so that it can be treated numerically. For that reason, a linear system of equations must be formed so as:

$$Ax=b,\ A \in \mathbb{R}^{nxm}, x \in \mathbb{R}^{n},\ b \in \mathbb{R}^{m} \quad (3)$$

where, matrix A is a discrete representation of kernel $K(s, t)$ from equation 2, **b** is a discrete vector that represents $g(s)$, and **x** is the solution vector that describes the desired solution $f(t)$. Vector **b** corresponds to the measured signal i.e., is the stress relaxation modulus. However, in linear discrete ill-posed problems that occur in measurements the right hand side **b** is contaminated by noise (error, **e**), which originates from discretization of the kernel and measurement errors i.e., $\mathbf{b} = \mathbf{b}' + \mathbf{e}$ where $\mathbf{b}'$ is the error free "true" measurement that is not accessible. The objective of the analysis is to determine the "true" solution $\mathbf{x}'$ in the error-free system $A\mathbf{x}' = \mathbf{b}'$. Since it is not possible to measure $\mathbf{b}'$ we need to determine an approximation of $\mathbf{x}'$ by computing an approximate solution **x**. Regularization is a mathematical technique that calculates a stable approximate solution in discrete ill-posed problems by replacing the linear system (3) by a closely related system that is less sensitive to perturbations (Engl et al., 1996; Groetsch, 1984; Hansen, 1998). Various regularization methods have been described in the literature (Hansen, 1998; Istratov and Vyvenko, 1999) with Tikhonov regularization being one of the popular techniques involved in such calculations. Tikhonov regularization replaces the linear system (3) by the regularized system:

$$(A^T A + \lambda I)\mathbf{x} = A^T\mathbf{b} \tag{4}$$

where λ is the regularization parameter, I the identity matrix and A^T the transpose of A. For $\lambda > 0$, the minimization problem has a unique solution $\mathbf{x}_\lambda = (A^T A + \lambda I)^{-1} A^T\mathbf{b}$ where $\mathbf{x}_\lambda$ is the regularized solution that corresponds to the approximate solution $\mathbf{x}$. Tikhonov regularization calculates a regularized solution $\mathbf{x}_\lambda$ by solving the following minimization problem:

$$x_\lambda = \min_{x \in \mathbb{R}} \left\{ ||Ax - b||^2 + \lambda ||x||^2 \right\} \tag{5}$$

where $||.||$ is the Euclidian norm of the matrices or vectors. The regularization parameter controls the sensitivity of the regularized solution $\mathbf{x}_\lambda$ to perturbations in A and $\mathbf{b}$ and also how much weight is given in the minimization of the solution norm $||\mathbf{x}||^2$ relative to the minimization of the residual $||A\mathbf{x} - \mathbf{b}||^2$. Therefore, calculation of regularization parameter is an important task because it ultimately controls the solution $\mathbf{x}_\lambda$ as it will be described later. The first step in the calculation of the spectrum is to discretize the kernel $K(s,t)$ from equation 2. Discretization is the transformation of continuous functions to discrete, in other words, the partition of continuous attributes into a finite set of adjacent intervals in order to generate attributes with a small number of distinct values (Tsai et al., 2008). This operation is necessary to transform the problem into the corresponding matrix form for numerical analysis with the aid of computers. To discretize kernel $K(s, t) = \exp(-t/s)$ MATLAB routine *discr* can be used that is published elsewhere and the reader can easily reproduce the MATLAB code in a personal computer (Kontogiorgos et al., 2009). In this algorithm the user sets the high and low boundaries of the relaxation spectrum and the algorithm creates a discrete space where the solution $\mathbf{x}_\lambda$ ranges using the real experimental time as one of its parameters. The high and low boundaries of the spectrum should be set using a heuristic approach (trial and error). The boundaries were relaxation times ranges are loosely defined depending on the system and material under investigation. Therefore the user should set various boundaries until a stable well resolved spectrum is obtained. Following this step, the singular value decomposition (SVD) of matrix A must be computed which is an essential intermediate step in the calculations of the solution of discrete ill-posed problems. A linear algebra theorem states that a rectangular matrix A can be broken down into the product of three matrices - an orthogonal matrix U (left singular vector), a diagonal matrix σ with nonnegative diagonal elements that decay gradually to zero (singular values of A), and the transpose of an orthogonal matrix V (right singular vector) such that:

$$A = U\sigma V^T = \begin{pmatrix} u_{11} & \cdots & u_{n1} \\ \vdots & \ddots & \vdots \\ u_{m1} & \cdots & u_{mm} \end{pmatrix} \cdot \begin{pmatrix} \sigma_{11} & \cdots & 0 \\ \vdots & \ddots & \vdots \\ 0 & \cdots & \sigma_{mn} \end{pmatrix} \cdot \begin{pmatrix} v_{11} & \cdots & v_{1n} \\ \vdots & \ddots & \vdots \\ v_{n1} & \cdots & v_{nn} \end{pmatrix}^T \tag{6}$$

The routine *csvd* that is available in Hansen's package can be used for this purpose and produces the matrices σ, U and V that satisfy the condition $A = U\sigma V^T$. These matrices will be used in the subsequent calculations and do not play any immediate role in the interpretation of the relaxation spectrum. For an advanced treatment of the theory and the role of the singular value decomposition the reader is referred to the appropriate literature (Hansen, 1998). As mentioned earlier, in the formulation of such problems a regularization parameter (λ) is involved that controls the trade-off between a small error in the solution (low λ values) and a smooth solution (large λ values) (Fig. 2). Clearly, the success of the regularization method depends on the appropriate choice of parameter λ (Hansen, 1992a). The L-curve is a plot for all valid regularization parameters of the semi-norm of the regularized solution *versus* the corresponding residual norm. In double logarithmic plots, the curve resembles the capital letter "L" and an essential feature is that the optimal regularization parameter corresponds to the corner of the L-curve (Hansen, 1992a). Consequently, after calculation of the singular value decomposition of matrix A, it is essential to optimize the regularization parameter λ. This can be performed with the routine *lcurve*. The script *lcurve* provides several methods for calculation of λ depending on the problem under investigation. In the present case, the "Tikhonov" method was used to calculate both the L-curve and the regularized solution in the following step. This routine requires the U and σ matrices that were produced in the previous step as well as the experimentally obtained storage modulus data. The routine *lcurve* changes the regularization parameter until a compromise is achieved between the minimization of the residual and solution norms. The vertical part of the curve (Fig. 2) corresponds to data where the solution norm is very sensitive to changes in the regularization parameter because the error in measurements dominates the regularized solution $\mathbf{x}_\lambda$.

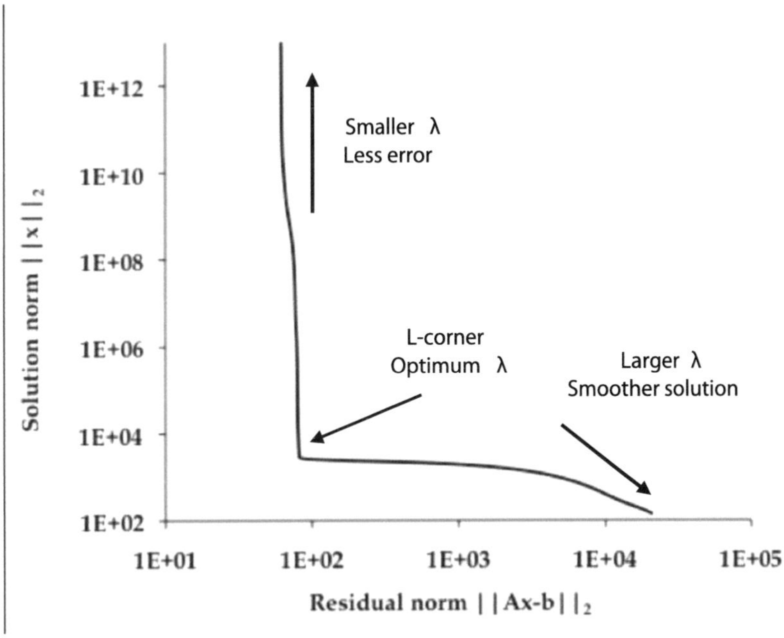

Fig. 2. Typical L-curve plot with the optimum regularization parameter located at the corner of the curve.

The horizontal part corresponds to solutions where the residual norm is the most sensitive to the regularization parameter because $\mathbf{x}_\lambda$ is now dominated by the regularization error (Hansen, 1992a). The optimum regularization parameter is located at the corner of this curve and this gives the most stable approximate solution $\mathbf{x}_\lambda$ to the problem. If one selects a regularization parameter that is smaller than the optimum, the resulting spectrum will be noisy resulting in significant loss of information and most of the times artificial peaks. On the contrary, if λ is greater than the optimum then the solution is over smoothed resulting in loss of resolution. Following calculation of the optimum regularization parameter the spectrum can be calculated with the use of *tiknonov* script that calculates the solution. This algorithm requires to input the matrices U, σ, V that were calculated previously, the vector **b** that contains the experimental values of storage modulus and the value of the regularization parameter that was calculated in the previous step.
It must be noted that the algorithm will calculate both positive and negative solutions to the problem. Obviously the negative solutions do not have physical meaning in the present case and must be dismissed. This can be done either manually by discarding the solutions when plotting the spectrum or modifying slightly the algorithm so it returns only the positive solutions. After the algorithm finishes the calculations the command *semilogx* (sp, X) can be used to plot the spectrum in a semi logarithmic plot. The parameter *sp* is the relaxation space as was calculated in the first step during discretization of the integral whereas X is the calculated $\mathbf{x}_\lambda$. Fig. 3 summarizes the procedure that was described and the algorithms that

STRESS RELAXATION MEASUREMENT
(At the LVR of the material)

↓

DISCRETIZATION OF THE INTEGRAL
(*discr*, formation of matrix A)

↓

SINGULAR VALUE DECOMPOSITION OF A
(*csvd*, calculation of matrices U, s, V)

↓

L-CURVE
(*lcurve*, calculation of optimum λ)

↓

CALCULATION OF THE SOLUTION
(*tikhonov*, calculation of spectrum)

↓

PLOT
(*semilogx*, plotting of spectrum)

Fig. 3. Schematic diagram with the steps involved in the calculation of relaxation spectra with MATLAB. In the brackets are shown the respective algorithms that should be used at each step of calculation.

should be used in each step. In the following section, we describe the application of the procedure in the calculations of the relaxation spectrum of hydrated gluten networks.

4. Case study: hydrated gluten networks

In the present section we reproduce part of the results of our previous work on the calculation of the relaxation spectrum of highly hydrated gluten networks (Kontogiorgos et al., 2009; Kontogiorgos and Kasapis, 2010). Gluten was chosen as model biopolymer system owning to its well-characterized viscoelastic properties and compositional complexity. Thus this model system is highly related to the complicated character found in biopolymer research, as opposed to the "simpler" synthetic polymers. Furthermore, the relaxation modes of gluten and flour have been previously studied and therefore we could compare our results with results from the literature (Bohlin and Carlson, 1981; Li et al., 2003; Phan-Thien and Safari-Ardi, 1998; Rao et al., 2000; Singh and MacRitchie, 2001).

Specimens of hydrated gluten (40% w/w protein solids, 60% w/w deionized water) were prepared as described previously (Kontogiorgos et al., 2007) and left to hydrate at 4 °C for 30 min. Stress relaxation measurements were performed with the Gemini HRnano (Malvern Instruments, UK) employing parallel serrated plate geometry of 40 mm diameter and 2 mm gap. At the end of the hydration period, samples were loaded onto the preheated platen of the rheometer (0, 30, or 50 °C). The samples were left to relax for 10 min prior to measurement as determined by preliminary time sweeps in dynamic oscillation on shear. The LVR of the material was then identified at each temperature with strain sweeps at an angular frequency of 1 rad/s and stress relaxation tests were carried out using the % instantaneous strain at each different temperature as calculated in the previous step. A solvent trap was also applied to minimize moisture loss during the course of stress relaxation testing which is particularly important at high temperatures. Thirty minutes relaxation following application of the instantaneous strain was found to be appropriate to obtain reproducible and highly resolved relaxation spectra with this particular system (Kontogiorgos et al., 2009). It should be noted that data of stress relaxation modulus ($G'(t)$) were collected in a logarithmic mode with respect to the timescale of observation so as to be able to observe the pseudo-equilibrium plateau of $G'(t)$.

Fig. 4 illustrates the stress relaxation measurements obtained with the methodology described above at various temperatures (0, 30 and 50 °C). As temperature increases the relaxation behavior of the specimen changes and in particular, storage modulus delays to establish the apparent pseudo equilibrium plateau. This indicates changes in the relaxation modes of the gluten network something that may affect the resolution and interpretation of the spectrum. However, as it was emphasized earlier, establishing a "true" molecular equilibrium with biological samples may not be possible since temperature often alters the interactions at molecular level and consequently the relaxation behavior of macromolecules. In the present case study, for example, gluten proteins may have different relaxation mechanisms at 0 °C compared to 50 °C. This could occur due to modification in the interaction forces between the protein chains i.e., hydrophobic or hydrogen interactions that play important role on the relaxation properties of gluten (Belton, 1999). Thus prolonged relaxation times may not be possible if time dependent structure-changes occur in the sample. Therefore, the necessary relaxation time must be determined experimentally for different biopolymeric materials as well as different temperatures so as to establish

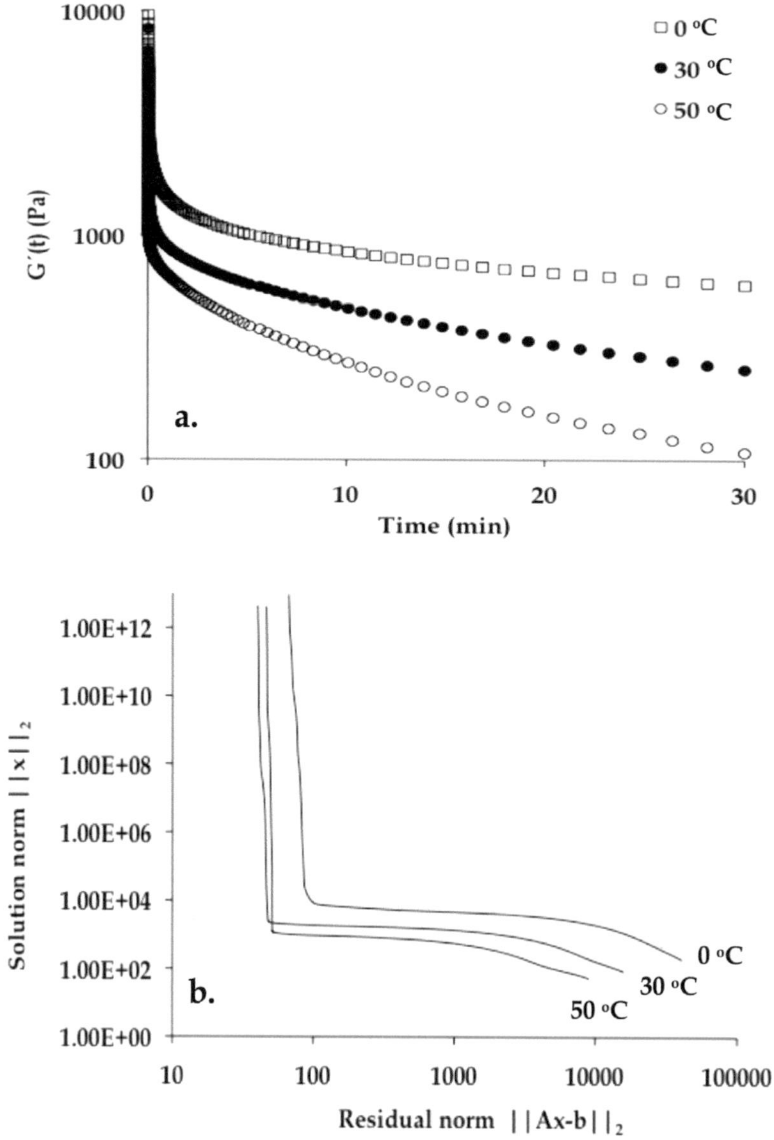

Fig. 4. (a) Stress relaxation measurements for hydrated gluten networks at different temperatures and (b) L-curves for the relaxation curves obtained from the measurements.

the optimum experimental conditions for reliable data collection. In the present work 10 min rest time (Fig. 1) followed by 30 min of relaxation were deemed to be adequate to give reproducible and highly resolved spectra. Following data collection, the optimum regularization parameter was calculated. As it is evident (Fig. 4b) depending on the temperature of measurement the optimum λ varies indicating that accurate determination of λ is essential for optimum calculation of the spectrum. Finally, after optimization of λ the relaxation spectrum was calculated (Fig. 5). Two relaxation regimes can be identified for this material at all temperatures, one at short (< 1 s) and another at long relaxation times (>1 s), which are qualitatively in agreement with previously reported relaxation spectra for various gluten samples (Bellido and Hatcher, 2009; Bohlin and Carlson, 1981; Li et al., 2003; Rao et al., 2000). Similar multimodal spectra have been observed in other biopolymers revealing the complex character of these materials (Mao et al., 2000; Ptaszek et al., 2009; Ptaszek and Grzesik, 2007; Singh Sodhi et al., 2010).

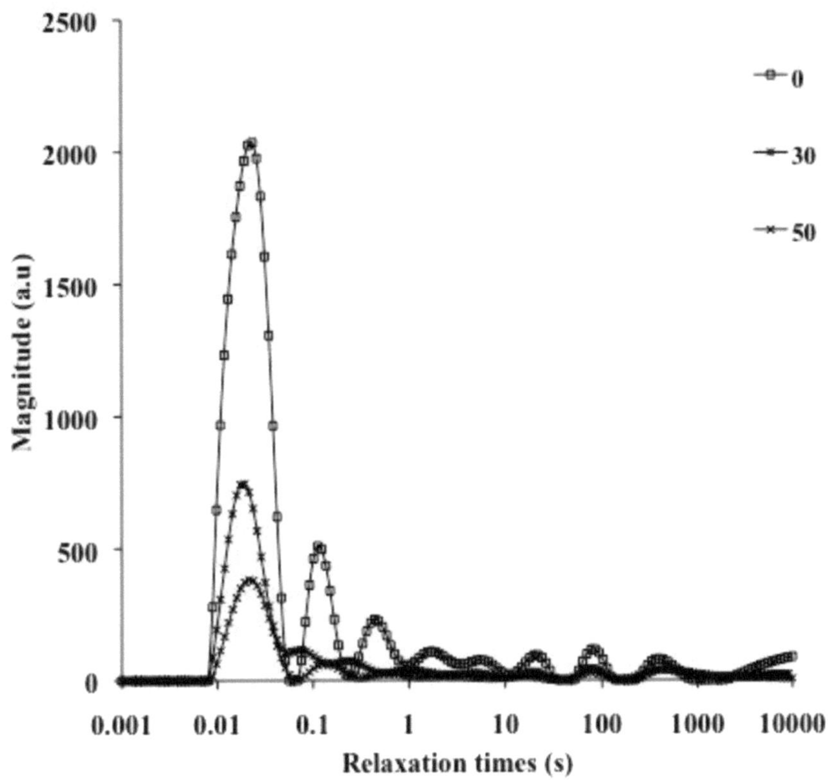

Fig. 5. Relaxation spectra of hydrated gluten at three different temperatures.

Elements with fast relaxation times return quickly to equilibrium after the initial excitation whereas those with slow relaxation times require longer times to reach the state before the application of stress. In the short relaxation times regime three relaxation processes can be distinguished with the fast relaxing species exhibiting strong temperature dependence. As the temperature increases their intensity decreases corresponding to a decrease in the amount of species that can be excited by the applied stress. This strong temperature dependence (one log cycle from 0 to 50 ^{o}C, Fig. 4a) suggests that fewer protein chains respond to the application of stress as the temperature increases or, in other words, more chains remain at the equilibrium state (Kontogiorgos and Kasapis, 2010). At long times, three more peaks can be observed that are mostly unaffected by temperature changes. These peaks correspond to elements with slow molecular rearrangements and require long time to return to equilibrium. Materials with high polydispersity and network structure like gluten, usually exhibit long relaxation events (Ferry, 1980). The fact that the long relaxation regime is not sensitive to changes in temperature suggests that rheological behavior could be controlled by the elements in the short relaxation times with respect to temperature changes

(Kontogiorgos and Kasapis, 2010). In both cases, these events are important for the optimum mixing behavior of flours if one considers that stress relaxation phenomena directly correspond to their mixing characteristics (Rao et al., 2000).
However, the lack of chemical identification of the resulting peaks of the spectra is an issue that limits further fundamental discussion and exploitation of the spectrum in practice. This is very important for highly inhomogeneous and polydisperse systems such as biopolymers and their mixtures in order to find links between the chemical nature and mechanical behavior of the samples. This will allow controlling the properties and designing new materials with superior mechanical performance. For example, if we could assign to a specific protein fraction the peaks at the short and long relaxation regimes then we could modify the mechanical properties of gluten by changing the ratio of the respective protein fractions depending on the desired application. Therefore, future research should focus in the identification of the molecular species that are responsible for the resulting peaks in the spectra. Finally, a further development of the present methodology would be to modify it accordingly so as to obtain the relaxation spectrum from dynamic oscillation measurements. This would be beneficial in order to extend the time (frequency) window of observation and obtain information at the limits of the viscoelastic behavior of the material. This will require development of discretization algorithms for the $G'(\omega)$, $G''(\omega)$ and then the problem can be treated as described previously.

5. Conclusions

A methodology to analyze stress relaxation data as a means to calculate the relaxation spectrum of biopolymeric materials has been described. Data collection should proceed after adequate resting time of the sample on the geometry of the instrument. Resting time should be measured every time when the experimental conditions change in order to ensure collection of reproducible data. Following data collection, Hansen's numerical algorithms that employ Tikhonov regularization can be used to calculate the spectrum using MATLAB as analytical software. In order to calculate the optimum spectrum the regularization parameter λ should be calculated first. For that purpose the L-curve criterion was used where the optimum λ is located at the corner of the L-curve. The presented methodology that returned highly resolved spectra with baseline resolution when applied to hydrated gluten networks. This methodology can be also used to treat exponential decays that are encountered in other physical phenomena as long as a discretization algorithm for the kernel that describes mathematically the system is available.

6. References

Bellido, G.G., Hatcher, D.W., 2009. Stress relaxation behaviour of yellow alkaline noodles: Effect of deformation history. Journal of Food Engineering 93, 460-467.

Belton, P.S., 1999. On the elasticity of wheat gluten. Journal of Cereal Science 29, 103-107.

Bohlin, L., Carlson, T.L.G., 1981. Shear stress relaxation of wheat flour dough and gluten. Colloids and Surfaces 2, 59-69.

Brabec, J.C., Rögl, H., Schausberger, A., 1997. Investigation of relaxation properties of polymer melts by comparison of relaxation time spectra calculated with different algorithms. Rheologica Acta 36, 667-676.

Correia, T., Gibson, A., Schweiger, M., Hebden, J., 2009. Selection of regularization parameter for optical topography. Journal of Biomedical Optics 14, 11.

Dahl, J., Hansen, P.C., Jensen, S.H., Jensen, T.L., 2009. Algorithms and software for total variation image reconstruction via first-order methods. Numerical Algorithms 53, 67-92.

Engl, H.W., Hanke, M., Neubauer, A., 1996. Regularization of inverse problems Kluwer, Dordrecht.

Ferry, J.D., 1980. Viscoelastic properties of polymers, 3rd ed. Wiley, New York.

Friedrich, C., Honerkamp, J., Weese, J., 1996. New ill-posed problems in rheology. Rheologica Acta 35, 186-193.

Groetsch, C.W., 1984. The theory of Tikhonov regularization for Fredholm equations of the first kind Pitman, Boston.

Hadamard, J., 1923. Lectures on Cauchy's problem in linear Partial differential equations Yale University Press, New Haven.

Hansen, P.-C., 1992a. Analysis of discrete ill-posed problems by means of the L-curve. SIAM Review 34, 561-580.

Hansen, P.-C., 1992b. Numerical tools for analysis and solution of Fredholm integral equations of the first kind. Inverse Problems 8, 849-872.

Hansen, P.-C., 1994. REGULARIZATION TOOLS: A MATLAB package for analysis and solution of discrete ill-posed problems. Numerical Algorithms 6, 1-35.

Hansen, P.-C., 2002. Regtools. http://www.mathworks.com/matlabcentral/fileexchange/loadFile.do?objectId=52&objectType=file

Hansen, P.C., 1998. Rank-deficient and discrete ill-posed problems SIAM, Philadelphia.

Hollingsworth, K.G., Johns, M.L., 2003. Measurement of emulsion droplet sizes using PFG NMR and regularization methods. Journal of Colloid and Interface Science 258, 383-389.

Honerkamp, J., 1989. Ill-posed problems in rheology. Rheologica Acta 28, 363-371.

Istratov, A.A., Vyvenko, O.F., 1999. Exponential analysis in physical phenomena. Review of Scientific Instruments 70, 1233-1257.

Kasapis, S., Norton, I., Ubbink, J., 2009. Modern biopolymer science, First edition ed. Academic Press, San Diego.

Kontogiorgos, V., Goff, H.D., Kasapis, S., 2007. Effect of aging and ice structuring proteins on the morphology of frozen hydrated gluten networks. Biomacromolecules 8, 1293-1299.

Kontogiorgos, V., Jiang, B., Kasapis, S., 2009. Numerical computation of relaxation spectra from mechanical measurements in biopolymers. Foods Research International 42, 130-136.

Kontogiorgos, V., Kasapis, S., 2010. Temperature dependence of relaxation spectra for highly hydrated gluten networks. Journal of Cereal Science 52, 100-105.

Li, W., Dobraszczyk, B.J., Schofield, J.D., 2003. Stress relaxation behavior of wheat dough, gluten and gluten protein fractions. Cereal Chemistry 80, 333-338.

Malkin, Y.A., 2006. Continuous relaxation spectrum - its advantages and methods of calculations. International Journal of Applied Mechanics and Engineering 11, 235-243.

Mao, R., Tang, J., Swanson, B.G., 2000. Relaxation time spectrum of hydrogels by CONTIN analysis. Journal of Food Science 65, 374-381.

Marino, I.-G., 2007. Rilt http://www.mathworks.com/matlabcentral/fileexchange/loadFile.do?objectId=6523&objectType=file.

Phan-Thien, N., Safari-Ardi, M., 1998. Linear viscoelastic properties of flour-water doughs at different water concentrations. Journal of Non-Newtonian Fluid Mechanics 74, 137-150.

Provencher, S.W., 1982a. A Constrained Regularization Method for Inverting Data Represented by Linear Algebraic or Integral-Equations. Computer Physics Communications 27, 213-227.

Provencher, S.W., 1982b. Contin - a General-Purpose Constrained Regularization Program for Inverting Noisy Linear Algebraic and Integral-Equations. Computer Physics Communications 27, 229-242.

Ptaszek, A., Berski, W., Ptaszek, P., Witczak, T., Repelewicz, U., Grzesik, M., 2009. Viscoelastic properties of waxy maize starch and selected non-starch hydrocolloid gels. Carbohydrate Polymers 76, 567-577.

Ptaszek, P., Grzesik, M., 2007. Viscoelastic properties of maize starch and guar gum gels. Journal of Food Engineering 82, 227-237.

Rao, V.K., Mulvaney, S.J., Dexter, J.E., 2000. Rheological characterization of long- and short mixing flours based on stress-relaxation. Journal of Cereal Science 31, 159-171.

Singh, H., MacRitchie, F., 2001. Application of polymer science to properties of gluten. Journal of Cereal Science 33, 231-243.

Singh Sodhi, N., Sasaki, T., Lu, Z.-H., Kohyama, K., 2010. Phenomenological viscoelasticity of some rice starch gels. Food Hydrocolloids 24, 512-517.

Tikhonov, A.N., Arsenin, V.Y., 1977. Solutions of ill-posed problems Wiley, New York.

Tsai, C.-J., Lee, C.-I., Yang, W.-P., 2008. A discretization algorithm based on class-attribute contigency coefficient. Information Sciences 178, 714-731.

Tschoegl, W.N., 1989. The phenomenological theory of linear viscoelastic behavior Springer-Verlag, Berlin.

van Beek, J.D., Meier, H.B., Schäfer, H., 2003. Inverse methods in two-dimensional NMR spectral analysis. Journal of Magnetic Resonance 162, 141-157.

Wendlandt, M., 2005. NLCSmoothReg http://www.mathworks.com/matlabcentral/fileexchange/7712.

26

Fluctuations of Stiff Polymers and Cell Mechanics

Jens Glaser and Klaus Kroy
Leipzig University
Germany

Indeed, the vista of the biochemist is one with an infinite horizon. And yet, this program of explaining the simple through the complex smacks suspiciously of the program of explaining atoms in terms of complex mechanical models.

Max Delbrück

1. Introduction

Understanding complex systems through the study of minimal models that capture their underlying universal principles has always been the tradition of physics. This reductionist approach is challenged by the vast complexity of life and the accumulating knowledge in molecular biology. Biological sciences have always laid an emphasis on diversity rather than on simplicity and universality. And rightly so, since diversity is a sine qua non of evolutionary robustness and adaptability (Kirschner & Gerhart, 1998). Bearing in mind this tension, the insight that cellular functions can be attributed to functional modules (Hartwell et al., 1999) as a higher level of biological organization offers a new perspective on a possible unification of the two seemingly contradictory paradigms. It stimulated the emergence of "bottom-up approaches" (Bausch & Kroy, 2006; Schwille & Diez, 2009; Liu & Fletcher, 2009) aiming at the reconstitution of functional modules of cell biology *in-vitro*. The reconstitution of a simplified biological system with a reduced number of components mutually depends on a detailed level of physical understanding, reveals how evolved biological systems work and provides insight into how new biological functions could be engineered.

Cellular mechanics represents an important example for the application of this idea (Bao & Suresh, 2003; Discher et al., 2009; Fletcher & Mullins, 2010). The bottom-up approach to cell mechanics has revealed the basic mechanisms underlying the complex mechanical behavior of the eukaryotic cytoskeleton (Fig. 1, left) by reconstituting self-assembling networks of biopolymers *in-vitro* in an attempt to balance the mutually conflicting demands for simplicity and complexity (Fig. 1, middle) (Bausch & Kroy, 2006).

The present contribution adopts the coarse-graining approach tested in polymer physics and explores how far it takes us in the task of understanding the functional modules responsible for cellular mechanics. We progress from a minimal model for single semiflexible polymers to a theoretical description of their complex networks. It turns out that on this basis many crucial features found in experimental studies of cellular mechanics can be understood qualitatively if not quantitatively. Semiflexible polymers are characterized by their persistence length ℓ_p, which is a mesoscopic length scale, much larger than the microscopic

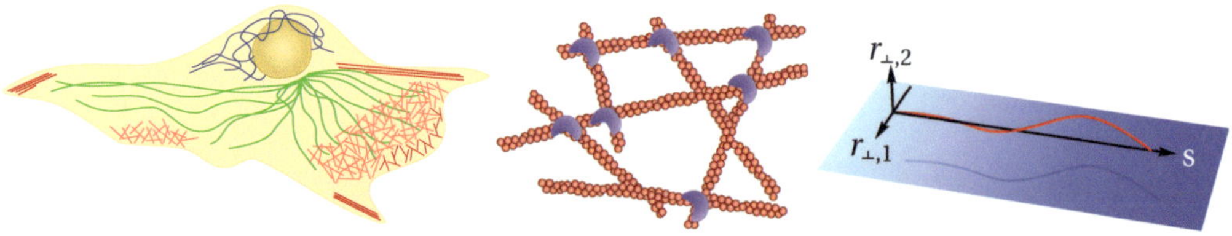

Fig. 1. Bottom-up approach to cell mechanics. *Left:* Schematic view of the cytoskeleton of the eukaryotic cell, showing microtubules (green), actin stress fibers and networks of the cortex and lamellipodium (red), and intermediate filaments (blue). *Middle:* A reconstituted actin network crosslinked by actin-binding proteins. *Right:* A single semiflexible filament described by a mathematical minimal model, the wormlike chain.

monomer length. It indicates the backbone length over which thermal fluctuations bend the polymer significantly, and microscopically, it arises from the backbone's finite bending stiffness, as described mathematically by the wormlike chain (WLC) model (Fig. 1, right).

Double-stranded DNA is a prototypical semiflexible polymer (Bustamante et al., 2003) with a persistence length of $\ell_p \approx$ 50nm that has been measured by single-molecule stretching experiments. The protein machinery for transcription and replication of DNA is highly adapted to the mechanical stiffness of DNA. Also, mechanical bending energy is required to wind DNA into a tightly packed conformation in the nucleosome.

As another important example, the semiflexible protein filaments of the cytoskeleton provide the structural basis of cellular mechanics. The cytoskeleton of the eukaryotic cell consists of three major classes of semiflexible filaments: microtubules, F-actin and intermediate filaments (see Fig. 1, left). In the cell, these filaments form self-assembling networks.

Microtubules are the most rigid of the cytoskeletal polymers with persistence lengths on the order of millimeters, and they are capable of bearing significant compressive load. They form a star-like network that spans the cell, which allows them to act as rails for intracellular transport. During cell division, this network transforms into a bipolar structure (the mitotic spindle) separating the DNA into two identical sets.

Filamentous (F-)actin is a biopolymer protein with $\ell_p \approx 10\mu$m (Isambert et al., 1995) assembled from globular (G)-actin monomers, which are of macromolecular size themselves. The actin cortex is a thin, membrane-bound F-actin network that is employed to maintain and transform the cell's shape. Lamellipodia, filopodia and microvilii are actin-rich structures, and polymerization-dependent forces push these cellular protrusions out of the cell. In muscles, actin provides tracks along which myosin motors walk to generate contractility.

The third type of cytoskeletal polymers, rope-like intermediate filaments, comprises a group of different biopolymer families, which are relatively flexible ($\ell_p \approx 1\mu$m) (Schopferer et al., 2009; Lin et al., 2010) and much less is known about their role in cell mechanics than for actin filaments and microtubules. Intermediate filaments lend mechanical support to the nuclear envelope. In the cytoplasm, a network of intermediate filaments helps the cell to resist shear stress.

In the cell, all three types of protein networks intertwine and interact. For example, the buckling resistance of microtubuli is enhanced by the lateral constraints provided by the surrounding actin and intermediate filament meshworks (Brangwynne et al., 2006), providing a natural paradigm for fiber-reinforced materials, which are also very popular in engineering.

In the following, we review the WLC model and its properties in thermal equilibrium and we infer salient predictions for the dynamics of single semiflexible polymers (Sec. 2). Recent results for their non-equilibrium dynamic response to stretching forces are briefly summarized. Subsequently, we address biopolymer networks *in-vivo* and *in-vitro* and review experimental results that were obtained using the bottom-up approach to cell mechanics (Sec. 3). Theoretical concepts for the description of semidilute solutions of WLCs are introduced. We review theories of the tube and its heterogeneities (Sec. 4), and models of crosslinked biopolymer networks (Sec. 5). Finally we provide a brief overview of models of the viscoelastic and inelastic dynamics of stiff polymer solutions and networks (Sec. 6).

2. Fluctuations and response of wormlike chains

We begin by introducing the mathematical minimal model of a semiflexible polymer, the wormlike chain (WLC). Historically, the concept of a semiflexible polymer that bends only on scales much larger than the monomer size was introduced to explain scattering experiments on thread-like molecules (Kratky & Porod, 1949). The description of a semiflexible polymer as a finitely extensible, differentiable space curve with a curvature energy was introduced in the framework of statistical mechanics by Saitô et al. (1967). We will henceforth refer to it as the WLC. This has become a standard model of polymer physics (Yamakawa, 1971; Doi & Edwards, 1988), and, in the context of biopolymers, it has been useful for the analysis of dynamic light scattering data of F-actin solutions (Farge & Maggs, 1993; Kroy & Frey, 1997). In particular, the simple analytical interpolation formula for the non-linear force-extension relation of a WLC proposed by Marko and Siggia explains force spectroscopy experiments with DNA (Bustamante et al., 1994; Marko & Siggia, 1995) and has led to a surge of applications in single-molecule experiments. Moreover, the WLC enters theories for polymers in confinement (Odijk, 1983; Semenov, 1986; Morse, 2001), under the application of forces (MacKintosh et al., 1995; Kroy & Frey, 1996; Seifert et al., 1996; Hallatschek et al., 2005), compressive load (Baczynski et al., 2007; Emanuel et al., 2007), under shear (Gittes et al., 1997; Morse, 1998c) or in flow fields (Morse, 1998b; Munk et al., 2006), for their bundles (Heussinger et al., 2007) or rings (Alim & Frey, 2007; Ostermeir et al., 2010). The WLC model has been used to characterize a wide range of other biological macromolecules besides DNA and cytoskeletal polymers, including muscle proteins (Tskhovrebova et al., 1997), RNA (Caliskan et al., 2005) or polysaccharides (Vincent et al., 2007).

In the following, we first concentrate on the fluctuations of single wormlike chains and their response to stretching forces, then we extend the picture to include the equilibrium and non-equilibrium dynamics.

2.1 Equilibrium properties of the WLC

2.1.1 Definition and basic properties

The WLC model represents the semiflexible polymer of contour length L by a differentiable space curve $\mathbf{r}(s)$ (see Fig. 1, right) with a curvature energy

$$\mathcal{H}_{WLC} = \frac{\kappa}{2}\int_0^L \mathrm{d}s\left[\mathbf{r}''(s)\right]^2, \tag{1}$$

where κ denotes the bending rigidity, together with the (local) constraint of inextensibility

$$|\mathbf{r}'(s)|=1. \tag{2}$$

Thermal averages are defined with respect to this Hamiltonian via a functional integral

$$\langle \ldots \rangle \equiv \int \mathcal{D}\mathbf{r}(s)\Psi[\mathbf{r}(s)]\ldots,$$

where

$$\Psi[\mathbf{r}(s)] \propto \delta\left\{\left[\mathbf{r}'(s)\right]^2 - 1\right\}\exp\left(-\frac{\mathcal{H}_{WLC}}{k_B T}\right)$$

is the statistical weight associated with the WLC Hamiltonian. The persistence length $\ell_p = \kappa/k_B T$ (in d=3) emerges as the correlation length of the exponential decay of contour tangents in thermal equilibrium, i.e.,

$$\langle \mathbf{t}(s)\mathbf{t}(s')\rangle = \exp\left(-\frac{|s-s'|}{\ell_p}\right), \tag{3}$$

with $\mathbf{t}(s) \equiv \mathbf{r}'(s)$. Eq. (3) follows from the formal equivalence between the statistical weight $\Psi[\mathbf{r}(s)]$ of a WLC conformation, expressed in terms of the tangent orientation $\mathbf{t}(s)$, and the Wiener measure for diffusion on the surface of the unit sphere $|\mathbf{t}| = 1$ (Landau & Lifshitz, 1980; Doi & Edwards, 1988). As a direct consequence of the tangent-tangent correlations, the mean-square end-to-end distance $\langle R^2\rangle$ of a WLC approaches the following asymptotic limiting cases, depending on the ratio of L to ℓ_p:

$$\langle R^2\rangle \to \begin{cases} L^2 & \ell_p \gg L \quad \text{(rigid rod)} \\ 2\ell_p L & \ell_p \ll L \quad \text{(flexible polymer).} \end{cases}$$

Thus, the persistence length demarcates cross-over from rigid rod behavior on short scales to flexible phantom chain (or random walk) behavior on large scales, where the effective step size or "Kuhn length" is $2\ell_p$.

2.1.2 Transverse fluctuations - the weakly bending rod

In many applications, semiflexible polymers are almost straight over the length scales of interest, either because of their intrinsic stiffness or because they are stretched by external forces. Thus, loops and overhangs of the contour are unlikely. In the weakly-bending rod (WBR) approximation the contour is parametrized by two-dimensional excursions $\mathbf{r}_\perp(s)$ transverse to a preferred axis lying along the longitudinal or $\|$-direction (as shown in Fig. 1, right),

$$\mathbf{r}(s) = [\mathbf{r}_\perp(s), s - r_\|(s)].$$

Here, $s-r_\|(s)$ is the coordinate along the preferred axis, and the quantity $r_\|(s)$ with $r_\|'(s) \ll 1$ is called the projected (or stored) length, referring to the contour length stored in the transverse undulations. Thus, for a stiff polymer the local arc-length constraint $|\mathbf{r}'(s)| = 1$ (Eq. (2)) is expanded as

$$r_{\|}'(s) \approx \frac{1}{2}\left[\mathbf{r}_{\perp}'(s)\right]^2 + \mathcal{O}[(\mathbf{r}_{\perp}')^4], \tag{4}$$

to leading order in the small transverse components $\mathbf{r}_{\perp}'$ of the contour tangent.

For a WBR, the exponential decay of the tangent correlations of a free WLC noted in Eq. (3) amounts to a diffusive growth of the mean-square displacement (MSD) of the transverse tangent vector as a function of the arc-length separation, $\langle[(\mathbf{r}_{\perp}'(s)-\mathbf{r}_{\perp}'(0)]^2\rangle = 2|s|/\ell_p$. For a WLC grafted at one end (s = 0) with $\mathbf{r}_{\perp}(0)$ = $\mathbf{r}_{\perp}'(0)$ = 0, the mean square transverse fluctuations are therefore calculated as

$$\langle \mathbf{r}_{\perp}^2(s)\rangle = \frac{2s^3}{3\ell_p}, \tag{5}$$

which can be interpreted as a roughness relation for the (asymptotically) self-affine contour fluctuations of the thermally agitated WBR.

2.1.3 Asymptotic distribution of end-to-end distances

An important quantity distinguishing a stiff polymer from a flexible one is the probability distribution $P(\mathbf{r}) \equiv \langle \delta[\mathbf{r} - \mathbf{R}]\rangle$ of the end-to-end-vector $\mathbf{R} \equiv \mathbf{r}(L) - \mathbf{r}(0)$. For flexible chains such as the freely-jointed chain, it is exactly known and for many purposes approximated well by a Gaussian centered around $\mathbf{r} = \mathbf{0}$ (Yamakawa, 1971). Stiff polymers behave drastically different, since their distribution $P(\mathbf{r})$ exhibits a peak near full extension. We quote here the exact asymptotic result for the $P(\mathbf{r})$ of a WBR from Wilhelm & Frey (1996).

$$P(\mathbf{r}) \sim \frac{\mathcal{N}}{x^{3/2}}\left(\frac{1}{x} - 2\right)\exp\left(-\frac{1}{4x}\right), \qquad x = \frac{\ell_p}{L}(1 - r/L). \tag{6}$$

This is the leading term of an infinite series for $P(\mathbf{r})$ and it is valid near full extension, i.e. for $1 - r/L \ll L/\ell_p$. The radial distribution function, obtained from $P(\mathbf{r})$ by multiplying with an additional measure factor $4\pi r^2$ in three dimensions, is compared with Monte-Carlo data for a WLC in Fig. 2 for several values of ℓ_p/L.

2.1.4 WLC under a strong stretching force

We calculate the nonlinear response of the WBR to a strong stretching force f acting at the ends. In the WBR parametrization, the Hamiltonian of a chain stretched by the force f reads

$$\mathcal{H}_f = \mathcal{H}_{WLC} + \mathcal{H}_{ext} \approx \frac{\kappa}{2}\int_0^L ds\left[\mathbf{r}_{\perp}''(s)\right]^2 + \frac{f}{2}\int_0^L ds\left[\mathbf{r}_{\perp}'(s)\right]^2, \tag{7}$$

where the last term $\mathcal{H}_{ext}$ is the work done by the external stretching force f, which is calculated from $\mathcal{H}_{ext} \equiv -\mathbf{fR} = f\, r_{\|}(L) + const.$ using the approximate local arc-length constraint, Eq. (4). Since we are primarily interested in a qualitative discussion (rather than in numerically exact prefactors), we employ scaling arguments to find the asymptotic force-extension relation for the WLC. First we observe the occurrence of a characteristic length scale $\ell_f \equiv \sqrt{\kappa/f}$ of the force, which is obtained by equating the two contributions to the

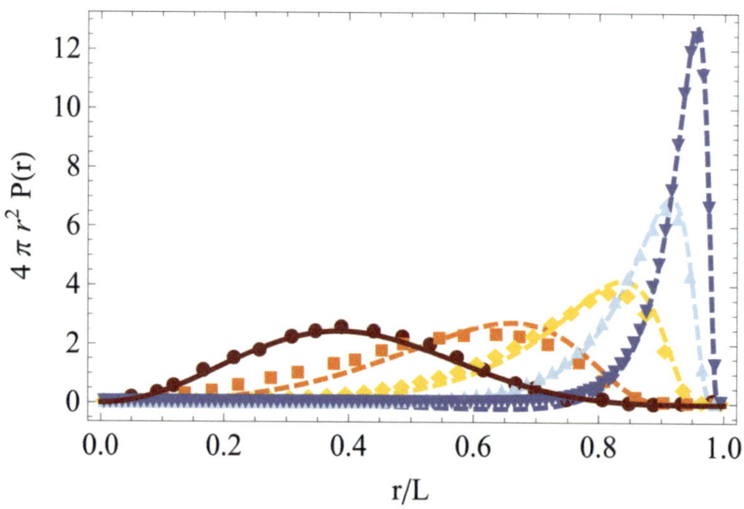

Fig. 2. Isotropic radial distribution function for a WLC of different $\ell_p/L = 0.1, 0.2, 0.5, 1, 2$ from left to right. Shown is the asymptotic formula Eq. (6) (dashed lines) and the Daniels approximation (Daniels, 1950; Yamakawa, 1971) (solid line), compared with Monte Carlo simulation data (from Wilhelm & Frey (1996) for $\ell_p/L = 0.1, 0.2$ and kindly provided by Sebastian Schöbl for $\ell_p/L = 0.5, 1, 2$; symbols).

Hamiltonian, Eq. (7). It indicates the length of unperturbed chain sections with a stored length $r_{\|}(\ell_f) \simeq \ell_f^2 / \ell_p$. For strong stretching forces, the chain may thus be viewed as a taut string of a number L/ℓ_f of subsections of length ℓ_f, and the asymptotic end-to-end distance follows as $R(f) = L - r_{\|,f}(L)$ from the total contraction $r_{\|,f}(L)$ of the chain,

$$r_{\|,f}(L) \sim \frac{L r_{\|}(\ell_f)}{\ell_f} \simeq \frac{L\ell_f}{\ell_p} = L\sqrt{\frac{k_B T}{\ell_p f}}. \tag{8}$$

This estimate differs from the exact asymptotic result merely by a factor of 1/2 (Fixman & Kovac, 1973; Marko & Siggia, 1995).

2.2 Dynamics of the WBR

2.2.1 Equation of motion of the WBR and the fluctuation-dissipation theorem

We formulate the linearized Langevin equations of motion of an overdamped WLC in a viscous solvent. They derive from the WLC Hamiltonian, Eq. (1), and we note that the elastic force per unit length is given by $\mathbf{f}_{el} = -\delta\mathcal{H}/\delta\mathbf{r}$, where $\mathcal{H}$ is a sum of two contributions:

$$\mathcal{H} = \mathcal{H}_{WLC} + \frac{1}{2}\int_0^L ds\, f(s,t)\left[\mathbf{r}'(s)\right]^2.$$

Here, a Lagrange multiplier force $f\,(s,\,t)$ enforces the local inextensibility constraint of the WLC, Eq. (2) (Goldstein & Langer, 1995). It has the physical interpretation of a local backbone tension. The friction force per unit length $\mathbf{f}_{visc} = -\zeta\dot{\mathbf{r}}$ is in the free-draining approximation mediated by a friction tensor

$$\zeta \equiv [\zeta_{\perp}(1-\mathbf{r}'\otimes\mathbf{r}')+\zeta_{\|}\mathbf{r}'\otimes\mathbf{r}'],$$

that reflects the anisotropic hydrodynamic interactions to leading order in two distinct friction coefficients $\zeta_{\|}$ and $\zeta_{\perp}$ for longitudinal and transverse motion, respectively (Doi & Edwards, 1988). To distinguish between contour length and time derivatives, we use primes and dots, respectively. Improved approximations for the viscous drag lead to logarithmic corrections to the linearized dynamics of a WBR (Granek, 1997; Glaser et al., 2008). The external and the stochastic thermal force per unit length are denoted by $\mathbf{g}$ and ξ, respectively. Then, the linearized projected Langevin equations of motion follow from a balance of forces $\mathbf{f}_{visc}+\mathbf{f}_{el}+\mathbf{g}+\xi=\mathbf{0}$ as

$$\begin{aligned} \zeta_{\perp}\dot{\mathbf{r}}_{\perp} &= -\kappa\mathbf{r}_{\perp}'''' + f\mathbf{r}_{\perp}'' + \mathbf{g}_{\perp} + \xi_{\perp}, \\ f' &= \mathbf{g}_{\|}. \end{aligned} \tag{9}$$

In order to arrive at Eq. (9), we expanded the equations to linear order in $\mathbf{r}_{\perp}$ using Eq. (2) (Hallatschek et al., 2007a). We also approximated $f\approx const.$ in the WBR-limit. Its leading (s,t)-dependence is however accessible via a dedicated perturbation scheme (see Sec. 2.2.3 below). The equations are completed by the correlations of the Gaussian distributed stochastic force density ξ,

$$\begin{aligned} &\langle \xi_{\perp,i}(s,t)\rangle = 0, \\ &\langle \xi_{\perp,i}(s,t)\xi_{\perp,j}(s',t')\rangle = 2k_BT\zeta_{\perp,ij}\delta(t-t')\delta(s-s'). \end{aligned}$$

These correlations are dictated by the fluctuation-dissipation theorem, which establishes a relation between the linear response of the chain $\langle \mathbf{r}(s,t)\rangle_{\mathbf{g}}$ and its equilibrium conformational correlations,

$$\frac{\delta\langle r_i(s,t)\rangle_{\mathbf{g}}}{\delta g_j(s',t')} = -\frac{\theta(t-t')}{k_BT}\frac{\mathrm{d}}{\mathrm{d}t}\langle r_i(s,t-t')r_j(s',0)\rangle \qquad \text{(FDT).} \tag{10}$$

The FDT can be formulated for all systems in thermal equilibrium (Chaikin & Lubensky, 1995).

2.2.2 Linear response of a WBR to a transverse force

We employ scaling arguments again to find an approximate solution to Eq. (9) for the linear *dynamic* response of a WBR to a transverse step force $\mathbf{G}_{\perp}$, acting for times $t>0$ at $s=s'$, i.e. $\mathbf{g}_{\perp}(s,t)=\mathbf{G}_{\perp}\delta(s-s')\theta(t)$ with $f=0$. It causes a growing indentation of width $\ell_{\perp}(t)$ and depth $\langle \mathbf{r}_{\perp}(s',t)\rangle_{\mathbf{g}_{\perp}}$, both of which are to be determined. The width $\ell_{\perp}(t)$ is inferred from the thermally averaged Eq. (9), which reads $\zeta_{\perp}\mathbf{r}_{\perp}/t \simeq \kappa\mathbf{r}_{\perp}/\ell_{\perp}^4(t)$ on the scaling level for $s\neq s'$, yielding $\ell_{\perp}(t)\simeq(\kappa t/\zeta_{\perp})^{1/4}$. To estimate $\langle \mathbf{r}_{\perp}\rangle_{\mathbf{g}_{\perp}}$, we carry out the ensemble average of Eq. (9) again and integrate over the spatial coordinate s, which gives:

$$\zeta_{\perp}\ell_{\perp}(t)\langle \dot{\mathbf{r}}_{\perp}\rangle_{\mathbf{g}_{\perp}} = \mathbf{G}_{\perp}, \tag{11}$$

The dynamics can therefore be understood in terms of a Stokes formula with a friction coefficient $\zeta_{\perp}(t)=\zeta_{\perp}\ell_{\perp}(t)$ that grows in time, corresponding to the increasing subsection of length $\ell_\perp(t)$ of the chain that is set into motion by the force $\mathbf{G}_\perp$. Eq. (11) then implies for the linear response $\langle \mathbf{r}_\perp \rangle_{\mathbf{g}_\perp}$ to the external force

$$\langle \mathbf{r}_\perp \rangle_{\mathbf{g}_\perp} \sim \mathbf{G}_\perp \frac{t^{3/4}}{\zeta_\perp^{3/4} \kappa^{1/4}}.$$

The above assumption of a purely transverse friction implied by the linearized Eq. (9) ceases to hold when further growth of the transverse indentation $\langle \mathbf{r}_\perp \rangle_{\mathbf{g}_\perp}$ is hindered by the limited availability of stored length, i.e. if additional contour length needs to be pulled in against longitudinal friction from the tails of the WBR. Thus transverse motion couples to longitudinal motion via a growing tension f and slows down at (sufficiently) long times (Obermayer & Hallatschek, 2007).

2.2.3 WBR under tension

The longitudinal dynamic response of a WBR to a stretching force $G_\parallel = f$ acting on the ends for $t > 0$ can analogously be inferred from scaling arguments. On the scaling level, the averaged Eq. (9) leads to $\zeta_\perp / t \simeq \kappa / \ell_\perp^4(t) + f / \ell_\perp^2(t)$, which implies $\ell_\perp(t) \simeq (\kappa t / \zeta_\perp)^{1/4}$ for $t \ll t_f \equiv \kappa\zeta_\perp / f^2$ and $\ell_\perp(t) \simeq (ft / \zeta_\perp)^{1/2}$ for $t \gg t_f$. The distinction between short and long times $t \lessgtr t_f$ is equivalent to the one between weak and strong forces, $f \lessgtr (\kappa\zeta_\perp / t)^{1/2}$. The former, linear response may be calculated from the corresponding longitudinal fluctuations using the FDT, Eq. (10). The result is that both quantities scale with time as $t^{3/4}$ (Granek, 1997; Everaers et al., 1999), similar to the transverse response (see Sec. 2.2.2). The long-time longitudinal response is estimated by observing that the chain consists of $L/\ell_\perp(t)$ subsections of length $\ell_\perp(t)$, which, by definition, have equilibrated at time t, i.e. they have been pulled essentially straight by the external force. The condition of straight subsegments implies that their elongation is equal to their initial equilibrium contraction $r_\parallel(\ell_\perp) \simeq \ell_\perp^2(t) / \ell_p$ (see Sec. 2.1.2), but with the opposite sign. The total change in end-to-end distance of the polymer follows as

$$\Delta R(t) \equiv r_\parallel(L, t=0) - \mathbf{r}_\parallel(L,t) \sim \frac{L\ell_\perp(t)}{\ell_p} \simeq \frac{Lf^{1/2}t^{1/2}}{\zeta_\perp^{1/2}\ell_p}, \tag{12}$$

for $\ell_\perp(t) \ll L$. This quantity saturates at its equilibrium value $r_\parallel(L) - r_{\parallel,f}(L)$ (see Sec. 2.1.4) when $\ell_\perp(t) \simeq L$.

If longitudinal friction was generated along the whole filament length L, the corresponding drag force $\zeta_\parallel L\Delta\dot{R}$ would exceed the external driving force f for times $t \lesssim t^* \simeq \zeta_\parallel^2 L^4 / \zeta_\perp \ell_p^2 f$ (Seifert et al., 1996; Ajdari et al., 1997; Everaers et al., 1999). This apparent contradiction indicates the breakdown of Eq. (12) at short times. It is avoided by considering that longitudinal friction is only generated inside a boundary layer of width $\ell_\parallel(t)$ growing with time, where the polymer contour is set into longitudinal motion. For the longitudinal motion, the length $\ell_\parallel(t)$ thus plays a role analogous to $\ell_\perp(t)$ for the perpendicular motion. Accordingly, $\ell_\parallel(t)$ can be defined by postulating an effective longitudinal Stokes equation, analogous to Eq. 11 for the transverse motion,

$$\zeta_\parallel \ell_\parallel(t) \Delta\dot{R}(t) \simeq f, \tag{13}$$

with a longitudinal friction coefficient $\zeta_{\|}\ell_{\|}(t)$. Comparison of Eq. (13) with Eq. (12) after replacing L by $\ell_{\|}(t)$ yields $\ell_{\|}(t) \simeq (ft\zeta_{\perp})^{1/4}(\ell_p / \zeta_{\|})^{1/2}$ for strong stretching forces (Seifert et al., 1996). The 'longitudinal equilibration length' is larger than the transverse one by a factor $\ell_{\|} / \ell_{\perp} \approx [\ell_p / \ell_{\perp}(t)]^{1/2} \gtrsim (\ell_p / L)^{1/2} \gg 1$ (for a sufficiently stiff polymer), and thus grows faster. Propagation of the tension $f(s, t)$ therefore only needs to be taken into account at short times for which $\ell_{\|}(t) \ll L$, i.e. when the tension has not yet equilibrated.

A systematic analysis of the phenomenon of tension propagation builds on this strong separation of length scales $\ell_{\|}(t) \gg \ell_{\perp}(t)$ (Hallatschek et al., 2005; 2007a;b). Via a (stochastic) multiple scale perturbation theory one can establish a coarse-grained deterministic theory for the polymer dynamics under strong tension. The spatially varying deterministic tension $f(s, t)$ is extracted by averaging over the transverse thermal fluctuations on short length scales. Previously known scaling results were recovered from this systematic theory as intermediate asymptotic regimes for specific scenarios, including pulling on the polymer, release of tension, and sudden temperature quench. The practically relevant case of pulling on a pre-stressed (Obermayer et al., 2007) or even pre-straightened (Obermayer et al., 2009) polymer, such as a polymer held in an optical trap, has been shown to lead to a wealth of new dynamic regimes and to depend sensitively on the initial conditions. Based on these results, the complex rheological modulus $G^{\star}(\omega) = G'(\omega) + iG''(\omega)$ for the response of a single WLC to an oscillatory longitudinal force could be calculated, and was shown to exhibit a $G^{\star}(\omega) \propto \omega^{7/8}$-regime for high frequencies (Hiraiwa & Ohta, 2008; 2009). Similar results were obtained for a chain with slightly extensible bonds (Obermayer & Frey, 2009). In affine shear flow, tension propagation can be neglected and the high-frequency modulus scales as $G^{\star}(\omega) \propto \omega^{3/4}$ (Gittes & MacKintosh, 1998; Morse, 1998c; Pasquali et al., 2001; Hiraiwa & Ohta, 2009).

3. Cells and gels

3.1 The bottom-up strategy and basic mechanics of the actin cytoskeleton

In this section, we review recent progress in the study of semiflexible polymer networks as simplified model systems for the actin cytoskeleton of the living cell (Kroy, 2006), focusing on their material properties (Kasza et al., 2007) and highlighting analogies (and differences) between both systems. We thus evaluate the usefulness of the bottom-up approach to cell mechanics by considering concrete examples of the linear and nonlinear rheology of cells and gels.

The mechanical properties of cells are considerably influenced by the cell cortex as a thin membrane-bound F-actin network capable of bearing substantial load (Stricker et al., 2010), although the contribution due to other intracellular compartments cannot be neglected (Hoffman & Crocker, 2009). A multitude of rheological techniques has been developed to characterize the response of the cell to mechanical perturbations (see Fig. 3). One may distinguish between passive and active techniques, which correspond to observing the the spontaneous motion of embedded tracer particles or to probing the deformation in response to an applied force, respectively. Only in equilibrium materials these methods yield the same results, whereas in the cell, this is in general not the case (see Sec. 3.5 below). The linear mechanical behavior of the cell, as characterized by the frequency-dependent complex shear modulus $G^{\star}(\omega)$, can be measured by passive methods under suitable conditions e.g. of ATP depletion (Bursac et al., 2005; Hoffman et al., 2006) or by active methods. It is intermediate between that of a solid and a liquid, and it is thus called viscoelastic.

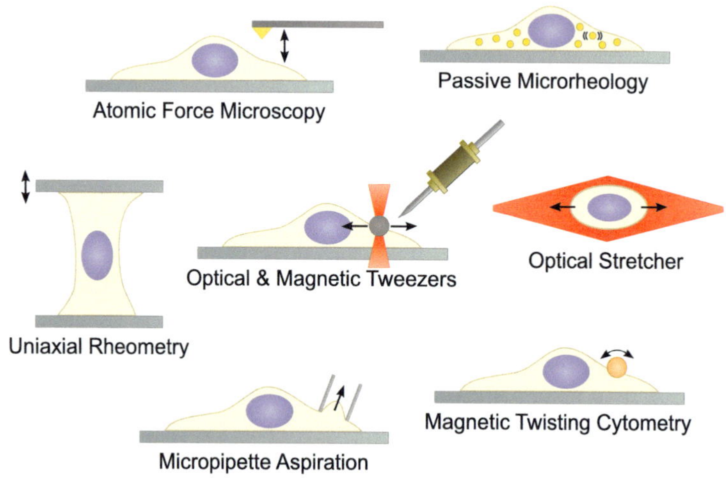

Fig. 3. Cells probed by common experimental methods, based on Kollmannsberger (2009) and Hoffman & Crocker (2009).

3.2 Importance of crosslinkers

The high-frequency shear-modulus of actin solutions and gels, a power-law resulting from single-filament dynamics (see Sec. 2.2.2 and Sections 4&5 below), crosses over to a rubber elastic plateau $G'(\omega) \sim G_0$ at low frequencies (Hinner et al., 1998; Gardel et al., 2004). This plateau is estimated by the tube model for entangled solutions (see Sec. 4 below) or the affine network model for gels (see Sec. 5 below). The difference between the absolute value of G_0 (on the order of $\approx$ 1Pa) and that of the weakly frequency-dependent shear modulus $G^{\star}(\omega)$ of cells (see Sec. 3.3 below; on the order of $\approx$ 1kPa) is now understood as a consequence of the different network elasticity in the absence or presence of crosslinkers and prestress. Networks of F-actin can be crosslinked using specific actin-binding-proteins (ABPs), and increasingly sophisticated studies have demonstrated that the relative crosslinker concentration, the type of crosslinker and even its molecular details provide fine-grained control over elastic and structural properties of the network (Gardel et al., 2004; 2006; Lieleg et al., 2010). Network elasticity is also determined by the flexibility of crosslinkers (Wagner et al., 2006). In addition, the network may be set under prestress, which can be externally applied (see Sec. 3.4 below) or internally generated, e.g. by molecular motors (see Sec. 3.5 below), which further increases the elasticity.

Cells choose between a multitude of different and partially redundant ABPs for crosslinking, but a large number of *in-vitro* studies concentrates on isolating the physical properties of networks crosslinked by a single type of molecule. This is justified by the fact that the rheological properties of composite networks containing different crosslinkers are largely determined by the crosslinker which outnumbers the other (Schmoller et al., 2008).

3.3 Plateau modulus vs. power-law rheology

The rheology of cells is well described by a power-law shear modulus $G^{\star}(\omega) \propto \omega^{\beta}$ at low frequencies with exponent $\beta = 0.1\text{–}0.25$, extending over up to three decades (Hoffman et al., 2006; Hoffman & Crocker, 2009). This has been interpreted as a sign of "glassy" dynamics (Fabry et al., 2001), as described by the generic model of "soft glassy rheology" (Sollich et al.,

1997). However, such slow dynamics was also recently observed in reconstituted biopolymer networks. High-precision dynamic light scattering studies have demonstrated that the plateau in the tracer particle MSD of actin solutions, which is related to the linear response function $G^{\star}(\omega)$ via the FDT (see Sec. 2.2.1), is in reality slanted, corresponding to a dynamic structure factor $S(q, t)$ that exhibits slow logarithmic decay of density fluctuations over several decades in time (Semmrich et al., 2007). This has been parametrized with high precision by the "glassy wormlike chain" model (see Sec. 6 below). Experiments on filamin crosslinked networks, on the other hand, have shown that power-law rheology may readily arise in these systems (Gardel et al., 2006), where it might be due to the flexibility of crosslinkers (DiDonna & Levine, 2006). Given the variety of crosslinker types available to the cell, power-law rheology could also result from the superposition of different crosslinker binding/unbinding rates (Lieleg et al., 2008). Common to these explanations for the power-law rheology of cells is a broad distribution of length, time or energy scales, which is supposed to have its origin in the physics of the stiff polymers and their crosslinkers rather than in the genuinely biological cell dynamics. Therefore, its study should be possible not only *in-vivo*, but also in *in-vitro* reconstituted functional modules.

3.4 Nonlinear strain-softening and stiffening

However, the nonlinear rheology of cells differs from that of reconstituted gels at first sight. Cells have been reported to become stiffer *or* softer with increasing strain, depending on the applied deformation protocol, whereas reconstituted gels usually strain-stiffen. More specifically, cells under uniaxial loading displayed an elastic stiffening response (Fernández et al., 2006; Fernández & Ott, 2008), but when subjected to a transient stretch, the opposite response, i.e. fluidization and subsequent recovery emerged, as shown in Fig. 4, left (Trepat et al., 2007). The first response can be seen as analogous to observations of stiffening in actin gels (Gardel et al., 2004; 2006; Tharmann et al., 2007) or other biopolymer gels (Storm et al., 2005), for which it was explained by the affine network model (see Sec. 5 below), nonlinear crosslink flexibility (Broedersz et al., 2008) or network geometry (Onck et al., 2005). By contrast, the fluidization of cells and similar observations of viscoplasticity in the living cell (Fernández & Ott, 2008) have been suggested to arise from the breaking of cytoskeletal bonds (Trepat et al., 2007; Wolff et al., 2010) (see also Sec. 6 below).

A similar phenomenology has also been demonstrated for actin solutions undergoing a transition from strain softening to stiffening upon changing the solvent or ambient parameters or the deformation rate (Semmrich et al., 2007; 2008; Lieleg & Bausch, 2007) (Fig. 4, right). This behavior, resembling a glass transition, might arise due to weak sticky interactions between filaments. Thus, the coexistence of a fluidiziation and a reinforcement response in cells, analogous to the continuous transition from strain-softening to stiffening observed in actin gels, could be interpreted as a common feature of the material properties of both systems.

3.5 Towards active materials

The living cell operates far from thermodynamic equilibrium, and in the cytoskeleton, energy stored in the form of ATP is constantly transformed into mechanical energy e.g. through the activity of motor proteins such as myosin, or through the ATP-dependent polymerization of actin filaments. The cytoskeleton might therefore be characterized as an active material (Fletcher & Geissler, 2009). The effect of active processes on cell rheology is

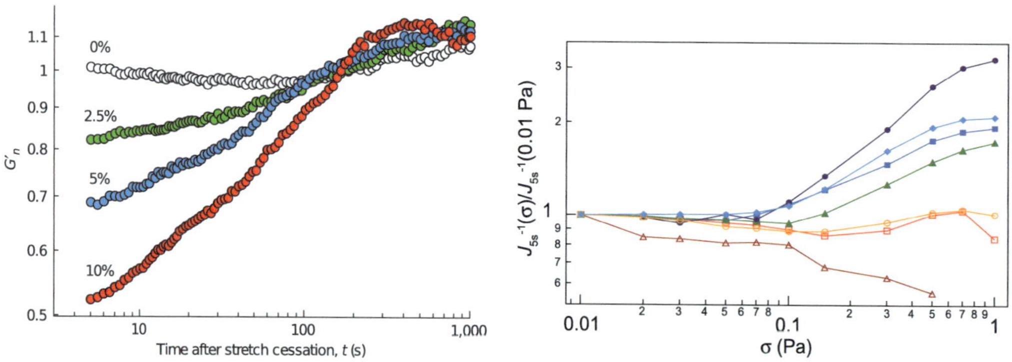

Fig. 4. *Left:* Fluidization and recovery of human airway smooth muscle cells after a single transient stretch, as measured by the normalized stiffness G_n' for different stretch amplitudes. Adapted from Trepat et al. (2007). Copyright © 2007 by Macmillan Publishers Ltd. *Right:* Temperature-induced transition from strain softening to stiffening in entangled F-actin solutions. The inverse of the normalized creep compliance J as a function of the applied stress σ for various temperatures from $T = 27 - 18°C$ (bottom to top). Adapted from Semmrich et al. (2007). Copyright © 2007 by The National Academy of Sciences of the USA.

signalled by the breakdown of the fluctuation-dissipation theorem (FDT) (see Sec. 2.2.1) (Lau et al., 2003; Bursac et al., 2005). Thus, in contrast to equilibrium materials, the linear response function $G^{\star}(\omega)$ can not be inferred from looking at the spontaneous fluctuations, e.g. of the position of a probe particle embedded in the cell, and in the presence of non-equilibrium fluctuations, it can only be obtained by active measurements. On the other hand, by combining passive and active methods in the same experiment, non-equilibrium contributions to the spontaneous fluctuations can be measured. By this method, a breakdown of the FDT at frequencies below 10Hz due to active motor-induced forces has been demonstrated in actin-myosin gels (Mizuno et al., 2007) and similarly in cells (Mizuno et al., 2009). Active processes driven by molecular motors lead to a variety of new and interesting phenomena in reconstituted systems (MacKintosh & Schmidt, 2010), including fluidization of actin solutions (Humphrey et al., 2002), active fluctuations of stiff microtubules embedded in the actin cytoskeleton (Brangwynne et al., 2008), or stiffening of crosslinked gels due to contractile tension generated by motors (Mizuno et al., 2007; Koenderink et al., 2009). Since many cytoskeletal structures involve contractile elements, active reconstituted gels are highly relevant for a more complete understanding of cellular mechanics.

3.6 Conclusion

In conclusion, reconstituted cytoskeletal systems exhibit many of the salient features of cell mechanics and they seem ideally suited to further study the intriguing viscoelastic, non-linear and viscoplastic properties of the living cell. It has been suggested that cell mechanics may be understood in terms of a small number of "laws" (Trepat et al., 2008). Here we have presented evidence that biopolymer gels exhibit mechanical properties of comparable robustness and universality. Their structural basis consists of scaffolding fibers, such as F-actin, and these may be combined with a variety of crosslinkers and ultimately with active components. This demonstrates that using simple, polymer-based model systems it is possible to explain a large number of cell mechanical observations by minimal assumptions.

Theoretical descriptions are therefore useful to establish the link between the rheological properties at a higher level to the underlying microscopic structures.

4. Tube model

In this section, we address theories for entangled solutions of stiff polymers (see Sec. 3). In the absence of crosslinkers, their equilibrium properties are successfully and quantitatively described by models of topological interactions. Macroscopic resistance against shear arises from the mutual impenetrability of the polymers - to deform a test filament, surrounding filaments need to be pushed out of the way, as familiar from knotted strings. The mathematical problem posed by highly entangled solutions is sufficiently complicated that it eludes a rigorous solution (Edwards, 1967; Everaers et al., 2004), yet, the tube model provides a simple and successful phenomenological description of their macroscopic properties.

4.1 Tube model for flexible polymers

The tube model was introduced by S. F. Edwards for melts and solutions of flexible polymers (Edwards, 1967; Doi & Edwards, 1988). The idea is to circumvent the explicit discussion of the complicated topological constraints and to represent them, effectively, by a harmonic potential for a test polymer, which is thus confined to a narrow tube-like cage. The polymer may escape its cage very slowly by a snake-like diffusive motion called reptation (de Gennes, 1971; Doi & Edwards, 1988; Schweizer et al., 1997). Another mechanism consists in the sudden release of constraints, when the ends of confining polymers slide past the test polymer. The reptation time τ_d is very sensitive on the contour length L. From the longitudinal diffusion coefficient $D_{\parallel} \propto L^{-1}$ one estimates it as $\tau_d \simeq L^2/D_{\parallel} \propto L^3$ for large L (de Gennes, 1971).

4.2 Tube model for semiflexible polymers

The principle of the tube model also applies to semiflexible polymers, with the main difference being the presence of an additional scale, the persistence length ℓ_p. It gives rise to a tightly entangled regime (Morse, 1998a; Uchida et al., 2008), where ℓ_p is larger than the mesh size ξ defined as $\xi \equiv \rho^{-1/2}$. Here, $\rho \equiv c_p L$ denotes the line concentration and c_p the polymer concentration. In a tightly entangled solution, the transverse bending undulations of all polymers are confined. The typical time scale τ_e on which they equilibrate and thus constitute the tube is on the order of milliseconds in actin solutions (Semmrich et al., 2007). On the other hand, typical semiflexible actin filaments have tube renewal times τ_d on the order of hours (Käs et al., 1994; He et al., 2007). Thus τ_d is much longer than the time needed by the transverse bending undulations to explore the cage and one can treat the polymer solution as an effective equilibrium system with a quenched network topology on intermediate time scales $\tau_e \ll t \ll \tau_d$.

The success of the tube model relies on the fact that it requires only a small set of parameters to predict most properties of the rheology and the dynamics quantitatively. These parameters can be inferred from an analysis of the topological problem in simulations (Everaers et al., 2004) or from simple self-consistent scaling arguments. More specifically, the tightly entangled state of semiflexible polymer solutions is characterized by the mean-field tube radius R and entanglement length $L_e \simeq R^{2/3}\ell_p^{1/3}$ of a test polymer, referring to the

magnitude of the confined transverse undulations of a WLC (see Eq. (5)) and the contour length between collisions, respectively (see Fig. 5, left) (Odijk, 1983; Semenov, 1986). The dependence of the values of R and L_e on monomer concentration $c \propto \rho$ is readily estimated by considering the binary collision of a test polymer with an obstacle filament. The average number of collisions per entanglement segment is on the order of one, which gives $c_p L L_e R \simeq 1$. Defining L_e as above, one then self-consistently obtains the tube radius $R \simeq \rho^{-3/5} \ell_p^{-1/5}$ and the entanglement length $L_e \simeq \rho^{-2/5} \ell_p^{1/5}$ (Semenov, 1986).
Macroscopic properties such as the plateau shear modulus G_0 or the osmotic compression modulus $\Pi(\rho)$ are expressed in terms of these fundamental parameters, and from the latter they inherit their characteristic concentration-dependence. Assuming that each collision of the test polymer with its tube contributes a free energy $k_B T$, the confinement free energy per filament scales as (Odijk, 1983; Isambert & Maggs, 1996; Burkhardt, 1995)

$$F \simeq \frac{k_B T L}{L_e} \sim \rho^{2/5}.$$

From this, the plateau shear modulus and the osmotic compression modulus are estimated as $G_0(\rho) \simeq \Pi(\rho) \equiv -N\partial F / \partial V \simeq k_B T \rho / L_e \propto c^{7/5}$ (Isambert & Maggs (1996); Morse (1998b)). This characteristic dependence on concentration has indeed been observed experimentally (Hinner et al., 1998; Tassieri et al., 2008; Vincent et al., 2007).
The tube model predicts a high-frequency limiting form of the complex, frequency-dependent shear modulus, resulting from the dynamic response of longitudinal fluctuations (see Sec. 2.2.3) (Morse, 1998c; Gittes & MacKintosh, 1998),

$$G^{\star}(\omega) = \frac{1}{15} \kappa \rho \ell_p (-2i\zeta / \kappa)^{3/4} \omega^{3/4}. \tag{14}$$

This high-frequency modulus has been confirmed experimentally (Gittes et al., 1997; Gisler & Weitz, 1999; Koenderink et al., 2006).

4.2.1 Microscopic models for the tube

The intuition associated with the tube model has been confirmed in single-molecule experiments. Because of the relatively larger dimensions of F-actin compared to flexible polymers, the tube around a single actin filament in entangled solution can directly be visualized microscopically using fluorescence labeling techniques (Käs et al., 1994). The possibility to measure the tube radius directly has spurred the development of quantitative tube models, based on a self-consistent binary collision approximation (BCA) (Morse, 2001; Hinsch et al., 2007) and an effective medium approximation (EMA) (Morse, 2001). These models predict the value of the tube radius relying on an analysis of the entanglement topology. For example, the BCA considers pairwise collisions of two weakly bending rods in a simple binary topology - "above" or "below". In a self-consistent calculation, the tube's strength is calculated as the cumulative effect of pair collisions in all possible configurations. The EMA, on the other hand, only discusses the average effect of the topological interactions. In this approximation, the polymer is coupled to an effective elastic background medium. Both theories give rise to conflicting predictions for the concentration-dependence of R and L_e, opening a debate on the appropriate theoretical description of semiflexible polymer solutions, and they have therefore been challenged by experiments

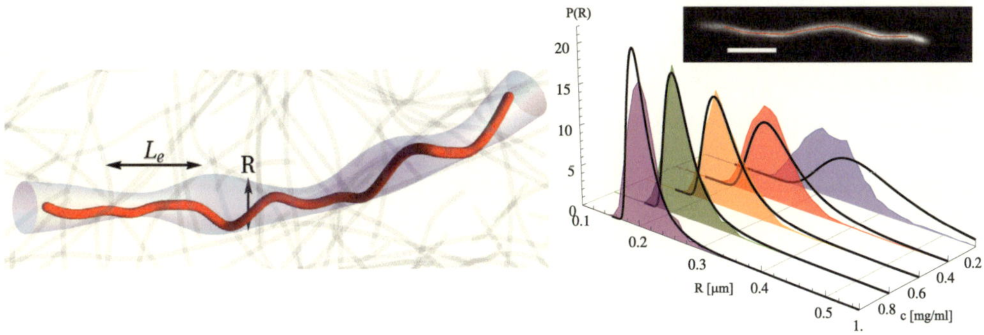

Fig. 5. *Left:* A stiff polymer confined in a tube of spatially varying radius R; L_e: entanglement length. Background polymers are depicted in gray. *Right:* Tube radius distribution $P(R)$ measured in entangled solutions of F-actin at different concentrations (shaded areas). Solid lines represent a global fit by a segment fluid theory. *Inset*: Superimposed confocal microscopy images of a fluorescent actin filament in a background solution and a spline representing the tube backbone; scale bar: 5μm. The width of the tube is inferred from Gaussian fits to the transverse intensity profile. From Glaser et al. (2010). Copyright © 2010 by The American Physical Society.

and simulations, which have either been in favor of the BCA (Romanowska et al., 2009; Wang et al., 2010; Ramanathan & Morse, 2007) or of the EMA (Tassieri et al., 2008). This controversy results in part from the close match of the scaling exponents in the BCA ($R \propto c^{-3/5}$) and the EMA ($R \propto c^{-1/2}$), suggesting that an unambiguous distinction between the two power-laws is extremely difficult to establish experimentally (Tassieri et al., 2008). In addition, respective conclusions must be drawn with care, since experiments typically yield skewed distributions of the tube radius, rendering an interpretation in terms of mere average values problematic (Wang et al., 2010; Glaser et al., 2010).

4.2.2 Tube width fluctuations

Indeed, experiments (Käs et al., 1994; Dichtl & Sackmann, 1999; Romanowska et al., 2009; Wang et al., 2010) and simulations (Hinsch et al., 2007) indicated that the assumption of a uniform tube width is a severe approximation. Pronounced tube fluctuations along a single actin filament have been measured by fluorescence microscopy (Glaser et al., 2010), as sketched in Fig. 5, left, and they can be analyzed by a systematic BCA-based theory. This theory extends D. Morse's mean-field approach and allows for a comprehensive characterization of the microstructure of entangled solutions. It predicts a tube radius distribution $P(R)$ quantifying the observed heterogeneities and compares favorably with experiments (see Fig. 5, right). An analysis of small remaining discrepancies in this comparison attributes them to collective modes of the effective medium. This suggests a way how to combine the BCA and the EMA to achieve a practically perfect explanation of the experimental data in the future (Glaser, 2010).

4.3 Recent developments and open problems

We now discuss experimental findings which, at first sight, seem to contradict the predictions of the tube model. For example, in experiments, the curvature distribution of the tube's primitive path has been measured and large curvatures occurred with a higher

frequency than expected for a free polymer (Romanowska et al., 2009). This was corroborated by computer simulations, where the same effect has been shown to occur when filament ends were allowed to move freely and to slide past fixed obstacles (Hinsch & Frey, 2009), suggesting that one may interpret it as a consequence of finite filament length.
A further observation is the slow logarithmic increase of the tube radius with time (see Sec. 3.3 and Sec. 6 below), which leads to an effective "softening" of the tube. In practice, however, this time dependence is weak, and, depending on the application, it may be accounted for by measuring an effectively saturated value of the tube radius.
Finally, let us comment on the nonlinear rheology. The response of physically entangled solutions to nonlinear strains is predicted by a nonlinear tube model, in which the tube is allowed to compress or expand (Morse, 1999; Fernández et al., 2009). The predicted universal strain-softening response is in contrast with the gradual transition from softening to stiffening in actin solutions (see Sec. 3.4 and Fig. 4, right), which has been interpreted as a consequence of weak crosslinking. On the other hand, the parameter changes that controlled this transition had little or no effect on the linear shear modulus (Semmrich et al., 2007; 2008). One might interpret this and the results of this section in the sense that the tube model has been validated for the linear response regime, while its predictions are overshadowed by (spurious) adhesion and crosslinking effects in most non-linear measurements at finite shear rates.
In summary, the tube model provides a detailed quantitative explanation of the mechanical properties of entangled polymer solutions. It reveals the important role of topological interactions in simple reconstituted cytoskeletal systems and serves as a well-defined reference description for studies of nontrivial dynamic and nonlinear effects. In a further step towards complexity, crosslinkers may be added.

5. Affine network model

The phenomenology of cytoskeletal networks with crosslinkers is broad and depends on a multitude of parameters, such as crosslinker type (rigid or flexible), crosslinker time scale (on/off-rate) and crosslinker/filament ratio (weakly or strongly crosslinked). After discussing elementary affine and non-affine deformation mechanisms, we give an outline of the affine network model as a simple zeroth order explanation for many of the observed effects in crosslinked networks, and we review progress on their understanding beyond the simplifying assumption of affine deformations.
Crosslinkers mediate local interactions between polymers. For the protein filament meshworks that make up the cytoskeleton, their action relies on specific binding sites on the polymers, in other words, they induce short-ranged, "patchy" attractions. In cases of extremely long bond lifetimes, these attractions may be modeled as geometric constraints (in addition to the above mentioned topological constraints due to the impenetrability of the polymer backbones). One commonly accounts for the presence of crosslinkers by introducing a new characteristic length scale L_c, representative of the mean distance between crosslinking sites along a single filament (MacKintosh, 2006), which is distinct from the entanglement length L_e and the geometrical mesh size ξ. On the level of a single filament, two different modes of deformation exist: longitudinal stretching/compression and transverse bending. Since, for rod-like networks, simple shear amounts to rotation and stretching/compression of filaments, only these latter modes should be relevant in a purely affine deformation. On the other hand, cooperative deformation mechanisms may result in

non-affine deformations involving bending of filaments. If the response to shear was dominated by non-affine filament bending, the plateau modulus of a densely crosslinked network approximated as a simple cubic mesh of semiflexible filaments with $\xi \simeq L_c$ would be expected to scale as (Kroy & Frey, 1996)

$$G_0 \simeq \frac{k_B T \ell_p}{\xi^4} \propto k_B T \ell_p c^2. \tag{15}$$

In the case of purely affine stretching, the modulus of the network originates from the mechanical response of a single WLC of length L_c. Stretching it an amount $\delta\ell \equiv \gamma L_c$ in the linear regime requires a force $f \simeq k_B T \ell_p^2 \delta\ell / L_c^4$ (MacKintosh et al., 1995). Defining the stress as $\sigma = f / \xi^2$, the shear modulus follows as

$$G_0 \equiv \frac{\sigma}{\gamma} \simeq \frac{k_B T \ell_p^2}{L_c^3 \xi^2}. \tag{15}$$

This is the rubber elasticity modulus of the affine semiflexible network model, which depends on filament and crosslinker concentration via the mesh size ξ and the cross-linker distance L_c. Using a plausible (ad-hoc) parametrization of the latter, agreement between Eq. (15) and experimental data for actin (Gardel et al., 2004; Tharmann et al., 2007) and intermediate filament networks (Lin et al., 2010) can be obtained.

The corresponding high-frequency modulus $G^\star(\omega)$ may be estimated from the rubber elastic modulus Eq. (15) in close analogy to the single-polymer results of Sec. 2.2.3, by replacing L_c with the dynamic equilibration length $\ell_\perp(t)$ for weak forces, evaluated at $t = i\omega$, yielding $G^\star(\omega) \simeq k_B T \ell_p^2 / \xi^2 \ell_\perp^3(t = i\omega) \propto \omega^{3/4}$. The exact asymptotic form of this complex viscoelastic high-frequency modulus is again given by Eq. (14), which includes prefactors and which constitutes a universal asymptotic result independent of the crosslinker or affinity length scale L_c. It has been verified for crosslinked networks and entangled solutions of F-actin (Koenderink et al., 2006).

The affine network model provides a simple natural explanation for the observed strain stiffening response observed for *in-vitro* networks and cells (see Sec.3.4) in terms of the nonlinear force-extension relation of a single WLC (see Sec. 2.1.4). Since it follows from Eq. (8) that $f \propto [L - R(f)]^{-2}$, the nonlinear differential modulus is thus obtained as (see Sec. 2.1.4) (MacKintosh et al., 1995; MacKintosh, 2006)

$$K' \equiv \frac{d\sigma}{d\gamma} \propto \frac{df}{dR} \propto f^{3/2} \propto \sigma^{3/2}. \tag{16}$$

Let us now turn to the question under which conditions the affine or non-affine deformation mechanisms prevail. This problem has been studied by simulations of two-dimensional crosslinked random fiber networks in the mechanical limit (Wilhelm & Frey, 2003; Head et al., 2003), and it was found that the transition from the non-affine bending to the affine stretching regime occurs for high crosslink densities and/or long fibers. The afine to non-affine transition of mechanical fibers has also been observed in regular networks based on a triangular geometry (Das et al., 2007). Strain field visualization in F-actin networks indeed seems to provide some evidence for a cross-over between distinct deformation modes,

depending on the ratio of polymer length to a characteristic non-affinity length scale (Liu et al., 2007). Heussinger and Frey have however shown by taking into account thermal fluctuations that affine deformations are unstable and non-affine intermediate asymptotic scaling regimes of the shear modulus dominate the mechanical response instead (Heussinger & Frey, 2006a;b). These are characterized by filament-filament correlations, which are not taken into account in the affine network model. Further simulation studies show that the strain field is in general non-affine (Onck et al., 2005), and that homogeneously crosslinked networks are softer in the linear regime and stiffen at higher strains than predicted by the affine network model (Huisman et al., 2008).
Moreover, the stiffening exponent of Eq. (16) is not universal (see also Sec. 3.4), as for example in filamin-crosslinked networks, strain-stiffening with an exponent close to one is observed (Gardel et al., 2006). Thus, models for flexible crosslinkers such as filamin have also been discussed. These may either be represented as a series of domains capable of unfolding (DiDonna & Levine, 2006), or as inextensible wormlike chains (Broedersz et al., 2008), leading to a softening or stiffening response, respectively. Recently theoretically studied problems also include thermodynamic properties and complex phase diagrams of polymer networks (Borukhov et al., 2005; Benetatos & Zippelius, 2007). An interesting interplay between single polymer and crosslinker dynamics arises in transiently connected networks, where the kinetics of the crosslinker leads to an additional viscous dissipation mechanism (Lieleg et al., 2008; Wolff et al., 2010) (see Sec. 6 below).

6. Towards viscoelastic and inelastic dynamics - the glassy wormlike chain

A study of semiflexible polymer dynamics in purely entangled solutions revealed that instead of a strict tube constraint, an additional relaxation mechanism exists at long times. More specifically, simulations have shown that the tube potential softens with time (Zhou & Larson, 2006; Ramanathan & Morse, 2007), implying a time-dependent tube radius. Analogous observations in flexible polymer melts have been attributed to constraint release (Vaca Chávez & Saalwächter, 2010). The dynamic structure factor of entangled F-actin solutions exhibited slow logarithmic decay beyond the cross-over time from free to confined polymer modes, extending over five decades in time (see Fig. 6, right, black curves) (Semmrich et al., 2007). The glassy wormlike chain (GWLC) model interprets this by an exponentially stretched relaxation time spectrum of the normal modes of an ordinary WLC on time scales longer than the equilibration time $\tau_\Lambda \simeq \tau_e$ (Kroy & Glaser, 2007; Kroy, 2008). This is accomplished by prescribing a modified relaxation time τ_λ for collective excitations of wavelength λ by

$$\tau_\lambda \to \tilde{\tau}_\lambda = \begin{cases} \tau_\lambda, & \lambda < \Lambda \\ \tau_\lambda \exp[\varepsilon(\lambda/\Lambda - 1)], & \lambda > \Lambda. \end{cases}$$

All relaxation times of bending modes of a wavelength longer than the characteristic interaction or bond length Λ - which may be different from the entanglement length (Glaser et al., 2008) - are thus multiplied by an Arrhenius factor involving the energy $\mathcal{E}k_BT$, and the number $\lambda/\Lambda{-}1$ of interactions. The Arrhenius factors have been attributed to free energy barriers arising either from finite free energy costs for tube deformations or from sticky interactions. Using this prescription in explicit expressions for the dynamic structure factor

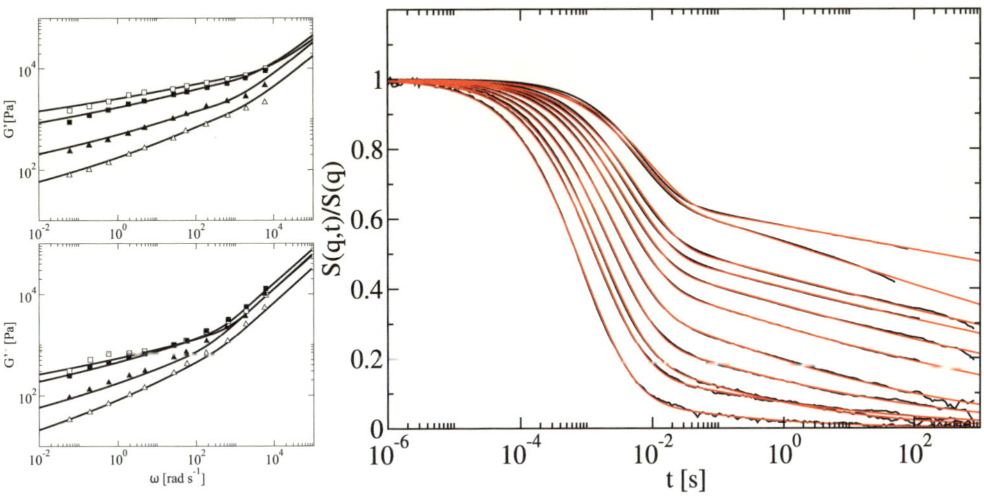

Fig. 6. *Left:* Storage and loss modulus $G'(\omega)$ and $G''(\omega)$ of human airway smooth muscle cells (symbols), for different pharmacological treatments. Solid lines are fits to the GWLC model described in Sec. 6. From Kroy & Glaser (2009). Copyright © 2009 by American Institute of Physics. Experimental data from Fabry et al. (2001). *Right*: Dynamic structure factor $S(q, t)$ of entangled F-actin solutions at various wave vectors q (black curves) fitted by the GWLC model (red curves) proposed in Sec. 6. Adapted from Semmrich et al. (2007). Copyright © 2007 by The National Academy of Sciences of the USA.

of stiff polymer solutions (Kroy & Frey, 1997; Glaser et al., 2008) yields excellent agreement of the model predictions with the data (Fig. 6, right, red curves). The "macrorheological modulus" $G^{\star}(\omega)$ of a GWLC is obtained by combining the affine network model of Sec. 5 with the stretched relaxation times of the GWLC. It exhibits (near) power-law rheology, in very good agreement with rheological data on live cells, see Fig. 6, left (Kroy & Glaser, 2009). Concerning the nonlinear response, the model accounts for the mutually opposite influences of stiffening due to backbone tension and the lowering of the energy barrier height $\mathcal{E}$ under the influence of an external force. Hence, the nonlinear differential shear modulus of a GWLC shows a variable degree of stiffening depending on the value of $\mathcal{E}$, followed by a softening response at high strains, which is in qualitative agreement with the experimental observations shown in Fig. 4, right.

The GWLC thus provides an efficient phenomenological description of slow dynamics in entangled solutions of stiff polymers, and conceptually, it might also be applied to weakly crosslinked networks. Indeed, the model has recently been extended to account for bond kinetics with defined rates and under the influence of a force (Wolff et al., 2010). The corresponding model for inelastic deformations in sticky polymer solutions evaluates the macroscopic shear modulus of the GWLC using a dynamic inter-bond distance $\Lambda(t)$ that accounts for the slow inelastic network evolution. In this scheme, it is straightforward to include the effect of prestress or backbone tension. It is then possible to account for the experimentally observed fluidiziation and recovery of live cells after transient shear deformations (see Fig. 4, left).

7. Summary and conclusion

The study of single biopolymers provides the basic knowledge necessary to describe their complex collective effects in a "bottom-up approach". The starting point of a description of stiff polymers is the wormlike chain model for a single semiflexible polymer, which already exhibits a rich phenomenology. Entangled solutions of stiff polymers are appropriately described by the tube model. In particular, we discussed heterogeneities of the tube width. The affine network model provides a simple explanation for the mechanics of densely crosslinked networks. Models of the viscoelastic and inelastic response of weakly and transiently crosslinked networks substantially extend prevailing theoretical approaches for cytoskeletal networks. Clearly, the development of mathematical toy models and systematic theories will remain a crucial element in the approach to a microscopic understanding of the origins of the remarkable mechanical properties of living matter.

8. Acknowledgments

We would like to thank Andrea Kramer for preparing Figs. 1 and 3, and Sebastian Schöbl for providing the data in Fig. 2. We acknowledge support by the Deutsche Forschungsgemeinschaft (DFG) through FOR 877 and the Leipzig School of Natural Sciences - Building with Molecules and Nano-objects.

9. References

Ajdari, A., Jülicher, F. & Maggs, A. (1997). Pulling on a Filament, *J. Phys. (France)* 7(7): 823–826.

Alim, K. & Frey, E. (2007). Shapes of Semiflexible Polymer Rings, *Phys. Rev. Lett.* 99(19): 198102.

Baczynski, K., Lipowsky, R. & Kierfeld, J. (2007). Stretching of buckled filaments by thermal fluctuations, *Phys. Rev. E: Stat., Nonlinear, Soft Matter Phys.* 76(6): 061914.

Bao, G. & Suresh, S. (2003). Cell and molecular mechanics of biological materials, *Nat. Mater.* 2(11): 715–25.

Bausch, A. & Kroy, K. (2006). A bottom-up approach to cell mechanics, *Nat. Phys.* 2(4): 231–238.

Benetatos, P. & Zippelius, A. (2007). Anisotropic Random Networks of Semiflexible Polymers, *Phys. Rev. Lett.* 99(19): 198301.

Borukhov, I., Bruinsma, R. F., Gelbart, W. M. & Liu, A. J. (2005). Structural polymorphism of the cytoskeleton: a model of linker-assisted filament aggregation., *Proc. Natl. Acad. Sci. U. S. A.* 102(10): 3673–8.

Brangwynne, C., Koenderink, G., MacKintosh, F. & Weitz, D. (2008). Nonequilibrium Microtubule Fluctuations in a Model Cytoskeleton, *Phys. Rev. Lett.* 100(11): 118104.

Brangwynne, C. P., MacKintosh, F. C., Kumar, S., Geisse, N. A., Talbot, J., Mahadevan, L., Parker, K. K., Ingber, D. E. & Weitz, D. A. (2006). Microtubules can bear enhanced compressive loads in living cells because of lateral reinforcement, *J. Cell. Biol.* 173(5): 733–41.

Broedersz, C., Storm, C. & MacKintosh, F. (2008). Nonlinear Elasticity of Composite Networks of Stiff Biopolymers with Flexible Linkers, *Phys. Rev. Lett.* 101(11): 118103.

Burkhardt, T. (1995). Free energy of a semiflexible polymer confined along an axis, *J. Phys. A.: Math. Gen.* 28: L629–635.

Bursac, P., Lenormand, G., Fabry, B., Oliver, M., Weitz, D. A., Viasnoff, V., Butler, J. P. & Fredberg, J. J. (2005). Cytoskeletal remodelling and slow dynamics in the living cell, *Nat. Mater.* 4(7): 557–61.

Bustamante, C., Bryant, Z. & Smith, S. (2003). Ten years of tension: single-molecule DNA mechanics, *Nature* 421(6921): 423–427.

Bustamante, C., Marko, J., Siggia, E. & Smith, S. (1994). Entropic Elasticity of lambda-Phage DNA, *Science* 265(5178): 1599–1600.

Caliskan, G., Hyeon, C., Perez-Salas, U., Briber, R., Woodson, S. & Thirumalai, D. (2005). Persistence Length Changes Dramatically as RNA Folds, *Phys. Rev. Lett.* 95(26): 268303.

Chaikin, P. & Lubensky, T. (1995). *Principles of condensed matter physics*, Cambridge University Press, Cambridge.

Daniels, H. (1950). The statistical theory of stiff chains, *Proc. Roy. Soc. Edinburgh* 63: 290–311.

Das, M., MacKintosh, F. & Levine, A. (2007). Effective Medium Theory of Semiflexible Filamentous Networks, *Phys. Rev. Lett.* 99(3): 038101.

de Gennes, P. G. (1971). Reptation of a Polymer Chain in the Presence of Fixed Obstacles, *J. Chem. Phys.* 55(2): 572-579.

Dichtl, M. & Sackmann, E. (1999). Colloidal probe study of short time local and long time reptational motion of semiflexible macromolecules in entangled networks, *New J. Phys.* 1: 18.

DiDonna, B. & Levine, A. (2006). Filamin Cross-Linked Semiflexible Networks: Fragility under Strain, *Phys. Rev. Lett.* 97(6): 068104.

Discher, D., Dong, C., Fredberg, J. J., Guilak, F., Ingber, D., Janmey, P., Kamm, R. D., Schmid-Schönbein, G. W. & Weinbaum, S. (2009). Biomechanics: cell research and applications for the next decade, *Ann. Biomed. Eng.* 37(5): 847–59.

Doi, M. & Edwards, S. F. (1988). *The Theory of Polymer Dynamics*, Oxford University Press, Oxford.

Edwards, S. F. (1967). The statistical mechanics of polymerized material, *Proc. Phys. Soc. London* 92(1): 9–16.

Emanuel, M., Mohrbach, H., Sayar, M., Schiessel, H. & Kulić, I. (2007). Buckling of stiff polymers: Influence of thermal fluctuations, *Phys. Rev. E: Stat., Nonlinear, Soft Matter Phys.* 76(6): 061907.

Everaers, R., Jülicher, F., Ajdari, A. & Maggs, A. (1999). Dynamic Fluctuations of Semiflexible Filaments, *Phys. Rev. Lett.* 82(18): 3717–3720.

Everaers, R., Sukumaran, S. K., Grest, G. S., Svaneborg, C., Sivasubramanian, A. & Kremer, K. (2004). Rheology and microscopic topology of entangled polymeric liquids, *Science* 303(5659): 823–6.

Fabry, B., Maksym, G., Butler, J., Glogauer, M., Navajas, D. & Fredberg, J. (2001). Scaling the Microrheology of Living Cells, *Phys. Rev. Lett.* 87(14): 148102.

Farge, E. & Maggs, A. C. (1993). Dynamic scattering from semiflexible polymers, *Macromolecules* 26(19): 5041–5044.

Fernández, P., Grosser, S. & Kroy, K. (2009). A unit-cell approach to the nonlinear rheology of biopolymer solutions, *Soft Matter* 5(10): 2047-2056.

Fernández, P. & Ott, A. (2008). Single Cell Mechanics: Stress Stiffening and Kinematic Hardening, *Phys. Rev. Lett.* 100(23): 238102.

Fernández, P., Pullarkat, P. A. & Ott, A. (2006). A master relation defines the nonlinear viscoelasticity of single fibroblasts, *Biophys. J.* 90(10): 3796–805.

Fixman, M. & Kovac, J. (1973). *Polymer conformational statistics: III. Modified Gaussian models of stiff chains, J. Chem. Phys.* 58 (4):1564-1568.

Fletcher, D. A. & Geissler, P. L. (2009). Active biological materials, *Annu. Rev. Phys. Chem.* 60: 469–86.

Fletcher, D. A. & Mullins, R. D. (2010). Cell mechanics and the cytoskeleton, *Nature* 463(7280): 485–92.

Gardel, M. L., Nakamura, F., Hartwig, J. H., Crocker, J. C., Stossel, T. P. & Weitz, D. a. (2006). Prestressed F-actin networks cross-linked by hinged filamins replicate mechanical properties of cells., *Proc. Natl. Acad. Sci. U. S. A.* 103(6): 1762–7.

Gardel, M. L., Shin, J. H., MacKintosh, F. C., Mahadevan, L., Matsudaira, P. & Weitz, D. A. (2004). Elastic behavior of cross-linked and bundled actin networks, *Science* 304(5675): 1301–5.

Gisler, T. & Weitz, D. (1999). Scaling of the Microrheology of Semidilute F-Actin Solutions, *Phys. Rev. Lett.* 82(7): 1606–1609.

Gittes, F. & MacKintosh, F. (1998). Dynamic shear modulus of a semiflexible polymer network, *Phys. Rev. E: Stat., Nonlinear, Soft Matter Phys.* 58(2): R1241–R1244.

Gittes, F., Schnurr, B., Olmsted, P., MacKintosh, F. & Schmidt, C. (1997). Microscopic Viscoelasticity: Shear Moduli of Soft Materials Determined from Thermal Fluctuations, *Phys. Rev. Lett.* 79(17): 3286–3289.

Glaser, J. (2010). Unpublished.

Glaser, J., Chakraborty, D., Kroy, K., Lauter, I., Degawa, M., Kirchgeßner, N., Hoffmann, B., Merkel, R. & Giesen, M. (2010). Tube Width Fluctuations in F-Actin Solutions, *Phys. Rev. Lett.* 105(3): 037801.

Glaser, J., Hallatschek, O. & Kroy, K. (2008). Dynamic structure factor of a stiff polymer in a glassy solution., *Eur. Phys. J. E, Soft Matter* 26(1-2): 123–36.

Goldstein, R. & Langer, S. (1995). Nonlinear Dynamics of Stiff Polymers, *Phys. Rev. Lett.* 75(6): 1094–1097.

Granek, R. (1997). From Semi-Flexible Polymers to Membranes: Anomalous Diffusion and Reptation, *J. Phys. II* 7(12): 1761–1788.

Hallatschek, O., Frey, E. & Kroy, K. (2005). Propagation and Relaxation of Tension in Stiff Polymers, *Phys. Rev. Lett.* 94(7): 077804.

Hallatschek, O., Frey, E. & Kroy, K. (2007a). Tension dynamics in semiflexible polymers. I. Coarse-grained equations of motion, *Phys. Rev. E: Stat., Nonlinear, Soft Matter Phys.* 75(3): 031905.

Hallatschek, O., Frey, E. & Kroy, K. (2007b). Tension dynamics in semiflexible polymers. II. Scaling solutions and applications, *Phys. Rev. E: Stat., Nonlinear, Soft Matter Phys.* 75(3): 031906.

Hartwell, L. H., Hopfield, J. J., Leibler, S. & Murray, A. W. (1999). From molecular to modular cell biology, *Nature* 402(6761 Suppl): C47–52.

He, J., Viamontes, J. & Tang, J. (2007). Counterion-Induced Abnormal Slowdown of F-Actin Diffusion across the Isotropic-to-Nematic Phase Transition, *Phys. Rev. Lett.* 99(6): 068103.

Head, D., Levine, A. & MacKintosh, F. (2003). Deformation of Cross-Linked Semiflexible Polymer Networks, *Phys. Rev. Lett.* 91(10): 108102.

Heussinger, C., Bathe, M. & Frey, E. (2007). Statistical Mechanics of Semiflexible Bundles of Wormlike Polymer Chains, *Phys. Rev. Lett.* 99(4): 048101.

Heussinger, C. & Frey, E. (2006a). Floppy Modes and Nonaffine Deformations in Random Fiber Networks, *Phys. Rev. Lett.* 97(10): 105501.

Heussinger, C. & Frey, E. (2006b). Stiff Polymers, Foams, and Fiber Networks, *Phys. Rev. Lett.* 96(1): 017802.

Hinner, B., Tempel, M., Sackmann, E., Kroy, K. & Frey, E. (1998). Entanglement, Elasticity, and Viscous Relaxation of Actin Solutions, *Phys. Rev. Lett.* 81(12): 2614–2617.

Hinsch, H. & Frey, E. (2009). Conformations of entangled semiflexible polymers: entropic trapping and transient non-equilibrium distributions, *ChemPhysChem* 10(16): 2891–9.

Hinsch, H., Wilhelm, J. & Frey, E. (2007). Quantitative tube model for semiflexible polymer solutions, *Eur. Phys. J. E, Soft Matter* 24(1): 35–46.

Hiraiwa, T. & Ohta, T. (2008). Viscoelastic Behavior of a Single Semiflexible Polymer Chain, *J. Phys. Soc. Jpn.* 77(2): 023001.

Hiraiwa, T. & Ohta, T. (2009). Viscoelasticity of a Single Semiflexible Polymer Chain, *Macromolecules* 42(19): 7553–7562.

Hoffman, B. D. & Crocker, J. C. (2009). Cell mechanics: dissecting the physical responses of cells to force, *Annu. Rev. Biomed. Eng.* 11: 259–88.

Hoffman, B. D., Massiera, G., Van Citters, K. M. & Crocker, J. C. (2006). The consensus mechanics of cultured mammalian cells, *Proc. Natl. Acad. Sci. U. S. A.* 103(27): 10259–64.

Huisman, E., Storm, C. & Barkema, G. (2008). Monte Carlo study of multiply crosslinked semiflexible polymer networks, *Phys. Rev. E: Stat., Nonlinear, Soft Matter Phys.* 78(5): 051801.

Humphrey, D., Duggan, C., Saha, D., Smith, D. & Käs, J. (2002). Active fluidization of polymer networks through molecular motors, *Nature* 416(6879): 413–6.

Isambert, H. & Maggs, A. C. (1996). Dynamics and Rheology of Actin Solutions, *Macromolecules* 29(3): 1036–1040.

Isambert, H., Venier, P., Maggs, A., Fattoum, A., Kassab, R., Pantaloni, D. & Carlier, M. (1995). Flexibility of Actin Filaments Derived From Thermal Fluctuations, *J. Biol. Chem.* 270(19): 11437-11444.

Käs, J., Strey, H. & Sackmann, E. (1994). Direct imaging of reptation for semiflexible actin filaments, *Nature* 368: 226–229.

Kasza, K. E., Rowat, A. C., Liu, J., Angelini, T. E., Brangwynne, C. P., Koenderink, G. H. & Weitz, D. A. (2007). The cell as a material, *Curr. Opin. Cell Biol.* 19(1): 101–7.

Kirschner, M. & Gerhart, J. (1998). Evolvability, *Proc. Natl. Acad. Sci. USA* 95: 8420–8427.

Koenderink, G., Atakhorrami, M., MacKintosh, F. & Schmidt, C. (2006). High-Frequency Stress Relaxation in Semiflexible Polymer Solutions and Networks, *Phys. Rev. Lett.* 96(13): 138307.

Koenderink, G. H., Dogic, Z., Nakamura, F., Bendix, P. M., MacKintosh, F. C., Hartwig, J. H., Stossel, T. P. & Weitz, D. A. (2009). An active biopolymer network controlled by molecular motors, *Proc. Natl. Acad. Sci. U. S. A.* 106(36): 15192–7.

Kollmannsberger, P. (2009). *Nonlinear microrheology of living cells*, PhD thesis.

Kratky, O. & Porod, G. (1949). Röntgenuntersuchung gelöster Fadenmoleküle, *Rec. Trav. Chim. Pays-Bas* 68: 1105–1123.

Kroy, K. (2006). Elasticity, dynamics and relaxation in biopolymer networks, *Curr. Opin. Coll. Interface Sci.* 11(1): 56–64.

Kroy, K. (2008). Dynamics of wormlike and glassy wormlike chains, *Soft Matter* 4(12): 2323–2330.
Kroy, K. & Frey, E. (1996). Force-Extension Relation and Plateau Modulus for Wormlike Chains, *Phys. Rev. Lett.* 77(2): 306–309.
Kroy, K. & Frey, E. (1997). Dynamic scattering from solutions of semiflexible polymers, *Phys. Rev. E: Stat., Nonlinear, Soft Matter Phys.* 55(3): 3092–3101.
Kroy, K. & Glaser, J. (2007). The glassy wormlike chain, *New J. Phys.* 9(11): 416–416.
Kroy, K. & Glaser, J. (2009). Rheological redundancy - from polymers to living cells, *AIP Conf. Proc.* 1151: 52–55.
Landau, L. & Lifshitz, E. (1980). *Statistical Physics, Third Edition, Part 1: Volume 5 (Course of Theoretical Physics, Volume 5)*, Butterworth-Heinemann, Oxford.
Lau, A., Hoffman, B., Davies, A., Crocker, J. & Lubensky, T. (2003). Microrheology, Stress Fluctuations, and Active Behavior of Living Cells, *Phys. Rev. Lett.* 91(19):198101
Lieleg, O. & Bausch, A. (2007). Cross-linker unbinding and self-similarity in bundled cytoskeletal networks, *Phys. Rev. Lett.* 99(15): 158105.
Lieleg, O., Claessens, M., Luan, Y. & Bausch, A. (2008). Transient binding and dissipation in cross-linked actin networks, *Phys. Rev. Lett.* 101(10): 108101.
Lieleg, O., Claessens, M. M. A. E. & Bausch, A. R. (2010). Structure and dynamics of crosslinked actin networks, *Soft Matter* 6(2): 218-225.
Lin, Y., Yao, N., Broedersz, C., Herrmann, H., MacKintosh, F. & Weitz, D. (2010). Origins of Elasticity in Intermediate Filament Networks, *Phys. Rev. Lett.* 104(5): 58101.
Liu, A. P. & Fletcher, D. A. (2009). Biology under construction: in vitro reconstitution of cellular function, *Nat. Rev. Mol. Cell. Biol.* 10(9): 644–50.
Liu, J., Koenderink, G., Kasza, K., MacKintosh, F. & Weitz, D. (2007). Visualizing the Strain Field in Semiflexible Polymer Networks: Strain Fluctuations and Nonlinear Rheology of F-Actin Gels, *Phys. Rev. Lett.* 98(19): 198304.
MacKintosh, F. C. (2006). *Elasticity and dynamics of cytoskeletal filaments and their networks*, Taylor & Francis, London, chapter 8, pp. 139–155.
MacKintosh, F. C., Käs, J. & Janmey, P. A. (1995). Elasticity of Semiflexible Biopolymer Networks, *Phys. Rev. Lett.* 75(24): 4425–4428.
MacKintosh, F. C. & Schmidt, C. F. (2010). Active cellular materials, *Curr. Opin. Cell Biol.* 22(1): 29–35.
Marko, J. & Siggia, E. (1995). Stretching DNA, *Macromolecules* 28(26): 8759–8770.
Mizuno, D., Bacabac, R., Tardin, C., Head, D. & Schmidt, C. (2009). High-Resolution Probing of Cellular Force Transmission, *Phys. Rev. Lett.* 102(16): 168102.
Mizuno, D., Tardin, C., Schmidt, C. F. & Mackintosh, F. C. (2007). Nonequilibrium mechanics of active cytoskeletal networks, *Science* 315(5810): 370–3.
Morse, D. C. (1998a). Viscoelasticity of concentrated isotropic solutions of semiflexible polymers. 1. Model and stress tensor, *Macromolecules* 31(20): 7030–7043.
Morse, D. C. (1998b). Viscoelasticity of Concentrated Isotropic Solutions of Semiflexible Polymers. 2. Linear Response, *Macromolecules* 31(20): 7044–7067.
Morse, D. C. (1998c). Viscoelasticity of tightly entangled solutions of semiflexible polymers, *Phys. Rev. E: Stat., Nonlinear, Soft Matter Phys.* 58(2): R1237–R1240.
Morse, D. C. (1999). Viscoelasticity of Concentrated Isotropic Solutions of Semiflexible Polymers. 3. Nonlinear Rheology, *Macromolecules* 32: 5934–5943.
Morse, D. C. (2001). Tube diameter in tightly entangled solutions of semiflexible polymers, *Phys. Rev. E: Stat., Nonlinear, Soft Matter Phys.* 63(3): 031502.

Munk, T., Hallatschek, O., Wiggins, C. & Frey, E. (2006). Dynamics of semiflexible polymers in a flow field, *Phys. Rev. E: Stat., Nonlinear, Soft Matter Phys.* 74(4): 041911.

Obermayer, B. & Frey, E. (2009). Tension dynamics and viscoelasticity of extensible wormlike chains, *Phys. Rev. E: Stat., Nonlinear, Soft Matter Phys.* 80(4): 040801(R).

Obermayer, B. & Hallatschek, O. (2007). Coupling of Transverse and Longitudinal Response in Stiff Polymers, *Phys. Rev. Lett.* 99(9): 098302.

Obermayer, B., Hallatschek, O., Frey, E. & Kroy, K. (2007). Stretching dynamics of semiflexible polymers, *Eur. Phys. J. E, Soft Matter* 23(4): 375–88.

Obermayer, B., Möbius, W., Hallatschek, O., Frey, E. & Kroy, K. (2009). Freely relaxing polymers remember how they were straightened, *Phys. Rev. E: Stat., Nonlinear, Soft Matter Phys.* 79(2): 021804.

Odijk, T. (1983). On the Statistics and Dynamics of Confined or Entangled Stiff Polymers, *Macromolecules* 1344(16): 1340–1344.

Onck, P., Koeman, T., van Dillen, T. & van Der Giessen, E. (2005). Alternative Explanation of Stiffening in Cross-Linked Semiflexible Networks, *Phys. Rev. Lett.* 95(17): 178102.

Ostermeir, K., Alim, K. & Frey, E. (2010). Buckling of stiff polymer rings in weak spherical confinement, *Phys. Rev. E: Stat., Nonlinear, Soft Matter Phys.* 81(6): 061802.

Pasquali, M., Shankar, V. & Morse, D. (2001). Viscoelasticity of dilute solutions of semiflexible polymers, *Phys. Rev. E: Stat., Nonlinear, Soft Matter Phys.* 64(2): 020802(R).

Ramanathan, S. & Morse, D. C. (2007). Simulations of dynamics and viscoelasticity in highly entangled solutions of semiflexible rods, *Phys. Rev. E: Stat., Nonlinear, Soft Matter Phys.* 76(1): 010501(R).

Romanowska, M., Hinsch, H., Kirchgeßner, N., Giesen, M., Degawa, M., Hoffmann, B., Frey, E. & Merkel, R. (2009). Direct observation of the tube model in F-actin solutions: Tube dimensions and curvatures, *Europhys. Lett.* 86(2): 26003.

Saitô, N., Takahashi, K. & Yunoki, Y. (1967). The Statistical Mechanical Theory of Stiff Chains, *J. Phys. Soc. Jpn.* 22(1): 219–226.

Schmoller, K., Lieleg, O. & Bausch, A. (2008). Cross-Linking Molecules Modify Composite Actin Networks Independently, *Phys. Rev. Lett.* 101(11): 118102.

Schopferer, M., Bär, H., Hochstein, B., Sharma, S., Mücke, N., Herrmann, H. & Willenbacher, N. (2009). Desmin and vimentin intermediate filament networks: their viscoelastic properties investigated by mechanical rheometry, *J. Mol. Biol.* 388(1): 133–43.

Schweizer, K., Fuchs, M., Szamel, G., Guenza, M. & Tang, H. (1997). Polymer-mode-coupling theory of the slow dynamics of entangled macromolecular fluids, *Macromol. Theory Simul.* 6: 1037–1117.

Schwille, P. & Diez, S. (2009). Synthetic biology of minimal systems, *Crit. Rev. Biochem. Mol. Biol.* 44(4): 223.242.

Seifert, U., Wintz, W. & Nelson, P. (1996). Straightening of Thermal Fluctuations in Semiflexible Polymers by Applied Tension, *Phys. Rev. Lett.* 77(27): 5389–5392.

Semenov, A. N. (1986). Dynamics of Concentrated Solutions of Rigid-chain Polymers Part 1. -Brownian Motion of Persistent Macromolecules in Isotropic Solution, *J. Chem. Soc., Faraday Trans.* 82(2): 317-329.

Semmrich, C., Larsen, R. J. & Bausch, A. R. (2008). Nonlinear mechanics of entangled F-actin solutions, *Soft Matter* 4(8): 1675-1680.

Semmrich, C., Storz, T., Glaser, J., Merkel, R., Bausch, A. R. & Kroy, K. (2007). Glass transition and rheological redundancy in F-actin solutions, *Proc. Natl. Acad. Sci. U. S. A.* 104(51): 20199–203.

Sollich, P., Lequeux, F., Hébraud, P. & Cates, M. (1997). Rheology of Soft Glassy Materials, *Phys. Rev. Lett.* 78(10): 2020–2023.

Storm, C., Pastore, J., MacKintosh, F., Lubensky, T. & Janmey, P. (2005). Nonlinear elasticity in biological gels, *Nature* 435(7039): 191.194.

Stricker, J., Falzone, T. & Gardel, M. L. (2010). Mechanics of the F-actin cytoskeleton, *J. Biomech.* 43(1): 9–14.

Tassieri, M., Evans, R. M. L., Barbu-Tudoran, L., Khan, G. N., Trinick, J. & Waigh, T. A. (2008). Dynamics of semi-flexible polymer solutions in the highly entangled regime, *Phys. Rev. Lett.* 101: 198301.

Tharmann, R., Claessens, M. & Bausch, A. (2007). Viscoelasticity of Isotropically Cross-Linked Actin Networks, *Phys. Rev. Lett.* 98(8): 088103.

Trepat, X., Deng, L., An, S. S., Navajas, D., Tschumperlin, D. J., Gerthoffer,W. T., Butler, J. P. & Fredberg, J. J. (2007). Universal physical responses to stretch in the living cell, *Nature* 447(7144): 592–5.

Trepat, X., Lenormand, G. & Fredberg, J. J. (2008). Universality in cell mechanics, *Soft Matter* 4(9): 1750.

Tskhovrebova, L., Trinick, J., Sleep, J. & Simmons, R. (1997). Elasticity and unfolding of single molecules of the giant muscle protein titin, *Nature* 387: 308–312.

Uchida, N., Grest, G. S. & Everaers, R. (2008). Viscoelasticity and primitive path analysis of entangled polymer liquids: from F-actin to polyethylene, *J. Chem. Phys.* 128(4): 044902.

Vaca Chávez, F. & Saalwächter, K. (2010). NMR Observation of Entangled Polymer Dynamics: Tube Model Predictions and Constraint Release, *Phys. Rev. Lett.* 104(19): 198305.

Vincent, R., Pinder, D., Hemar, Y. & Williams, M. (2007). Microrheological studies reveal semiflexible networks in gels of a ubiquitous cell wall polysaccharide, *Phys. Rev. E: Stat., Nonlinear, Soft Matter Phys.* 76(3): 031909.

Wagner, B., Tharmann, R., Haase, I., Fischer, M. & Bausch, A. R. (2006). Cytoskeletal polymer networks: the molecular structure of cross-linkers determines macroscopic properties, *Proc. Natl. Acad. Sci. U. S. A.* 103(38): 13974–8.

Wang, B., Guan, J., Anthony, S. M., Bae, S. C., Schweizer, K. S. & Granick, S. (2010). Confining Potential when a Biopolymer Filament Reptates, *Phys. Rev. Lett.* 104(11): 118301.

Wilhelm, J. & Frey, E. (1996). Radial Distribution Function of Semiflexible Polymers, *Phys. Rev. Lett.* 77(12): 2581–2584.

Wilhelm, J. & Frey, E. (2003). Elasticity of Stiff Polymer Networks, *Phys. Rev. Lett.* 91(10): 108103.

Wolff, L., Fernandez, P. & Kroy, K. (2010). Inelastic mechanics of sticky biopolymer networks, *New J. Phys.* 12(5): 053024.

Yamakawa, H. (1971). *Modern Theory of Polymer Solutions*, Harper & Row, New York.

Zhou, Q. & Larson, R. (2006). Direct calculation of the tube potential confining entangled polymers, *Macromolecules* 39(19): 6737–6743.

Section C

27

Detect Structural Features of Asymmetric and Symmetric CH_2 and CH_3 Functional Groups and Their Ratio of Biopolymers Within Intact Tissue in Complex Plant System Using Synchrotron FTIRM and DRIFT Molecular Spectroscopy

Peiqiang Yu, PhD.
College of Agriculture and Bioresources, University of Saskatchewan
51 Campus Drive, Saskatoon, S7N 5A8
Canada

1. Introduction

1.1 Bright synchrotron radiation

Synchrotron light is photon beam, which is million times brighter than sunlight. A synchrotron facility is a giant particle accelerator that turns electrons into light (Wikipedia, 2010). How a Synchrotron Works? Electric Gun (EG: *Synchrotron Component 1*) in a synchrotron blasts out electrons. The electrons are accelerated to nearly speed of light by Linear Accelerator (LA: *Synchrotron Component* 2). A small synchrotron orbiting Booster Ring (BR: *Synchrotron Component* 3) increase their energy. Finally electron bunches enter Storage Ring (SR: *Synchrotron Component* 4). The bunches orbit ring controlled by serial magnets. Special magnet called "Undulator" causes electrons to rapidly change course and vast emitting energy in form of photons. Then photons enter Beamline (BL: *Synchrotron Component* 5) and directed to device called "monochromator" which selects specific wavelength required for each experiment. Finally, light focus onto sample at Experimental End Station (ES: *Synchrotron Component* 6), where scientists collect data to determine molecular structure (Dumas, 2003; Marinkovic et al., 2002; Miller, 2009; CLS, 2010; Wikipedia, 2010). The extremely bright synchrotron light makes it possible to detect biomaterial structure (make-up or conformation) at both molecular and cellular levels (Marinkovic & Chance, 2006; Miller & Dumas, 2006).

1.2 Probe molecular structure of biopolymer using synchrotron-based and DRIFT molecular spectroscopy

Snchrotron FTIR microspectroscopy has been developed as a rapid, direct, non-destructive and bioanalytical technique. This cutting-edge analytical technology is capable of exploring the molecular chemistry or structure of a biological tissue without destruction inherent structures at ultra-spatial resolutions (Dumas, 2003; Yu, 2004; Marinkovic & Chance, 2006; Miller & Dumas, 2006; Yu et al., 2008). This technique has been used for molecular imaging and mapping in plant/feed/food tissue (Yu et al., 2004; Marinkovic & Chance 2006; Miller &

Dumas 2006; Yu et al., 2007). The diffused reflectance infrared Fourier transform spectroscopy (DRIFT) molecular spectroscopy is capable of exploring the biopolymer conformation through molecular and functional group spectral analyses (Doiron et al., 2009a; Liu & Yu, 2010). It can be used to detect molecular structure changes in oil seeds affect by heat processing (Doiron et al., 2009b) and cereal grains after processing (eg bio-ethanol processing).

1.3 Objective of this article

In this article, a novel approach was introduced to study the molecular structural features of the asymmetric and symmetric CH_2 and CH_3 and their ratios of biopolymers in complex intact plant system. Recent progress and applications were reported on using synchrotron-based bioanalytical technique (SFTIRM) and DRIFT molecular spectroscopy to image ultraspatial distribution of intensities of the asymmetric and symmetric CH_2 and CH_3 functional groups in complex intact plant tissue and detect the effects of plant seed variety and bioethanol processing on plant-based biopolymer structure changes on a molecular basis in terms of the asymmetric and symmetric CH_2 and CH_3 functional groups.

2. Molecular structure and spectral features of biopolymer in complex pant system

2.1 Spectral characteristics of biopolymers in complex plant system

In complex plant system, the biopolymers which are highly related to nutrient availability, including protein, lipid, structure and non-structural carbohydrates (cellulosic compounds, starch). These biopolymers have very unique structures, therefore resulting in unique spectral bands for each biopolymer. Fig. 1. showed a typical ultra-spatially resolved synchrotron-based FTIRM spectrum in complex plant tissue (aleurone layer tissue of sorghum with pixel size: 10 μm × 10 μm) in the mid-IR region ca. 4000-800 cm^{-1}.

The function groups of biopolymers in complex plant system included N-H and O-H stretch, C-H groups stretch, amide I and II, C=O carbonyl, CHO and cellulosic compounds. The protein amide I frequency band centered at 1650 cm^{-1} and amide II centered at ca. 1550 cm^{-1}, both of which arise from specific stretching and bending vibrations within the protein backbone (Jackson et al., 1989; Harris & Chapman, 1992; Jackson & Mantsch, 1995, 1996). The cellulosic compound band is centered at ca. 1245 cm^{-1}. The non-structural carbohydrate of starch is at ca. 1025 cm^{-1} (Wetzel et al., 1998, 2003).

2.2 Asymmetric and symmetric CH_2 and CH_3 groups and their ratio in complex plant system

In Fig. 1, the functional groups include C-H stretching group bands. These groups include (1) Asymmetric CH_3 stretching band; (2) Asymmetric CH_2 stretching band; (3) Symmetric CH_3 stretching band; and (4) Symmetric CH_2 stretching band. Fig. 2 shows the CH_3 and CH_2 asymmetric and symmetric stretching band features in typical ultra-spatially resolved synchrotron-based FTIRM spectrum in complex plant tissue (aleurone tissue of sorghum, pixel size: 10 μm × 10 μm) in the C-H stretch region ca. 3015-2780 cm^{-1} showing C-H function groups of biopolymers: CH_3 asymmetric stretch at ca. 2958 cm^{-1}, CH_2 asymmetric stretch at ca. 2934 cm^{-1}, CH_3 symmetric stretch at ca. 2874 cm^{-1} and CH_2 symmetric stretch of acyl chains at ca. 2851 cm^{-1} (Wetzel & LeVine 1999; Jackson & Mantsch, 2000). The ratios of the infrared absorbance intensity peak height or area are obtained by the height or area under one functional group band divided by the height or area under another functional group band (Yu et al., 2005).

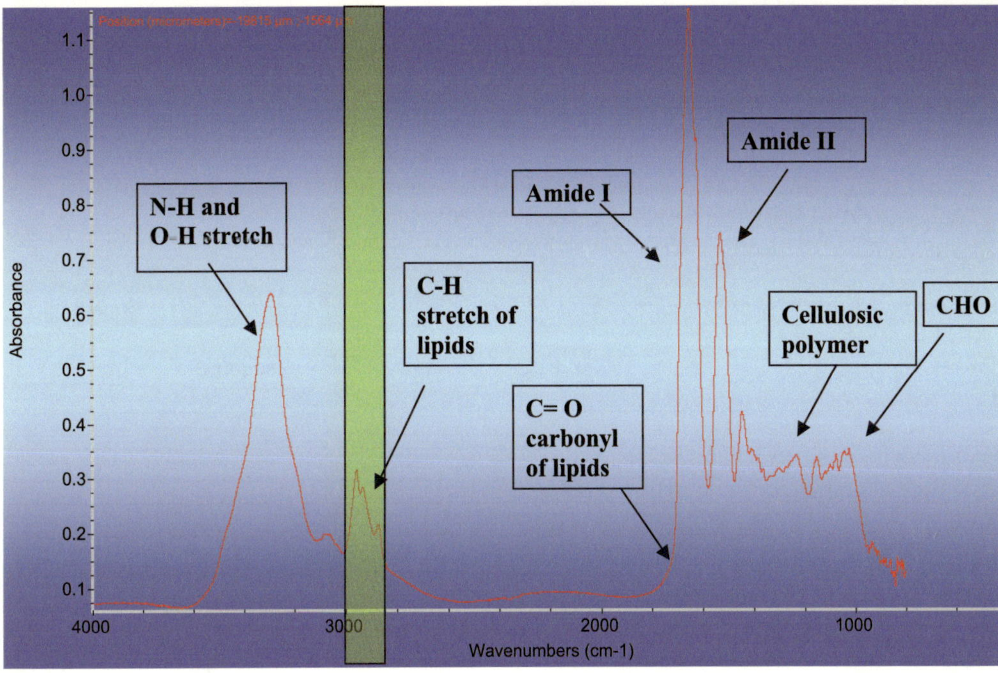

Fig. 1. Typical ultraspatially resolved synchrotron-based FTIRM spectrum in complex plant tissue (aleurone layer tissue of sorghum, pixel size: 10 μm × 10 μm) in the mid-IR region ca. 4000-800 cm^{-1} showed the function groups of biopolymers in complex plant system: N-H and O-H stretch, C-H groups stretch, amide I and II, C=O carbonyl, CHO and cellulosic compounds.

2.3 Molecular structure of a biopolymer in relation to nutrient availability

Nutrient availability of a biopolymer in animal and human is highly related to it molecular structure features (Walker et al., 2009). In an animal study (Doiron et al., 2009b), the protein molecular structure such as alpha-helix and beta-sheet ratio (Yu, 2006) highly associated with truly absorbed protein supply (metabolizable protein) in the small intestine to dairy cattle. Not only protein molecular structure but also amide I to II ratio in the new co-products of bioethanol processing also associated with true protein supply to dairy cattle.

3. Molecular imaging and molecular spectral analyses to discriminate and classify plant-based molecular structures in asymmetric and symmetric CH_2 and CH_2 groups

3.1 Imaging and mapping of asymmetric and symmetric CH groups in complex intact plant tissue

Using the synchrotron FTIR microspectroscopy, functional group images can be generated by plotting the intensity of synchrotron IR absorption bands as a function of *xy* position

(Budevska, 2002; Yu, 2005; Yu et al., 2007). Different coverage of the plant tissue with measurements could also be achieved by varying the step size and the dimensions of the image mask (aperture size). The big advantage of using synchrotron beam with FTIR spectroscopy to map the tissues is that diffraction limits and spatial resolution are achievable (Budevska, 2002; Wetzel et al., 1998). The false color maps were used (colors representing band intensities), which were derived from the area under particular spectral features.

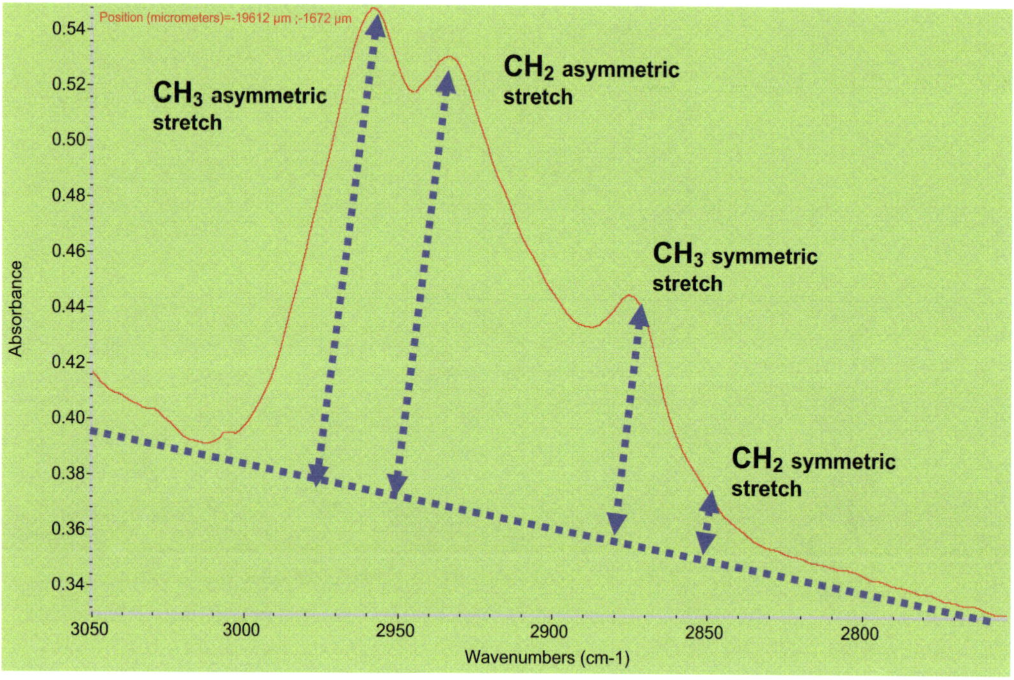

Fig. 2. The CH_2 and CH_3 asymmetric and symmetric stretching band features in typical ultra-spatially resolved synchrotron-based FTIRM spectrum in plant tissue (aleurone tissue of sorghum, pixel size: 10 µm × 10 µm) in the C-H stretch region ca. 3015-2780 cm^{-1} showing C-H function groups of biopolymers in complex plant tissue system: CH_3 asymmetric stretch at ca. 2958 cm^{-1}, CH_2 asymmetric stretch at ca. 2934 cm^{-1}, CH_3 symmetric stretch at ca. 2874 cm^{-1}, and CH_2 symmetric stretch at ca. 2851 cm^{-1}.

3.2 Agglomerative hierarchical molecular spectral cluster analysis

Agglomerative hierarchical cluster analysis (AHCA) is used to determine the main sources of variation in asymmetric and symmetric CH_2 and CH_3 groups spectra and the results are displayed as dendrograms using Ward's method (Cytospec Software). Doiron et al. (2009b) used AHCA analysis to detect the structure difference between the raw and heated treatment of flaxseed (cv. *Vimy*)

3.3 Spectral principle components analysis

Spectral Principle Components Analysis (PCA) used to determine major sources of variation in the asymmetric and symmetric CH_2 and CH_3 groups spectra. The purpose of PCA is to derive a small number of independent linear combinations (PCs) for a set of variables, while retaining as much information in the original variables as possible. This analysis allows the global study of the relationships between a set of quantitative characters p (eg. chemical functional groups) observed for a set of n samples (eg. spectra). The outcome of such an analysis can be presented either as 2D (two PCs) or 3D (three PCs) scatter plots (Sockalingum et al., 1998).

4. Recent progress in asymmetric and symmetric CH_3 and CH_2 research in complex plant system

4.1 Application I: Imaging of spatial distribution of asymmetric and symmetric CH_2 and CH_3 groups with advanced synchrotron FTIR microspectroscopy

Using the synchrotron FTIR microspectroscopy, the functional group images of asymmetric and symmetric CH_2 and CH_3 in complex plant system can be generated within cellular and subcellular dimensions. Fig. 3 shows a visible image of the plant sorghum seed tissue. Fig. 4 shows plant CH functional group images (a) CH_3 asymmetric stretch at ca. 2960 cm^{-1}; (b) CH_2 asymmetric stretch at ca. 2929 cm^{-1}; (c) CH_3 symmetric stretch at ca. 2877 cm^{-1}; (d) CH_2 asymmetric stretch at ca. 2848 cm^{-1} of the plant sorghum seed tissue from the pericarp (outside), seed coat, aleurone layer and endosperm.

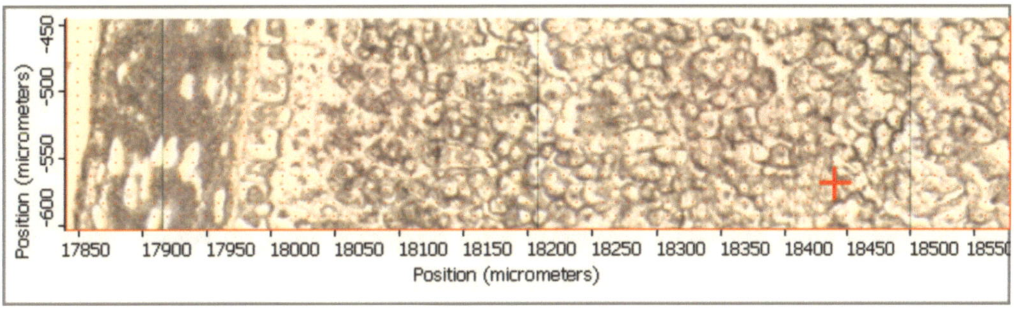

Fig. 3a. A visible image of the plant sorghum seed tissue

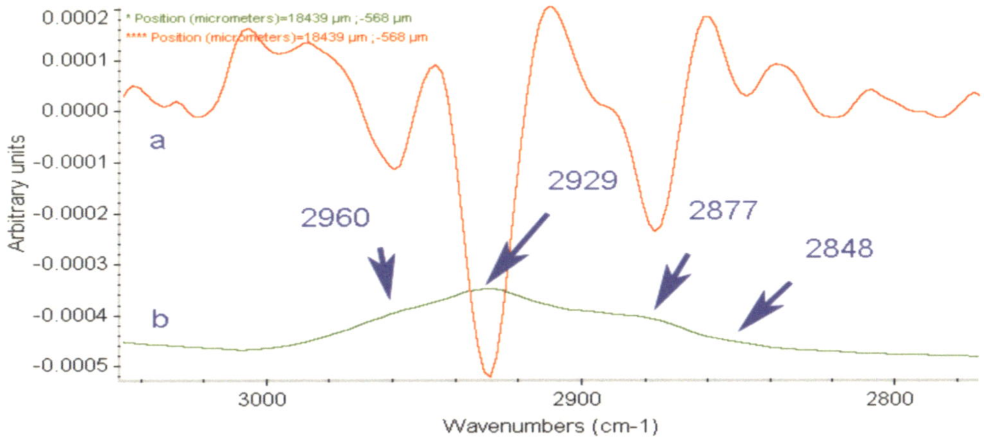

Fig. 3b. Spectra corresponding to the pixel at the cross-hair in the visible image (spectrum pixel size 10 ×10 µm).

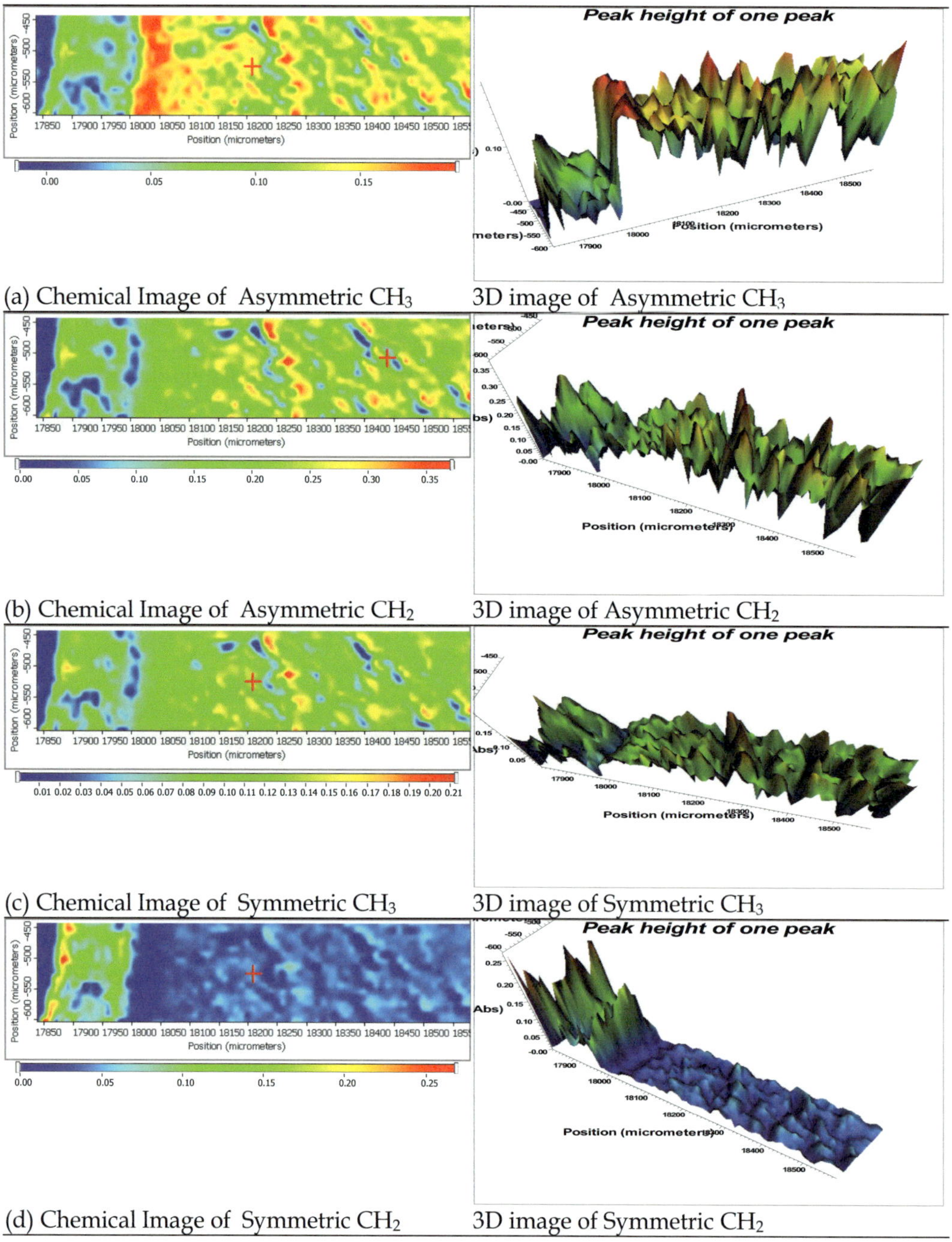

Fig. 4. CH functional group images [(a) CH_3 asymmetric stretch at ca. 2960 cm^{-1}; (b) CH_2 asymmetric stretch at ca. 2929 cm^{-1}; (c) CH_3 symmetric stretch at ca. 2877 cm^{-1}; (d) CH_2 asymmetric stretch at ca. 2848 cm^{-1}] of the sorghum seed tissue from the pericarp (outside), seed coat, aleurone layer and endosperm, using synchrotron-based FTIR microspectroscopy (1. Visible image; 2 Chemical image; 3. Spectra corresponding to the pixel at the cross-hair in the visible image) (spectrum pixel size 10 ×10 μm).

4.2 Application II: comparison of yellow- (*Brassica Rapa*) and brown-seeded (*Brassica Napus*) canola within cellular dimensions in structural features of asymmetric and symmetric CH_2 and CH_3, explored with advanced synchrotron FTIR microspectroscopy

This study aimed to use advanced synchrotron technique- SRFTIRM to compare two types of canola seed (yellow- (*Brassica Rapa*) vs. Brown-Seeded (*Brassica Napus*) Canola) at a cellular level in structural features of asymmetric and symmetric CH_2 and CH_3. The results showed the average absorbed peak intensities of CH functional groups: CH_3-asymmetric; CH_2-asymmetric; CH_3-symmetric; and CH_2-symmetric stretching vibrations of the yellow- and brown-seeded canola tissues, but no differences were found between the yellow- (*Brassica* rapa) and brown-seeded (*Brassica* napus) canola tissue, indicating similar CH_2 and CH_3 content between the two seed types. The peak height ratios of total CH_2:CH_3; CH_3-asymmetric: CH_3-symmetric; CH_2-asymmetric: CH_2-symmetric and total CH-asymmetric: CH-symmetric were 1.06 and 1.13, 1.28 and 1.26, 2.90 and 3.08, 1.82 and 1.78, for the yellow- and brown-seeded canola, respectively. Based on average data from each canola tissue, no significant differences were found, indicating that lipid chain length and branching is similar between the two seed types (Yu et al., 2005).

4.3 Application III: effect of bio-ethanol processing on structure changes of asymmetric and symmetric CH_2 and CH_3 on a molecular basis

Recently there is a dramatic increase in the bioethanol production which results in different types of co-products (Nuez-Ortín & Yu, 2009, 2010). This study aimed to reveal molecular structures of the new co-products of bioethanol production affected by bio-ethanol processing, identify the differences in asymmetric and symmetric CH_2 and CH_3 on a molecular basis between grains and new co-products and between different types of the bioethanol co-products. Fig. 4 (D. Damiran, personal communication) showed that the structural asymmetric and symmetric CH_2 and CH_3 groups were different between the grains and the new co-products of bio-ethanol production.

Cluster analyses (Fig. 5) indicated that the spectra of asymmetric and symmetric CH_2 and CH_3 between the grain (corn) and co-products of bioethnaol production (corn DDGS) were fully discriminated. This indicates that bioethanol processing change molecular structure of asymmetric and symmetric CH_2 and CH_3 conformation of the grain.

Fig. 6 show results from principal component analysis of the spectrum of asymmetric and symmetric CH_2 and CH_3. The first two PCs (obtained after data reduction) were plotted and show that the original grain (corn) and co-products of corn DDGS from bioethanol production can be grouped in separate ellipses. The first two PCs explain 99.98%, and 0.01 of the variation in the structural molecular spectrum data set between the corn and corn DDGS. Therefore principal component analysis show the distinguished difference between the corn and corn DDGS. These results indicate that the structural make-up in terms of asymmetric and symmetric CH_2 and CH_3 conformation between the corn and corn DDGS is fully distinguished.

5. Conclusions and implications

Using synchrotron-based SFTIRM and DRIFT molecular spectral spectroscopy, the structural feature and conformation of functional groups of asymmetric and symmetric CH_2

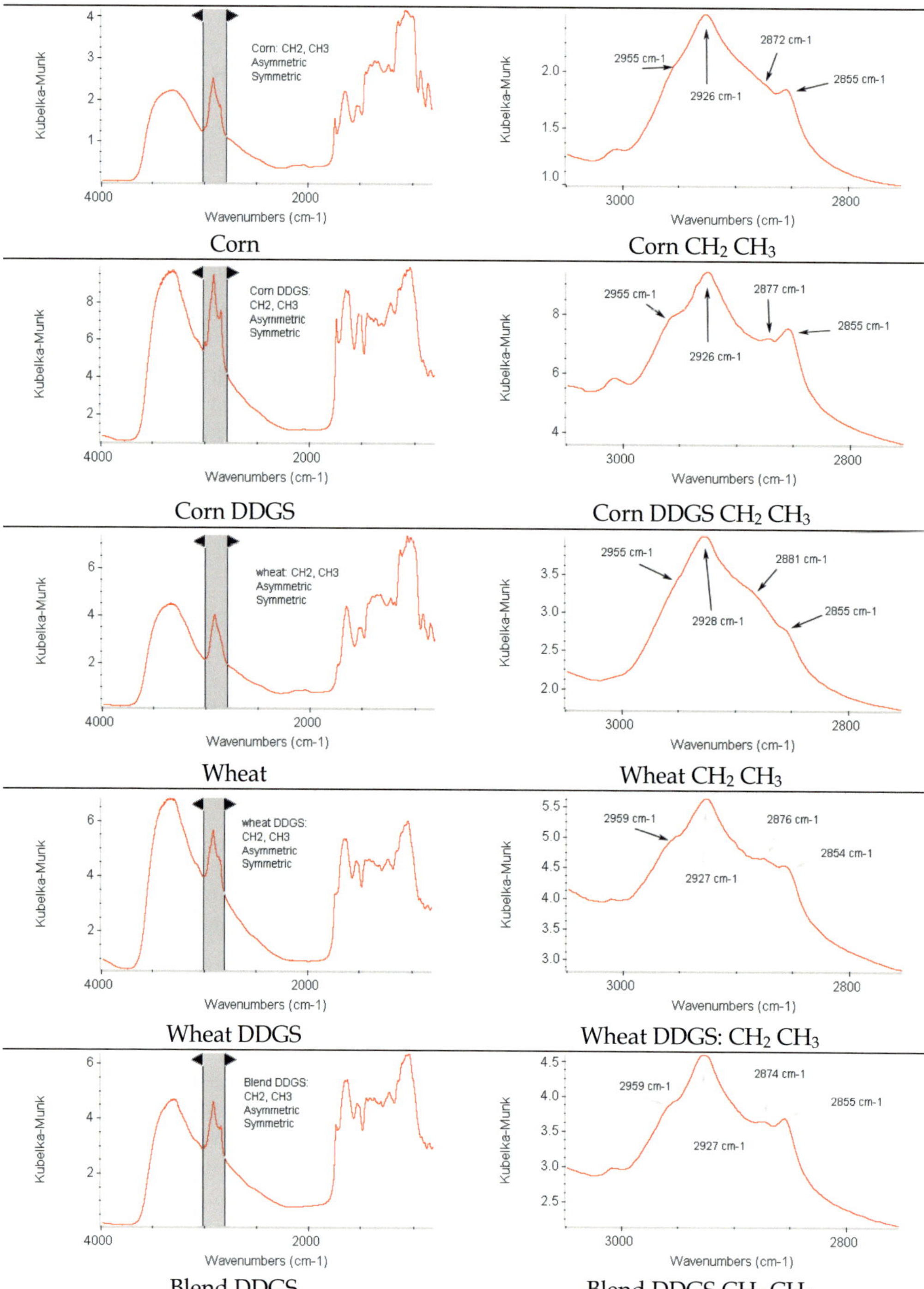

Fig. 4. Comparison of different types of DDGS and original feedstock in CH_3 and CH_2 asymmetric and symmetric stretching bands profiles using the DRIFT molecular spectroscopy: Effect of Bio-Ethanol Processing

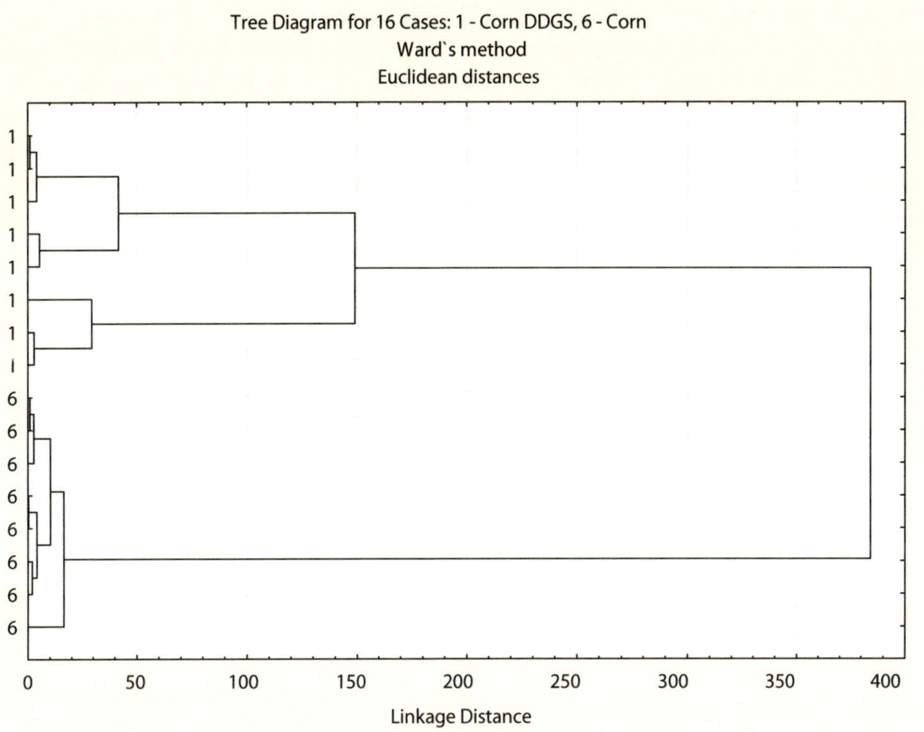

Fig. 5. Cluster analysis of spectrum detected with DRIFTS (acquired at 4 cm^{-1} resolution in the reflectance mode with subtraction of the KBr background, accumulated 256 times; range of wavenumbers, 3027-2813 cm^{-1}, IR reflectance Kubelka-Munk units) obtained from corn DDGS (1) vs. corn grain (6) from all original spot spectrum [CLA: (a) region of CH groups (CH_2, CH_3) ca. 3027-2813 cm^{-1}; (b) distance method, Euclidean; (c) cluster method, Ward's algorithm] (source: D. Damiran)

and CH_3 and their ratios (asymmetric CH_2 to symmetric CH_2 ratio; asymmetric CH_3 to symmetric CH_3 ratio; asymmetric CH_2 to asymmetric CH_3 ratio; symmetric CH_2 to symmetric CH_3 ratio) on a molecular basis could be detected in complex plant system. The molecular structure changes induced by various processing (eg. heating and bio-ethanol processing) or molecular structure difference between plant varieties could be revealed. The ultra-spatial distribution of the intensities of the asymmetric and symmetric CH_2 and CH_3 and their ratios could be imaged. The molecular structure difference between plant varieties or structure change induced by processing in terms of asymmetric and symmetric CH_2 and CH_3 and their ratios will affect functionality and nutrient availability and biodegradation in human and animals.

6. Acknowledgments

This research was supported by grants from the Natural Sciences and Engineering Research Council of Canada (NSERC- Individual Discovery Grant) and the Saskatchewan Agricultural Development Fund (ADF) and Ministry of Agriculture Strategic Research Chair Fund. The National Synchrotron Light Source in Brookhaven National Laboratory (NSLS-

BNL, New York, USA) is supported by the U.S. Department of Energy contract DE-AC02-98CH10886. The Center for Synchrotron Biosciences (U2B), Case Western Reserve University, was supported by the National Institute for Biomedical Imaging and Bioengineering under P41-EB-01979. The Canadian Light Sources (CLS) is supported by various federal and provincial funding agencies in Canada. The author is grateful to Megan Bourassa, Jennifer Bohon, Randy Smith Nebojsa Marinkovic and Lisa Miller at U10B and U2B (NSLS-BNL, New York), and Tim May, Tor Pederson and Luca Quaroni (CLS, University of Saskatchewan) for helpful data collection 01B1-1 (Mid IR) experimental station and Daal Damiran for providing one Figure for this review article.

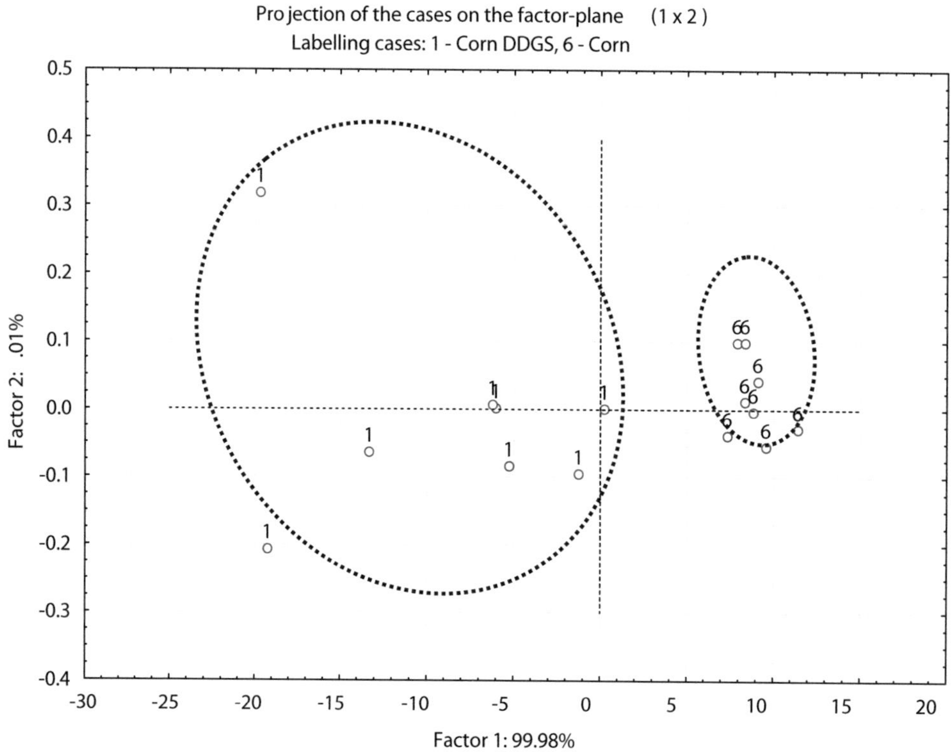

Fig. 6. Scatter plot of the 1st principal component vs. the 2nd principal component (a) of PCA analysis of spectrum detected with DRIFTS (acquired at 4 cm^{-1} resolution in the reflectance mode with subtraction of the KBr background, accumulated 256 times; range of wavenumbers, 3027–2813 cm^{-1}, IR reflectance Kubelka-Munk units) obtained from corn DDGS (1) vs. corn grain (6) from all original spot spectrum (CH groups (CH_2, CH_3) 3027-2813 cm^{-1}): The 1st and 2nd principal component explains 99.98, and 0.01% of the total variance, respectively (source: D. Damiran)

7. References

Budevska, B. O. (2002). Applications of Vibrational Spectroscopy in Life, Pharmaceutical and Natural Sciences. In: *Handbook of Vibrational Spectroscopy*, Vol. 5, Chalmers, J. M.

and P.R. Griffiths, P.R. (Ed.), pp 3720-3732. John Wiley and Sons, Inc., New York, NY, USA

CLS. (2010). Synchrotron facts. Available: http://www.lightsource.ca/education/whatis.php Accessed Mar, 2010.

Cytospec. Software for Infrared Spectral Imaging. V. 1.1.01

Damiran, D. & Yu, P. (2010). Structural make-up, biopolymer conformation and biodegradation characteristics of newly developed super genotype of oats (CDC SO-I vs. conventional varieties): A novel approach. *J Agric Food Chem.* 58: 2377-2387

Doiron, K. J. & Yu, P., Christensen, C. R., Christensen, D. A. & McKinnon, J. J. (2009a). Detecting molecular changes in *Vimy* flaxseed protein structure using synchrotron FTIRM and DRIFT spectroscopic techniques: structural and biochemical characterization. *Spectroscopy* 23: 307 322

Doiron, K. J., Yu, P., McKinnon, J. J., & Christensen, D. A. (2009b). Heat-induced protein structures and protein subfractions in relation to protein degradation kinetics and intestinal availability in dairy cattle. *J Dairy Sci* 92:3319-3330

Dumas, P. (2003). Synchrotron IR Microspectroscopy: A Multidisciplinary Analytical Technique. The 6th Annual Synchrotron CLS Users' Meeting and Associated Synchrotron Workshops University of Saskatchewan, Canada. Nov 13-15.

Harris, P. I. & Chapman, D. (1992). Does Fourier-transform infrared spectroscopy provide useful information on protein structures? *Trend Biochem Sci* 17: 328-333.

Jackson, M., Haris, P. I. & Chapman, D. (1989) Conformational transitions in poly-L-lysine: an FT-IR spectroscopic study. *Biochim Biophys Acta* 998:75.

Jackson, M. & Mantsch, H. H. (1995). The use and misuse of FTIR spectroscopy in the determination of protein structure. *Biochemistry and Molecular Biology* 30: 95-120.

Jackson, M. & Mantsch, H. H. (1996). Biomedical Infrared Spectroscopy, Pages 311-340 in Infrared Spectroscopy of Biomolecules, eds. HH. Mantsch, D. Chapman, Wiley-Liss, New York.

Liu, N. & Yu, P. (2010). Using DRIFT molecular spectroscopy with uni- and multivariate molecular spectral techniques to detect plant protein molecular structure difference among different genotypes of barley. *J Agric Food Chem.* In Review

Marinkovic, N. S. & Chance, M. R. (2006) Synchrotron Infrared Microspectroscopy. In: (Meyers R, ed) *Encyclopedia of Molecular Cell Biology and Molecular Medicine,* 2nd ed., Vol 13, Wiley Inc. pp 671-708.

Marinkovic, N. S., Huang, R., Bromberg, P., Sullivan, M., Toomey, J., Miller, L. M., Sperber, E., Moshe, S., Jones, K. W., Chouparova, E., Lappi, S., Franzen, S. & Chance, M. R. (2002). Center for Synchrotron Biosciences' U2B beamline: an international resource for biological infrared spectroscopy. *J Synchrotron Rad* 9:189-197.

Miller, L. M. (2009). Infrared Microsperctroscopy and Imaging. Available: http://nslsweb.nsls.bnl.gov/nsls/pubs/nslspubs/imaging0502/irxrayworkshopintroduction.ht Accessed Oct.

Miller, L. M. & Dumas, P. (2006). Chemical imaging of biological tissue with synchrotron infrared light. *Biochim Biophys Acta* 1758: 846-857.

Nuez-Ortín, W. G. & Yu, P. (2009). Nutrient variation and availability of wheat DDGS, corn DDGS and blend DDGS from bioethanol plants. *J Sci Food Agric.* 89: 1754–1761

Nuez-Ortín, W. G. & Yu, P. (2010). Effects of Bioethanol Plant and Co-products Type on the Metabolic Characteristics of the Proteins. *J Dairy Sci.* In press.

Sockalingum, G. D., Bouhedja, W., Pina, P., Allouch, P., Bloy, C. & Manfait, M. (1998). FT-IT spectroscopy as an emerging method for rapid characterization of microorganisms. *Cell Mol Bio* 44: 261-269.

Walker, A. M., Yu, P., Christensen, C. R., Christensen, D. A. & McKinnon, J. J. (2009). Fourier Transform infrared microspectroscopic analysis of the effects of cereal type and variety within a type of grain on molecular structural make-up in relation to rumen degradation kinetics. *J Agric Food Chem* 57: 6871–6878

Wetzel, D. L. & LeVine, S. M. (1999). Imaging molecular chemistry with infrared microscopy. *Science.* 285: 1224-1225.

Wetzel, D. L., Srivarin, P. & Finney, J. R. (2003). Revealing protein infrared specrtral detail in a heterogeneous matrix dominated by starch. *Vibrational Spectroscopy* 31: 109-114.

Wikipedia. (2010). Synchrotron light source. http://en.wikipedia.org/wiki/Synchrotron_light_source. Accessed April.

Yu, P. (2004). Application of advanced synchrotron-based Fourier transform infrared microspectroscopy (SR-FTIR) to animal nutrition and feed science: a novel approach. British J Nutri 92: 869-885.

Yu, P., McKinnon, J. J., Christensen, C. R. & Christensen, D. A. (2004). Imaging Molecular Chemistry of Pioneer Corn. *J Agric Food Chem* 52:7345-7352

Yu, P., Christensen, C. R., Christensen, D. A. & McKinnon, J. J. (2005). Ultrastructural-chemical makeup of yellow- (*Brassica Rapa*) and brown-seeded (*Brassica Napus*) canola within cellular dimensions, explored with synchrotron reflection FTIR microspectroscopy. *Can J Plant Sci* 85: 533-541.

Yu, P., Block, H., Niu, Z. & Doiron, K. J. (2007). Rapid characterization of molecular chemistry and nutrient make-up and microlocalization of internal seed Tissue. *J Synchrotron Rad* 14: 382-390.

Yu, P., Doiron, K. & Liu, D. (2008) Shining light on the difference in molecular structural chemical make-up and the cause of distinct biodegradation behavior between malting- and feed-type barley: A novel approach. *J Agric Food Chem (Molecular Nutrition Section)* 56: 3417-3426.

Section D

28

Molecularly Imprinted Polymers (MIPs) in Biomedical Applications

Francesco Puoci, Giuseppe Cirillo, Manuela Curcio, Francesca Iemma, Ortensia Ilaria Parisi, Umile Gianfranco Spizzirri and Nevio Picci
Department of Pharmaceutical Sciences, University of Calabria
I-87036, Rende (CS) - Italy

1. Introduction

Generations of scientists have been intrigued by the binding phenomena involved in interactions that occur between natural molecular species, and over the years, numerous approaches have been used to mimic these interactions. Complex formation between a host molecule and the guest involves recognition, which is the additive result of a number of binding forces (Figure 1).

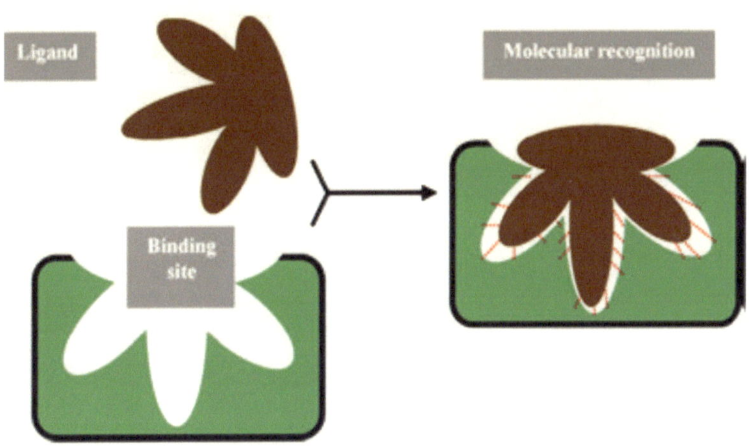

Fig. 1. Schematic representation of molecular recognition process. Adapted from Hillberg & Tabrizian, 2008.

Within biological systems, these are usually dynamic and are the result of a mass of non-covalent interactions, which act collectively to form a very stable system. Molecular imprinting is a relatively new and rapidly evolving technique used to create synthetic receptors, having recognition properties comparable to the biological systems and it also possesses great potential in a number of applications in the life Sciences. Primarily, molecular imprinting aims to create artificial recognition cavities within synthetic polymers (Alvarez-Lorenzo & Concheiro, 2004; Ramström & Ansell, 1998; Mosbach & Ramström, 1996). It is a relatively simple concept, which involves the construction of sites of specific

recognition, in synthetic polymers (Owens et al., 1999; Wulff, 1995; Caro et al., 2002; Joshi et al., 1998). The template of choice is entrapped in a pre-polymerization complex, consisting of functional monomers with good functionality, which chemically interact with the template. Polymerization in the presence of crosslinker serves to freeze these template-monomer interactions and subsequent removal of the template results in the formation of a molecularly imprinted polymer matrix (Figure 2).
Enormous interest has also been shown in imprinted materials as they mime biological receptors for the screening of new substances with potential pharmacological activity or to specifically detect drugs in biological fluids in screening assays for drugs of abuse. Such specificity is comparable with monoclonal antibodies used in immunoassay techniques (Pap et al., 2002; Chapuis et al., 2003; Caro et al., 2003; Vandevelde et al., 2007). Molecular imprinting is a well-developed tool in the analytical field, mainly for separating and quantifying very different substances, including drugs and bio-active molecules contained in relatively complex matrices. Moreover, the information generated about polymer synthesis procedures and the properties outlined for optimum performance in separation-based technologies may be a good starting point to create imprinted polymers useful in biomedical applications such as drug delivery systems, polymeric traps for toxic metabolites, etc. (Cunliffe et al., 2005). The chapter will focus on the most representative applications of MIPs in the biomedical field.

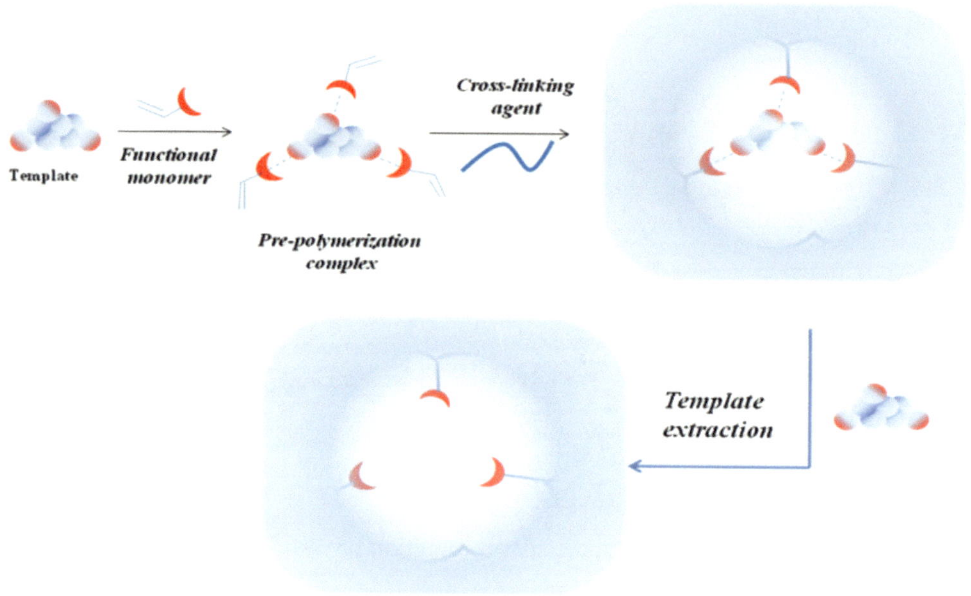

Fig. 2. Schematic representation of MIP synthesis.

2. Synthesis of MIP

Molecular imprinting is a very useful technique to incorporate specific substrate recognition sites into polymers. Molecular recognition characteristics of these polymers are attributed to complementary size, shape, and binding sites imparted to the polymers by the template molecules. The specific binding properties of MIP must be attributed to specific interactions between the template and the functional groups in the polymeric network, thus the choice of the functional monomers is of primary importance to obtain performing imprinted materials (Puoci et al., 2005; Curcio et al, 2009).

MIPs can be synthesized following three different imprinting approaches (Caro et al., 2002), as follows:

1. The non-covalent procedure is the most widely used because it is relatively simple experimentally and the complexation step during the synthesis is achieved by mixing the template with an appropriate functional monomer, or monomers, in a suitable porogen (solvent) (Joshi et al., 1998). After synthesis, the template is removed from the resultant polymer simply by washing it with a solvent or a mixture of solvents. Then, the rebinding step of the template by the MIP exploits non-covalent interactions.
2. The covalent protocol, which requires the formation of covalent bonds between the template and the functional monomer prior to polymerization. To remove the template from the polymer matrix after synthesis, it is necessary to cleave the covalent bonds. To this end, the polymer is then refluxed in a Soxhlet extraction or treated with reagents in solution (Ikegami et al., 2004).
3. The semi-covalent approach is a hybrid of the two previous methods. Thus, covalent bonds are established between the template and the functional monomers before polymerization, while, once the template has been removed from the polymer matrix, the subsequent re-binding of the analyte to the MIP exploits non-covalent interactions, as the non-covalent imprinting protocol.

The binding sites obtained by molecular imprinting show different characteristics, depending on the interactions established during polymerization. The average affinity of binding site prepared using bonding by non-covalent forces is generally weaker than those prepared using covalent methods because electrostatic, hydrogen bonding, π-π and hydrophobic interactions, between the template and the functional monomers, are used exclusively in forming the molecular assemblies (Hwang & Lee, 2002). Moreover, an excess of functional monomer relative to the template is usually required to favor template-functional monomer complex formation and to maintain its integrity during polymerization. As a result, a fraction of the functional monomers is randomly incorporated into the polymer matrix to form non-selective binding sites.

However, when covalent bonds are established between the template and the functional monomer prior to polymerization, this gives rise to better defined and more homogeneous binding sites than the non-covalent approach, since the template-functional monomer interactions are far more stable and defined during the imprinting process.

Nevertheless, non covalent imprinting protocol is still the most widely used method to prepare MIP because of the advantages that it offers over the covalent approach from the point of view of synthesis.

In some polymers prepared by the non-covalent procedure, it has been observed that the binding of the template to the polymer can sometimes be so strong that it is difficult to remove the last traces of template, even after washing the polymer several times (Martin et al., 2003; Andersson et al., 1997).

When the MIP is used, small amounts of residual template can be eluted. This bleeding is a problem mainly when the MIP has to be applied to extract trace levels of the target analyte.

To overcome this drawback, some authors have synthesized MIP using an analogue of the target molecule as a template (the template-analogue approach) (Dirion et al., 2002). In this way, if the MIP bleeds template, then the elution of the template does not interfere in the quantification of the target analyte. Andersson was the first author to synthesize a MIP using a template analogue. In this case, a MIP selective for sameridine was prepared using as a template a close structural analogue of sameridine. However, it should be pointed out that the use of template analogues is not always the solution, because sometimes is not possible to identify and to source a suitable analogue. For this reason, other methods, such as thermal annihilation, microwave-assisted extraction (MAE) and desorption of the template with supercritical fluids have also been developed to remove the template from the MIP (Ellwanger et al., 2001).

It should also be mentioned that, as a control in each polymerization, a non-imprinted polymer (NIP) is also synthesised in the same way as the MIP but in absence of the template. To evaluate the imprinting effect, the selectivities of the NIP and MIP are then compared.

It is important to state that MIP can be obtained in different formats, depending on the preparation method followed. To date, the most common polymerizations for preparing MIPs involve conventional solution, suspension, precipitation, multi-step swelling and emulsion core-shell. There are also other methods, such as aerosol or surface rearrangement of latex particles, but they are not used routinely.

When a MIP is obtained by conventional solution polymerization, the resultant polymer is a monolith, which has to be crushed before use, except when the MIP is prepared in situ. However, suspension polymerization (in fluorocarbons or water) and precipitation polymerization allow MIPs to be prepared in the form of spherical polymer particulates.

Conventional solution polymerization is the most common method because of its simplicity and universality. It does have some drawbacks as the processes of grinding and sieving not only are wasteful and time consuming, but also may produce irregularly sized particles.

Another important parameter to be considered in the synthesis of MIP is the type of initiator system.

The widespread use of traditional free radical polymerization methods for the preparation of molecularly imprinted polymers can be attributed to a tolerance for a wide range of functional groups and template structures. In essence, the free radicals generated during the addition polymerization do not interfere with the intermolecular interactions critical for the non-covalent imprinting system.

Generally, in the synthesis of MIP, the free radicals are generated by decomposition of azo-compounds, peroxides and thermal iniferters which require relatively high polymerization temperature to ensure their rapid decomposition.

The polymerization temperature is also an important parameter to be considered in order to obtain performing MIP. A high temperature, indeed, is expected to drive the equilibrium away from the template-functional monomer complex toward the unassociated species, resulting in a decrease in the number of imprinted cavities. Thus, several strategies have been planned to create a stable pre-polymerization complex by decreasing the kinetic energy of the system, a parameter that strongly depends on the polymerization temperature. For example, UV induced polymerization processes were successfully employed in the synthesis of MIP selective for different kinds of template (Puoci et al., 2008a; Puoci et al., 2007a). Moreover, even if conventional initiator systems have been applied in polymerization and copolymerization with the convenience of working at a lower temperature, they show the disadvantage of the possible introduction of harmful and toxic chemical side products.

In a recent work, (Cirillo et al., 2010a) $FeCl_2/H_2O_2$ redox initiator system was employed to synthesize a theophilline imprinted polymer. Hydroxyl radical is the active species that is generated from the reduction of hydrogen peroxide at the expense of Fe^{2+} ions.
A great number of studies have investigated the use of Fenton reactions for water remediation through pollutant degradation. Fenton reagents have been used as radical initiator in vinylic polymerization or grafting for more than 50 years. However, almost no reference has been made to its use to initiate molecularly imprinted polymerization.
The advantages of this kind of initiator system consist of the low working temperature, the absence of any kind of toxic reaction products, that is desiderable for materials to be employed in biomedical field, and the possibility to decrease the polymerization time (2h for the synthesis of Redox MIP *vs* 24 h for the synthesis of conventional MIP synthesized by azo-initiators). The whole of these aspects contributes to preserve the stability of the pre-polymerization complex, thus improving the imprinting efficiency of the obtained materials.

3. Applications of MIP

Molecular imprinting has now become an established method and has also been applied in the areas of biomedical and analytical chemistry. MIP have been used as chromatographic stationary phases (Turiel & Martin-Esteban, 2004) for enantiomeric separations (Bruggemann et al., 2004), solid-phase extraction (Haupt et al., 1998), catalysis (Ye & Mosbach, 2001a) and sensor design (Mosbach, 2001), as well as for protein separation (Hansen, 2007), as receptor (Haupt, 2003), antibody (Ye & Mosbach, 2008) and enzyme mimics (Yu et al., 2002), and most recently as drug delivery systems (DDS) (Alvarez-Lorenzo & Concheiro, 2008).

3.1 MIP as basis of Drug Delivery Systems

In the last few years, a number of significant advances have been made in the development of new technologies for optimizing drug delivery (Schmaljohann, 2006). To maximize the efficacy and safety of medicines, drug delivery systems (DDS) must be capable of regulating the rate of release (delayed- or extended-release systems) and/or targeting the drug to a specific site. Efficient DDS should provide a desired rate of delivery of the therapeutic dose, at the most appropriate place in the body, in order to prolong the duration of pharmacological action and reduce the adverse effects, minimize the dosing frequency and enhance patient compliance. To control the moment at which delivery should begin and the drug release rate, the three following approaches have been developed (Chien & Lin, 2002):

a. ***rate-programmed drug delivery***: drug diffusion from the system has to follow a specific rate profile;
b. ***activation-modulated drug delivery***: the release is activated by some physical, chemical or biochemical processes;
c. ***feedback-regulated drug delivery***: the rate of drug release is regulated by the concentration of a triggering agent, such as a biochemical substance, concentration of which is itself dependent on the drug concentration in the body.

When the triggering agent is above a certain level, the release is activated. This induces a decrease in the level of the triggering agent and, finally, the drug release is stopped. The sensor embedded in the DDS tries to imitate the recognition role of enzymes, membrane receptors and antibodies in living organisms for regulation of chemical reactions and for maintenance of the homeostatic equilibrium.

Molecular imprinting technology can provide efficient polymer systems with the ability to recognize specific bioactive molecules and a sorption capacity dependent on the properties and template concentration of the surrounding medium; therefore, although imprinted DDS have not reached clinical application yet, this technology has an enormous potential for creating satisfactory dosage forms.

The following aspects should be taken into account:

a. *Compromise between rigidity and flexibility.*

The structure of the imprinted cavities should be stable enough to maintain the conformation in the absence of the template, but somehow flexible enough to facilitate the attainment of a fast equilibrium between the release and re-uptake of the template in the cavity. This will be particularly important if the device is used as a diagnostic sensor or as a trap of toxic substances. In this sense, non-covalent imprinting usually provides faster equilibrium kinetics than the covalent imprinting approach (Allender et al., 2005). The mechanical properties of the polymer and the conformation of the imprinted cavities depend to a great extent on the proportion of the cross-linker. Mostly imprinted systems for analytical applications require around 25-90% of cross-linker agent. These cross-linking levels increase the hydrophobicity of the network and prevent the polymer network from changing the conformation obtained during synthesis. As a consequence, the affinity for the template is not dependent on external variables and it is not foreseen that the device will have regulatory or switching capabilities. The lack of response capability to the alterations of the physico-chemical properties of the medium or to the presence of a specific substance limits their potential uses as *activation-* or *feedback-modulated* DDS. A high cross-linker proportion also considerably increases the stiffness of the network making it difficult to adapt the shape of the administration site and causing mechanical friction with the surrounding tissues (especially when administered topically, ocularly or as implants).

b. *High chemical stability.*

MIP for drug delivery should be stable enough to resist enzymatic and chemical attack and mechanical stress. The device will enter into contact with biological fluids of complex composition and different pH, in which the enzymatic activity is intense. Ethylene glycol dimethacrylate (EGDMA) and related cross-linkers, which are the most usual ones, have been proved to provide stable networks in a wide range of pHs and temperatures under *in vitro* conditions (Svenson & Nicholls, 2001). However, additional research should be carried out to obtain information about its behaviour in vivo environments, where esterases and extreme pHs seem to be able to catalyse its hydrolysis (Yourtee et al., 2001). Additionally, it has to be taken into account that the adaptability of molecular imprinting technology for drug delivery also requires the consideration of *safety* and *toxicological* concerns. The device is going to enter into contact with sensitive tissues; therefore, it should not be toxic, neither should its components, residual monomers, impurities or possible products of degradation (Aydin et al., 2002). Therefore, to ensure biocompatibility it might be more appropriate to try to adapt the imprinting technique to already tested materials instead of creating a completely new polymeric system. On the other hand, most classical MIP are created in organic solvents to be used in these media, taking advantage of electrostatic and hydrogen bonding interactions. The presence of residual organic solvents may cause cellular damage and should be the object of a precise control. In consequence, hydrophilic polymer networks that can be synthesised and purified in water are preferable to those that require organic solvents. A hydrophilic surface also enhances biocompatibility and avoids adsorption of

proteins and microorganisms (Anderson, 1994). Additionally, many drugs, peptides, oligonucleotides and sugars are also incompatible with organic media.
A wide range of cross-linked hydrogels have been proved to be useful as drug delivery platforms (Davis & Anseth, 2002). Molecular imprinting in water is still under development and difficulties arise due to the considerable weakness of electrostatic and hydrogen-bonding interactions in this polar medium, which decrease the affinity and selectivity of MIP for the ligand (Komiyama et al., 2003). Nevertheless, hydrophobic and metal co-ordination interactions are proving to be very promising to enhance template and functional monomer association in water (Piletsky et al., 1999).
It is clear that the polymer composition and solvent are key parameters in the achievement of a good imprinting and that, in consequence, a compromise between functionality and biocompatibility is needed.
To date, several MIP based drug delivery devices were prepared for the sustained/controlled release of anticancer, antibiotic and anti-inflammatory drugs, obtaining a great efficiency in the release modulation.
One of the most relevant challenges in this field is intelligent drug delivery combined with molecular recognition. Intelligent drug release refers to the release, in a predictable way, of a therapeutic agent in response to specific stimuli such as the presence of another specific molecule or small changes in temperature, pH, solvent composition, ionic strength, electric field, or incident light (Gil & Hudson, 2004; Peppas & Leobandung, 2004). The ability of polymers to reversibly respond to small environmental changes mainly depends on different interactions between functional segments of the polymer network (Puoci et al., 2008b).

3.1.1 pH responsive MIP

pH-responsive polymers are characterized by swelling/shrinking structural changes in response to environmental pH changes (Morikawa et al., 2008; Oh & Lee, 2008; Pérez-Aòvarez et al., 2008). Such a polymeric network, containing ionizable groups, is able to accept or donate protons at a specific pH, thereby undergoing a volume phase transition from a collapsed state to an expanded state. Weak polyacids and weak polybases represent two types of pH-sensitive polyelectrolyte. To date, there have been a number of papers in the literature describing the synthesis and applications of pH-sensitive polymer hydrogels based on molecular imprinting technology to be applied as base excipients for drug delivery formulations. (Gil & Hudson, 2004).
As reported, in the synthesis of an efficient imprinted polymers, the first parameter to be considered is the choice of the suitable functional monomer, and for this scope, a screening of different functional monomers should be made. In a recent work (Cirillo et al., 2010b), three different MIP for the selective release of glycyrrhizic acid were synthesized employing methacrylic acid (MAA) as acidic, 2-(dimethylamino)ethyl methacrilate (DMAEMA) as basic, and 2-hydroxyethylmetacrylate (HEMA) as neutral functional monomer, in order to evaluate the effect of the different monomer to the recognition properties of the resulting materials. The most promising matrix to be applied as glycyrrhizic acid controlled delivery device in gastrointestinal was found to be the MAA-containing MIP, while the DMAEMA-MIP was not effective in this direction because of the high non-specific hydrophobically driven interaction between polymeric matrices and template. The HEMA-containing MIP was found to be less effective as a result of the lower capacity of HEMA to form hydrogen bond comparing to MAA.

Another work (Puoci et al., 2007b) reports on the synthesis of MIP for the sustained release of this molecule in gastro-intestinal simulating fluids. The imprinted polymers were found to have a better ability to control drug release compared with non-imprinted polymers due to the presence of specific binding sites in the polymeric network that are able to release the drug much more slowly: the drug release from NIP was indeed remarkably faster than that observed from MIP. These remarkable differences depend on the different recognition properties of the two polymeric matrices (Figure 3).

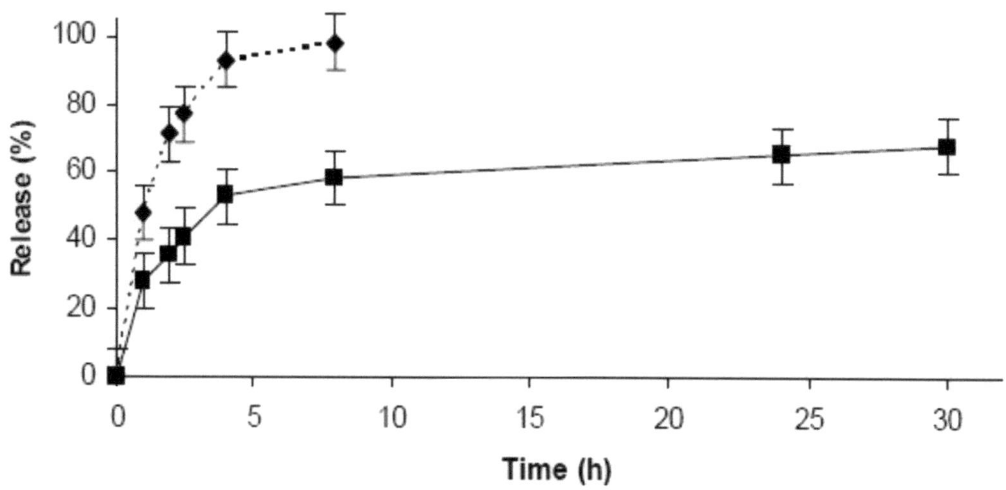

Fig. 3. Gastrointesinal release profile of 5-FU by MIP (_■_) and NIP (- -♦- -). Adapted from Puoci et al., 2007a.

The non-imprinted polymers, indeed, do not have specific binding cavities for the drug, while the MIP samples, because of their specific structure, strongly bound the drug by non-covalent interactions in the cavities formed during the polymerization procedure in the presence of the analyte. This observation supports a model of retention mechanism, which assumes that the acid groups of the selective sites have stronger interaction with the drug than the non-selective sites. At low pH (1.0) values, the carboxylic groups are not ionized and there is a good interaction with the template. These results might help us to understand the behavior of these matrices when the pH increases. Under these conditions, that simulate the intestinal fluid, in the non-imprinted polymers the antioxidant is bound with non-covalent interactions on the surface of the matrices. At pH 6.8, the diffusion rate of the buffer on the polymer surface is fast, the carboxylic groups are ionized, and the drug is rapidly released. Instead, in the MIP case, the diffusion rate of the buffer into specific cavities of imprinted polymers is slower, and the functional groups are ionized more slowly, resulted in well controlled release.

Similar results were obtained for the release of antioxidant molecules such as tocopherol (Puoci et al., 2008c), and phytic acid (Cirillo et al., 2009a) confirming that MIPs represent a very useful polymeric device for the selective and controlled release of a therapeutic agent in gastrointestinal fluids.

However, the reported synthetic approaches (bulk polymerization) yields particles with limited control on particle size and shape. In literature, several attempts have been applied to produce monodispersed molecularly imprinted polymeric particles using methods such

as suspension polymerization in water (Lai et al., 2001) , dispersion polymerization (Say et al., 2003), liquid perfluorocarbon (Mayes & Mosbach, 1996), and via aqueous two-step swelling polymerization (Piscopo et al., 2002). However, during the polymerization procedure, these techniques require water or highly polar organic solvents, which frequently decrease specific interactions between functional monomers and template molecules. Precipitation technique not only allows to avoid these disadvantages, but also to obtain monodispersed molecularly imprinted micro- and nanospheres, without the integrity and stability of recognition sites compromised (Wei et al., 2006). Moreover, spherical shape should be advisable in order to avoid swelling anisotropic behavior associated with other geometries (Iemma et al., 2008).
Based on these considerations, micro- and nano-spherical imprinted polymers (Figure 4) were prepare for the sustained release of sulfasalezine in gastrointestinal simulating fluids (Puoci et al., 2004) and 5-FU in plasma simulating fluids (Cirillo et al., 2009b), respectively. A better control on drug delivery was obtained, the spherical shape, indeed, allows to eliminate the anisotropic swelling normally associated with others geometries.

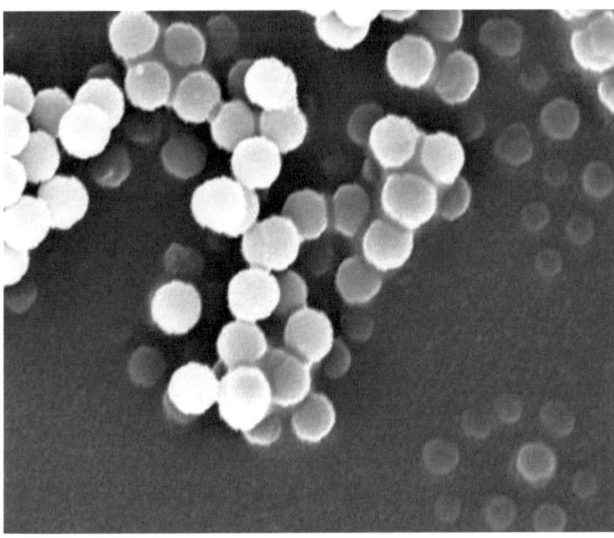

Fig. 4. SEM image of 5-FU molecularly imprinted nanospheres. Adapted from Cirillo et al., 2009b.

Recently, furthermore, a different approach was used for the synthesis of imprinted microspheres to be applied in the sustained release of paracetamol. Most of the developed imprinting protocols, indeed, can be successfully used to produce MIP for recognition of a large range of guest molecules predominantly in organic solvent-based media, while they often fail to generate MIP for use in pure aqueous environments (Benito-Peña et al., 2009). This depends on the non-specific hydrophobically driven bonds between template and surface of materials. In addition, biological sample components, such as proteins and lipids, are strongly adsorbed to the polymeric surfaces, negatively interfering with their recognition properties (Boos & Fleischer, 2001). Thus, in order to obtain MIP able to work in aqueous media, such as biological fluids or environmental waters, a considerable reduction of these non-specific interactions is required (Bures et al., 2001). For this purpose, different methodologies were developed (Mullet & Pawliszyn, 2003; Sambe et al., 2007). A widely used approach is the insertion of a hydrophilic monomer such as 2-hydroxyethyl methacrylate (HEMA) in the pre-polymerization mixture. This compound is known to impart water compatibility in a number of unrelated systems, but it is also able to interfere with the formation of the pre-polymerization complex interacting with several analytes by hydrogen bonds formation (Tunc et al., 2006). Another approach, involving a two step polymerization procedure, is the hydrophilic modification of MIP surface using glycerol monomethacrylate (GMMA) and glycerol dimethacrylate (GDMA). This materials avoid the destructive deposition of biomacromolecules on the polymeric surface, allowing an enhanced imprinting effect, especially in SPE protocols (Sanbe & Haginaka, 2003; Haginaka et al., 1999). A more promising approach is to use a monomer that less interferes in the pre-polymerization complex formation, but able, at the same time, after a post-polymerization straight forward modification, to impart water compatibility to the system. Glycidilmethacrylate (GMA) is useful for this purpose because its oxygen atom, bounded to two carbons, has lower capacity to form hydrogen bonds than a free hydroxy group. Furthermore, the epoxide ring opening carried out to the formation of a hydrophilic external layer on the polymeric surface. With this reaction, it is possible to modify hydrophobic matrices in more water compatible ones, more suitable to be employed in biological media because of the reduction of non-specific hydrophobic interactions (Puoci et al., 2009; Parisi et al. 2009).

3.1.2 Thermo-responsive MIP

A great number of synthetic, naturally occurring, and semisynthetic polymers display discrete, rapid, and reversible phase transformations as a result of conformational changes in response to temperature (Curcio et al., 2010). Polymers can exhibit either a lower critical solution temperature (LCST), below which they are soluble in deionized water, or an upper critical solution temperature (UCST), above which they are soluble. A balance of hydrophilic/ hydrophobic groups in the network determines the onset of the response that switches these "smart" materials in a controlled manner by adjusting the temperature. The responsive behavior of polymers with LCST properties is characterized by interactions between the hydrophobic groups, such as methyl, ethyl, and propyl groups, which become stronger than the hydrogen bonds with increasing temperature. On the other hand, in materials with UCST properties, the opposite is true and they swell at high temperature and shrink at low temperature. Poly(*N*-isopropylacrylamide) (PNIPAM) is the polymer most widely studied in this context because of its low critical solution temperature (LCST) in the range of 25-32 °C, i.e. close to the temperature of the human body (Iemma et al., 2009). In recent years, MIPs exhibiting thermoresponsive behavior have also been studied. One of the first reports concerned temperature-sensitive imprinted polymeric gels based on *N*-isopropylacrylamide (NIPAM), acrylic acid, and *N,N'*-methylenebis(acrylamide) (BIS), which were prepared in the presence of a template such as DL-norephedrine hydrochloride or DL-adrenaline hydrochloride (Watanabe et al., 1998). The imprinted and non-imprinted gels prepared in 1,4-dioxane showed a volume change in aqueous solution as a function of temperature. However, when the guest molecule was present in a saturated solution, the polymers exhibited another phase ("molecular recognition phase"), the volume of which was responsive to the concentration of the guest molecule. An interesting study (Alvarez-Lorenzo et al., 2001) reported on temperature-sensitive polymeric gels based on NIPAM, methacrylic monomers, and *N,N'*- methylenebis(acrylamide) as cross-linker, which were capable of reversibly adsorbing and releasing divalent ions. The effects of various methacrylate salts on the binding of divalent ions were reported. Imprinted gels prepared with calcium methacrylate or lead methacrylate showed higher affinity for target molecules as compared to randomly polymerized gels containing methacrylic acid (MAA) or lithium methacrylate as adsorbing monomers. The affinity decreased in the swollen state but was recovered upon shrinking. This suggests that the imprinted gels possessed a reversible adsorption ability, which was controlled by the folding and unfolding of the polymer, i.e. the volume phase transition. A different procedure was employed to prepare temperature responsive imprinted polymers without using a template (D'Oleo et al., 2001). These polymers were based on NIPAM and *N,N*-cystaminebis(acrylamide) weakly cross-linked with *N,N*-methylenebis(acrylamide). After polymerization, the disulfur bridges in the pendant cystamine groups were cleaved and oxidized to form a pair of sulfonic functions capable of interacting with divalent cations.

3.1.3 Photo-responsive MIP

The interaction between light and a material may be used to modulate drug delivery. This can be accomplished by using a material that absorbs light at a specific wavelength and then uses the energy from the absorbed light to modulate drug delivery (Suzuki & Tanaka, 1990). Since light stimulus can be imposed instantly and delivered in specific amounts with high accuracy, light-sensitive hydrogels may have special advantages over systems that rely on other stimuli. The capacity for instantaneous delivery of the stimulus makes the development of light-sensitive materials important for various applications in both the engineering and biochemical fields (Yui et al., 1993). For example, molecularly imprinted membranes, based on a polymerizable derivative of azobenzene, *p*-phenylazoacrylanilide (PhAAAn), with photoregulated ability to interact reversibly with a predetermined compound such as dansylamide, were synthesized (Minoura et al., 2003). A mixture of ethylene glycol dimethacrylate and tetraethylene glycol diacrylate was used to prepare PhAAAn-containing membranes in the presence of the template. PhAAAn serves not only as a photoresponsive monomer but also as a functional monomer. Upon UV irradiation of these membranes, PhAAAn undergoes *trans*-to-*cis* isomerization and upon visible light irradiation, *cis*-to-*trans* isomerization occurs. Correspondingly, the shape, intensity, and positions of the absorption bands change.

3.3 MIP as chemo/biosensors

A sensor is a device that responds to a physical or chemical stimulus by producing a signal, usually electrical. As highlighted by Hillberg et al., 2005, although this is often the case for physical effectors, such as temperature, light, or weight, this is less commonly the case when a sensor's target is a particular molecule, ion or atom. In these situations an "effect" can either be specific or non-specific, can be informative or misleading. For instance the absorbance or emission spectra of an excited metal atom in a flame can be diagnostic of a particular metal whilst the UV absorption spectra of a mystery solution can be indicative but is seldom specific. And of course an additional, and often overriding complication, is that it is unusual, in "real" samples, for there to be a single species present. More commonly an analyte of interest is accompanied by a number of different species, all present at different concentrations and all adding to the complexity of the analytical problem.

All over the world, billions of dollars are spent annually on chemical/biological detections related to medical diagnosis, environmental monitoring, public security and food safety because lab analysis using expensive equipment is usually cumbersome and time-consuming. Therefore, there has been a pressing societal need for the development of chemo/biosensors for the detection of various analytes in solution and atmosphere, which are both less expensive and simpler to construct and operate. Although considerable progress was made in the past several decades, the chemo/biosensor field remains underdeveloped and at a low level of commercialization because of the lack of alternative strategies and multidisciplinary approaches (Guan et al., 2008).

The standard approach to the analytical analysis of complex matrices is the separation of the different components. Typically, therefore, before a sensor can be used to perceive and quantify one component in a mixed solution, the various components of the complex mixture must be separated, usually by a chromatographic process, so that some form of non-selective sensor, e.g. UV absorbance measurements, can be used to detect and quantify each individual component.

In order to improve the performance of chemical sensors, an improvement of their selectivity is required, so that a particular chemical species can be detected and assayed without the need for a possibly lengthy separation stage. In this direction, a technological approach is the development of the biosensor (Updike & Hicks, 1967). A biosensor is a sensor that uses biological selectivity to limit perception to the specific molecule of interest. A typical biosensor consists of two main components: the chemosensory materials (receptors) that can selectively bind target analytes and the efficient transducer that can transform the binding events into a readable signal output related to the analyte concentration in the Sample (Eggins, 2002). The efficiency of chemosensors is largely dependent on the selectivity and sensitivity of the used sensory materials to a target species. The traditional approaches are to immobilize a biological or biologically derived sensing element acting as receptor on the surface of a physical transducer to provide selective binding of analytes (Figure 5) (Orellana & Moreno-Bondi, 2005; Jiang & Ju, 2007). As sensing element, it is possible to use either biological macromolecules (e.g. antibodies, enzymes, receptors and ion channel proteins, nucleic acids, aptamers and peptide nucleic acids) or biological systems (e.g. ex vivo tissue, microorganisms, isolated whole cells and organelles). However, the small surface area and non-tunable surface properties of transducers greatly limit the efficiency of chemosensors, especially for the detection of ultratrace analytes.

Recently, nanomaterials have found a wide range of applications as a material foundation of chemosensors, and have exhibited various degrees of success in the improvement of detection sensitivity and selectivity (Gao et al., 2007; Xie et al., 2006; Banholzer et al., 2008). Nanomaterials themselves can also form a novel platform of chemical/biological detections due to their unique electrical, optical, catalytic or magnetic properties (Chen et al., 2004). Moreover, the large surface-to-volume ratio and good dispersivity of nanomaterials provide a huge adsorptive surface for enriching target species (Xie et al., 2008). Although biological receptors have specific molecular affinity and have been widely used in diagnostic bioassays and chemo/biosensors, they are often produced via complex protocols with a high cost and require specific handling conditions because of their poor stability, and the natural receptors for many detected analytes don't exist (Whitcombe et al., 2000; Wulff, 2002; Haupt & Mosbach, 2000; Ye & Haupt, 2004). Thus, there has been a strong driving force in synthesizing artificial recognition receptors. Molecular imprinting is one of the most efficient strategies that offer a synthetic route to artificial recognition systems by a template polymerization technique (Ye & Mosbach, 2001b; Spivak, 2005; Zhang et al., 2006). In this direction a recent review (Hillberg et al., 2005) highlight the importance of the concept of "engineerability" of MIP, defining as "engineerability" the materials ability to be integrated into an electro-mechanical device (Adhikari & Majumdar 2004). To date molecularly imprinted polymers have been successfully used with most types of transduction platforms and a range of methods have been used to bring about close integration of the platform with the polymer.

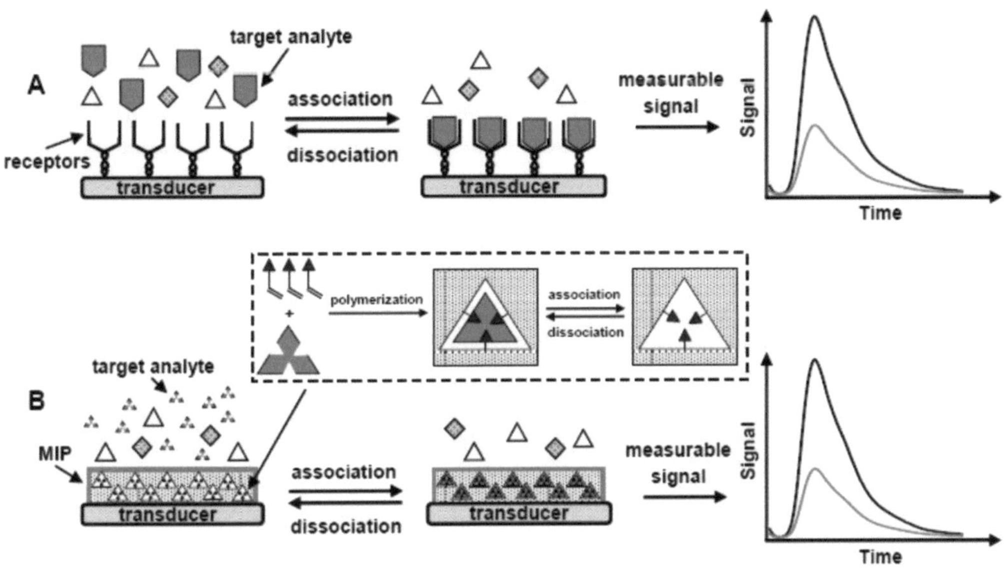

Fig. 5. Schematic representation of Biosensor. Adapted from Guan at al., 2008.

During the past ten years, the literatures on the development of MIP-based sensors, in particularly electrochemical (Riskin et al., 2008; Kan et al., 2008a) and optical (McDonagh et al., 2008; Basabe-Desmonts et al., 2007; Li et al., 2007; Feng et al., 2008) sensors, have been dramatically growing.

3.3.1 Electrochemical sensors

MIP-based electrochemical sensors were first reported in the early 1990s by Mosbach's group (Andersson et al., 1990). They described the integration of a phenylalanine anilide imprinted polymer into a field effect capacitance sensor and reported a significant reduction in the overall capacitance of the system when the sensor was exposed to the template (Hedborg et al., 1993). It was also observed that no such effect was observed when the sensor was exposed to the potential cross-reactants tyrosine anilide and phenylalaninol. The capacitance sensors based on MIPs were also fabricated and used to detect many other analytes such as amino acid derivatives with a detection limit of 500 ppm (panasyuk et al., 1999), and barbituric acid with a detection limit of 3.5 ppm (Mirsky et al., 1999). During the past decade, remarkable progress in MIP-based electrochemical sensors have been achieved by the use of conductometric/potentiometric measurements and MIP nanomaterials, greatly extending the range of detected targets and improving the sensitivity, selectivity and simplicity of electrochemical sensors (Zhou et al., 2003). Different MIP sensing device designed with therapeutic application were prepared, by employing amperometric and/or voltammetry measurements and using several different templates, such as morphine, (Kriz & Mosbach, 1995), atrazine (Kim et al., 2007), benzyltriphenylphosphonium chloride (Kriz & Mosbach, 1995), thiophenol (Kröger et al., 1999), glutamic acid (Ouyang et al., 2007). Recently, the electrochemical sensors are fabricated by installing MIP nanomaterials, as recognition elements, onto the surface of electrode. The changes of current and peak voltage at cyclic voltammetry upon the analyte binding can sensitively respond to the concentration and kind of analytes, respectively, because of the oxidation or reduction of analytes at the MIP-modified electrode. In this direction, sensor for the detection of several analytes were developed: (Prasad et al., 2010a; Prasad et al., 2010b), tolazoline (Zhang et al., 2010a), tryptophan (Prasad et al., 2010c; Kong et al., 2010), clindamycin (Zhang et al., 2010b), 2,4-dichlorophenoxy acetic acid (Xie et al, 2010), histamine (Bongaers et al., 2010), caffeine (Alizadeh et al., 2010; Vinjamuri et al., 2008), uracil and 5-fluorouracil (Prasad et al., 2009a), salicylic acid (Kang et al., 2009), uric acid (Patel et al., 2009), resveratrol (Xiang & Li, 2009), hydroquinone (Kan et al., 2009; Kan et al., 2008a), bisphenol (Kolarz & Jakubiak, 2008), dopamine (Kan et al., 2008b).

3.3.2 Optical sensors

Of various signal transducers, optically addressable sensors based on fluorescent "turn-on" or "turn-off" mechanism have been demonstrated to be highly desirable for a variety of small molecular analytes in many challenging environments, due to their high signal output and feasible measurements (Holthoff & Bright, 2007a; Holthoff & Bright, 2007b). One of the first earliest MIP sensors studies described an optical device for sensing l-dansyl phenylalanine (Kriz et al., 1995). In this simple study polymer particles, imprinted with the fluorescent template l-dansyl phenylalanine, were sealed beneath a quartz window and re-exposed to the template. The fluorescence response of the systems was shown to be related to the concentration of template and importantly that this response was stereoselective. recent progress in the covalent linkage of MIPs to optical transducers has allowed for the realisation of highly efficient and robust optical MIP-based molecular recognition sensors (Henry et al., 2005). Most of the strategies involve in the design and use of fluorescent ligands and fluorotag-ligand conjugates in the preparation of the fluorescent sensors. Fluorescent functional monomers are coupled with imprinted sites, exhibiting fluorescence enhancement or quenching upon the analyte binding. In this direction, a 2-acrylamidoquinoline as a fluorescent functional monomer with a polymerizable acrylate moiety and a fluorescent hydrogen-bonding moiety was designed and synthesized (Kubo et al., 2005). The template cyclobarbital was imprinted into a polymer matrix by using the fluorescent functional monomer, in which the remarkable fluorescent enhancement upon the hydrogen bonding of the target into the imprinted sites was observed. The fluorescent sensor demonstrated the ability to signal the presence and concentration of the analyte with a detection range of 0.1-2.0 mM.

Wang et al. developed a system that responded to the binding event with a significant fluorescence intensity change without the use of an external quencher (Wang et al., 1999; Gao et al., 2001). The key to this was the use of a fluorescent, anthracene containing monomer that was substituted with a boronic acid containing group. When the template, d-fructose, was re-introduced into the system a large change in fluorescence was observed.

This was attributed to the reformation of the boronic ester with the cis-diol of the fructose. In a different approach, Ye et al. incorporated a fluorescent scintillant into polymer microspheres imprinted with (S)-propranolol (Ye et al., 2002). When the MIP was used in scintillation proximity assays the specific binding of the radio-labelled template resulted in a transfer of energy from template to scintillant resulting in the generation of a fluorescence signal. Furthermore, Detection and quantification of Dextromethorphan is a pharmacological important marker drug used to identify the activity of the CYP2D6 class of p450 monoxygenases, is achieved by measuring the refractive index changes of multiple surface plasmons resulting from the binding to template pockets within the thin layer imprintd β-cyclodextrin polymer.
To date, MIP based optical sensors were successfully prepared for the selective recognition of different templates, such as, digoxin (Gonzalez et al., 2009), monoamine naphthalenes (Valero-Navarro et al., 2009), atrazine (Wu et al., 2008), aflatoxin B1 (Mosbach, 2006), dopamine (Kan et al. 2008b), In addition to the abovementioned fluorescence enhancement, photoinduced electron transfer (Leung et al., 2001), quencher-analyte competition adsorption (Liao et al., 1999) and chemiluminescene (Lin & Yamada, 2000) have been extensively explored to signal the analyte binding events. Photoinduced electron transfer has been a very popular mode of sensing in fluorescent molecular recognition in recent years (Basabe-Desmonts et al., 2007). It was demonstrated that the use of electron transfer mechanism as a means of signal transduction is feasible for the fluorescent detection of non-fluorescent analyte. A sol-gel molecularly imprinted luminescent sensor was fabricated by using a tailor-made organosilane as fluorescent functional monomer and 2,4-D as template molecule. Luminescence of the templatewas greatly enhanced by the formation of acid-base ion pairs with 2,4-D, because of the suppression of photoinduced electron transfer quenching on the anthryl fluorophore emission. Therefore, the imprinted sol-gel materials exhibited a selective fluorescent response to 2,4-D by the significant enhancement of fluorescence. A gradually rising trend in luminescent intensity was observed with increasing 2,4-D concentration from 10 to 166.6 μg mL^{-1}, while the control materials showed negligible response in luminescent intensity (Leung et al., 2001).

3.3.3 Mass sensitive devices

In principle, the measurement of mass is the most general method suitable for the detection of any analyte since the mass is a universal property of matter. Piezoelectric devices such as a quartz crystal microbalance (QCM) can provide an extremely sensitive measurement to the mass of the analyte binding at the surface of piezoelectric materials. When the mass of a piezoelectric material (e.g. quartz) changes there is an accompanying change in the resonant frequency and this change can be measured very precisely. A general rule of thumb being that for a system resonating at 10 MHz a change in mass of 1ng results in a 1 Hz chance in resonant frequency. In practice this means that when an analyte binds to the surface of a piezoelectric device, such as a quartz crystal microbalance (QCM), its presence is detected through a change in the resonant frequency of the system. When a molecular imprinted polymer layer is attached to the surface of a QCM the system can be used to measure template specific binding with high degree of sensitivity. MIP-based piezoelectric sensors has increased at a relatively slow rate compared with electrochemical and optical sensors, the synergetic advantages of the selectivity provided by MIP with the sensitivity provided by piezoelectric sensing makes the sensors almost universally applicable with good limits of detection, low cost and the possibility of easy miniaturization and automation (Haupt et al., 1999; Tanaka, 2007; Ayela et al., 2007). The applications of MIP nanomaterials in piezoelectric sensors extends from small molecules to biomacromolecules and to bulky analytes such as microorganisms and cells. In 1996 Dickert and Thierer coated QCM surfaces with cross-linked polyurethanes molecularly imprinted with different solvents (Dickert & Thierer, 1996). The resulting sensor was shown to be selective for the template solvent. This is particularly interesting since it suggested that polymer selectivity could be achieved for small and poorly functional molecules such as tetrahydrofuran and chloroform. Krozer (Reimhult et al., 2008) reported the QCM sensor with dissipation (QCM-D) by coating the sensor surface with pre-made molecularly imprinted nanoparticles. The nanoparticles were physically entrapped into a thin poly(ethylene terephthalate) (PET) layer spin-coated on the transducer surface. By controlling the deposition conditions, a high nanoparticle loading can be gained in the stable PET layer, allowing the recognition sites in nanoparticles to be easily accessed by the test analytes. The highest uptake of the nanoparticle film to propranolol corresponded to approximately 2 nmol cm^{-2} or about 1x1015 molecules cm^{-2}. The detection limit of the MIP-QCM sensor was about 10 μM, and the chiral recognition and discrimination between *R*- and *S*-propranolol can also be achieved.

Mass sensitive MIP sensors have also been viewed as good candidates for use in therapeutic monitoring and a number of therapeutically interesting targets have been studied. Liang et al. developed a highly selective and sensitive caffeine sensor which performed well in both serum and urine samples (Liang et al., 1999), the same group also used a similar bulk acoustic wave mass sensitive techniques to prepare MIP sensors for the direct determination of epinephrine (Liang et al., 2000), the antimicrobial agent pyrimethamine (Peng et al., 2000a), Phenobarbital (Peng et al., 2000b), (Yao et al., 2000), atropine (Peng et al., 2000c), and dopamine (Prasad et al., 2009b).

3.4 MIP as artificial receptors and antibodies

The design and synthesis of biomimetic receptor systems capable of binding a target molecule with similar affinities and specificities to their natural counterparts has long been a goal of bioorganic chemistry. Due to their unique binding characteristics (in terms of affinity and specificity), their high chemical and physical stability, ease availability and low cost, molecularly imprinted polymers are sometimes referred to as artificial antibodies and are considered an alternative to antibodies. (Ye & Haupt, 2004)

Molecularly imprinted polymers are certainly very different from antibodies; they are large, rigid and insoluble, whereas antibodies are small, flexible and soluble. However, as before mentioned, MIPs share with antibodies one of their most important features: the ability to selectively bind a target molecule.

To be used as receptor or antibodies and potentially as a drug, a MIP should be water-compatible and be synthesized from biocompatible building blocks.

At the time being, the majority of reports on molecularly imprinted polymers describe organic polymers synthesized from vinyl or acrylic monomers by radical polymerisation, and using non-covalent interactions. This can be attributed to the rather straightforward synthesis of these materials, and to the vast choice of available monomers with different functional groups. These can be basic (*e.g.* vinylpyridine) or acid (*e.g.* methacrylic acid), permanently charged (*e.g.* 3-acrylamidopropyltrimethylammonium chloride), hydrogen bonding (*e.g.* acrylamide), hydrophobic (*e.g.* styrene), metal coordinating, *etc.* These functional monomers are sometimes considered analogous to the 20 amino acids that constitute the building blocks of proteins. These simple monomers have association constants with the template that are too low for the formation of a stable complex (although in the final polymer, the formation of several simultaneous interactions and a favourable entropy term normally assure tight binding of the target molecule). During non-covalent imprinting, functional monomers have to be used in excess to shift the equilibrium towards complex formation, resulting in some functional groups being randomly distributed throughout the polymer, which in turn is one of the reasons for non-specific binding. Compared to proteins that nature has selected for the required recognition and binding properties through evolution or, in the case of antibodies, clonal selection, this is a considerable drawback. Therefore, somewhat more sophisticated monomers are being designed that form more stable interactions with the template molecule or substructures thereof, and that can be used in a stoichiometric ratio. Other organic polymers are sometimes used for imprinting that are either better suited for a specific application or easier to synthesise in the desired form, for example poly(phenylene diamine), overoxidised polypyrrole, or polyurethanes. Imprinting is also possible in inorganic matrices, in particular sol- gels of silica or titanium dioxide. The molecular imprinting technique can be applied to different kinds of target molecules, ranging from small, organic molecules (*e.g.* pharmaceuticals, pesticides, amino acids and peptides, nucleotide bases, steroids and sugars) to peptides and proteins.

The first paper in this application area being a report by Mosbach's group on the development of a MIP-based immunoassay against theophylline and diazepam (Vlatakis et al., 1993). In this and other examples, MIPs have been used as substitutes for antibodies in radioimmunoassays (RIA) for drugs, showing strong binding to the target analytes and cross-reactivity profiles similar to those of antibodies. The dissociation constants that have been measured by some authors were found to be in the nanomolar to micromolar range (Andersson et al., 1995; Ramstrom et al., 1996). This is in the same range as the average antibody, although antibodies exist that have an affinity for their antigen several orders of magnitude higher. In a competitive radioimmunoassay based on MIP, the radioisotope-labeled target analyte is incubated with increasing amounts of non-labeled target to compete for binding to a limited amount of MIP. After the equilibrium is reached, the amount of label bound to the MIP, which is inversely related to the concentration of nonlabeled analyte, is quantified by radioactivity measurements.

A plot of bound label against the concentration of non-labeled analyte gives a typical sigmoidal calibration curve, which can be used to calculate the MIP's binding affinity and site population. The concentration of non-labeled analyte displacing 50% of the label is defined as the $IC50$ value. The same experiment can be repeated using other related drugs as the competing ligand, which gives displacement curves (and $IC'50$) shifted to a higher concentration range. The MIP's cross-reactivity for the new ligand is defined as the percentage of $IC50/IC'50$.

By incorporating an appropriate scintillation reporter element, MIPs can be designed to directly generate a specific physicochemical signal upon binding of an analyte. In Figure 6, the principle of of using a "universal" scintillation reporter embedded in molecularly imprinted microspheres is reported (Ye & Mosbach, 2001c; Ye et al., 2002).

The MIP containing the scintillation reporter is imprinted against a β-adrenergic antagonist, *S*-propranolol. When tritium-labeled *S*-propranolol binds to the MIP, its β-radiation triggers the nearby reporter to emit long wavelength fluorescence that can be directly quantified.

When used in competitive-assay mode, the fluorescence signal decreases due to the non-labeled analyte competing for the limited number of binding sites. This MIP-based scintillation proximity assay (SPA) has the potential to provide a very high sample throughput, since it is a quasi-homogeneous assay that does not require washing steps to separate unbound radioligand from its bound fraction before quantification.

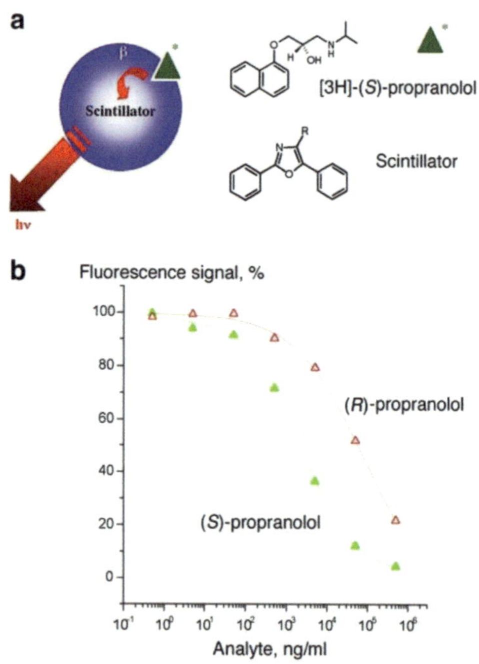

Fig. 6. MIP-based proximity scintillation assay. **a** The *S*-propranolol- imprinted microspheres contain a scintillation reporter located in proximity to the specific binding site. Binding of [3H]*S*-propranolol makes the β-electron from the radioisotope decay stimulate the reporter to generate long wavelength fluorescence. **b** Calibration curve. In competitive mode, the non-labeled *S*-propranolol displaces the [3H]*S*-propranolol, and so reduces the fluorescence signal. Adapted from Ye et al., 2002.

Imprinted-polymer-based assays are conveniently performed using radiolabels, because the labelled analyte has the same structure as the original template. However, this involves the handling of radioactive materials and produces radioactive waste, which is sometimes undesirable. Interest is therefore increasing in the development of alternative assay formats based on other detection methods that could use, just like immunoassays, an enzyme reaction or fluorescence for detection. Several years ago competitive immunoassays that use a fluorescent probe (Haupt et al., 1998) or an electroactive probe (Kroger et al., 1999) for detection were proposed. These assays were based on a polymer imprinted with the herbicide 2,4-dichlorophenoxyacetic acid, and the probes were not related to the analyte but had some structural similarity with it. It was shown that although binding of the probes to the polymer was only a few percent as compared to the analyte, specificity and selectivity of the assay were on a par with a competitive radioligand binding assay using the same polymer and the radiolabelled analyte. The fluorescent assay could be performed in aqueous buffer as well as in organic solvents such as, acetonitrile. The real challenge, however, has always been to use enzyme labels. Although most common with immunoassays, enzymes seemed to be less practical in MIPs assays for two reasons: first, they often only work in aqueous buffers, whereas the use of many imprinted polymers used to be restricted to organic solvents. Second, the rather hydrophobic nature and highly cross-linked structure of the polymer limits the access of the imprinted binding sites by the large protein molecules.

However, during the last few years, MIPs that perform well in aqueous solvents have been developed, and Haupt et al. have shown that the problem of binding site accessibility might be circumvented by using, instead of large porous MIP particles, imprinted microspheres that have binding sites at or close to their surface (Andersson, 1996). They have developed ELISA-type assays where the analyte was labelled with the enzyme peroxidase. Thus, colorimetry or chemiluminescence could be used for detection. A colorimetric assay has also been reported by Piletsky and colleagues. They have developed a method where the polymer is *in situ* synthesised in the wells of a polystyrene microtiter plate. Aminophenylboronic acid was polymerised in the presence of epinephrine (the target analyte) using oxidation of the monomer by ammonium persulfate. This process resulted in the grafting of a thin polymer layer onto the polystyrene surface (Piletsky et al., 2000). The polymer was then used in a competitive enzyme-linked assay with a conjugate of horseradish peroxidase and norepinephrine.

4. References

Adhikari, B. & Majumdar, S. (2004) Polymers in sensor applications. *Progress in Polymer Science*, 29, 699– 766.

Alizadeh, T.; Ganjali, M.R.; Zare, M. & Norouzi, P. (2010). Development of a voltammetric sensor based on a molecularly imprinted polymer (MIP) for caffeine measurement. *Electrochimica Acta*, 55, 1568-1574.

Allender, C.J.; Brain, K.R. & Heard, C.M. (2005). Molecularly imprinted polymers-preparation, biomedical applications and technical challenges. *Progress in Medicinal Chemistry*, 36, 235-291.

Alvarez-Lorenzo, C. & Concheiro, A. (2004). Molecularly imprinted polymers for drug delivery. *Journal of Chromatography B*, 804, 231-245.

Alvarez-Lorenzo, C. & Concheiro, A. (2008). Intelligent drug delivery systems: Polymeric micelles and hydrogels. *Mini-Reviews in Medicinal Chemistry*, 8, 1065-1074.

Alvarez-Lorenzo, C.; Guney, O.; Oya, T.; Sakai, Y.; Kobayashi, M.; Enoki, T.; Takeoka, Y.; Ishibashi, T.; Kuroda, K.; Tanaka, K.; Wang, G.; Grosberg, A.Y.; Masamune, S. & Tanaka, T. (2001). Reversible adsorption of calcium ions by imprinted temperature sensitive gels. *Journal of Chemical Physics*, 114, 2812-2816.

Anderson, J.M. (1994). In vivo biocompatibility of implantable delivery systems and biomaterials. *European Journal of Pharmaceutics and Biopharmaceutics*, 40, 1-8.

Andersson, L.I. (1996) Application of molecular imprinting to the development of aqueous buffer and organic solvent based radioligand binding assays for (S)-propranolol. *Analytical Chemistry*, 68, 111–117.

Andersson, L.I.; Miyabayashi, A.; O'Shannessy, D.J. & Mosbach, K. (1990). Enantiomeric resolution of amino acid derivatives on molecularly imprinted polymers as monitored by potentiometric measurements. *Journal of Chromatography*, 516, 323-331

Andersson, L.I.; Muller, R.; Vlatakis, G. & Mosbach, K. (1995). Mimics of the binding sites of opioid receptors obtained by molecular imprinting of enkephalin and morphine. *Proceedings of the National Academy of Sciences of the United States of America*, 92, 4788-4792.

Andersson, L.I.; Paprica, A. & Arvidsson, T. (1997). A highly selective solid phase extraction sorbent for preconcentration of sameridine made by molecular imprinting. *Chromatographia*, 46, 57-66.

Aydin, O.; Attila, G.; Dogan, A.; Aydin, M.V.; Canacankatan, N. & Kanik, A. (2002). The effects of methyl methacrylate on nasal cavity, lung, and antioxidant system (An Experimental Inhalation Study). *Toxicologic Patholoty*, 30, 350-356.

Ayela, C.; Vandevelde, F.; Lagrange, D.; Haupt, K. & Nicu, L. (2007). Combining resonant piezoelectric micromembranes with molecularly imprinted polymers. *Angewandte Chemie - International Edition*, 46, 9271-9274.

Banholzer, M.J.; Millstone, J.E.; Qin, L.D. & Mirkin, C.A. (2008). Rationally designed nanostructures for surface-enhanced raman spectroscopy. *Chemical Society Reviews*, 37, 885-897.

Basabe-Desmonts, L.; Reinhoudt, D.N. & Crego-Calama, M. (2007). Design of fluorescent materials for chemical sensing. *Chemical Society Reviews*, 36, 993-1017.

Benito-Peña, E.; Martins, S.; Orellana, G. & Moreno-Bondi, M.C. (2009). Watercompatible molecularly imprinted polymer for the selective recognition of fluoroquinolone antibiotics in biological samples. *Analytical and Bioanalytical Chemistry*, 393, 235–245.

Bongaers, E.; Alenus, J.; Horemans, F.; Weustenraed, A.; Lutsen, L.; Vanderzande, D.; Cleij, T.J.; Troost, G.J.; Brummer, R.-J. & Wagner, P. (2010). A MIP-based biomimetic sensor for the impedimetric detection of histamine in different pH environments. *Physica Status Solidi (A) Applications and Materials*, 207, 837-843.

Boos, K.S. & Fleischer, C.T. (2001). Multidimensional on-line solid-phase extraction (SPE) using restricted access materials (RAM) in combination with molecular imprinted polymers (MIP). *Journal of Analytical Chemistry*, 371, 16–20.

Brüggemann, O.; Visnjevski, A.; Burch, R. & Patel., P. (2004). Selective extraction of antioxidants with molecularly imprinted polymers. *Analytica Chimica Acta*, 504, 81-88.

Bures, P.; Huang, Y.; Oral, E. & Peppas, N.A. (2001). Surface modifications and molecular imprinting of polymers in medical and pharmaceutical applications. *Journal of Controlled Release*, 72, 25–33.

Caro, E.; Marcè, R.M.; Cormack, P.A.G.; Sherrington, D.C. & Borrull, F. (2003). On-line solid-phase extraction with molecularly imprinted polymers to selectively extract substituted 4-chlorophenols and 4-nitrophenol from water. *Journal of Chromatogaphy A*, 995, 233-240.

Caro, E.; Masquè, N.; Marcè, R.M.; Borrull, F.; Cormack, P.A.G. & Sherrington, D.C. (2002). Non-covalent and semi-covalent molecularly imprinted polymers for selective on-line solid-phase extraction of 4-nitrophenol from water samples. *Journal of Chromatography A*, 963, 169-178.

Chapuis, F.; Pichon, V.; Lanza, F.; Sellergren, B. & Hennion, M.-C. (2003). Optimization of the class-selective extraction of triazines from aqueous samples using a molecularly imprinted polymer by a comprehensive approach of the retention mechanism. *Journal of Chromatography A*, 999, 23-30.

Chen, J.R.; Miao, Y.Q.; He, N.Y.; Wu, X.H. & Li, S.J. (2004). Nanotechnology and biosensors. *Biotechnology Advances*, 22, 505-518.

Chien, Y.W. & Lin, S. (2002). Optimisation of treatment by applying programmable rate-controlled drug delivery technology. *Clinical Pharmacokinetics*, 41, 1267-1299.

Cirillo, G.; Curcio, M.; Parisi, O.I.; Puoci, F.; Iemma, F.; Spizzirri, U.G. & Picci, N. (2009a). Gastro-intestinal sustained release of phytic acid by molecularly imprinted microparticles. *Pharmaceutical Development and Technology*, DOI: 10.3109/10837450903397602.

Cirillo, G.; Iemma, F.; Puoci, F.; Parisi, O.I.; Curcio, M.; Spizzirri, U.G. & Picci, N. (2009b). Imprinted hydrophilic nanospheres as drug delivery systems for 5-fluorouracil sustained release. *Journal of Drug Targeting*, 17, 72-77

Cirillo, G.; Parisi, O.I.; Curcio, M.; Puoci, F.; Iemma, F.; Spizzirri, U.G. & Picci, N. (2010b). Molecularly Imprinted Polymers as Drug Delivery Systems for the Sustained Release of Glycyrrhizic Acid. *Journal of Pharmacy and Pharmacology*, 62, 577-582.

Cirillo, G.; Puoci, F.; Curcio, M.; Parisi, O.I.; Iemma, F.; Spizzirri, U.G. & Picci, N. (2010a). Molecular imprinting polymerization by Fenton reaction. *Colloid and Polymer Science*, 288, 689-693.

Cunliffe, D.; Kirby, A. & Alexander, C. (2005). Molecularly imprinted drug delivery systems. *Advanced Drug Delivery Reviews*, 57, 1836-1853.

Curcio, M.; Puoci, F.; Spizzirri, U.G.; Iemma, F.; Cirillo, G.; Parisi, O.I. & Picci, N. (2010). Negative Thermo-responsive Microspheres Based on Hydrolyzed Gelatin as Drug Delivery Device. *AAPS PharmSciTech*, DOI:10.1208/s12249-010-9429-5.

Curcio, M.; Parisi, O.I.; Cirillo, G.; Spizzirri, U.G.; Puoci, F.; Iemma, F. & Picci N. (2009) Selective Recognition of Methotrexate by Molecularly Imprinted Polymers" *E-polymers, 78*, 1-7.

D'Oleo, R.; Alvarez-Lorenzo, C. & Sun, G. (2001). A new approach to design imprinted polymer gels without using a template. *Macromolecules*, 34, 4965-4971.

Davis, K.A. & Anseth, K.S. (2002). Controlled release from crosslinked degradable networks. *Critical Reviews in Therapeutic Drug Carrier Systems*, 19, 385-423.

Dickert, F.L. &. Thierer S. (1996). Molecularly imprinted polymers for optochemical sensor. *Advanced Materials*, 8, 987–990.

Dirion, B.; Lanza, F.; Sellergren, B.; Chassaing, C.; Venn, R. & Berggren, C. (2002). Selective solid phase extraction of a drug lead compound using molecularly imprinted polymers prepared by the target analogue approach. *Chromatographia*, 56, 237-241.

Eggins, B.R. (2002). *Chemical sensors and biosensors*. John Wiley, Chichester, UK.

Ellwanger, A.; Berggren, C.; Bayoudh, S.; Crecenzi, C.; Karlsson, L.; Owens, P.K.; Ensing, K.; Cormack, P.A.G.; Sherrington, D.C. & Sellergren, B. (2001). Evaluation of methods aimed at complete removal of template from molecularly imprinted polymers. *Analyst*, 126 784-792.

Feng, F.D.; He, F.; An, L.L.; Wang, S.; Li, Y.H. & Zhu, D.B. (2008). Fluorescent conjugated polyelectrolytes for biomacromolecule detection. *Advanced Materials*, 20, 2959-2964.

Gao, D.; Zhang, Z.; Wu, M.; Xie, C.; Guan, G. & Wang, D. (2007). A surface functional monomer-directing strategy for highly dense imprinting of TNT at surface of silica nanoparticles. *Journal of the American Chemical Society*, 129, 7859-7866.

Gao, S.H.; Wang, W. & Wang, B.H. (2001). Building fluorescent sensors for carbohydrates using template-directed polymerizations. *Bioorganic Chemistry*, 29, 308– 320.

Gil, E.S. & Hudson, S.M. (2004). Stimuli-reponsive polymers and their bioconjugates. *Progress in Polymer Science (Oxford)*, 29, 1173-1222.

González, G.P.; Hernando, P.F. & Alegría, J.S.D. (2009). An optical sensor for the determination of digoxin in serum samples based on a molecularly imprinted polymer membrane. *Analytica Chimica Acta*, 638, 209-212.

Guan, G.; Liu, B.; Wang, Z. & Zhang, Z. (2008). Imprinting of Molecular Recognition Sites on Nanostructures and its Applications in Chemosensors. *Sensors*, 8, 8291-8320.

Haginaka, J.; Takehira, H.; Hosoya, K. & Tanaka, N. (1999). Uniform-sized molecularly imprinted polymer for (S)-naproxen selectively modified with hydrophilic external layer. *Journal of Chromatography A*, 849, 331–339.

Hansen, D.E. (2007). Recent developments in the molecular imprinting of proteins. *Biomaterials*, 28, 4178-4191.

Haupt, K.; Mayes, A.G. & Mosbach, K. (1998). Herbicide assay using an imprinted polymer-based system analogous to competitive fluoroimmunoassays. *Analytical Chemistry*, 70, 3936-3939.

Haupt, K. & Mosbach, K. (2000). Molecularly imprinted polymers and their use in biomimetic sensors. *Chemical Revies* 100, 2495-2504.

Haupt, K. (2003). Imprinted polymers - Tailor-made mimics of antibodies and receptors. *Chemical Communications*, 9, 171-177.

Haupt, K.; Dzgoev, A. & Mosbach, K. (1998). Assay system for the herbicide 2,4-dichlorophenoxyacetic acid using a molecularly imprinted polymer as an artificial recognition element. *Analytical Chemistry*, 70, 628-631.

Haupt, K.; Noworyta, K. & Kutner, W. (1999). Imprinted polymer-based enantioselective acoustic sensor using a quartz crystal microbalance. *Analytical Communications*, 36, 391-393.

Hedborg, E.; Winquist, F.; Lundström, I.; Andersson, L.I. & Mosbach, K. (1993). Some studies of molecularly-imprinted polymer membranes in combination with field-effect devices. *Sensors and Actuators, A: Physical* 37-8, 796–799.

Henry, O.Y.F.; Cullen, D.C. & Piletsky, S.A. (2005). Optical interrogation of molecularly imprinted polymers and development of MIP sensors: A review. *Analytical and Bioanalytical Chemistry*, 382, 947-956.

Hillberg, A.L.; Brain, K.R. & Allender, C.J. (2005). Molecular imprinted polymer sensors: Implications for therapeutics. *Advanced Drug Delivery Reviews*, 57, 1875–1889.

Hillberg, A.L. & Tabrizian, M. (2008). Biomolecule imprinting: Developments in mimicking dynamic natural recognition systems. ITBM-RBM 29, 89–104.

Holthoff, E.L. & Bright, F.V. (2007a). Molecularly templated materials in chemical sensing. *Analytica Chimica Acta*, 594, 147-161.

Holthoff, E.L. & Bright, F.V. (2007b). Molecularly imprinted xerogels as platforms for sensing. *Accounts of Chemical Research*, 40, 756-767.

Hwang, C.C. & Lee, W.C. (2002). Chromatographic characteristics of cholesterol-imprinted polymers prepared by covalent and non-covalent imprinting methods. *Journal of Chromatography A*, 962, 69-78.

Iemma, F.; Cirillo, G.; Spizzirri, U.G.; Puoci, F.; Parisi, O.I. & Picci, N. (2008). Removal of metal ions from aqueous solution by chelating polymeric microspheres bearing phytic acid derivatives. *European Polymer Journal*, 44, 1183-1190.

Iemma, F.; Spizzirri, U.G.; Puoci, F.; Cirillo, G.; Curcio, M.; Parisi, O.I. & Picci, N. (2009). Synthesis and release profile analysis of thermo-sensitive albumin hydrogels. *Colloid and Polymer Science*, 287, 779-787.

Ikegami, T.; Mukawa, T.; Nariai, H. & Takeuchi, T. (2004). Bisphenol A-recognition polymers prepared by covalent molecular imprinting. *Analytica Chimica Acta*, 504, 131-135.

Jiang, H. & Ju, H.X. (2007). Enzyme-quantum dots architecture for highly sensitive electrochemiluminescenece biosensing of oxidase substrates. *Chemical Communications*, 4, 404-406.

Joshi, V.P.; Karode, S.K.; Kulkarni, M.G. & Mashelkar, R.A. (1998). Novel separation strategies based on molecularly imprinted adsorbents. *Chemical Engineering Science*, 53, 2271-2284.

Kan, X.; Geng, Z.; Wang, Z. & Zhu, J.-J. (2009). Core-Shell molecularly imprinted polymer nanospheres for the recognition and determination of hydroquinone. *Journal of Nanoscience and Nanotechnology*, 9, 2008-2013.

Kan, X.; Zhao, Q.; Zhang, Z.; Wang, Z. & Zhu, J.-J. (2008a). Molecularly imprinted polymers microsphere prepared by precipitation polymerization for hydroquinone recognition. *Talanta*, 75, 22-26.

Kan, X.; Zhao, Y.; Geng, Z.; Wang, Z. & Zhu, J.-J. (2008b). Composites of multiwalled carbon nanotubes and molecularly imprinted polymers for dopamine recognition. *Journal of Physical Chemistry C*, 112, 4849-4854.

Kang, J.; Zhang, H.; Wang, Z.; Wu, G. & Lu, X. (2009). A novel amperometric sensor for salicylic acid based on molecularly imprinted polymer-modified electrodes. *Polymer - Plastics Technology and Engineering*, 48, 639-645.

Kim, S.N.; Rusling, J.F. & Papadimitrakopoulos, F. (2007). Carbon nanotubes for electronic and electrochemical detection of biomolecules. *Advanced Materials*, 19, 3214-3228.

Kolarz, B.N. & Jakubiak, A. (2008). Catalytic activity of molecular imprinted vinylpyridine/acrylonitrile/ divinylbenzene terpolymers with guanidyl ligands-Cu(II) inside the active centres. *Polimery/Polymers*, 53, 848-853.

Komiyama, M.; Takeuchi, T.; Mukawa, T. & Asanuma, H. (2003). *Molecular Imprinting*, Wiley–VCH, Weinheim.

Kong, Y.; Zhao, W.; Yao, S.; Xu, J.; Wang, W. & Chen, Z.(2010). Molecularly imprinted polypyrrole prepared by electrodeposition for the selective recognition of tryptophan enantiomers. *Journal of Applied Polymer Science*, 115, 1952-1957.

Kriz, D. & Mosbach, K. (1995). Competitive amperometric morphine sensor-based on an agarose immobilized molecularly imprinted polymer. *Analytica Chimica Acta*, 300, 71- 75.

Kriz, D.; Ramström, O.; Svensson, A. & Mosbach, K. (1995). Introducing biomimetic sensors based on molecularly imprinted polymers as recognition elements. *Analytical Chemistry*, 67, 2142–2144.

Kröger, S.; Turner, A.P.F.; Mosbach, K. & Haupt, K. (1999). Imprinted polymer based sensor system for herbicides using differential pulse voltammetry on screen printed electrodes. *Analytical Chemistry*, 71, 3698– 3702.

Kubo, H.; Yoshioka, N. & Takeuchi, T. (2005). Fluorescent imprinted polymers prepared with 2-acrylamidoquinoline as a signaling monomer. *Organic Letters*, 7, 359-362.

Lai, J.P.; Lu, X.Y.; Lu, C.Y.; Ju, H.F. & He, X.W. (2001). Preparation and evaluation of molecularly imprinted polymeric microspheres by aqueous suspension polymerization for use as a high-performance liquid chromatography stationary phase. Analytica Chimica Acta, 442, 105–111.

Leung, M. K-P.; Choe, C-F. & Lam, M. H-W. (2001). A sol-gel derived molecular imprinted luminescent PET sensing material for 2, 4-dichlorophenoxyacetic acid. *Journal of Materials Chemistry*, 11, 2985-2991.

Li, J.; Kendig, C.E. & Nesterov, E.E. (2007). Chemosensory performance of molecularly imprinted fluorescent conjugated polymer materials. *Journal of the American Chemical Society*, 129, 15911-15918.

Liang, C.D.; Peng, H.; Bao, X.Y.; Nie, L.H. & Yao, S.Z. (1999). Study of a molecular imprinting polymer coated BAW bio-mimic sensor and its application to the determination of caffeine in human serum and urine. *Analyst*, 124, 1781–1785.

Liang, C.D.; Peng, H.; Zhou, A.H.; Nie, L.H. & Yao, S.Z. (2000). Molecular imprinting polymer coated BAW bio-mimic sensor for direct determination of epinephrine. *Analytica Chimica Acta*, 415, 135–141.

Liao, Y.; Wang, W. & Wang, B.H. (1999) Building fluorescent sensors by template polymerization: the preparation of a fluorescent sensor for L-tryptophan. *Bioorganic Chemisty*, 27, 463-476.

Lin, J.M. & Yamada, M. (2000).Chemiluminescent reaction of fluorescent organic compounds with KHSO5 using cobalt(II) as catalyst and its first application to molecular imprinting. *Analytical Chemistry*, 72, 1148-1155.

Martin, P.; Jones, G.R.; Stringer, F. & Wilson, I.D. (2003). Comparison of normal and reversed-phase solid phase extraction methods for extraction of β-blockers from plasma using molecularly imprinted polymers. *Analyst*, 128, 345-350.

Mayes, A.G. & Mosbach, K. (1996). Molecularly imprinted polymer beads: Suspension polymerization using a liquid perfluorocarbon as the dispersing phase. *Analytical Chemistry*, 68, 769–3774.

McDonagh, C.; Burke, C.S. & MacCraith, B.D. (2008). Optical chemical sensors. *Chemical Reviews*, 108, 400-422.

Minoura, N.; Idei, K.; Rachkov, A.; Uzawa, H. & Matsuda, K. (2003). Molecularly Imprinted Polymer Membranes with Photoregulated Template Binding *Chemistry of Materials*, 15, 4703-4704.

Mirsky, V.M.; Hirsch, T.; Piletsky, S.A. & Wolfbeis, O.S.(1999). A spreader-bar approach to architecture: formation of stable artificial chemoreceptors. *Angewandte Chemie - International Edition*, 38, 1108-1110.

Morikawa, H.; Koike, S.; Saiki, M. & Saegusa, Y. (2008). Synthesis and characterization of the PEG-basecl nonionic surfactants endowed with carboxylic acid moiety at the hydrophobic terminal. *Journal of Polymer Science, Part A: Polymer Chemistry*, 46, 8206-8212.

Mosbach, K. & Ramström, O. (1996). The emerging technique of molecular imprinting and its future impact on biotechnology. *Bio/Technology*, 14, 163-170.

Mosbach, K. (2001). Towards the development of artificial recognition elements in sensor technology with emphasis on molecular imprinting (E), *Electrochemistry*, 69, 919-.

Mosbach, K. (2006). The promise of molecular Imprinting. *Scientific American*, 295, 86-91.

Mullett, W.M. & Pawliszyn, J. (2003). The development of selective and biocompatible coatings for solid phase microextraction. *Journal of Separation Science*, 26, 251–260.

Oh, K.T. & Lee, E.S. (2008). Cancer-associated pH-responsive tetracopolymeric micelles composed of poly(ethylene glycol)-b-poly(L-histidine)-b-poly(L-lactic acid)-b-poly(ethylene glycol). *Polymers for Advanced Technologies*, 19, 1907-1913.

Orellana, G. & Moreno-Bondi, M.C. *Frontiers in chemical sensors: novel principles and techniques*. Springer-Verlag, Berlin, Heidelberg, New York, NY, 2005.

Ouyang, R.; Lei, J.; Ju, H. & Xue, Y. (2007). A molecularly imprinted copolymer designed for enantioselective recognition of glutamic acid. *Advanced Functional Materials* 17, 3223-3230.

Owens, P.K.; Karlsson, L.; Lutz, E.S.M. & Andersson, L.I. (1999). Molecular imprinting for bio- and pharmaceutical analysis. *Trends in Analytical Chemistry*, 18, 146-155.

Panasyuk, T.L.; Mirsky, V.M.; Piletsky, S.A. & Wolfbeis, O.S. (1999). Electropolymerized molecularly imprinted polymers as receptor layers in a capacitive chemical sensors. *Analytical Chemistry*, 71, 4609-4613.

Pap, T.; Horvath, V.; Tolokàn, A.; Horvai, G. & Sellergren, B. (2002). Effect of solvents on the selectivity of terbutylazine imprinted polymer sorbents used in solid-phase extraction. *Journal of Chromatography A*, 973, 1-8.

Parisi, O.I.; Cirillo, G.; Curcio, M.; Puoci, F.; Iemma, F.; Spizzirri, U.G. & Picci, N. (2009) Surface modifications of molecularly imprinted polymers for improved template recognition in water media. *Journal of Polymer Research*, DOI 10.1007/s10965-009-9322-7.

Patel, A.K.; Sharma, P.S. & Prasad, B.B. (2009). Electrochemical sensor for uric acid based on a molecularly imprinted polymer brush grafted to tetraethoxysilane derived sol-gel thin film graphite electrode. *Materials Science and Engineering C*, 29,1545-1553.

Peng, H.; Liang, C.D.; He, D.L.; Nie, L.H. & Yao, S.Z. (2000a). Bulk acoustic wave sensor using molecularly imprinted polymers as recognition elements for the determination of pyrimethamine. *Talanta*, 52, 441- 448.

Peng, H.; Liang, C.D.; He, D.L.; Nie, L.H. & Yao, S.Z. (2000b). Nonaqueous assay system for phenobarbital using biomimetic bulk acoustic wave sensor based on a molecularly imprinted polymers. *Analytical Letters*, 33, 793–808.

Peng, H.; Liang, C.D.; Zhou, A.H.; Zhang, Y.Y.; Xie, Q.J. & Yao, S.Z. (2000c). Development of a new atropine sulfate bulk acoustic wave sensor based on a molecularly imprinted electrosynthesized copolymer of aniline with o-phenylenediamine. *Analytica Chimica Acta*, 423, 221–228.

Peppas, N.A. & Leobandung, W. (2004). Stimuli-sensitive hydrogels: Ideal carriers for chronobiology and chronotherapy. *Journal of Biomaterials Science, Polymer Edition*, 15, 125-144.

Pérez-Álvarez, L.; Sáez-Martínez, V.; Hernáez, E. & Katime, I. (2008). Synthesis and characterization of pH-sensitive microgels by derivatization of npa-based reactive copolymers. *Materials Chemistry and Physics*, 112), 516-524.

Piletsky, S.A.; Andersson, H.S. & Nicholls. I.A. (1999). Combined Hydrophobic and Electrostatic Interaction-Based Recognition in Molecularly Imprinted Polymers. *Macromolecules*, 32, 633-636.

Piletsky, S.A.; Piletska, E.V.; Chen, B.; Karim, K.; Weston, D.; Barrett, G.; Lowe, P. & Turner A.P.F. (2000). Chemical grafting of molecularly imprinted homopolymers to the surface of microplates. Application of artificial adrenergic receptor in enzyme-linked assay for β-agonists determination. *Analytical Chemistry*, 72, 4381-4385.

Piscopo, L.; Prandi, C.; Coppa, M.; Sparnacci, K.; Laus, M.; Lagana, A.; Curini, R. & D'Ascenzo, G. (2002). Uniformly sized molecularly imprinted polymers (MIPs) for 17 β-estradiol. *Macromolecular Chemistry and Physics*, 203:532–1538.

Prasad, B.B.; Madhuri, R.; Tiwari, M.P. & Sharma, P.S. (2010a). Imprinted polymer-carbon consolidated composite fiber sensor for substrate-selective electrochemical sensing of folic acid. *Biosensors and Bioelectronics*, 25, 2140-2148.

Prasad, B.B.; Madhuri, R.; Tiwari, M.P. & Sharma, P.S. (2010b). Electrochemical sensor for folic acid based on a hyperbranched molecularly imprinted polymer-immobilized sol-gel-modified pencil graphite electrode. *Sensors and Actuators, B: Chemical*, 146, 321-330.

Prasad, B.B.; Madhuri, R.; Tiwari, M.P. & Sharma, P.S. (2010c). Enantioselective recognition of d- and l-tryptophan by imprinted polymer-carbon composite fiber sensor. *Talanta*, 81, 187-196.

Prasad, B.B.; Srivastava, S.; Tiwari, K. & Sharma, P.S. (2009a). Development of uracil and 5-fluorouracil sensors based on molecularly imprinted polymer-modified hanging mercury drop electrode. *Sensors and Materials*, 21, 291-306.

Prasad, B.B.; Srivastava, S.; Tiwari, K. & Sharma, P.S. (2009b). Trace-level sensing of dopamine in real samples using molecularly imprinted polymer-sensor. *Biochemical Engineering Journal*, 44, 232-239.

Puoci, F.; Garreffa, C.; Iemma, F.; Muzzalupo, R.; Spizzirri, U.G. & Picci, N. (2005). Molecularly imprinted solid phase extraction for detection of sudan I in food matrices. *Food Chemistry*, 93, 349-353.

Puoci, F.; Cirillo, G.; Curcio, M.; Iemma, F.; Parisi, O.I.; Castiglione, M. & Picci, N. (2008c). Molecularly imprinted polymers for α-tocopherol delivery. *Drug Delivery*, 15, 253-258.

Puoci, F.; Cirillo, G.; Curcio, M.; Iemma, F.; Spizzirri, U.G. & Picci, N. (2007a). Molecularly Imprinted Solid Phase Extraction for the selective HPLC determination of α-tocopherol in Bay Leaves. *Analytica Chimica Acta*, 593, 164-170.

Puoci, F.; Curcio, M.; Cirillo, G.; Iemma, F.; Spizzirri, U.G. & Picci, N. (2008a). Molecularly imprinted solid-phase extraction for cholesterol determination in cheese products. *Food Chemistry*, 106, 836-842.

Puoci, F.; Iemma, F.; Cirillo, G.; Curcio, M.; Parisi, O.I.;Spizzirri, U.G. & Picci, N. (2009). New Restricted Access Materials combined to Molecularly Imprinted Polymers for Selective Recognition/Release in Water Media. *European Polymer Journal*. 45, 1634-1640.

Puoci, F.; Iemma, F. & Picci, N. (2008b). Stimuli-responsive molecularly imprinted polymers for drug delivery: A review. *Current Drug Delivery*, 5, 85-96.

Puoci, F.; Iemma, F.; Cirillo, G.; Picci, N.; Matricardi, P. & Alhaique, F. (2007b). Molecularly imprinted polymers for 5-fluorouracil release in biological fluids. *Molecules*, 12, 805-814.

Puoci, F.; Iemma, F.; Muzzalupo, R.; Spizzirri, U.G.; Trombino, S.; Cassano, R. & Picci, N. (2004). Spherical Molecularly Imprinted Polymers (SMIPs) via a Novel Precipitation Polymerization in the Controlled Delivery of Sulfasalazine. *Macromolecular Bioscience*, 4, 22-26.

Ramström, O. & Ansell, R.J. (1998). Molecular imprinting technology: Challenges and prospects for the future. *Chirality*, 10, 195-209.

Ramström, O.; Ye, L. & Mosbach, K. (1996). Artificial antibodies to corticosteroids prepared by molecular imprinting. *Chemistry and Biology*, 3, 471-477.

Reimhult, K.; Yoshimatsu, K.; Risveden, K.; Chen, S.; Ye, L. & Krozer, A. (2008). Characterization of QCM sensor surface coated with molecularly imprinted nanoparticles. *Biosensors and Bioelectronics*, 23, 1908-1914.

Riskin, M.; Tel-Vered, R.; Bourenko, T.; Granot, E. & Willner, I. (2008). Imprinting of molecular recognition sites through electropolymerization of functionalized Au nanoparticles: development of an electrochemical TNT sensor based on π-donor-acceptor interactions. *Journal of the American Chemical Society*, 130, 15911-15918.

Sambe, H.; Hoshina, K. & Haginaka, J. (2007). Molecularly imprinted polymers for triazine herbicides prepared by multi-step swelling and polymerization method. Their application to the determination of methylthiotriazine herbicides in river water. *Journal of Chromatography A*, 1152, 130–137.

Sanbe, H. & Haginaka J. (2003). Restricted access media-molecularly imprinted polymer for propranolol and its application to direct injection analysis of β-blockers in biological fluids. *Analyst*, 128, 593–597.

Say, R.; Birlik, E.; Ersoz, A.; Yilmaz, F.; Gedikbey, T. & Denizli, A. (2003). Preconcentration of copper on ion-selective imprinted polymer microbeads. *Analytica Chimica Acta*, 480, 251–258.

Schmaljohann, D. (2006). Thermo- and pH-responsive polymers in drug delivery. *Advanced Drug Delivery Reviews*, 58, 1655-1670.

Spivak, D.A. (2005). Optimization, evaluation, and characterization of molecularly imprinted polymers. *Advanced Drug Delivery Reviws*, 57, 1779-1794.

Suzuki, A. & Tanaka, T. (1990). Phase transition in polymer gels induced by visible light. *Nature*, 346, 345-347.

Svenson, J. & Nicholls, I.A. (2001). On the thermal and chemical stability of molecularly imprinted polymers. *Analytica Chimica Acta*, 435, 19-24.

Tanaka, M. (2007). An industrial and applied review of new MEMS devices features. *Microelectronic Engineering*, 84, 1341-1344.

Tunc, Y.; Hasirci, N.; Yesilada, A. & Ulubayram, K. (2006). Comonomer effects on binding performances and morphology of acrylate-based imprinted polymers. *Polymer*, 47, 6931–6940.

Turiel, E. & Martın-Esteban, A. (2004). Molecularly imprinted polymers: Towards highly selective stationary phases in liquid chromatography and capillary electrophoresis. *Analytical and Bioanalytical Chemistry*, 378, 1876-1886.

Updike, S.J & Hicks, G.P. (1967). The enzyme electrode. *Nature*, 214, 986–988.

Valero-Navarro, A.; Salinas-Castillo, A.; Fernández-Sánchez, J.F.; Segura-Carretero, A.; Mallavia, R. & Fernández-Gutiérrez, A. (2009). The development of a MIP-optosensor for the detection of monoamine naphthalenes in drinking water. *Biosensors and Bioelectronics*, 24, 2305-2311.

Vandevelde, F.; Leïchlé, T.; Ayela, C.; Bergaud, C.; Nicu, L. & Haupt, K. (2007). Direct patterning of molecularly imprinted microdot arrays for sensors and biochips. *Langmuir* 23, 6490-6493.

Vinjamuri, A.K.; Burns, S.C. & Dahl, D.B. (2008). Caffeine and Theobromine Selectivity Using Molecularly Imprinted Polypyrrole Modified Electrodes. *ECS Transactions*, 13, 9-20.

Vlatakis, G. ; Andersson L.I.; Muller, R. & Mosbach K. (1993). Drug assay using antibody mimics made by molecular imprinting. *Nature*, 36, 645-647.

Wang, W.; Gao, S.H. & Wang, B.H. (1999). Building fluorescent sensors by template polymerization: the preparation of a fluorescent sensor for d-fructose, *Organic Letters*, 1, 1209- 1212.

Watanabe, M.; Akahoshi, T.; Tabata, Y. & Nakayama, D. (1998). Molecular specific swelling change of hydrogels in accordance with the concentration of guest molecules. *Journal of the American Chemical Society*, 120, 5577-5578.

Wei, S.; Molinelli, A. & Mizaikoff, B. (2006). Molecularly imprinted micro and nanospheres for the selective recognition of 17β-estradiol. *Biosensors and Bioelectronics*, 21, 1943–1951.

Whitcombe, M.J.; Alexander, C. & Vulfson, E.N. (2000). Imprinted polymers: Versatile new tools in synthesis. *Synlett*, 911-923.

Wu, Z.; Tao, C.-A.; Lin, C.; Shen, D. & Li, G. (2008). Label-free colorimetric detection of trace atrazine in aqueous solution by using molecularly imprinted photonic polymers. *Chemistry - A European Journal*, 14, pp. 11358-11368.

Wulff, G. (1995). Molecular imprinting in cross-linked materials with the aid of molecular templates - A way towards artificial antibodies. *Angewandte Chemie - International Edition*, 34, 1812-1832.

Wulff, G. (2002). Enzyme-like catalysis by molecularly imprinted polymers. *Chemical Reviews*, 102, 1-27.

Xiang, H.Y. & Li, W.G. (2009). Electrochemical sensor for trans-resveratrol determination based on indium tin oxide electrode modified with molecularly imprinted self-assembled films. *Electroanalysis*, 21,1207-1210.

Xie, C.; Gao, S.; Guo, Q. & Xu, K. (2010) Electrochemical sensor for 2,4-dichlorophenoxy acetic acid using molecularly imprinted polypyrrole membrane as recognition element. *Microchimica Acta*, 169, 145-152.

Xie, C.; Liu, B.; Wang, Z.; Gao, D.; Guan, G. & Zhang, Z. (2008). Molecular imprinting at walls of silica nanotubes for TNT recognition. *Analytical Chemistry*, 80, 437-443.

Xie, C.; Zhang, Z.; Wang, D.; Guan, G.; Gao, D. & Liu, J. (2006). Surface molecular self-assembly strategy for TNT imprinting of polymer nanowire/nanotube arrays. *Analytical Chemistry*, 78, 8339-8346.

Yao, S.Z.; Peng, H.; Liang, C.D.; Wu, Y. & Nie, L.H. (2000). Biomimetic bulk acoustic wave sensor for determination of trimethoprim in the organic phase based on a molecular imprinting polymer. *Analytical Sciences*, 16, 211 - 215.

Ye, L. & Haupt, K. (2004) Molecularly imprinted polymers as antibody and receptor mimics for assays, sensors and drug discovery. *Analytical and Bioanalytical Chemistry*, 378, 1887–1897.

Ye, L. & Mosbach, K. (2001b). Molecularly imprinted microspheres as antibody binding mimics. *Reactive and Functional Polymers*, 48, 149-157.

Ye, L. & Mosbach, K. (2001c) Generation of new enzyme inhibitors using imprinted binding sites: The anti-idiotypic approach, a step toward the next generation of molecular imprinting. *Journal of American Chemical Society*, 123, 2901-2902.

Ye, L. & Mosbach, K. (2001a). The technique of molecular imprinting - Principle, state of the art, and future aspects. *Journal of Inclusion Phenomena*, 41, 107-113.

Ye, L. & Mosbach, K. (2008). Molecular imprinting: Synthetic materials as substitutes for biological antibodies and receptors. *Chemistry of Materials*, 20, 859-868.

Ye, L.; Surugia, I. & Haupt, K. (2002). Scintilation proximity assay using molecularly imprinted polymers. *Analytical Chemistry*, 74, 959- 964.

Yourtee, D.M.; Smith, R.E.; Russo, K.A.; Burmaster, S.; Cannon, J.M.; Eick, J.D. & Kostoryz, E.L. (2001). The stability of methacrylate biomaterials when enzyme challenged: Kinetic and systematic evaluations. *Journal of Biomedical Materials Research*, 57, 522-531.

Yu, Y.; Ye, L.; Haupt, K. & Mosbach, K. (2002). Formation of a class of enzyme inhibitors (drugs), including a chiral compound, by using imprinted polymers or biomolecules as molecular-scale reaction vessels. *Angewandte Chemie - International Edition*, 41, 4459-4463.

Yui, N.; Okano, T. & Skurai, Y. (1993). Photo-responsive degradation of heterogeneous hydrogels comprising crosslinked hyaluronic acid and lipid microspheres for temporal drug delivery. *Journal of Controlled Release*, 26, 141-145.

Zhang, H.Q.; Ye, L & Mosbach, K. (2006). Non-covalent molecular imprinting with emphasis on its application in separation and drug development. *Journal of Molecular Recognition*, 19, 248-259.

Zhang, J.; Wang, Y.; Lv, R. & Xu, L. (2010a). Electrochemical tolazoline sensor based on gold nanoparticles and imprinted poly-o-aminothiophenol film. *Electrochimica Acta*, 55, 4039-4044.

Zhang, Z.; Hu, Y.; Zhang, H. & Yao, S. (2010b). Novel layer-by-layer assembly molecularly imprinted sol-gel sensor for selective recognition of clindamycin based on Au electrode decorated by multi-wall carbon nano tube. *Journal of Colloid and Interface Science*, 344, 158-164.

Zhou, Y.X.; Yu, B. & Levon, K. (2003). Potentiometric sensing of chiral amino acids. *Chemistry of Materials*, 15, 2774-2779.

29

Hydrogels as Potential Nano-Scale Drug Delivery Systems

Mohammad Reza Saboktakin, Ph. D of Nano Chemistry
Islamic Republic
Iran

1. Introduction

Man has always been plagued with ailments and diseases of both the body and the mind. However, dedicated research from scientists all over the world has made it possible to treat, prevent and eradicate many of these diseases that plague man[1]. The field of pharmaceutical science has been developing steadily over the years, and has today become invaluable in helping to keep us healthy and prevent disease. An avenue of research that has progressed a great deal in the past few decades is the treatment of diseases via biomolecules such as drugs, proteins etc. Initially, these could only be administered in limited manner, due to limitations of drug delivery through harmful environments in the body[2]. Thus mobility reduced the effectiveness of administered drugs. Progress came with the development of biomaterial carriers which could be encapsulated, or immobilized with drugs, allowing the drug to safely reach the required site without harm. These carriers allowed for the release of drug in sites which were previously inaccessible. The nature of these carriers progressed over the years from ceramics, to natural, to synthetic materials. Factors such as integrity, biocompatibility and flexibility were considered, and lead to the use of hydrophilic three dimensional matrics as carrier materials[3]. These are a class of materials known as *Hyrogels.*

These three dimensional polymer matrices are capable of imbibing large amounts of water, and biological fluids. This property of hydrogels is the reason behind its varied applications ranging from food additives to pharmaceuticals and clinical applications. Synthetic hydrogels prepared from a varied range of monomers have found many applications especially in tissue-engineering scaffolds, as carriers for implantable devices, and drug delivery devices. Synthetic hydrogels provide an effective and controlled way in which to administer protein and peptide based drugs for treatment of a number of diseases. A successful drug delivery device relies not only on competent network design, but also on accurate mathematical modeling of drug release profiles. Hydrogels have ordered polymer networks, with well-defined chemistries yielding well-defined physicochemical properties and easily reproducible drug release profiles[4]. In order to accurately understand and model drug release profiles from a material, it becomes essential to have a quantitative mathematical understanding of material properties, interaction parameters, kinetics, and transport phenomena within the material in question. The network structure also plays a key role in diffusional behavior, mesh size and stability of incorporated drug[9]. It is this well-defined order that enables accurate network design by identifying key parameters and mechanisms that govern the rate and extent of drug release.

Hydrogels have thus become a premier materials used for drug delivery formulations and biomedical implants, due to its biocompatibility, network structure, and molecular stability of the incorporated bioactive agent[5].

Hydrogels, The swellable polymeric materials, have been widely investigated as the carrier for drug delivery systems. These biomaterials have gained attention owing to their peculiar characteristics like swelling in aqueous medium, pH and temperature sensitivity or sensitivity towards other stimuli. Hydrogels being biocompatible materials have been recognized to function as drug protectors, especially for peptides and proteins, from in vivo environment. Also these swollen polymers are helpful as targetable carriers for bioactive drugs with tissue specificity.

During the past decade, novel polymeric micropheres, polymer micelles, and hydrogel-type materials have been shown to be effective in enhancing drug targeting specificity, lowering systemic drug toxicity, improving treatment absorption rates, and providing protection for pharmaceuticals against biochemical degradation. These are all goals of drug delivery. In addition, several other experimental drug delivery systems show signs of promise, including those composed of biodegradable polymers, dendrimers(so-called star polymers), electroactive polymers, and modified C-60 fullerense(Also known as "Buckyballs")[6].

Polymer drug delivery nano-scale systems are based on " *Nano carriers*" which are composed of mixing polymeric chemical compounds with drugs to forms complex, large molecules, which " *carry*" the drug across physiological barriers. Illustrative examples of these polymeric compounds are poly(rhylene-glycol)-poly(alpha, beta-asparic acid), carboxylates, and heterobifunctional polyethylene glycol, in addition to others.

Another type of nanotechnology revolves around the use of "*Hydrogel*" as nano carriers of drugs. The principle behind this technology is to use a chemical compound which traps a drug and then releases the active compound by "*Swelling*" or expanding inside of specific tissues, thus allowing a higher concentration of the drug in a biodegradable format. Hydrogels are very specialized systems and are generally formulated to meet specific needs for the delivery of individual drugs.

During the past two decades, research into hydrogel delivery systems in nano - scale has focused primarily on systems containing polyacrylic acid(PAA) backbones. PAA hydrogels are known for their super-absorbency and ability to form extended polymer networks through hydrogen bonding [7]. In addition, they are excellent bioadhesives, which means that they can adhere to mucosal linings within the gastrointestinal tract for extended periods, releasing their encapsulated medications slowly over time .

One example of the complexity of these systems is a glucose-sensitive hydrogel that could be used to deliver insulin to diabetic patients using an internal pH trigger. This system features an insulin-containing " *Reservoir*" formed by a poly[methacrylic acid -g-poly(ethylene glycol)]hydrogel membrane into which glucose oxidase has been immobilized. The membrane itself is housed between nonswelling, porous "*Molecular fences*".

2. Properties of hydrogels

Hydrogels ate water swellen polymer matrics, with a tendency to imbibe water when placed in aqueous environment. This ability to swell, under biological conditions, makes it an ideal material for use in drug delivery and immobilization of protein, peptides, and other biological compounds. Due to their high water content, these gels resemble natural living tissue more than any other type of synthetic biomaterials . These networks, have a three dimensional structure, crosslinked together either physically (entanglements, crystallites), or chemically(tie-points, junctions). This insoluble crosslinked structure allows immobilization of active agents, biomolecules effectively, and allows for its release in well-defined specific manner. Thus the hydrogels biocompatibility and crosslinked structure are responsible for its varied applications[8].

3. Mechanical properties

For non biodegradable applications, it is essential that the carrier gel matrix maintain physical and mechanical integrity. mechanical stability of the gel is, therefore, an important consideration when designing a therapeutic system. For example, drugs and other biomolecules must be protected from the harmful environments in the body such as, extreme pH environment before it is released at the required site. To this end, the carrier gel must be able to maintain its physical integrity and mechanical strength in order to prove an effective biomaterial. The strength of the material can be increased by incorporating crosslinking agents, comonomers, and increasing degree of crosslinking . There is however an optimum degree of crosslink, as a higher degree of crosslinking also leads to brittleness and less elasticity. Elasticity of the gel is important to give flexibility to the crosslinked chains, to facilitate movement of incorporated bioactive agent. Thus a compromise between mechanical strength and flexibility is necessary for appropriate use of these materials.

4. Biocompatible properties

It is important for synthetic materials, such as hydrogels, to be biocompatible and nontoxic in order for it to be a useful biomedical polymer. Most polymers used for biomedical application must pass a cytotoxicity and in-vivo toxicity tests. Most toxicity problems associated with hydrogels arise due to unreacted monomers, oligomers and initiators that leach out during application. Thus an assessment of the potential toxicity of all materials used for fabrication of gel is an integral part of determining sutiability of the gel for biological applications . To lower chances of toxic effects, the use of initiators is being eliminated, with the advent of gamma irradiation as polymerization technique. Steps are also taken to eliminate contaminants from hydrogels by repeated washing and treatment. Also, kinetics of polymerization has been studied, so as to achieve higher conversion rates, and avoid unreacted minomers and side products.

5. Classification of hydrogels

Hydrogels can be classified as neutral or ionic, based on the nature of sode groups . In neutral hydrogels, the driving force for swelling is due to the water-polymer thermodynamic mixing contribution to the overall free energy, along with elastic polymer contribution. The swelling of ionic hydrogels is also affects by the ionic interactions between charged polymers and free ions. Ionic hydrogels containing ionic groups, such as carboxylic acid, imbibe larger amount of water, because of its increased hydrophilicity. Examples of such gels are poly(acrylic acid), and polyamines. Hydrogels are also classified as homopolymers or copolymers, based on the method of preparation[9]. Hydrogels can be classified based on the physical structure of the network as amorphous, semicrystalline, hydrogen bonded structures, supermolecular structures and hydrocolloidal aggregates. An important class of hydrogels are the stimuli responsive gels. These gels show swelling behavior dependent radiation. These properties allow for usage in a number of applications, such as seprartion membranes, biosensors, artificial muscles, and drug delivery devise.

6. Types of hydrogels

6.1 pH sensitive or ion sensitive hydrogels

These hydrogels respond to changes in pH of the external environment. These gels have ionic groups (which are readily ionizable side groups)attached to impart peculiar characteristics. Some of the pH sensitive polymers used in hydrogels preparations are polymethyacrylic acid(PAA), polymethyl methacrylate (PMMA), polyacrylamide (PAAm), polydimethylaminoethylmethacrylate(PDEAEMA) and polyethylene glycol. These polymers though in nature are hydrophobic but swells in water depending upon the pH prevalent in the external environment. Any change in pH of the biological environment causes changes in the swelling behavior, for example, the hydrogel of caffeine is prepared with polymer PDEAEMA at pH below 6. 6. As the polymer shows high swellability but when pH changes to higher side, the polymer showed shrinkage leading to drug release. The other pH sensitive hydrogels are copolymer of PMMA and polyhydroxyethyl methyl acrylate(PHEMA) which are anionic copolymers, swell high in neutral or high pH but do not swell in acidic medium. It was also observed that pH and ionic strength determines kinetics of swelling of PHEMA and guar gum. Other drugs that have been delivered through pH sensitive hydrogels are Tri polymer of N-vinyl-2-pyrrolidone methacrylamide and itaconic acid, polydimethylaminoethylmethacrylate, polyethyleneglycol, copolymer of poly methacrylic acid and polyethylene glycol, copolymer of cationic guar gum and acrylic acid monomer. pH sensitive hydrogels have also been used to encapsulate proteins in acrylamide polymer cross-linked with bisacrylamide acetal cross-linkers. At pH of around 5, the pore size of the acetal cross-linked hydrogels increases leading to release of protein. However at neutral pH, the acetal groups remain inact as cross linkers and protein do not diffuse out easily[10].

6.2 Temperature sensitive hydrogels

The hydrogels being cross-linked polymers are temperature sensitive. These hydrogels are pharmaceutically well accepted owing to large number of temperature sensitive drugs being delivered in these dosage forms . The release as well as mechanical characteristics of drug and hydrogels are altered with the change in the temperature of external environment. Negative thermo-sensitive hydrogels contract upon heating above their low critical solution temperature. Positive thermo-sensitive hydrogels contract upon cooling above their upper critical solution temperature. In general, these hydrogels are hydrophobic polmers which show variable network in response to temperature thus modulate the drug release. These thermo- sensitive gels are specific, controllable and biocompatible drug delivery devices. They could be biodegradable also. The drugs which are widely been explored for such devices are usually from category of anticancer, antidiabetic, hormones or proteins and peptides. Sometimes these gels are formed within the system and are particularly beneficial for tissue targeting to inflamed or diseased areas. Drugs like insulin, heparin and indometacin have delivered using these types of hydrogels. Tanaka developed the thermo-sensitive hydrogels of PNIPAAm(polyisopropylacrylamide). The crosss-linked polymers containing 75% NIPAAm(N-isopropyl acrylic amide)and rest of MAA(methacrylic acid) showed temperature dependent swelling . However, the combined effect of temperature and pH controls the drug release only when hydrogel gets swollen.

Thermo-sensitive macrocapsules of nanoparticles have been developed recently where the matrix consists of temperature sensitive ethylcellulose polymer being coated with thermo-sensitive membrane prepared by cross-linking poly NIPAAm hydrogel. The drug release is expected to be high when high temperature causes collapsing of the membrane leading to large void formation. These polymers exhibit phase separation at lower critical solution temperature of about 32°C in aqueous solution. A novel thermo-sensitive hydrogel of PNIPBAm{poly(N-isopropyl-3-butenamide)}was synthesized by Xu et al. . The synthesized gel showed smaller pore size with the increase in concentration of cross-linker but the swelling ratio was high with a gel containing low concentration of cross-linker. Another thermo-sensitive hydrogel of polyorganophosphazene polymers bearing alpha - amino-omega-methyl-poly ethylene glycol (AMPEG) and hydrophobic L-isoleucine ethyl ester side groups so synthesized showed variable physical appearance from transparent sol to translucent gel depending upon temperature and was utilized to entrap natural insulin source for prolonged release. The same polymer has been used for controlled release of an anticancer drug-doxorubicin. The drug showed sustained release over a period of 20 days with no effect on gel characteristics like viscosity or gel strength and thus could be injected for its depot therapy. Some of temperature sensitive hydrogels having pharmaceutical applicability include : poly organophosphazene with α-amino omegamethylpolyethylene glycol, copolymer of gelatin and PVA, Co-polymer of poly-PNIPA and poly-PNIPA-Co-AA, NIPAAm-Co. AAm, NIPAAm-Co-AAm, polyepsilon caprolactone-co-lactide-polyethylene glycol, chitosan. These hydrogels are not having thermo-sensitive response only, but they are biodegradable also. That is why they are preferred for oral drug delivery. One such example is novel biodegradable, thermo-sensitive hydrogel of poly(epsilon-caprolactone-co-lactide-polyethylene glycol) copolymer used for iniectable drug delivery systems for proteins and peptides[11].

6.3 Glucose sensitive hydrogels

These hydrogels are sugar sensitive and show variability in response depending upon the presence of glucose. One of such pharmaceutical hydrogel system is the cross- linked poly(methacrylamido phenylboronic acid)-co-acrylmide hydrogel which liberates the drug in a controlled manner only when the concentration of glucose is high in the surrounding environment causing swelling of hydrogel. Usually, glucose sensitive hydrogels are based on implantable sensor which is sensitized to glucose concentration from 0-20 mM. Insulin loaded hydrogels of cross-linked co-polymers of polyethyleneglycol and methacrylic acid have been prepared by partitioning the insulin concentration. The micro particles of hydrogels showed no leakage under acidic conditions while the release was highest at pH 7. 4 . A similar glucosesensitive hydrogel was prepared by photopolymerization of 2-hydroxyethyl methacrylate and 3-acrylamido phenyl boronic acid. The liberation of insulin was glucose concentration dependent. The hydrogels based on sulfonamide chemistry, where the hydrogel showed maximum swelling at pH 7. 4 in a local glucose environment of 0-300 mg/dl for delivery of insulin, was an enzymatic approach. The hydrogel of poly(2-hydroxyethyl methacrylate-co-N, N-dimethylaminoethyl methacrylate or poly HEMA-co-DMAEMA polymer entrapped the insulin, glucose oxidase and catalase enzymes. Under the environment of glucose, the glucose diffuses in the hydrogel from blood and gets converted to gluconic acid which raises the pH thus causes swelling of the hydrogel. Swelling of hydrogel leads to liberation of insulin which controls the glucose level in the blood. Through such controlled release devices not only the insulin release is controlled (by varying the concentration of cross linking agent)but also, the morphology of hydrogel is regulated by oxygen uptake.

Based on a similar approach of stimuli sensitive hydrogels, a conjugated polymer of monomethoxy poly(ethylene glycol) with glucose containing polymer showed reversible gel to sol phases depending on the concentration of glucose in the external environment. The viscosity of the hydrogel decreased with the addition of glucose. A part from gel to sol approach for glucose sensitive sensitive hydrogels, the other approach is competitively binding insulin to concanavalin A, which is a lectin protein that reacts with specific sugar residues present at terminals so that in the presence of glucose, insulin is displaced. Thus, in general, glucose sensitive hydrogels are formed by immobilizing glucose oxidase enzyme which catalyses beta D-glucose to gluconic acid and hydrogen peroxide. The release of gluconic acid decreases pH of the external environment causing decrease in swelling behavior. This enzyme can be present in bound form or it could be attached to the polymer chain. The conducting behavior of gels which gives the idea of swelling vary with the ions liberated due to formation of gluconic acid or by ionization of amines present in the polymer(usually acrylates) used for preparation of hydrogels. Therefore, these smart biomaterials show controlled delivery of solute usually proteins like insulin, lysozyme or BSA(Bovine serum albumin)in response to external environment.

Apart from temperature, pH, glucose sensitive hydrogels, other stimuli like light, electric field, chemicals and ions have been utilized in formation of responsive hydrogels. But these have not gained considerable attention in the field of drug delivery[12].

6.4 Nanohydrogels

Nanohydrogels are the hydrogels which are prepared in water by self aggregation of polymers of natural origin like dextran. These types of hydrogels are formed from natural polysaccharides like dextran, pullulan, or cholesterol-containing polysaccharide. the cholesterol-containing polysaccharide is stirred at 50°C for 12 h in aqueous buffer which leads to swelling of the cholesterol-containing polysaccharide. After sonication at 25°C for 10 min, nanoparticles of hydrogelsare formed. The size and density of hydrogel nanoparticles can be controlled by changing the degree of substitution of cholesterol groups of such polysaccharides. These hydrogels are of nano dimensions usually of 20-30 nm and are used for cell targeting as they release the entrapped drug by swelling caused by change in the pH of the surrounding environment. Drugs like adriamycin has been deliverd to tumor cells and the drug showed pH dependent release and the highest release was when pH was below 6. 8 . These nanoparticles of hydrogels have been used for controlled release of proteins like lysozyme, albumin, immunoglobulin. The amount of protein released is dependent on the square root of time. Hyrogels especially of dextran are made biodegradable by encapsulation of enzyme dextranase. The hydrogels of pullulan nanoparticles have been used for cell targeting by encapsulating active drug in aqueous core of Aerosol OT/n-hexane[13].

Similar hydrogels have been made by self-assembling nanoparticles of linoleic acid modified chitosan. 1. 8% linoleic acid substituted chitosan has structural integrity and shows loading capacity of 19. 85 to 37. 57% of bovine srum albumin. These nanoparticles and are ideal for tissue targeting. the nanohydrogel of polysaccharide-mannose from saccharomyces cerevisiae have been prepared or capsulating insulin or BSA. the incorporation of calcium phosphate prevents the initial burst release thus these hydrogels are used for controlled drug delivery.

7. Preparation methods of hydrogels

Hydrogels are polymeric networks. This implies that crosslinks have to be present in order to avoid dissolution of the hydrophilic polymer chian in aqueous solution. Hydrogels are most frequently used for controlled release of bioactive agents and for encapsulation of cell and biomolecules. In many of these cases the three dimensional structure of the hydrogels have to disintegrate into harmless non toxic products to ensure biocompatibility of the gel. The nature of the degradation productions can be tailored by a proper selection of the hydrogel building blocks. Keeping this consideration in mind, various chemical and physical crosslinking gels have ionic or covalent bonds between polymer chains . Even though this leads to more mechanical stability, some of the crosslinking agents used can be toxic, and give unwanted reactions, thus rendering the hydrogel unsuitable for biological use. These adverse effects can be removed with the use of physically crosslinked gels. In physically crosslinked gels, dissolution I prevented by physical interations between different polymer chains. Both of these methods are used today for preparation of synthetic hydrogels and discussed in detail.

8. Chemically crosslinked gels

As stated earlier, chemically crosslinked gels are mechanically quite stable due to the ionic and covalent bond which comprises these gels. However the addition of crosslinking agent leads to adverse effects if the compound is toxic, which on liberation in the body becomes quite harmful. the various methods for chemical crosslinking are as follows:

- **Crosslinking of Polymers**

In this method chemically crosslinked gels are formed by radical polymerization of low molecular weight monomers, or homopolymers, or copolymers in the presence of crosslinking agent. This reaction is mostly carried out in solution for biomedical applications. Most hydrophilic polmers have pendant hydroxyl group, thus agents such as aldehydes, maleic and oxalic acid, dimethylurea, diisocyanates etc that condense when organic hydroxyl groups are used as crosslinking agents. the solvent used for these reactions is usually water, but methanol, ethanol and benzyl alcohol have also been used. these solvents can be used only if after formation of network structure, the solvent can be exchanged with water.

A typical reaction scheme for this type of crosslinking is shown:

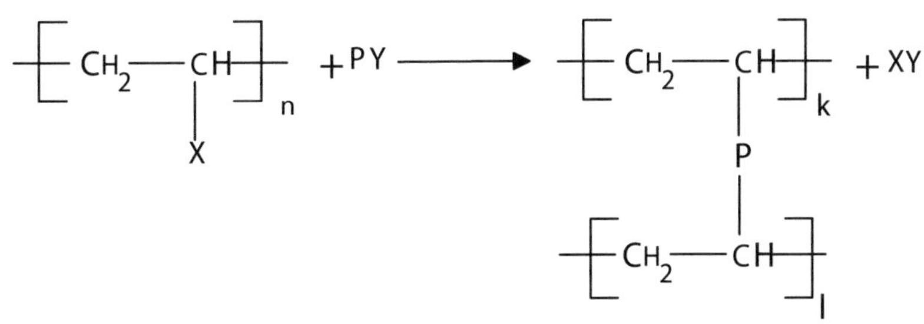

9. Typical reaction scheme for Flory type crosslinked structure

End-linking and cross-linking reactions may also occur in the absence of cross-linking agents if a free radical initiator can be used which forms free radicals in the backbone chain.

- **Copolymerization/Crosslinking Reactions**

Copolymerization reactions are used to produce polymer gels, many hydrogels are produced in this fashion, for example poly(hydroxyalkyl methacrylates). Initiarors used in these reactions are radical and anionic initiators. Various initiators are used, such as Azobisisobutyronitrile(AIBN), benzyl peroxide etc. Solvents can be added during the reaction to decrease the viscosity of the solution.

- *Kinetic mechanism*

The whole crosslinking mechanism consists of four steps: initiation, propagation, crosslinking, and termination. Termination can occur by combination, disproportionation, and chain transfer to monomer. An example of a representative reaction scheme follows:

Initiation

$$I \xrightarrow{k_d} 2A$$

$$A + M_1 \xrightarrow{k_{i1}} P_{1,0,0}$$

Propagation and cross-linking

$$P_{p,q,r} + P_{x,y,z} \longrightarrow Q_{p,q,r} + P_{x,y-1,z+1}$$

Termination by combination

$$P_{p,q,r} + P_{x,y,z} \xrightarrow{k_{tc11}} M_{p+x,q+y,r+z}$$

Termination by disproportionation

$$P_{p,q,r} + P_{x,y,z} \xrightarrow{k_{d111}} M_{p,q,r} + M_{x,y,z}$$

Chain Transfer to monomer

$$P_{p,q,r} + M_1 \xrightarrow{k_{f11}} M_{p,q,r} + M_{p,q,r} + P_{1,0,0}$$

$$P_{p,q,r} + P_{x,y,z} \xrightarrow{k_{f12}} M_{p.q.r} + P_{x,y-1,z+1} + Q_{0,0,1}$$

HEMA as the monomethacryl monomer and EGDMA as dimethylacryl monomer, I is the initiator, and A is a molecule with initiated radical. Here P and Q represent living polymer chains with monomethylacyl and dimethylacyl monomer terminal groups, respectively and M is dead polymer chain. The subscripts p, q, r are used to describe primary chain, they refer to monomethylacryl units, pendant methylacyl groups, and cross-links per chain respectively[14] .

- **Cross-linking by High Energy Radiation**

High energy radiation, such as gamma and electron beam radiation can be used to polymerize unsaturated compounds. Water soluble polymers derivatized with vinyl groups can be converted into hydrogels using high energy radiation. For example, PEG derivatized to PEGDA can form hydrogels once irradiated with UV radiations. Polymers without additional vinyl groups can also be cross-linked via radiation. On exposure to gamma or electron beam radiation, aqueous solutions of polymers form radicals on the polymer chains (e. g by the hemolytic scission of C-H bonds). Also the radiolysis of water molecules generates the formation of hydroxyl groups which can attack polymer chains also resulting in the formation of microradicals. Recombination of these microradicals on different chains results in the formation of covalent bonds and finally in a crosslinked structure. the swelling and permeability characteristics of the gel depend on the extent of polymerization, a function of polymer and radiation dose (in general crosslinking density increases with increasing radiation dose). The advantage of using this process for gel formation is that in can be done in water under mild conditions without the use of a crosslinking agent. However there are some drawbacks to using this method, the bioactive material has to be loaded after gel formation, as irradiation might damage the agent. Also in some gels like PEG and PVA, the crosslinks consist of C-C bonds, which are not biodegradable.

- **Cross-linking Using Enzymes**

Recently a new method was published using an enzyme to synthesize PEG-based hydrogels. A tetrahydroxy PEG was functionlized with addition of glutaminyl groups and networks were formed by addition of transglutaminase into solution of PEG and poly(lysine-co-phenylalanine). This enzyme catalyzed reaction between γ-carboxamide group of PEG and the ε-amine group of lysine to obtain an amide linkage between polymers. The gel properties can be tailored by changing ratios of PEG and lysine.

10. Physically cross-linked gels

Chemically crosslinked gels imply use of a cross-linking agent, which is often toxic. This requires that the cross-linking agent be removed from gel, which can affect the gel integrity. For these reasons, physically cross-linked gels are now coming into prominence. Several methods have been investigated exploring preparation of physically cross-linked gels. Below are mentioned some of the most widely used methods and their areas of application.

- **Cross-linking by Ionic interactions**

An example of cross-linking via ionic interactions is cross-linking of Alginate consists of glucuronic acid residues and mannuronic residues and can be cross-linked by calcium ions. Cross-linking can be carried out at normal temperature and pH. These gels are used as matrix for encapsulation of cells and for release of proteins. Also Chitosan based hydrogels, as well as dextran based hydrogels, cross-linked with potassium ions also other gels synthesized with ionic interactions. In addition to anionic polymers being cross-linked with methalic ions, hydrogels can also be obtained by complexation of polyanions and polycations[15].

- **Cross-linking by Crystallization**

An aqueous solution of PVA that undergoes a freeze-thaw process yiels a strong highly elastic gel. Gel formation is attributed to the formation of PVA crystallites which act as physical cross-linking sites in the network. The gel properties could be modified by varying polymer concentration, temperature, and freezing and thawing cycle times. These gels have been shown to be useful for drug release.

- **Cross-linking by hydrogen Bonds**

Poly(acrylic acid) and poly(methacrylic acid) form complexes with poly(ethylene glycol) by hydrogen bonding the oxygen of the poly(ethylene glycol) and the carboxylic acid group of poly(methacrylic acid). Also hydrogen bonding has been observed in poly(methacrylic acid-g-ethylene glycol). The hydrogen bonds are only formed when the carboxylic acid groups are protonated. This also implies that the swelling of gels is pH dependent. Recently a hydrogen system was developed using the principle of DNA hybridization via hydrogen bonding . In this approach, oligodeoxyribounucleotides were coupled to a water soluble polymer. Hydrogels were prepared by addition of a complementary oligodinucleotide (ODN) either conjugated to the same water soluble polymer or, in its free form, to an aqueous solution of the ODN derivatized water soluble copolymer.

- **Cross-linking by Protein Interaction**

Genetic engineering has also been used for the preparation of hydrogels. The major advantage is that the sequence of peptides and, therefore its physical and chemical properties can be precisely controlled by the proper design of the genetic code in synthetic DNA sequences. Cappello and colleagues prepared sequential block copolymers containing a repetition of silk-like and elastine-like blocks, in which the insoluble silk segments are associated in the form of aligated hydrogen bonded beta strands or sheets. These hydrogels can also be used for drug delivery with drug delivery release influenced by concentration, polymer composition, and temperature. Crosslinking by antigen-antibody interaction was also performed, in which an antigen (rabbit IgG) was grafted to chemically crosslinked polyacrylamide in the presence of an additional crosslinker. Additionally hydrogels have been prepared by immobilizing both the antigen and the antibody in the form of an interpenetrating network polymer network . This approach might permit drug delivery in response to specific antigen[16].

11. Monomers used for fabrication of hydrogels

The monomers used for fabrication of these biocaompatible hydrogels have expanded from a handful of choices, to several novel materials with tailor-made proerties suited to particular applications. The first synthesis of hydrogel was that of Wichterle and lin using PHEMA(poly)hydroxyethyl methacrylate)) as monomer. Depending upon the application, hydrogel monomers are chosen according to their properties, ease of delivery or encapsulations, as well as cost and availability . One of the most traditional monomers used for drug delivery of proteins is biodegradable PLGA(polymers of lactic and glycolic acid). However, these hydrophobic materials have a tendency to denature protein as well as cause inflammation due to degradation, These problems were overcome when researchers turned towards hydrophilic monomers. Monomers such as acrylic acid, polyethylene glycol, and methacrylic acid are all materials used in therapeutic applications. Researchers are today trying to custom - make materials to suit specific applications. PNIPAAm (poly(N-isopropylacrylamide), PVA(polyvinyl alcohol)are all synthesized by new preparation techniques, for distinct applications. Table 1 provides monomers used in the synthesis of hydrogels for pharmaceutical applications.

12. PEG(polyethylene glycol) as suitable material

It is known that hydrophilic monomers provide a distinct advantage in both fabrication and application of hydrogels. the premier material used today for both drug delivery, cell encapsulation and as adhesive promoters is poly(ethylene glycol)hydrogels. PEG has many unique properties which make it an ideal choice. PEG and its "stealth" properties, that is once its attached to certain formulations, it allows slow release of the formulation, thus

Monomer Abbreviation	Monomer
HEMA	Hydroxyethyl methacrylate
HEEMA	Hydroxyethoxyethyl methacrylate
HDEEMA	Hydroxydiethoxyethyl methacrylate
MEMA	Methoxyethyl meacrylate
MEEMA	Methoxyethoxyethyl methacrylate
MDEEMA	Methoxydiethoxyethyl methacrylate
EGDMA	Ethylene glycol dimethacrylate
NVP	N-vinyl-2-pyrrolidone
NIPAAm	N-isopropyl AAm
VAc	Vinyl acetate
AA	Acrylic acid
MAA	Methacrylic acid
HPMA	N-(2-hydroxypropyl methacrylamide)
EG	Ethylene glycol
PEGA	PEG acrylate
PEGMA	PEG methacrylate
PEGDA	PEG diacrylate
PEGDMA	PEG dimethacrylate

Table 1.

enabling controlled release, as well as reduce uptake of harmful immunoglobins. This allows longer dosage and reduces immunogenicity of substances such as adenosine deaminase(ADA) and asparaginase. PEG is non toxic, thus ideal for biological applications, and can be injected into the body without adverse effects. It is also an FDA approved materials for use in humans. PEGylation is an important technique being developed for drug delivery, involves attachment of PEG to proteins and drugs, and has great potential for improving pharmokinetic and pharmodynamic properties of delivery drugs. Thus PEG has varied used in the medical field, including drug delivery(e. g. ; treatment of hepatitis C), laxatives, cell immobilization, (as adhesion promoters), biosensor materials, and encapsulation of islets of langerhans for treatment of diabetes. It is also used as carrier material for encapsulated cells for tissue engineering purposes. Thus PEG, with its biocompatibility, flexibility and stealth properties is an ideal material for use in pharmaceutical applications[17].

13. Applications of hydrogels

Water-swollen cross-linked hydrogels have varied applications in fields such as food additives, pharmaceutical as well as biomedicine. The pioneering work on cross-linked HEMA hydrogels was done by Wichterle and Lim in 1954. From their reseach and discovery of the hydrophilic and biocompatible properties of hydrogels, there emerged a new class of hydrogel technologies based on biomaterial application. Lim and Sun in 1980 demonstrated the successful use of calcium alginate microcapsulates for cell encapsulation. Later natural polymers such as collagen, and shark cartilage were incorporated into hydrogels as wound dressings. Natural and synthetic polymers are used encapsulation of cells, as well as encapsulation of islets in a semipermeable membrane. Hydrogels have been used to prevent adhesions and prevent thrombosis after surgery, and as cell adhesion resistant surfaces. Microfabricated hydrogel arrays are also used for biosensing . Hydrogels now play an important role in tissue engineering scaffolds, biosensor and bioMEMS devices and drug carriers.

Among these applications, hydrogel-based drug delivery devices have become a major area of study, and several commercially available producs are already in the market. Proteins, peptides, DNA based drugs can all be delivered via hydrogel carrier devices. The various properties of hydrogels such as biocompatibility, hydrophilicity, flexibility all make it ideal for use as drug delivery matrix.

Hydrogels show good compatibility with blood and other body fluids, thus are used as materials for contact lenses, burn wound dressings, membranes, and as coating applied to living surfaces. Natural and synthetic polymers have applications as wound dressings, encapsulation of cells, and recently are being used in the new field of tissue engineering as matrices for repairing and regenerating a wide variety of tissues and organs. When parts, or whole tissues, organs fail, treatments include repair, replacement with a natural or synthetic substitute, or regeneration. Implants have been reasonably successful;however tissue engineering holds great promise for regeneration. Hydrogels are now being considered as ideal matrices for tissue engineering[18].

14. Drug delivery

The current growth of hydrogel applications in drug delivery and biosensors is ascribed in part to the biocompatibility of hydrogels, and in part to fast and reversible volume changes in response to external stimuli such as temperature, pH, electric and magnetic fields, or analyte concentration. Thus these hydrogels are sometimes called " Stimulus" responsive polymers.

One approach is to use pH sensitivity to mediate changes in swelling . A pH - sensitive hydrogel undergoes very large and reversible volume changes in response to pH changes within the hydrogel. Two main types of pH -sensitive hydrogels are acidic hydrogels and basic hydrogels. Acidic hydrogels by definition will be ionized and hence swollen at high pH, and uncharged and unswollen at low pH. Swelling behavior of a basic hydrogel has the opposite dependence on pH. The pH sensitivity is caused by pendant acidic and basic groups such as carboxylic acids, Sulfonic acids, primary amines, and quaternary ammonium salts. Carboxylic acid groups for example are charged at high pH and uncharged at low pH, whereas the reverse is true for primary amine groups and quaternary ammonium salts.

In a biosensor, the swelling and shrinking of the hydrogel is usually made to be responsive to charges in the level of a biological indicator or molecule of interest. This is generally achieved by incorporating into the hydrogel an enzyme, receptor, antibody, or other agent which binds the molecule of interest. Oxioreductase enzymes are one category of such agents, which find particular use in biosensors. The characteristics of oxidoreductase enzymes of particular value in sensor applications is the production of oxygen by the enzyme reaction.

Among the oxidoreductase currently being investigated for use in biosensors are glucose oxidase(for sensing blood sugar levels), cholesterase(for sensing cholesterol levels), alcohol dehydrogenase(for sensing alcohol levels), and penicillinase(for sensing penicillin levels). Besides those named here, there are over 100 known oxidoreductase enzymes, and at least some of these are likely to find future use in biosensors[19].

A pH - sensitive hydrogel containing glucose oxidase (GOx) enzyme is called a glucose-sensitive hydrogel(GSH) due to its responsiveness to environmental glucose concentrations. Thermally stable GOx is a flavin-containing glycoprotein which catalyzes a reaction which is very specific for glucose, and which produces gluconic acid and hydrogen peroxide in the presence of glucose and oxygen as shown below. Therefoere, increases in the environmental glucose concentration lower the pH value within the GSH.

Several attempts have been made to utilize this catalytic reaction in glucose biosensors. Glucose biosensors based on amperometric methods are the most highly developed. In the amperometric method, an electrode is used which produces a current proportional to the diffusional flux of hydrogen peroxide to the electrode surface, or, alternatively, proportional to the diffusional flux of oxygen to the electrode surface. At steady state, the diffusional flux of hydrogen peroxide to the electrode surface equals the rate at which hydrogen peroxide is produced by the GOx reaction in the hydrogel adjacent to the electrode. However, unlike the hydrogels considered here, the hydrogels in amperometric glucose biodensors do not swell in response to pH changes.

An important physical property of pH-sensitive GSHs is the ability to change volume in response to changes in environmental glucose concentrations, due to changes in pH within the hydrogel caused by the reaction of the GOx enzyme. This physical phenomenon has been applied in insuline delivery devices to control insulin permeability through GSHs.

pH-sensitive glucose hydrogels are useful in devices using either amperometric means or pressure transducers to detect glucose concentrations. For such applications, two major problems with the GOx enzymatic process have been identified: insufficient oxygen supply for the reaction, and the decay of the GOx activity with time due to peroxide inducted degradation[20].

For both insulin delivery devices and glucose biosensors, GOx stability is essential for long term use *in vivo*. For insulin delivery devices and the pressure based glucose biosensors, a rapid swelling kinetic is also important, to provide the best performance. The use of hydrogels containing oxidoreductase enzymes in biosensors and controlled drug delivery systems and more particularly to the inclusion of catalase in such biosensors and drug delivery systems has been reported. The invention comprises a hydrogels containing an analyte-sensitive enzyme which, with the catalase being present in amounts ranging from about 100 units/ml to about 1000 units/ml. The term " *Hydrogel*" is intended to encompass any polymer matrix suitable for use in hydrated conditions. In one embodiment, the analyte is glucose and the analyte-sensitivie enzyme is glucose oxidase. In addition to glucose oxidase, the invention is applicable any analyte-sensitive enzyme which generates hydrogen peroxide as part of the reaction. these include monoamine oxidase as well as many oxidoreductases. The invention further encompasses biosensors incorporating these hydrogels . The hydrogels may preferably be formulated such that swelling of the gel permits flow of a drug such as insulin out of the gel. Thus, in a further embodiment the invention encompasses analyte responsive drug delivery devices containing hydrogels which meet the above description. The hydrogels may be used with biosensors or drug - delivery devices which use pressure transducers or amperometric means to register analyte concentration. Hydrogels according the invention may also be used with devices employing gas reservoirs or semipermeable membranes. The invention further includes methods for using catalase in hydrogels, biosensors and analyte-responsive drug delivery devices[21].

Drug delivery has been a subject of intense studies over recent years. The goal is to achieve sustained(or slow) and /or controlled drug release and thereby improve efficacy, safety, and/or patient comfort. A sustained and /or controlled release of the drug agents is achieved by the retardation of drug diffusion by and/or gradual disintegration of the polymer matrix following application.

In-situ gelation is a process of gel formation at the site of application after the composition or formulation has been applied to the site. In the field of human and animal medicine, the sites of application refers to various injection sites, topical application sites, and others where the agents are brought into contact with tissues or body fluids. As a drug delivery agent, the *in-situ* gel has an advantage related to the gel or polymer network being formed *in-situ* providing sustained release of the drug agent. At the same time, it permits the drug to be delivered in a liquid form. The *in-situ* gelation compositions using ionic polysaccharides have been reported, which consist of a drug, a polymer and a gel forming ionic polysaccharide which consist of two components, an ionic polysaccharide and a cross-linking ion capable of cross-linking the former. The *in-situ* formation is induced by the application of the cross-linking ions.

Thus, a great need exists for a simpler and more efficient in-situ gelling composition that employs only a low polymer concentration for the purposes of drug delivery. Pectin is a biodegradable acidic carbohydrate polymer. Pectin is commonly found in plant cell walls. The cell wall of a plant is devided into three layers consisting of the middle lamella, the primary wall and the secondary cell wall. The middle lamella is richest in pectin. The chemistry and biology of pectin have been extensively reviewed.

Current commercial pectins are mainly from citrus and apples. However, besides cirtrus and applies pectins can also be isolated from many other plants. All vetetacble and fruits that have been examined contain pectins. Pectins from sugar beets, sunflowers, potatoes, and grapefruits are just a few other well known examples. A pectic substance to provide a biodegradable *in-situ* gelling composition for animal and human use was prepared. The composition transformed from a liquid into a gel following administration to the target site. Preferably the pectic substance was *Aloe* pectin. This composition could control, or sustain, the release of a physiologically active agent in the body of an animal . It provided a transparent polymer solution wherein no dramatic increase in gel cloudiness was created beyond certain concentration ranges. Preferably the composition was capable of creating an in-situ gel at low concentrations. The polymer solution was transparent wherein a thickener is added. Preferably the composition is capable of creating an in-situ gel at low concentrations to be delivered in the liquid form and provide a composition for drug delivery a therapeutic or diagnostic agent incorporated into the formulation or composition. These agents can be small molecules as well as large ones such as proteins like insulin. Preferably the composition is capable of forming an in-situ gel at low concentrations[22].

Two classes of polymers that are currently receiving widespread attention in biosensor development are hydrogels and conducting electroactive polymers. the integration of two materials to produce electroactive hydrogel composites are reported that physically entrap enzymes within their matrics for biosensor construction and chemically stimulated controlled release. Enhanced biosensing capabilities of these membranes have been demonstrated in the fabrication of glucose, cholesterol and glucose amperometric biosensors. All biosensors displayed extended linear response ranges(10^{-5} -10^{-2} M), rapid response times (< 60 sec.), retained storage stabilities of up to 1 year, and excellent screening of the physiological interferents ascorbic acid, uric acid, and acetaminophen. When the cross-linked hydrogel components of these composite membranes were prepared with the amine containing dimethylaminoethyl methacrylate monomer the result was polymeric devices that swelled in response to pH changes(neutral to acidic). Entrapment of glucose oxidase within these materials made them glucose responsive through the formation of gluconic acid. When insulin was co-loaded with glucose oxidase into these "*bio smart*" devices, there was a twofold increase in insulin release rate when the devices were immersed in glucose solutions. This demonstrates the potential of such systems to function as a chemically-synthesized artificial pancreas.

Glucose-sensitive hydrogel membranes have been synthesized and characterized for their rate - of delivery of macromolecules. The mechanism for changing this rate is based on variable displacement of the affinity interaction between dextran and concanavalin A(con A). Membranes were constructed from cross-linked dextrans to which con A was coupled via a spacer arm. Changes in the porosity of the resulting hydrogel in the presence of glucose led to changes in the diffusion rate observed for a range of proteins. Gels of specified thickness were cast around to nylon gauze support (pore size, 0. 1 mm) to improve mechanical strength. Diffusion of proteins through the gel membrane was determined using a twin-chamber diffusion cell with the concentrations being continuously minitored using a UV spectrophotometer. Changes in the transport properties of the membranes in response to glucose were explored and it was found that, while 0. 1M D-glucose caused a substantial, but saturatrable, increase in the rates of diffusion of both insulin and lysozyme, controls using glycerolor L-glucose(0. 1M) had no significant effect. Sequential addtition and removal of external glucose in a stepwise manner showed that permeability changes were reversible. As expected, diffusion rates were inversely proportional to membrane thickness. A maximum increase in permeability was observed at pH 7. 4 at 37 degrees C. The results demonstrate that this hydrogel membrane functions as a smart material allowing control of solute delivery in response to specific changes in its external environment[23].

A novel UV polymerized glucose-response mixture containing concanavalin A (con A) and dextran was synthesized and characterized as a " smart" biomaterial to form the basis of a closed-loop delivery device. Dextran and con A precursors were modified with acrylic side groups and then UV polymerized to produce covalently bonded mixtures which were examined by FTIR. The viscoelastic properties of these polymerized mixtures containing glucose concentrations between 0% and 5% w/w were also examined using oscillatory rheometry within the linear viscoelstic range across a frequency range of 0. 01-50 Hz. As the formulation glucose concentration was raised, a graded decrease in storage modulus, loss modulus and complex viscosity when compared at 1 Hz was observed. Increasing the mixture irradiation time produced viscosity profiles at higher values throughout the glucose concentration range. The subsequent testing of such formulations in in vitro diffusion experiments revealed that the leaching of the mixture components is formulation dependent and is restricted significantly in the covalently bonded mixtures. Insulin delivery in response to glucose in the physiologically relevant glucose concentration range was demonstrated using the novel polymerized mixture at 37 degrees C. The performance of this covalently cross-linked glucose-responsive biomaterial has been improved in terms of increased mixture stability with reduced component leaching . This could, therefore be used as the basis of the design of a closed-loop drug delivery device for therapeutic agents used for the management of diabetes mellitus[24].

Treatment of diseases has always been a major issue for researchers for as long as mankind has existed. As technology has advanced, proteins, peptides, and other materials have been identified as " drugs" which can be used to treat physiological life processes, pain, and discomfort. Drugs can vary in their characteristics to the extent that drugs used to treat the same symptoms might differ in characteristics such as hydrophilicity, chemical composition, size and effectiveness. An increasing understanding of cellular at the biology at the molecular level and breakthroughs in proteomics have led to concept of gene delivery. Drugs have to reach the site of action following administration(oral intravenous, transdermal etc.) in a specific manner and in specific quantity. This is the basis of the drug delivery field. Drug delivery aims at delivering the right drug at the right place, at right concentration for the period of time. Sometimes direct delivery of such drugs is difficult, due to the treacherous route of delivery or discomfort caused to the patient. for such cases, strategies have been developed for delivering drug with a carrier. the drug carrier, whether it be an implantable device, or long chain polymer must be biocompatible with the drug and the body. Drug delivery systems alter the biodistribution and pharmakinetics of the drug . Therefore one must take into account obstacles such as drug solubility, enzyme degradation, toxicity, inability to cross biological barriers as well as adverse environmental conditions. In order to make the delivery of the drug effective without causing an immune response in the body, proper design and engineering of the drug delivery system is essential[25].

15. Hydrogels in drug delivery

Localized drug delivery can be achieved by introducing the drug directly at the target site. the major class of biomaterials considered as implantable drug delivery systems are hydrogels. These hydrophilic networks are capable of absorbing great amounts of water while maintaining structural integrity. Their structural similarity to the extracellular matrix makes it biocompatible. These synthetic polymers have generated wide interests and are now at the forefront of drug delivery research.

In order to incorporate a performed gel into the body, an opening must be created, with at least the same dimension as that of the gel. This leads to potential risk and discomfort to the patient. Thus focus has shifted to developing injectable materials with ability to form three dimensional matrices under physiological matrices. This in situ formation can be achieved through specific chemical cross-linking reactions. Gel structuring is triggered by environmental stimuli(pH, temperature, solvent exchange etc.). Synthetic hydrogels, with their ability to imbibe water, flexibility, and biocompatibility, are ideal carriers for the development for novel pharmaceutical formulations and for the delivery of drugs, proteins, and as targeting agents for drug delivery . The network structure and the nature of components play a key role in the diffusional behavior, molecular mesh size changes, and stability of the incorporated bioactive agent[26]. The use of hydrogels allows not only delivery of drugs, but also controlled release, in the manner required by the pharmaceutical scientists. For example, drugs can be delivered only when needed, may be directed to specific site, and can be delivered at specific rates required by the body. In the last 20 tears, advanced drug delivery formulations have been examined in great detail. Reviews related to the various applications of hydrogels in drug delivery and various sites available in the body for such are readily available.

16. Properties useful in drug delivery

Hydrigels posses several properties that make them an ideal material for drug delivery. First, hydrogels can be tailored ro respond to a number of stimuli. This enables sustained drug delivery corresponding to external stimuli such as pH or temperature. These pH sensitive gels are useful in oral drug delivery as they can protect proteins in the digestive track. pH responsiveness is also useful for lysosomal escape during gene delivery. Second, hydrogels can also be synthesized to exhibit bioadhesiveness to facilitate drug targeting, especially through mucus membranes, for non-invasive drug administration. finally, hydrogels also have a " *stealth*"characteristic in vivo circulation time of delivery device by evading the host immune response and decreasing phagocytic activity[27].

17. Applications of Hydrogels in drug delivery

Advances in recombinant protein technology have identified several protein and peptide therapeutics for disease treatment. However, the problem which plagued researchers as how to effectively deliver these biomolecules. Due to their large molecular weight, and three dimensional structure, the most commomly used route for drug administration is by intravenous or subcutaneous injection. Unfortunately proteins and peptides are prone to proteolytic degradation, thus they experience short plasma circulation times and rapid renal clearance, leading to multiple daily injections or increased dosage in order to maintain the required drug therapeutic levels. Multiple injections are difficult for the patient, while high dises might be toxic, and induce serious immune response. Hydrophobic polymeric controlled release formulations, such as PLGA, offer a sustained release mechanism in which drug release rates can manipulated by changing polymer molecular weight and composition. These polymers however induce adverse effects to the encapsulated proteins or peptides during network preparation and delivery, as well as trigger the immune response. Hydrophilic hydrogels, on the other hand, provide relatively mild network fabrication technique and drug encapsulation conditions, making them the ideal material for use indrug delivery. Thus hydrogels are primarily used for encapsulation of bioactive materials and their subsequent controlled release. If designed properly, hydrogels can be used in a variety of applications such as sustained, targeted, or stealth biomolecule delivery. Hydrogel based delivery devices can be used for oral. ocular, epidermal and subcutaneous application[28].

- **drug Delivery in GI Tract**

The ease of administration of drugs, and the large surface area for absorption makes the GI tract most popular rote for drug delivery. It is however, also a very complex rote, so that versatile approaches are needed to deliver drugs for effective therapy. Hydrogel-based devices can be designed to deliver drugs locally to specific sites in the GI tract. For example, Patel and Amiji proposed stomach-speicific antibiotic drug delivery systems for the treatment of *Helicobacter pylori* unfection in peptic ulcer disease. They developed cationic hydrogels with pH sensitive swelling and drug release properties for antibiotic delivery in the acidic environment of the stomach. There are still many drawbacks for peroral delivery of peptides and proteins to GI tract, like protein inactivation by digestive enzymes in the GI tract and poor epithelial permeability of the drugs. However, certain hydrogels may overcome some of these problems by appropriate molecular design of formulation, for example Akiyama reported novel peroral dosage forms of hydrogel formulations with protease inhibitory activities[29].

Recently oral insulin delivery using pH responsive complexation hydrogels was reported. the hydrogels used were crosslinked copolymers of PMMA with graft chains of polyethylene glycol . These hydrogels protect the insulin in the harsh, acidic environment of the stomach before releasing the drug in the small intestine.

The colonic region has also been considered as a possible absorption site for orally administered proteins and peptides, mostly due to a lower proteolytic activity in comparison to that in the small intestine. Several hydrogels are currently being investigated as potential devices for colon-specific drug delivery. These include chemically or physically cross-linked polysaccharides such as dextran, guar gumand insulin. They are designed to be highlyswollen or degraded in the presence of colonic enzymes or microflora, providing colon-specificity in drug **delivery[30].**

- **Rectal Delivery**

This route has been used to deliver many types of drugs for treatment of diseases associated with the rectum, such as hemorrhoids. This route is an ideal way to administer drugs suufering heavy first-pass metabolism. There are however, some drawbacks associated with rectal delivery. For example, due to discomfort arising from given dosage forms, there is substantial variability in patient's acceptance of treatment. Also, if drugs diffusing out of the suppositories are delivered in an uncontrolled manner, they are unable to be retained at a specific position in the rectum, and tend to migrate upwards to the colon. This leads to variation of availability of drugs, especially those that undergo extensive first-pass elimination. Hydrogels offer a way in which to overcome these limitations, provided that the hydrogels show bioadhesive properties. It was reported that increased bioavailability of propanol subject to extensive first-pass metabolism was observed by adding certain mucoadhesive polymeric compounds to poloxamer-based thermally gelling suppositories. the polymeric compounds tested were polycarbophil and sodium alginate. Miyazaki et al. investigated the potential application of xyloglucan gels with a thermal gelling property as matrices for drug delivery. Another important issue in rectal drug delivery is to avoid rectal irritation. The products discussed above, indicated no such mucosal irritation after drug administration[31].

- **Ocular delivery**

Drug delivery to the eye is difficult due to its protective mechanisms, such as effective tear drainage, blinking, and low permeability of the cornea. Thus, eye drops containing drug solution tends to be eliminated rapidly from the eye and the drugs show limited absorption, leading to poor aphthalmic bioavailability. Due to the short retention time, a frequent dosing regimen is necessary for required therapeutic efficacy. These challenges have motivated researchers to develop drug delivery systems that provide prolonged residence time.

The earlier dosage forms, such as suspension and ointments colud be retained in the eye, but sometimes gave patients an unpleasant feeling because of the nature of solids and semi-solids. Hydrogels, because of their elastic properties can represent an ocular drainage-reisitant device. In-situ forming hydrogels are attractive as an ocular drug delivery system because of their facility in dosing as a liquid, and long term retention property as a gel after dosing[32-35].

Cohen et al. developed an in-situ gelling system of alginate with high gluronic acid contents for the ophthalmic delivery of pilocarpine. This system extended the duration of the pilocarpine to 10 hr, compared to 3 hr when pilocarpine nitrate was dosed as a solution. Chetoni et al. reported silicone rubber hydrogel composite ophthalmic inserts. An in-vivo study using rabbits showed a prolonged release of oxytetracycline from the inserts for several days.

Vascular endothelial growth factor(VEGF) has been identified as akey regulator of angiogenesis. It can act as an endothelial cell mitogen and increase vascular permeability along with angiogenesis. Elevated VEGF level has been correlated with several ocular diseases, such as age-related macular degeneration and diabetic retinopathy. On the basis of these findings, in the past several years, considerable progress has been made in the treatment of the wet form of age - related macular degeneration and diabetic retinopathy by using anti-VEGF therapy. Several clinical trials employing ranibizumab, including ANCHOR and MARINA, have demonstrated the success of anti-VEGF therapy.

Although intravitreal anti-VEGF therapy is a very promising treatment, the major drawback is that the treatment must be repeatedevery 4 to 6 weeks. This is not a desirable method of delivery for several reasons: patient discomfort;the need for repetitive injections with inherent complications, including endophthalmitis, retinal tear and detachment, intraocular hemorrhage, and cataract formation;and bolus administration of the agent. Currently, there is no alternative method for delivery of the anti-VEGF agent into the eye;hence, there is a great need and desire to develop a relatively noninvasive delivery method that is more effective and longer lasting than the current clinical regimen. Since the development of hydrogels in 1960, they have been of great interest to biomaterial scientists and tissue engineers.

- **Transdermal Delivery**

Drug delivery to the skin has been generally used to treat skin diseases or for disinfection of the skin. In recent years, however a transdermal route for the delivery of drugs has been investigated. Swollen hydrogels can be delivered for long duration and can be easily removed. These hydrogels can also bypass hepatic first-class metabolism, and are more comfortable for the patient. Hydrogel based delivery devices have been proposed by Sun et al., such as composite membranes of crosslinked PHEMA with a woven polester support. Also hydrogels have been reported which have been obtained by the copolymerization of bovine serum albumin(BSA) and PEG. These hydrogels can be used as controlled release devices in the field of wound dressing . Hubbell has also carried out extensive research on in-situ photopolymerization made from terminally diacrylated ABA block copolymers of lactic acid oligomers(A) and PEG(B) for barriers and local drug delivery in the control of wound healing .

Current research in this field is now focused on electrically –assisted delivery using iontophoresis and electroporation. Hydrogel-based formulations are being look at for transdermal ionphoresis to obtain enhanced permeation of products in question such as, hormones and nicotine[36-38].

- **Subcutaneous Delivery**

Among the varied possible pharmaceutical applications of hydrogels, the most substantial application is probably in implantable therapeutics. Implantable devices that are subcutaneously inserted tend to illicit immune response of the body, leading to inflammation, carcinogenicity and immunogenicity. Thus biocompatibility becomes a major issue, and all implantable materials must be compatible with the body. Hydrogels are an ideal candidate for implantable materials. They also have other properties which make them a vaiable choice, (1) minimal mechanical irritation upon in-vivo implantation due to their soft, elastic properties(2) prevention of protein absorption and cell adhesion arising from the low interfacial tension between water and hydrogels (3)broad acceptability for individual drugs with swelling for release of incorporated drug in specific manner. Thus, hydrogels are an ideal material to be used for delivery of proteins and peptides[40].

Hydrogel formulations for subcutaneous delivery of anticancer drugs have been proposed. For example, crosslinked PHEMA was applied to cyratabine(Ara-C). Current studies on implantable hydrogels are leading towards the development of biodegradable systems, which don not require surgical removal once the drug has been administered. Biodegradable PEG hydrogels are now at the forefront of this research, and several novel systems have been developed. One type is synthesized via a polycondensation reaction between functional PEG acids and brabched PEG polyols. Another type is PEG based hydrogels having functional groups in which the protein drugs can be covalently attached to the gel network via ester linkages. In this case, the release of the immobilized proteins would be controlled by the hydrolysis of ester linkage between the gel and protein, followed by diffusion of protein, and degradation of gel[41-45].

18. References

[1] Furda I. ;(1993);"Aminopolysaccharides-their potential as dietary fiber. In: Furda I, ed. ;*Unconventional Sources of Dietary Fiber, Physiological and In vitro Functional Properties*, Washington, DC:American Chemical Society ; 105-122.

[2] Chourasia M. K., Jain S., K. ;(2003);"Pharmaceutical approaches to colon targeted drug delivery systems", *Journal of Pharmaceutical sciences*;6:33-66.

[3] Davaran S., Hanaee J., Khosravi A., (1999);"Release of 5-aminosalicylic acid from acrylic type polymeric prodrugs designed for colon-specific drug delivery", *Journal of Control Release*;58:279-287.

[4] Schacht E., Gevaert A., Kenawy E., R. ;(1996);" Polymers for colon specific drug delivery ", *Journal of Control Release*;58:327-338.

[5] Chung K., T., Stevens S., E., Cerniglia C., E. ;(1992);"The reduction of azo dyes by the intestinal microflora";*Critical. Review Microbiol*ology;18:175-190.

[6] Yamaoka T., Makita Y., Sasatani H., Kim S. I., Kimura Y. ;(2000);" Linear type azo-containing polyurethane as drug-coating material for colon-specific delivery : its properties degradation behavior and utilization for drug formulation"; *Journal of Control Release*;66:187-197.

[7] Shantha K. L., Ravichandran P., Rao K. P. ;(1995);" Azo polymeric hydrogels for colon targeted drug delivery";*Biomaterials*;16:1313-1318.

[8] Van den Mooter G., Samyn C., Kinget R. ;(1992);"Azo polymers for colon-spicific drug delivery";*International Journal of Pharmceutical*;87:37-46.

[9] Ghandehari H., Kopeckova P., Kopecek J. ;(1997);"In vitro degradation of pH sensitive hydrogels containing aromatic azo bonds";*Biomaterials*;18:861-872.

[10] Kakoulides E. P., Smart J. D., Tsibouklis J. ;(2000);"Azo crosslinked poly(acrylic acid) for colonic delivery and adhesion specificity synthesis and characterization";*Journal of Control Release*;52;291-300.

[11] Yamoto A., Tozaki H., Okada N., Fujita T. ;(2000);"Colon specific delivery of peptide drugs and anti-inflammatory drugs using chitosan capsules";*STP Pharma Science*;10:23-43.

[12] Agnihotri S. A., Mallikarjuna N. N., Aminabhavi T. M. ;(2004);"Recent advances on chitosan-based micro- and nanoparticles in drug delivery"; *Journal of Control Release*;100:5-28.

[13] Calvo P., Remunan-Lopez C., Vila-Jato J. L., Alonso M. J. ;(1997);"Novel hydrophilic chitosan-polyethylene oxide nanoparticles as protein carriers";*Journal of Applied Polymer Science*;63:125-132.

[14] De S., Robinson D. ;(2003);"Polymer realationships during preparation of chitosan-alginate and poly-l-lisyne-alginate nanospheres";*Journal of Control Release* ;89:101-112.

[15] Chen Y., Mohanraj V., Parkin J. ;(2003);"Chitosan-dextransulfate nanoparticles for delivery of an anti-angiogenesis peptide"; *International Journal of Peptide Research Therapy*;10:621-629.
[16] Ma Z., Yeoh H. H., Lim L. Y. ;(2002);"Formulation pH modulates the interaction of insulin with chitosan nanoparticles";*Journal of Pharmaceutical Sciences*;91:1396-1404.
[17] Tiyaboonchai W., Woiszwillo J., Sims R. C., Middaugh C. R. ;(2003);"Insulin containing polyethylenimine-dextran sulfate nanoparticles";*Journal of Pharmaceutical Sciences*;255:139-151.
[18] Davis S, S;(1990);"Assessment of gastrointestinal transit and drug absorption", In:Prescott LF, Nimmo WS, eds. *Novel Drug Delivery and Its Therapeutic Application*, Chichester, UK:Wiley;89-101.
[19] Chourasia M. K., Jain S., K. ;(2003);"Pharmaceutical approaches to colon targeted drug delivery systems", *Journal of Pharmaceutical Science*;6:33-66.
[20] Davaran S., Hanaee J., Khosravi A., (1999);"Release of 5-aminosalicylic acid from acrylic type polymeric prodrugs designed for colon-specific drug delivery", *Journal of Control Release*;58:279-287.
[21] Schacht E., Gevaert A., Kenawy E., R. ;(1996);" Polymers for colon specific drug delivery ", *Journal of Control Release*;58:327-338.
[22] Chung K., T., Stevens S., E., Cerniglia C., E. ;(1992);"The reduction of azo dyes by the intestinal microflora";*Critical Review Microbiol*ology;18:175-190.
[23] Yamaoka T., Makita Y., Sasatani H., Kim S. I., Kimura Y. ;(2000);" Linear type azo-containing polyurethane as drug-coating material for colon-specific delivery : its properties degradation behavior and utilization for drug formulation"; *Journal of Control Release*;66:187-197.
[24] Shantha K. L., Ravichandran P., Rao K. P. ;(1995);" Azo polymeric hydrogels for colon targeted drug delivery";*Biomaterials.* ;16:1313-1318.
[25] Van den Mooter G., Samyn C., Kinget R. ;(1992);"Azo polymers for colon-spicific drug delivery";*International Journal of Pharmaceutical*;87:37-46.
[26] Ghandehari H., Kopeckova P., Kopecek J. ;(1997);"In vitro degradation of pH sensitive hydrogels containing aromatic azo bonds";*Biomaterials*;18:861-872.
[27] Kakoulides E. P., Smart J. D., Tsibouklis J. ;(1998);"Azo crosslinked poly(acrylic acid) for colonic delivery and adhesion specificity synthesis and characterization";*Journal of Control Release*;52"291-300.
[28] Furda I., (1998);"Aminopolysaccharides-their potential as dietary fiber. In: Furda I, ed. *;Unconventional Sources of Dietary Fiber, Physiological and In vitro Functional Properties*, Washington, DC:American Chemical Society;;105-122.
[29] Ormrod D. J., Holmes C. C., Miller T. E. ;(1998);"Dietary Chitosan inhibits hypercholesterolaemia and atherogenesis in the apolipoprotein E-deficient mouse model of atherosclerosis";*Atherosclerosis*;138;329-334.
[30] Yamamoto A., Tozaki H., Okada N., Fujita T. ;(*2000*);"Colon specific delivery of peptide drugs and anti-inflammatory drugs using chitosan capsules";*STP Pharma Science*;10:23-43.
[31] Agnihotri S. A., Mallikarjuna N. N., Aminabhavi T. M. ;(2004);"Recent advances on chitosan-based micro- and nanoparticles in drug delivery"; *Journal of Control Release*;100:5-28.
[32] Calvo P., Remunan-Lopez C., Vila-Jato J. L., Alonso M. J. ;(1997);"Novel hydrophilic chitosan-polyethylene oxide nanoparticles as protein carriers";*Journal of Applied polymer science*;63:125-132.

[33] De S., Robinson D. ;(*2003*);"Polymer realationships during preparation of chitosan-alginate and poly-l-lisyne-alginate nanospheres";*Journal of Control Release* ;89:101-112.

[34] Chen Y., Mohanraj V., Parkin J. ;(2003);"Chitosan-dextran sulfate nanoparticles for delivery of an anti-angiogenesis peptide"; *International Journa lof Peptideand protein Research*;10:621-629.

[35] Ma Z., Yeoh H. H., Lim L. Y. ;(2002);"Formulation pH modulates the interaction of insulin with chitosan nanoparticles"; *Journal of Pharmaceutical S*ciences;91:1396-1404.

[36] Tiyaboonchai W., Woiszwillo J., Sims R. C., Middaugh C. R. ;(2003);"Insulin containing polyethylenimine-dextran sulfate nanoparticles";*Journal of Pharmaceutical S*ciences;255:139-151.

[37] Pan Y., Li j., Zhao H. ;(2002);"Bioadhesive polysaccharide in protein delivery system:chitosan nanoparticles improve the intestinal absorption of insulin in vivo"; *International Journal of Pharmaceutical*;24:139-147.

[38] Saboktakin, M. R. ;Tabatabaie, R. ;Maharramov, A. ; Ramazanov, M. A. ;(2010);" A Synthetic Macromolecule as MRI Drug Carriers : Amino Dextran - coated Iron Oxide Nanoparticles", *Journal of Carbohydrate Polymers*; 80(3), 695-698.

[39] Saboktakin, M. R. ;Tabatabaie, R. ;Maharramov, A. ; Ramazanov, M. A. ;(2010);" Synthesis and Characterization of New electrorheological Fluids by Carboxymethyl Starch Nanocomposites " ; *Journal of Carbohydrate Polymers*;79, 4, 1113-1116.

[40] Saboktakin, M. R. ;Tabatabaie, R. ;Maharramov, A. ; Ramazanov, M. A. ;(2009); "pH-sensitive starch hydrogels via free radical graft copolymerization, synthesis and properties", *Journal of Carbohydrate Polymers* ;77(3), 634-638.

[41] Saboktakin, M. R. ;Tabatabaie, R. ;Maharramov, A. ; Ramazanov, M. A. ;(2009); "Synthesis and Characterization of Superparamagnetic Nanoparticles Coated with Carboxymethylstarch(CMS) for Magnetic Resonance Imaging Technique", *Journal of Carbohydrate Polymers*;78, 292-295.

[42] Saboktakin, M. R. ;Tabatabaie, R. ;Maharramov, A. ; Ramazanov, M. A. ;(2010); " Synthesis and Characterization of Chitosan Hydrogels Containing 5-Aminosalicylic Acid Nano Pendents for Colon - Specific Drug Delivery ", *Journal of Pharmaceutical Sciences*, 2010, Accepted & in Press .

[43] Saboktakin, M. R. ;Tabatabaie, R. ;Maharramov, A. ; Ramazanov, M. A. ;(2010); " Synthesis and Characterization of Superparamagnetic Chitosan-Dextran Sulfate Hydrogels for Targeted Drug Delivery to the Colon ", *Journal of Carbohydrate Polymers* ; Accepted and in Press.

[44] Saboktakin, M. R. ;Tabatabaie, R. ;Maharramov, A. ; Ramazanov, M. A. ;(2010); " Synthesis and Characterization of Superparamagnetic Chitosan-Dextran Sulfate Hydrogels for Targeted Drug Delivery to the Colon ", *Journal of Carbohydrate Polymers*, Accepted and in Press.

[45] Saboktakin, M. R. ;Tabatabaie, R. ;Maharramov, A. ; Ramazanov, M. A. ;(2010); "" Synthesis and Characterization of Biodegradable Chitosan Beads as Nano Carriers for Local Delivery of Satranidazole" ; *Journal of Carbohydrate Polymers*, Accepted & in press .

30

Biopolymers for Military Use: Opportunities and Environment Implications – A Review

Teodora Zecheru
NBC Defense and Ecology Scientific Research Center
Romania

All of us are responsible for the protection of our environment.

1. Introduction

When reading this title, everyone tends to wonder why is there need to talk about biopolymers in the defense field. Isn't the same subject as for the civil field? Or, are we really debating about bio-polymers?

Allow me to recall two of the most known and used definitions of biopolymers:

- chemicals that preserve their efficacy while maintaining low toxicity;
- bricks of human organs or bones reconstruction.

But aren't they also important puzzle pieces when considering the environment?

In this regard, another definition should be added: biopolymers are substances produced without toxic byproducts; they are biodegradable or, at least, biocompatible, having a minimal impact on the environment.

The governments finance military forces, whose duties are the preparation and conduction of different operations, in order to defend people and land. While maintaining continuous improvement of infrastructure and equipment, military forces must assume a growing significance to environmental compliance and restoration efficiency.

Military purposes include utilization of explosive devices and protection equipment. We may number here also biopolymers used as drug delivery matrices. Or as matrices for drug entrapment, being known that every organic molecule has a particular structure that may imprint in a biopolymer matrix.

It is known that the material and its use are inseparable, and that not every material is suitable for any kind of application. Some of them perform successfully under static conditions, but fail or perform undesirably during real situations. Therefore, they must be evaluated together.

In this context, the obtaining and use of environment-friendly compositions, or biopolymers, in order to minimize the risks of potential accidents, while improving the performances, is mandatory.

2. State-of-the-art

Different environments imply different threats. Regarding the military, the hazards have always existed, either at the work place or in the theatres of operation, conducting to health threat to personnel. Even from its early history, military medicine focused on the reduction or, more, on the elimination of injuries caused by nourishment, drinks, wastes, insects, climate, equipment worn, etc. (Fig. 1). Though, it was demonstrated recently that risks from occupational and environmental exposures to toxic chemicals, hazardous materials, and common military chemical compounds, should decrease.

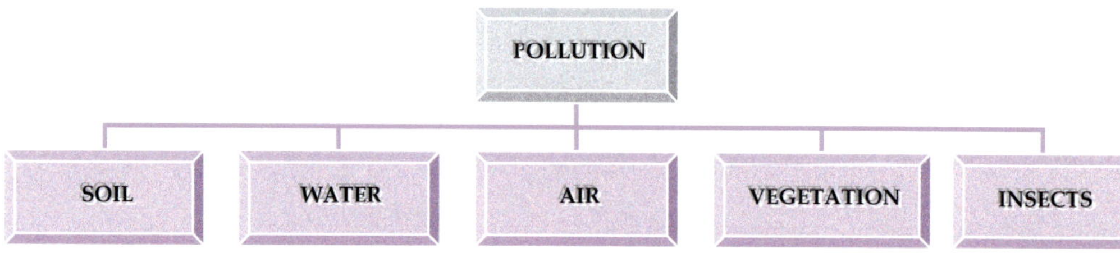

Fig. 1. Indirect influence of natural factors on the humans' health

In order to accomplish their missions, the staff has to be well trained and healthy. Therefore, occupational and environmental health hazards must be identified and their effects on the health diminished.

Still wondering why biopolymers in military field, then? Due to the actual legislation and concerns, the armies must consider the modification of ammunition compositions, in order to decrease or to eliminate the toxicity ratio to human health and to the environment.

The fields taken into consideration in the present paper are exposed in Fig. 2.

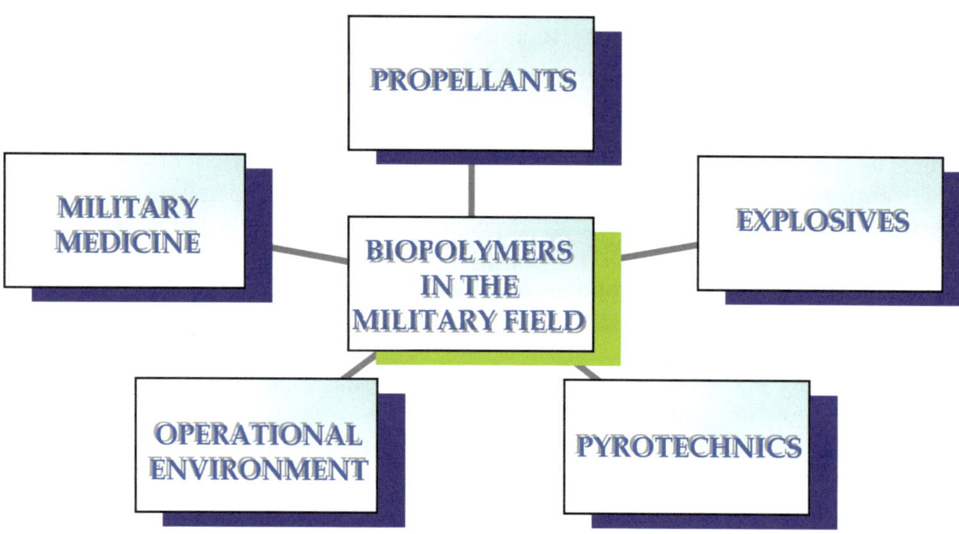

Fig. 2. Fields of use of biopolymers in the army

Firing and munitions testing inevitably conduct to the dispersion of energetic materials and solid by-products in the environment, where energetic materials represent the generic name for any propellant, explosive and pyrotechnics used in weapons or munitions.

Consequently, soil or groundwater contamination by energetic materials, either at the production sites or in military training or testing area occurs. These energetic materials imply a particular environmental concern due to their different chemical, physical and

toxicological properties. Further, every national army has to implement environmental-friendly activities, ensuring that no adverse effect upon the surrounding environment is performed.
All aspects of the life cycle of munitions imply an impact on the environment or human health. Munitions production and firing, demolition procedures and destruction of out-dated ammunition conduct to dispersion of energetic compounds into the environment.
Environmental protection should, therefore, be considered versus its effect on the natural environment. Moreover, it is of a great importance to understand the linkages between the effect on the environment and the impact in relation with staff safety, force protection, and force health protection.
Considering the environmental effects of training, operations, and logistics activities, the environmental damage and costs would decrease. A continuous protection of the environment involves availability of the land to conduct real training operations, without encountering environmental issues.
Environmental protection should be recognized in material acquisition procedures, training, and facilities operations. In brief, it should become an institutional and personal ethic, and military staff should become aware that pollution prevention is a proactive measure rather than just a mere compliance or reaction to laws and regulations.
Moreover, the need for strong environment legislation, especially within the context of military units' closure and demilitarization, conducts to important technical studies on the environment impact of munitions and composition modification.

3. Green military chemistry

The concept of Green Chemistry is not new, dating back in the 60's, but only refined and implemented in the 90's, focusing on minimization of the environmental impact of manufacturing processes through the management of energy, wastes, and products.
Its 12 principles, as described by Anastas and Warner (1998), are as follows:

1. It is better to prevent waste than to treat or clean up waste after it is formed.
2. Synthetic methods should be designed to maximize the incorporation of all materials used in the process into the final product.
3. Wherever practicable, synthetic methodologies should be designed to use and generate substances that possess little or no toxicity to human health and the environment.
4. Chemical products should be designed to preserve efficacy of function while reducing toxicity.
5. The use of auxiliary substances (solvents, separation agents, etc.) should be made unnecessary whenever possible and, innocuous when used.
6. Energy requirements should be recognized for their environmental and economic impacts and should be minimized. Synthetic methods should be conducted at ambient temperature and pressure.
7. A raw material feedstock should be renewable rather than depleting whenever technically and economically practical.
8. Unnecessary derivatization (blocking group, protection/de-protection, temporary modification of physical/chemical processes) should be avoided whenever possible.
9. Catalytic reagents (as selective as possible) are superior to stoichiometric reagents.
10. Chemical products should be designed so that at the end of their function they do not persist in the environment and break down into innocuous degradation products.

11. Analytical methodologies need to be further developed to allow real-time in-process monitoring and control prior to the formation of hazardous substances.
12. Substances and the form of a substance used in a chemical process should be chosen so as to minimize the potential for chemical accidents, including releases, explosions, and fires.

These principles present a multidimensional matrix to guide the design of individual components of any manufacturing process. Life-cycle analysis provides a means to measure how well the principles are being employed, and economics determines whether the process can be commercialized (Slater et al., 2003).

Therefore, these disciplines provide a means to design, monitor, assess, and improve the sustainability of biopolymer production.

4. Propellants

Propellants are designed to produce high temperatures and pressures in a closed chamber for the acceleration of projectiles, rockets, or missiles through the intermediary of the resulting propulsive force. Generally, important burning rates and high temperatures are prerequisites for a propellant (Singh et al., 2006).

In function of their composition, there are two major categories of solid propellants: homogeneous (single-, double-, and triple-base propellants) and heterogeneous propellants (composite and granulated propellants).

Single-base propellants use nitrocellulose (NC) as the main constituent (NC content: 85–96%). Other constituents are chemical stabilizers, such as diphenyl amine and inert or energetic plasticizers, such as dibutyl phthalate, dibutyl sebacate, camphor, or an isomeric mixture of 2,4-dinitrotoluene and 2,6-dinitrotoluene, respectively. In addition to the improvement of the mechanical properties of the propellant, inert plasticizers as well as flame suppressors (such as potassium sulfate or nitrate) act as retardants and coolants, while energetic plasticizers contribute to the total energy output. Single-base propellants are used for all kinds of ammunition for small arms: cannons, tanks, aircraft, and anti-aircraft weaponry.

Double-base (or "smokeless") propellants were developed for long distance firing with large caliber cannons needing higher bullet speeds and more energetic propellants. Therefore these propellants use NC and nitroglycerine (NG) or other liquid nitrate esters. Because of problems arising from the high freezing point of pure NG, modern mixtures of NG with glycol nitrates, such as diethylene glycol, trimethylene glycol, or other alcohols are used, or the NG is completely replaced. Smokeless double-base propellants consist in 50–60% NC and 30–49% NG or the alternative nitrates.

Triple-base propellants contain nitroguanidine (NQ) as a third constituent in addition to NC and NG. NQ has a low flame temperature, but high nitrogen content. Thus, a large volume of gas is produced upon ignition of these "cold powders". Their use prevents damage to the barrel of large-caliber weapons.

Long distance solid-rockets and missiles use heterogeneous propellants. These composite propellants consist in an oxidizer, such as, ammonium perchlorate or ammonium nitrate, and a fuel, such as aluminum. The ignition of such a rocket motor produces a huge cloud of smoke particles and hydrochloric acid.

In general, the aim is to produce a large amount of gaseous combustion products with low molecular mass and high combustion temperature, in order to maximize the specific

impulse. In contrast to solid propellants, which are placed directly in the combustion chamber, liquid propellants are injected from external tank into the chamber at the time of ignition (TM 9-1300-214, 1984).
There are two classes of liquid propellants: monopropellants (liquid propellants consisting in a single substance or a homogeneous mixture of substances, such as 80-99% hydrogen peroxide and hydrazine) and bipropellants: a fuel and an oxidizer injected simultaneously. Examples of such fuels are hydrocarbons, alcohols, amines, or hydrazines, as oxidizers hydrogen peroxide, concentrated nitric acid, or nitrogen dioxide. The two components are stored in separate tanks and their simultaneous injection into the combustion chamber initiates the reaction.
When talking about rockets, the characteristics refer to their performances, reliability and, of course, the costs. In the present, the environment impact is also considered, especially when taking into consideration the rocket propellants.
In solid-propellant rockets, the oxidizer is the ammonium perchlorate (NH_4ClO_4) and the fuel is the aluminum (Al). Consequently to the combustion, hydrogen chloride (HCl) and alumina (Al_2O_3) are generated. HCl is known to have a hazardous influence to the environment, and, although Al_2O_3 is not basically harmful, its release as small particles of aerosol represents a potential toxic effect to the human body.
Compounds such as lead (II) salicylate, stearate or, 2-ethylhexanonate are utilized as burn rate catalysts. Military propellants based on beryllium fuels disperse potent carcinogenic beryllium aerosols and can be regarded as severe sources of pollution, despite of their good performance. In other cases, lead oxides, such as PbO, PbO_2, and Pb_3O_4 or lead (II) chromate or nitrate are, or have been used, as oxidizers in pyrotechnics, especially in "electric matches" (non-explosive fuses). They are utilized for all kinds of pyrotechnic and technical initiation purposes. Furthermore, these lead oxides are occasionally constituents of crackling fireworks and millisecond-delay blasting caps.
Recently, efforts at making lead-free electric matches have increased, through the utilization of nanoscale thermite materials, so-called metastable intermolecular composites (MIC). Apart from avoiding lead compounds, these MICs are characterized by a decreased sensitivity to friction, impact, and heat, coupled with good performance (high combustion temperature).
In this respect, an important issue arises: is it possible to completely replace the solid-propellants?
Researches were performed in this view, and it was found to be very difficult to achieve. Therefore, the idea of a hybrid propellant for the replacement of the solid rocket propellant was embraced.
The conditions to be achieved for the fuel are: high energetic potential and low impact on the environment. In order to meet these needs, organic polymers containing energetic bonds must be considered, such as the azide polymers: glycidyl azide polymer, allyl azide polymer, etc.

5. Explosives

Explosives are materials that detonate. In explosives, a fast reaction produces a very high pressure shock in the surrounding medium. They may be used independent of, or form a part of, ammunition. For military management purposes, the two are controlled as one category of weapons or armaments.

The variation in the properties of explosives is put to practical use in armaments by an arrangement called an explosive train. An explosive train consists in elements arranged upon their decreasing sensitivity and increasing potency. The first element, the initiator, consists in a small quantity of highly sensitive material. This one consists in a primary explosive and other ingredients.

5.1 Military primary explosives

Primary explosives are easily detonated by heat, spark, impact, or friction. In large quantities these materials are extremely hazardous because of their great sensitivity. The other ingredients in the priming composition increase the sensitivity of the mixture to the desired property, such as percussion or heat.

The second element, the booster, contains a larger quantity of less sensitive but more powerful material called a secondary or high explosive. The booster is used either as an intermediate stage to detonate material that is too insensitive to be detonated by the relatively weak initiator or to ensure complete detonation of the main charge. The main or bursting sensitive material but comprises the bulk of the explosive charge (Meyer et al., 2002; TM 9-1300-214, 1984).

Examples of the most common primary explosives are: lead azide, $Pb(N_3)_2$, mercury fulminate, $Hg(ONC)_2$, diazodinitrophenol (DDNP) (Fig. 3-a), lead styphnate (LSty) (Fig. 3-b), lead mononitroresorcinate (LMNR) (Fig. 3-c), potassium dinitrobenzofuroxane (KDNBF) (Fig. 3-d).

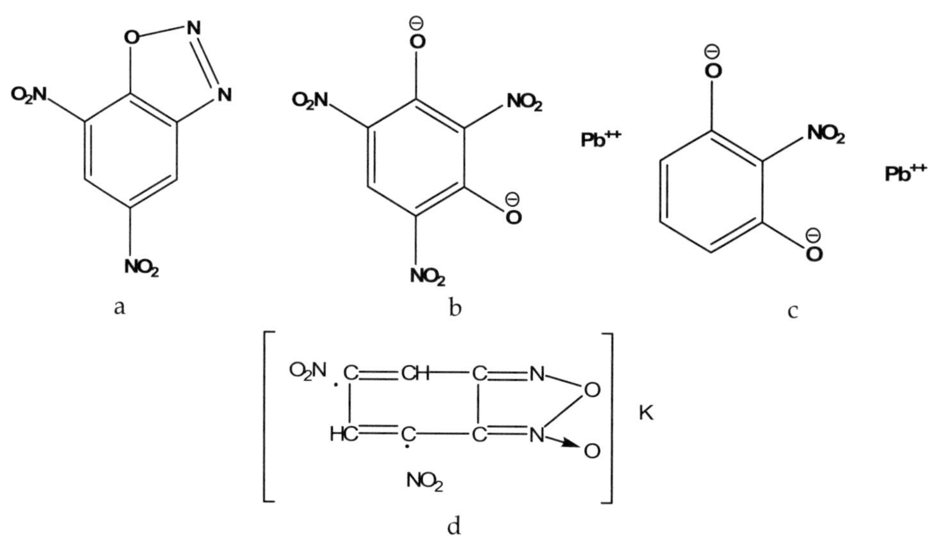

Fig. 3. Structural formulas for: a - DDNP; b - LSty; c - LMNR; d - KDNBF.

5.2 High explosives

Secondary explosives differ from primary explosives in several major characteristics. Small, unconfined charges (one to two grams) of secondary explosives, even though ignited, do not transfer easily from a burning or deflagration reaction to a detonation. Except in the case of dust clouds, ignition by electrostatic spark is difficult. The shock required for ignition is greater for a secondary explosive than for a primary explosive. In some cases, two other elements, a delay and a relay, may be added to the explosive train between the initiator and booster. The delay is calibrated to prevent detonation of the booster for a specified length of time. A delay is considered a pyrotechnic device. A relay may be required to strengthen the relatively weak output of the delay to detonate the booster.

Aliphatic nitrate esters are compounds prepared by O-type nitration in which a nitro group is attached to an oxygen atom of the compound being nitrated. Some of them are: 1,2,4-butanetriol trinitrate (BTN) (Fig. 4-a); diethyleneglycol dinitrate (DEGN) (Fig. 4-b); nitrocellulose (NC); nitroglycerin (NG); nitroguanidine (NQ); nitrostarch (NS); triethylene glycoldinitrate (TEGN) (Fig. 4-c); pentaerythritol tetranitrate (PETN) (Fig. 4-d); 1,1,1 trimethylolethane trinitrate (TMETN) (Fig. 4-e); cyclotetramethylenetetranitramine (HMX)

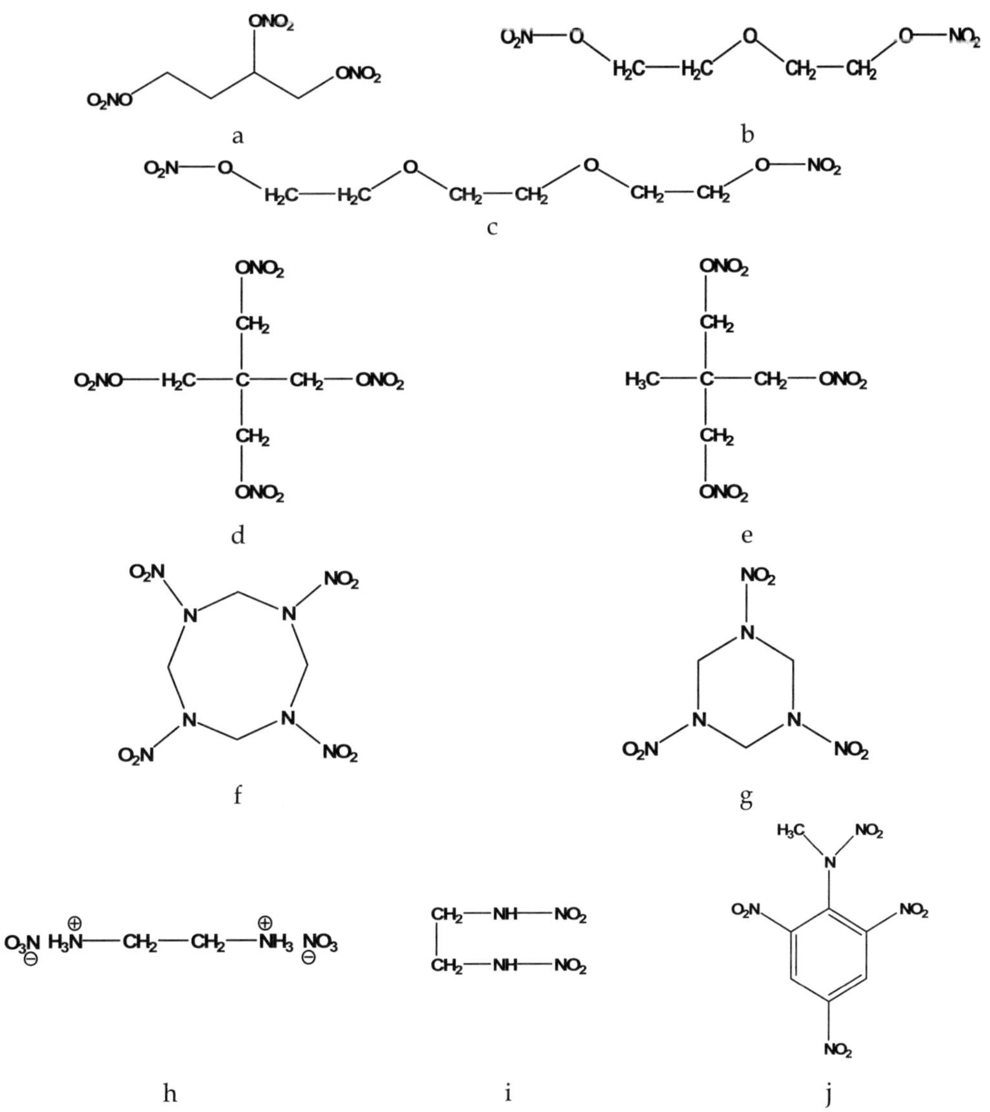

Fig. 4. Structural formulas of high explosives - aliphatic nitrate esters

(Fig. 4-f); cyclotrimethylenetrinitramine (RDX) (Fig. 4-g); ethylenediamine dinitrate (EDDN) (Fig. 4-h); ethylenedinitramine (Haleite) (Fig. 4-i); 2,4,6 trinitrophenylmethylnitramine (Tetryl) (Fig. 4-j).

Nitroaromatics are compounds prepared by C-type nitration in which a nitrogroup is attached to a carbon atom of the compound being nitrated.

In fig. 5 there are given structural formulas of the most known such explosives: ammonium picrate (Fig. 5-a); 1,3-diamino-2,4,6-trinitrobenzene (DATB) (Fig. 5-b); 2,2'4,4',6,6'-hexanitroazobenzene (HNAB) (Fig. 5-c); hexanitrostilbene (HNS) (Fig. 5-d); 1,3,5 triamino-2,4,6 trinitrobenzene (TATB) (Fig. 5-e); 2,4,6 trinitrotoluene (TNT) (Fig. 5-f).

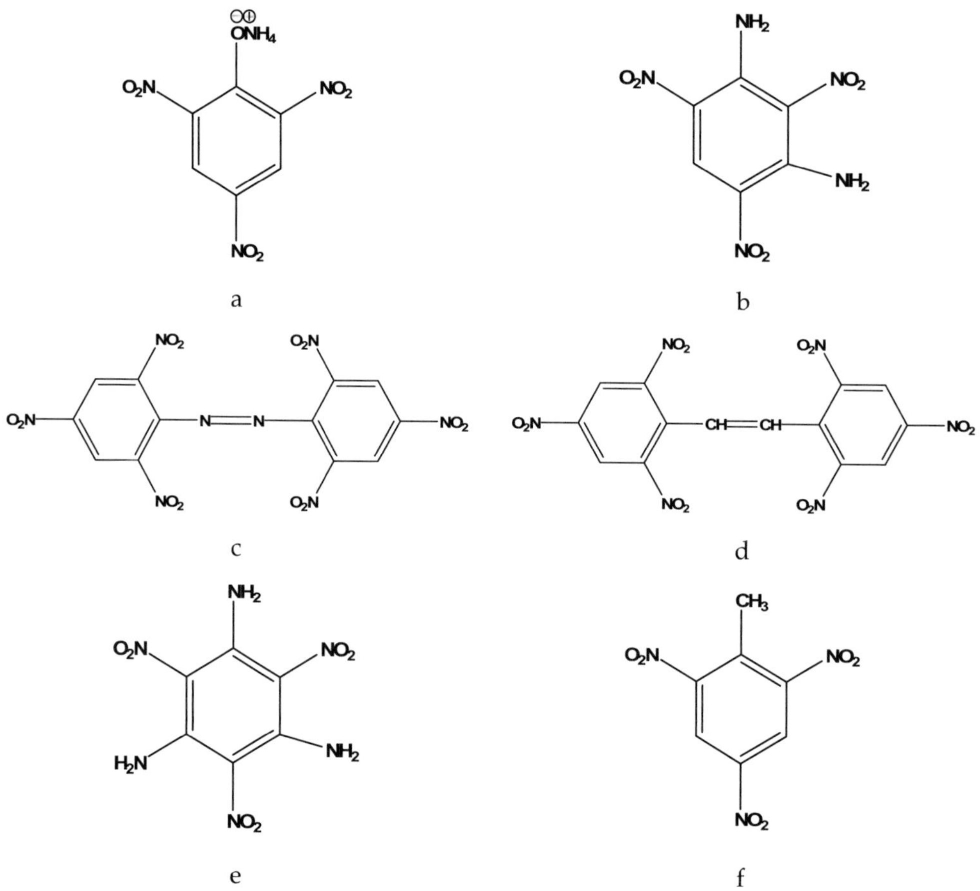

Fig. 5. Structural formulas of high explosives - nitroaromatics

Compositions are explosives in which two or more explosive compounds are mixed to produce an explosive with more suitable characteristics for a particular application. There are:

- binary mixtures, such as amatols - binary mixtures of ammonium nitrate and TNT; ednatols - mixtures of haleite and TNT; octols - mixtures of HMX and TNT; pentolites - castable explosive mixtures of PETN and TNT; tetrytols - mixtures of TNT and tetryl.
- ternary mixtures: amatex (RDX, TNT and ammonium nitrate); ammonals (ammonium nitrate and powdered aluminum incorporated with high explosives such as TNT, DNT,

and RDX); minols are mixtures of TNT, ammonium nitrate, and aluminum; torpex (RDX, TNT, aluminum powder, and wax).

- quaternary mixtures: depth bomb explosive (DBX) consisting in TNT, RDX, ammonium nitrate, and aluminum.
- Plastic Bonded Explosives (PBX) contain a large percentage of basic explosives such as RDX, HMX, HNS, or PETN in intimate mixture with a polymeric binder such as polyester, polyurethane, nylon, polystyrene, various types of rubbers, nitrocellulose, or teflon. In some instances a plasticizer such as dioctylphthalate (DOP), DPA, or butyldinitrophenylamine (BDNPA) is included in the ingredients as well as a fuel such as powdered aluminum or iron (Meyer et al., 2002; TM 9-1300-214, 1984).

6. Pyrotechnics

The massive use of polymer materials in the field is driven by their remarkable combination of properties, low weight and ease of processing.

Flame and incendiary compositions represent the oldest chemical weapons, dating back to the caveman's use of flames and burning coals to discourage animals and enemies. While screening smoke was employed in early conflicts, the results of isolated incidents were always uncertain to justify the adoption of smoke as a warfare agent. Dense smoke clouds generated by the black powder became of a great importance. The clouds obscured the field of vision, interfering with aiming and firing, and hampered the movement of troops. More recently, these characteristics have been exploited tactically by the planned employment of screening smoke munitions.

Meanwhile, the development of flame retardant materials requires scientific expertise in the field (Laoutid et al., 2009). Due to their chemical structure, mainly consisting in carbon and hydrogen, polymers used are highly flammable, often being accompanied by the production of corrosive or toxic gases and smoke during combustion. Consequently, improving the fire retardant behavior of polymers is a major challenge for extending their use to most applications (Pal & Macskasy, 1991). Safety requirements are currently becoming more and more drastic in terms of polymers' reaction to fire and their fire resistance performances, while various flame retardant additives, such as halogenated additives, are being phased out for their proven or suspected adverse effects on the environment. The combined challenge thus consists in developing effective and environmentally friendly flame retardant systems for polymer materials (Troitzsch, 2004; Woycheshin & Sobolev 1975).

Also, fireworks, though spectacular and entertaining, are a source of concern because of environmental pollution. A firework disperses the pollutants over a large area. In other cases, such as handheld military flares, however, inhalation of toxic combustion products exhibits a severe health threat.

The substances released upon pyrotechnics decomposition are harmful to the environment. This problem was identified decades ago; however, efforts in this regard are rather recent.

The development of green pyrotechnics is under considerable cost pressure, since it is difficult for new products to compete with the low cost of traditional formulations. Green pyrotechnics therefore need governmental or other external support to succeed (Steinhauser & Klapotke, 2008).

All pyrotechnic compositions contain at least an oxidizer and a fuel. Additional ingredients include binding agents, retardants, waterproofing agents, smoke dyes or color intensifiers.

a. Oxidizers are substances in which an oxidizing agent is liberated at the high temperatures of the chemical reaction involved. Two oxidizing agents and a fluorine-containing compound are currently used in pyrotechnic compositions. Oxygen is provided by different nitrates: of barium, strontium, sodium, and potassium, perchlorates: of ammonium, potassium, or peroxides: of barium, strontium, and lead. Use of a specific oxidant is determined by the desired burning rate, luminosity, and color. Fluorine is usually provided by polytetrafluoroethylene or chlorotrifluoroethylene.
b. Fuels include finely powdered aluminum, magnesium, metal hydrides, red phosphorus, sulfur, charcoal, boron, silicon, etc. The most frequently used are powdered aluminum and magnesium. Additional materials such as binding agents, waterproofing agents, and color intensifiers also act as fuels.
c. Binding agents make the oxidizer and the fuel adhere when compressed into pyrotechnic mixtures. They also prevent segregation and ensure a more uniformly blended composition. Binders also help to obtain maximum density and increased burning efficiency. Non-energetic binders can either be water-soluble (dextrin, polyvinyl alcohol, or arabic rubber) or organic-solvents-based (vinyl alcohol acetate resin, polymethyl methacrylate, or other organic polymers), or solvent-free materials, such as some epoxy binders or Laminac (an unsaturated polyester with styrene cross-links), which binds upon addition of a catalyst.
d. Retardants are used to reduce the burning rate of the fuel-oxidizing agent mixture, with a minimum effect on the color intensity of the composition. Some retardants act only as inert diluents, while others take part in the combustion reaction at a much slower rate than the metallic fuels. Calcium carbonate, sodium oxalate, strontium resinate, titanium dioxide, polyvinyl chloride, ethyl cellulose, paraffin, linseed oil, castor oil, asphaltum, and sulfur are the most important retardants used. Certain of these serve other purposes as well. For example, sodium oxalate and polyvinyl chloride act also as color intensifiers, titanium dioxide is a source of incandescent solid particles in the flame produced by the composition, and the waxes and oils act as fuels, binding agents, and waterproofing agents, as well as retardants.
e. Waterproofing agents are applied as a coating on metallic fuels, such as a coating of dried linseed oil on magnesium, or as an ingredient uniformly distributed throughout the rest of the composition. In some cases, the metallic fuel is given a coating by treatment with a solution of acidic or acidified potassium dichromate. Waxes, metal resinates, and natural and synthetic resins are used for distribution throughout the composition. Many of these serve also as binding agents.
f. Color intensifiers are chlorinated organic compounds, such as hexachloroethane (C_2Cl_6), hexachlorobenzene (C_6Cl_6), polyvinyl chloride, and dechlorane ($C_{10}Cl_{12}$). Since they are less reactive than metallic fuels, color intensifiers act to some extent as combustion retardants. Following their physical characteristics, certain intensifiers, such as polyvinyl chloride, can serve also as binding agents.
g. Smoke dyes are azo and anthraquinone dyes. These dyes provide the color in smokes used for signaling, marking, and spotting.

In the respect of environment protection, green pyrotechnics should primarily avoid perchlorates and heavy metals, taking into account that, for example, the contribution of fireworks to the total annual emission and deposition of lead was estimated to be almost 0.8%.

Chromium was found in scalp hair of factory workers and was identified as the cause of headaches and dizziness. The toxicity of chromium is very much dependent on the oxidation state of the ion, Cr(VI) compounds being the most toxic and carcinogenic compounds (in contrast to metallic chromium, Cr(II), and Cr(III) compounds). For unknown reasons, the chromate cations play an important role in their toxicity.

Although strontium compounds are generally not very toxic, strontium chromate is one of the most potent animal carcinogens ever identified.

Traces of heavy metals (As, Cd, Hg) are found in fireworks as a consequence of contaminated raw materials, but they do not contribute significantly to the total emission and deposition of these elements within the area (TM 9-1300-214, 1984).

Pyrotechnic mixtures are not limited to those using oxygen as the oxidizing species. In terms of green chemistry, compositions based on metals and halogenated organic compounds or polymers, such as magnesium-teflon-Viton or MTV and Al-Teflon/Viton mixtures or similar compositions, are superior to other pyrotechnics because of the poor solubility in water and resulting low bioavailability of both the components and the combustion products (MgF_2, AlF_3). The environmental compatibility of the Berger mixture using zinc or zinc oxide and hexachloroethane, however, is poor. These mixtures cause the formation of zinc chloride (more bioavailable, though sensitive to hydrolysis) and several organic chlorides.

Derivatives of these compositions have also to be investigated for the achievement of environment protection targets.

7. Operational environment. Disposal, destruction and demilitarization

7.1 Staff protection equipment

Operational shields with overhead and frontal protection are designed for the staff protection. Employees must wear protection equipment when explosive materials are destroyed by detonation or when explosive materials which may detonate are being burned. Also, the wearing of protective equipment is required during exposure or threat of exposure to chemical, biological, radiological or nuclear agents, or unknown agent or agents, when the route of exposure may be inferred from the presence of contaminant on the clothing and skin of victims, while enabling these people to accomplish their assigned missions.

Military protection equipment refers to (EMS; Iacobidze et al., 2003; SLAC, 2009; Wiener, 1996):

- protection respiratory devices: rubbery masks for the protection of the respiratory tract, eyes, and mucous membranes, providing protection against vapors involved in chemical warfare and against biological warfare particles;
- garment ensembles, giving protection against chemical and biological warfare agents and radioactive alpha and beta particles;
- wraps, chemical- and biological-protective for casualties in contaminated environments in which personnel are unable to wear battledress overgarments, the top of the garment having a charcoal lining similar to the BDO, while the bottom being constructed of impermeable rubber;
- individual ballistic protection equipment, against different fragments, splinters and small caliber projectiles, against shock waves from different explosive devices, shells and projectiles; fire-resistant garment;

- gloves, consisting in protective outer glove pairs made out of butyl rubber and an inner glove for absorption of perspiration;
- safety glasses, with permanently attached side shields, required whenever there is a potential for projectile objects in the work area;
- safety goggles and face shields, required when handling any chemical or process that can create fine dust, fumes, mists, and sprays;
- footwear covers, butyl rubber or vinyl overboots for the protection of the combat boots against all agents.

The materials used for those equipments are:

- natural fibers and fabrics, consisting in bulk fibers, yarns, or woven cloth manufactured from plants, such as cotton, wool, linen (flax), sisal, jute, hemp, or silk;
- synthetic materials, consisting in bulk fibers, yarns, woven cloth or other textile products manufactured from polymer-based materials, such as polyethylene, polypropylene, polyamide (nylon), polyester, polyurethane, aramid, or other thermoplastics (butyl and halobutyl rubbers); for example, bullet-proof vests are made of ultra-high molecular weight polyethylene (UHMWPE), with Dyneema and Spectra trade marks of polyethylene-based fibres; shields are made of polycarbonate or polymethyl methacrylate;
- aramid fibers and aramid fabrics consisting in bulk, chopped fibers, continuous strands or woven cloth forms of aromatic polyamide thermoplastic for reinforcing polymer matrix composites and other applications. Nomex, Kevlar, Twaron and Tyvek are trade marks for the most known and used para-aramid synthetic fibers as safety materials, Nomex presenting also a very important flame-resistant potential.

All the materials used are both protective and human-compatible (tested against human skin, eyes, hair), they are practical to wear, light, flexible and cover the vulnerable areas for their particular application.

7.2 Materials disposal

Military polymers and materials cannot be safely disposed by simply dissolving in solution and eliminating, due to the fact that they are water-insoluble, they conduct to toxic by-products, and are hazardous to the environment. Therefore, disposal would consist in burning, detonation, or chemical decomposition.

Destruction methods depend on the quantity and nature of materials to be destroyed, the facilities available, and the land topography. Destruction of explosive materials will be accomplished by burning or detonation. The only exception is made in the case of small quantities of explosives, which can be destroyed by chemical means (TM 9-1300-214, 1984).

It is absolutely prohibited to bury energetic materials in the ground or to dump them in waste places, rivers, or sea. Site selection, physical security, personnel training, emergency equipment, and procedures are governed by environmental regulations, particularly applicable hazardous waste regulations, and by the safety considerations which follow.

Open burning and open detonation operations will be conducted in accordance with applicable air, hazardous waste, and other environmental allowances (Fig. 6). The site selected for the destruction of explosives and other energetic materials shall be located at the maximum practicable distance from buildings or public roads, in order to avoid dispersion of fragments and debris. In all disposal and destruction activities, the quantity of explosives to be safely destroyed at one time shall be determined carefully by starting with a limited quantity and then gradually increasing that quantity until the optimum amount consistent with safe and efficient operation is determined.

During disposal and destruction operations, the possibility that the mass detonates must be considered and protective measures for the staff and the endowments should be taken. The number of people in the area exposed to the hazard must be kept to a minimum, and the personnel engaged in burning explosives should be provided with flame resistant clothing.

7.3 Demilitarization

The problem of what to do with outdated ammunition to prevent both pollution and hazardous situations from occurring is an enormous one which dates back more than a century. At first glance, two answers become apparent: complete disposal or partial disposal with some recycling. The various chemical constituents of the warhead, propellant, or pyrotechnic are not only explosively hazardous, but are frequently toxic. Disposal by dumping into the oceans, incineration, or detonation have been shown to be not only dangerous but an addition to world pollution and as such, a persistent universal health hazard. Further, the problem of pollution from all military and non military sources became so acute that both state and private facilities are set up to handle pollution abatement and waste recycling. To this end, efforts are developed for safe, efficient, and non-polluting methods of disposal or recycling of outdated ammunition, in particular, their energetic material content.

Fig. 6. Burning of pyrotechnical compositions in the field

Certain energetic materials such as relatively stable high explosives and pyrotechnics can be easily reclaimed and reloaded, but solid propellants, with a limited storage life, require significant degrees of processing before they can be recycled or converted to other products. In many instances, energetic material component recovery was not economically feasible. Energy conservation requirements strongly favor the recycling of energetic materials over pollution-free disposal (or the wasting) of these materials (TM 9-1300-214, 1984).
Preliminary removal of the high explosive charge from a mine, projectile, or shell usually involves the use of hot water or steam to liquefy the explosive, which is then separated from the water by gravity, or contour drilling followed by high pressure water erosion to remove the high explosive residue.

8. Biopolymers and their use in military medicine

Cellulose is the most abundant and most widely used biopolymer. This and related polymers have been produced in the past century and are used in applications including films and injection-molded products.
Like cellulose, chitin is a natural polysaccharide that is extremely abundant. Highly purified chitin, called chitosan, has a variety of medical and food uses and has an important potential of plastic-use. However, there are currently no commercial plastics processes that use chitin as a raw material.
Starch-based plastics find an increased interest in the polymer markets, as biodegradable materials and resins for injection molding. Starch has also been incorporated with non-biodegradable polymers in an effort to have them partially decompose. In this case, only a small portion of the polymer is actually biodegradable (Slater et al.).
Modern medicine has begun to understand and use the benefits provided by biotechnology in health care and casualty care. Practical knowledge of the causes of human disease, biological targets for new drugs, genetic markers, and sophisticated diagnostic tests would increase the effectiveness of medicine (EMS).
As regarding the highly specialized necessities of military medicine, it may provide unique opportunities to absorb these advances at a rapid rate. The developments significantly enhance the capabilities in war-zone medicine. And an increasing number of new capabilities will emerge for carrying out health operations other than war.
A key contributor to this revolution has become biomaterials, and, in particular, biopolymers.
Medical applications of biopolymers include three directions:

- extracorporeal uses: catheters, tubing, and fluid lines; dialysis membranes/artificial kidneys; ocular devices; and wound dressings and artificial skin;
- permanently implanted devices, such as sensory devices; cardiovascular devices;
- temporary implants, such as degradable sutures, implantable drug delivery systems, scaffolds for cell or tissue transplants, temporary vascular grafts and arterial stents.

In this context, biopolymers represent for military medicine an important field of continuous research, in order to improve the characteristics required for the personnel in the theatre.

9. Conclusion

As environmental concerns and the availability of finite resources become increasingly important, the defense industry must begin to focus on the next generation of raw materials and manufacturing technologies. Manufacturing based on the use of renewable resources as primary feedstocks will increasingly be required to achieve economic and environmental sustainability.

To have a safe, effective, and efficient program of ammunition recycling, ammunition that can be readily recycled must be available. To this end, a new concept in ammunition design has to be conceived. Biopolymers represent a potential alternative to old energetic materials. Therefore, ammunition items (explosives, propellants, and pyrotechnics) will be designed and fabricated so that to be easily and safely recycled.

Meanwhile, each one of us should be aware that terrorists do not care about environment, laws and regulations and the national armies have to be prepared to face properly any kind of attack, using specific protection equipment and endowment, continuously improved.

10. References

Anastas, P.T. & Warner, J.C. (1998). *Green Chemistry. Theory and Practice.* Oxford University Press Inc., ISBN: 0-19-850234-6, Oxford

Emergency Medical Services (EMS). Personal protection and safety principles. In: *Managing Hazardous Materials Incidents. Emergency Medical Services.* Vol. 1, p. 15-19

Emergency Medical Services (EMS). Respiratory protection. In: *Managing Hazardous Materials Incidents. Emergency Medical Services.* Vol. 1, p. 20-22

Iakobidze, N.; Tavadze, G. & Khvadagiani, A. (2003). Individual Ballistic Protection Equipment, *RTO HFM Symposium on "NATO Force Health Protection Requirements from Pre- to Post-Deployment: Population Health for the Military"*, p. 39-1 - 39-8, Antalya, Turkey, April 2003, RTO-MP-109.

Laoutid, F. ; Bonnaud, L. ; Alexandre, M. ; Lopez-Cuesta, J.-M. & Dubois, Ph. (2009). New prospects in flame retardant polymer materials: From fundamentals to nanocomposites. *Materials Science and Engineering R: Reports*, Vol. 63, No. 3 (January 2009), p. 100–125, ISSN: 0927-796X

Meyer, R.; Kohler, J. & Homburg, A. (2002). *Explosives* (5th ed.). Wiley-VCH Verlag GmbH & Co. KGaA, ISBN: 3-527-60051-5 (electronic), New York

Pal, G. & Macskasy, H. (1991). *Plastics: Their Behavior in Fires*, Elsevier, ISBN: 0-444-98766-5, New York

Singh, R.P.; Verma, R.D.; Meshri, D.T. & Shreeve, J.M. (2006). Recent Developments in High Nitrogen Energetic Salts Including Ionic Liquids, *Angew. Chem. Int. Ed.*, Vol. 45, p. 3584-3601, ISSN: 1521-3773.

SLAC Environment, Safety, and Health Manual (SLAC-I-720-0A29Z-001) (2009). Chapter 19, "Personal Protective Equipment"

Slater, S.; Glassner, D.; Vink, E. & Gerngross, T. (2003). Evaluating the Environmental Impact of Biopolymers, *Biopolymers, Vol. 10: General Aspects and Special Applications*, p. 473-480, ISBN-10: 3-527-30229-8, ISBN-13: 978-3-527-30229-1, WILEY-VCH, Weinheim

Steinhauser, G. & Klapotke, T.M. (2008). "Green" Pyrotechnics: A Chemists' Challenge. *Angew. Chem. Int. Ed.*, Vol. 47, p. 3330 - 3347, ISSN: 1521-3773.

Technical Manual TM 9-1300-214 (1984). Military Explosives

Troitzsch, J. (2004). *Plastics Flammability Handbook: Principles, regulations, testing, and approval* (3rd ed.), Hanser Publishers, Munich/Hanser Gardner Publications, Inc., ISBN: 3-446-21308-2, Cincinnati

Wiener, S.L. (1996). Strategies for the prevention of a successful biological warfare aerosol attack. *Mil. Med.*, Vol. 161, No. 5 (May 1996), p. 251-256, ISSN: 0026-4075

Woycheshin, E.A. & Sobolev, I. (1975). Effect of Particle Size on the Performance of Alumina Hydrate in Glass-Reinforced Polyesters. *J. Fire Flammability/Fire Retard. Chem.*, Vol. 2, p. 224-241, ISSN: 0362-1693